D1496084

Tropical Forests of the Guiana Shield

Ancient Forests in a Modern World

Tropical Forests of the Guiana Shield

Ancient Forests in a Modern World

Edited by

D.S. Hammond

Formerly of Iwokrama International Centre for Rain Forest Conservation and Development Georgetown, Guyana

CABI Publishing

CABI Publishing is a division of CAB International

CABI Publishing
CAB International
Wallingford
Oxfordshire OX10 8DE
UK

Tel: +44 (0)1491 832111
Fax: +44 (0)1491 833508
E-mail: cabi@cabi.org
Web site: www.cabi-publishing.org

CABI Publishing
875 Massachusetts Avenue
7th Floor
Cambridge, MA 02139
USA

Tel: +1 617 395 4056
Fax: +1 617 354 6875
E-mail: cabi-nao@cabi.org

A catalogue record for this book is available from the British Library, London, UK.

Library of Congress Cataloging-in-Publication Data

Tropical forests of the Guiana Shield: ancient forests of the modern world / edited by D. Hammond

p. cm.

Includes bibliographical references and index.
ISBN 0-85199-536-5 (alk. paper)
1. Rain forest ecology--Guyana Shield. 2. Rain forests--Guyana Shield. 3. Natural History--Guyana Shield. 4. Guyana Shield. I. Hammond, D. (David S.) II. Title.

QH111.T76 2005
578.734′098--dc22

2004021133

ISBN 0 85199 536 5

Typeset by MRM Graphics Ltd, Winslow, Bucks
Printed and bound in the UK by Biddles Ltd, King's Lynn

Contents

Contributors

Alexander, E.E., *Planning and Research Development Division, Guyana Forestry Commission, 1 Water Street, Kingston, Georgetown, Guyana.*

Basset, Y., *Smithsonian Tropical Research Institute, Apartado 2072, Balboa, Ancon, Panama.* (e-mail: bassety@tivoli.si.edu)

Charles, E., *Faculty of Agriculture/Forestry, University of Guyana, Turkeyen, Georgetown, Guyana.* (e-mail: elroy_c@yahoo.co.uk)

De Dijn, B.P.E., *National Zoological Collection, University of Suriname, University Complex, Leysweg, PO Box 9212, Paramaribo, Suriname.* (e-mail: nzcs@cq-link.sr)

Forget, P.-M., *Département Ecologie et Gestion de la Biodiversité, Museum National d'Histoire Naturelle, UMR 5176, CNRS-MNHN, 4 Av. Du Petit Chateau, F.91800 Brunoy, France.* (e-mail: forget@mnhn.fr)

Hammond, D.S., *Iwokrama International Centre for Rain Forest Conservation and Development, Georgetown, Guyana.* Current Address: *NWFS Consulting, 15595 NW Oak Hill Dr., Beaverton, OR 97006, USA.* (e-mail: dhammond@nwfs.biz)

Houter, N.C., *Department of Plant Ecophysiology, Utrecht University, Sorbonnelaan 16, 3584 CA Utrecht, The Netherlands.*

Pons, T.L., *Department of Plant Ecophysiology, Utrecht University, Sorbonnelaan 16, 3584 CA Utrecht, The Netherlands.* (e-mail: t.l.pons@bio.uu.nl)

Rijkers, T., *Department of Forest Ecology and Forest Management, Wageningen University, PO Box 47, 6700 AA Wageningen, The Netherlands.*

Rose, S.A., *Planning and Research Development Division, Guyana Forestry Commission, 1 Water Street, Kingston, Georgetown, Guyana.*

Springate, N.D., *Department of Entomology, The Natural History Museum, Cromwell Road, London SW7 5BD, UK.* (e-mail: nds@nhm.ac.uk)

Acknowledgements

Any work is the product of numerous contributions at many different moments in many different ways. This book is no different and the individuals who have contributed through their patient and persistent collection of data in the field, organization and compilation of references, time series and spatial information and administrative support are numerous. I would particularly like to thank again Kate Lance, Kwasie Crandon, Roxroy Bollers, George Roberts, Catherine Clarebrough, Luc van Tienen, Mariska Weijerman, Arnoud Schouten and Dexter Angoy for their efforts in collecting data presented or referred to in various chapters of the book. Several UK and Guyanese volunteer and training groups did a marvellous job with data collection. Twydale Martinborough deserves special thanks for her efforts in meticulously compiling much of the data on production statistics and Chanchal Prashad for her excellent work in digitizing and compiling many of the geological and historical GIS coverages presented in figures throughout the book. Thanks are extended to all of the scientists and resource management professionals in French Guiana, Guyana and Suriname for sharing information and literature early in the book's development. I also owe a debt of gratitude for the efficient and timely administrative support of Juliet Dos Santos and Jean Bacchus in making the compilation of 'grey' literature and statistics that much easier. The amazing efficiency and speed through which high quality data were made available by governmental institutions in Guyana, Venezuela, Suriname, France, Brazil, the USA and the UK, often free-of-charge and through ftp downloading, has enhanced in no small measure the quality of content and avoided the labyrinth typically encountered en route to acquiring such types of environmental data. I would like to acknowledge their effort here as a means of registering support for the growing trend in making these types of data available to the international community. These include ANEEL (Brazil), CDC (USA), CNRS (France), CPRM (Brazil), DAAC (USA), DANE (Colombia), GEBCO (UK), IBAMA (Brazil), IBGE (Brazil), INMET (Brazil), IRD (France), INSEE (France), MARNR (Venezuela), NOAA (USA), NODC (USA), OCEI (Venezuela), ODP (USA), SCOPE (USA), UEA-CRU (UK), USGS (USA), and Woods Hole Oceanographic Institute – LBA (USA). Similarly, I wish to thank agencies and commissions under the UN umbrella – the FAO, UNEP, UNESCO, UNDP, IPCC and WCD, for making data and literature they have generated widely and easily available.

Administrative, logistical, data-sharing and library support were provided to the book at various stages by the Tropenbos-Guyana Programme, Tropenbos Foundation, Guyana Forestry Commission, Geology and Mines Commission, Imperial College – Silwood Park, UK Natural History Museum, CABI Bioscience, and the Iwokrama International Centre for Rain Forest Conservation and Development. In particular, thank you for the great cooperation and exchange with various institutions in Guyana, including the Guyana Forestry

Commission, University of Guyana, National Agricultural Research Institute and Environmental Protection Agency.

Without the financial support provided to the editor and for project development, this book would not have been possible. To this end, generous support provided at various early stages by the Department for International Development-UK, the Tropenbos Foundation and the European Commission (to the Iwokrama Centre) is gratefully acknowledged. Support during the final stages was provided solely by the editor.

I wish to acknowledge the professional and motivational support provided at various times through the development of this book by Professor Val Brown (now at University of Reading), David Cassells (Iwokrama/World Bank), Ben ter Welle (Tropenbos/Utrecht University/GTZ), my friend and colleague Pierre-Michel Forget (MNHN – Paris) and Hans ter Steege (National Herbarium – Utrecht). I also would like to thank the many scientific colleagues who agreed to review chapters at various stages and often at short notice and the contributing authors for their unprecedented patience while this volume went through various changes.

The editor would like to dedicate his effort on this book to the memory of Timothy C. Whitmore, one of the great tropical botanists, ecologists and foresters of the 20th century and a scientific guiding light during my time living and working in Guyana.

Most of all, thanks to my family for their endless patience.

About the Editor

David S. Hammond has been researching and working to help conserve and sustainably develop neotropical forests since 1987. He received his BSc in Botany–Environmental Science from Miami University, USA, and a PhD in Environmental Sciences from the University of East Anglia, UK. He currently resides in Portland, Oregon.

Acronyms and Abbreviations

ANEEL	Agencia Nacional de Energia Eletrica
AVHRR	Advanced very high resolution radiometry
BP	Before present (normally 1950)
CAM	Crassulacean acid metabolism
COADS	Coupled ocean–atmospheric data set, NOAA
CVG	Corporacion Venezolana de Guayana
DAAC	Distributed Active Archive Center, NASA
DANE	Departamento Administrativo Nacional de Estadistica, Colombia
DNPM	Departamento Nacional de Produção Mineral, Brazil
DOC	Dissolved oxygen concentration
DWIC	Dutch West Indies Company
ENSO	El Niño – southern oscillation
ETR	Electron transport rate
FAO	Food and Agriculture Organization of the United Nations
GDP	Gross domestic product
GFC	Guyana Forestry Commission
GGMC	Guyana Geological and Mines Commission
GOES	Geostationary operational environmental satellite
HIV	Human immuno-deficiency virus
Hydromet	Guyana Hydrometeorological Office
IBAMA	Instituto Brasileiro do Meio Ambiente e dos Recursos Naturais Renováveis
IBGE	Instituto Brasileiro de Geografia y Estatistica
INDERENA	Instituto de Desarrollo de los Recursos Naturales, Colombia
INPARQUES	Instituto Nacional de Parques, Venezuela
INRA	Institut National de Recherche Agronomique
ISR	Incoming short-wave radiation
ITCZ	Inter-tropical convergence zone
JERS	Japanese earth resources satellite
LAPD	Latin American Pollen database
LAR	Leaf area ratio
LGM	Last Glacial Maximum
LMA	Leaf mass per unit area
LMF	Leaf mass fraction
NAR	Net assimilation rate

NCAR	National Center for Atmospheric Research
NCEP	National Center for Environmental Prediction, NOAA
NEP	Net ecosystem productivity
NOAA	National Oceanic and Atmospheric Administration
NPV	Net present value
NTFP	Non-timber forest product
OCEI	Oficina Central de Estadistica e Informatica, Venezuela
OLR	Outgoing long-wave radiation
ONF	Office National de Forêt, France
ORSTOM	Institut de Recherche Scientifique pour le Développement en Coopération
PFD	Photon flux density
PS II	Photosystem II
RGR	Relative growth rate
RIL	Reduced impact logging
SAR	Synthetic aperture radar
SLP	Sea-level atmospheric pressure
SOI	Southern oscillation index
SPC	Spare productive capacity
SST	Sea-surface temperature
TATE	Trans-Amazonian tectonothermal episode
TRMM	Tropical rainfall measuring mission
TSS	Total suspended solids
TZ+	Total dissolved cation concentration
USGS	United States Geological Survey
VPD	Vapour pressure deficit
WBR	World base reference
WCMC	World Conservation Monitoring Centre
WMSSC	World Monthly Surface Station Climatology, NCAR
WPA	World petroleum assessment
WRI	World Resources Institute
WUE	Water use efficiency

1 Ancient Land in a Modern World

David S. Hammond
Iwokrama International Centre for Rain Forest Conservation and Development, Georgetown, Guyana. Currently: NWFS Consulting, Beaverton, Oregon, USA

Introduction

The Guiana Shield could be described as a land of old rock, poor soils, much water, extensive forest and few people. These five attributes, perhaps better than any other, lay down a foundation for much of the geographic and historic variation that has shaped the shield, its forests and the way these have and will be conserved and used.

Few tropical forest regions of the world can trace their geological origins directly back to the earliest days of life itself. Even fewer can be recognized as having largely escaped thus far the sweep of modern human society as it appropriates an ever increasing share of the global natural resource base. Covering much of the northeast corner of South America between the Orinoco and Amazon Rivers, the contemporary forest landscape of this region is a complex of ancient and recent geological features, many of these showing few signs of recent human modification. Today, forests of the shield rest upon a silent geological landscape. Billions of years of repeated deformation have all but exhausted the geological processes that catalyse the natural transfer of nutrients, soil and biomass across other landscapes. During the Precambrian and throughout the Phanerozoic, the shield landscape has been shaped by an incredible sequence of repeated fracturing, volcanism, erosion and deposition that was compounded by millennial-scale effects of regional climate change. In recent times, the silence of the geological landscape has been broken by a growth in industrial human activity. Placed in perspective, the forests of the Guiana Shield that are used and conserved today represent a mere snapshot of the evolutionary process extending back to a Cretaceous Gondwanaland, more than 120 million years ago (Romero, 1993) (see Chapters 2 and 7). From all available evidence, the distribution of closed forests in the region over the last several million years has continually expanded and contracted in response to climate change and its impact on sea level, rainfall and temperature (see Chapter 2). Compared with the natural dynamism of the much younger geological landscapes of western Amazonia and Central America, however, recent surface and internal changes to the lithosphere ('crust') are absent or indiscernible in the modern Guianan landscape (see Chapter 2). The mountains have been weathered to their roots and the resulting sediments reworked to such an extent that, in many areas, the nutrient and water holding capacity has largely disappeared (see 'Soil and soil fertility', Chapter 2). The shaping influence of this environment on many short- and long-term biological processes is increasingly becoming clear as we accrue information on contemporary and historical events in the region. The biological evidence, however, does not in all cases reinforce the view that an exhausted edaphic environment leads to a proportional decline in the variety of life, as plants and animals find novel means to cope with diminishing opportunities for survival and growth (see Chapters 3–5) and other processes continue to play a more important role in sustaining diversity (Chapter 7). The implications of the geological age of the shield to the debate on human use and conservation of these forests is profound, but often diluted due to the scale and duration of its shaping influence. There is no factor, apart from the current decisions made by human society, that wields greater overarching influence on the size and location of resource use and conservation opportunities in the region and the land's future ability to sustain economic growth than the twin foundational forces of geology and climate (see Chapters 2, 8, and 9).

The (sparse) archaeological and historical record suggests that society and culture in the shield region have come and gone and recycled again through repeated waves of colonization and abandonment (see 'Human Prehistory of the Guiana Shield', Chapter 8). Today, the region remains one of the least populated areas of the world (Fig. 1.1 and Chapter 8). In fact, population densities in most parts of the shield are more comparable to the coldest and least productive (shield) areas of the planet, such as Nunuvut (northern Canada) and Siberia (Russia), than to other tropical forest regions. Yet the Guiana Shield is covered by vast stretches of forest and savanna that, at first glimpse, appear no less productive and accommodating to human life than any other tropical environment. Other tropical countries, such as Indonesia, southern India, Bangladesh and Nigeria, have population densities comparable to the USA, Great Britain or Germany. How can such a vast area of tropical forest remain so sparsely populated despite a long history of human colonization (see 'Colonial History', Chapter 8)? Is the paradox of a tropical forest landscape both verdant and impoverished an adequate explanation? Or have other factors worked surreptitiously to shape the relationship between people and the environment of the Guiana Shield?

The answer to this and other large-scale questions is rarely addressed adequately by any single perspective. The 'forêt dynamique' can follow a trajectory

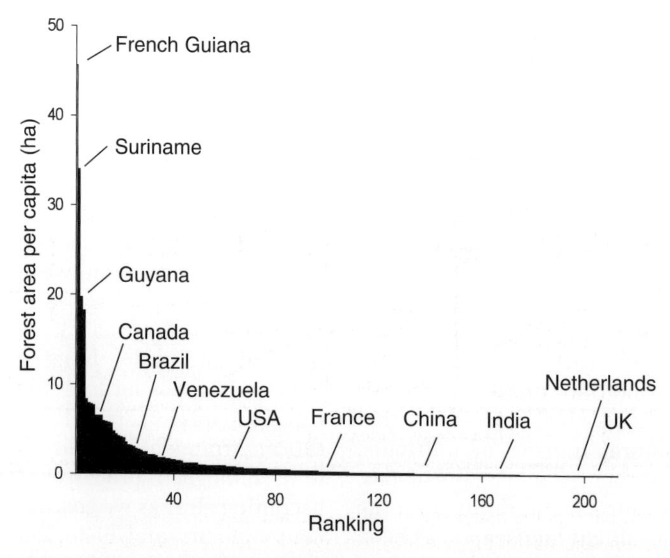

Fig. 1.1. Two-hundred and thirteen countries ranked by forest area per capita in 2000. Rankings of countries are influenced by both the natural extent of forest cover and losses due to deforestation. Data source: FAO (2001). France (Metropole) and French Guiana (DOM) are considered separately here for comparative purposes.

that is not necessarily the sole product of understood biophysical mechanisms or contemporary social processes. The sum total of numerous anomalies, the relicts of past events unrecorded and poorly understood, may plausibly constrain the course of natural processes, and thus drive forest change along a course that is otherwise assigned wholly to environmental forces (in statistical parlance, an 'aggregated' type II error). Human history and the social dynamics of decision-making can interweave with a wide range of biological and physical processes to create a unique and unreplicable forest filled with its own special mix of life. The forces that work to fashion the forest we see today and that we will see in the future are not easily attributed to any single approach or perspective. A simplified depiction of these broad interdisciplinary relationships and their scale of operation is offered in Fig. 1.2. Management of forest cover and composition based on a static snapshot of either biological or social structure, composition and function overlooks the essence of that which constitutes a forest ecosystem: change. The thematic approach taken in this book is aimed at drawing together what we know about these forces, the forest components and processes they influence, and how these are linked to forest composition, function and change in the Guiana Shield region. Do forest-use planning horizons adequately account for both slow-moving, low-frequency and short-term, high-frequency forces and reflect the proportional influence attached to their level of resonance through the forest ecosystem? And can these system attributes be sensibly considered in practical management action in the field? Many approaches already work to this end. Yet important facets of forest use remain largely outside the scope of resource management, confounding long-term sustainability. The interaction of socio-economic and biophysical foundations vary across the neotropics. The applicability of practices across this variance towards a common goal, thus may require further thought (see Chapter 9).

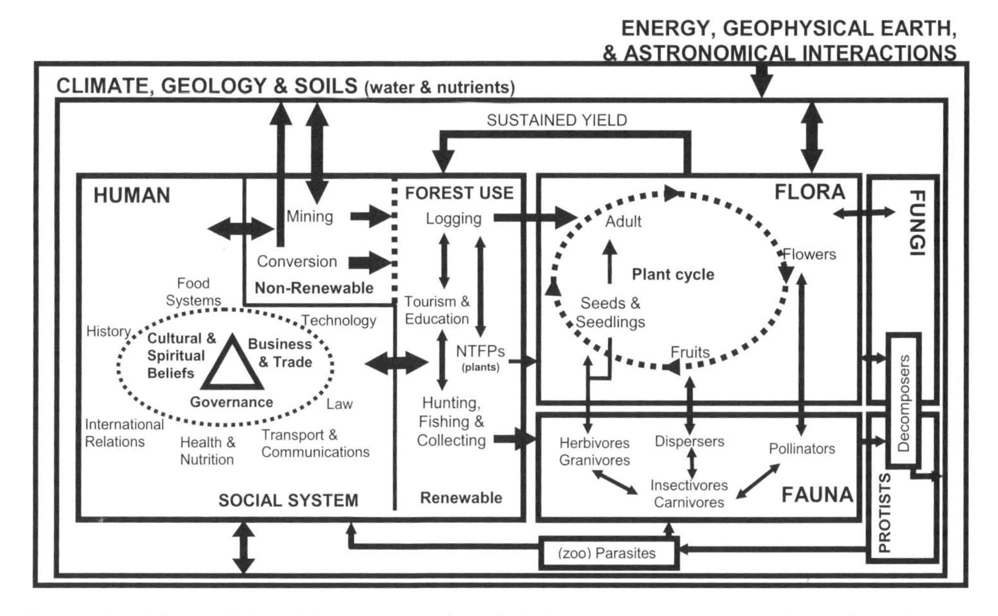

Fig. 1.2. Simplified depiction of the major social, biological and physical forces working to shape the tropical forest ecosystem. The size of the line increases with a putative increase in magnitude of control, magnitude defined as a combination of spatial extent and rapidity of change. Dual arrowheads represent direct feedback. See Chapters 2 and 7 for details of energy, geophysical and astronomical forces and geology, climate and soils; Chapters 7–9 for social forces; Chapters 3, 4, 7 and 9 for plant systems and Chapters 4–7 for animal systems.

Contents of this Book

Thus, the objective of this volume is to provide a wide base of information compiled in a single source so the broadest range of influences can be considered by the reader and assessed for their relevance across the Guiana Shield region (and perhaps, the larger neotropics). The book also aims to provide a glimpse at the inner workings of the Guiana Shield as a regional landscape, rather than only as a series of political entities or as a poorly studied part of the Amazon Basin. Both biophysical and social (Chapters 8, 9) features of the region are covered, but the emphasis on geological (Chapter 2), pedological (Chapter 2), climatic (Chapter 2), hydrological (Chapter 2), botanical (Chapters 3–5), ecological (Chapters 4, 5, 8, 9) and zoological (Chapters 5–7) research is apparent.

The book content is wide-ranging. More effort was expended on trying to broadly triangulate the complex of attributes that combine to shape forests of the shield region than diving to the deepest level of detail in each of the various fields covered. As a consequence, many key concepts, processes and attributes may appear inadequately characterized in the view of experts and for this the editor apologizes. It is my hope that the effort to bring a broad cross-disciplinary approach to theories and explanations that are often otherwise delimited and isolated by a specific scientific focus can compensate in some small measure for the lack of technical refinement. A solid reference base underpinned by an extensive bibliography should be used to explore in greater detail the many areas only briefly addressed in this single volume. Contributors have made a special effort to review a growing volume of research conducted within the region and organize this into a series of sectional topics that improve access to specific areas of interest. In many instances, very little research has been conducted and this condition is noted in each of the chapters explicitly or as a set of testable hypotheses. Elsewhere, comparisons with other tropical forest regions are constructed as a way of characterizing the different environments within the Guiana Shield by their socio-economic ranking (Chapters 8, 9) or position along a larger-scale biophysical gradient (Chapters 2 and 7).

Chapter 2 does much of the latter. It attempts to bring together information from many sources to define the spatial distribution and properties of rock, soils, rivers and lakes, rainfall and temperature and compare these with other neotropical regions. It also describes the main regional landforms and landscape macro-features. Biophysical features are continually changing and this chapter also briefly describes many, but certainly not all, of the main factors that govern the change trajectory. In many instances, the specific effects of these factors on the Guiana Shield are poorly understood, often due to sparse research and monitoring. In other cases, the importance and/or relevance of specific processes have ebbed and waned over the course of the region's history. Climate change mechanisms in particular are summarily described in an effort to highlight the linkages between terrestrial, marine and atmospheric systems and how these have shaped forest evolutionary development, distribution and composition across the Guiana Shield.

In Chapter 3, Thijs Pons and colleagues provide an interesting look at the ecophysiological function of forest plants in the Guiana Shield. They present a wide range of experimental results and address these within a broader context defined by other studies conducted within the region and elsewhere. The chapter addresses most of the main plant physiological pathways: photosynthesis, growth, transpiration, water and nutrient uptake and discusses changes in these in relation to canopy structure, openness and water availability, among others. The chapter also briefly explores the role of symbionts in catalysing and sustaining nutrient uptake pathways in a highly limited growth environment. The authors conclude that regional forest tree species can be described by wide-ranging combinations of different ecophysiological attributes.

Chapter 4 focuses on reviewing the large body of literature published on the terrestrial fauna known to reside in the region's forests and how these animals make use of the plant resources. The authors review what is known about dietary habits of the main forest fauna taxa, but with an emphasis on mammals. The chapter also discusses the cross-regulating effects that plants and animals can exert on each other. Dispersal and pollination are perhaps of greatest importance to this exchange and these important components of plant–animal relationships are discussed in relation to the type and diversity of animal food preferences, along with other activities such as seed and seedling predation, folivory and nectar/pollen-feeding. An accompanying checklist of known vertebrates (less fish) from the region represents a simple taxonomic snapshot of the diversity, but provides a good point source for comparison with other regions and across the shield itself.

Chapter 5 by Yves Basset *et al.* provides a broad review of the work done with above-ground, leaf-feeding insects in the region, including a historical account of insect collecting and curation in the three Guianas. The authors briefly address several important topics relevant to tropical forest folivory and folivore diversity, such as host-specificity, stand (mono)dominance and spatial resource distribution. The authors emphasize the relatively sparse entomological studies carried out in the region.

Chapter 6 addresses the specific role of insects as flower visitors. It examines the role of visitors as both pollinators and nectar-robbers, providing examples drawn from the very few studies carried out at locations within the Guiana Shield. De Dijn briefly examines trends in pollinator and plant diversity, but highlights the paucity of concrete information from the region. The chapter concludes with suggestions for future pollination research in the Guiana Shield.

Chapter 7 explores the ecological and evolutionary responses of forests to changes characterized in Chapter 2. Forest structure and composition are interlocked with individual recruitment, growth and survivorship along a reaction function. Variation in these attributes in many ways better describes a forest than traditional static measures of standing diversity and biomass because these quantities represent a response to past forces without first understanding how and for how long these resonate through a forest system. The chapter draws on a large volume of published research to explore regional differences in many of the common disturbances believed to drive forest change. It also compares several plant life-history attributes and geographical ranges believed to be important in distinguishing system contrasts. It concludes with a discussion of the relative role of geomorphic and geographical controls in shaping forest trajectories and an eclecticistic proposal for observed differences in forest composition and diversity.

Chapter 8 approaches the description of the regional forests from a (pre-)historical and socio-economic perspective. It summarizes the relatively small amount of archaeological and anthropological knowledge of pre-Columbian Amerindian societies and the dynamics of settlement, migration and forest use within the region. The chapter does not attempt to cover the diverse and unique cultural drivers of these societies (e.g. cosmology), although the role of ritual, religion, kinship and other cornerstones of social structure are recognized as important factors influencing both Amerindian and non-Amerindian forest use patterns and purposes. The chapter also provides a brief chronology of post-Columbian colonization and forest use within the region and the historical establishment and development of forestry, mining and agriculture. The chapter finishes with a brief description of the current social and economic condition of the different countries/provinces/states found within the Guiana Shield with an emphasis on the historical transition from industrial uses of NTFPs, timber and minerals.

Chapter 9 draws upon preceding chapters to explore forest conservation and management direction in the Guiana Shield since its earliest inception through to recent initiatives. This section attempts to highlight relationships between regional landscape conservation, patterns of biological diversity, economic use of forestland resources and global economic drivers, drawing upon the growing volume of studies published on these topics recently.

Ancient Land, Modern Name

The Guiana Shield draws its name from the fusion of two terms – one ancient, one modern. 'Guiana', 'Guayana' and 'Guyana' are believed to be linguistic variants on a traditional Amerindian word widely interpreted to mean 'water' or 'much water' (but see Williams, 1923; Cummings, 1963).[1] The word 'Shield' in this context refers to a modern scientific term used by geologists to describe a large region of very old, exposed Precambrian basement rock that is not affected to any large extent by modern volcanic or tectonic activity (Press and Siever, 1982; Gibbs and Barron, 1993; Goodwin, 1996). The Guiana Shield region, however, was not commonly referred to as a single geological province until the late 1950s, when techniques for isotopic dating of rock improved and evidence supporting plate tectonic theory began to accrue. The area is also described in regional languages of French, Spanish and Portuguese as the 'Bouclier Guyanais', 'Escudo de Guayana' and 'Escudo Guyanense', respectively (see Gibbs and Barron, 1993). The names Guiana and Guayana are interchangeable and applied equally in describing the specific shield region (e.g. Maguire, 1970; see http://www.guayanashield.org).

The region was defined as a discrete geopolitical entity in Western literature long before it was recognized as a geological province. This was at first based on the mythical view that the lands were part of a vast Amerindian empire, centred on the fabled city of Manoa and akin to those encountered by the Spanish in Peru and Mexico (Raleigh, 1596). Later, the geopolitics of European imperial ambitions divided South America into three broad areas, viz. Brazil (Portugal), New Grenada (Spain) and Guiana (England, France and The Netherlands). On this basis, Guiana was considered at the time by some to extend between the Amazon, Orinoco and Negro Rivers (Bancroft, 1769), though the locations of international boundaries within the region were not established until almost a century later (Schomburgk, 1840, 1848), and even then returning to the search for El Dorado as the root basis of colonial claims to the region (see Burnett, 2000).[2]

Guiana Shield, Tropical Forest Cover and the Global Distribution of Precambrian Landscapes

In part the significance of the Guiana Shield forests rests with the underlying geological landscape and the relative importance of the region to the global stock of tropical forests growing on surfaces dominated by the Precambrian. There are nine main areas of exposed Precambrian rock, belonging to ten geological provinces (after Osmonson *et al.*, 2000), that are subject to climatic conditions capable of sustaining tropical moist forests worldwide (after Olson *et al.*, 2001) (Fig. 1.3). Of these, three are located in South America, four in Africa and two in India. Indian and Madagascan forests are located at the northern and southern maximum of tropical climates, respectively, and would typically classify as sub-tropical, while the remaining areas consist of a wide variety of tropical forest assemblages, including both open and closed canopy formations. The total global area of Precambrian rock naturally covered by tropical forests is estimated at 5.08×10^6 km^2. The Guiana Shield is by far the largest contiguous area of exposed Precambrian rock naturally covered by tropical forests in the world, accounting for one-third of this total (Fig. 1.4). When further considering the levels of deforestation that have affected Precambrian areas in India, Madagascar, the Atlantic coast of Brazil and West Africa over the last 50 years, the relatively large area of forest remaining in the Guiana Shield accounts for more than half of the remaining tropical moist forests growing on

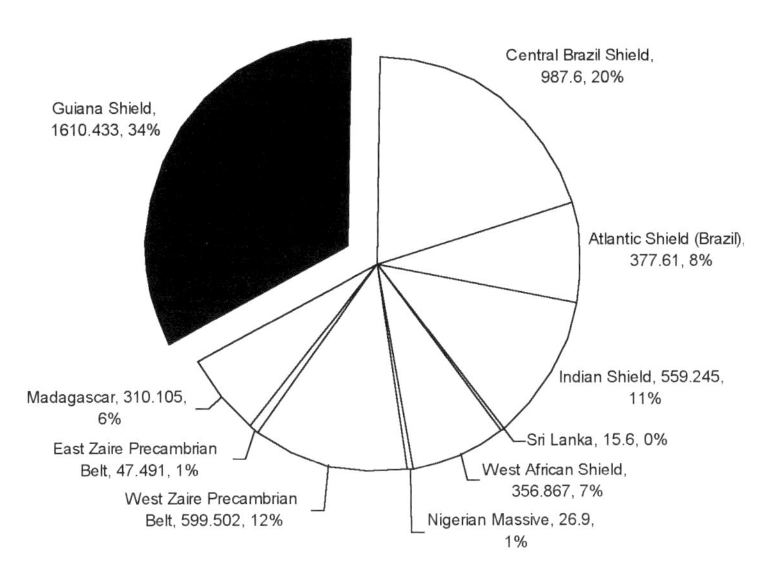

Fig. 1.3. The estimated global area (× 1000 km^2) of tropical moist forest naturally occurring on Precambrian landscapes grouped by geological province. Forest areas were estimated by spatially intersecting a modified subset of Precambrian geological provinces derived from Osmonson *et al.* (2000) and a subset of tropical moist forest ecoregions derived from Olson *et al.* (2001). See these references for details of positional accuracy and resolution.

Precambrian geological landscapes worldwide.

The significance of this statistic rests in the biophysical and socio-economic control that underlying geology exerts upon forest ecological and evolutionary processes, the way in which forestlands are used and managed, and the opportunities and limits that this region's forest resources can offer to local rural communities, national economies and the global storage of biological diversity, carbon and water. The voyage of the shield area atop its underlying crustal plate has combined with movement of the other continental land masses to place this Precambrian region in a unique geographical position along the meteorological equator, a position not held by any other area of similar geological antiquity (Chapter 2). Its current position along the western rim of the Atlantic basin also exposes the ancient geology of this region to a tropical climate that is responsive to a wide range of forcing factors, ranging from El Niño events through to periodic fluctuation in solar sunspot activity and planetary orientation (see Chapter 2). While many regions currently accommodating tropical forests formed part of the sea floor over most of our planet's geological history, the landscape of the Guiana Shield was actively evolving in response to climatic fluctuations (see 'Sea-level change', Chapter 2). This long history of exposure is now strongly reflected in the spatial distribution of soils, water and minerals (see Chapter 2), the ways in which forest plants and animals interact to survive and reproduce (see Chapters 3, 4, 5 and 7), and in our own (pre-)history of occupation (Chapter 8).

Location and Size of the Guiana Shield

Location

The Guiana Shield is a vast expanse of lowland forest, mountains, wetlands and savanna in northern South America, wedged between the llanos of northern Venezuela and eastern Colombia and the Amazon River in northern Brazil. The size, shape and location of the shield have been outlined since the early 1950s by numerous

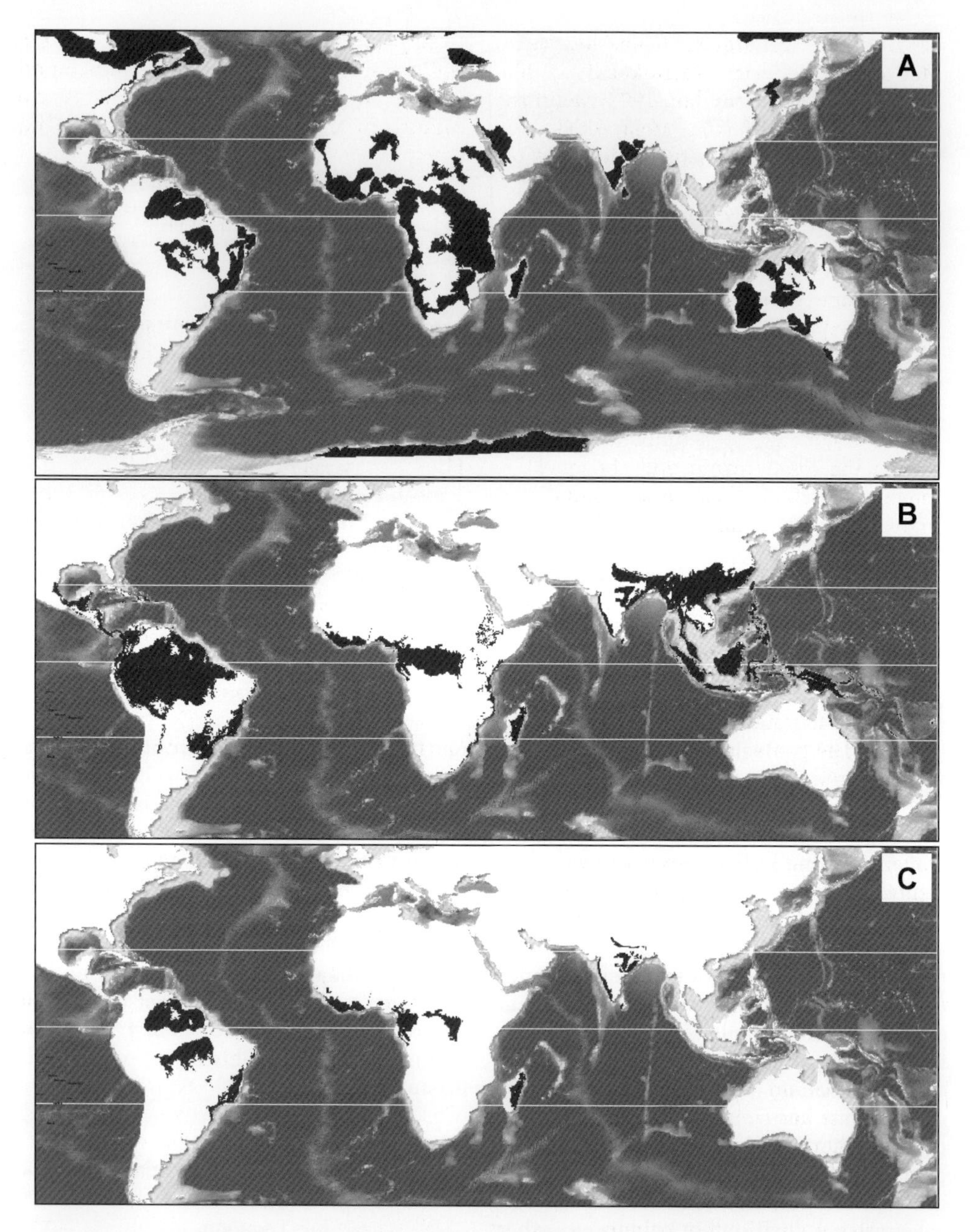

Fig. 1.4. (A) Distribution of the major areas of exposed Precambrian rock, (B) maximum extent of modern (sub)tropical forest cover (not including losses to deforestation) and (C) major tropical forest areas situated on Precambrian formations, all in black. Distribution of Precambrian rock based on USGS WPA-derived coverage of global geological provinces (Osmonson *et al.*, 2000) and cross-checked with Goodwin (1996) and references contained therein. Distribution of tropical forests based on WWF Ecoregion classification system coverage (Olson *et al.*, 2001) and cross-checked with FAO (2001), Collins *et al.* (1991), Sayer *et al.* (1992) and Harcourt and Sayer (1996). Note that the Mercator Projection stretches graticules at high latitudes. This super-sizes areas of Precambrian rock occurring above the 20th parallel in (A). Background image shows ocean floor topography with major tectonic joints and platforms as white lines and surfaces. Courtesy of ESRI.

geographers, geologists (Gansser, 1954; McConnell, 1961; Kalliokoski, 1965; Blancaneaux and Pouyllau, 1977; Mendoza, 1977; Gibbs and Barron, 1993; USGS and T.M. Corporacion Venezolana de Guayana, 1993; Goodwin, 1996) and more recently, conservation groups (e.g. http://www.guayanashield.org). All of the South American craton consists of Precambrian rock of one form or another, but not all remains exposed, principally due to downwarping of the crust along faults, geosynclines and other points of subsidence that criss-cross the continent. Many of these areas are covered by sediments deposited during the Phanerozoic and the question remains whether these areas should be included when defining the location and area of the shield.

The most detailed and extensive depiction of the Guiana Shield, as a geological entity, has been drawn up by Gibbs and Barron (1993). Considering only those areas largely occupied by Precambrian rock formations in the map presented in Gibbs and Barron (1993), the shield area consists of northwestern and southeastern lobes, separated by two central sedimentary depressions extending from Atlantic and Amazonian sediment margins towards the centre of the region (Fig. 1.5A). Including Atlantic and central areas of Quaternary sediment as depicted by Gibbs and Barron extends the area considerably (Fig. 1.5B). Adding the large area along the south-central margin of the shield located between the Japurá and Solimões Rivers extends this to the maximum area delimited by Gibbs and Barron and others (Fig. 1.5C). In this case, the northern and southern boundaries can be defined by the maximum width of the Orinoco and Amazon River floodplains, respectively, while the Atlantic coast of South America and the central Chiribiquete Plateau in southeastern Colombia could be considered as the eastern and western boundaries. On this basis the shield ranges more than 14 degrees in latitude between its most northerly point at the Orinoco Delta (10°N, 62°W) and southerly point approximately 75 km north of Coari, Brazil (4°S, 63°W), on the south bank of the Solimões. The most easterly point is found in the coastal swamplands north of the mouth of the Amazon near Sucuriju, Amapá, Brazil (2°N, 50°W). The shield reaches its most westerly point at Colima Cumare (0°40′N, 74°W), north of the Caquetá River in Colombia.

Other depictions have generally agreed with Gibbs and Barron's demarcation, though the inclusion/exclusion of the large Atlantic and Amazonian margins of shield-derived sediments repeatedly acts as the main filter in defining the areal extent of the shield. Williams and others depicted the shield as an area that extended from the Atlantic seaboard westward to the Rio Negro between the Orinoco in the north and band of Palaeozoic sediment running immediately north and in parallel with the Amazon River. As part of a World Petroleum Assessment (Schenk *et al.*, 1999), the US Geological Survey delimited the Guiana Shield area as a geological province that excludes Quaternary sedimentary lowland areas around the confluence of the Negro and Branco Rivers and in the Berbice Basin (Fig. 1.5D). This depiction of the shield area has a western margin along the Brazil–Colombia border. The area has been similarly delimited in Goodwin (1996), Berrangé (1977), Putzer (1984) and Räsänen (1993). Pouyllau (1976, cited in Blancaneaux and Pouyllau, 1977) similarly excluded areas covered by deep sediments along the Atlantic and Amazon margins, but also placed the western margin fully east of the Rio Negro. Dosso (1990, cited in Chareyre, 1994, Fig. 2) depicted the Guiana Shield as limited to an area even further east and south of the Orinoco and Rio Negro Rivers, but including sedimentary margins along the Atlantic, North Pará and Amapá. In the process of defining regional conservation priorities at a 2002 meeting in Suriname, participants identified a shield area very similar in size and shape to Gibbs and Barron's depiction including Phanerozoic sedimentary cover (http://www.guayanashield.org) (Fig. 1.5E).

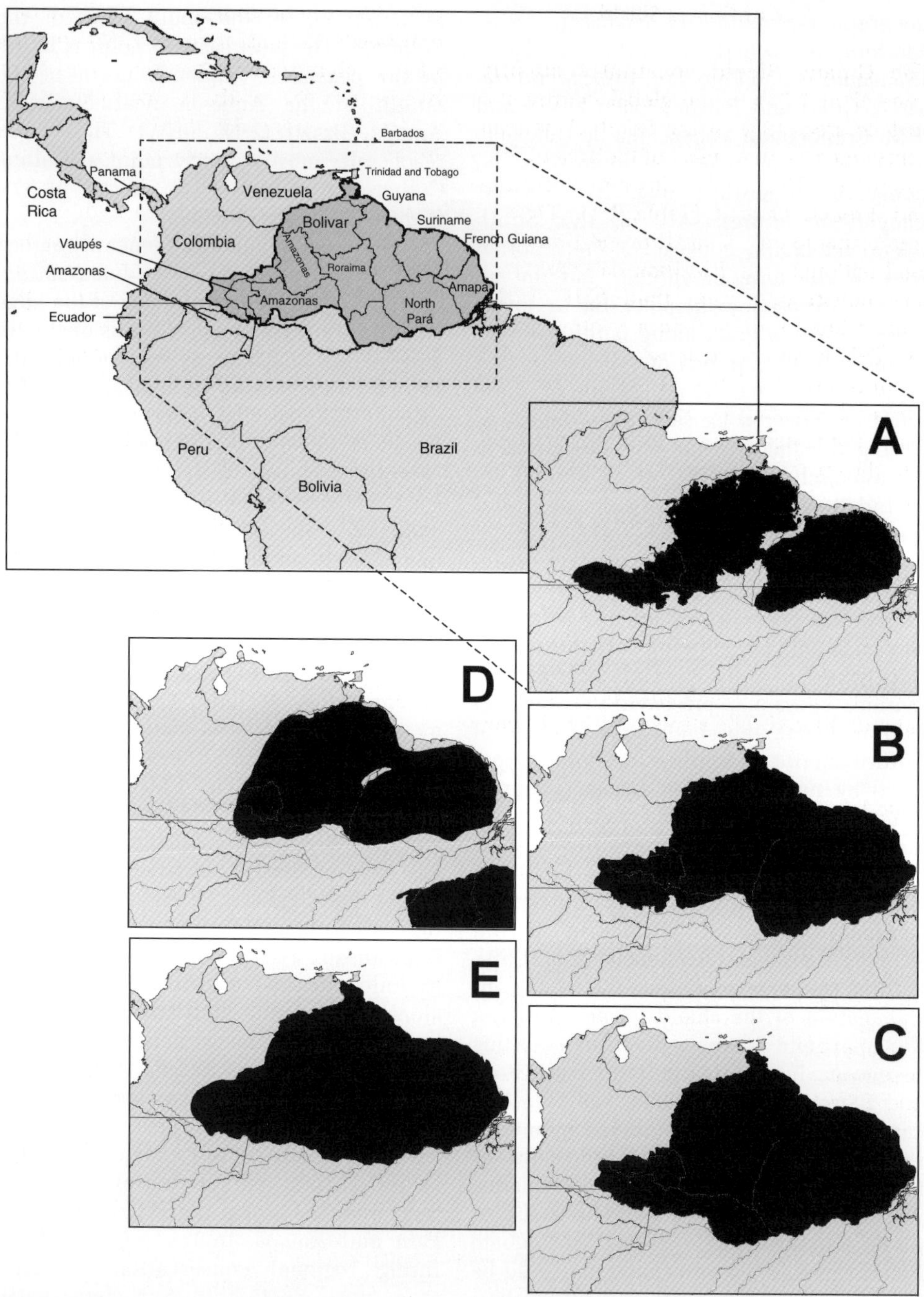

Fig. 1.5. Location and areal extent of the Guiana Shield showing several possible configurations defined as: (A) Precambrian zones+deep Phanerozoic sedimentary cover principally of shield origin (adapted from Gibbs and Barron, 1993); (B) same as (A) but excluding deep Putomayo–Solimoes sedimentary plains; (C) Precambrian zones only, based on rock types dated >550 Ma as assigned in Gibbs and Barron (1993); (E) Precambrian+shallower parts of central sedimentary basins (Osmonson *et al.*, 2000; Berrangé, 1977). The Guiana Shield region as defined in (D) is based on the results of a conservation priority-setting workshop for the region held in 2002 (http://www.guayanashield.org).

Size of Guiana Shield

The Guiana Shield constitutes slightly more than 1.5% of the global continental surface area, 13% of the South American continent and all or most of the area of four countries: Venezuela, Guyana, Suriname and French Guiana (Table 1.1). Though amounting to only a small proportion of its total national area, Brazilian national territory constitutes more than half of the Guiana Shield, more than any other country. This large area falls within four states in northern Brazil – Amapá, Roraima, Pará and Amazonas – and accounts for virtually the entire area of the shield draining into the greater Amazon Basin (see Chapter 2).

The balance of each nation's interest in the Guiana Shield depends on many factors, but the relationship between the proportion of a country's area in the shield and its relative contribution to the shield (Table 1.1) suggests that Brazil's relationship with the region is at the same time both substantial and diluted. The structuring of policies and allocation of resources in the Brazilian case needs to consider a national interest that is significantly broader and more diverse than the relatively small part of the country that is found within the shield area. But at the same time, Brazil occupies more than half of the shield area, and these same broad policies can affect the region as a whole more than those implemented by any other country. France, too, has a relationship with the region that considers national domestic needs extending outside the Guiana Shield. Other countries whose land area is entirely or predominantly located within the shield area have an interest that is more attenuated to the prevailing shield environment, but have a relatively modest influence on how this region as a whole is conserved and developed. Despite this asymmetry, considerable spatial variation in forest composition, diversity and accessibility across the shield area suggests that each country has an important role to play in maintaining regional forest cover, productivity and conservation quality.

The important interplay of socio-

Table 1.1. Estimated area of the Guiana Shield (GS) and its breakdown by constituent country.

	Land area (km^2)			
Country	In GS[a]	Entire country[b]	% country area in GS	% contribution to GS
Brazil	1,204,279	8,456,510	14.2	52.6
Colombia	170,500	1,038,710	16.4	7.5
French Guiana	88,150	88,150[c]	100.0	3.9
Guyana	214,980	214,980	100.0	9.4
Suriname	156,000	156,000	100.0	6.8
Venezuela	453,950	882,060	51.5	19.8
Guiana Shield	2,287,859			100.0

[a]Based on (1) Brazil – area figures for munícipios in Amapá, Roraima, Pará and Amazonas falling within the adopted Guiana Shield boundary provided by IBGE (2000). The shield area of administrative units that extend outside the perimeter was calculated from available statistics through geoprocessing in ArcView 3.2. (2) Colombia – area figures based on values provided through anonymous ftp by DANE (2000) and cross-corrected by geoprocessing of Japurá, Negro and Orinoco watershed units with shield and international boundary coverages using ArcView 3.2. (3) Venezuela – based on area figures for municipios in Delta Amacuro, Bolivar and Amazonas states published by OCEI (2000).
[b]National land area figures are consistent with those published in Harcourt and Sayer (1996), FAO (2001), but figures for French Guiana and Suriname are less than those published by FAO (1993) and commonly cited in various encyclopaedias, almanacs and other statistical compilations that most likely adopted figures from previous citations. The sum of watershed area calculations for French Guiana and Suriname generally support figures stated here (Amatali, 1993; Hiez, 1964).
[c]Only Départment Otre Mer (DOM) du Guyane – not including metropolitan France.

economic and biophysical factors indicates that the size and shape of the shield used for the purpose of describing regional features may vary depending on the factor under consideration. While the definition of the shield as a geological province undoubtedly underpins all others, the term 'Guiana Shield' is increasingly used as a means of demarcating and defining a region of interest for many non-geological attributes (e.g. Bryant *et al.*, 1997; Haden, 1999). Expanding the definition of the Guiana Shield as an area of biogeographical (e.g. floristic province; Maguire, 1966; Mori, 1991) and conservation significance (e.g. http://www.guayanashield.org) influences the shape and area of interest, one that is not necessarily identical to a shield defined in the strictest sense by the presence of underlying exposed Precambrian rock (Fig. 1.5A,D vs. 1.5C,E). In this context, defining an area that includes the main sedimentary depressions traversing the area would be more appropriate since the upland and lowland landscapes are strongly interlinked.

Impinging uncertainties

The location, size and shape of the Guiana Shield presented here is based on a series of geological and topographical thresholds. While the shield is considered to be a distinct geological entity, the fine-scale resolution of its boundaries is little more than a good approximation for a number of reasons. First, it forms one part of a much larger biogeochemical region and the processes that have shaped the shield environment are shared, to varying extent, with other areas of South America and other continents (especially West Africa). The area of mineralogical affinity that defines the modern-day shield area expands as one moves further back in time through the Precambrian. Secondly, the area remains remote and many parts have been poorly surveyed. Our knowledge of more specific small-scale landscape features of the region have in the past been largely interpolated at larger scales, though improvements in remotely sensed data acquisition are improving this knowledge base. This coverage also varies by country due to differences in national technical capability. Suriname and Guyana are particularly constrained in this regard in comparison with their French, Brazilian and Venezuelan neighbours.

Whether the vast areas of Phanerozoic cover along the Atlantic coast and in northern Pará, Amazonas and Roraima states should be included or excluded from the Guiana Shield has largely favoured their exclusion from a strict geological perspective. However, many included areas also have significant sedimentary cover formed under the same conditions and through the same processes as those excluded (Gibbs and Barron, 1993). Much of the cover is thought to have originated from the erosion of older sedimentary structures (e.g. the Roraima Group) and basement rock with significant redeposition and remixing through oscillating periods of marine flooding, retreat and evaporation. Precambrian rock rests below all of these sedimentary basins and its presence/absence does not qualify as a selective factor alone. The threshold depth and mineralogical provenance of Phanerozoic cover that should be included as part of the Guiana Shield is, however, a major criterion playing on the size and shape of the shield and has yet to be objectively established.

There are also complications with generalizing the extent of the shield area when considering the eastern and western limits. The eastern edge can vary depending on whether current or past sea level is considered as a limit to the eastern boundary. The western edge of the shield is even more difficult to define unambiguously, since significant geological and floristic elements typical of central shield locations can extend as far west as southeastern Colombia north of the Solimões River (Gibbs and Barron, 1993; Duivenvoorden and Lips, 1995; Cortès and Franco, 1997; Cortès *et al.*, 1998). Information from these areas suggests that flora, fauna, soils and hydrological attributes are equally or more typical of central and western Amazonian forestlands (Duivenvoorden and Lips, 1995) as biogeographical and geological influences of the Andes to the west and north intermingled with the much older shield contributions centred to the east and south. From an anthropological perspective, the Amerindian languages spoken in southern Colombia are also considered part of the Tukano/Ticuna linguistic groups rather than the shield-centred Carib/Arawakan/Yanomami groups (Chapter 8). Southern parts of the geological shield are largely inhabited by Tupi/Guarani-speaking peoples traditionally found living south of the Amazon River but not in any other part of the Guiana Shield. These and other attributes that together help in creating a distinct picture of the Guiana Shield region, and through this a better sense of its size as it relates to many different attributes, are described in much greater detail later.

Notes

[1] The etymological origins of 'Guiana' are debatable, though the modern translation is widely accepted to mean 'water' or 'many waters'. Schomburgk (1840) suggested that the name is derived from a small tributary of the Orinoco, while Codazzi (1841) argued that the name originated with an Amerindian tribe, the Guayanos, and that the term for pale or white, 'uayana', was used in describing the first Europeans they encountered. Certainly a vast area extending inland from the Atlantic was referred to as Guiana since the earliest written accounts of exploration by Raleigh (1596), Keymis (1596) and Harcourt (1613), though they used this name without qualifying its origins. A succinct account of some of the possibilities and variations is also given by Berry *et al.* (1995).

[2] Segments of the international boundaries within the shield are still disputed more than 150 years later (Braveboy-Wagner, 1984; Burnett, 2000).

References

Amatali, M. (1993) Climate and surface water hydrology. In: Ouboter, P.E. (ed.) *The Freshwater Ecosystems of Suriname*. Kluwer Academic, Dordrecht, The Netherlands, pp. 29–52.

Bancroft, E. (1769) *An Essay on the Natural History of Guiana in South America*. T. Becket and P.A. De Hondt, London.

Berrangé, J.P. (1977) *The Geology of Southern Guyana, South America*. Overseas Memoir 4. HMSO, London.

Berry, P.E., Holst, B.K. and Yatskievych, K. (1995) *Flora of the Venezuelan Guayana. Volume 1: Introduction*. Timber Press, Portland, Oregon.

Blancaneaux, P. and Pouyllau, M. (1977) Les relations géomorpho-pédologiques et la retombée nord-occidentale du massif guyanais (Vénézuela). *Cahiers ORSTOM, Série Pédologique* 15, 437–448.

Braveboy-Wagner, J.A. (1984) *The Venezuela–Guyana Border Dispute: Britain's Colonial Legacy in Latin America*. Westview Press, Boulder, Colorado.

Bryant, D., Nielsen, D. and Tangley, L. (1997) *The Last Frontier Forests – Ecosystems and Economies on the Edge*. World Resources Institute, Washington, DC.

Burnett, D.G. (2000) *Masters of all they Surveyed – Exploration, Geography, and a British El Dorado*. University of Chicago Press, Chicago, Illinois.

Chareyre, P. (1994) *Régénération Naturelle en Guyane Française. Distribution Spatiale de Quelques Espèces Ligneuses dans une Forêt Secondaire de 18 Ans*. UFR de Sciences, Université Paris XII Val de Marne, Paris.

Codazzi, A. (1841) *Resumen de la Geografia de Venezuela*. H. Fournier, Paris.

Collins, N.M., Sayer, J.A. and Whitmore, T.C. (1991) *The Conservation Atlas of Tropical Forests: Asia*. Macmillan, London.

Cortès, R. and Franco, P. (1997) Análisis panbiogeográfico de la flora de Chiribiquete, Colombia. *Caldasia* 19, 465–478.

Cortès, R., Franco, P. and Rangel, O. (1998) La flora Vascular de la Sierra de Chiribiquete, Colombia. *Caldasia* 20, 103–141.

Cummings, L.P. (1963) The name Guiana: its origin and meaning. *Journal of the British Guiana Museum and Zoo* 38, 51–53.

DANE (2000) Informacion geoestadistico. http://www.dane.gov.co
Duivenvoorden, J.F. and Lips, J.M. (1995) *A Land–Ecological Study of Soils, Vegetation and Plant Diversity in Colombian Amazonia.* Tropenbos Foundation, Wageningen, The Netherlands.
FAO (1993) *Forest Resources Assessment 1990: Tropical Countries.* FAO Forestry Paper 112. FAO, Rome.
FAO (2001) *Global Forest Resources Assessment 2000.* FAO Forestry Paper 140. FAO, Rome.
Gansser, A. (1954) Observations on the Guiana Shield (South America). *Ecologae Geologicae Helvetiae* 47, 77–112.
Gibbs, A.K. and Barron, C.N. (1993) *The Geology of the Guiana Shield.* Oxford University Press, Oxford, UK.
Goodwin, A. (1996) *Principles of Precambrian Geology.* Academic Press, New York.
Haden, P. (1999) *Forestry Issues in the Guiana Shield Region: A Perspective on Guyana and Suriname.* EU Tropical Forestry Paper 3. Overseas Development Institute, London.
Harcourt, C.S. and Sayer, J.A. (1996) *The Conservation Atlas of Tropical Forests: the Americas.* Simon and Schuster, London.
Harcourt, R. (1613) *A Relation of a Voyage to Guiana.* John Beale, London.
Hiez, G. and Dubreuil, P. (1964) *Les Régimes Hydrologiques en Guyane Francaise.* ORSTOM, Paris.
IBGE (2000) Cidades@. http://www.ibge.gov.br
Kalliokoski, J. (1965) Geology of north-central Guyana Shield, Venezuela. *Geological Society of America Bulletin* 76, 1027–1050.
Keymis, L. (1596) *A Relation of the Second Voyage to Guiana.* Thomas Dawson, London.
Maguire, B. (1966) Contributions to the botany of the Guianas. *Memoirs of the New York Botanical Gardens* 15, 50–128.
Maguire, B. (1970) On the flora of the Guayana Highland. *Biotropica* 2, 85–100.
McConnell, R.B. (1961) The Precambrian rocks of British Guiana. *Timehri* 40, 77–91.
Mendoza, V. (1977) Evolucion tectonica del Escudo de Guayana. *Boletín de Geologia (Publicacion Especial)* 7, 2237–2270.
Mori, S.A. (1991) The Guayana lowland floristic province. *Comptes Rendu de la Société de Biogéographie* 67, 67–75.
OCEI (INE) (2000) Aspectos fisicos. http://www.ocei.gov.ve
Olson, D.M., Dinerstein, E., Wikramanayake, E.D., Burgess, N.D., Powell, G.V.N., Underwood, E.C., D'Amico, J.A., Itoua, I., Strand, H.E., Morrison, J.C., Loucks, C.J., Allnutt, T.F., Ricketts, T.H., Kura, Y., Lamoreux, J.F., Wettengel, W.W., Hedao, P. and Kassem, K.R. (2001) Terrestrial ecoregions of the world: a new map of life on Earth. *Bioscience* 51, 933–938.
Osmonson, L.M., Persits, F.M., Steinhouer, D.W. and Klett, T.R. (2000) *Geologic Provinces of the World – wrld_prvg.* USGS, Denver, Colorado.
Press, F. and Siever, R. (1982) *Earth.* W.H. Freeman and Co., San Francisco, California.
Putzer, H. (1984) The geological evolution of the Amazon basin and its mineral resources. In: Sioli, H. (ed.) *The Amazon: Limnology and Landscape Ecology of a Mighty Tropical River and its Basin.* Dr W. Junk, Dordrecht, The Netherlands, pp. 15–46.
Raleigh, W. (1596) *The Discoverie of the large, rich, and bewtiful empyre of Guiana, with a relation of the great and golden citie of Manoa (which the Spanyards call El Dorado) and of the provinces of Emeria, Arromaia, Amapaia, and othe countries, with their rivers, adjoyning.* Robert Robinson, London.
Räsänen, M. (1993) La geohistoria y geologia de la Amazonia Peruana. In: Danjoy, W. (ed.) *Amazonia Peruana. Vegetacion Humeda Tropical en el Llano Subandino.* PAUT/ONERN, Jyväskylä, Finland, pp. 43–65.
Romero, E.J. (1993) South American paleofloras. In: Goldblatt, P. (ed.) *Biological Relationships Between Africa and South America.* Yale University Press, New Haven, Connecticut, pp. 62–85.
Sayer, J.A., Harcourt, C.S. and Collins, N.M. (1992) *The Conservation Atlas of Tropical Forests: Africa.* Macmillan, London.
Schenk, C.J., Viger, R.J. and Anderson, C.P. (1999) *Maps Showing Geology, Oil and Gas Fields and Geologic Provinces of South America.* USGS Open File Report. 97-470D. USGS, Denver, Colorado.
Schomburgk, R.H. (1840) *A Description of British Guiana.* Simpkin, Marshall and Co., London.
Schomburgk, R.H. (1848) *Reisen in British Guiana den Jahren 1840–1844.* J.J. Weber, Leipzig.
USGS and T.M. Corporacion Venezolana de Guayana (1993) *Geology and Mineral Resource Assessment of the Venezuelan Guayana Shield.* USGS Bulletin 2062. US GPO, Washington, DC.
Williams, J. (1923) The name Guiana. *Société des Américanistes de Paris – New Series* 15, 19–34.

2 Biophysical Features of the Guiana Shield

David S. Hammond
Iwokrama International Centre for Rain Forest Conservation and Development, Georgetown, Guyana. Currently: NWFS Consulting, Beaverton, Oregon, USA

Geology – a Precambrian Conundrum

The physical appearance of the Guiana Shield is as much a reflection of early global tectonic evolution as it is the consequence of localized weathering and deformation. The mixing of many different geological events and processes at different spatial scales over a considerable period of time has created a modern landscape that is both diverse and patchy. From ancient rifting and volcanism through to the erosional retreat of the massive sediment forming the Guayana Highlands and its subsequent redeposition during interglacial sea transgressions, the Guiana Shield embraces a vast geological history that has substantially influenced both the type and extent of modern plant and animal life, as well as the pace and scale of human development. The Guiana Shield owes its modern physical landscape to the same set of geological processes that have affected other tropical forestlands. The relative influence, however, of these different processes in shaping global tropical forestlands, both past and present, emphasizes the important role that rock formation and deformation processes can play in constraining the trajectory of forest evolution. This chapter attempts to describe the salient geological features of the Guiana Shield and illustrate some of the similarities and contrasts between this region and other parts of the neotropics and the more extensive, global Precambrian. Taken together, a picture emerges of the Guiana Shield as a canvas, sewn together from different materials on which a unique collage of forest consociations has developed and evolved.

The South American craton

The Guiana Shield can be viewed as a cluster of three bulges separated by a series of depressions in the centre of the South American plate. This plate extends in the east many hundreds of kilometres beneath the waters of the equatorial Atlantic to the mid-Atlantic ridge, a long snake-like zone of sea-floor spreading that has separated and gradually distanced the South American from the African craton since the mid-Jurassic. To the west, the South American plate abruptly ends off the western Pacific coast along the 'trailing' boundary of the plate where the giant Pacific plate rock moves beneath, or subducts the South American plate. The continental crust riding atop the South American plate can be variously divided according to the age of the dominant superficial rock formations. The mainly metamorphosed rock of the South American continental crust, or craton, generally consists of (after Goodwin, 1996):

1. Three major areas of exposed Precambrian rock, the Guiana, Central

 Tropical Forests of the Guiana Shield (ed. D.S. Hammond)

Brazilian (aka Guaporé; Gibbs and Barron, 1993) and Atlantic (aka Uruguay) Shield (4.6–0.57 Ga BP);

2. The surrounding regions that are predominantly covered with often deep, reworked and consolidated Phanerozoic sediment (0.57–0.01 Ga BP); and

3. Mountain-building areas, such as the Andes, that continue to develop along the edges of the South American and neighbouring plates in response to subduction of the lithosphere (0.245 Ga–present).

Precambrian rock can also be found underlying many of the sedimentary and mountainous regions, but it is in the shield areas where these represent the main surface formations.

Areas of exposed Precambrian crust are generally sub-classified according to the age of the formations, ranging from the oldest rocks on the planet, formed around 4.6 billion years ago, to those developing in the Late Proterozoic, around 570 million years before present (BP). Precambrian rock is not difficult to find. Given that the geological history of the planet is largely Precambrian, it is not surprising that approximately 72% of the global continental crust is thought to have been formed during this era (Poldevaart, 1955). Thus at the core of each modern continent is a large, Precambrian craton that has been relatively stable, in geological terms, for at least 500 million years.

Goodwin (1996) estimates that there are 30.3 million square kilometres of exposed Precambrian crust variously distributed over approximately 80 shields, belts, blocks, uplifts and sub-cratons on all eight continents (Fig. 2.1, also Fig. 1.3, Chapter 1). Africa, South America and North America alone account for nearly three-quarters of all exposed Precambrian rock, with the vast Canadian Shield representing the largest and oldest contiguous area of this type in the world. The South American shield areas, in contrast, consist mainly of younger Precambrian formations of the middle to late Proterozoic (0.6–1.7 Ga). Of all the continents, the South American craton is the 'youngest'. It has the least amount of rock formed during the Archean eon (>2.5 Ga), representing 16% of the exposed Precambrian formations and only 5% of the total craton (exposed + buried Precambrian rock) (Goodwin, 1996) (Fig. 2.1).

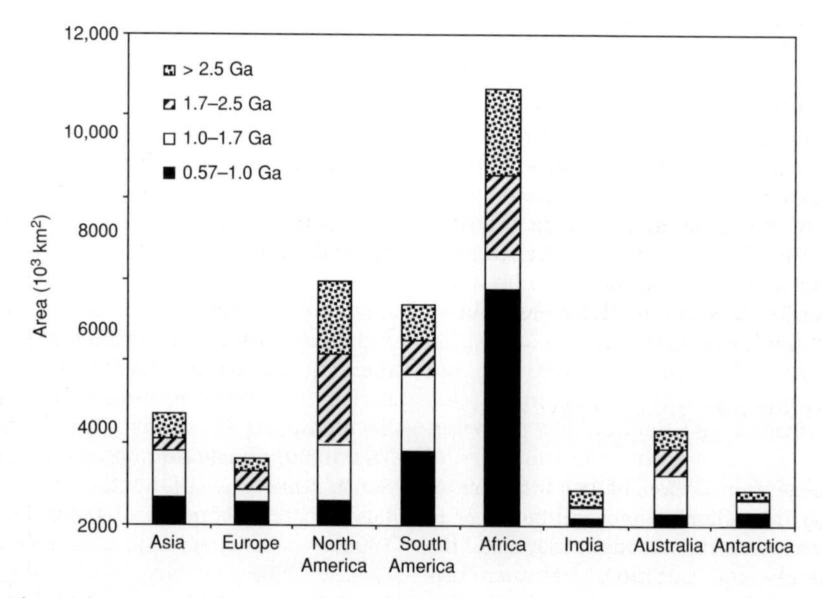

Fig. 2.1. The areal extent and relative age of Precambrian formations on the eight continents. Data from Goodwin (1996).

The Guiana Shield

The Guiana Shield accounts for more than 50% of the exposed Precambrian surface area in South America (Goodwin, 1996) (Fig. 1.2, Chapter 1), nearly 98% of this area consisting of rock formed during the Proterozoic (2.4 to 0.57 Ga BP). While most of the major rock formations in the shield were being formed and modified during this period, relatively few changes were taking place to the older rock making up other well known Precambrian areas of the world (Gibbs and Barron, 1993). This makes the Guiana Shield the largest area of exposed rock in the world that was formed during the Proterozoic and of particular interest to Precambrian geologists. The age and extent of the main basement complex exposed in the shield should also be, however, of concern to those focused on the biology, conservation and use of tropical forests. The geological foundation has been changed over the last 570 million years through uplifting, subsidence, faulting, sea-level change, erosion and substantial weathering. Interacting with climate, these evolving geological features have strongly influenced the modern-day distribution of plants, animals and minerals of the Guiana Shield through their effects on hydrology, weather, soils and topography.

To understand the relative uniqueness of this geological legacy and the extent of its impact on tropical forests, we need to further resolve the distribution of Precambrian formations across the major tropical forest regions of the planet and more specifically dissect the development of the Guiana Shield through time and examine how the resulting features are spatially distributed across the shield area today.

Comparison with geology of other tropical forest regions

The global distribution of tropical forests generally is believed to have shifted continuously in response to changes in global and regional climate, particularly temperature and rainfall. Processes that form the underlying substrate of these regions have played an equally important role in determining the composition, structure, extent and persistence of tropical forest cover. The geological processes and events that shaped the large-scale landscapes of modern-day tropical forestlands have arguably been the primary factor responsible for calibrating the long-standing trajectory of forest evolution in these regions (see Chapter 7).

The relative importance of different geological processes, such as faulting, uplifting, volcanism, sedimentation and erosion, has not remained constant across all regions. As a result, the Guiana Shield has stronger geological affinities with some tropical forest regions compared to others. This is thought to have led, in some instances, to convergence in the observed attributes of modern-day forest plant species and forest stands as they respond to edaphic conditions derived from similar geological foundations (e.g. low pH and high aluminium and iron mobility). Tectonics, the partnering progenitor of change, along with climate, is a good place to start in drawing out these differences.

PLUTONIC VS. VOLCANIC A good primary dichotomy distinguishing the geology of tropical forestlands is based on the relative importance of sub-surface, plutonic and above-surface, volcanic rock-forming processes. The dominance of one over the other is largely related to the distance of a forestland area from the perimeter of its parenting tectonic plate and how rock has been principally formed over time. The tropical forestlands of South America, west-central Africa, India and Australia are currently all centrally located on their respective plates and principally reflect ancient diastrophic deformation of a mainly plutonic Precambrian crust (Goodwin, 1996), while those of Central America, the Caribbean, the Pacific Islands and virtually all of South-east Asia are located on plate edges/hotspots and are formed primarily through more recent, and more dynamic, volcanic and sedimentary processes. Inevitably, some overlap exists across this dichotomy. Many of the prominent geological formations in the Guiana Shield owe their uniqueness to ancient flow of volcanic material (e.g. of the Uatuma Supergroup) (Gibbs and Barron, 1993), while Precambrian crust underlies much of southern New Guinea (Drexel *et al.*, 1993). The 2000 km long Grenville Belt, considered part of the Canadian Shield, extends along a narrow strip as far south as the Pacific *selva baja* in Mexico's Oaxaca state (Hoffmann, 1989). The vast majority of tropical forestland in South-east Asia, the Pacific Islands, Central America and the Caribbean rests, however, upon a landscape shaped principally by *modern* extrusive volcanic and sedimentary deposition of fine-grained rocks and their subsequent gradation. Those in eastern South America, west-central Africa, southern India and northern Australia share a common foundation of coarse-grained plutonic rocks metamorphosed, often repeatedly, then exposed and reburied, more passively, through selective gradation, uplifting and subsidence.

Volcanic landscapes are typically renewed and reshaped more frequently than shield areas and have a shorter, and often more tumultuous, cycle of gradation. Plutonic landscapes are more rigidly structured, comparably quiescent, are subject to a much longer cycle of gradation, and exposed in modern times to change shaped almost exclusively by fluvio-deltaic and marine sedimentation. The consequences of these large-scale differences cut down to the smallest scales. For example, they have substantial impacts on the rate, type and spatial scale of gradation, the chemistry of soil formation (e.g. feldspar decay products) and their subsequent influence on plant–water–nutrient relationships. Further aspects of these differences as they are linked to soil–plant relationships within the Guiana Shield are covered in more detail later (see 'Main Soils of the Guiana Shield').

EXTENT OF PRECAMBRIAN COVERAGE A second dichotomy useful in differentiating the geological condition of the Guiana Shield from other tropical forest regions is based on the relative contribution of Precambrian crust to the surface area of each craton. Formations in Africa and South America account for a much greater fraction of the Precambrian surface – between 2.5 and 10 times greater – than in either Australia or India (Fig. 2.2). The latter India craton, while formed around a series of Precambrian formations, is largely covered in the north by recent fluvial sediment flowing from the Himalayas through the Ganges and Indus Rivers and the marine-derived Deccan Traps (Naqvi and Rogers, 1987). Southern areas are edged by marine deposits. The major geological formation of Australia is due to folding of Palaeozoic marine and platform sediments deposited following subsidence in the east-central region. The oldest and most exposed Precambrian formations dominate Western Australia's dry tropical forestlands (Drexel *et al.*, 1993).

EXTENT OF EXPOSURE Africa is closer to South America in extent, age and composition of its Precambrian than any other tropical continent. This is hardly surprising given their fusion as part of Gondwana prior to rifting in the late Jurassic. However, the exposed formations account for a much higher percentage of the total Precambrian in Africa than in South America. In fact, the figure is more than twice that of South America (Goodwin, 1996), translating into an additional area of over 10 million square kilometres. Most of this supports dry forest, scrub and desert, with a much smaller fraction covered by tropical moist forests in comparison to South America, where the Guiana Shield accounts for the largest amount (see Chapter 1). The difference in amount of exposed Precambrian basement provides another useful means of contrasting the geology of the Guiana Shield from other tropical regions. The implications of this to tropical forest dynamics at the cratonic level are discussed further in Chapter 7.

PHANEROZOIC COVER Across the South American craton, only the Guiana, Central Brazilian and Atlantic Shields exhibit sig-

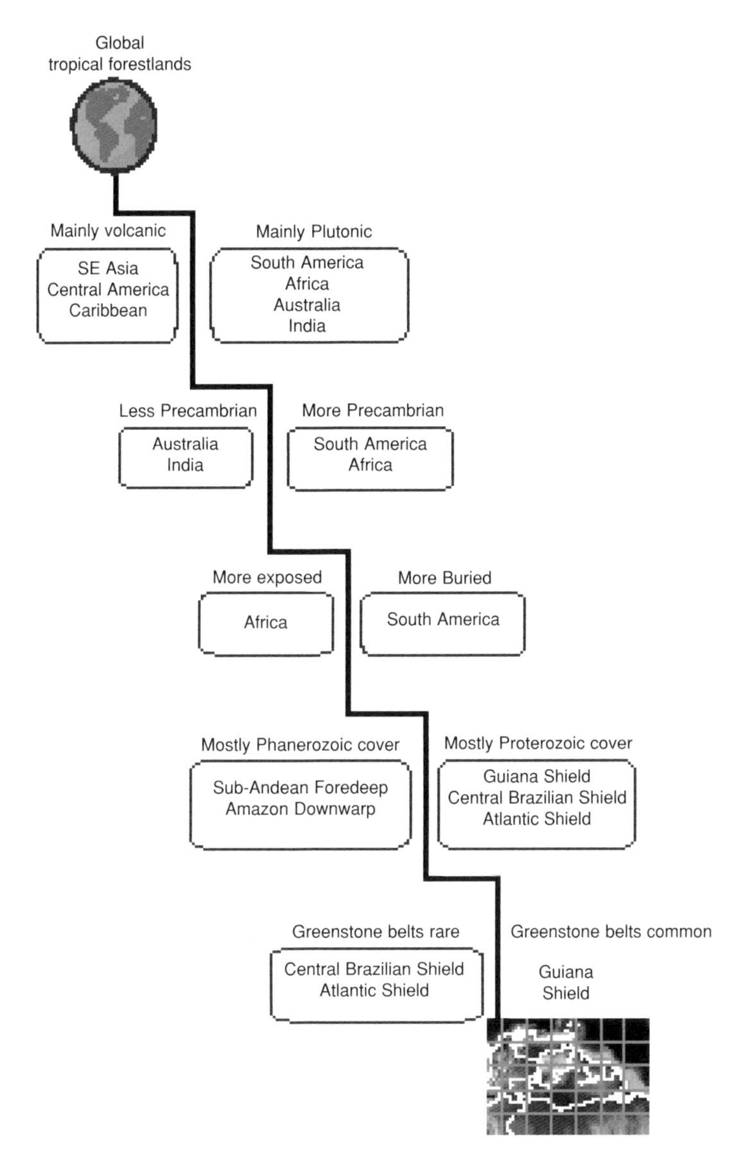

Fig. 2.2. Distinguishing the Guiana Shield from other tropical geological regions in South America and the world. Based on information in Goodwin (1996) and Gibbs and Barron (1993).

nificant areas of exposed Precambrian crust. The remaining cratonic area is covered almost entirely by sediments of Phanerozoic age. The division between Proterozoic and Phanerozoic cover within the South American craton constitutes a fourth dichotomy in the geological distinction of the Guiana Shield. The development of the Andes along the western and northern rim of the craton created a significant geosynclinal subsidence between the forming western cordillera and well-established east-central shield areas (Gibbs and Barron, 1993). While still formed of Precambrian crust, this Sub-Andean Foredeep received significant volumes of sediment during the Phanerozoic through the accelerated gradation brought on by the rise of the Andes (Klammer, 1984; Goodwin, 1996). Rapid deposition only served to promote further subsidence of the trough and increase sediment aggradation through isostasy, resulting in modern sediment depths exceeding several kilometres in some areas. Development of the Amazon Downwarp followed a similar path as sediment was pushed eastward along the newly formed Amazon River after significant uplifting in the west and eustatic subsidence in the east created a major east-to-west incline.

GREENSTONE BELTS The Guiana Shield is the geological sibling of the Central Brazilian and Atlantic Shield areas. They share a similar distribution of igneous, metamorphic and sedimentary formations, many of which formed concomitantly across a Precambrian West Gondwanan super-shield that was later bisected by the Amazon Downwarp. One of two main geological features distinguishing the Guiana Shield from its southern hemispheric siblings is the relative abundance of greenstone belts (*roches vertes*). The other is the thick, Proterozoic sedimentaries forming the Guayana Highlands. Greenstone complexes cover extensive areas along the north-eastern rim of the Guiana Shield, but are scarcely discernible in either Central Brazilian or Atlantic Shield areas (Gibbs and Barron, 1993; Goodwin, 1996). The occurrence of these metamorphosed basalts is thought to be coincident with the axis of Atlantic rifting (Gibbs and Barron, 1993), a feature that would have been in much closer proximity to the eastern Guiana Shield than the Central Brazilian Shield during the early Cretaceous initiation of the rifting process and subject to more pronounced compression as the African plate rotated counter-clockwise in relation to South America (e.g. see Rabinowitz and LaBrecque, 1979), giving rise to the South Atlantic first (Pindell and Dewey, 1982). The relative abundance of greenstone belts is a fifth dichotomy useful in distinguishing the geology of the Guiana Shield from that of other neotropical forest-lands.

Formative events shaping the Guiana Shield

The timeline of geological development in the Guiana Shield is both long and complex. Often sequences linking modern-day features to specific geological eras, periods or epochs are tentative, uncertain, or simply unknown to Precambrian geologists. The establishment of the main formative events and arrangement here of features according to their proposed geological age is principally based on Kalliokoski (1965), Groeneweg and Bosma (1969), McConnell and Williams (1969), Berrangé (1977), RADAMBRASIL (1973–1978), Mendoza (1977), Cordani and de Brito Neves (1982), Teixeira *et al.* (1989) and in particular, more recent treatments by Gibbs and Barron (1993) and USGS and TM Corporacion Venezolana de Guayana (CVG-TM) (1993).

During the earliest stages, rock formation and deformation were believed to reflect large-scale processes, such as regional metamorphism, that commonly affected many of the modern-day continental components of the former Gondwanaland. These were largely associated with the creation, reactivation and deformation of a crystalline basement complex dominated by felsic granulites (remelted and reshaped silica-rich rocks) (Gibbs and Barron, 1993). After the final break-up of the southern supercontinent began about 135 million years ago, South America was in effect an island continent (Press and Siever, 1982). Processes that earlier influenced several modern-day continents collectively, began to increasingly affect each independently as the parent crustal plates carried the other continents away from South America, while South America itself remained close to its modern-day position (Goldblatt, 1993).

The Guiana Shield became isolated from its sibling shield areas in South America as regional uplift of the Guayana Highlands and later, the Andes, altered the direction and number of grading bevels (large erosion slopes) across the continent (Gibbs and Barron, 1993). Further faulting, downwarping due to sediment loading and more localized uplifting continued to alter the type and extent of gradational processes. Later, deposition of sediments from the Andes, uplifting of the shield highlands and continuing sea transgression into the surrounding basins, served to change the topography of the region even further. The consequent modern-day landscape is diverse and patchy, particularly in the northwest of the shield (Guayana Highlands and vicinity).

In an effort to organize the chronology of geological events shaping South America, and more specifically the Guiana Shield, geologists have assigned modern formations of similar composition and conformation to ranked associations. Each association has then been assigned a relative position within a geochronology. Not all of these formations are fully or even partially visible (or readily accessible) at the surface. The extent and conformation of these have been more completely assessed through drilling and using geophysical techniques (e.g. radar, gravity and aeromagnetic) to characterize their lithology (e.g. USGS and CVG-TM, 1993). Analysing the relative abundance of different (non) radioactive isotopes in rock samples has been used to establish a position for each association within the geochronology. Where rocks have been seen as a complex product of many different events occurring at different points in time, it often has not been possible to assign a reliable date (Gibbs and Barron, 1993). At the same time, the accuracy and precision of an isotopic date, like that of many other techniques for reconstructing palaeohistories, diminishes with antiquity, often leading to reported confidence intervals attached to Precambrian dates greater than the entire Cenozoic (70 million years) (see radioisotopic dates presented in Gibbs and Barron (1993)). This fact has prevented geologists from establishing a pinpointed cause-and-effect chronosequence for rock products since the calculated age intervals of consecutive events often will overlap. Unconformity in the stratification of rocks can further complicate the development of a detailed chronology of events. This leads to overlap in the assignment of formative events in the geology of the Guiana Shield. Table 2.1 presents a chronosequence of the major geological associations recognized for the Guiana Shield and the names given to these in each of the six countries within the shield region.

TECTONIC PROCESSES These processes include *diastrophic* activities that deform the surface through folding, twisting, warping, shearing or compressing rock and *volcanic* activities that lead to the distribution of heated magma as lava or pyroclastics from volcanoes and fissures (rifts). Diastrophic events, leading to rock metamorphism and reactivation (remelting), have affected nearly all geological formations in the Guiana Shield (the Roraima Supergroup/Avanavero Suite complex and Mesozoic sediments being the exceptions). Volcanic events have had a more punctuated influence on the shield landscape as a consequence of these larger diastrophic events. Greenstone belts, the swarms of 'younger' dykes and sills, the Uatuma Supergroup volcanics and the Apoteri Suite underlying the main Takutu Graben Fill have all developed through volcanic ejection emanating from tensional forces fissuring the surrounding rock.

The main diastrophic episodes known or expected to have played a prominent role in shaping the modern geological landscape of the Guiana Shield include the following tentative assignments and dates based on correlation of a number of different sources listed with each episode – they are not all considered definitive. Their occurrence relative to other selected geological and climatological events shaping South America can be seen in Fig. 2.3.

1. Proto-Shield Period (>3.4 Ga BP) (also 'Old Crystalline Basement' (Berrangé, 1977), Proto-Imataca (Montgomery, 1979)). Submarine formation of early planetary crust and Imataca protolith.

2. Imataca Episode (3.4–2.7 Ga BP) (also referred to as the Gurian Orogenesis (Huber, 1995a)). Submarine formation of Imataca granulites and gneisses, intermixed with manganese-rich metasediments iron forma-

Table 2.1. Chronology of rock formation in the Guiana Shield. Based on information from Gibbs and Barron (1993) and references therein.

Age (Ga BP)	Geologic time interval episode	Parent association	Component association					
			Venezuela	Guyana	Suriname	Guyane	Brazil	Colombia
>2.7	Archean	Imataca Belt	Imataca Complex					
2.7–2.0	Early Proterozoic	Central Granulite Belt	Kanku Complex		Falawatra Group		Apiaú, Camanaú-Curiuaú, Jauaperi Complexes	
		Tumucumaque Belt				Alitany Orthogneiss Suite	Ananaí, Tartarugal Grande Metamorphic Suites	
		Uraniqúera Belt					Urariqúera Metamorphic Suite	
		Guaspati Belt	Carichapo Group					
		Caroni-Paragua Belt	Santa Barbara, Caroni, Chiguao Groups					
		Botanamo Belt	Botanamo Group					
		Barama Belt		Arakaka, Matthews Ridge Tenapu Formations				
		Mazaruni Belt		Issineru, Haimaraka Formations				
		Cuyuni Belt		Cuyuni Group				
		Marowijne Supergroup			Matapi, Paramaka, Armina Groups			
		Maroni Supergroup				Paramaca, Bonidoro, Orapu Groups		
		Amapa Supergroup					Villa Nova, Cauarane, Parima, Kwitaro, Coeroeni Groups	
2.2–1.9	*Trans-Amazonian Tectnothermal Episode*	Metadolerites	Dikes I–III	Older Basic Intrusives Metadolerites	undesc?	Malipapane metadolerite dikes	undesc?	
		Large Ultrabasics		Omai horblendites, Kauramembu ultrabasics Itaki gabbros				
		Badidku Suite		Appinite Suite, Late Kanematic Suite	De Goeje Gabbro Suite	Tampoc meta/ultrabasics	Tapuruquara Association	
		Granitoids	Supamo Complex, El Manteco-Guri Young Granites	Bartica Assemblage, N. Guyana Younger Granites, Southern Guyana Granite Complex, Essequibo-Courantyne Granite Complex, Makarapan Alkali Granite	Diapiric intrusions, Shallow-level granites, Deep-level granites	Granites Caraibe, Granites Guyanais, Granites Galibi	Agua Branca Adamellite (Rio Novo Granodiorite)	
1.9–1.2	Middle Proterozoic	Uatuma Supergroup	Cuchivero Group	Iwokrama Formation, Kuyuwini Group, Muruwa Formation	Dalbana Formation, Ston Formation		Surumu, Iricoumé Formations, Mapuera Intrusive Suite	
		Roraima Supergroup	Uairen, Kukuenan, Uaimapué Mataui Formations	Upper, Middle Lower Members			Mataui, Uailan, Arai	

continued

Table 2.1. *continued.*

Age (Ga BP)	Geologic Time interval/episode	Parent association	Component association: Venezuela	Guyana	Suriname	Guyane	Brazil	Colombia
1.9–1.2	Middle Proterozoic	Quasi-Roraima Mesas	Sipapo Block, Cerro Paru	Makari Mountain	Tafelberg, Emma Range		Tepequém, Urutanim, Uafaranda, Surucucu, Rio Novo, Araca, Neblina Padre Massif, Urupi Massif, Uneuixi	
		Avanavero Suites (Roraima Intrusives)	Cano Roja, Meseta de Boro Sipapo, Duida, Paru dikes	Tumutamari Dike, Ebini Sill, Kopinang Sill, Waracabra Norite Sill	Avanavero, Bakhuis Dikes	undesc.	Pedra Preta, Arai, Uaicas, Quarente Islas, Pixilinga Sills	
		Vaupés/Tunuí Supergroup	Cineruco Group				Tunuí, Caparro Groups	La Pedrera, Piraparana, Guainia, Naquém Groups
		Parguaza Granites	Parguaza	Post-Roraima acid intrusives (Amatuk Dike, Kanaima Felsites)			Surucucus, Abonari Intrusive Suites, Madeira, Aqua Boa Plutons	Parguaza, Mitú Complex
		Alkali Intrusives	Cerro Impacto	Muri Mountain			Catrimani, Apiau syenites, Mutum Mt., Seis Lagos, Mapari, Maraconai, Maicuru, Apupariu	
1.3–1.2	*Nickerie Episode*	Mylonitized faults	Orinoquean Event	Kmudku Episode	Nickerian Episode		Jari-Falsino Episode	
1.2–0.57	Late Proterozoic	Cachoeira Seca Suite					Seringa Suite	
0.57–0.22	Palaeozoic	PAPA Dikes		southern Guyana swarms			Roraima swarms	
		Margin Sediments	Hato Viejo Formation				Prosperança, Trombetas Formations	(west) Araracuara Formation
0.22–0.07	Mesozoic	Apatoe Dike Suite (Basic Dike Suite)		southern Guyana swarms	western Suriname swarms		Roraima swarms	
		Rewa Group		Takutu, Rupununi, Pirara, Manari Formations, Apoteri Suite			undesc.	
		Amazon Margin Sediment					Alter de Chão	
		Guyana Basin		Stabroek, Potoco, Canje, New Amsterdam Formations				
0.07-0.053	Cenozoic (Palaeocene)	Coastal Sedimentary Plains		Georgetown Formation	Onverdacht Formation			
0.053-0.037	Eocene			Pomeroon Formation				
0.037-0.026	Oligocene		Bauxite Hiatus	Bauxite Hiatus	Bauxite Hiatus			
0.026-0.007	Miocene			Courentyne Formation	Courentyne Formation		Para Group (Piraracu, Solimoes Formations)	
0.007–0.002	Pliocene		Mesa Formation	Berbice Formation	Coeswijne Formation		Barreira Formation	Pebas Formation
0.002–0.0001	Pleistocene			North Rupununi Savannas Formation, Berbice Formation	Zanderij Formation	Sables Blancs	Iça, Boa Vista Formations	
<0.0001	Recent		River silts and clays	Demerara Formation, River silts and clays	Coropina Formation, River silts and clays	Coastal, River silts and clays	Coastal, River silts and clays	River silts and clays

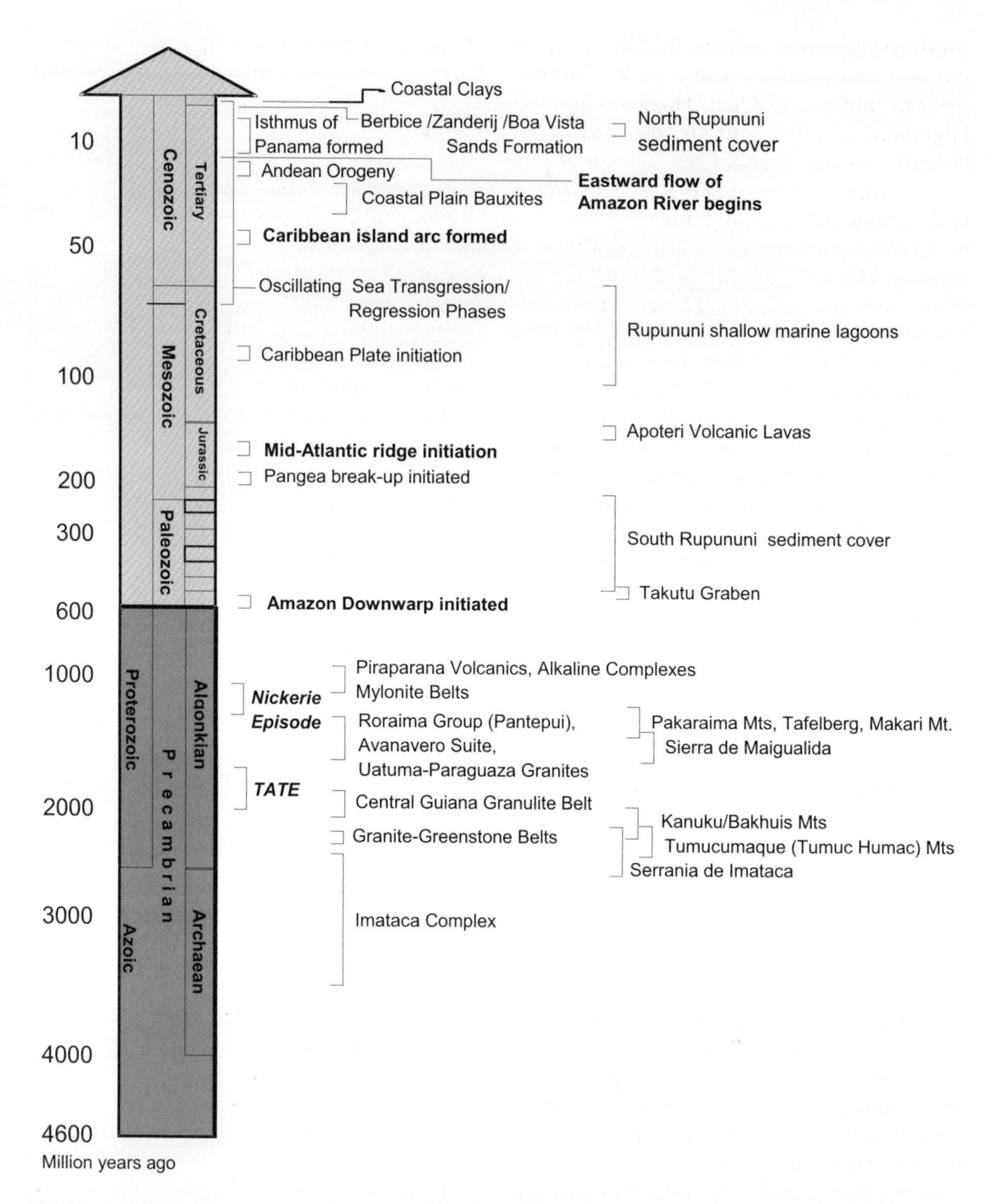

Fig. 2.3. Selected tectonic and gradational events shaping the geological cycle of the Guiana Shield. Selected South American, Guiana Shield and formation-specific events are aligned to the left, centre and right, respectively. Prominent events are in bold. TATE=Trans-Amazonian Tectonothermal Episode.

tions, dolomites and anorthosites. Subsequently reactivated and metamorphosed.

3. Pre-Transamazonian Orogenesis (2.6–2.1 Ga BP) (Mendoza, 1977; Huber, 1995a; also referred to as the Early Proterozoic Arc stage (Gibbs and Barron, 1993)). Submarine formation of Kanuku, Bakhuis granulites and Falawtra gneisses of the Central Granulite Belt and the Tumucumaque granulites and orthogneisses. Widespread greenstone (metamorphosed basalts) belts produced by considerable folding of early volcanic and sedimentary rocks formed prior to continental crust. Their modern NE alignment is attributed to the later mid-Atlantic rifting. Formation of non-greenstone supracrustals of the Kwitaro Group and Parima Suite, among others.

4. Trans-Amazonian Tectonothermal Episode (TATE) (2.4–2.1 to 2.0–1.7 Ga BP) (also referred to as Trans-Amazonian Orogenic Cycle, Akawaian Episode (Berrangé, 1977), Trans-Amazonian Orogenesis (Mendoza, 1977; Huber, 1995a)). The most important event affecting nearly all of the South American proto-craton. Submarine formation of feldspar-rich granitoids that today account for most of the widely distributed crystalline basement exposed, most notably, in southern Guyana and Venezuela and over most of Suriname, French Guiana (Granites Caraibe and Guyanais), Amapá and North Pará. Thought to have created an early island arc, not unlike the Caribbean of today (Gibbs and Barron, 1993) that was later in-filled and uplifted.

5. Uatuma Episode. The main volcanic event to affect the modern shield landscape occurred after the TATE during the Middle Proterozoic (1.9–1.7 Ga BP) and consistent in timing with the western Parguaza episode. The development of the Uatuma Supergroup of acid to intermediate volcanics and subvolcanic intrusives preceded the later depositional episode forming the Roraima Supergroup (Fig. 2.3). Acid (silica-rich) rocks formed during this period are sandwiched between an underlying basement of granites and gneisses and overlying sedimentary cover associated with the Roraima Group and more recent deposition (Gibbs and Barron, 1993).

6. Parguaza Tectonothermal Episode (1.9?–1.5 Ga BP). Correlated with the hypothesized Rio Negro-Juruena Episode (Tassinari (1984), cited in Gibbs and Barron (1993). Also referred to as the Parguaza Event (Priem *et al.*, 1982)). An uncertain submarine event believed to have led to the formation of the Mitú complex and, importantly, the formation of NW–SE rift valleys embracing the Parguaza granitic batholith along the upper Rio Negro and Atabapo Rivers in western Venezuela/Colombia and along the Guaniamo and upper Ventuari Rivers.

7. Nickerian Episode (1.3–1.1 Ga BP) (correlated with the Orinoquean Event (Venezuelan Guayana), the Kmudku Episode (south Guyana), Jari-Falsion Episode (Amapá) (see Gibbs and Barron, 1993)). Also referred to as the Nickerian Compressional Episode and Orinocan Orogenesis (Mendoza, 1977; Huber, 1995a). Another important episode leading to cratonic formation by shearing and in-filling of block-fault fractures through mechanical transport of angular fragments of sedimentary rock and subsequent lithification of these deposits through fault movement (mylonite formation).

8. Dolerite dyke intrusion (0.49–0.14 Ga BP) (Basic dyke suite (Berrangé, 1977), Apatoe dyke suite (Gibbs and Barron, 1993)). A series of silica-poor dyke suites intruded upon the surrounding metamorphosed rocks throughout much of the shield area during the Palaeozoic and Mesozoic eras as a consequence of localized tensile fracturing of the surrounding rock due perhaps to eastern plate boundary initiation (Caribbean) and ocean-floor spreading (Mid-Atlantic). In southern Guyana, these long, sinuous structures can run 50+ km and consist mainly of calcium, magnesium and iron amphibolites, pyroxenes and olivines (Berrangé, 1977). The last of the main tectonic episodes to influence the modern-day landscape of the Guiana Shield.

GRADATIONAL PROCESSES These processes include degradation through mechanical and chemical weathering and erosion as well as aggradation through the lithification of sediments deposited by marine, fluvial or Aeolian (wind) action. In many instances, the resulting sedimentary rock deposits rest unconformably atop other strata.

1. Roraima Episode (1.7–1.6 Ga BP). The principal sedimentary event shaping the Guiana Shield. The widespread deposition of sediment in the uplifted back-arc basin, further uplifting and block-faulting, and subsequent weathering and erosion of the Roraima Group consolidated the region as a cratonic feature and created the Guayana Pantepui and Pakaraima Mountains (Fig. 2.3).

2. Takutu Episode (0.18–0.13 Ga BP). The Rewa Group forms a series of littoral mudstone, siltstone, shale and sandstone deposits distributed across the North Rupununi Savannas and parts of Roraima state, Brazil (Fig. 2.3). They are thought to be derived from the volcanic and subvolcanics of the Iwokrama Formation, adjacent to the Takutu Graben (Berrangé, 1977) and lying unconformably beneath more recent laterite platforms (Nappi Laterite Formation).

3. 'Courentyne Episode' (<0.10 Ga BP) (after Courentyne Group (Berrangé, 1977)). While not an episode recognized *per se* by shield geologists, the series of (mainly) sedimentary events dominating the past 100 million years have without doubt substantively altered the landscape of the region. In time, as plate movements continue, it is likely that this run of sedimentation will be eclipsed by other more formative tectonic upheavals, compartmentalizing the late Mesozoic/Cenozoic as an episode. Formations identified along the Atlantic coast, in the Takutu Graben, on the northern rim of the Amazon River, and along the Rio Branco and Rio Orinoco show concordance in relation to depositional dynamics, though not perhaps sediment provenance. The series of depositional cycles over the period were strongly linked to the rise and fall of sea level concomitant with changes in climate (glaciation) and the effects that inland sea advance had on eustatic subsidence, and with sea retreat, uplifting of the foreshore basins. Marine ingress along lower rivers and floodplains has mixed with and been reworked by more recent river transport to create the modern-day sand and silt plateaus common throughout many parts of the Guiana Shield. Sedimentation shaped the surface along erosion bevels formed by the combination of erosion and depositional forces and structural changes in the planation surface (e.g. faulting, subsidence and uplifting). The uplift of the Andes during the Tertiary represents the most significant of these structural changes affecting sedimentation.

Prominent geological regions of the Guiana Shield

Rock associations in the Guiana Shield have generally been referred to as supergroups, groups, complexes, belts, suites and/or formations. Akin to the relationship between plants and taxonomists, these associations have been periodically dismembered, regrouped, promoted and relegated by shield geologists as new evidence, or (re-)interpretation of existing evidence, leading to new proposals (Gibbs and Barron, 1993). Typically, supergroups are composed of groups or complexes (depending on degree of conformation) and these, in turn, of formations, each designating a closer lithological association. Belts are formations of conformable lithology that normally distribute anisotropically and are principally formed of granulite or greenstone (*roches vertes*) in the shield, typical products of large-scale regional metamorphism. The Guiana Shield has been fractured into tentative geological *provinces*, based on a number of distinguishing features (e.g. age, metamorphic grade, exposure) and major rock associations present in each sector of the region and described by various geologists (Mendoza, 1980; Cordani and de Brito Neves, 1982; Gibbs and Barron, 1993) (Fig. 2.4). All the provinces rest upon a basement of crystalline rock, but subsequent metamorphism, sedimentation, volcanism and mylonitization have altered the superficial layers extensively. Provinces can be described according to the predominant source of these layers and the main rock associations found within each contiguous area (see Fig. 2.4 for distribution of rock associations).

MAINLY METAMORPHOSED CRYSTALLINE AND METAVOLCANIC ROCKS (GRANULITES, GRANITOIDS, GRANITES, GREENSCHISTS) These are the oldest rocks in the shield. They can be found predominantly in five areas, consistent with five of seven geological provinces con-

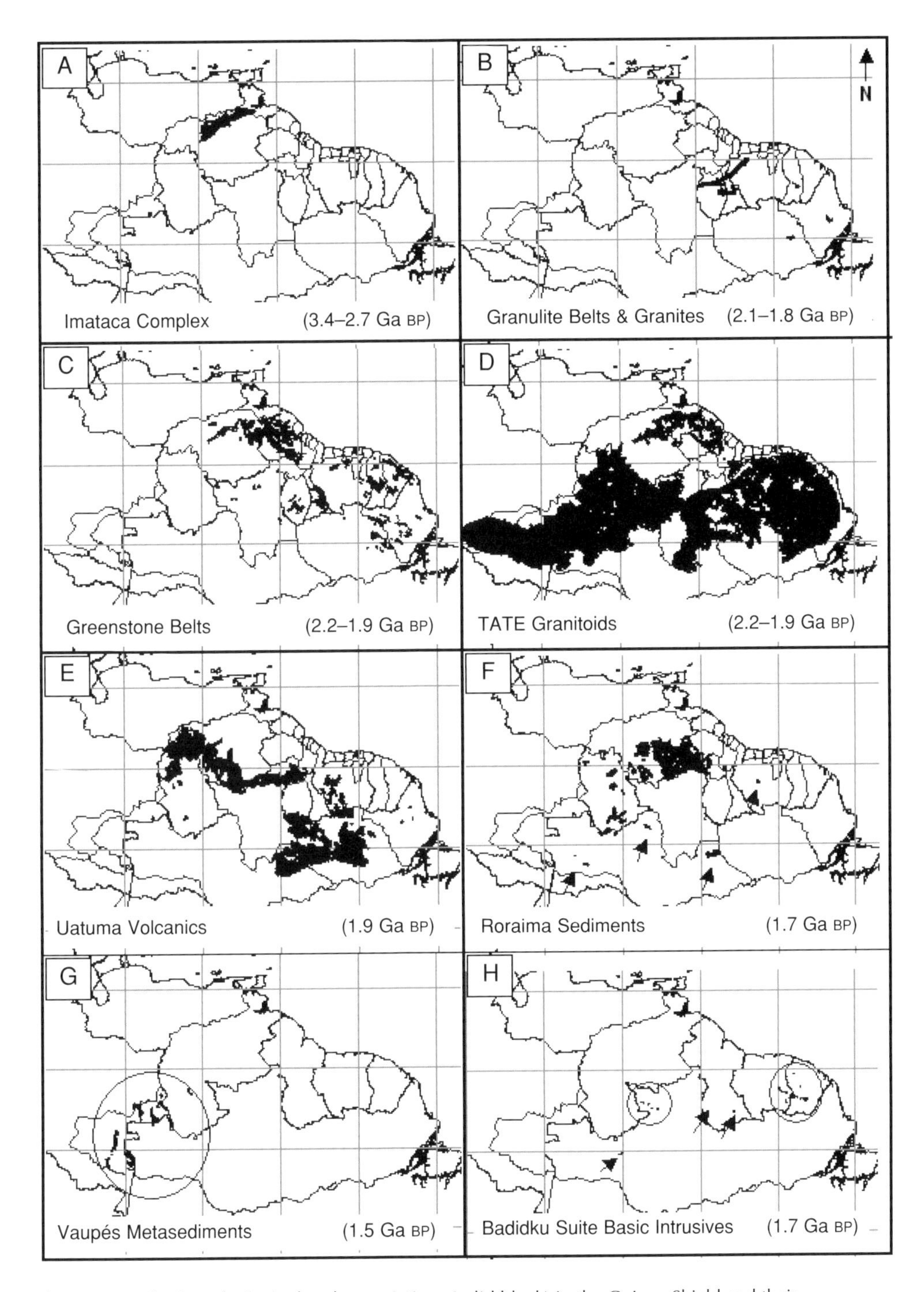

Fig. 2.4. Distribution of principal rock associations (solid black) in the Guiana Shield and their approximate age of formation over the last 3.4 billion years. Adapted from appendix map in Gibbs and Barron (1993). TATE=Trans-Amazonian Tectonothermal Episode. *continued*

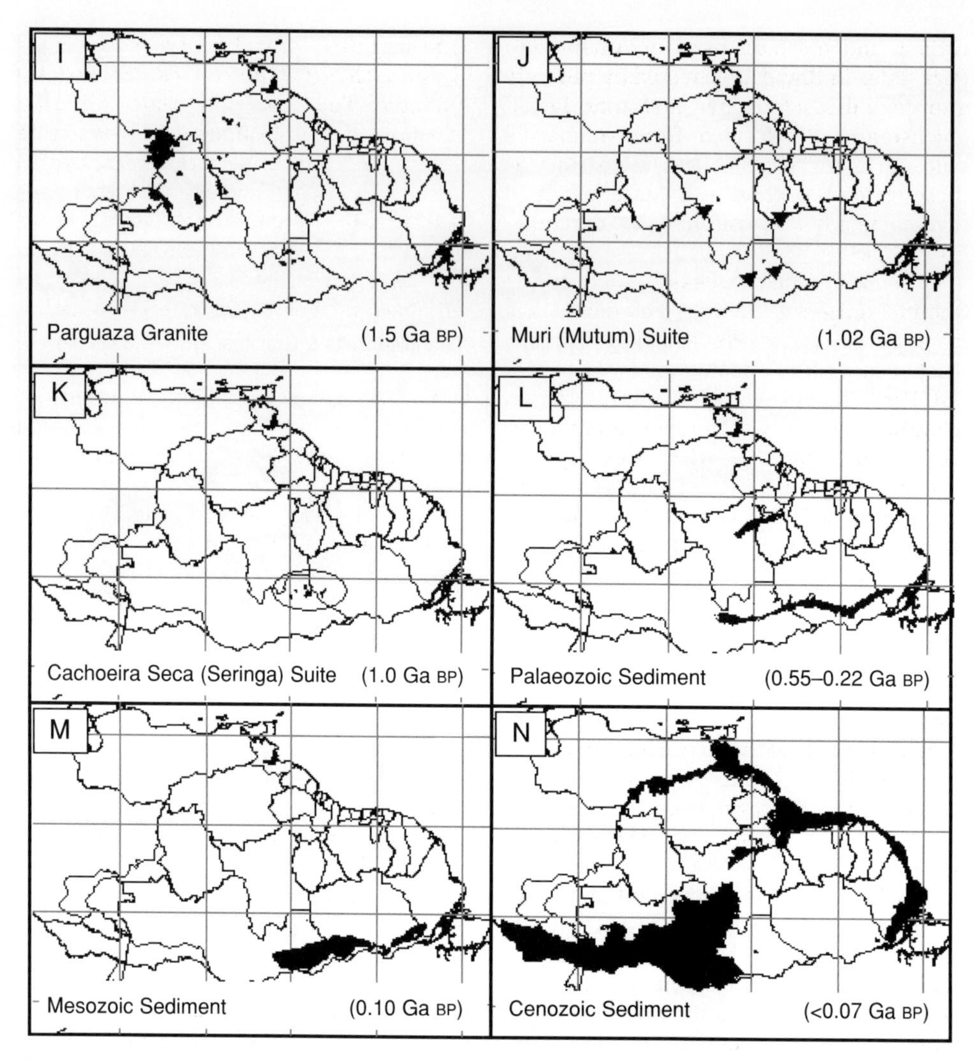

Fig. 2.4. *Continued.*

sidered by Gibbs and Barron (1993) as making up the Guiana Shield:

1. Imataca Province. The first and oldest formation is the Imataca complex, a compact, contiguous area of Archaean gneiss and granulite that is exposed directly south of the lower Orinoco River floodplain (Fig. 2.4A). Only approximately 2% of the shield surface is currently considered to have been formed during the Archaean (>2.5 Ga BP), and nearly all of this area exposed today is found within the Imataca complex (Gibbs and Barron, 1993).

2. Pastora–Amapá Province. A long fault (Guri) separates the southern border of this area from the second group that consists of early Proterozoic granite, greenstone and gneiss formations running in a northwest to southeast direction through much of northern Bolivar state and the Cuyuni/Mazaruni regions of northwest Guyana (left section of Fig. 2.4C).

3. Pastora–Vila Nova Province. The third area is separated from these areas by the Berbice Basin/Takutu Graben/Central Guiana Granulite Belt complex. Southeastern Suriname, most of French

Guiana and northern parts of Amapá and Pará states in Brazil are covered by plutonic granites dissected by metamorphosed greenstones that formed from submarine volcanic activity along fracture seams as these were opened up as a consequence of regional tectonic stresses (cluster on right of Fig. 2.4C).

4. Western side – east section of Roraima–Uatuma Province. The fourth area of exposed early Precambrian rock is located in eastern Roraima and northwestern Pará. It consists of central Guianan granulites and early Proterozoic granite and gneiss interspersed with the middle Proterozoic volcanics, granites and metasediments that make up the Wassarai, Kamoa and Acarai mountains along the southern border of Guyana and highlands further south along the eastern border of Roraima and Amazonas states with Pará (Fig. 2.4C, D, E).

5. Western side – west section of Roraima–Uatuma Province and north Vaupés–Parguaza Province. The fifth area is found in the Territorio Federal Amazonas of Venezuela, Roraima and Amazonas states in Brazil and Vaupés state in Colombia and covers much of the western Gran Sabana, upper Orinoco and upper Rio Negro basins. This area is dominated by exposed granitoids associated with the Trans-Amazonian Tectonothermal Episode, the massive Parguaza granite batholith surrounding Puerto Ayacucho upstream from the Meta-Orinoco confluence and localized sedimentary cover (Fig. 2.4D, E). Further south these formations are increasingly covered by Phanerozoic platform sediments deposited as water flow direction and volume within the Amazon Basin varied with eustatic and orogenic activity over the last 570 million years.

MAINLY ACID VOLCANICS, SUBVOLCANICS AND ALKALI BASALTS These rocks originate from volcanic and subsequent plutonic activity during the Middle Proterozoic. Volcanic rocks are principally of extrusive, feldspar-rich rhyodacites formed in pyroclastic, lava and ash flows exposed along rivers (Berrangé, 1977; Gibbs and Barron, 1993). Subvolcanics are largely plagioclase-biotite-rich, granitic intrusives forming uplands. The small exposure of akali basalts in the south (Cachoeira Seca suite) is localized. Exposed formations can be found in two of Gibbs' seven provinces, located along a central northwest to southeast axis. Kloosterman (1973) suggested that these centrally located deposits are the remnants of three supercalderas (large volcanoes).

1. Eastern side – east section of Roraima–Uatuma Province. The first area runs from a southern buried margin running parallel to the Amazon River from Manaus to Obidos northward into southwest Guyana and Suriname.

Highly metamorphosed and weathered volcanics of the Uatuma Supergroup, forming in part the Kuyuwini and Burro-Burro groups, are exposed along the northern rim of the Kanuku Graben in central Guyana (Iwokrama Formation) and in the south of Guyana along the Brazil border (Berrangé, 1977). They extend into southwest Suriname along the Zuid River and upper Orenoque Rivers through to the Dalibana Creek area (Fig. 2.4E).

Subvolcanic intrusives of metamorphosed granitics form a number of important highlands in southern Guyana, including the Kamoa, Amuku, Acarai and Wassarai Mountains (Berrangé, 1977). Volcanic materials also contribute to their formation to varying degrees. The Wilhelmina Massif of south-central Suriname is principally formed from metamorphosed quartz porphyries, rhyolite pumice and rhyodacites with subvolcanic granitoid intrusives (Verhofstad, 1971; Gibbs and Barron, 1993).

2. Central strip – west section of Roraima–Uatuma Province. The second area runs westward from the Makari Mountain in the upper Berbice of Guyana, through the Siparuni area and southern Pakaraimas along the southern rim of the main Roraima Group forming the tepui of the Guayana Highlands. The Uatuma Group in fact runs beneath nearly the entire Roraima sedimentary group (Gibbs and

Barron, 1993). This area of Proterozoic volcanic and subvolcanic formations continues westward between the Ventauri and lower Caura Rivers, terminating at the eastern margin of the Orinoco (Fig. 2.4E). They consist mainly of metamorphosed tuffaceous lavas, pumice and ignimbrites (lithified volcanic ash and breccia). Highland centres forming the Sierra Maigualida, Sierra de Guampi and Sierra Guanay ranges are granitic intrusives formed from the Santa Rosalia and San Pedro batholiths.

MAINLY SEDIMENTARY While recognizing that sediments are continuously eroded and deposited through both mechanical and chemical action, the main sedimentary formations seen today in the Guiana Shield can be attributed to three main gradation phases (adapted from Gibbs and Barron, 1993): (i) submarine sedimentary episodes following Archaen/Early Proterozoic crustal formation that were subsequently metamorphosed and degraded, often leading to unconformity with early, mainly metamorphosed, igneous rock (e.g. Kwitaro, Parima Groups); (ii) wide-ranging and deep (900+ m) sediments deposited as epicontinental (shallow marine), fluviodeltaic and Aeolian strata during the Middle Proterozoic and subsequently lightly metamorphosed and degraded (Roraima Supergroup and Quasi-Roraima formations, Vaupés Supergroup); (iii) extensive Phanerozoic sedimentary episodes associated with climate-induced changes in sea level, uplifting, subsidence, faulting and mountain-building (see Table 2.1). Significant sedimentary formations can be found in three provinces delimited by Gibbs and Barron (1993). Localized and often not insignificant formations can be found, however, throughout the Guiana Shield interspersed with metamorphosed volcanic, subvolcanic and plutonic groups.

1. Eastern side – west section of Roraima–Uatuma Province. The remnants of the massive Roraima Group cover sediments are found mainly in central Bolivar state and form the Serra Pacaraima that extends eastward into central Guyana and southward into northern Roraima state in Brazil (Fig. 2.4F). The little-metamorphosed deposits of the Roraima formation unconformably overlie the volcanics of the Uatuma Supergroup and are highly intercalated with intrusive, silica-poor dolerite and gabbro dykes, sills and sheets of the Avanavero Suite (Gibbs and Barron, 1993). These are most visible as rounded hills and ridges between sedimentary tepui tablelands and in the Gran Sabana of Venezuela, but are most commonly exposed among the younger Roraima deposits in Guyana's Pakaraima Mountains.

The Roraima Supergroup includes many notable flat-topped peaks (mesas, cerros, tepuis) and compacted highland (massif) areas within the northern shield area, including Mount Roraima (at the junction with Guyana and Brazil), Kukenán (Kukenaam), Los Testigos massif, Jaua massif, Auyán massif and Chimantá massif, among others in Venezuela (see Huber, 1995a). Merumé, Ayanganna, Kurungiku, Wokomung, Ayanguik are prominent components of the Roraima in Guyana, and Serra Telequén on the Brazil side of the frontier with Venezuela.

Members of the Quasi-Roraima Formation (after Gibbs and Barron, 1993) are distinct table top massifs that form an archipelago of outlying, isolated peaks (arrows in Fig. 2.4F). It has been suggested that these were connected with the main Roraima group in a single sedimentary cover with a size eightfold that of its current surface area (Gansser, 1954). Subsequent selective degradation of weaker substrate elements over the past 1.6 billion years has left the remaining highland areas fragmented. Notable among these outlying peaks and massifs is the Makari Mountain above Canister Falls in Guyana, the Tafelberg and Emma Range in east-central Suriname, the Urupi Massif in the northeastern corner of Amazonas state, Brazil, Serra Araca (Jauri), Padre and Pico de Neblina (transnational) near the Venezuelan frontier, and Uneuxi between the Negro and Japurá Rivers. In Venezuela, a series of outlying massifs extend in a north–south alignment near the Casiquiare rift valley in Amazonas state. These include from north to south, Yutajé massif, Parú massif, the Jaua massif, the Duida-Marahuaka massif and the Neblina-Aracamuni massif (again see Huber, 1995a).

2. Southwestern Vaupés–Parguaza Province. This area is covered by metamorphosed sediment ridges of the Middle Proterozoic Vaupés (Tunui) Supergroup and much younger, largely unconsolidated sheet sediments of Tertiary/Quaternary age (Mariñame and Pebas formations) (Gibbs and Barron, 1993).

The Vaupés formations extend between the Vaupés and Caquetá Rivers in eastern Colombia and along the upper Rio Negro (Guainia) in the vicinity of the Casiquiare. These quartz-rich structures are thought to have originated in an epicontinental environment, having been deposited unconformably upon a metamorphosed basement complex of gneiss reactivated during the Parguazan Tectonothermal Episode, approx. 1.5 Ga BP (Priem *et al.*, 1982). The northern group runs parallel to the proposed Casiquiare rift valley that structurally separates the upper reaches of the Negro and Orinoco Rivers (Fig. 2.4G).

3. Atlantic coast section of Berbice–Boa Vista Province. The Berbice Basin spreads fan-like towards the Atlantic coast between the Essequibo (Guyana) and Maroni (Suriname/Guyana) Rivers from a narrowing inland reach extending as far south as the upper Berbice River (upper band of Fig. 2.4N). Bouger gravity measurements indicate deep bedded sediments and a significant dip that extends southwestward towards the Takutu Graben – an area of crustal downwarping that has led to subsidence and block faulting along its margins. Subsidence is inclined towards the Atlantic. The combination of continued subsidence, periodic uplifting, oscillating sea-level change and river action, particularly during the Tertiary/Quaternary period, created a sequence of depositional environments centred on a Berbice–Courentyne Rivers axis that led to deep layering of white, quartzite sands atop kaolins and other chemically weathered products (such as bauxite) of the underlying basement complex. Bauxite and kaolin deposits formed through the chemical weathering of feldspar-rich basement complex rocks prior to sedimentation and are thought to trace the former coastline during the Eocene/Oligocene period, 25–35 million years ago (Van der Hammen and Wymstra, 1964). The initial deposits (Stabroek, Onverdacht, Nickerie Formations) of conglomeratic sandstones, siltstones and shales were deposited in the upper Aptian epoch (late Cretaceous), around 90 million years ago and were followed by limestones, shales and carbonates (Canje, New Amsterdam, Georgetown and Pomeroon Formations) through to the mid-Miocene (18 Ga BP) (Gibbs and Barron, 1993). Subsequent depositional phases during the Plio-Pleistocene (the Courentyne, Berbice, Coeswijne, Zanderij Formations) consisted principally of the quartzite white sands. Up until the Pleistocene, these sediments were deposited in a marine environment. Local uplifting during the Pleistocene tilted the Atlantic edge of the basin upward and brought the white sand formation above sea level. These sediments attain a depth of more than 2000 m near the mouth of the Courentyne River and are largely responsible for the capture and inclusion of organic material offshore that has subsequently led to the formation of commercially attractive oil deposits (Bleakley, 1957; Van der Hammen and Wymstra, 1964). Uplifting created a new erosion bevel along the Atlantic coast as fluvial action reworked earlier submarine deposits. Unconformable deposition of silts and clays along the uplifted basin shore (Demerara and Coropina Formations) has occurred during the Quaternary, yielding the coastal strip historically underpinning cash-crop cultivation in Guyana and Suriname (see Chapter 8).

4. Central Takutu Graben section of Berbice–Boa Vista Province. The inland connection with the downwarping Berbice Basin runs southwestward for 300 km from the upper Berbice through to the Rio Branco south of Boa Vista as a 30–50 km wide graben, or rift valley, created through tensional block faulting in the late

Jurassic–early Cretaceous. The Rupununi Savanna north of the Kanuku Mountains and south of the Pakaraimas is principally a product of this development and the subsequent in-filling (Manari, Pirara, Rupununi, Takutu Formations) and lithification under shallow marine conditions produced a series of shales, siltstones, sandstones and carbonates to a depth of up to 7 km (Berrangé, 1977; Crawford *et al.*, 1984). These rest atop lavas of the Apoteri Formation that run along the eastern margin of the graben and surface in the north to form the Toucan Hills. Fluviatile and Aeolian deposits of silt and sand accumulating since the late Miocene (12 million years BP) constitute the main surface structure (North Savanna Formation) down to a depth of approx. 25 m (Crawford *et al.*, 1984) (upper band in Fig. 2.4L).

5. Southwestern section of Berbice–Boa Vista Province. The southwestern end of the Cenozoic sediment filling the downwarp-graben structure bisecting the Guiana Shield (Fig. 2.4N). Running southward along the Rio Branco from Caracarai, the mainly quartzite sand cover is the product of sea transgressions backing up the Amazon River during the Plio-Pleistocene and later, uplifting and subsequent fluvial action of the Rio Branco and Rio Negro (but for alternative origin of recent sediments, see Leenheer and Santos (1980)). Referred to as the Içа Formation, it gradates in the south with the older Pirarucu and Solimões Formations of the Miocene–Holocene Pará Group and the younger top sediments distributed in relation to the alternating Quaternary glacial and inter-glacial periods.

Rock chemistry of principal formations

The main geological formations recognized by shield geologists have been formed through a sequence of oscillating intrusive and extrusive events consistent with the main tectonothermal episodes. This is reflected in the chemistry of the various rock types that make up each formation. Molecular oxides, the building blocks of rock, are found throughout the Earth's crust. They combine to characterize the main minerals that are then mixed under a wide range of temperature and pressure conditions to form rock types. Rock types subjected to similar development and deformation at the same or similar time are found in each formation. The chemical composition of non-sedimentary rocks falls into two main categories that are grouped variously into the geological regions described above. Felsic rocks in the Guiana Shield contain a very high relative abundance of silica dioxide (SiO_2) compared to its overall crustal abundance and relatively low contributions from the other main oxides of iron (Fe), magnesium (Mg), calcium (Ca), potassium (K) and sodium (Na) (see Fig. 2.5 – groups with grey columns). Mafic and ultra-mafic rock types are, in contrast, highly deficient in SiO_2 compared to its overall crustal abundance. These rocks form under much higher temperature and pressure conditions that accompany subterranean intrusion into subsurface fractures in the surrounding country rock. They typically contain elevated contributions from minerals that readily precipitate (harden) first at higher temperatures, such as MgO and CaO, and depending on the type of intrusion, K_2O, Na_2O, FeO and Al_2O_3 (Fig. 2.5 – groups with black columns). These minerals then consolidate first, leaving the siliceous materials to solidify when temperatures finally drop below their much lower melting point. A third rock group consists entirely of lithified and unlithified sedimentary deposits, but these are formed initially through surficial weathering processes. The very high SiO_2 content of many of these deposits in the Guiana Shield reflects the past dominance of felsic weathering surfaces and a protracted life dominated by repeated episodes of weathering, deposition, lithification, erosion and leaching.

Chemical profiles of a wide range of rock samples presented in Gibbs and Barron (1993) and presented in Fig. 2.5 strongly illustrate the distinction between the two main non-sedimentary rock groups. The most extensive formations in the shield, commonly referred to as the

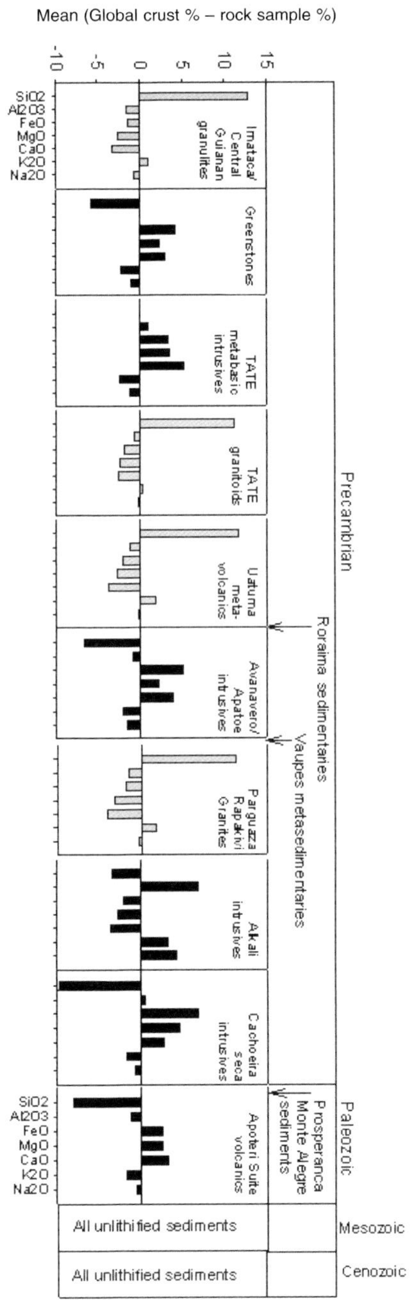

Fig. 2.5. A chronology of oxide formation based on sample data presented in Gibbs and Barron (1993). The compositions of rock samples are given for the main oxides that contribute through weathering to soil formation and nutrient influx from parent rock, viz. silicon, iron, magnesium, calcium, potassium and sodium. Their contribution to average composition of rock within each group is presented as the difference from their overall contribution to global crust (represented by the zero axis). Groups of grey columns represent mainly extrusive formations that are dominated by felsic (silica-rich) rock types and groups of black columns represent those intrusives dominated by mafic (silica-poor) rock types. Episodes leading to sedimentary formations are indicated by arrows and labels.

'crystalline basement complex', consist principally of felsic rock types. These include the Imataca and Central Guiana granulites (Fig. 2.5, Fig. 2.4A, B), granitoids formed during the Trans-Amazonian Tectonothermal Episode (Fig. 2.5, Fig. 2.4D), the Uatuma metavolcanics (Fig. 2.5, Fig. 2.4E) and the Parguaza/Surucucu granite group (Fig. 2.5, Fig. 2.4I). Mafic-dominated rock types are limited to exposure of intrusive formation surfaces that are often very limited in extent (e.g. Badidku Suite (Fig. 2.4H) and Muri Suite (Fig. 2.4J)). The extensive, and much older, low- to medium-grade Greenstone belts along the northeastern rim of the shield area (Fig. 2.4C) represent the most important exception to this rule and one of the defining geological attributes of the shield area (see above, 'Comparison with geology of other tropical forest regions').

The significance of the relatively high contributions of SiO_2 to the modern forest landscape of the Guiana Shield when compared to other tropical regions of the world is compelling. The influence of this ubiquitous compound in all of its mineral forms has fundamentally shaped the structure and function of plant life (see Chapter 3), and the opportunities and challenges faced by people in their efforts to make a living from these resources. At the same time, the geographical distribution of contrasting mafic and ultramafic rock dominants has influenced the landscape-level diversity of the shield's forests and played a controlling role in the mineral exploitation and conservation of these lands (see Chapter 9).

Tectonics, Topography and Landforms

Regional tectonic features

The Guiana Shield rests in the north-central quadrant of the South American tectonic plate and just south of the tri-juncture with the very active eastern margin of the Caribbean Plate and southernmost extension of the North American Plate (Fig. 2.6A). At some distance away from the Atlantic coastline of the shield, near the major axis of the Atlantic Basin, the African and South American plate margins meet, forming an important sea-floor spreading zone.

The modern shield region has been devoid of volcanic activity since the Mesozoic. A number of important volcanic belts, however, are located north and west of the area along regional plate subduction zones (Fig. 2.6A, solid triangles). Former tectonic events have created a series of much smaller fault lines across South and Central America, but concentrated along the Andes and the Guiana and Brazilian Shield areas (Fig. 2.6A, parallel lines).

Topography

The broad-scale topography of the Guiana Shield and its periphery can be dissected into three distinct 'highland islands' separated through crustal subsidence along a number of distinct structural faults, warps and geosynclines (Fig. 2.6A, B). Many of these features are immediately recognizable as forming significant landscapes in their own right. The formative features of these areas have been described variously by others (RADAMBRASIL, 1972; PRORADAM, 1979; Sombroek, 1990; Huber, 1995a). Most geologists believe that the interstitial lowlands acted as depositories of sediment eroded from the highland areas and, in part, pushed inland as sea level rose and transgressed during interglacial stages of the Quaternary period (Irion, 1976; Gibbs and Barron, 1993). As sediment-laden waters naturally flowed in and through these troughs, the weight of deposited sediment caused the crust in turn to further subside. Subsidence caused the deposition zone to deepen even further, maintaining an erosion bevel that kept these areas as the principal zone of deposition. This phenomenon extended the aggrading phase of a geomorphic cycle that forms much of the shield's modern appearance. The resulting surfaces are vast and encompass some of the heaviest sediment deposits found in the region, often reaching several kilometres deep. A number of these low-lying areas are in fact

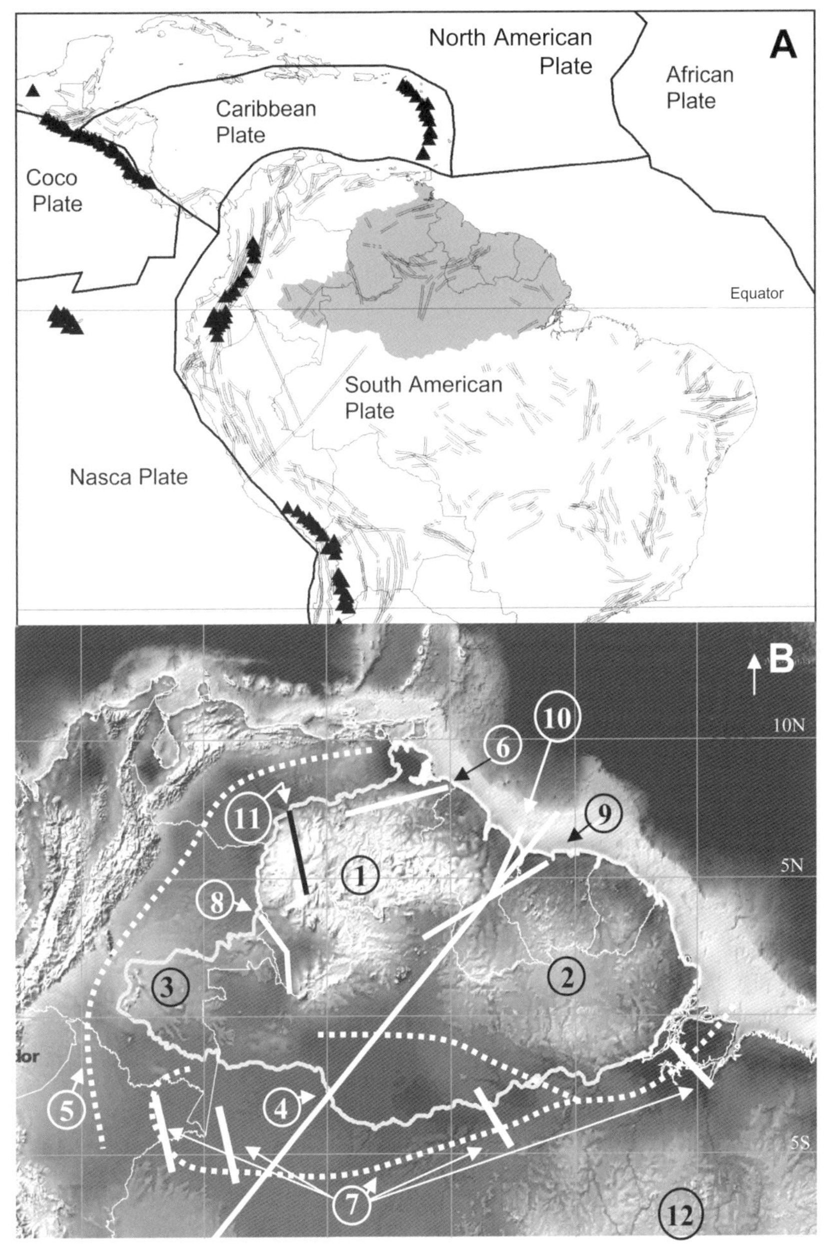

Fig. 2.6. (A) Tectonic features of South America and the Guiana Shield, including crustal plate margins (single lines), volcanoes (solid triangles) and major continental stress fault lines (double lines) as depicted in Data+ and ESRI (1996), Gibbs and Barron (1993) and Putzer (1984). (B) Schematic of topographical macro-features shaping the Guiana Shield. (1) Guayana Highlands, (2) Serra Tumucumaque Uplands and Rolling Hills, (3) Chiribiquete Plateau, (4) Pisco-Jura Megafault (incorporating the Takutu Graben and Berbice Basin), (5) Sub-Andean Trough, (6) Guri Fault System, (7) Amazon Downwarp (and uplifted arches), (8) Casiquiare Rift (including proposed graben and deep-shear structures), (9) Takutu Graben-Kanuku/Bakhuis Horst, (10) Berbice Basin, (11) Suapure-Mavaca/Ventauri Rift, (12) Brazilian Shield adapted from USGS and CVG-TM (1993), Gibbs and Barron (1993), Klammer (1984) and Kalliokoski (1965).

not considered to be within the Guiana Shield as defined *sensu stricto* due to their more recent Quaternary ages (e.g. see maps in Räsänen, 1993; Huber, 1995a; Goodwin, 1996). But their role in delimiting and fractionating the highland Precambrian core of the shield and the important influence they have had on the geological, biological, economic and social history of the region warrant inclusion here.

The average (±1 SD) elevation for the entire shield area is estimated at 270 (±341) m above sea level (asl).[1] The relatively unique deposition of conglomeritic sediments en masse that have come to be known as the Roraima Supergroup has created an unusually wide topographic range for an area dominated by Precambrian geology. Compared to the other two main shield areas in the Americas, the Brazilian and Canadian Shields, the elevational range found within the Guiana Shield is noticeably greater, varying between a maximum of 3014 m near the centre of the shield area (No. 1 in Fig. 2.6B) down to a minimum of 15 m below current sea level in the reclaimed coastlands of the Berbice along the northern shield perimeter (No. 10, Fig. 2.6B, also see Fig. 2.8).

The highest elevations and most rugged topography are found between 3° and 6°. The aggregation of peaks and high valleys of the Guayana Highlands rapidly give way southwards to flat low-lying valleys barely above sea level (Fig. 2.7). The Orinoco River cuts deeply along the northern margin of the shield, barely ascending more than 100 m asl at its upper juncture with the Casiquiare (Fig. 2.7, 3° N). The Rio Branco and Rio Negro start from a relatively narrow valley sandwiched between the steep southern edge of the Guayana Highlands (Fig. 2.7, 3° N to Equator) and the more gradual rise associated with the western margin of the southeastern 'island', referred to here as the Tumucumaque Uplands (Fig. 2.6B, No. 2). Southward the Rio Branco basin fans outwards to meet up with the lower Rio Negro, occupying an area nearly 800 km wide (Fig. 2.7, 1° N to Equator). Only isolated remnants associated with the Roraima formation and various intrusive rocks break up this peneplain (Fig. 2.7, Equator). The exposed crystalline basement that occupies much of the Tumucumaque Upland area (Fig. 2.6B, No. 2) also loses much of its relatively modest topographic variation southward, as downwarping associated with the Amazon Trough led to deep and extensive sedimentation across much of the area (Fig. 2.7) (Gibbs and Barron, 1993).

Shield macro-features

GUAYANA HIGHLANDS The Guayana Highlands dominate Venezuela east of the Orinoco and extend across much of west-central Guyana and northern Roraima state in Brazil. This is the largest of the three 'islands' of shield crust that rise above the surrounding basins (*cuencas*) formed through downwarping processes. It represents the northernmost part of the Guiana Shield and the most rugged topography in the region. Born from a mixture of ancient sedimentary rock deposited atop the crystalline basement and intruded by igneous rock of the Avanavero Suite, the Guayana Highlands are dominated by the high flat-topped peaks of the Roraima Supergroup and Quasi-Roraima formation and the rounded granite peaks of the Parguaza and Imataca complexes to the north and south-western edges of the area (Figs 2.6B, 2.7). The northern perimeter of this region is distinctly and drastically defined by the current course of the Orinoco River, as it separates the rapidly rising hills of the Parguaza and Uatuma formations from the flat, low-lying llanos (Fig. 2.7, 7° N, 6° N). While this major river system drains nearly the entire highland area (and adjacent lands in eastern Colombia and the Venezuelan llanos), the main stem follows a relatively weak hydraulic gradient, rising less than 0.1 m for every 1 km length travelled up to its headwater tributaries. The main stem barely reaches more than 100 m asl up to this point. In contrast, many of the main waterways feeding into the Orinoco from within the Guayana Highlands area, such as the Caura and Ventauri Rivers, have main stems that run in valleys extending well

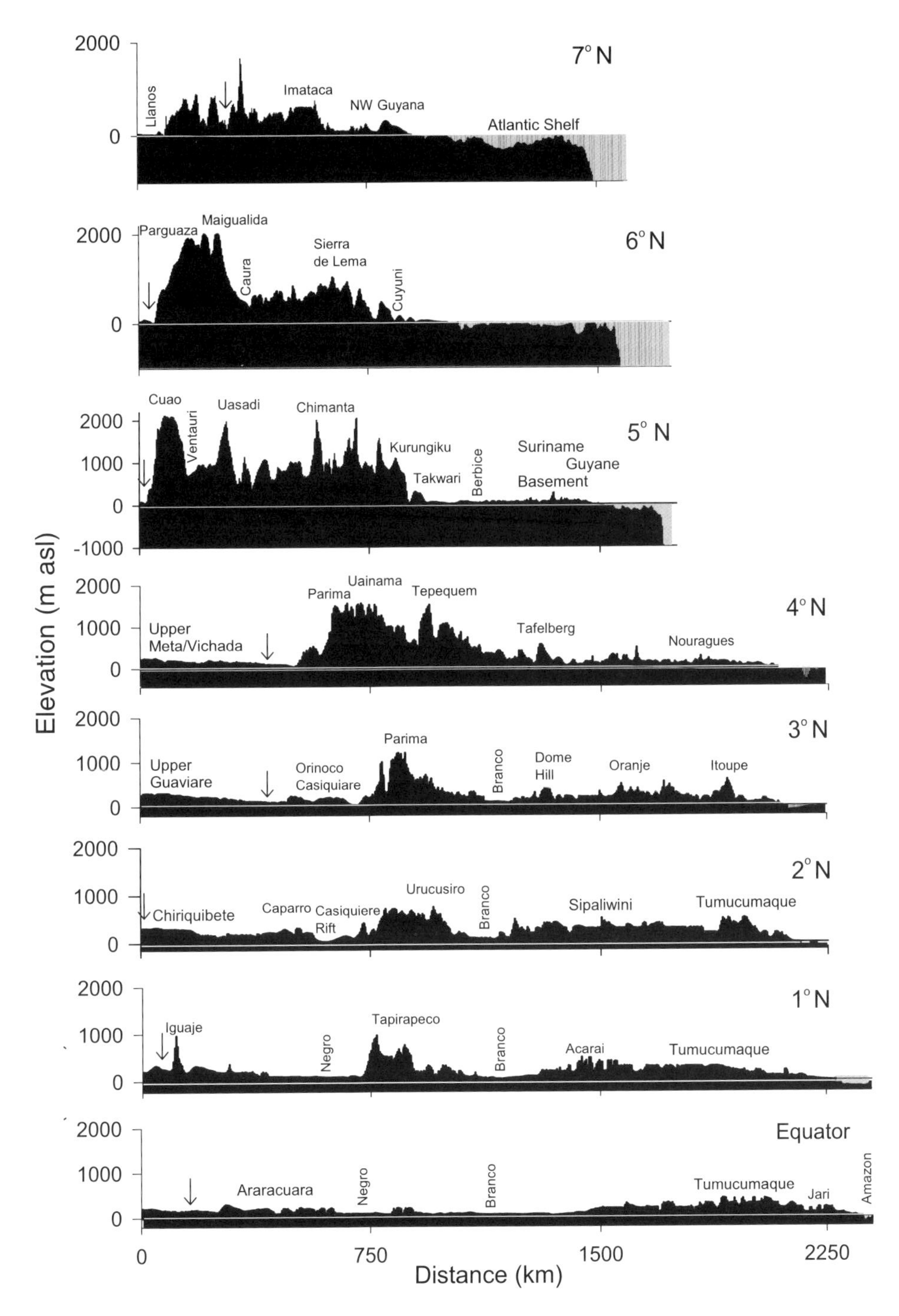

Fig. 2.7. Two-dimensional west to east topographic profiles across the Guiana Shield.

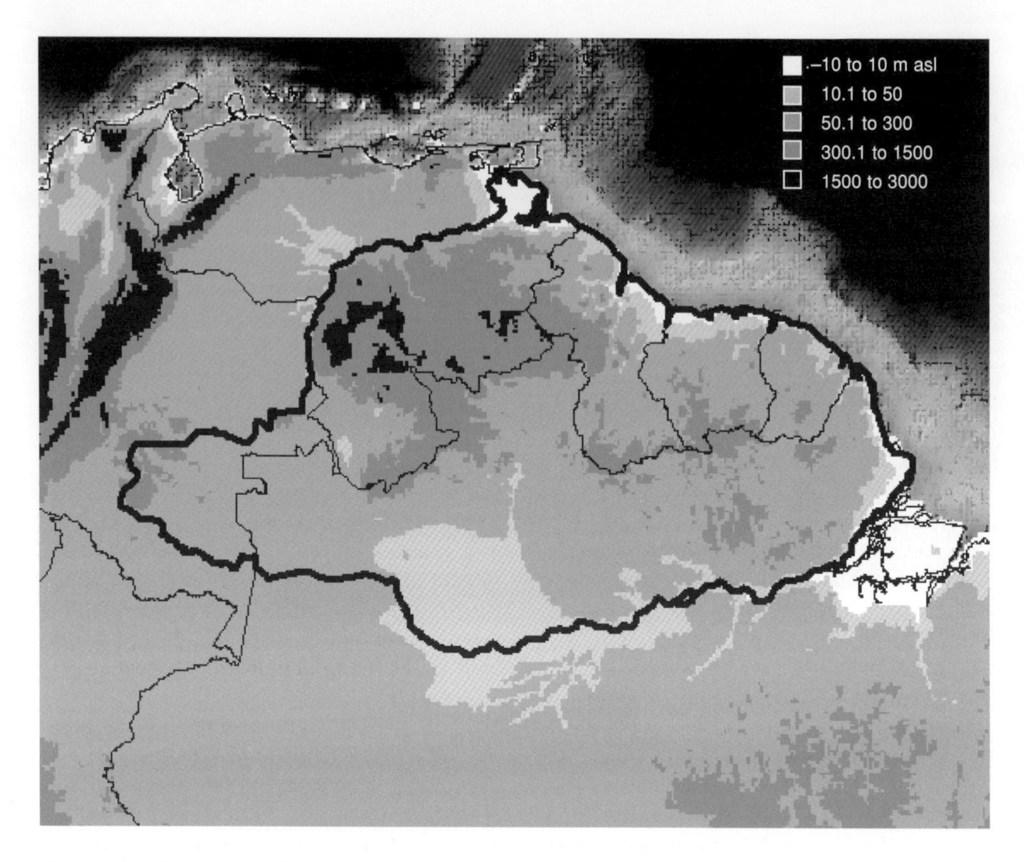

Fig. 2.8. Landforms in the Guiana Shield based on elevation intervals adapted from Daniel (1984), Sombroek (1990) and Huber (1995a). Intervals derived from the ETOPO5′ gridded world elevation coverage with a maximum resolution of 5 arcseconds (9 km^2) (Haxby *et al.*, 1983; Edwards, 1986). Areas in black outside the delimited shield area (thick black line) are >3000 m asl (Andes) or ≤ 200 m asl (Atlantic Ocean).

above sea level and achieve hydraulic gradients exceeding 1 m/km (Peña, 1996) (Fig. 2.7, 6° N). The Gran Sabana, a patch of grassland that extends into Guyana (Rupununi savanna) and Brazil, attains its highest elevation in southeastern Venezuela as a shelf-like plateau between 600 and 1000 m asl.

TUMUCUMAQUE UPLANDS The topography and elevational range of the Tumucumaque Uplands pales in comparison with the Guayana Highlands area. It represents the heart of the Precambrian Guiana Shield environment, where the crystalline basement complex is exposed more than in any other area. Formed principally of granitoid and metavolcanic formations dissected by greenstone belts and, to a much lesser extent, granulite belts, the area culminates in a series of central 'highland' massifs forming an arc from south-central Suriname (Wilhelmina Mts) and then along the southern boundary separating the three Guianas from the Brazilian states of Roraima (Acarai Mts), Pará and Amapá (Tumuc-Humac Mts). From this arc, both the southern and northern uplands slope gently downwards towards the Amazon River and Atlantic sea shelf, respectively (Figs 2.7, 2.8).

CHIRIBIQUETE PLATEAU The Chiribiquete Plateau of south-central Colombia rises to an elevation of 900 m asl and forms the westernmost surficial landscape included in the Guiana Shield (Fig. 2.6B, No. 3) and includes the La Chorrera massif. The western margin of the mainly quartzic, sandstone plateau is separated from the eastern Andes by the deep Tertiary sediments of the Sub-Andean Trough that runs along the northern rim of the Guiana Shield (Fig. 2.6B, No. 5). The sandstones forming the plateau are commonly referred to as the Araracuara Formation. They are of comparable composition to those forming the Roraima highlands, but are considerably younger, being of Palaeozoic, rather than Precambrian, age (Estrada and Fuertes, 1993). They rest unconformably atop metamorphosed rocks forming the Mitú Complex, a series of mainly gneisses thought to have been formed as a consequence of a second, recently postulated, tectonothermal event metamorphosing and remobilizing the basement complex in the western part of the shield (Gibbs and Barron, 1993). The original age over most of this basement has been obliterated by this event, but isotopic dating indicates metamorphic formation around 1.5 Ga, about the time that the Parguaza Granites were forming in western Venezuela/eastern Colombia (Table 2.1) (Fig. 2.7, 6° N).

KANUKU/BAKHUIS HORST The Kanuku Mountains define the northeastern edge of the Tumucumaque Uplands where this meets the North Savannas Rift Valley, a section of the much longer Takutu Graben (Fig. 2.6B, No. 9) (Berrangé, 1977). This massif forms part of the Kanuku Complex, a formation of mainly high-grade metamorphic rocks, paragneisses, migmatites and granulites, which extends over 22,000 km^2 of southern Guyana and represents the western leg of the much longer Central Guiana Granulite Belt (Fig. 2.4B). The origin of the various rock units forming the Kanuku complex is not entirely clear, but a number of pathways have been proposed describing the transition from sedimentary (Berrangé, 1977) and crystalline basement precursors to the present-day metamorphics. The eastern leg of this belt, the Falawatra Group, runs through eastern Suriname to form the Wilhelmina and Bakhuis Gebergte (Mountains) (Gibbs and Barron, 1993). These horsts were formed by the lowering of adjacent lands, the North Rupununi and Berbice basins, through block faulting within the crystalline basement (Berrangé, 1977), rather than an uplifting due to other regional metamorphic processes. The Kanuku horst extends eastward from the savanna strip embracing the Takutu River between the Serra do Tucano (Toucan Hills) in Brazil and Yamat Mountain in Guyana. The horst is dissected by the northward-flowing Rupununi, Kwitaro and Rewa Rivers in the east and reaches its highest elevation, 960 m asl, in the central segment (Mt Naraikipang). The Wilhelmina and Bakhuis ranges extend north from the Lucie River near to the border with Guyana to the uplands forming the headwaters of the Kabalebo and Nickerie Rivers to the west and the Coppename to the east.

ATLANTIC COASTAL SHELF Stretching more than 1550 km between Trinidad (Paria Shelf) and the mouth of the Amazon River (Amazon Shelf) and extending between 20 and 80 km offshore, the build-up of Phanerozoic sediment atop the submarine extension of the Precambrian basement has formed a shallow marine environment. Throughout this zone it is not uncommon to encounter patches extending several square kilometres in size that are covered with less than 2 m of water during low tide. The shelf is particularly wide at the northern end near to the mouth of the Orinoco River. The rapid increase in depth offshore from the mouth of the Courentyne River forms part of the larger downwarping known as the Berbice Basin and represents one of the narrowest points in the shallow marine shelf environment along with the rapid downslope near the Oyapock River on the French Guiana–Amapá border. The modern coastal shelf deposits are believed to have been formed from two sources contributing material over the Cenozoic: (i) the bulk of coarser sandy sediments ejected from rivers spilling directly on to the coastal shelf (Pujos *et al.*, 1990) with some localized longshore redistribution (Krook,

1979); and (ii) lighter silt and mud from the Amazon River transported and deposited along the Guiana coast (Eisma and van der Marel, 1971). Virtually all of the materials contributed via the Amazon are believed to be Andean in origin, with little contribution from the rivers draining the western shield area (Gibbs, 1967; Mario and Pujos, 1998). Nearly 10% (1.5×10^8 t/year) of the total suspended matter ejected currently into the Atlantic from the Amazon is believed to be transported directly northwestward by the Guiana Current (Muller Karger *et al.*, 1988). Finer silt and clay are also transported through more passive longshore drift and ocean swells from the Amazon Cone along the Guiana coast (Wells and Coleman, 1981; Pujos and Froidefond, 1995). These fluid muds form the characteristic intertidal banks occupying most of the near coastline. The relative contribution of deposits from each of these two sources has changed over time in response to geological and climatic events. Oscillating transgression and regression of marine waters due to climate change, changes in forest cover and, over a longer timescale, changes in the erosion bevel of the Amazon and Guiana Shield regions due to regional uplifting and downwarping have all affected the dynamics of coastal sedimentation along the Guiana coast over the course of the last 100 million years (Bleakley, 1957; McConnell, 1969; Nota, 1969; Pujos *et al.*, 1990; Mario and Pujos, 1998).

Major faults, downwarps, rift valleys and geosynclines

Geomagnetic, gravity and seismic surveys show that the Guiana Shield is fractionated by hundreds of faults and rift-like structures (compiled by A.K. Gibbs in Fig. 1.7 of Gibbs and Barron (1993)). By far the majority of these structures contained within the crystalline basement core of the shield find their origins in the Precambrian, though many have been reactivated later. Most of these range from 1 to 100 km in length, but a few major faults extend for hundreds of kilometres. Downwarping involves faulting and deformation at a much larger scale and in response to craton-altering mechanisms, such as the uplift of the Andes or eustatic tilting of the continental crust during periods of sea transgression into the major river basins within and surrounding the shield. Several of the largest and most important of these identified in Fig. 2.6 are described briefly below.

Pisco-Juruá megafault

This major fault is thought to run north-northeast from the western Cordillera of Peru, across the upper Amazon (Solimões) Basin along the alignment of its northeastward flowing tributary, the Juruá River, and ends at the Berbice Basin on the northern rim of the shield area (Fig. 2.6A and 2.6B, No. 4). The megafault has been proposed as a tensional feature running roughly perpendicular to the main mid-Atlantic rift axis and remaining active until the Mesozoic, though its original activation may have been as much as 1 billion years earlier, during the Nickerian Episode (Gibbs and Barron, 1993). It has been implicated in the development of the Takutu Graben as the interaction between tensional and torsional forces between the Kanuku/Bakhuis horst formations and the megafault 'twisted' the crust in the graben vicinity resulting in a rhombochastic orientation of the horst-graben association (Szatmari, 1983, cited in Gibbs and Barron, 1993).

Sub-Andean Trough

A massive sedimentary depression lies directly east-southeast of the Andean Cordillera as it wraps around the northern and western margin of the South American craton. It is wedged between the Andes and the Guiana Shield in the north, the Palaeozoic Amazon Downwarp in the middle and the western margin of the Brazilian Shield in the south (Fig. 2.6B, No. 5). The area accounts for a large proportion of the lowland tropics in western Brazil, Peru, Ecuador, Colombia and, to a lesser extent, Bolivia. The area is typically separated from the Amazon Downwarp at the Iquitos Arch near the confluence of the Solimões with the Napo, Ucayali and Javari Rivers. The Sub-Andean Trough (after Devol and Hedges, 2001) has been variously referred to as the Andean Geosyncline (Williams *et al.*, 1967), pre-Andean region (Fittkau, 1971), Sub-Andean Foreland (Putzer, 1984), Amazon Foreland Basin (Cuenca de antearco de la Amazonia) (Räsänen, 1993), and Cis-Andean Plains (Sternberg, 1995), among others. The crystalline basement rock beneath the trough is entirely covered with relatively recent, but very deep, sedimentary deposits eroded from the eastern face of the Andean highlands. The relatively compressed timeline of deposition and change that has affected this and the Amazon Downwarp areas has produced one of the largest Tertiary sedimentary basins in the world. Sediments found in this depression are thought to have been deposited over a series of episodes dating back to the Palaeocene, 70 million years ago. These are variously represented by the Ipururo and Pebas (Solimões) formations of the Plio-Pleistocene, the Curaray, Chambira, Arajuno, Chalcana (Pozo), Cachiyacu (Rio Branco) and Tiyuyacu (Contamana) formations of the Mio-Eocene, and the Cruzeiro Formation of the Palaeocene (Beurlen, 1970; Campos and Bacoccoli, 1973; Galvis *et al.*, 1979). Consisting mainly of river alluvium, sands, conglomerates, ferruginous clays, silts and sideritic ironstones, the deposits can attain depths between 4000 and 10,000 m (Putzer, 1984; Gibbs and Barron, 1993; Räsänen, 1993).

Guri fault system

The 400 km long Guri or Guri-El Piar fault is a system or zone of multiple transcurrent faults, sheared and partially mylonitized, striking west-southwest to form the southern boundary of the Imataca Complex (Fig. 2.6A and 2.6B, No. 6) (Kalliokoski, 1965; Gibbs and Barron, 1993; Goodwin, 1996). Outliers of the Imataca Complex occur south of the fault system (USGS and CVG-TM, 1993), but it represents a general line of contact between Archean rock in the north and the (predominantly) Proterozoic granite-greenstone rock formations to the south (Gibbs and Barron, 1993). The fault system runs parallel to the El Pao Fault to the north and is believed to have developed approximately 1.2–1.3 Ga during the Nickerian Episode (Gibbs and Barron, 1993). Both the Guri-El Piar and the El Pao strike parallel to the projected alignment of the Pisco-Juruá Megafault (similarly of Nickerian age) and roughly perpendicular to the hypothesized orientation of the Atlantic fracture zone during its formative stages of development prior to mid-Atlantic rifting.

Amazon Downwarp

The lowland area of sedimentary cover running along the main stem of the Amazon for 3500 km between the southern rim of the Guiana Shield and the northern rim of the Brazilian Shield is referred to here as the Amazon Downwarp (Fig. 2.6B, No. 7) and is somewhat narrower than the geochemical province described as Central Amazonia by Fittkau (1971). This area accounts for almost half of the entire Amazon Basin (Stallard and Edmond, 1987). It is separated from the western Sub-Andean Trough by the ridge of uplifted and exposed Precambrian rock of the Iquitos Arch running north–south across the upper Solimões in eastern Peru/western Brazil (Putzer, 1984). The Amazon Downwarp is dissected by three additional uplifted arches all running roughly parallel to one another (Fig. 2.6B, No. 7). The Purús Arch runs north-northwest from the lower Madeira and across the Purús and Amazon Rivers just upstream from their confluence. The Carauarí Arch extends between the lower Putamayo River in Colombia southward to the upland region separating the Juruá and Purús Rivers. The Gurupá Arch runs along a similar alignment east of the Rio Xingu mouth and along the eastern flank of Marajó Island, at the mouth of the Amazon. The arches are thought to have played a prominent role in regulating the extent of sea transgression into the downwarped area and subsequent deposition of sediments. The Iquitos Arch is believed to have been the easternmost upland edge of a palaeo-watershed draining westward during the Tertiary prior to the uplifting of the Ecuadorian Andes and closing of the Pacific outlet (Putzer, 1984).

The Amazon Downwarp interfaces with the southern rim of the Guiana Shield through a series of sedimentary formations running parallel to the main stem of the Amazon River. The oldest, and furthest from the river, date back to the Palaeozoic. They consist of the Prosperança, Trombetas, Jatapu, Lontra, Ereré, Curua, Monte Alegre, Itaituba and Nova Olinda Formations in order of decreasing age between the Cambrian and upper Permian periods of the Palaeozoic (Fig. 2.4L) (Putzer, 1984; Gibbs and Barron, 1993). These bands of silts, sands, conglomerates, evaporates, sandstones, shales and ironstones grade into the upper Cretaceous (Mesozoic) red clays, silts and sandstones of the Alter De Chaõ Formation (Fig. 2.4M). These are partly covered by more recent sand and silt deposits of the Barreiros, Piracurú, Solimões, Içа, Tucunaré and Boa Vista formations dating from the Miocene to the Plio-Pleistocene (Fig. 2.4N). Recent silts and clays are deposited along and across the main river channel floodplains of the Amazon.

The distribution of many of the older series of sediments reflects oscillation and variation in sea levels over the past 500 million years and changes in the angle of the South American cratonic planation surface as deformation and tectonism changed the tilt of the continent (Klammer, 1984). The distribution and unique composition of many of these sedimentary events, such as the Monte Alegre greenish sands and Belterra Clays, has fundamentally influenced forest life in the downwarp, from the green waters of the Tapajós to the Babaçu forests near Santarém.

The classical volume edited by Sioli (1984b) gives an extensive and comprehensive overview of the main features of the Amazon Downwarp and its linkages with the peripheral landscape.

Casiquiare Rift

Bellizia (1972, cited in Blancaneaux and Pouyllau, 1977) described this structure as one of three major fracture zones running parallel from north to south through the Venezuelan Amazon (the others being the Supare-Mavaca and Ventauri Rifts, Fig. 2.6A and 2.6B, No. 11). It has also been referred to as the Parguaza Subduction and Atabapo-Negro suture zone (Gibbs and Barron, 1993). This suture-like structure running north-northwest in southwestern Amazonas state of Venezuela effectively separates and deflects the upper reaches of the Negro (Guianía) and Orinoco Rivers, leaving the Casiquiare River (sometimes 'canal') to act as one of only two modern connections between the Atlantic-draining and Amazon-draining waterways of the shield (Fig. 2.9A). Simple Bouger gravity anomalies calculated for the area underscore the existence of a high density body beneath the suture running from 3° N and extending to a point just south of Puerto Ayacucho near the confluence of the Guayapo and Autana Rivers to the east of the Orinoco (USGS and CVG-TM, 1993). It is postulated by Tassinari (in Gibbs and Barron, 1993) to be the subduction zone of a tectonic 'palaeoplate' linked to a second Post Trans-Amazonian tectonothermal episode, around 1.9–1.76 Ga, that reactivated basement rock during the Parguaza Episode (also Rio Negro–Juruena Episode) around 1.5 Ga. Formations to the west of the suture zone, such as the gneisses of the Mitú Complex in Colombia, show concordant ages around 1.5 Ga, while those basement granitoids east of the zone are consistent with the TATE around 2.1–1.8 Ga. Others interpret the zone as simply a mobile belt formed of high-grade (granulite) metamorphics (Bridger, 1984, cited in Gibbs and Barron, 1993). Regardless of its exact origins, the 300 km wide, localized subsidence created in the zone has produced a hydrological peneplain with stagnant drainage culminating in the Rebalse de Macavacape along the upper reaches of the Rio Atabapo (Fig. 2.9A). The famous bi-directional flow of the Casiquiare Rivers is in part affected by the lack of bevel within the rift and the rise and fall of flood stage within the basin. This can be seen clearly in a SAR image taken as part of the May–July, 1996 JERS-1 mission (Fig. 2.9A).

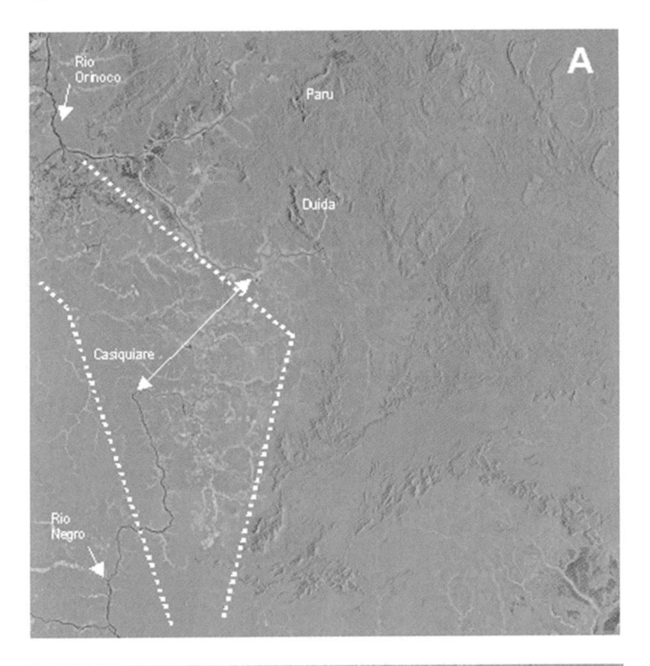

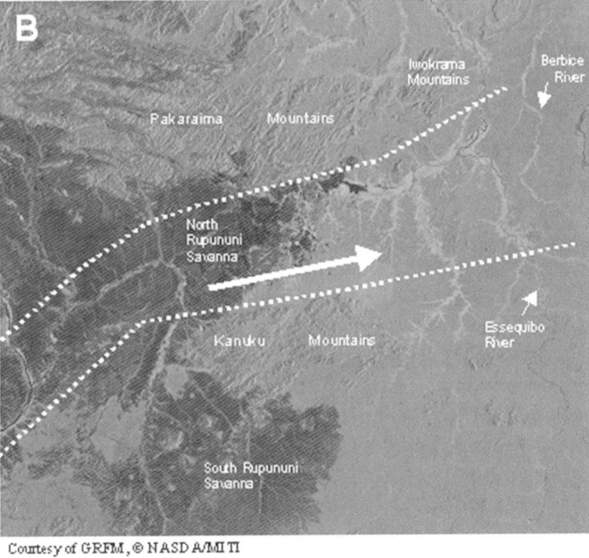

Courtesy of GRFM, © NASDA/MITI

Fig. 2.9. JERS-1 radar image of two major rift valleys in the Guiana Shield. (A) The Casiquiare Rift area during the May wet season with rift sutures suggested by Bellizzia (in Blancaneaux and Pouyllau, 1977). (B) The Takutu Rift area with north and south rift boundaries as described by Berrangé (1977) and Crawford *et al.* (1984). Images were taken during the May wet season, 1996. Flooded forest along the main rivers, creeks and swamplands is identified by a double bounce reflectance signature seen here in white against contrasting upland terra firme forests (in grey) and open, cleared savanna or aquatic habitats (in black).

Takutu Graben

The savanna valley that runs along the northern flank of the Kanuku Mountains and the southern flank of the Pakaraima Mountains in southern Guyana and the northeastern Brazilian state of Roraima is the product of two parallel major faults with a series of intervening minor faults (Berrangé, 1977; Crawford *et al.*, 1984). The (re-)activation of these faults created a cascading series of downthrown blocks of Precambrian basement rock to a depth of over 7 km (Crawford *et al.*, 1984). The graben itself extends between its northernmost point located upstream from the village of Apoteri, near the confluence of the Essequibo and Rupununi Rivers and southern limit on the Rio Branco (Fig. 2.9B). Both Boa Vista, the state capital of Roraima, and Lethem, the regional capital of southern Guyana, rest atop the sedimentary fill of the rift valley. The South Rupununi savannas do not fall within the graben, resting atop much older Kanuku Complex paragneisses, granulites and granites formed during the Precambrian (Berrangé, 1977).

Landforms

Considerable effort has been made by geologists, geographers, soil scientists and botanists to characterize various subregions of the Guiana Shield by a series of landforms linking topography, geological history, soil and hydrology (Sombroek, 1990; Huber, 1995a). Here, these features have been combined into five large landform categories, each encompassing different subsets of the shield macro-features described above. Elevation is considered as the principal factor, given its strong relationship with geology, soil formation and hydrology (through erosion bevels) in landscapes dominated by the Precambrian (Gibbs and Barron, 1993).

Recent Coastal Plains (–10 to 10 m asl)

Along the Atlantic coast between the mouths of the Orinoco and Amazon Rivers, a narrow plain of flat, low-lying sediments extends up to 80 km inland from the present-day coastline (Fig. 2.8). This belt accounts for approximately 5% of the shield area, reaching its greatest width near the mouth of the four largest rivers draining the area, the Orinoco, Essequibo, Courantyne/Berbice and Amazon (Fig. 2.8). These areas are subject to daily tidal oscillations extending many kilometres upstream. In fact, the Recent Coastal Plains rarely rise higher than 10 m asl and the vast majority of this landform that has not been empoldered is subject to tidal submersion.

The coastal belt consists of a series of unconsolidated clay and silt layers deposited over the past 10,000 years. Fluid muds, a dense suspension of solid clay and silt particles, can extend for kilometres along the seaward margin of the more stable coastal mud banks (often anchored by mangrove). These represent the current land-building front between the built-up coastal plain backlands and open marine environment of the coastal shelf. Silt and clay forming the fluid muds have largely arrived along the coast through the conveying action of the Guiana Current running northwestward along the Atlantic coast (e.g. Ryther *et al.*, 1967; Wells and Coleman, 1981; Muller Karger *et al.*, 1988). These currents deflect the massive Amazon sediment plume northward, transporting about 10% of this material from the Amazon River mouth and depositing this along the Guiana coast (Muller Karger *et al.*, 1988). The ejection of sediment from the many rivers draining the shield interior and spilling their sediment load directly into the Atlantic has compounded deposition, creating localized sediment ‘bulges’ on the seaward side of the major rivers. The extent and distribution of coastal sediments has been shaped over time by variations in sea level, coastal downwarping, tidal strength, rate of sediment injection, and most recently, agricultural development and the installation of coastline defences. Despite dramatic increases in the rate and area of deforestation in the Amazon Basin, the impact of this phenomenon on existing nutrients and sediment levels in the mainstem Amazon is thought to remain negligible in comparison to the contribution made through natural weathering of the Andean Highlands (DeMaster and Aller, 2001).

Efforts in Guyana to 'hold back the sea' have created the most heavily populated agricultural zone along the Guiana Atlantic seaboard, but only after considerable cost and effort. Up until the early 1700s, the Atlantic coastline largely remained a mangrove- and palm-dominated tidal wetland interspersed with upland patches of forest. Extensive coastal defence works in the 1700s opened up a large empoldered landscape ideal for lowland sugar cultivation. Since 1850, however, the reclaimed coastline has slowly moved inland as sediment accrued along the artificially abrupt coastline (Case, 1943). In many areas of Guyana, coastal defence and land reclamation works have retained a polder-like zone of land that is up to 15 m below sea level. This area is repeatedly affected by breaches in sea defences and many parts have returned to their brackish condition. Where the solid rock of the Precambrian Rolling Hills landform extends close to the coastline, particularly in central French Guiana and northern Amapá, the coastal plains are largely restricted to the offshore tidal flats consisting of mud banks and fluid muds up to several metres deep.

The Recent Coastal Plains landform encompasses both the Young Coastal Plain (Demerara Formation) and Old Coastal Plain (Coropina Formation) zones typically used to describe the Cenozoic deposits circumscribing the Precambrian shield centre (Fig. 2.4K) (Bleakley, 1957; Daniel, 1984; Gibbs and Barron, 1993).

Tertiary Sandy Plains (10–50 m asl)

Rising above the Recent Coastal Plains and the riverine floodplains of the Amazon and its main tributaries, the Tertiary Sandy Plains cover approximately 30% of the Guiana Shield land area. This landform roughly approximates the *sandy Plains* proposed by Sombroek (1990) and is included in Huber's (1995a) *Lowlands* category. Formed of heavily reworked and leached sands, they cover a large belt running parallel to the Recent Coastal Plains along the Atlantic and covering much of the lowlands surrounding the Amazon, lower Negro, Japurá/Caquetá and lower Branco Rivers in northern Brazil/southern Colombia. Much of the topographic relief of the underlying Precambrian rock is buried beneath the numerous stratified layers of sand and loam that can reach several thousand metres thick in regions of significant crustal downwarping along the lower Berbice/ Courentyne Rivers.

The Tertiary Sandy Plains include sedimentary formations associated with periodic sea level change and eustatic movement of the shield area since the late Cretaceous/early Tertiary and include: the New Amsterdam, Georgetown, Pomeroon, Courantyne and Berbice Formations in Guyana, the Nickerie, Onverdacht, Coeswijne and Zanderij Formations in Suriname, the Alter De Chaõ, Pirarucú, Solimões, Iça and Boa Vista Formations in Brazil, Mesa Formation in Venezuela and Pebas Formation in Colombia. Subsequent erosion of these plains has led to a series of moderately sloped valleys that are subject to widespread vertical infiltration where sands are deep and largely devoid of clay and loam.

Precambrian Rolling Hills (50–300 m asl)

By far the vast majority of the shield landscape rests atop granitoids formed during the massive Trans-Amazonian Tectonothermal Episode (Fig. 2.4D). Combined with Precambrian greenstones in the north and south and metavolcanics of the central shield region, a gently undulating terrain has been created through synclinal folding and differential weathering of these and surrounding associations over the last 2 billion years. Today, this can be seen as a series of (flat-topped) hills, ridges and valleys with an elevation of between approx. 80 and 300 m.

The Precambrian Rolling Hills and the stratigraphic units underlying them account for more than 50% of the shield area. In Guyana, this landform has been included in the *Precambrian Lowlands* of Daniel (1984), along with a number of other structures described independently below. Sombroek (1990) described this area as the *crystalline shield Uplands.* They include much of the Imataca terrain of northeastern Venezuelan Guayana, extending eastward through the Northwest region of Guyana to the lower Essequibo River. Dissected by the Berbice Basin and Falawtra Group in Suriname, the Rolling Hills continue eastward across most of Suriname south of the Zanderij, across virtually all of (French) Guyane and into west-central Amapá. Above the 2nd parallel, this landform covers most of Pará, Roraima and Amazonas states, only intersected by the Iça Formation running between the Rio Branco and Rio Negro. The Rolling Hills continue westward into eastern Colombia and most of the southern half of Amazonas state in Venezuela.

Guiana Uplands (300–1500 m) and Guayana Highlands (1500–3000 m asl)

The second most important landform found in the Guiana Shield, referred to here as the Guiana Uplands, is dominated by structures forming the Roraima, Vaupés and Uatuma Supergroups and intrusive rock suites of varying age (see Figs 2.4 and 2.8). The *Inselberg complexes,* and parts of the *Sandstone table lands* and *crystalline shield Uplands* of Sombroek (1990) are included here in the Guiana Uplands. A good part of the Guayana Highland and many inselbergs of the Tumucumaque Upland macro-features are included in the Uplands landform, along with the Chiribiquete Plateau in Colombia and isolated massifs and peaks rising above 300 m that are scattered throughout the surrounding lowlands (Figs 2.6B, 2.7). Together, they are represented by a number of mountainous areas ranging from 300 to 3000 m in elevation and covering approximately 15% of the shield area (the Roraima Supergroup covers 10% of the shield area or 163,000 km^2 (McConnell, 1959), 45% of this area formed by the Pakaraima Highland massif alone (Gibbs and Barron, 1993)). The elevation of these areas has been created through region-wide weathering, depositional and volcanic processes and more localized uplifting along block fault margins. This is in part referred to in Guyana by Daniel (1984) as the Pakaraima Mountain Region and would incorporate both the Upland and Highland classifications provided by Huber (1995a) for the Venezuelan Guayana portion of the shield. Many lower, outlying and isolated peaks and hills (300–500 m asl) not considered in these classifications are also included here.

At 3014 m asl, Sierra de la Neblina on the Venezuelan Amazonas–Brazilian Roraima border reaches the highest elevation in the Guiana Shield. An isolated and dramatic chain of clustered table top escarpments, Neblina is one of more than 70 tablelands, often referred to as tepuis, mesas, tafelbergs or chapadões, found throughout eastern Venezuela (50), western Guyana (12), Suriname (1) and Brazil (7) that together form the Guayana Highlands. They are the uppermost remnants of a once widespread, +3000 m thick, sedimentary plain that is believed to have covered as much as one-third of the modern-day area of the Guiana Shield (Gansser, 1954). The second highest peak in the shield, Roraima, rests at the international boundary between Brazil, Venezuela and Guyana, some 800 km northeast of Neblina. Maximum elevations drop off rapidly from the twin highland centres formed by the main Pakaraima Mountain massif occupying eastern Venezuela and western Guyana, and the string of clustered tepuis running southwest across southern Bolivar and Amazonas states in Venezuela. Juliana Top in Suriname reaches a height of only 1230 m, while Mont Itoupe at 840 m, the highest peak in French Guiana, is barely one-third the height of Neblina. Contrasting soils, climate and vegetation further underscore the distinction between the Upland and Highland landforms. The highest peaks in Colombian Guayana and the shield area forming part of the Brazilian states of Amazonas and Pará are also of modest elevation and represented by isolated massifs within

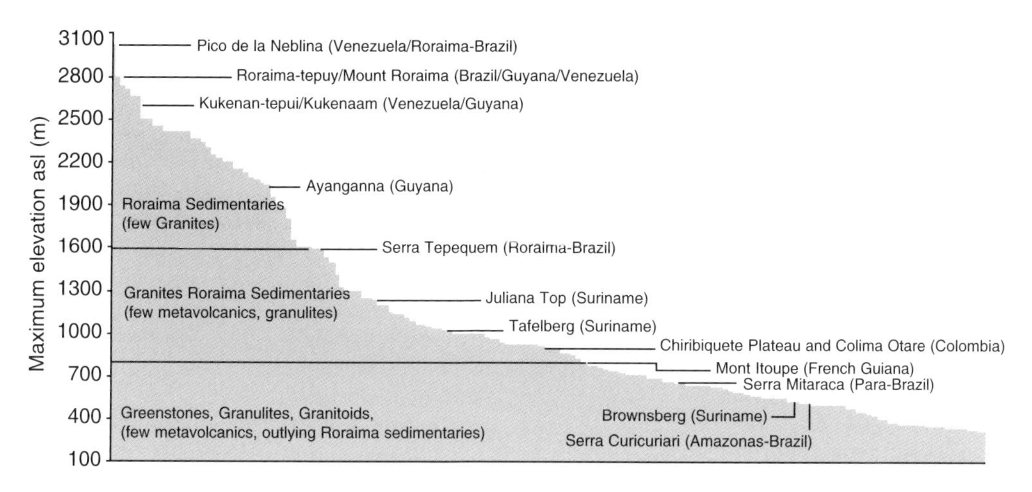

Fig. 2.10. The known summit elevation of 185 notable tepuis/mesas, horsts, inselbergs/lajas, mountains, hills and plateaus found in the Guiana Shield. The name of the peak representing the highest elevation in each of the countries forming the shield region is in brackets.

an otherwise flat, low-lying landscape (Fig. 2.10).

There is a strong relationship between the geological origins of the highland areas and their current summit elevations with a transition from sedimentary rock-dominated structures at the higher elevations through to igneous-dominated structures at lower highland elevations (Fig. 2.10). Generally speaking, areas in the Guiana Shield attaining summits in excess of 1600 m are sedimentary in origin and belong to the Roraima Supergroup. This would include virtually all of the *altiplanicies* and *picos* found in the Venezuelan Guayana, northern Roraima State and the Pakaraima Mountains of Guyana and is roughly consistent with the Highlands classification given by Huber (1995a). There are several exceptions, however (e.g. Cerro Coro Coro, C. Yavi), linked to the intrusive granitic batholith forming large parts of the Parguaza Granite (Vaupés Supergroup) in northwestern Amazonas state of Venezuela and adjacent Colombia (Gibbs and Barron, 1993).

Highland areas ranging from 800 to 1600 m originate from a variety of geological structures, including those formed through lithified sediments (of the Roraima Supergroup), granitic intrusives, granulitic horsts and ancient metamorphosed volcanics (Uatuma volcanics). This elevational band is consistent with the Uplands (500–1500 m) of Venezuelan Guayana (Huber, 1995a), but can be extended at the shield level to include many areas of the Pakaraima, Wassarai, Acarai and Kamoa Mountains in Guyana, the Roraima Highlands in Brazil, the Chiribiquete Plateau, Colima Otare, C. Main Hanari and C. Campana in Colombia as well as horsts and other lower massifs scattered in Guyana (Kanuku Mts) and Suriname (Bakhuis Mts, Hendrik Top, Tafelberg, Wilhelmina Mts).

Hills and mountains of sedimentary origin are virtually absent at an elevational range of 300–800 m. Nearly all of these landscape features owe their elevation to a combination of regional tectonic uplifting, changes in the erosion bevel (caused by eustatic tilting) and more localized differential weathering processes that have eliminated surrounding country rock to expose hard, silica-rich granites and granulites and laterite-capped dolerites and other silica-poor rock types as isolated massifs and ridgelines. These structures account for most of the highland areas found in French Guiana, Suriname and the Brazilian states of Amapá, Pará and Amazonas.

Weathering rates

Rock weathers through two main processes: one driven by mechanical agents, such as wind, ice or water, and another working through a slow chemical breakdown, dissolution and export of materials through groundwater flows.

An abundant forest cover and extended weathering history has effectively restricted natural physical weathering to the steepest slopes of the Guayana Highlands. Rock formations naturally prone to erosion were removed from the surface long ago, leaving only the most resistant rock types. The combination of these factors has virtually eliminated physical weathering as a collective force degrading rock surfaces in the region (but see Chapter 8 on mining).

The estimated chemical weathering rate for the shield region (10 mm/ka after Edmond *et al.*, 1995) is one of the lowest in the world and 8–70 times lower than rates estimated for Andean drainages using either a dissolution rate or cationic concentration method as the basis for calculation (Stallard, 1988; Edmond *et al.*, 1996; Mortatti and Probst, 2003).

In part, the low modern chemical weathering rate estimated for the shield is due to a high quartz (SiO_2) abundance. Quartz is the most common siliceous mineral found in the Guiana Shield (distribution of A, B, D and E in Fig. 2.4 and mineral composition in Fig. 2.5) but is also notoriously impervious to dissolution, registering one of the lowest rates among common silicate minerals (Table 2.3 in Lerman, 1994). Rock types with much lower or no silicate content (e.g. carbonates, evaporites) tend to have much higher dissolution, and therefore estimated weathering, rates regardless of acidity. Without persistent high soil acidity (e.g. under more arid and/or cooler climatic conditions), it is not unreasonable to suggest that the weathering rate of the shield's crystalline basement would be even lower than current estimates, although wind-driven weathering would in all likelihood increase. Wind-driven deposition has been implicated in the development of parabolic dunes and sand sheet features of the (Pleistocene) Boa Vista Formation in Roraima, Brazil and southern Guyana (Latrubesse and Nelson, 2001), along the Rio Negro (Filho *et al.*, 2002) and in the Colombian–Venezuelan llanos north of the shield region.

This combination of geological, climatic and biological conditions promotes highly acidic (low pH) soil conditions (through microbial/root respiration of CO_2 that combines with water to form carbonic acid, H_2CO_3) and thus a predominance of chemical (low pH) over physical (high pH) weathering of rock. The consequence of an environment dominated by chemical weathering is an *in situ* accumulation of these weathering products in the form of soil. Environments dominated by physical weathering forces export these products, leading to a decrease in soil thickness. In the Guiana Shield, soil profiles are believed to be thickening (Mortatti and Probst, 2003). It is the considerable depth of many soil types in the region, combined with climate-driven changes to their physical properties, that leads to their strongly controlling effect on forest composition, productivity and structure.

Soils and Soil Fertility – Poorest of the Poor

Generally speaking, soils of the wet tropics are acidic (pH 3.5–6), poor in available nutrients (especially phosphorus: Schulz, 1960; Vitousek, 1984) and calcium (Schulz, 1960; Jordan and Herrera, 1981), often contain high aluminium and iron levels and have a low cation exchange capacity (Sanchez, 1976). It is generally accepted that the main determinants of modern-day wet tropical soil properties have to do with the mineral composition of the original parent material, (historic) water table dynamics, acid chemistry, biological action, human modification and weathering age. Each soil 'type' is the product of its own unique life history, changing continuously (but often indiscernibly) in response to its surrounding environment. Modern soil taxonomy attempts to simplify an otherwise continuous variation in soil characteristics by boxing these into general soil groups that exhibit certain properties at a particular point in time (FAO, 1998). Importantly, the distribution of these soil types, and the impact of their wide-ranging properties upon tropical forest growth, composition and distribution, vary considerably from one region to another (Sanchez, 1976; Richter and Babbar, 1991).

Soil properties exert an important direct or indirect control function over tropical forest vegetation composition, structure and function at a wide range of scales (e.g. Richards and Davis, 1933; Richards, 1952; Schulz, 1960; Ogden, 1966; Hall and Swaine, 1976; Ashton, 1977; Huston, 1980; Lescure and Boulet, 1985; Vitousek and Denslow, 1987; Gentry, 1990; Terborgh, 1992; ter Steege *et al.*, 1993; Clark, 1994). As more data have accrued and more interdisciplinary research has been conducted in areas previously considered inaccessible, it has been possible to further evaluate the specific mechanisms driving tropical forest soil–vegetation relationships in light of other intervening factors, such as climate and disturbance (e.g. Huston, 1994; Clinebell *et al.*, 1995; Duivenvoorden and Lips, 1995; ter Steege and Hammond, 2001). Soils are an important medium through which underlying differences in geology, hydrology and topography influence the spatial transitions in composition and structure of tropical forests. Their value in explaining forest patterns across the landscape mosaic of the Guiana Shield is of particular significance (see Chapter 7).

This section aims to provide a general overview of the classification, distribution and characteristics of soils in the Guiana Shield, often drawing comparisons with other tropical forest regions to emphasize the spatial contrasts that typify this tropical pedosphere.

Soil classifications systems used in the Guiana Shield

There are four major soil classification schemes that are currently used to categorize soil properties across the Guiana Shield:

1. FAO/UNESCO system and its offspring, the WBR (World Base Reference for Soil Resources) international system are used increasingly throughout the region, but particularly in Guyana and Suriname (FAO-UNESCO, 1974; FAO, 1988, 1998).
2. US Soil Taxonomy classification scheme that has been used widely in Guyana, Suriname, Colombia and Venezuela (USDA, 1996).
3. Sistema Brasileiro de Classificação de Solos (CiBCS) used throughout the Brazilian Legal Amazon (EMBRAPA-CNPS, 1999).
4. Référéntial Pédologique (Baize and Girard, 1995) and its predecessors (e.g. Leveque, 1961; Aubert, 1965) used in French Guiana.

Guyana (Stark *et al.*, 1959; Khan *et al.*, 1980) and Suriname (van der Eyk, 1957) also have used rudimentary soil classification series based on the more common soil forms mainly encountered along the Atlantic coast and sedimentary plains. In Guyana, this consists of several 'series' covering soils derived from igneous intrusives (Prosperity Creek), crystalline basement and other silica-rich igneous rock formations (Durban) and unconsolidated Tertiary sediments (Tiwiwid and Kasarama) (e.g. Stark *et al.*, 1959).

The reference base of the FAO WBR consists of 153 soil units falling variously into 30 main soil groups. Thirteen of these main groups occur in the Guiana Shield, but only five account for approx. 90% of the soils in the area and two widely distributed lowland soil groups, the Ferralsols and Acrisols, represent more than 65% of the shield soil cover.

Classifying soils in the Guiana Shield using the 12 Soil Orders of the US Soil Taxonomy also pinpoints five of these that cover over 90% of the area. Again, two of these, the Oxisols and Ultisols, account for three-quarters of the soils known in the region (e.g. Westin, 1962).

The Brazilian system of soil classification consists of 70 soil groups. Five of these account for the majority of the soil cover in the southern half of the shield where they have been used. Two of these, the Latossolos and Podzólicos, are believed to account for more than 70% of the soils in the entire Brazilian Legal Amazon (Volkaff, 1984; Prado, 1996).

Soil cover in Guyana can be assigned mainly to four large groups within the French system of soil classification, in part due to the small area and rather uniform Precambrian landscape relative to the other parts of the shield region. Over three-quarters of the area can be included under the broad groups headed by the Sol Ferrallitiques, Sols ferrugineux, Sols podzoliques and Sols hydromorphes (e.g. Leveque, 1961; Blancaneaux, 1973).

There is some complementarity between the various soil classification systems, but generally speaking, less group-by-group equivalency than one would hope for. These asymmetric relationships are expressed through inconsistencies in the assignment of field samples across systems. For example, van Kekem *et al.* (1996) relate acri-haplic and acri-xanthic Ferralsols of central Guyana to the Kandiudults (Ultisols) of the US Soil Taxonomy system, while Dubroeucq *et al.* (1999) pair haplic and xanthic Ferralsols with Haploperoxes and Acroperoxes (Oxisols). Relationships between the main soil groups in one system and those of another are often discordant, with elements found within various major groups in one system often falling within a single group under another.[2]

The FAO/WBR system has been adopted here to describe and compare soils found within the Guiana Shield and contrast these with the soils found in other tropical forestland areas. Reference to other soil names and descriptors is retained where this will assist in linking local knowledge with the broader classificatory framework.

Main soils in the Guiana Shield

At least 43 of the 153 global soil units recognized under the FAO WBR system (FAO, 1988) are currently known to occur in the Guiana Shield area (Schulz, 1960; Leveque, 1961; Westin, 1962; FAO, 1965; Blancaneaux, 1973; RADAMBRASIL, 1973–1978; PRORADAM, 1979; Gavaud *et al.*, 1986; Sombroek, 1990; Duivenvoorden and Lips, 1995; UNEP *et al.*, 1995; van Kekem *et al.*, 1996; Ramos and Blanco, 1997; Dubroeucq *et al.*, 1999). More than 65% of the surface area, however, is covered by only nine soil units falling in two of the main reference groups: the Ferralsols and Acrisols. The remaining 35% of the area consists of approximately 34 units within 11 other reference groups reflecting a wide range of soil conditions (Fig. 2.11). Soils are strongly, but not exclusively, linked to topography and landform (Sombroek, 1990), and the main soils in the Guiana Shield have been grouped below according to their relative topographic position.

Highland soils

Highland soils are best described as thin, young and weakly developed formations distributed in relatively small patches with widely ranging properties strictly influenced by localized biogeochemical conditions. They are often intermixed with other soil groups typically associated with more extensive lowland formations. Variation in biogeochemical conditions is strongly associated with small-scale topographic relief.

CAMBISOLS These soils are patchily associated with valleys and pediments of the Pakaraima Mountain area in eastern Venezuela–western Guyana and the foothills of the Kanuku Mountains in Guyana (Fig. 2.11A, striped areas). They can vary in depth, have a relatively high cation exchange capacity (CEC) (FAO, 1998) and are composed of the younger weathering products of mainly basic intrusive formations (FAO, 1965). In the Pakaraima Mountains, these are largely associated with the Avanavero Suite (Gibbs and Barron, 1993). Cambisols are not restricted to highland regions alone. The mottled clays characterizing Gleyic Cambisols typify many lowland alluvial

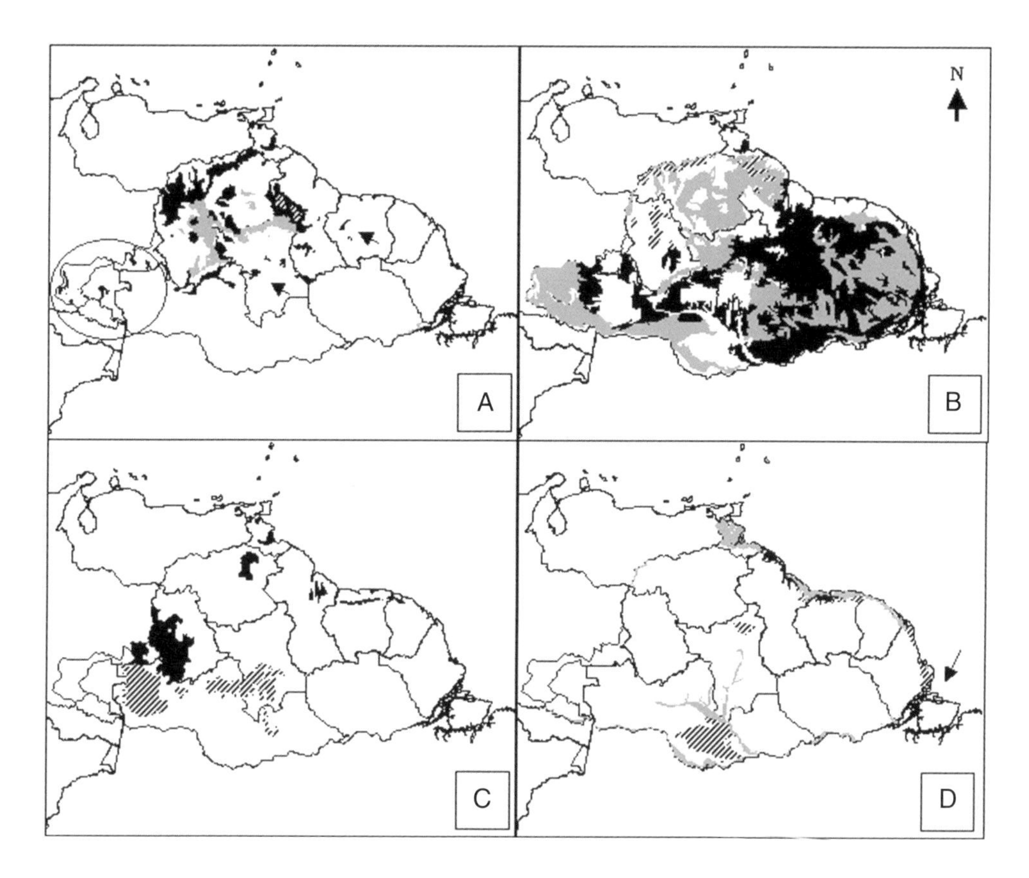

Fig. 2.11. The 12 soil reference groups dominating the Guiana Shield pedosphere presented here in four sets approximating an altitudinal gradient from highest (A) to lowest (D) elevations. (A) Leptosols (black, arrows and inside circle), Regosols (grey), Cambisols (striped). (B) Ferralsols (black), Acrisols (grey), Lixisols (striped). (C) Arenosols (black), Podzols (striped). (D) Histosols (black), Gleysols (grey), Plinthosols (striped), Solonchaks (arrow). Soil distributions presented here are based on the combined FAO soil (FAO, 1988) and landform coverages produced at a 1:1 million scale through the SOTER project (UNEP *et al.*, 1995). These soil type distributions are not applicable at smaller scales (see text). Eutric and dystric Fluvisols are distributed at smaller spatial scales throughout the area and are not presented here.

plains, while Ferralic Cambisols, particularly those in a rudic (stony) phase, are more commonly encountered on mountainous slopes and high hills.

LEPTOSOLS AND REGOSOLS Very shallow, mineral soils recently formed through weathering (e.g. exfoliation) of exposed rocks along the summits and upper slopes of the Guayana Highland massifs of the Roraima Formation are typically classified as Leptosols (aka Lithosols) (Sombroek, 1990), but variously intermix with other upland Regosol and Cambisol soils in relation to localized changes in topography (Fig. 2.11A). They are characterized by their shallow depth, abundance of coarse rock fragments and, consequently, virtual absence of diagnostic horizons. Shallow and weakly formed soils associated with the slopes and summits of the larger Precambrian granitic batholiths and outlying inselbergs (lajas) are grouped as Regosols. According to SOTER (UNEP *et al.*, 1995), these soils are broadly associated with the highland areas in Venezuela, Guyana, northern Brazil (Roraima) and Colombia (Fig. 2.11A, black). Topography and elevation were most likely of overriding importance in assigning Leptosol and Regosol soils to these regions using the SOTER approach and their true distribution is certainly more interspersed with other soils than appears in Fig. 2.11. They are associated with the Parguaza Formation, Imataca Complex, (quasi-)Roraima Formation, Avanavero Suite, Uatuma Volcanics and Mitú complex.

The small-scale impact of localized changes in topography on highland soil properties cannot be overstated. For example, soil samples collected on the table summit (*altiplanicie*) of the Sierra de Maigualida within the Caura basin region of Venezuelan Guayana have been described as Lithic Tropofibrists and Tropohemists (Histosols) in the USDA Taxonomy (Ramos, 1997). These soils are often associated with localized depressions of poorly drained, stagnating organic matter with little mineral horizon formation and are more akin to lowland swamp bogs formed in localized rock depressions or areas of occluded drainage at much lower elevations (e.g. van Kekem *et al.*, 1996). At lower elevations in the Cerro Mani–Cerro Chanaro area of the Guayana Highlands, soils on high hills were, in contrast, identified as Typic Kanhapludults (haplic Ferralsols, skeletic?) intermixed with bare rock (Ramos and Blanco, 1997).

Upland and sedimentary plain soils

The undulating hills and valleys and Tertiary sedimentary plains of the Guiana Shield are dominated by soils that are more extensively distributed, tend to form larger patch sizes and do not embrace the almost indescribable spatial transitions in soil properties that typify the younger soil surfaces of the highlands. These broad soil units cover many of the flat, peneplanic lowland troughs, basins and depressions surrounding and dissecting the much older exposed Precambrian basement complex. These include large areas between major rivers in the shield, including between the Negro and Branco, Mazaruni and Essequibo, Berbice and Courentyne and Orinoco and Negro Rivers (FAO, 1965; Sombroek, 1990). They also extend over much of the exposed basement complex, most notably across the Tumucumaque Uplands where silica-rich rocks have been chemically degraded to form deep silt and clay-containing soil horizons (Fig. 2.6B, No. 2).

FERRALSOLS AND ACRISOLS Ferralsols are deep soils typified by a highly weathered subsurface dominated by low-activity clays (saprolite, pipe clays), such as kaolinite with a relatively high iron content (Baillie, 1996). They are largely the product of extensive *in situ* weathering and decomposition of silica-rich parent material over a prolonged period. Ferralsols dominate most of the Guiana Shield and are only widespread in tropical South America and Africa (Sanchez, 1976; Richter and Babbar, 1991), where tropical forest vegetation rests atop ancient crystalline basement rock, often covered by deep sedimentary deposits of Precambrian provenance. They are variously intermixed with soil types belonging to the Acrisol group throughout the Guiana Shield and tropical South America (Sanchez, 1981). Acrisols are also deep soils characterized by a dense clay subhorizon, but are generally associated with younger, less highly weathered parent material, and a clay subhorizon developed largely through downward clay migration (illuviation) of deposited materials, rather than those formed from rock decomposition. They typically contain higher concentrations of weatherable minerals and active clay particles than the Ferralsols (UNEP *et al.*, 1995; FAO, 1998), but specific chemical properties of soil samples attributed to the two groups overlap extensively and these soils types are often tightly interwoven at smaller spatial scales (Cuevas, 2001) and often separated only by slope position (e.g. Hapludoxs vs. Kandiudults in Dubroeucq and Volkaff, 1998). Duivenvoorden and Lips (1995) recognized this relationship by referring to the dominating lowland soil units as Ali-Acrisols (Alisols × Acrisols) and Acri-Ferralsols (Acrisols × Ferralsols).

Ferralsols are widespread throughout

the upper Berbice Basin and Takutu Graben and across the Tumucumaque Uplands. They are noticeably less dominant over large parts of the western and southern sedimentary plains and hills in Brazil and Colombia where Acrisols become more common (PRORADAM, 1979; IGAC, 1993; Duivenvoorden and Lips, 1995) (Fig. 2.11B, grey shading), principally due to the declining influence of stable Precambrian rocks structures on soil development dynamics (Fig. 2.11B, black shading). Ferralsols (mixture of Udults and Orthoxs in USDA Taxonomy system) have been assigned to lowland slopes, plateaus and terraces within Venezuelan Guayana (Westin, 1962; MARNR, 1985; Gavaud *et al.*, 1986), but SOTER assign a dominant role to Acrisols soils over most of the Venezuelan Guayana region (UNEP *et al.*, 1995) (Fig. 2.11B, grey shading). Fuentes and Madero (1996) indicate that the most extensively distributed soils of the large Caura Basin in Venezuelan Guayana are Ultisols (Acrisols) with a relatively minor occurrence of Oxisols (Ferralsols). In the Caquetá region of Colombia, Duivenvoorden and Lips (1995) suggest that the increase in Ali-Acrisol cover is largely due to the contribution of deposits from younger Andean parent material, while Acri-Ferrasol soils are derived largely from more extensively weathered shield parent rock. Both groups, however, show some of the lowest standing nutrient levels and storage capacities relative to most other mesic soil types.

LIXISOLS According to the SOTER database (UNEP *et al.*, 1995), soils of the Lixisol group are generally found in the lowland basins of western Bolivar and Amazonas states along the northern rim of the Guiana Shield (Fig. 2.11B). These are generally associated with the floodplain areas embracing the upper Ventauri, upper Cuyuni and confluences of the Orinoco with the Caura, Aro and other large tributaries. They are generally classified as a mixture of Psamments and Ustoxs in the USDA Taxonomy (MARNR, 1985; Gavaud *et al.*, 1986).

The physical properties of these soils are very similar to those of the Acrisols and Luvisols and are difficult to distinguish in the field (FAO, 1998). Generally, they are considered to be younger, and more fertile, than Acrisols.

ARENOSOLS AND PODZOLS – WHITE SANDS Large areas within the sedimentary plains are dominated by soils that largely consist of coarse sand, with varying, but always relatively small, amounts of clay, silt or loam and no rock fragments. The extreme situation is exemplified by the very deep albic Arenosols, sometimes called giant Podzols (Baillie, 1996). The upper horizons of these white, or bleached, sands (*sable blancs*) can extend to depths of several metres, and show some of the lowest CECs known, but also very low aluminium saturation (van Kekem *et al.*, 1996). The pure sand matrix has very little, if any, water-holding capacity and dry seasons in the region bring a period of dry down that results in virtually no moisture remaining in the top horizons of deep deposits (e.g. Jetten, 1994). Other arenosolic soils on slopes and in valleys often show an increase in clay content with depth (e.g. gleyic Arenosols), but remain highly infertile. These can also vary in their hydrological properties, particularly the effects of fluctuating water tables on clay illuviation activity and hardpan development.

Podzols represent another variation on a sand-dominated theme. They tend to contain more silt than Arenosols in a patchwork of bleached sand and silt pockets overlying a relatively thin subhorizon of cemented organic matter and aluminium or iron oxides (hard pan, humic pan or *ortstein*) (Sombroek, 1990; Baillie, 1996; FAO, 1998). It is thought that this hard pan is formed through the action of fluctuating water tables that gradually leach organic matter and clay constituents from the upper profile down to a level consistent with the minimum water level, where these become concentrated (podzolization) (Cooper, 1979). This cemented layer occludes vertical drainage to varying degrees and all Podzols remain saturated during (part of) the year, but may briefly dry down in the upper horizon to levels achieved by excessively drained Arenosols during dry periods (Jetten, 1994; Coomes and Grubb, 1996). The depth and horizontal extent of the hard pan may vary and this influences the degree of drainage and duration of flooding (van der Eyk, 1957; Klinge, 1968). Some Podzols remain saturated throughout most of the year due to the presence of a hard pan at shallower depths or a topographic location susceptible to regular riverine flooding (groundwater Podzols) (Bleakley and Khan, 1963). This seasonal oscillation between flood and drought conditions combined with low nutrient retention make these soils some of the poorest known in the tropics and represent one of the most unique challenges to plant life (Heyligers, 1963).

Arenosols and Podzols are commonly encountered throughout the sedimentary plains of the Guiana Shield (Fig. 2.11C), but tend to be found in large, relatively isolated patches surrounded by sandy soils with pronounced clay subhorizons, such as the Ferralsols and Acrisols. The transition from these sand-dominated soil patches to neighbouring types with higher clay and silt content is often abrupt. Their distribution is strongly shaped by those of the Tertiary sedimentary formations covering the region, but usually represent a subsection of this cover (compare Fig. 2.11C with Fig. 2.4N). For example, Schultz (1960) estimated that albic Arenosols accounted for only 17% of the Zanderij Formation in Suriname, the remaining area dominated by clay and silt-containing ferrasolic soil types. In Guyana, they are thought to occupy only 26% of the sedimentary cover attributed to the Berbice Formation, about 8% of the national land area (FAO, 1965).

PLINTHOSOLS – LOWLAND NEO-LATERITES Like Podzols, the Plinthosols ('groundwater laterites, *lateritas hydromórficas, schol soils*') can also be characterized by a hardpan at their more advanced stages of development. This, however, is typically found at or near to the surface (<100 cm depth) and is formed from the cementation of iron or aluminum-rich kaolinite (plinthite), rather than organic matter (FAO, 1998). This hardpan, better known as laterite, and also variously called petroplinthite, ferrite, *ripio*, *arrecife*, ferricrete, *cuirass*, ironpan or ironstone (Huber, 1995a; Baillie, 1996; van Kekem *et al.*, 1996; Tardy, 1997; FAO, 1998; EMBRAPA-CNPS, 1999) is exclusively linked to clay and iron-rich soils that have been exposed over a period of time to fluctuating water tables within a low-lying sedimentary plain.

Plinthosols are more commonly associated with soil series in the Amazon Basin than in any other tropical region, though their distribution is pan (sub)tropical (Richter and Babbar, 1991; FAO, 1998). The largest fraction of their mapped area in South America is located in the basin below 100 m asl that is west of Manaus. Extensive flooding and drying was concentrated in this section of the Amazon and Sub-Andean basins during the Pleistocene (Irion, 1984). Within the shield area, Plinthosols can be found to occur in association with other soil types across large areas of the Rio Negro–Amazon sedimentary plains, along the Atlantic coastline and in the Takutu Graben region where flooding conditions continue to support their formation on iron–aluminium rich substrates (Fig. 2.11D). Plinthosols are estimated to cover about 7% of the Brazilian Amazon (Richter and Babbar, 1991) and between 8% and 12% of Guyana (FAO, 1965).

Upland laterite soils – Palaeo-plinthosols and inverse topography

Plinthosols are the contemporary equivalent of much older iron hardpans now found covering many ridges, benches and hill tops (*cuirasses*) in the upland and highland areas. They are believed to have been formed during the Eocene-Oligocene (Berrangé, 1977) or Plio-Pleistocene (Sinha, 1968), but with possible episodes dating as far back as the early Palaeozoic (Gibbs and Barron, 1993). In some instances, older forms found in lowland settings have degraded and then recemented later, making it difficult to distinguish them from neo-laterites. Neo-laterites are generally considered to have formed within the last 10,000 years (Sinha, 1968).

Anyone who has travelled through parts of the forested or savanna interior of the shield has probably encountered laterite in one of its various forms. However, they may not have fully understood how strongly petroplinthic formation, or laterization, has shaped modern topography of the Guiana Shield. As you travel across one of the many boulder-strewn hills covered with pot-holed and pea-gravelled red rock, you may have in fact been moving on what was once a flooded valley floor. At one time these palaeo-valleys were subject to similar hydrological conditions seen in areas occupied by Plinthosols today. As erosion proceeded to reshape the former landscape, the iron hardpan that had formed under fluctuating water table conditions in the valleys reduced the rate of valley weathering relative to the surrounding uplands. This left erosion of the surrounding highlands to outpace the valleys. Eventually, the former upland areas eroded to a lower level, forming new valleys. The old valleys, now capped with laterite and relatively impervious to degradation through normal weathering processes, transformed into strings of upland hills and ridges. Later regional uplifting further emphasized the erosional differences between laterite uplands and catalysed erosion in valleys as new bevels reinvigorated the deformation process. This pronounced inverse topography is seen throughout much of the Guiana Shield landscape today, but particularly within and along the major depressions that were subjected to staggered periods of flooding throughout the Cenozoic. Repetition of the laterization process in 'new' valleys over a much longer period of time has created a series of planation surface relicts (Choubert, 1957; McConnell, 1968).

The formation of Plinthosols and laterites requires a relatively high concentration of iron in the parent material (Tardy, 1997). Plutonic (ultra)mafic intrusives typically present higher concentrations of iron (and magnesium) (see Fig. 2.5) and it is not surprising that most laterite-capped hills and mountains of the Guiana Shield are formed (but not exclusively) atop exposed elements of those intrusive rock suites that have a high ferro-magnesium (and aluminium) content (e.g. van der Eyk, 1957). The most prevalent of these are the unmetamorphosed dolerites of the Basic Dyke Suite emplaced during the Permian, Triassic and Jurassic periods (Table 2.1) (Priem *et al.*, 1968; Berrangé, 1977).

Upland palaeo-plinthosols are classified differently from their modern counterparts because weathering over time has broken down, dissected, dissolved and occasionally reconsolidated the ironpan in various ways. Where the hardpan remains relatively solid, continuous and close to the surface, the soils have been classified as being Leptosols in a petroferric phase (FAO, 1988), or more recently reclassified as dystric Plinthosols (FAO, 1998). Where hardpan has degraded, laterization has produced smaller aggregates rather than sheets, or deformation has led to mixing in with subhorizons of more silicic content, the soils can be classified as Ferralsols in a skeletic phase (e.g. van Kekem *et al.*, 1996).

Subterranean laterites – calcine bauxite

In some instances, the plinthite or hardpan is covered by very deep sedimentary cover (e.g. Berbice Formation in Guyana) that has been washed clean of any organic or mineral coatings. When the overlying sedimentary cover is substantial, buried plinthic or petroplinthic horizons are no longer considered as forming part of the pedosphere but rather as mineral deposits. Under these conditions, chemical weathering proceeds at a low pH in a reducing, rather than oxidizing, environment and the iron oxides that would normally bind to form Plinthosols at the surface are instead dissolved or kept in solution as free iron. In many cases, high concentrations of alumina remain, attached only to the surrounding clay matrix and relatively free of iron (Krook, 1969; Gibbs and Barron, 1993). These form most of the major high-grade bauxite deposits in the Guiana Shield, although lower grade deposits (e.g. higher iron content) have been formed through other bauxitization pathways (e.g. Bakhuys Mts in Suriname; Krook and de Roever, 1975). (See also Chapter 8).

Soils of the coastal and riverine floodplains

The youngest soils in the shield area are those currently and continuously being shaped by fluvial, tidal, stagnating and anthropogenic processes operating within the long, narrow coastal and riverine floodplains that embrace and dissect the older sedimentary plains, uplands and highlands. They are often collectively referred to as alluvial soils, but include a number of different soil types recognized under the WBR. They are typically located at the lowest elevations within each landform wherever flooding is permanent or seasonal. The most extensive cover is restricted to the margins of the Guiana Shield in the form of three narrow strips, one running parallel to the Atlantic seashore between the Orinoco and Amazon and another two mirroring the course of the lower Branco and Negro Rivers in Brazil, respectively (Fig. 2.11D). Floodplain soils properties are also found associated with smaller waterways throughout the shield and with numerous natural and artificial swamplands formed through localized depressions. Water is the overwhelming factor shaping all of the soils typically encountered in the coastal and riverine floodplains and as a consequence, most floodplain soil types exhibit relatively few diagnostic horizons due to continuous action of the fluctuating tides and river stages.

ALLUVIAL SOILS – FLUVISOLS AND GLEYSOLS

Fluvisols have upper layers that are continuously changing as fresh materials are regularly deposited and removed through tidal and fluvial action and are typified by the presence of young organic matter within the upper horizons (USDA, 1996; FAO, 1998). Floodplain areas that are saturated for most of the year are characterized by Gleysols. These soils are characterized by reducing conditions and a low pH that bring iron into solution in a manner similar to that forming calcine bauxite strata, but much closer to the surface. Fluvisols and Gleysols are among some of the youngest soils typically found in tropical forest areas and as a consequence of their age and mode of development are found as a minor soil type throughout the tropical forest regions of the world (Richter and Babbar, 1991). They arguably represent the youngest soils in the Guiana Shield.

The constant reworking of the upper horizon of fluvisolic soils principally involves the addition of silts and clays through flocculation (colloidal suspension as aggregates), but transfer of larger particles through saltation (bouncing along the channel floor) and organic matter can also take place at higher current speeds. The result of this action is a regular recharging of the nutrients in the upper horizon. The magnitude and quality of the recharge, however, depends on the nutrient source (Sombroek, 1991). Rivers draining the Andean cordillera and Central American sierras transport and deposit much larger volumes of relatively high activity clays compared to black and clear water rivers draining the Guiana Shield, creating a striking contrast in the concentrations of nutrients in the upper horizon(s) of soils within and outside the shield (Fig. 2.12) (Sollins *et al.*, 1994; van Kekem *et al.*, 1996; Zarin, 1999). Even differences in mineral provenance within relatively young geological landscapes, such as the Andes, can lead to significant differences in the concentrations of exchangeable nutrients transported and deposited to form Fluvisols and other alluvial soils (Zarin, 1999). At both scales, differences can begin to affect vegetation patterns.

Fluvisols are thought to cover approximately 2% of tropical South America (Richter and Babbar, 1991), but more than 85% of this area is found within the Sub-Andean Trough and Amazon Downwarp areas and mostly outside of the Guiana Shield region (RADAMBRASIL, 1973–1978; IGAC, 1993). Gleysols are more common, covering nearly 4% of tropical South America and again, largely concentrated in the major sedimentary basins and depressions criss-crossing the Amazon. Gleysols

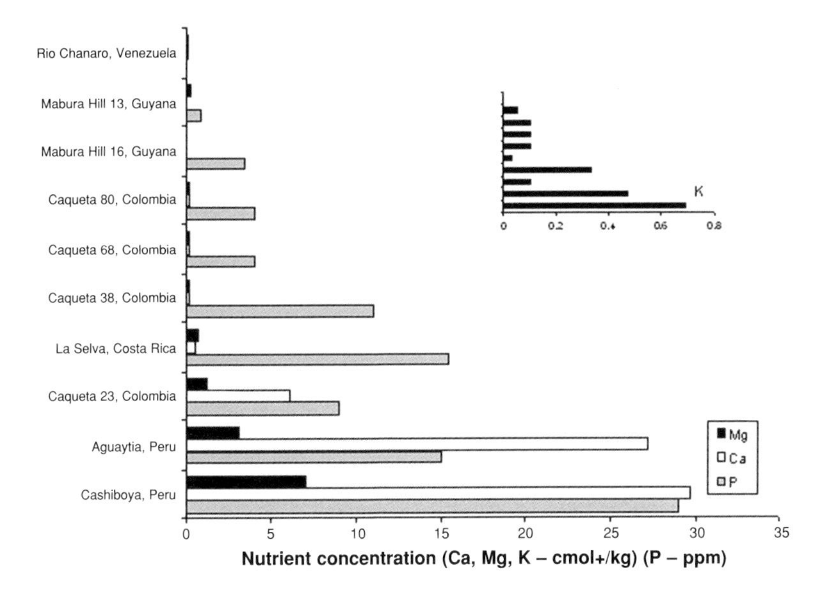

Fig. 2.12. Concentrations of exchangeable and available nutrients measured in various Fluvisol soil profiles sampled in the Guiana Shield, Central America and western Amazonia. Values are for uppermost horizon(s) only (<35 cm). Data sources: Venezuela (Ramos and Blanco, 1997), Mabura Hill (van Kekem *et al.*, 1996), Colombia (Duivenvoorden and Lips, 1995), Costa Rica (Sollins *et al.*, 1994), Peru (Zarin, 1999). Values were interpolated from two consecutive depths, when uppermost depth presented was less than 15 cm.

probably range more extensively than Fluvisols along the Atlantic coast and major river systems in the Guiana Shield (Fig. 2.11D). This is not surprising given the widespread exposure to saturating water tables across the peneplanic regions of the Guiana Shield. At larger scales, these soils are rarely found alone and tend to be intermixed with Vertisols, Histosols, Plinthosols, Acrisols and other gleyic soils as localized topographic and hydrological conditions change (Sombroek, 1991). For example, in the Caura Basin sector of the Venezuelan Guayana, floodplains typically are covered by a variety of alluvial soil types classified as Ultisols (USDA) (Ramos and Blanco, 1997). Alluvial soils are expected to cover between 10% and 15% of Guyana, but again most are intermixed with a wide range of hydromorphic and sedimentary types (FAO, 1965).

HISTOSOLS Gleysols that accumulate organic matter to form a humus-rich (peat, *pegasse*, muck) upper horizon are typically classified as Histosols (*sols à couche de pegasse*). This organic accumulation normally occurs under the highly acidic conditions that hinder decomposition in backswamp environments typically developing on fluvio-marine sediments. Despite the predominance of these sediments, Histosols are not common in tropical South America relative to other soil types, accounting for less than 1% of the area (Richter and Babbar, 1991). A large fraction of this area is concentrated in two zones. The first is a 60,000 km^2 Pleistocene/ Holocene alluvial fan in the Pastaza-Marañon region of the Sub-Andean Trough (Sombroek, 1991). This lowland depression is situated between the Andean foothills to the west and the Iquitos Arch to the east. As a consequence, seasonal saturating conditions are widespread, making this basin one of the largest flooded areas in Amazonia and ideal for Histosol development (Räsänen, 1993). The second zone is located within the Guiana Shield along a ribbon of Atlantic coastal backswamp that can range up to 65 km inland between the Orinoco and Berbice Rivers (Fig. 2.11D) (FAO, 1965). This band continues to a lesser extent along the coast of Suriname (van der Eyk, 1957), French Guiana (Leveque, 1961), Amapá and a portion of the Orinoco delta (Westin, 1962; MARNR, 1985). This soil type has been estimated at 4% of the land area in Guyana (FAO, 1965), but is less than 1% in other shield countries (van der Eyk, 1957; Westin, 1962). Histosols can also be found in patches at more local scales. Subsidence caused by subterranean decay of surface laterites, drainage impediments caused by human activity and perched water tables can all create the conditions needed for localized accumulation of organic matter and development of histic soil horizons.

Histosols can also be found at localized depressions in large Podzol plateaus in the Rio Negro Basin where a spodic horizon occludes drainage and sparks a process of lateral transformation from Podzol to Histosol as the occluding spodic horizon moves upward, raising the perched water table and stimulating further organic deposition and retention (Dubroeucq and Volkaff, 1998). These local peat swamps situated within white sand plains are also known to occur in the Burro-Burro River and Demerara regions of Guyana (D. Hammond, personal observation; Hawkes and Wall, 1993; van Kekem *et al.*, 1996), but are normally classified as carbic Podzols due to the thinner organic matter horizon.

The accumulation of organic matter in the histic horizon can exceed 10 m along the area of greatest development between the Amakura and Pomeroon Rivers in the northeastern corner of the shield (FAO, 1965), declining to 2–3 m in Suriname (van der Eyk, 1957) and normally to depths less than 2 m at inland lowland forest locations (e.g. Duivenvoorden and Lips, 1995; Fuentes and Madero, 1996; van Kekem *et al.*, 1996; Dubroeucq *et al.*, 1999). Histosols are also found on the exposed summits of highland tepuis, although these develop under much lower ambient temperatures and are in many ways more closely related to temperate peat bog soils (Gavaud *et al.*, 1986; Huber, 1995a).

MANGROVE AND ESTUARINE SOILS – SOLONCHAKS Solonchaks are Gleysol-like coastal foreland soils dominating estuarine and tidal mangrove areas. They are principally characterized by their high electrical conductivity due to saturation with a brackish solution of readily soluble salts derived from seawater (FAO, 1998). They are not widely distributed in South America, account for less than 0.1% of the Brazilian Legal Amazon (Richter and Babbar, 1991) and are largely restricted to a very thin ribbon of land directly in contact with the tidal belt area of the Atlantic coastal floodplain. According to SOTER, a thin strip of coastal mangrove in southeastern Amapá, part of the Amazon estuary, is dominated by Solonchak-type soils (UNEP *et al.*, 1995) (Fig. 2.11D). These are known to extend across Marajó Island (Sombroek, 1991). Many areas of Solonchak soils have been modified through human activity, particularly where agricultural empoldering has replaced brackish seawater with freshwater and their former distribution was in all likelihood more extensive, as in the case of much of the Atlantic coastal perimeter of the shield area. In places, Solonchaks are found to incorporate an acid sulphate layer ('cat clays') that developed from sulphide-bearing strata oxidizing after drainage (Baillie, 1996). These areas, where they develop, are highly toxic to plant life, showing one of the lowest soil pH conditions known to occur in the tropics (less than 3.0).

HUMAN SOILS – ANTHROSOLS A wide range of soils in the Guiana Shield have been modified through various agricultural activities (e.g. tilling, irrigating, draining, forest clearing and burning) that accompany food and livestock production. These are generally referred to as Anthrosols. Other human activities, such as mining, urban and village settlements, forestry, and other forms of infrastructural development have deposited a wide range of anthropogeomorphic material in soils and these have altered the physical and chemical properties of the pedosphere at deposition sites. Most Anthrosols and other human-altered soils are concentrated along the main river and coastal floodplains. The combination of accessibility and relatively high fertility (where these are not stagnic or salic) have made these the soils of choice for cultivation over the last 13,000 years (e.g. Roosevelt, 1991) (see Chapter 8). In riverine areas subject to strong and rapid deposition–erosion cycles, the pedological signature of human use may not last long enough to significantly change the dominant fluvic properties associated with continuous sedimentation. In contrast, upland (*terra firme*) soils may bear the imprint of former agrarian activity much longer. Balée (1989) estimates that around 12% of *terra firme* forests in the Amazon are anthropogenic and this figure may prove to be even higher. Perhaps the most well-known soils believed to be associated with former human inhabitation and land use are the nutrient-rich *terra prêta do índio* (Indian black land) soils from Ferralsol-dominated locations along the main lowland rivers in Brazil, Peru, Ecuador and Colombia (Eden *et al.*, 1984; Balée, 1989; Wood and McCann, 1999). An association of these soils with relatively fertile Nitisols (Alfisols, *terra roxa estruturada*) borne on mafic intrusives has been suggested for some locations in western Amazonia (Moran, 1995) and it is likely that geological control over pre-Columbian agricultural site success cannot be fully discounted despite views suggesting that highly infertile soils have been managed in the past to make them highly suitable to agriculture (Wood and McCann, 1999).

Formative factors in Guiana Shield soil development

Given the manifestly important role soil formation and function play in agricultural development, it is not surprising that of all the scientific disciplines, only the geological development and composition of the Guiana Shield, and its impact on commercial mineral extraction, has a more extensive published record of research within the region. The commercial consequences of this enquiry, however, have not proven necessarily to be entirely of social benefit. Moreover, the impacts of mineral and agricultural development have not often proven to be comfortably compatible with the understood prehistoric rate and course of ecosystem change within the shield (see Chapters 8 and 9). Understanding the evolution of soils, their present-day distribution and characteristic properties helps to shed light on one of many controls exerted by the physical landscape on Guiana Shield forests, their distribution and composition.

A review of the main soils of the Guiana Shield has shown a complex of ultra-weathered substrate that can be broadly characterized by relatively low pH, widespread and record-setting nutrient deficiencies (base saturations of <1% and cation exchange capacities of <0.1 cmol+/kg are common) and low organic carbon content (except histic swamp soils and indurated layers of podzols). As with all attempts to generalize, there are many exceptions and the most significant of these are characterized below. Many factors have influenced soil development in the region, as they have done elsewhere, and it is not the purpose here to review universally recognized soil-forming processes or the influence of local landscape attributes, such as slope position (Jenny, 1980). Rather, I hope to highlight and qualify a number of formative factors that to one extent or another combine in distinguishing the soil environments of the shield region from those found beneath other tropical forestlands (and savannas). Four aspects of the underlying geological and climatic history of the region in particular assist in understanding how certain broad soil types have come to dominate large areas throughout the Guiana Shield.

Age matters: Precambrian vs. Cenozoic

More than 90% of the modern Guiana Shield surface is shaped by underlying rock formations set in place either more than 600 million (Precambrian) or less than 70 million (Tertiary) years ago (Fig. 2.13). The record of diastrophic activity in the shield during the intervening Palaeozoic/lower Mesozoic appears limited to minor fault fracturing and depositional activity along its margins and in the Takutu graben (Gibbs and Barron, 1993) (Fig. 2.3). All but one of the major tectonothermal episodes shaping the mineral composition of the igneous and metamorphic parent rock were completed by the upper Proterozoic. Only swarms of dykes and sills forming the Apotoe Dyke Suite were emplaced during the early Phanerozoic. Weathering and sedimentation phases predominantly shaped the geomorphic evolution of the shield region throughout the Palaeozoic and Mesozoic eras (Fig. 2.4L), but the evaporates, ironstones, clastics and other sedimentary rocks formed from these deposits are now largely buried by the massive and extensive influx of unconsolidated sediments that took place during the Cenozoic, mainly due to the uplift of the Andes and frequent sea-level movements over the pleni-glacial (Fig. 2.4N).

Only the (sub)tropical forest soils atop the exposed Precambrian crust of the Man and Benin–Nigeria Shields of West Africa, the northern half of the Brazilian Shield and the Indian Craton compare in surface area with those of the Guiana Shield (Chapter 1, Fig. 1.2 (Goodwin, 1996)). The soils of all of the other major tropical forest regions in the world are principally derived from much younger, and active, parent rock associated with volcanic activity and orogenic uplift dating back no more than 30 million years BP (Borneo, most of Papua New Guinea, SE Asia, Polynesia, Central America, Caribbean) or have formed in large sedimentary basins (Amazon Downwarp, Sub-Andean Trough, Congo Basin, Deccan Traps). The characteristics of these 'young' soils borne from primary weathering of volcanic, as compared to plutonic, rock strike a significant contrast (e.g. Cambisols, Luvisols, Andosols, Nitisols, Kastanozems vs. Ferralsols, Arenosols, Podzols, Plinthosols). Even when considering only the soils formed atop Cenozoic sedimentary deposits, many of the soils

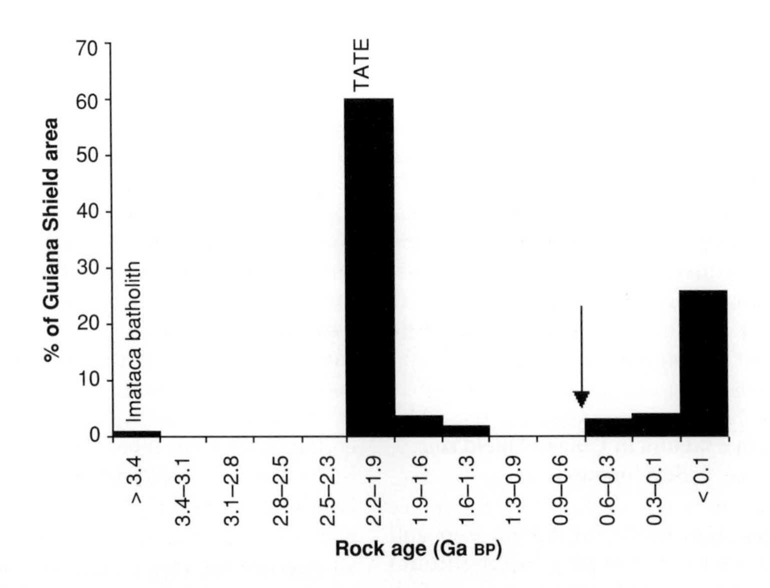

Fig. 2.13. Approximate age intervals (in Ga BP) of the different rock formations making up the present-day Guiana Shield landscape according to radioisotopic, palynological and stratigraphic dating of rock and sediment referenced in Gibbs and Barron (1993). Note that Precambrian ages (left of arrow) reflect both new rock formation and re-activation of previously formed rock, while Phanerozoic formations (right of arrow) are largely sedimentary (except the Apotoe Dyke Suite).

dominating Central America, the Caribbean, the western Amazonian foreland and SE Asia have been formed from primary weathering surfaces created through relatively recent tectonic and volcanic mountain-building processes. These sedimentary processes are active and continuously recharging depositional basins with new material from abundant weathering surfaces. In contrast, most Tertiary-aged sediments in the Guiana Shield have been redeposited from secondary or tertiary point sources as a consequence of more gradual episodic change, often associated with the rise and fall of sea-levels and tectonic uplifting. The long history of soil material formation is inextricably associated with the dynamics of soil material migration from the highland and upland 'islands' into subsiding depressions throughout the shield area.

Movement matters: autochthonous vs. allochthonous processes

The source–sink relationship between upland areas largely structured by Precambrian rock and the reworked sedimentary surfaces of the lowland plains has produced a major dichotomy in characterizing soil formations of the Guiana Shield. The breakdown of mainly Precambrian rocks *in situ* has created a number of primary soil groups. These display contrasting chemical and physical properties that mirror differences in their parent material (felsic vs. (ultra)mafic vs. sedimentary) and the palaeo-climatic conditions under which the uppermost portions of parent rock were broken down to form soil. Subsequent erosion and redeposition of these autochthonous materials has over an extended period of time created a new set of secondary, allochthonous soil types that are commonly associated with the Cenozoic sedimentary plains, but also savannas and more localized valleys in the upland and highland regions of the shield. The properties of these secondary soil groups in turn reflect their own relatively brief life history of changing water table conditions, biological activity and climate over the last 70 million years.

The main, and oldest, autochthonous soils in the Guiana Shield are represented by the Ferralsol, Acrisol and (to much lesser extent) Lixisol groups (FAO/WBR classification). The properties of these soils principally reflect chemical weathering processes dominating soil formation under warm and wet environmental conditions modulated by forest cover. These soils are recognizable as forming much of the Precambrian Rolling Hills, Guiana Upland and Guayana Highland landforms. Chemical weathering proceeds mainly through contact between water and rock minerals under closed forest cover, creating a heavy clay subhorizon (saprolite). Ambient temperature, mineral attributes of the rock, acidity of the water and the duration and frequency of their contact control the rate of chemical reactions driving the breakdown process.

The allochthonous soil materials in the Guiana Shield are represented by two main groups. The older of these two groups consists of the Arenosols and Podzols associated with highly leached Plio-Pleistocene sediments, and largely forming the Tertiary Sandy Plains. The second, younger group, consisting mainly of Histosols, (humic and eutric) Gleysols and to a lesser extent Fluvisols, is linked to recent and continuing deposition of Amazon silts and clays along the Atlantic coast and shield sediments along river floodplains under varying hydrological conditions. They fall mainly within the Recent Coastal Plains. The iron-sesquioxide rich Plinthosols can fall in either group, depending on their topographic position and concretionary stage, both in part reflecting on age of the soil.

Arenosols and Podzols have been affected by chemical weathering of their constituent minerals at numerous stages. Originating from *in situ* breakdown of metamorphosed Precambrian granites, volcanics and, most importantly, Roraima conglomerates and sandstones, the weathered products were then fluvio-deltaically transported in subsiding depressions during periods of inland sea movement (Bleakley, 1957). Once deposited, these in turn were increasingly weathered (Heyligers, 1963), but starting from a benchmark defined by the previous weathering, not rock-forming, processes. Dubroeucq and Volkaff (1998) suggest that the large Podzols of the upper Rio Negro have developed *in situ* as one stage in a lateral transformation process that entails the deformation and planation of Oxisol/Ultisol-dominated hills and subsequent depletion of clays from these soils (Ferralsols) through podzolization. Podzolization commences as changes in horizontal water table movements and filtration of humic materials spark the development of a spodic horizon and consequently further elevation of the water table. Both theories may be complementary in describing the evolution of these soils, only at different spatial and temporal scales. Water, as both a weathering and depositional agent, is implicated in both proposals and clearly has been the main catalyst in shaping the attributes of both autochthonous and allochthonous soils in the Guiana Shield.

Water matters: flooding, oscillating water tables and coastal migration

Water has arguably been the most important surficial agent affecting the physical landscape of the Guiana Shield. Its role in the formation of soils can be broadly attributed to past and present fluctuations in the marine and freshwater tables that have led to flooding, drought and coastal migration (sea transgression–regression). Water tables can be viewed as consisting of an amplitude (stage height), wavelength (stage duration) and frequency (seasonal, historical occurrence) (see 'River, Lake and Tidal Systems', below). Different combinations of these three parameters can describe most soil attributes found in the Guiana Shield because most areas have been significantly affected by changes in the water table in one way or another (e.g. hardpan formation), at one time or another (e.g. laterite planation surfaces).

Fluvisols form through flooding that occurs for several days or months on an annual or supra-annual basis. The delivery of new sediments at an average frequency, through large increases in stage height (high amplitude), but over a relatively brief duration (short wavelength) recharges the upper horizon of soils in these areas while importantly maintaining an oxidizing environment. Gleysols are also flooded frequently, but normally at lower stage heights (modest amplitude), for a greater duration (very long wavelength) and every year (very high frequency). Histosols (swamp and bog soils) have an even more pronounced flat-line profile. Akin to Fluvisols, Gleysols and Histosols, these soils are also recharged through the addition of silts, clays and organic matter. Due to the long wavelength and high frequency of flooding conditions, however, these soils are characterized by a reducing, rather than oxidizing, environment.

The impact of historic, mainly Pleistocene, fluctuations in sea level on soils in the Guiana Shield is most obvious at large spatial scales in the low-lying depressions, grabens and downwarpings found throughout the region. Extensive and prolonged coastal inundation combined with its impeding effects on freshwater drainages created deltaic, estuarine, lacustrine and littoral depositional conditions (McConnell, 1958; Bleakley and Khan, 1963; Sinha, 1968; Daniel, 1984) that have contributed to the extensive formation of Arenosol, Podzol and Ferralsol-dominated surfaces throughout the main lowland areas along the rim of the exposed basement 'islands'.

Fluctuating water tables have also been the main agent responsible for the formation of Plinthosols, and eventually hardened laterites, in many areas. Plinthosols develop from highly weathered soils that have relatively high iron contents, such as Ferralsols. When these soils are exposed to a series of water table fluctuations, the iron goes into solution, 'washes' through the soil matrix, and is deposited as a concentrated clay band, called plinthite, at the zone of intermittent saturation. Laterite forms when the plinthite precursor undergoes an 'irreversible drying' phase after the water table lowers during drier climatic intervals, often forming multiple layers as the lowering process occurs in stages (Bleakley, 1964). Iron cations within the plinthite then fully oxidize and concretize to form hardened ironstone layers of varying thickness and density. The importance of rock provenance in delivering the iron needed for laterite development is another factor interacting with water that has contributed to the modern soil landscape of the Guiana Shield.

Mineral matters: granitoids, basic intrusives and Roraima sedimentaries

The conditions that have led to rock formation and persistence vary widely within the Guiana Shield. This is reflected, for example, in the contrasting elemental composition of extrusive and intrusive rock formations in the region (see Fig. 2.5) and the relative contribution of sedimentary and metamorphic rock types in shaping the sharply contrasting topographic relief dominated by sedimentary plains, inselbergs and tepuis (see Fig. 2.8). Generally speaking, parent material plays an important role in shaping soil properties, but with time its impact can become one of soil horizon thickening, rather than soil horizon diversification (Hole, 1961, cited in Dubroeucq and Volkaff, 1998). Only where parent rock is close to the surface and covered by younger, thin-horizoned soils does horizon development proceed dynamically. The properties of deeper, more highly weathered soils formed in geologically stable regions are believed to eventually 'decouple' from the underlying rock strata (Burnham, 1989). Numerous authors have commented on the impacts of varying mineral content of parent material on soil type and properties in South America and other tropical regions (Harrison and Reid, 1911; Hardy and Follet-Smith, 1931; van der Eyk, 1957; Jenny, 1980; Burnham, 1989; Sombroek, 1991; Huber, 1995a; Baillie, 1996; Osher and Buol, 1998; Leigh, 1999). Importantly, this geological variation can be indirectly, but strongly, associated (via soil) with changes in forest composition and structure (Richards, 1952; Ogden, 1966; Huston, 1980; Duivenvoorden and Lips, 1995).

Mineral provenance in the Guiana Shield is most visibly expressed in soil properties through three main pathways: (i) *in situ* formation of clays; (ii) supply of sediments to lowland floodplain and sedimentary plain soils; and (iii) development of laterites.

TATE Granitoids dominate the geological landscape of the Guiana Shield (Fig. 2.4D). As a group, they consist of a complex mix of granites, syenites, diorites, gneisses and amphibolites, among others. All of these granitoids are composed principally of minerals, such as k-feldspar ($KAlSi_3O_8$) and quartz (SiO_2), which contain relatively high concentrations of silica and aluminium (Fig. 2.5) (Gibbs and Barron, 1993). K-feldspar decomposes into kaolinite ($Al_2Si_2O_5(OH_4)$) through the hydrolytic action of carbonic acid (H_2CO_3) – carbon dioxide gas dissolved in water. Potassium, chlorine and silica dioxide are dissolved and washed away, leaving only kaolinite. Under the extensive chemical weathering regime found under tropical forests, kaolinite can be broken down further, dissolving the remaining silica and leaving behind gibbsite ($Al(OH)_3$), the main material comprising bauxite. Quartz is virtually undissolvable and remains with the degraded clays (or bauxite) as sand, forming soil horizons in response to different hydrological conditions. Ferralsols, Acrisols and other soils with high aluminium contents are principally the product of feldspar weathering through hydrolysis and the retention of hard quartz (sand).

In contrast, many of the dykes, sills and ridges seen throughout the Guiana Shield are formed from intrusive and volcanic rocks with a quite different mineral make-up. Basic intrusives and greenstones formed from the precipitation of mafic minerals with higher melting points. As a group, the basic intrusives and greenstones consist of metamorphosed plutonic and volcanic rocks, such as andesite, basalt, gabbro and dunite (Gibbs and Barron, 1993). All of these basic (mafic) rocks contain very little, if any, k-feldspar and large quantities of plagioclase feldspars ($NaAlSi_3O_8$; $CaAlSi_2O_8$), olivine ($(Mg,Fe)_2SiO_4$) and/or pyroxene ($(Mg,Fe)SiO_3$) that consist mainly of silica, sodium, calcium, magnesium and, importantly, iron. The dissolution of silica leaves these minerals to form free cations (e.g. Mg^{2+}, Fe^{2+}). This cationic form of iron oxidizes readily in solution to form ferric iron and then precipitates to produce limonite, a ferric oxide that gives many red tropical soils their distinct colour. This reaction also takes place in granitoid-borne soils, but is less pronounced. Certain types of Ferralsols, Acrisols, Nitisols, Plinthosols, Alisols, Cambisols and Leptosols are often, but not exclusively, traced back to a basic intrusive/volcanic provenance in the Guiana Shield and elsewhere. The high iron content of products formed from decomposition of basic intrusive rock makes these ideal surfaces for laterite formation when exposed to the oscillating water table needed for plinthite development.

The Roraima sedimentaries form a large part of the Guayana Highlands and consist mainly of coarse-grained conglomerates, sandstones, breccia, arkose and greywacke with some finer-grained siltstones and shales (Gibbs and Barron, 1993). Most of the coarse sedimentary materials are made up almost entirely of quartz and k-feldspar. Finer materials often contain more mafic minerals. Quartz sands form the main contribution of these Precambrian sedimentary rocks to soil formation in the Guiana Shield. Most large sedimentary depressions covered by quartz sand Arenosols, Podzols and Ferralsols are believed to have formed from a Roraima sedimentary provenance (Bleakley, 1957; Gansser, 1974), either through direct source-to-sink transfer or through secondary transfer phases, such as interglacial sea transgressions. Soils with high sand contents in some parts of the western shield area, however, have originated from Palaeozoic sandstones of the Chiribiquete complex or weathered igneous and metamorphic formations of the Andes in Colombia (Duivenvoorden and Lips, 1995). Some areas may have also developed through podzolization of Acri-Ferralsol soils formed *in situ* from the underlying Precambrian granitoids (Dubroeucq and Volkaff, 1998; Dubroeucq *et al.*, 1999).

Distribution of soils in relation to other neotropical forest regions

The differences in relative occurrence of tropical soil types within regions have been emphasized at the global scale (Sanchez, 1981; Richter and Babbar, 1991; Huston, 1994) and between landforms and regions within continents (e.g. Sombroek, 1991; Moran, 1995). Within the Guiana Shield, soils have been mapped at national scales (van der Eyk, 1957; Leveque, 1961; FAO, 1965; CVG-TECMIN, 1991), but with few comparisons across the geological landscape. National soil surveys can be of tremendous use in managing forests for conservation and use, but artificial political boundaries also can unintentionally censure interpretation of soil effects on forest structure, composition and function when the full range of conditions are not considered (Huston, 1994, p. 515).

The geologically benign shield area is dominated by soil types that exhibit properties on one extreme end of the global geochemical spectrum. Ferralsols, Arenosols, Podzols and Histosols (not shown in Fig. 2.14) account for a larger portion of the shield area than in Central America, the Sub-Andean Trough (Amazon Foredeep) or the Amazon Downwarp (Fig. 2.14, see Fig. 2.6B for regions). These soils present some of the lowest levels of exchangeable nutrients and their corollaries, pH, CEC and high activity clay abundance. In contrast, the Central American zone houses as a proportion of its total area, more of the moderate to high fertility (but not necessarily arable) soil groups, such as the Cambisols, Nitisols, Regosols, Vertisols and (not shown) Fluvisols, Alisols, Luvisols and Phaeozems. Infertile Podzols, tropical Arenosols and Ferralsols are virtually absent from the Central American landscape. The western Amazon Foredeep and west-to-east

Amazon Downwarp are more similar to the Guiana Shield but show important differences in relative soil dominance. Acrisols (Ultisols) are more widely distributed compared to Ferralsols and two large groups of hydromorphic soils, the Gleysols and Plinthosols, are particularly widespread in comparison to other regions (Fig. 2.14). Fluvisols also account for a more significant fraction of these two regions than the Guiana Shield.

Interpreting the general nutrient status of broad soil groups based on field data also illustrates the contrasts between regional soil-forming pathways and provenances, even when comparing similarly classified samples. For example, the exchangeable nutrients contained in floodplain soils have been shown to reflect their rock provenances (Fig. 2.12) and those in western Amazonia are thought to contribute much higher concentrations of macro-nutrients essential to plant growth (Sanchez *et al.*, 1982; Gentry and Terborgh, 1991 cited in Linna, 1993), even when considering local variation in rock parentage (Zarin, 1999). Properties of Tertiary sedimentary land units in the uplands of Araracuara, Colombia can also be distinguished by their parent material provenance (Duiven-voorden and Lips, 1998). Yavitt (2000) concluded that small-scale variation in rock provenance had little discernible effect on levels of available phosphorus, nitrogen and sulphur in soils formed atop sedimentary and andesitic parent materials on Barro Colorado Island (BCI). However, as the author notes, nitrogen is not a good indicator of a rock provenance effect since this is rarely involved in mineral formation and cations that better reflect differences in phenocrystic andesite, foraminiferal limestone and sandstone mineral weathering, such as calcium, sodium and magnesium, were not analysed. Localized veins composed of sulphide minerals are commonly found emplaced in country rock of many different types and would strongly

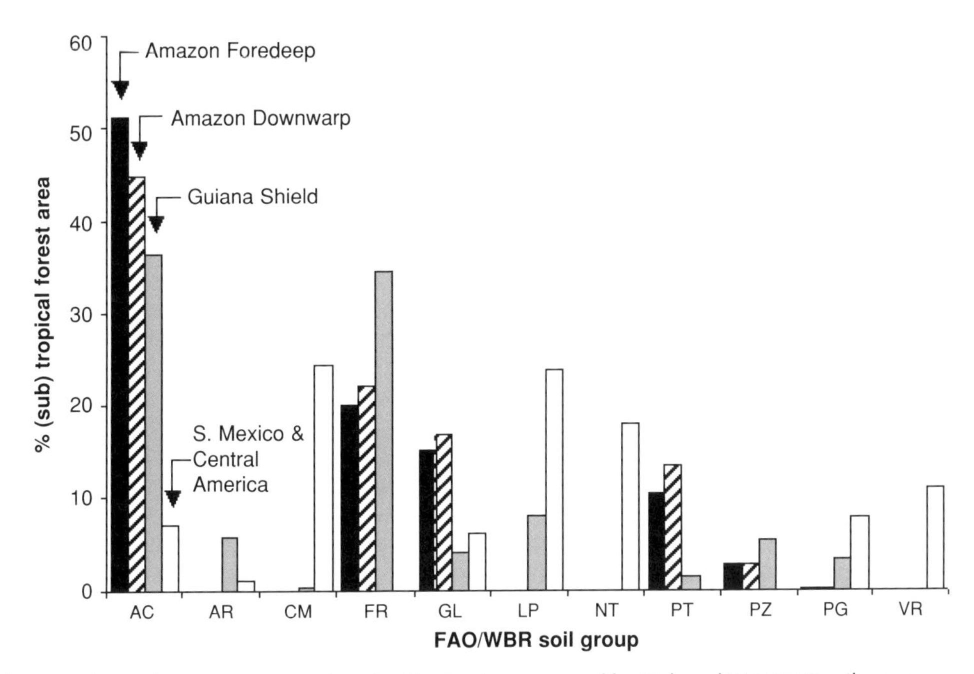

Fig. 2.14. Approximate percentage of tropical lowland area covered by 11 broad FAO/WBR soil groups within each of four neotropical regions (see Fig. 2.8 for locations). AC: Acrisols, AR: Arenosols, CM: Cambisols, FR: Ferralsols, GL: Gleysols, LP: Leptosols, NT: Nitisols, PT: Plinthisols, PZ: Podzols, RG: Regosols, VR: Vertisols. Distribution of soils based on SOTER global soil map (1:1 million) (UNEP *et al.*, 1995) and tropical lowlands on WWF Ecoregion map (1:1 million) (Olson *et al.*, 2001).

reflect size and distribution of sampling effort.

The rock provenance effect can also assist in comparing soil properties at the level of the neotropics. On the one end there is a predominance of soils in Central America (e.g. La Selva, Sollins *et al.*, 1994) and western Amazonia (e.g. WA floodplains, Zarin, 1999) that have relatively high levels of exchangeable nutrients such as calcium, magnesium, potassium and phosphorus (Fig. 2.15). These soils tend to have lower levels of dissolved aluminium and iron binding these elements in a relatively high soil pH environment. On the other end, there is a dominance of soil groups in the Guiana Shield (Caura, 24 Mile Reserve, Mabura Hill, Caquetá in Fig. 2.15) (Schulz, 1960; Jordan and Herrera, 1981; Lescure and Boulet, 1985b; Aymard *et al.*, 1998) that have very low standing levels of exchangeable and available nutrients either due to high ionic-binding with aluminium and iron sustained under very low pH conditions or simply a scarcity of ionically active clays, such as montmorillonite, vermiculite and illite (e.g. Ferralsol, Arenosol, Spodzol). In some cases, such as that of the Caquetá region in Colombia, the intermediate range in concentrations recorded may be the product of more dynamic mixing of materials from both Andean and Guianan origins (Duivenvoorden and Lips, 1998).

This general positioning of regions along a fertility gradient reflects the relative abundance of different soil types within each regional mosaic. All of the regions contain soils associated with both extremes of tropical fertility, but those covering Central America tend to extend across a greater range (e.g. Yavitt and Wieder, 1988; Sollins *et al.*, 1994) than those in the Guiana Shield, which tend to cluster at the lower end (e.g. Duivenvoorden and Lips, 1995; van Kekem *et al.*, 1996; Ramos, 1997) (Fig. 2.15). The Amazon and Sub-Andean sedimentary depressions tend to have a more finely mixed distribution of soil-building materials, with the well-drained soil types of Andean mafic or fine sedimentary provenance (e.g. Nitisols, Alisols, Luvisols, Fluvisols) generally exhibiting the highest concentrations of exchangeable cations. Western Amazon and Central American forests grow on soils with exchangeable nutrient concentrations similar to those found in the Guiana Shield – but young, steep and rapidly eroding planation surfaces created by the rising mountain slopes of Central America and western Amazonia increase the relative contribution of enriched sedimentary soil types to the adjacent lowland forest regions due to the action of more substantive erosional processes in the alpine highlands, such as glaciation and mass wasting of Cenozoic rock (see 'Physico-chemical transport', below). In contrast, highland contributions of lowland soil-building materials in the Guiana Shield are nominal and largely derived from Precambrian sedimentary formations.

Climate and Weather – a Simple Matter of Rainfall?

Weather is a collection of the many smaller-scale manifestations of larger climatic effects. Conversely, climate is often defined as the average condition associated with more variable weather. Weather delivers the impact generated by the interaction of global scale climate-regulating processes, such as planetary motion, solar sunspot activity, plate tectonics, volcanics and atmospheric and oceanic circulation, with regional land topography, hydrology and vegetation cover. More recently, human activity has become an important forcing factor at both global and regional scales.

Energy is the *lingua franca* of climate and weather and both are a way of viewing the same spatial and temporal variation in radiative heating and cooling, only at different scales. At the largest scale, global climate is shaped by a multitude of factors operating within a complex of interacting subsystems (Robinson and Henderson-Seller, 1999). A number of these subsystems exert a particularly pronounced effect on the climate and weather of tropical forestlands, but the impact of these does not resonate equally across all regions and is

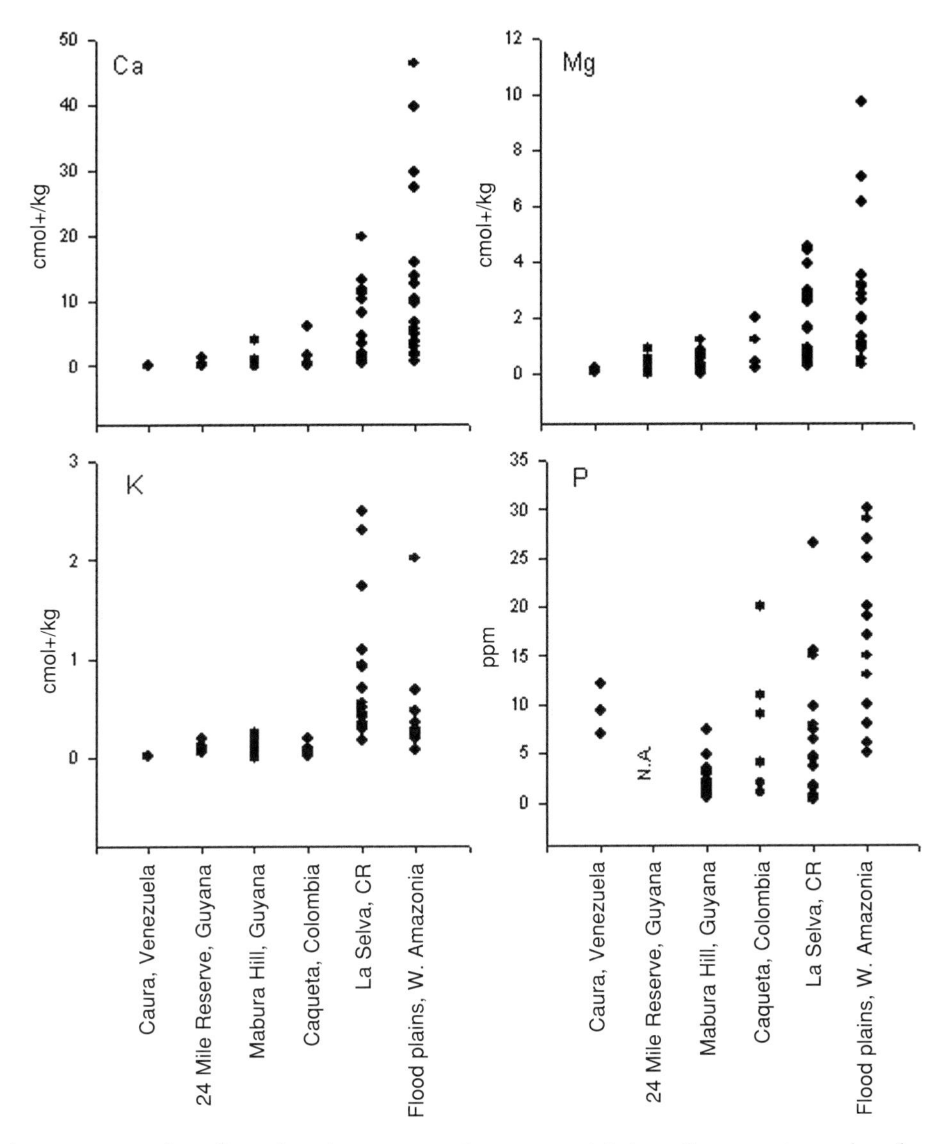

Fig. 2.15. Range of standing soil nutrient concentrations measured during soil surveys at a number of neotropical locations. All values are taken from upper profiles (most cases 20–35 cm depth). Values were interpolated from two consecutive depths (A+B horizons) when uppermost depth presented was less than 15 cm. Data sources: 24 Mile Reserve (Ogden, 1966), Mabura Hill (van Kekem *et al.*, 1996), Caquetá (Duivenvoorden and Lips, 1995), Caura (Fuentes and Madero, 1996; Briceño *et al.*, 1997; Ramos and Blanco, 1997), La Selva (Sollins *et al.*, 1994), western Amazon floodplains (Hoag, 1987 in Zarin, 1999). An outstanding phosphorus value of 145 ppm from the floodplains of western Amazonia was rejected as an outlier. Note that methods of extraction used are different depending on predominant soil attributes at different locations, but still may not yield perfectly proportional extraction efficiencies.

unlikely to have done so in the past, despite speculation otherwise. In the pantheon of factors shaping forests, climate stands next to geology as a giant and long-term variation in rainfall and temperature have proven to be of particular significance. The impacts of climate and geology are both long-standing and widespread, and as a consequence, often imperceptible or even misleading at the minor scales of time and space that define many scientific studies. As we will see in the case of the neotropics, scaling up from sample to system may be full of pitfalls when different climate mechanisms can bring similar rainfall patterns to different locations. Geographic location becomes an important element in defining not only the annual cycle of seasonal rainfall, but also how this cycle has and is likely to fluctuate in response to more expansive climate-driving forces. With recent rapid advances in climatological research techniques and technologies, an evolving picture of regional interconnections between land, sea and air that drives modern variability in tropical rainfall and *inter alia*, the evolving structure, function and composition of tropical forests, is beginning to emerge.

Rainfall variation, pace-setter of tropical life

Up until the mid-20th century, Western scientific understanding of tropical rainfall variation was limited. To 19th-century scientists and natural historians accustomed to seasonal changes in day-length and temperature (the two are of course bound together as cause and effect), the climate of tropical forestlands for all intents and purposes appeared uniform (e.g. Wallace, 1878). The account of Bates in his Amazon travelogue exemplifies this early perception:

> A little difference exists between the dry and wet seasons ... It results from this, that the periodical phenomena of plants and animals do not take place at about the same time in all species, or in the individuals of any given species.... Plants do not flower or shed their leaves, nor do birds moult, pair, or breed simultaneously ... With the day and night always of equal length, the atmospheric disturbances of each day neutralizing themselves before each succeeding morn; with the sun in its course proceeding midway across the sky, and the daily temperature the same within two or three degrees throughout the year – how grand in its perfect equilibrium and simplicity is the march of Nature under the equator! In some areas the convectional thunderstorms are so regular that appointments are made 'before' and 'after the rains'. (1879, Ch. II, p. 27)

The erroneous view that tropical forests were uniformly affected by an unchanging weather regime was in many ways symptomatic of the limited breadth and depth of scientific instrumentation and technology available to scientists at the time. Their impressive empirical powers unknowingly relegated the effect of many tropical climatic processes, both past and present, to the backbench of scientific enquiry. Starting from such an early misconception meant that substantial breakthroughs in our broader understanding of the large-scale mechanisms responsible for rainfall variation in the tropics gained little momentum until relatively recently. Even up until as late as the 1960s some climatologists had not fully accepted the notion of meteorological interconnections, though these were proposed for some tropical areas as early as the 1920s (Walker, 1923, 1924) and views of Amazonia as a stable environment persisted (e.g. Richards, 1952; Schwabe, 1969). Establishing a linkage between a number of important oceanic and atmospheric state variables, such as sea temperature and rainfall, helped to establish a mechanistic foothold in explaining global variability in rainfall and the key role that sea temperature plays in regulating climate in an otherwise equivalent barotropic atmosphere (e.g. Ichiye and Peterson, 1963; Bjerknes, 1966, 1969).

The mistaken belief that an absence of temperature-driven seasonality and little interannual variability somehow simplified the growth cycle in the wet tropics has faded with time and the mounting record of failure met in trying to harness this 'simplicity' through large-scale commercial forestry and agricultural practices and technologies (e.g. see Moran, 1982). The annual arrival and disappearance of heavy rainfall had already inculcated an implicit understanding of rainfall seasonality and its implications into the long-established agrarian cultures and societies of the lowland tropics long before the arrival of Western scientific theory. As a consequence, many of their customs and traditions reflected, and continue to reflect, a refined understanding of the effects of what otherwise may appear as relatively minor seasonal and interannual fluctuations in rainfall, the true pace-setter of contemporary tropical life (see Chapters 4 and 7).

About this section

This section aims to provide a synoptic account of the complex pool of global, regional and local factors shaping climate and weather in the Guiana Shield and to describe and categorize, albeit briefly, the spatial and temporal variation in these factors by drawing upon wide-ranging published research and presenting this in tandem with data available from selected weather stations in and adjacent to the region. In particular, it focuses, albeit only superficially, on the extent and nature of large-scale, modern variation in rainfall, temperature, cloud cover and radiative fluxes across the shield area and in relation to other neotropical regions. It is important to note, however, that variation in climate and weather of the Guiana Shield remains one of the most poorly monitored in the world (Folland *et al.*, 2001). The main processes that govern rainfall and temperature in the region and how their interactions (could) control variation are also briefly considered. This section also attempts to demarcate the current debate surrounding prehistoric climatic conditions across neotropical lowlands, the extent of evidence available for the Guiana Shield region and a rough compositing of this wide-ranging information in considering climate change over the Quaternary (as recent as 1000 ka BP). It is not intended to act as a substitute for more substantial texts or the wide range of published studies exploring tropical climate, and the reader is referred to the many references cited in this section for further clarification of the complex interrelationships only summarized briefly here.

Modern climatic regions

The Guiana Shield embraces a large part of the climatological variability that can be found anywhere in the lowland tropics. According to the Köppen climate classification system, the area is consistent with three major climate zones, *Af*, *Aw* and *Am*, based on temperature and precipitation (Walsh, 1996; Robinson and Henderson-Seller, 1999). These are roughly distributed along a longitudinal line with a corridor of relatively dry *Aw* climate separating the main western area classified as *Af* from the eastern rim classified by a mixture of *Am* and *Af* subtypes (Fig. 2.16). Tepuis and other peaks of restricted surface area in the Guayana Highlands are exposed to high amounts of precipitation and much lower temperatures and would generally be classified as *Cfb* in Köppen's system. These island-like areas of much cooler and wetter climate break the narrow corridor of *Aw* type climate that runs roughly north–south between the southern cerrados of the Brazilian Shield and the northern llanos of central Venezuela and northeastern Colombia.

The distribution of climatic regions in the Guiana Shield under Köppen's system is mirrored when classifying the area using the perhumidity index (PI) devised by Walsh (1996). The PI measures the depth and extent of seasonality in tropical rainfall. This method yields similar results for the region except in the way the climate of the western tongue of the shield is classified. Under Walsh's scheme, this area is classified as a *tropical superwet* (PI $\geq$20) with a zonal transition eastward across three other climate types, viz. *tropical wet* (PI = 10–19.5), *tropical wet seasonal* (PI = 5–9.5) and *seasonal* (PI <5).

According to a similar precipitation–temperature system devised by García (1987, cited in Huber, 1995a), the main western and eastern *Aw* and *Am* zones in the shield would classify as a *macrothermic* (>24°C) *ombrophilous* (>2 m rainfall per annum) climate. A *macrothermic tropophilous* (1>x>2 m rainfall per annum) climate type under García's system is con-

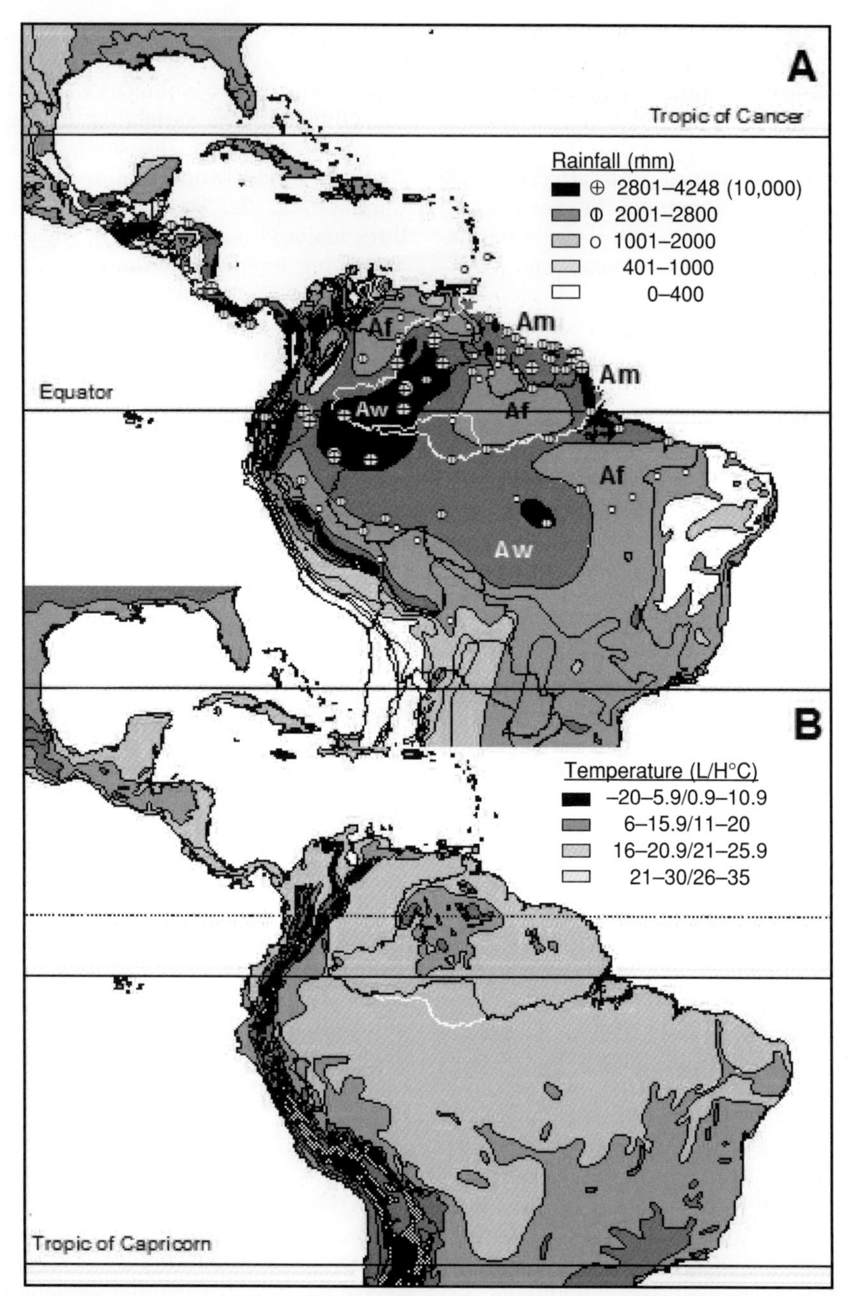

Fig. 2.16. The distribution of average annual rainfall for 87 neotropical locations expressed as standard deviates. Standard deviates range from –3 (smallest points) to +4 (largest points). Rainfall for all 87 locations ranges between 903 (Tegucigalpa, Honduras) and 4248 (La Selva, CR) with an average of 2221 mm. Each location average is based on at least 10 years of monthly totals. In most cases these records fall between 1950 and 2000. Note the position of the 'meteorological equator' (hatched line) at 5° N of the equator (solid line) and in relation to locations recording larger (3+) positive standard deviates (large circles) and the relatively dry southern rim of the Amazon Basin (–1/–2 SD; smaller circles). Stippled lines are the 2000 mm isohyet as mapped by Sombroek (1999). Aw<2000<Af /Am. See text for data sources references and definitions.

sistent with Köppen's *Aw* designation. Cooler highland types would include the *mesothermic* (12–18°C) *ombrophilous* and *submicrothermic* (8–12°C) *ombrophilous* classifications.

The Guiana Shield represents the most extensive Precambrian area classified under Köppen's wettest and warmest climate category, *Af* (Walsh's *tropical wet/superwet* and García's *macrothemic ombrophilous*), despite a large central swathe of the region experiencing a more severe seasonal decline in rainfall (*Aw*). The three other Precambrian areas with (sub)tropical forest, viz. the Brazilian Shield and Precambrian areas in Africa and India, also experience more severe drops in rainfall during the dry season months and are for the most part classified as *Aw*. The broad expanse of *Af* climate in the Guiana Shield correlates well with its principal rank as the world's largest repository of tropical moist forest on Precambrian rock (Chapter 1).

Climate classificatory systems help to understand large-scale spatial variation in average rainfall and temperature across the shield, but also mask longer-term fluctuations in the depth and breadth of seasonality and any anisotropies that occur across the region. This drawback of the PI (and other) system(s) is noted by Walsh (1996), but the system does fit well with the current (natural) distribution of recognized tropical forest formations. Given the importance of seasonal environmental fluctuations as phenological pace-setters in the tropics, interannual variation in rainfall status adds an important dimension to a region's contemporary climate regime and, *inter alia*, power to exploratory analyses attempting to unscramble the host of factors shaping standing forest diversity and productivity.

Modern centres of high rainfall in the Guiana Shield

The Guiana Shield embraces some of the wettest and driest locations found within the equatorial trough. As may be expected for such a vast, relatively unpopulated area, ground station data for the region are few and far between and principally located on the perimeter of the shield along the main stems of the largest rivers. Few stations, combined with erratic record-keeping and infrequent reporting at existing stations, makes this one of the most poorly monitored in the western hemisphere, particularly those areas where population densities are at their lowest (Folland *et al.*, 2001).

Despite this limitation, existing station data and interpolations derived from these have identified several centres of high annual precipitation (>3000 mm per annum) within the Guiana Shield (e.g. Sombroek, 1999). The easternmost of these is centred on northern French Guiana and extends southward outside the Guiana Shield as a narrow coastal band along the Atlantic seaboard of Amapá and across the mouth of the Amazon River (Fig. 2.16A) (Snow, 1976).

In the north-central region of the shield a second small centre extends across the Guayana Highlands between the Pakaraima Mountains of west-central Guyana and eastern Venezuela through to the northern coastal swamplands of northwest Guyana (Fig. 2.16A) (Snow, 1976; Huber, 1995a). This centre of high annual precipitation is isolated by a narrow corridor of relatively dry lowland climate, referred to here as the Savanna Trough, that extends between the Roraima massif along the Guyana–Venezuela–Brazil border and the north–south alignment of the upper Caroni River and extends southward across the Amazon Downwarp west of Santarem (Fig. 2.16A) (Soubiés, 1979; Sombroek, 1999). Northward, a reduced rainfall regime fans out across much of the lower Caroni, Paragua and Caura River basins and the upper Cuyuni River watershed. Rain data from stations in this corridor, such as Santa Elena de Uairén, Kamarata, La Paragua, Ciudad Bolívar, Anacoco and Tumeremo, among others, show a distinct suppression of the June–July wet season (Fig. 2.16A) compared to high precipitation centres immediately east and west of the corridor. Isolated tepuis found within the corridor, however, have recorded annual totals equal to or greater than the Pakaraima highland region (Huber, 1995a), almost entirely due to local orographic effects.

A third centre of high precipitation is both the largest and wettest of the three, extending from the upper Orinoco in Venezuela and across the western tongue of the Guiana Shield in northern Brazil and south-central Colombia. This centre forms the eastern flank of a much larger zone of exceptionally high rainfall that extends across most of the lowland forest areas of the Sub-Andean Foredeep and Amazon Downwarp regions of Colombia, Ecuador, western Brazil and northern Peru (see Figs 2.6B and 2.16A) (Snow, 1976; Salati and Marques, 1984, p. 106; Figueroa and Nobre, 1990; Walsh, 1996). The high annual rainfall total in this western zone is due to the virtual absence of any month receiving less than 60–100 mm of rain, a characteristic that normally defines tropical seasonality according to Köppen (1918), Walsh (1996) (a PI value of 20–24) and others. In comparison, eastern and central centres of high rainfall in the shield normally include at least one or two months with less than 100 mm of rainfall.

Seasonal variation in rainfall

Rainfall in the Guiana Shield shows seasonal variation that generally conforms to one of three general models referred to here as: (i) northern uni-modal; (ii) southern uni-modal; and (iii) Atlantic bi-modal. The northern uni-modal model is applicable to 70% of the shield area falling between 1° S and 5° N. Rainfall peaks between May and August, ranging from 100 to 600 mm per month for the period. Both forest and savanna locations, such as Boa Vista in Roraima and Araracuara, Colombia, conform to this model, but with very different annual totals (Fig. 2.17, centre and right columns).

The southern uni-modal model appears in 20% of the extended shield area found south of the equator between 1° and 3° S, but extends much further throughout most of southern Amazonia. The main peak in rainfall in this area occurs from November to January with slightly lower average monthly totals than occur in the northern uni-modal area (e.g. Manaus, Leticia in Fig. 2.17).

The Atlantic bi-modal model fits well with the seasonality of rainfall over approximately 10% of the shield area running parallel to the northeastern Atlantic coastline. The major peak of rainfall occurs at the same time as in the northern uni-modal areas to the west, but with a second much smaller peak occurring from November to January, the typical peak for southern uni-modal regions (e.g. Paramaribo, Georgetown in Fig. 2.17). In some parts of the shield where a bi-modal rainfall predominates, this second peak is much larger and the seasonality is in effect uni-modal. The large amount of rain falling annually along the coastal zone of French Guiana, Amapá and across the mouth of the Amazon in comparison to more northerly Atlantic coastline stations (e.g. Paramaribo, Georgetown, Port Kaituma) is a fine example of this effect (Fig. 2.17), which can be attributed to the interplay of regional convective and localized dynamic uplifting processes (Robinson and Henderson-Seller, 1999) and a surface roughness gradient from ocean to forest (Snow, 1976). These coastal effects along the Guiana Shield are the same as have promoted growth of the Brazilian Atlantic forest south of the –6th parallel (Morellato and Haddad, 2000).

Generally speaking, there are two continental-scale gradients of declining rainfall running perpendicular to one another, one meridional (longitudinal) and the other zonal (latitudinal). These gradients, however, are not apparent when examining total annual rainfall for sites within the area affected, largely due to the compounding effects of continental, local topographic and maritime influences. Anisotropies of declining rainfall at this scale are 'phase-dependent', occurring exclusively during the normal dry season months.

Rainfall differences in the zone affected by a northern uni-modal or bi-modal regime are the result of a strong continental-scale west to east decline in

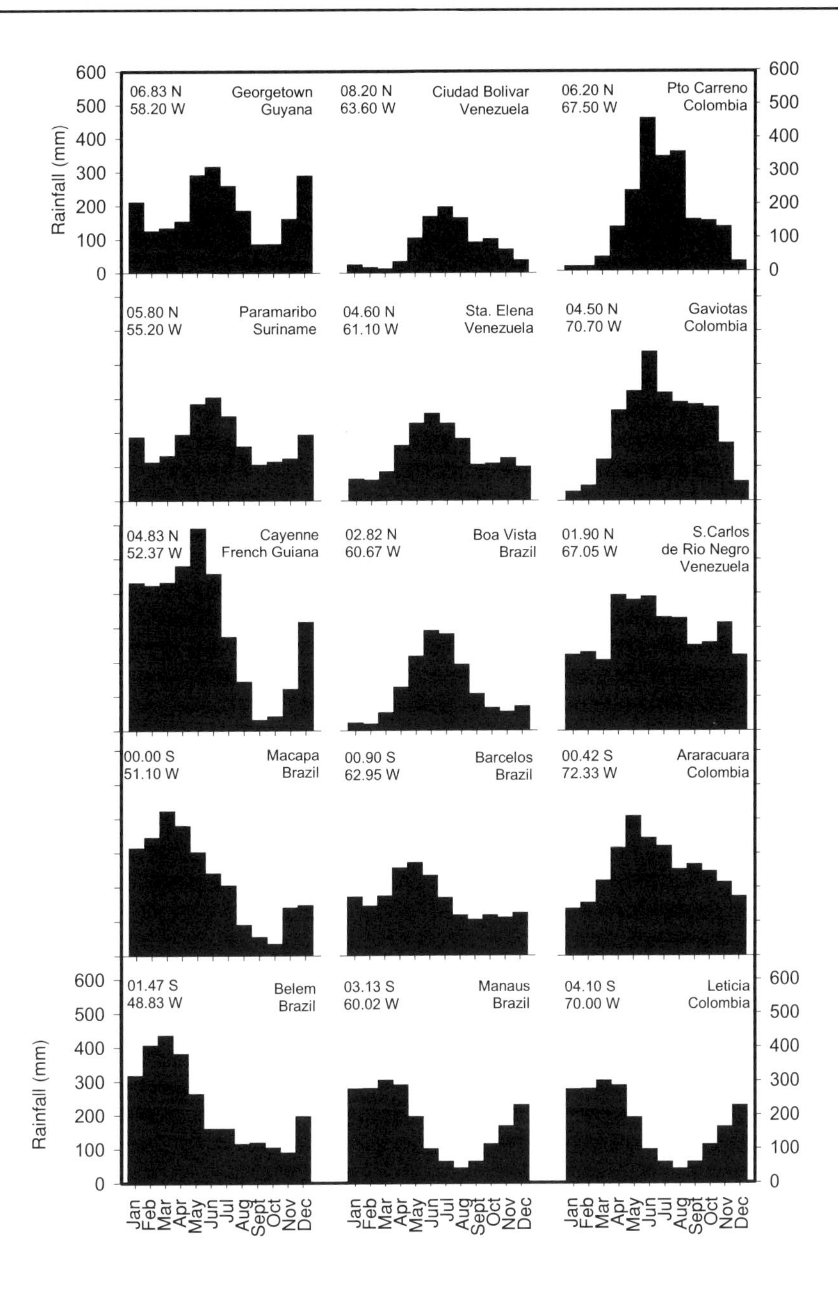

Fig. 2.17. Average monthly precipitation for 15 stations representing the eastern (left column), central (middle) and western (right column) climatic zones in the Guiana Shield. Station data are presented from north (top row) to south (bottom) and represent averages from 10+ years of data collected between 1950 and 2000. Rainfall data sources: see Notes.

precipitation during the months of September and October (–November) (Fig. 2.18A). During most other months of the year, rainfall variation is overwhelmingly shaped by the Inter-Tropical Convergence Zone (ITCZ) interacting with finer-scale landscape features (e.g. topography, drainage) throughout the area north of the equator. During this period, the meridional gradient disappears altogether (Fig. 2.18A). The decline, which varies in slope by year, is due almost entirely to a precipitous seasonal drop in rainfall across the eastern half of lowland South America, while locations in the western superwet zone north of the equator experience a much smaller seasonal reduction in the amount received (Fig. 2.17).

The second seasonal anisotropy in rainfall occurs as a north to south decline that for the most part extends outside the Guiana Shield. It forms across the area south of the shield (i.e. 2°–10° S) that conforms to a uni-modal regime with the area within the southern Guiana Shield forming the upper end of the 'phase-dependent' gradient. The north to south decline in this instance occurs for the dry season months of April to July (Fig. 2.18B). The trough that occurs during this dry phase of the southern uni-modal deepens southward and defines the gradient. Rainfall during the wet season phase from November to January is roughly uniform across the region, variation again being largely a function of smaller-scale differences. This gradient extends from just north of the Amazon River and southwards towards the central part of the Brazilian Shield (about 10° S) where the ranging ITCZ meets its southernmost limit across the South American continent (Snow, 1976; Critchfield, 1983). Localized convection also diminishes as the high atmospheric water vapour storage that characterizes the central Amazon Downwarp (Salati and Marques, 1984), Sub-Andean Foredeep and western parts of the Guiana Shield declines southward across the Brazilian Shield.

When combining annual and seasonal isohyet maps of 'normal' precipitation across the Amazon Basin, the gradient in seasonal rainfall decline from west to east is also evident (Figs 8 and 9 in Salati and Marques, 1984). A more recent map compiled by Sombroek (1999) depicts the variation in seasonality across the Amazon region. The map shows the (interpolated) spatial distribution of dry month frequency across tropical South America and identifies a large zone north of the equator where no month receives less than 50 mm of rain (drought conditions, after Walsh, 1996). This zone is bisected by the north–south corridor of drier climate (*Aw*) that coincides with much of the Rio Branco/Gran Sabana/Rupununi savanna (Fig. 2.16). The meridional gradient in dry season rainfall decline across northern South America is clearly more refined, with many locations not reaching physiological drought conditions as depicted in Sombroek's map, but still registering diminishing rainfall during the identified months. To a large extent, this is a consequence of ITCZ behaviour across the region and its impact on dry season timing. Westward locations tend to experience a (less severe) seasonal dip in rainfall two months later (Dec–Feb) than eastern locations (Sept–Oct) (Fig. 2.18). Despite the staggered arrival of the 'dry' season at western locations, the total number of dry days (<0.1 mm) occurring at sites during the September–October period remains relatively high, but occurs in much shorter consecutive periods compared with December–February, when periods of greater than six consecutive days are more common (e.g. Snow, 1976; Duivenvoorden and Lips, 1995). This high incidence of single dry days nested within longer wet periods contrasts with the more prolonged periods with and without rain that typify many sites in the eastern Guiana Shield (e.g. Frost, 1968) and constitute the difference in seasonal monthly and annual totals between the two regions (Snow, 1976) (see Chapter 7). Long-term series of daily and monthly rainfall records indicate that these differences are amplified during strong ENSO phases as a consequence of asymmetric response along the east–west gradient. A discussion and review of research assigning a plausible role of these asymme-

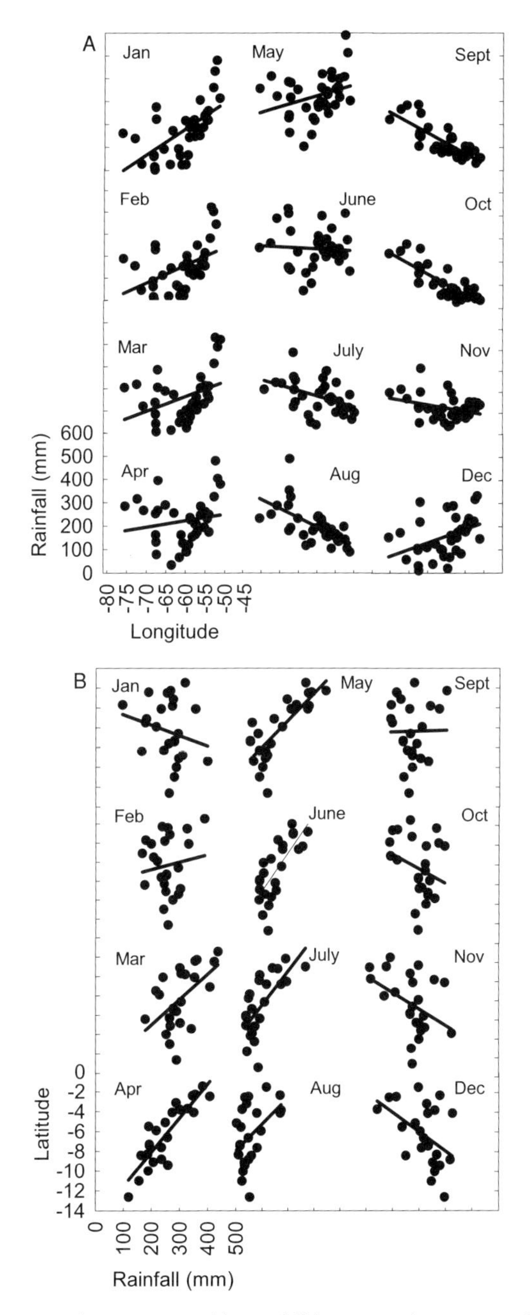

Fig. 2.18. Simple dissection of average monthly rainfall by geographic position for locations in tropical South America. (A) Monthly rainfall vs. longitude across the Guiana Shield. Only monthly data from lowland (<200 m asl) sites affected by the northern uni-modal rainfall pattern and falling between 9° N and 1° S are included. (B) Monthly rainfall vs. latitude across the southern Guiana Shield, Amazon Downwarp and Brazilian Shield. Only data from lowland sites affected by the southern uni-modal pattern are included. Monthly values represent averages for 10+ years. Lines represent best fit, least-squares regression. Rainfall data sources: see Notes.

tries in driving variation in forest distribution, productivity and perhaps, standing species diversity across the neotropics is discussed in Chapter 7.

Historic variation in rainfall

Rainfall across the Guiana Shield follows a fairly predictable seasonal pattern consistent with the three major models, with relatively little variation in each month's rank contribution to annual rainfall. The actual total annual or monthly amount of rainfall received, however, is subject to much greater variation, particularly where precipitation patterns are sensitive to oscillations in regional atmospheric–oceanic circulations, rather than more predictable and regular effects of convection driven by local topography or evapotranspiration.

Century-long records of monthly rainfall for stations within the Guiana Shield are limited to a few of the main population centres. Most of these stations are located on or near the perimeter of the shield area (as described in Chapter 1) and records from stations located in the interior of the shield are typically of much shorter duration and often missing periods ranging from one month to several years. Long-term, monthly time series maintained at locations in the Caribbean (e.g. Puerto Rico (since 1899), Barbados (1853), Trinidad (1862)) and eastern Brazil (e.g. Fortaleza (1849), Recifé (1875)) are of greatest duration and completion. One station directly on the shield perimeter, the Botanic Gardens, Georgetown, Guyana (1880) has maintained a record comparable with these other stations. Since 1950, a 'modern' record of monthly rainfall has been maintained at a much larger number of stations both within and adjacent to the region. Restricting comparisons among stations to this more recent period, a general picture of variation in rainfall over a semi-centennial scale within and surrounding the Guiana Shield can be assembled from the same historical station data employed in composite climatological analyses.[3]

Seasonally adjusted average annual rainfall shows little or no overall trend between 1958 and 2002 at station locations in or adjacent to the Guiana Shield (Fig. 2.19). Similar analyses carried out on 5 × 5° gridded averages as part of the IPCC WG1's assessment of global rainfall trends show a general decline of 4–12% over the eastern half of the Guiana Shield during the period 1946–1975 (Fig. 2.25i in Folland *et al.*, 2001). This contrasts with a positive trend in annual rainfall of 4–8% calculated for grid cells covering western Amazonia and Central America (Panama/Costa Rica) over the same period.[4] All stations examined, however, show irregular oscillatory variation with positive (wet) and negative (dry) phases lasting anywhere between 1 and 5 years. Long-term rainfall records at a number of stations show reasonable concordance in the timing of these multi-annual phases (Fig. 2.19), although less so when also considering phase duration and amplitude. This concordance is, logically, most common among adjacent stations, though zonal relationships appear more important than meridional ones. Only during significantly strong La Niña (e.g. 1989–90) and El Niño (e.g. 1982–83, 1997–98) phases of the Southern Oscillation does a wider regional spread of available station data show increasing concordance (Fig. 2.19). Again, however, the trend towards spatial concordance in rainfall variability is limited to an area east of Manaus, north of Salvador and south of Panama and is centred on the eastern half of the Guiana Shield. Rainfall variability at western and southern Amazonian stations (e.g. Leticia, Fig. 2.19) shows only weak or slightly negative correlations with easterly stations during these periods. Analyses employing empirical orthogonal functions (EOFs) to discriminate different sources of variation yield similar conclusions for eastern and southern areas in Brazil (Tanaka *et al.*, 1995), only at larger spatial scales. Generally speaking, locations west of the Savanna Trough show a declining long-term precipitation relationship with those found east of this dry zone.

The annual course of seasonal transitions in the Guiana Shield and elsewhere in the tropics follows a predictable periodi-

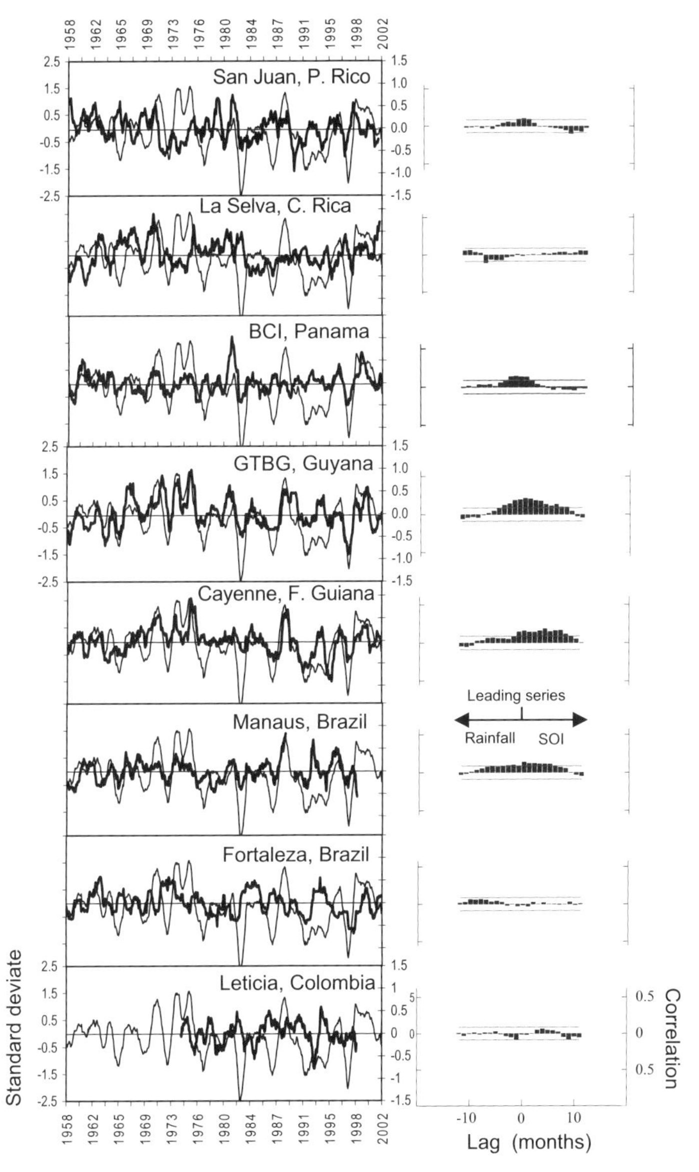

Fig. 2.19. Time series of SOI and rainfall between 1958 and 2002 for selected stations located on or near the perimeter of the Guiana Shield. Shown are series of standardized monthly rainfall and SOI values smoothed using a 12 month MA. Standardized rainfall values were based on mean and variance of all data available for the period. Standardized SOI values were taken directly from normalized values calculated by the Climatic Research Unit/University of East Anglia using methods described by Ropelewski and Jones (1987). Negative and positive trending in standardized values for the SOI are generally consistent with El Niño/La Niña conditions, respectively. Cross correlation plots for each rainfall station cover a 24-month lag between the standardized monthly SOI and rainfall values. Bars above or below horizontal lines depict P=0.05 level of significance of cross correlation at that lag. Positive and negative values represent correlations with rainfall leading and following SOI, respectively. Single missing values were interpolated using a local quadratic smoothing routine (SYSTAT 8.0). Cross correlations for series not spanning the full period between 1958 and 2002 were truncated prior to analysis. See Note 3 for rainfall data sources.

city. At the decadal scale, deviations in monthly rainfall over the Guiana Shield and the adjacent regions of the neotropics accumulate and the regular periodicity of rainfall associated with seasonality disappears at longer timescales when the seasonal component is removed (Figs 2.17, 2.19). Frequency, duration and amplitude of multi-annual dry and wet phases can vary considerably. Despite this less predictable behaviour, seasonally decomposed rainfall variation at the longer interannual timescale also fluctuates in a periodic manner. These fluctuations, however, reflect the composite action of a more varied series of periodic cycles shaping the global circulation.

Within the global circulation, the interaction between sea surface temperatures (SST) and atmospheric pressure (AP) in the tropical Atlantic and Pacific Oceans wields the largest influence on the annual and interannual variation in rainfall within the Guiana Shield (Fu *et al.*, 2001; Liebmann and Marengo, 2001). Three main compartments in the global circulation of moisture, viz. oceanic currents, atmospheric cells and terrestrial stores, interact to influence the movement and magnitude of SST and AP throughout the tropics (Philander, 1990; Robinson and Henderson-Seller, 1999). Energy exchange between atmosphere, ocean and land inextricably bind the behaviour of one to the other, but they are treated here as separate processes in order to focus on several key features of each compartment affecting rainfall and temperature in the Guiana Shield and adjacent tropical regions.

Modern centres of low temperature in the Guiana Shield

Spatial coverage and length of the temperature record lags available information on rainfall. This lack of data is most apparent at upper elevation locations of the central shield (e.g. Huber, 1995a). In part this reflects the fact that seasonal variation in temperature is relatively small in the barotropic environment of the equatorial trough, particularly in comparison to the daily range, and has therefore not ranked high among priorities for meteorological monitoring in the tropics. But it also reflects a declining capacity to maintain the more extensive monitoring systems established in the 1950s across a relatively remote region of the world.

The Guiana Shield embraces a larger, spatially weighted range in sub-alpine temperature than any other part of the neotropics. This range consists of one major and two minor highland centres of relatively low temperature (average minimum: 6–20°C) surrounded by an extensive lowland area of much higher temperatures (average minimum: 21–30°C) (Fig. 2.16B). Centres of relatively low temperature are located on the three main upland 'islands' found within the region, the Guayana Highlands, Tumucumaque Uplands and Chiribiquete Plateau (Fig. 2.6B; 'Shield macro-features', above). Lowest temperatures are found at the highest elevations on isolated tepuis and peaks formed from the Roraima sedimentaries, Avanavero intrusives and Parguaza granites of the Guayana and Tumucumaque regions and the Palaeozoic sedimentary rock of the Araracuara Formation forming the Chiribiquete Plateau (Fig. 2.16B).

Daily and seasonal variation in temperature

Lowest mean daily temperatures in the Guiana Shield are expected to occur on the summits of the highest tepuis of the Guayana Highlands during the peak season of rainfall and cloud cover. Highest maximum daily temperatures occur in the low-lying sections of the Savanna Trough during the peak dry season (Frost, 1968) when cloud cover (October) and moderating latent heat sources (March) are minimal. Daily temperatures can range as much as 20°C, but typically vary between 7°C and 15°C over a 24 h period at most monitoring stations within the Guiana Shield (Schulz, 1960; Ratisbona, 1976; Snow, 1976; Vargas and Rangel, 1996a). The width of the daily range also typically varies by as much as

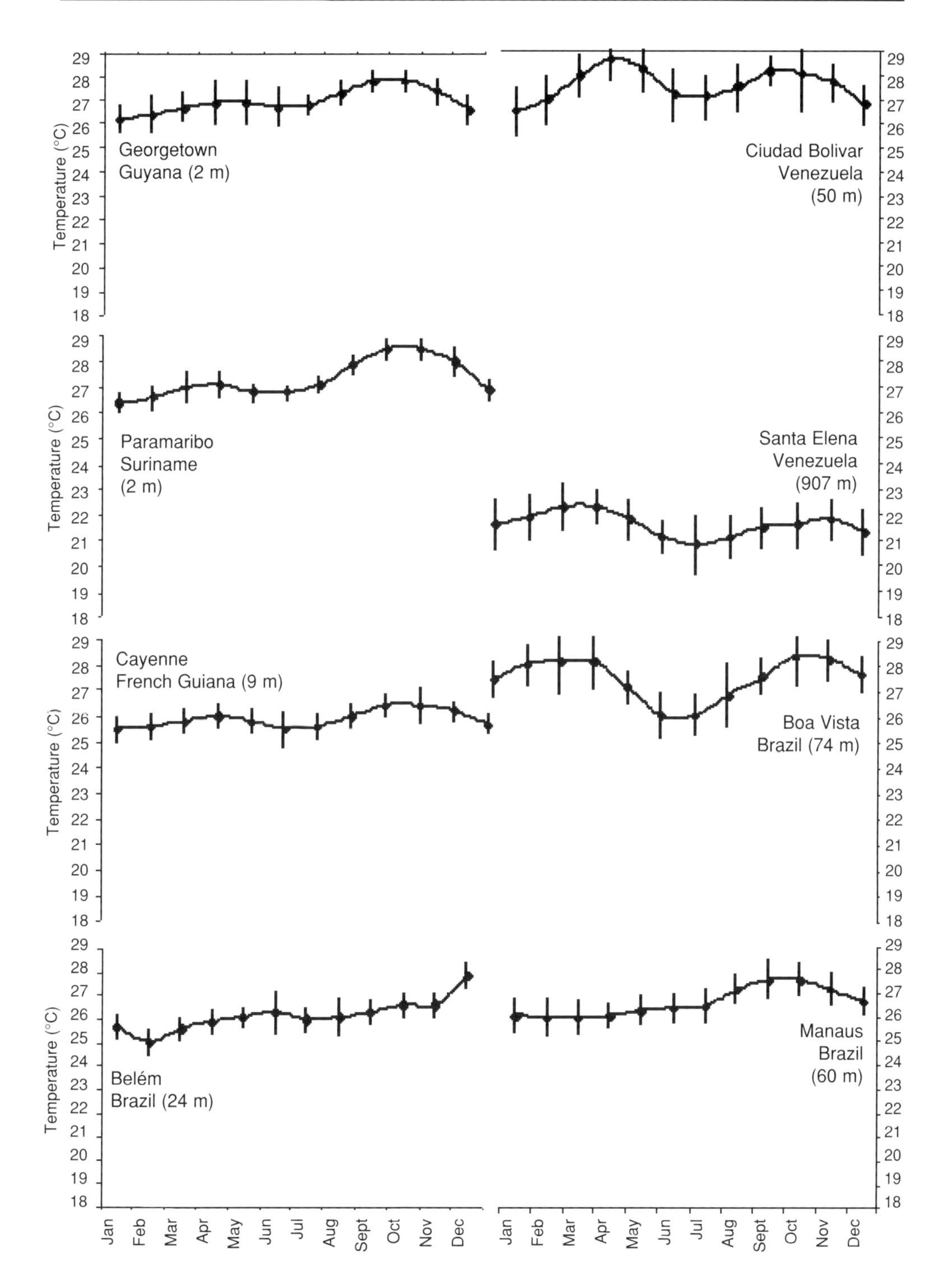

Fig. 2.20. Mean monthly surface temperatures at 8 stations representing eastern (left); and central (right) climatic zones in the Guiana Shield. Data are presented from north (top) to south (bottom) and represent averages from 10+ years of data collected between 1950 and 2000. Data sources: Snow (1976), Johnson (1976).

2–5°C between seasons based on available station data. Widest ranges occur in open savannas during dry season months when there is little cloud cover to modulate radiation fluxes. Conversely, coastal and highland forest locations across the region exhibit the lowest seasonal change in daily temperature range (Snow, 1976).

Seasonal variation, while small by comparison, is not uniform across the region and changes in range width are demonstrable. Available data suggest that intra-annual variation generally ranges between 2°C and 4°C (Fig. 2.20). Stations registering relatively uniform rainfall (e.g. Cayenne, French Guiana) experience the lowest seasonal variation in mean temperatures, while regions subject to more pronounced uni-modal rainfall in the Savanna Trough typically show the greatest seasonal change in mean temperature (e.g. Boa Vista, Roraima, Brazil) (Salati and Marques, 1984).

Historical variation in temperature

A sparse history of temperature data collection across the Guiana Shield and the south-central Amazon has created a noticeable gap in the global grid of historical temperature coverages (Fig. 2.9 in Folland *et al.*, 2001). This gap has been most persistent over the Brazilian Legal Amazon, including the southern, sedimentary lowlands of the Guiana Shield.

Despite the problem of spatial inconsistency in data coverage, some stations have been producing quality data for many decades. When combined, these data suggest that the Amazon region as a whole has been experiencing a trend of increasing temperature since the early 1900s, estimated as a rate of +0.56°C per century by Victoria *et al.* (1998). Spatially averaging available station data for the Guiana Shield region also indicates a slightly lower positive trend (+0.33°C per century) in temperature for the area between 1880 and 2002 (Jones *et al.*, 1999) (Fig. 2.21). Grid cells over the Guiana Shield that form part of the IPCC's gridded global depiction of historical temperature change indicate a similar rate of increase for most lowland areas since 1946, when adequate data coverage began (Folland *et al.*, 2001). However, gridded rates across the shield region, or the neotropics for that matter, do not appear uniform in this depiction. Rates within cells have not remained constant over the 20th century, nor have different cells necessarily shown similar rates of change, although autocorrelation between grid values at certain spatial scales is apparent. Most noticeable is the declining trend in temperatures (–0.2°C/decade) assigned to

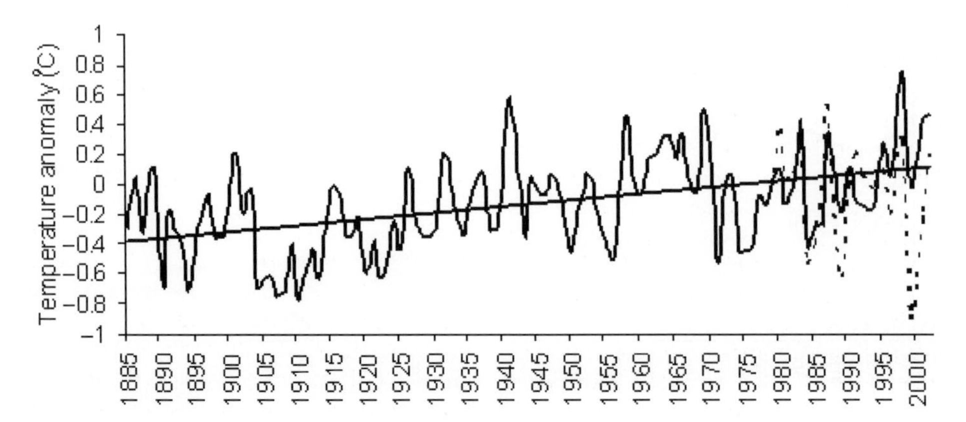

Fig. 2.21. Spatially averaged temperature anomalies for the Guiana Shield and adjacent regions falling between 10° N and 05° S and 50° W and 75° W. Anomalies from 1880 to 1998 are referenced to the mean for the period 1961–1990 with a spatial resolution of 5° × 5°. Line is best-fit trend line. See Jones *et al.* (1999) for details of parent data set construction and anomaly calculations.

cells over the Guayana Highlands, particularly from 1946 to 1975 (Fig. 2.9c,d in Folland *et al.*, 2001). Areas of the Guiana Shield covered by the grid analysis of Folland *et al.* generally appear to have warmed at a lower rate than adjacent western and southern regions of Amazonia. These areas show rates of increase that are two to four times greater than those of the Guiana Shield and Central American regions. Importantly, the analysis also indicates that western Amazonia may have experienced warming (of 0.2–0.4°C/decade) from 1946 to 1975 while the Guiana Shield (–0.4–0.2°C/decade) and Central America (–0.2–0°C/decade) generally experienced a decline in temperature. This could be important in framing regional variation in net ecosystem productivity of tropical forests, given the role of temperature and moisture in defining variation in both plant growth and microbial respiration across Amazonia, all else being equal (see Chapter 7).

Seasonal to multi-decadal rainfall and temperature-regulating mechanisms

Solar heating

The simplest and most direct influence shaping the seasonal rise and decline of rainfall in the tropics is the annual swing in the amount of solar radiation received in the northern (boreal) and southern (austral) hemispheres. This occurs as a consequence of Earth's axial tilt in relation to the Sun that swings the belt of maximum insolation between northern and southern hemispheres through the course of a single orbit (one year). In the higher latitudes, this swing strongly affects temperature and characterizes seasonality. While this seasonal swing in temperature does not directly affect the tropical belt (but see 'Guiana Shield rainfall and temperature regulation in relation to other neotropical regions' and 'Mid-latitude incursions', this chapter), it does affect seasonal rainfall in the tropics by forcing the meridional oscillation of the atmospheric convergence zones that bring strong rain across much of the tropics. It also passively directs the conveyance of warm and cold water from one hemisphere to another which attenuates the atmospheric response (especially at high latitudes) to the swing in solar seasons (see 'Oceanic components', this chapter).

If the annual cycle of differential solar heating was the only factor shaping seasonality, the amount and timing of rainfall in the tropics would be highly predictable at all but the largest timescales (see 'Longer-term climate forcing factors', below). As it stands, the amplitude of variation becomes less predictable over larger time intervals as the impact of different factors operating at different periodicities begins to accumulate. Other mechanisms must be acting to modify the seasonal trek established by the basic astronomical relationships between our planet and the Sun. These mechanisms, in the seas and atmosphere, are driven by and create a cascade of further adjustments to initial changes in the Sun–Earth interactions. They can be collectively considered as the global and regional circulation.

Global and regional circulation

Warm and cold currents, both in the air and sea, are continuously moving and shifting their geographic track and the strength and direction of flow. It is these movements and their degree of connectivity within and between regions that modulate variations in solar heating of the surface. The temporal variation of rainfall experienced in the Guiana Shield region is directly influenced by the cycling of moisture in the atmosphere through two very large and important meridional (parallel to longitude) and zonal (parallel to latitude) components – the Hadley and Walker Circulations, respectively. Equatorward movements of extratropical highs (that limit rainfall) and mid-latitude atmospheric Rossby waves (that generate rainfall), the movement of easterly (African) waves across the northern tropical Atlantic and the seasonal formation of their offspring, hurricanes, in the tropical mid-Atlantic, also influence rainfall in the Guiana Shield on a regular basis, although not in equal parts.

Rainfall variation is also influenced indirectly by the geographic distribution of warm and cold surface seawater in the tropical Atlantic and Pacific Oceans through their impacts on ocean–atmosphere exchange processes. The distribution of seawater temperature primarily reflects the push-and-pull of seasonal hemispheric heating, consequent changes in wind flow, and the effects this has on the track and strength of major surface currents operating in these tropical maritime regions (Tomczak and Godfrey, 1994). These shifts can occur as frequently as every year or as more gradual adjustments to changes in the global oceanic conveyor (Broecker, 1991). They can also be affected by more frequent adjustments in the form of wave-like packets of warm water sent zonally across oceanic basins, referred to as Kelvin (Kessler and McPhaden, 1995) and Rossby waves (Dickinson, 1978). Changes in wind stress, coastline geometries and the resonance dynamics of reflecting waves have been identified as key features of these phenomena affecting seawater temperatures (Philander, 1990).

Importantly, temperature changes also reflect the effects these shifts can have on deep seawater upwelling and the geometry of the thermocline within a basin over any particular period of time. Major oceanic currents are driven by wind, gradients of salinity and temperature created by fluxes of heat and freshwater (the thermohaline circulation) and inertia created through the global ocean conveyor. Currents play an important primary role in shaping rainfall variability over (parts of) the Guiana Shield since they convey warm water. The main currents in the Atlantic and Pacific directly involved in tropical water movements, include the massive pantropical North Equatorial current (surface), South Equatorial current (surface), Equatorial Undercurrent (thermoclinic) and North Equatorial Countercurrent (surface), and the much smaller Peru current (surface, Pacific), North Brazil current (surface, Atlantic), Brazilian Coastal Current (surface, Atlantic), Guiana current (surface, Atlantic) and Caribbean current (surface, Atlantic) (Richardson and Walsh, 1986).

Oceanic currents and the overlying atmosphere interact to create a continuous cascade of change in the geographic distribution of their main energy-transfer and storage zones. These zones can be characterized by a large number of (collinearly) related parameters. Sea-surface temperature (SST), surface windstress (SW), outgoing long-wave radiation (OLR), atmospheric pressure (AP) and sea current velocity (SCV) are particularly useful (leading, tracking and simultaneous) indicators. Cascading changes in the state conditions of these commonly monitored attributes are associated with rainfall variability at daily to multi-decadal scales within the South American tropics (e.g. Meisner and Arkin, 1987; Nobre and Shukla, 1996; Fu *et al.*, 2001; Liebmann and Marengo, 2001; Paegle and Mo, 2002). The main atmospheric pump contributing rainfall that sustains tropical forests in the Guiana Shield, however, is the thermally direct Hadley Circulation and its rainfall-producing centre, the Inter-Tropical Convergence Zone, or ITCZ.

ATMOSPHERIC COMPONENTS – THE HADLEY CIRCULATION The Hadley Circulation consists of two cells. Each of these cells is sustained by moisture-laden air rising within the ITCZ, moving poleward through the upper troposphere, subsiding (descending) in the sub-tropical latitudes to form the trade winds that then flow equatorward along an arc-like trajectory, shaped by the Coriolis effect, to once again rise within the ITCZ. Moisture gathered along this journey is trapped within the low-level trade wind flow as it travels equatorward by persistent higher atmospheric warm air trapping the cooler trade winds along the poleward end of the cycle, leaving the ITCZ as the only rainfall-producing phase of the cycle (Robinson and Henderson-Seller, 1999). The meridional migration of this airflow pattern largely influences the seasonal cycle of rainfall across the Guiana Shield in tandem with the main oceanic currents that seasonally redistribute energy and modulate the meridional shift of warm SSTs. Because heating of the Earth's surface is roughly equal throughout the year across the tropics, the effects of seasonal differences in heating and cooling that create differences between land and sea convection patterns at higher latitudes are minimized, decreasing the longitudinal differences in cloud migration over tropical continents compared to adjacent oceanic zones.

THE ITCZ The ITCZ is a belt of hydrostatic instability that occurs in a ring of low pressure circling the Earth called the equatorial trough (Fig. 2.22). It is maintained by the convergence and convective ascent of low-altitude trade winds as part of the Hadley Circulation. It is the main cause of heavy rainfall and thick seasonal cloud cover that typify many parts of the tropics. ITCZ-related cloud cover over the equatorial Atlantic has two components: (i) cloud blocks produced through localized convection; and (ii) cloud trains associated with the westward movement of African easterly waves.[5]

The ITCZ oscillates between a northern and southern limit that, on average, tends to reside for longer periods north of the equator, although it is capable of extending much further southward across South America (Fig. 2.22B) (Critchfield, 1983). This northern hemispheric asymmetry may be related to the meridional distribution and geometry of continental land masses and their consequent effects on large-scale wind patterns that sustain the ITCZ (Philander *et al.*, 1996). The lag between the seasonal shift of the main solar heating zone from one hemisphere to the other and movement of warm ocean water deflects the main residence time of the ITCZ northward. Wind patterns are strongly shaped by the distance and difference in atmospheric pressure centres (Fig. 2.22B). These pressure centres are directly affected by sea-surface temperatures (SSTs). SSTs, in turn, vary with changes in the action of ocean currents affected by winds to form a positive feedback loop. Atmospheric convergence zones, like the ITCZ, typically reside over the warmest surface waters and migrate with latitudinal movements of the warm oceanic currents. Warm ocean currents act as the main engine to hydrostatic instability and increased cloud formation in the ITCZ (Fig. 2.22A), although a high SST does not always lead to large-scale convective activity (Graham and Barnett, 1987). Nonetheless, a northward asymmetry in 'normal' meridional ranging of the ITCZ typically results in much higher rainfall amounts in locations between 2° and 8° N and the creation of a 'meteorological' equator at 5° N, where ITCZ-driven rainfall is commonly received in varying amounts during most, but not all, months of a 'normal' year (Snow, 1976). It is the extreme departures from the 'normal' seasonal timing of ITCZ migration, forced by a wide array of climate-driving factors discussed later, that leads to substantial amplification of seasonal rainfall patterns in the region.

Importantly, the latitudinal limits of the ITCZ vary zonally (Critchfield, 1983). The tropical forests of East Africa, India, SE Asia and Polynesia are subject to an annual migration that can range from 30° N to 15° S. The neotropics are affected by a much narrower ranging of the ITCZ. In an area extending from the eastern Pacific, over the Sub-Andean Trough and to the western rim of the Guiana Shield, this remains largely north of the equator with a seasonal migration of no more than 10°. The movement of the ITCZ dips southward over the eastern half of tropical South America during the boreal autumn–winter, passing over most of the Guiana and Brazilian Shield areas.

The annual cycle of seasonal rainfall over most of the Guiana Shield is established by the meridional migration of the ITCZ (Snow, 1976; Figs 2.17, 2.22A, 2.23). The (semi-)annual passover of the ITCZ sets up the normal periodicity of high rainfall across eastern South America and largely accounts for the latitudinal shifts in the calendric timing of wet and dry seasons encountered across the tropics. The deepest part of the dry season is established when the atmospheric instability produced by the ITCZ is at its furthest distance from any given location within the equatorial trough. Given the range limits of the ITCZ and its

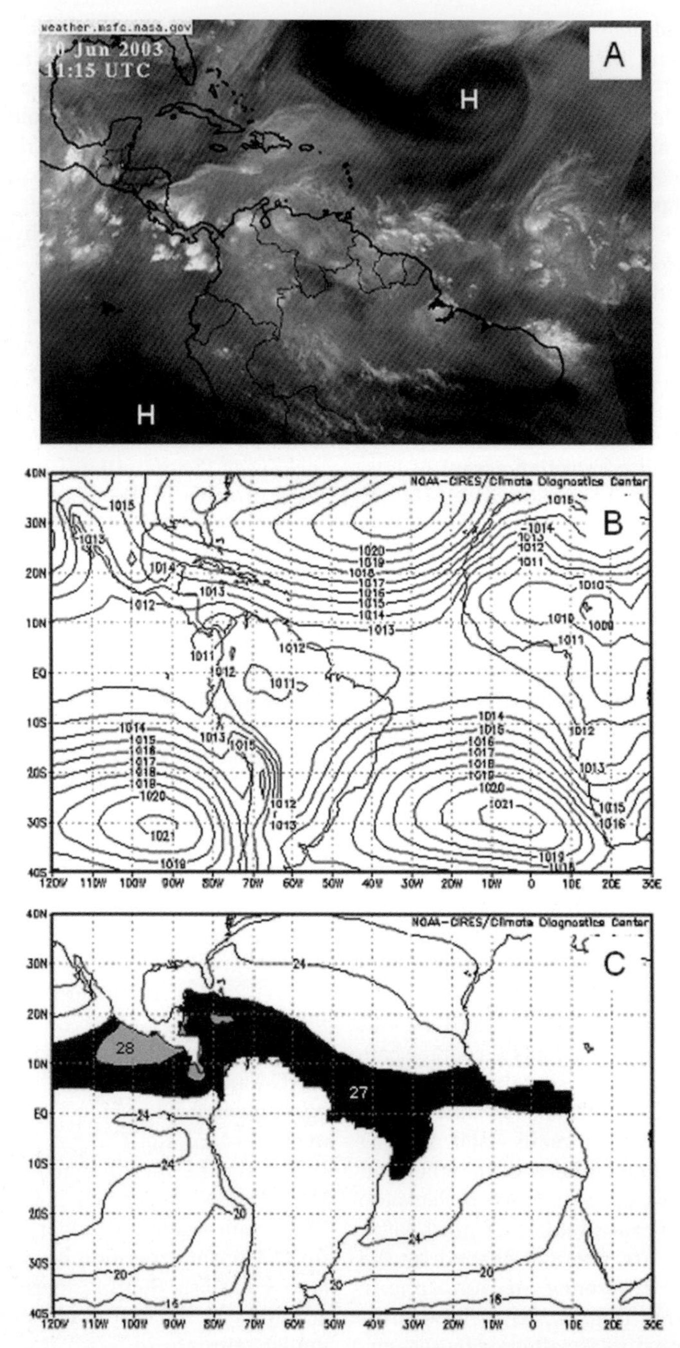

Fig. 2.22. (A) GOES-8 satellite image of ITCZ over the Guiana Shield during the wet season. The ITCZ is identified by the series of wave trains running north of the equator between the oceanic centres of high pressure (H). (B) Mean sea-level pressure (mb), 1960–1997, all months. Note high pressure centres over oceans. Stacked isobars identify location of the westward flowing trade winds to the north and south of the equatorial trough. (C) Mean sea surface temperatures (°C), 1960–1997, all months. Shaded zones are areas with mean temperature exceeding 27°C. SLP and SST distributions based on NCEP Reanalysis dataset and COADS 1°-enhanced, respectively. See Woodruff *et al.* (1987, 1998) for description of COADS.

disproportionate residence time just north of the equator, it is not surprising that stations in this area experience dry seasons of less severity and shorter duration over the course of a 'normal' year (Figs 2.17, 2.23).

The distribution and position of the Guiana Shield places virtually the entire area within this zone of higher ITCZ activity, making it the largest, contiguous area of tropical forest located within this pantropical belt of high ITCZ-driven rainfall. Only the tropical superwet SE Asian forests on Borneo and (previously) Sumatra and Peninsular Malaysia also fall largely within 3° of the meteorological equator (at 5° N). But rainfall variability in the western Pacific is most closely linked to changes in another atmospheric conveyor, the Walker Circulation.

ATMOSPHERIC COMPONENTS – THE WALKER CIRCULATION While the Hadley Circulation and its action on sustaining rainfall within the ITCZ has been recognized for centuries, the zonal compartment of circulation influencing rainfall in the Guiana Shield and other regions, the Walker Circulation, has only in the last 40 years come to be recognized for the important impact it has on neotropical rainfall and global climate. Like the Hadley Circulation, the Walker Circulation (WC) is thermally direct. It is also driven by a zone of hydrostatic instability. This zone in the Pacific is normally located over the far western Pacific and Indonesia with a high altitude eastern flow, subsidence of dry air along the eastern Pacific (western coast of South America) and the sea-level return of this air westward towards the western Pacific zone of convection (an equatorial extension of the trade winds) (Bjerknes, 1969; Philander, 1983; Rasmusson and Wallace, 1983). Some climatologists have also suggested a Walker cell for the Atlantic Basin, with the convective zone located over the Amazon and subsidence occurring over the zone of South Atlantic high pressure and northeastern Brazil (Kidson, 1975, cited in Philander, 1990).

THE PACIFIC ENSO The main convection centre of the Walker Circulation oscillates zonally (west to east), but unlike the meridional oscillation of the ITCZ, the circulatory path of the WC periodically fails altogether. This drives the main zone of convective instability (and rainfall) much further into the eastern Pacific after having amplified the seasonal southward migration of the ITCZ. This periodic event represents a phase in the Southern Oscillation (SO), a repeating pattern of reversal in the pressure gradient between the eastern (commonly tracked at Tahiti) and western (Darwin, Australia) Pacific (defined by the Southern Oscillation Index (SOI)) and is strongly correlated with heavy rainfall and drought events in many regions of the world.[6] Periods of heavy rainfall and drought in the eastern Pacific have become known as La Niña (LN) and El Niño (EN) events, respectively, although the circumstances defining these periods are, in many ways, subjective, since they simply represent phases in a larger circulatory process that may vary in its intensity and duration depending on the combined result of a number of different atmospheric and oceanographic conditions (Philander, 1990). Adopting one conservative measure[7] of EN and LN, at least 22 distinct periods consistent with a strong to severe El Niño phase of the SO have occurred since 1870. During the same period, at least 16 strong to severe La Niña events have taken place. Due to the greater frequency and socio-economic importance of EN periods and the absence of any recognized standard for delimiting these, the term EN is often combined with the SO to define the anomalous change in conditions across the equatorial Pacific as ENSO, the El Niño–Southern Oscillation. Simply stated, EN is a phase of ENSO when anomalously low surface pressure, heavy rainfall, warm SSTs and weakened trade winds prevail in the east-central Pacific. LN describes a phase of anomalously strong, and rapid, emergence of these conditions over the western tropical Pacific, often immediately following the relaxation of EN conditions. ENSO thus can be viewed as embodying all phases of a continuous oscillatory process.

The zonal migration of convective instability between the eastern and western

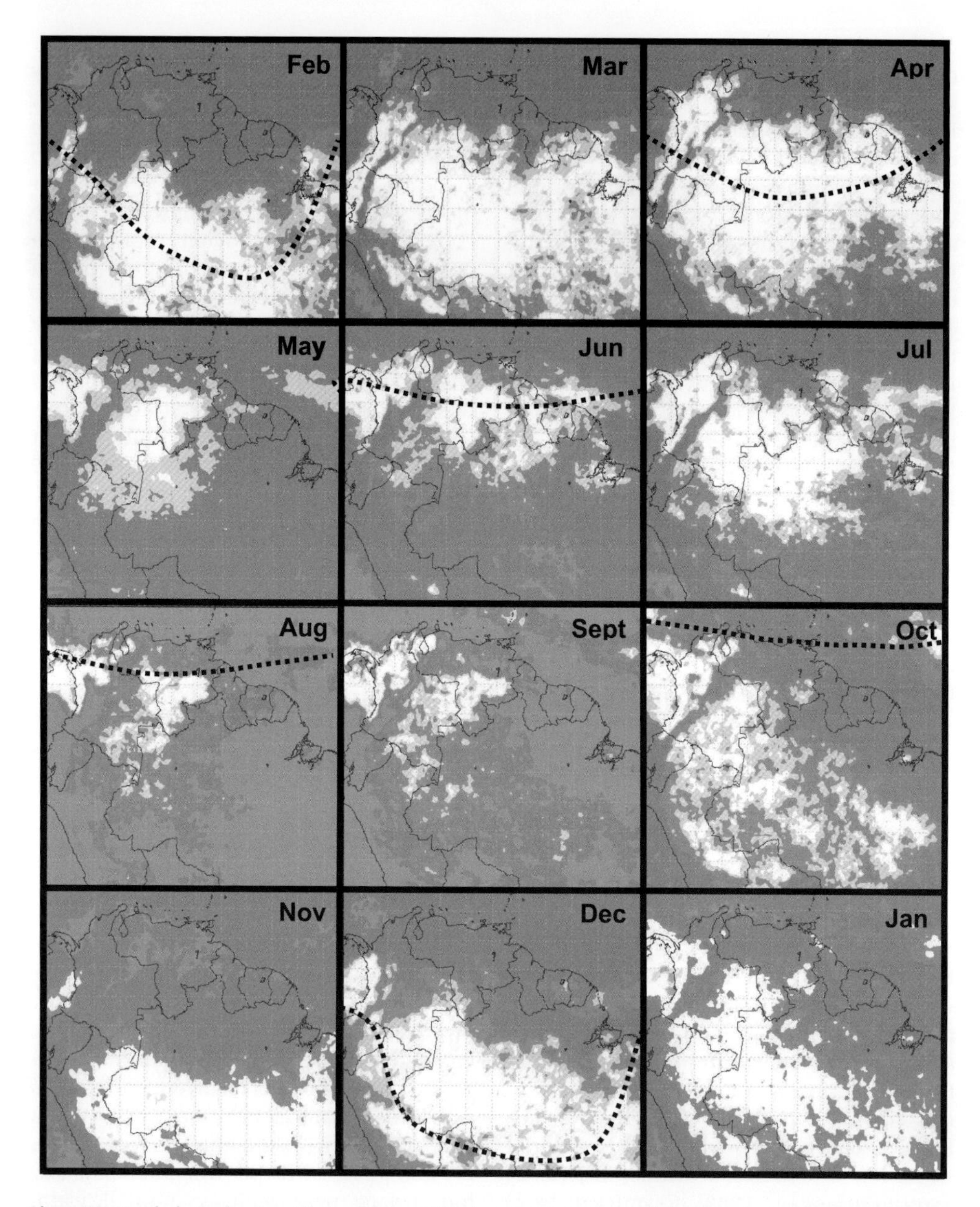

Fig. 2.23. Yearly latitudinal oscillation of the ITCZ depicted by estimated distribution (white cover) of total monthly precipitation >200 mm (>300 mm for Aug–Sept) across tropical South America from February 2002 to January 2003). Estimates derived from data collected using the precipitation radar sensor based on NASA's Tropical Rainfall Measuring Mission (TRMM) satellite. The deeper and more prolonged southward migration of the ITCZ along the eastern side of South America can be clearly seen affecting an east to west increase in rainfall within the Guiana Shield from October to January 2003 – a weak negative SOI year. Images are threshold grey-scale versions of those created by the GDAAC Hydrology Data Support Team – NASA.

Pacific that is directed by ENSO-related changes to the Walker Circulation is not directly related to rainfall over the Guiana Shield. An indirect connection, or teleconnection, between changes in a number of tracking parameters, including the SOI, that are used to identify warm (western rainfall) and cold (western drought) phases of the ENSO, and multidecadal variation in rainfall across much of the Guiana Shield has been established (Ropelewski and Halpert, 1987, 1989, 1996; Kiladis and Diaz, 1989; Dai and Wigley, 2000). Many, but not all, of the most severe historical flooding and droughts known to have occurred in eastern lowland South America have been tied phenomenologically to the magnitude of west to east change in the Pacific circulation (Fig. 2.19; Kiladis and Diaz, 1986). The correlation between the SOI and anomalous rainfall also appears to dampen immediately south and north of the Guiana Shield, although total rainfall also generally declines (Fig. 2.19) (Marengo and Hastenrath, 1993; Ropelewski and Halpert, 1996). For example, rainfall variation southeast of the Guiana Shield, in northeastern Brazil around Recifé, is significantly correlated with variation in the SOI, but the SOI only accounts for 10% of this variation (Kousky *et al.*, 1984). The absence of a significant leading or trailing association with the SOI at Fortaleza is also indicative of other factors, particularly disturbances to the meridional Atlantic SST gradient, overshadowing the minor effects zonal variation in the tropical Pacific climate generally exerts over the area (Fig. 2.19) (e.g. Rogers, 1988). This, of course, is not the case for the Guiana Shield region. In fact, the region has been identified by Ropelewski and Halpert (1987, 1989, 1996) as having the most consistent negative (positive) precipitation responses to EN (LN) events in the neotropics. Typically, the median deviation from 'base year' rainfall during the 18–24 months of an EN (LN) period is in the range of −50 to −300 mm (+100 to +400) (Ropelewski and Halpert, 1996). The results of Dai and Wigley's (2000, Figs 1 and 2) EOF analysis on rainfall data covering the period 1900–1998 confirm these results, estimating a −50 to −200 mm response to typical EN phases over the Guiana Shield.

This impact of ENSO on rainfall is to amplify the existing seasonal variation in rainfall rather than alter seasonality (Hastenrath, 1984a; Philander, 1990). Across NE South America, this has occurred as a consistent precipitation deficiency from July to March following the onset of ENSO conditions in the equatorial Pacific since the late 19th century (Ropelewski and Halpert, 1987). This can manifest itself in various ways. For example, during severe EN phases, the effect across the eastern third of the Guiana Shield is an attenuation of rainfall during the May–July wet season, a deepening of the subsequent September–October dry season and significantly, the January–March 'little' dry season. During LN phases, the opposite is true, with an amplification of rainfall during the wet season and attenuation of the dry season. At smaller timescales, EN phase effects are often seen as unusually long series of consecutive days without precipitation, creating drought-like conditions towards the end of these anomalously long rainless periods (Hammond and ter Steege, 1998; van Dam, 2001) (see Chapter 7). During strong LN phases, enhanced wet season rainfall is derived from an early onset and/or extension of the wet season with higher peak rainfall rates often leading to unusually severe flooding and mass wasting of exposed riverine soils. These often immediately follow severe EN events, such as occurred most recently in 1982–83 and 1997–98 (rapid (−) to (+) SOI in Fig. 2.19). It is important to note, however, that ENSO can interact with other local and regional climate forcing mechanisms to create considerable spatial variation in the timing and intensity of response (strength of cross correlations, Fig. 2.19) (e.g. Panama, Rand and Rand, 1982, p. 49; Costa Rica, Waylen *et al.*, 1996). In fact, given the wide range of modulating factors shaping station-specific rainfall values, the strong cross correlations of rainfall with SOI scores at several eastern Guiana Shield stations attests to the importance of this phenomenon in regulating inter-annual and decadal rainfall variation across the region (Fig. 2.19).

The spatial extent of ENSO-correlated variations in regional rainfall extends across the Guiana Shield, but with a clear epicentre spreading from the Pakaraima Mts region of west-central Guyana through to northern Amapá in Brazil (Ropelewski and Halpert, 1996). The analysis of Dai and Wigley (2000) extends this centre southward to cover most of northeastern Brazil. Negative precipitation anomalies during warm EN phases drop off rapidly from an average shortfall of over 300 mm of rain at the epicentre to less than 50 mm north and south of the shield region.[8] Ropelewski and Halpert (1996) identified a similar distribution of positive anomalies associated with the cold LN phases that are centred on the Guiana Shield and decline outside the region. As mentioned previously, shifts in the median precipitation received in the region during warm and cold phases are not symmetric, with LN-phase median values registering within the 60–70th percentile of base precipitation years (50% is the base median) and EN phase median values falling within the 30–40th percentile.

Evidence suggests that spatial variation in rainfall remains stationary during EN phase reductions across the Guiana Shield, but that stationarity is scale-dependent. In other words, the area of maximum rainfall expands and contracts around each stable centre when shifting into LN and EN phases, respectively, but the centres do not shift. Instead, there is a general landscape-scale decline in precipitation with high and low rainfall centres remaining stationary. For example, the reduction in rainfall distribution across French Guiana between a 'neutral' year (1956) and a strong EN year (1958) shows persistence of maximum rainfall south of Cayenne between the upper Sinnamary and Approuague Rivers (graphique 8 vs. 9 in Hiez and Dubreuil, 1964).

THE ATLANTIC ENSO Several authors have identified the existence of a Walker Circulation spanning the equatorial Atlantic/South America similar to the well-known Pacific cell (Covey and Hastenrath, 1978; Zebiak, 1993). The cell is defined by a zone of convective instability located over the Amazon (Kidson, 1975), a tropospheric eastward movement of dry air that subsides off the southwestern African coast and then is driven across the Atlantic and the dry zone of northeastern Brazil by strong southern trade winds towards the Amazon. Again, changes to the circulation largely occur as a consequence of the strengthening and relaxation of the trade winds north and south of the equator. Wang (2002) refers to variability in this circulation as the Atlantic zonal equatorial mode and suggests that the Hadley and Walker circulations over the Atlantic are affected by the Pacific ENSO through anomalously early warming of the tropical north Atlantic during the boreal spring following EN onset. However, unlike EN periods in the Pacific, migration of the westernmost convergence zone eastward during periods of trade wind relaxation is not known to occur in the Atlantic cell (Philander, 1990), possibly due in part to internal land surface effects (e.g. soil moisture variation) maintaining convectivity over the Amazon Basin during an event. Instead, SST in the western Atlantic respond alone to relaxation of the trade winds, creating a rainfall response gradient spanning the persistent western Amazon centre of convection and the belt of rainfall failure that is most consistent along the northeastern region of the Guiana Shield. The ENSO–precipitation relationships described by Ropelewski and Halpert (1996) show a gradient across tropical South America that is consistent with this west to east decline during Pacific EN events. The much higher rainfall received in northwest Amazonia further suggests that modest amplification of seasonality due to Atlantic/Pacific ENSO would have a relatively limited effect on the region's water balance. Rainfall anomalies over the Andean highlands to the west are more likely to influence surface water balances in the northern Sub-Andean Foredeep region (Vuille *et al.*, 2000).

Atmospheric components – mid-latitude incursions

Rainfall and temperature in the Amazon region are not solely the product of changes in Hadley and Walker Circulations. Regional incursions of polar fronts are known to affect tropical forestlands and cerrados of the southern Amazon and occasionally penetrate into the equatorial zone, particularly in western Amazonia (Ratisbona, 1976). It has also been noted that south polar incursions are capable of moving further northward into Venezuela (Riehl, 1977), connecting via the Orinoco Plains with the Atlantic high pressure (Azores High) system (Ratisbona, 1976) that seasonally flows southward across the llanos of Venezuela and northern Guiana Shield behind the southward-moving ITCZ. The path of this Atlantic high pressure incursion southward is in part shaped by a blocking effect of the Venezuelan/Colombian Andes and the southeastern trajectory of the Atlantic coastline forcing zonal movement of warm SSTs as they migrate with the swing in solar insolation between hemispheres.

FRIAGEMS Rapid incursions of cold air masses can bring temperature depressions of 8–10°C, and well below the 18°C minimum expected at most tropical forest locations. When a cold, polar front surges deep into the equatorial zone, it is locally referred to as a *friagem*, and brings brief periods of unusually low temperatures that typically last 5–6 days, with a 2–3 day depression in temperature and humidity, although longer events have been recorded (Ratisbona, 1976; Walsh, 1996; Marengo *et al.*, 1997). These incursions are relatively common from May to September (Marengo and Nobre, 2001), when the ITCZ is near its northernmost position and high pressure zones of the southern hemisphere trail northward.

Equatorial incursions typically emanate from the southwest, first affecting southwestern Amazonian forests as air is channelled northward along the eastern base of the Andes and western rim of the Brazilian Shield uplands. Only the strongest events, probably precipitated by anomalous northward extension and residence of the ITCZ, result in a *friagem* reaching areas of the Guiana Shield and, in this case, with effects largely restricted to southwestern areas (North Pará, Roraima, Colombian, Venezuelan and Brazilian states of Amazonas) (Fig. 13 in Ratisbona, 1976) and extreme minima dissipating as the cold air mass is modified during its move across Amazonia (Marengo *et al.*, 1997).

Oceanic components – Pacific currents and SST migration

While a strong ENSO–precipitation relationship clearly exists for the Guiana Shield, atmospheric behaviour of the Walker Circulation alone is insufficient to explain this phenomenon, although interbasin atmospheric connections can be important. The role of migrating Pacific sea-surface temperatures (SST) is crucial in understanding EN/LN development and thus the link between rainfall in the Guiana Shield and changes occurring in the equatorial Pacific as a consequence of ENSO. Migration of SST is largely controlled by changes to surface oceanic currents whose strength and direction are in turn mainly driven by surface wind direction, speed and persistence. The establishment of seawater density gradients due to spatial differences in temperature and salinity (the thermohaline circulation) also contributes to the migration of SST, but more slowly.

Seawater is steadily conveyed around the global oceans by the main oceanic currents, but also migrates across oceanic basins as a series of fast-moving waves that can act to intensify or even redirect current flow. These waves develop as a near-term oceanic adjustment to changes in atmospheric wind flow conditions (Gill, 1982; Philander, 1990). While numerous wave forms are known to develop and flow across all of the main ocean basins and along the main coastlines, two of the major waves known to specifically affect SST migration across the tropical Pacific are equatorial Kelvin waves and Rossby Waves.

Equatorial Kelvin waves are packets or pulses of relatively warm water that have propagated off the massive warm pool encompassing the Indonesian western Pacific (Glantz, 1998). They regularly form during intervening periods between a relaxation and subsequent restrengthening of surface winds associated with the Madden–Julian Oscillation, an important global atmospheric cycle with a 40–50 day periodicity affecting variability over the western Pacific (Madden and Julian, 1972). Once formed, Kelvin waves travel across the equatorial Pacific at a rate of around 13 km/h conveying warm water towards the normally cool eastern Pacific zone (Philander, 1990). Importantly, they have a downwelling effect, deepening the layer of warm water above the thermocline.

Deflection of Kelvin waves against the South American continent is one method of producing westward travelling Rossby waves (Gill, 1982; Philander, 1990). These waves travel at a much slower rate and are not restricted to the equatorial waters as are Kelvin waves. They can propagate in a poleward direction, the speed of the poleward ends declining as their distances from the leading centre increases, creating a fan-like configuration (Philander, 1990; Chelton and Schlax, 1996). Equatorial Rossby waves reaching the western rim of the Pacific are often redeflected westward, forming new Kelvin waves. This resonance of eastward and westward propagating waves sets up a delayed oscillatory mechanism that is believed to play an important role in regulating the phase-swings inherent to the ENSO (Schopf and Suarez, 1987). In particular, high-frequency propagation of large Kelvin waves foreshadows the onset of the eastward migration of warm SST associated with strong EN events. The excitation of the Pacific through an increase in Kelvin, Rossby and recycled Kelvin–Rossby waves acts to attenuate earlier zonal movement of the warm SST (and zone of convection) eastward and shift conditions towards a LN event (Philander, 1990).

Atmospheric convergence zones, such as the ITCZ, South Atlantic Convergence Zone (SACZ) or South Pacific Convergence Zone (SPCZ), occur exclusively over zones of warm SST. Conversely, subsidence of dry air occurs over cold waters. This wind–temperature relationship is at the heart of an ocean–atmosphere coupling that drives precipitation and its variation over many tropical regions. In the Pacific, the strong trade winds flowing (south)westward just north of the ITCZ push warm surface waters of the north equatorial current towards Austral-Indonesia. This centre of warm water in the western Pacific drives convection, supplies the tropospheric eastward flow of air that subsides in the eastern Pacific and, through the generation of westward surface trade winds moving towards the low pressure zone over warm water, continues to maintain the location of warm SST in the western Pacific by sustaining the westward flow of surface waters. The interaction between normal seasonal shifts of the ITCZ and the SPCZ as a consequence of extra-equatorial changes in radiative heating during the annual planetary orbit around the Sun modulates rainfall distribution in the Pacific.

Rapid changes occur to this condition when initial external forcing factors (e.g. MJO anomalies), that are not yet completely understood, cause the zonal trade winds to relax, allowing the warm waters in the western Pacific to move rapidly eastward. This movement alters the depth of the thermocline and the eastward advection of warm waters competes with upwelling of cold water in the eastern equatorial Pacific. Upwelling of deep, cold water normally 'sets' the low moisture retention qualities of the surface winds along the western coast of South America and along the equatorial Pacific as Ekman drift drives warmer waters offshore and poleward to produce a tongue of cold water (Philander, 1990). As a consequence, when this equatorial upwelling fails, the eastern region becomes hydrostatically unstable sooner and the convergence zone forms over the eastern, rather than the western half of the equatorial Pacific with concomitant changes in the position of atmospheric centres of high pressure.

A collapse in the westward-flowing trade winds during EN periods also leads to a southward displacement of the ITCZ towards the equator, and pulls the SPCZ northward so that all three of the main convection zones are centred in the eastern half of the equatorial Pacific when normally they would be distributed along the southern, western and northern flanks of the region (Rasmusson and Wallace, 1983; Philander, 1990). This movement, precipitated by the eastward shift in warm SST, also leads to a transgression of mid-latitude westerlies into the equatorial belt. This in turn heightens the flow of warm waters eastward. The shifting of warm SST zones also inhibits the normal seasonal migration of the ITCZ northwards as the stationary warm waters oppose the effects of seasonal changes in radiative heating of the planet's surface. The series of changes that occur in the Pacific as a consequence of ENSO are not limited to the Pacific. As the trade winds relax and the zone of warm SSTs shifts in the Pacific during strong EN periods, shifts in the oceanic and atmospheric conditions in the Atlantic basin are set in motion. Consequently, interannual variation in ITCZ-driven rainfall over the east-central region of the Guiana Shield, the Caribbean and Central America are variously affected through this teleconnection.

The migratory behaviour of the warm SST belt in the Pacific drives the vigour of the Walker Circulation. The relative importance of seasonal, meridional SST movements to the ITCZ in the Pacific still requires consideration (Philander, 1990), particularly in its potential role in terminating EN/LN events (Harrison and Vecchi, 1999), but changes to the equatorially contained Walker Circulation are currently regarded as the overriding climatological response to equatorial SST anomalies.

Oceanic components – Atlantic currents and SST migration

In the same way that a northerly ITCZ, presence of easterly trade winds and maintenance of a western zone of convective instability define atmospheric parallels between the Atlantic and Pacific Basins, they also share a number of common oceanic features (Xie *et al.*, 1999). The line of maximum SST, the thermal equator, rests north of the geographic equator between 3° and 10° N for both oceans (Fig. 2.22C) (Philander, 1990). The eastern Atlantic along the southwest African coast is driven by a similar current dynamic as that operating in the coastal waters of the eastern equatorial Pacific. Both zones are defined by a shoaling thermocline, cold-water upwelling, eastern tongue of equatorial cold water and advection of warmer surface water poleward, a process that maintains the westward flow of the trade winds towards the convergence zones of convection (Fig. 2.22C) (Xie *et al.*, 1999). Together these features predispose neotropical forests to the same ENSO-type amplification of seasonal rainfall patterns as have been observed in the Maritime region (Oceania) of the western Pacific, but as a consequence of different pathways. Differences in the relative size of the two ocean basins, the way that the currents function and perhaps most crucially, the nature of the ocean–atmosphere couplings, force different rainfall responses and variabilities in the Atlantic and Pacific.

Climatologists and oceanographers are only beginning to understand the various roles that the many different structures and attributes of the tropical Atlantic play in regulating basin-wide rainfall (including the Guiana Shield). As a consequence, the wide variety of analytical models currently being employed often generate conflicting, if not opposing, conclusions regarding the relative significance of both atmospheric and oceanic processes operating within both the Atlantic and Pacific. On many points, consensus remains elusive (e.g. existence of a true Atlantic 'dipole'). None the less, a solid core of key features regulating rainfall as part of the tropical Atlantic circulation is emerging. The most salient of these are briefly described below and their inter-relationships within and between oceanic basins are characterized. Among these, the dynamical heating and cooling of surface water in the tropical Atlantic, like that in the Pacific, exerts a pivotal influence on climate variation within and along the perimeter of the Atlantic basin.

The geographic location of anomalously warm SST is the simplest predictor of heavy rainfall in most coastal regions of the tropical Atlantic basin, including Central America, the Caribbean and northeast Brazil (Hastenrath, 1984a,b; Enfield, 1996). In the tropical Atlantic, the development of anomalously cold and warm SSTs is largely led by changes in trade wind flow (Déqué and Servain, 1989; Huang and Shukla, 1997; Dommenget and Latif, 2000). Composite analyses carried out by Nobre and Shukla (1996) indicate that this weakening (strengthening) of trade wind stress leads maximal warming (cooling) of SST in the eastern tropical Atlantic by an average of two months. This in part translates into a zonal shift in the thermocline depth on a seasonal basis, leading to a contraction of the cold-water upwelling zone in the southeastern tropical Atlantic and a weakening of the SST that concentrate as the tropical Western Hemisphere Warm Pool (WHWP) (after Wang and Enfield, 2003) across the eastern Pacific, Central America and the Caribbean during the boreal wet season, May–July (Fig. 2.22C). As this zone of warm SST moves southward, the heavy rainfall associated with the ITCZ moves in concert. It returns the following season when the WHWP expands over the boreal summer as a response to a seasonal increase in incoming solar radiation across the northern hemisphere (see section on Solar heating, above), restrengthening of the southeasterly trade winds and conveyance of warm water along the north Brazil/Guyana/Caribbean current chain. Changes to this annual pattern of meridional movement along an SST gradient defines the main oceanic influence on rainfall over most of eastern tropical South America, Central America and the Caribbean at interannual and decadal timescales.

NORTH AND SOUTH SEA-SURFACE TEMPERATURE ANOMALIES (SSTA) AND THE ATLANTIC DIPOLE

The meridional gradient in tropical SST is demarcated by the WHWP in the northwestern tropical Atlantic and a zone of relatively cold water off the (very dry) southwestern coast of Africa (Fig. 2.22C). A number of studies have suggested that decadal variation in SSTA in the tropical north and south Atlantic occurs across the gradient as a basin-wide dipole response to changes in wind-induced latent heat fluxes (Carton *et al.*, 1996; Chang *et al.*, 1997; Huang and Shukla, 1997). Other studies suggest that anomalies can appear dipole-like, but fluctuations on either end of the Atlantic are more dynamic and linked to independent processes operating within each hemisphere at different timescales (Houghton and Tourre, 1992; Enfield and Mayer, 1997; Mehta, 1998; Sutton *et al.*, 2000). In either case, the inter-hemispheric SST gradient has been found to oscillate noticeably at a timescale of about 13 years (Mehta and Delworth, 1995; Chang *et al.*, 1997), bringing changes to the seasonal movement of the ITCZ/SACZ and rainfall to eastern South America. An anomalously cold south tropical Atlantic (SSTA–) and warm north tropical Atlantic (SSTA+) set up conditions that strengthen southeasterly and weaken northeasterly trade winds, anomalously pushing the ITCZ northward and through this, promoting anomalously high rainfall over the eastern Guiana Shield, Caribbean and lower Central America. These conditions in the Atlantic have been known for some time to correlate well with drought across northeastern Brazil (Bahía) south of the Guiana Shield (Hastenrath and Heller, 1977).

EAST AND WEST SSTA AND THE ATLANTIC ENSO MODE

When trade winds are strengthened, the westward flow of surface waters in the equatorial Atlantic deepens warm waters at the western edge of the basin and shoals these in the east, promoting greater cold-water upwelling along parts of the West African coast and the equator. When trade winds relax, opposing conditions dominate and warm waters pooled in the northwestern Atlantic flow eastward, carried by westerly surface winds (Wang, 2002). Periodically, trade winds strengthen and relax more extensively than 'usual' and this variation creates zones of anomalously warm and cold SSTs. This scenario mirrors the well-known zonal shift in SSTAs that occurs in the equatorial Pacific. The Atlantic ENSO, however, is damped in comparison to the high amplitude fluctuations of the Pacific ENSO that are driven mainly by variation in surface wind stress (Philander, 1990). Surface wind stress is believed to play only a minor role in forcing interannual variability in SSTs in the equatorial and north tropical Atlantic (Zebiak, 1993). Instead, variation in wind-induced latent heat flux[9] has been assigned the dominant role (Chang *et al.*, 1997; Häkkinen and Mo, 2002). In the subtropical Atlantic, several modelling studies have suggested that changes in heat flux feedback between atmosphere and surface waters are important in explaining variability in tropical Atlantic SSTs (e.g. Carton *et al.*, 1996). Saravanan and Chang (1999) go further to suggest that this thermodynamic feedback is affected by the spatial variation in the mixed layer depth, and through this interaction, strongly influences SST variability in the subtropical seas, poleward of 10° latitude. The zonal shift in SST and SLP conditions ascribed to an ENSO in the tropical Atlantic is considered stable (Zebiak, 1993), Delecluse *et al.* (1994) suggested that ENSO in the Pacific may be remotely forcing unusual zonal shifts to the ocean–atmosphere exchange in the tropical Atlantic, but the complex of mechanisms that would link Pacific and Atlantic ENSO type variations still awaits clear deciphering.

The impact of the Atlantic ENSO variation on rainfall has focused largely on the eastern rim of the tropical Atlantic, where SSTAs, particularly in the south and northeastern tropical and subtropical Atlantic, have been associated with anomalous seasonal rainfall in the Sahel, along the Guinean and Angolan Coasts and in the Congo Basin (Hastenrath, 1984a; Mo *et al.*, 2001; Vizy and Cook, 2001; Kouadio *et al.*, 2003). While the impact of the Atlantic ENSO on equatorial Atlantic SST variability is apparently similar to, but less than, the Pacific ENSO's effects on SSTs in the Pacific, the resulting effects of this relatively weak relaxation of the Walker Circulation on rainfall variation in the Guiana Shield are less clear. This is particularly true in light of the other impacts that the Pacific ENSO, the meridional Atlantic SST gradient (aka Atlantic Dipole) and other remote forcing factors may exert in concert or during alternating seasonal phases. Marengo and Hastenrath (1993) indicate that the Amazon Basin may have become a zone of subsidence during the extreme 1983 EN event as the zone of convergence associated with the convective arm of the Atlantic Walker Circulation shifted to the west of the Andes, but generally western Amazon precipitation anomalies are not related to ENSO.

Remote forcing factors

The latitudinal oscillation of the ITCZ brings heavy seasonal rainfall to the eastern Guiana Shield, the Caribbean and southern Central America. Its seasonal trek is driven by movement of Atlantic warm waters that change position in response to seasonal shifts in hemispheric solar heating, the course of the oceanic currents, behaviour of the major atmospheric circulations and effect of basin geometry. Ocean–atmosphere models indicate that these internal processes driving change in the Atlantic basin would not sustain or achieve the observed variability in SSTAs without the action of remote forcing factors (e.g. Nobre *et al.*, 2003). Connections with remote atmospheric and oceanic circulation patterns in the tropical Pacific and extra-tropical Atlantic can act to attenuate or amplify the effects of seasonal changes to the tropical Atlantic SST gradient. By forcing anomalous changes in the location, extent and intensity of the main warm SST pool (via strengthening and relaxation of the trade winds), they force a shift in, amongst other phenomena, the normal seasonal movements of the ITCZ, the production and track of African easterly waves (and hurricanes) and rainfall across the region.

PACIFIC ENSO EFFECTS ENSO has been shown to play a significant role in shaping

SST variability in the tropical Atlantic at interannual to multi-decadal timescales (Hameed *et al.*, 1993; Mo and Häkkinen, 2001). A study by Enfield and Mayer (1997) (and later Ruiz-Barradas *et al.*, 2000; Mélice and Servain, 2003) showed that the development of Atlantic SST anomalies lagged those induced directly by ENSO in the Pacific by 4–5 months and that these were consistently most severe along the western edge of the tropical north Atlantic (i.e. the WHWP). Mélice and Servain identified a quasi-decadal (9.6 years) response of tropical north Atlantic SST anomalies to the SOI. This response was asynchronous with the (14 year) signal identified for a tropical south Atlantic SSTA–SOI relationship, further suggesting that northern and southern oceanic–atmospheric processes may respond independently to different remote forces and are not oscillating in a sustained north–south dipole.

Northwestern Atlantic rainfall responses to ENSO development in the Pacific are not necessarily proportional. Often a severe EN/LN phase fails to translate into equally severe rainfall anomalies at locations normally considered responsive to ENSO (see Fig. 2.19, SOI vs. rainfall). In part, this relates to the intensity of zonal shifts in ocean–atmosphere conditions in the Pacific that translate into EN/LN events. But other, longer-term oscillations that alter background SSTs in the eastern tropical Pacific can interact to enhance the impacts of the ENSO. Mestas-Nuñez and Enfield (2001) constructed a synthetic characterization of ENSO by canonically correlating various indicators and compared this with SST anomalies in the (Niño-3 zone) eastern tropical Pacific. They found that 40–50% of the amplitude in SSTs achieved during the most severe 1982–83 and 1997–98 EN events was not associated with the canonical ENSO, but longer-term oceanic adjustments sub-strengthening the impacts of a strong negative ENSO phase. This alone, however, does not result in a proportional amplification of rainfall in the western Atlantic sector, even in areas, such as the eastern Guiana Shield, that have shown a very consistent response to EN/LN events.

The most severe amplification of seasonal rainfall patterns in the Caribbean, NE Brazil and Central America has been attributed to periods when Pacific and Atlantic SST (or SLP) patterns achieve a certain conformation. Again, severe events do not occur regularly because changes in other atmospheric components of the global circulation affecting tropical Atlantic SSTs do not always oscillate in a way that reinforces the effects of ENSO. Spectral analyses suggest that oscillatory phases of different forcing factors are typically attenuating rainfall responses because their respective phases of development and dissipation are discordant, and this mitigates the development of strong SST anomalies. However, at certain periodicities, the oscillatory phases of different forcing factors that promote positive/negative SST anomalies conform, creating a composite effect that amplifies anomalous behaviour of SSTs in the tropical Atlantic (e.g. Uvo *et al.*, 1998; Enfield and Alfaro, 1999; Mo and Häkkinen, 2001).

In the tropical eastern Pacific, 79% of variation in SST anomalies is directly attributable to ENSO. The amount of SST variation in the tropical north Atlantic explained by ENSO in the Pacific is considerably less, about 25% according to Enfield and Mayer (1997). The dominance of the zonal Walker Circulation in the Pacific, differences in basin size, land distribution and the overriding importance of a meridional, rather than zonal, gradient as the foundation on which SSTAs develop in the Atlantic create conditions that are more difficult to attribute to any single influence, internal or remote. Other remote influences can act as important sources of variation at alternate periodicities. One of the main forcing factors interacting with the Atlantic meridional gradient and Pacific ENSO to shape the magnitude of SSTAs in the tropical North Atlantic is the North Atlantic Oscillation.

NAO EFFECTS In the Atlantic, research suggests that circulatory patterns other than those driven by the thermally direct Hadley

Circulation can also force more subtle changes in the SST gradient through variation in their own oscillatory behaviour. As a consequence, they may periodically alter the seasonal positioning of subsidence and convergence zones of the Hadley, driving multi-year to centennial-scale rainfall variation in parts of the region. The North Atlantic Oscillation (NAO) has been variously placed high on the list of co-contributors, along with ENSO. The influence that NAO has on rainfall in the basin is believed to be delivered indirectly through its forcing effects upon wind flow and oceanic heat transfer patterns. The ITCZ remains the main delivery vehicle for rainfall over much of the eastern half of the Guiana Shield and adjacent regions, but the cascade of adjustments forced through interconnecting atmospheric and oceanic circulatory features is ultimately driving the variation in rainfall that occurs between years, decades and centuries.

The NAO is defined by a large-scale fluctuation of sea-level pressure between centres near Iceland (low pressure) and the Azores islands (Portugal) (high pressure) (van Loon and Rogers, 1978; Rogers, 1984). The behaviour of NAO during the boreal winter is the dominant atmospheric dynamic in the North Atlantic and has proven to be an important source of long-term variability in global climate. NAO is known to vary seasonally, inter-annually and at decadal and multi-decadal scales. Extreme high index phases are characterized by a very strong Azores high pressure in the south-east and an intense Icelandic low pressure in the far north. They bring strong mid-latitude westerlies, anomalously low SST to the tropical Atlantic and much reduced rainfall over the Caribbean (Malmgren *et al.*, 1998; Robertson *et al.*, 2000; Giannini *et al.*, 2001). Its effect on rainfall in the Guiana Shield has not been specifically studied. However, results achieved using general circulation models (GCMs) suggest that inter-annual fluctuations in simulated NAO conditions are strongly associated with SST anomalies in the tropical and subtropical North Atlantic and variations in these are known to affect rainfall in the eastern Guiana Shield region (Nobre and Shukla, 1996; Fu *et al.*, 2001).

AMPLIFICATIONS Recent climatological analyses indicate that the most extreme precipitation events across northern South America, the Caribbean and lower Central America may occur when both the Pacific ENSO and NAO simultaneously reach extremes. Dry periods are characterized by strong warm ENSO phases (+SSTA in the eastern Pacific) and strong positive NAO phases (+SLP anomalies over the Azores) and exceptionally wet periods when signs are reversed. The western tropical Atlantic is the only region known to experience variations in SLP and SSTA associated with extreme phases of both the NAO and ENSO, establishing a unique spectrum of oscillating conditions. Several of the most severe droughts on record (1877–78, 1982–83, 1997–98) as having occurred in the Guiana Shield and Central America are coincident with simultaneous extremes in the ENSO and NAO (Rogers, 1984; Huang *et al.*, 1998; Malmgren *et al.*, 1998; Giannini *et al.*, 2001). During other phases, these systems oscillate in a manner that acts to attenuate the effects they have on SLP and SSTAs, and through these, the seasonal positioning of the ITCZ. These amplifying and mitigating interactions between ENSO, NAO and other remote forces help to explain why the variation in rainfall across the Guiana Shield is not consistently and proportionally responsive to the intensity of EN/LN events (Fig. 2.19).

Certain interannual or decadal-scale state conditions in other oceanic and atmospheric systems also amplify SSTAs and phases of the seasonal rainfall cycle over the Guiana Shield region. According to Enfield and Alfaro (1999), one of the most important of these, the Atlantic meridional gradient ('dipole'), interacts with negative (positive) conditions in the eastern tropical Pacific to further enhance (reduce) rainfall over the Caribbean, lower Central America and northern South America when the 'dipole' phase is characterized by anomalous northern tropical Atlantic conditions of opposite sign to those in the eastern

Pacific. Wang (2002) further concludes that ENSO can interact with both the Atlantic zonal equatorial mode (Atlantic Walker Circulation) and the Atlantic meridional gradient to enhance warm SSTAs in the tropical north Atlantic during the boreal spring following the mature phase of an EN event.

Land-borne components – topography, evapotranspirative recycling and surface albedo

Terrestrial-borne effects of varying topography, surface albedo and vegetation cover interface with local hydrological conditions and the larger global circulation to define further variation in rainfall at smaller spatial scales. In the western Guiana Shield and Sub-Andean Foredeep these are perhaps the dominant mechanisms regulating rainfall during all but the most extreme ocean–atmosphere phases, as the effects of varying oceanic conditions in the Atlantic dampen towards the continental centre (Fu *et al.*, 2001) and westward water vapour flux declines (e.g. Fig. 18 in Salati and Marques, 1984).

TOPOGRAPHY, OROGRAPHY AND SOIL MOISTURE

Variation in surface roughness (and therefore frictional force) can foster localized precipitation in a number of ways. Along coastal margins, moisture-laden surface winds travelling from smooth warm ocean surfaces across the mangrove-covered and forested coastline profiles can create conditions leading to rainfall. This effect is believed to enhance seasonal rainfall received along the Atlantic margin of the Guiana Shield as low-level easterlies move across the warm northward-flowing Brazil and Guiana Currents (Snow, 1976). The high centre of rainfall in southeastern French Guiana and Amapá is due in part to this effect, although other factors are also contributing to the high precipitation rates in the area.

Severe topographic transitions involving mountains create more substantive frictional forces, rapid uplift and formation of rain clouds. Orographic lifting plays a significant role in maintaining the relatively high local rainfall received on the windward slopes of massifs forming the Guayana Highland region. The long narrow corridor of relatively low rainfall dividing eastern and western parts of the shield is interrupted by the eastern extension of the Pakaraima Mts and surrounding tepui formations (Fig. 2.16A). Leeward of these mountain features, rainfall shadows predominate (Frost, 1968; Snow, 1976). This may assist in explaining the relatively narrow north–south belt of 'dry' climate bisecting the Guiana Shield. As moisture affecting the eastern Guiana Shield is driven westward into the Guayana Highlands by the trade winds and sea breezes, leeward slopes are starved of this addition to the main ITCZ source. Orographic lifting atop isolated horst formations, such as the Kanuku Mountains in southern Guyana or Bakhuys Mountains in Suriname, can also substantially amplify otherwise relatively modest local rainfall and water balances (Frost, 1968; Snow, 1976).

The topographic effect of the Guayana Highlands on rainfall nonetheless is slight in comparison with the orographic impact of the eastern Andean slopes. Rainfall along these slopes can reach above 4000 mm/year as the trade wind flows pick up moisture across the lowland basin and then are rapidly forced upward (Johnson, 1976). Orographic uplift typically leads to cloud types different from those typifying adjacent lowland systems. Orographic clouds tend to configure as stratus or altostratus types, while convective clouds over the lowlands are mainly cumulo-nimbus. Both deliver rainfall, but with differing intensities and rain rates. Rainfall from orographic clouds along the eastern Andean slopes occurs over longer periods with fewer extreme rainfall events and is substantially augmented at certain elevations by fog-delivered precipitation (e.g. Grubb and Whitmore, 1966). Downslope at lowland sites where rainfall is delivered mainly through convection, the opposite is true (Fig. 2.24). The significance of this difference rests in the impact of slope and rain

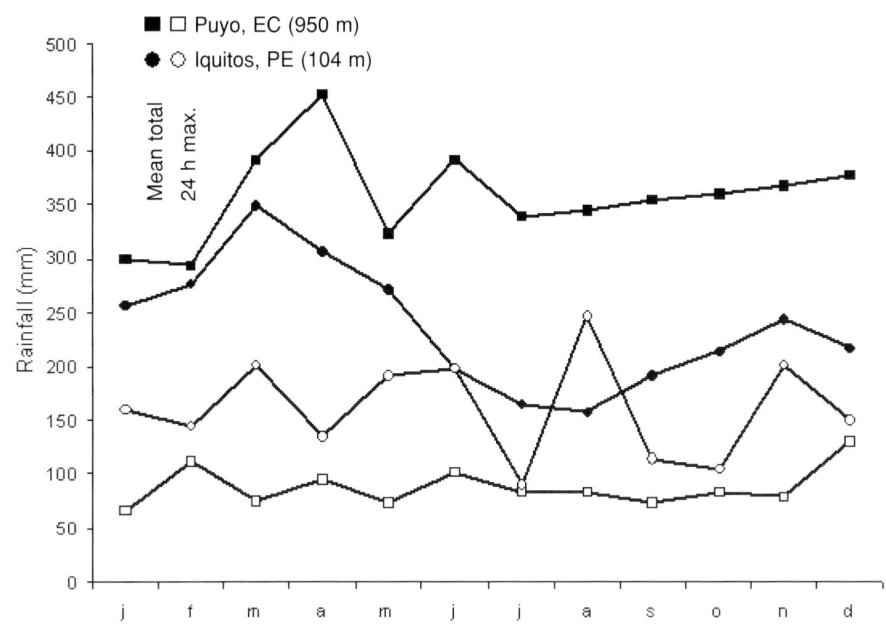

Fig. 2.24. Comparison of mean monthly total and 24 hour maximum rainfall for highland (Puyo, Ecuador) and lowland (Iquitos, Peru) locations in western Amazonia. Data for period 1949–1970. Source: Johnson (1976).

rate on surface run-off in these regions, as well as the effects of a more uniform cloud cover and rainfall on expected changes in temperature with varying elevation (environmental lapse rates). In the Guiana Shield, orographic effects reflect the modest elevation of a scattered and localized archipelago of ancient horst (Kanuku, Bakhuys), batholithic (Parguaza) and sedimentary (Roraima Supergroup) formations that are for the most part centrally placed atop the crystalline basement complex and fed by ocean-modulated trade winds and sea breezes. The contrast with orographic conditions in western Amazonia could not be more striking. Uplift is fed by substantial, water-laden northeasterlies impacting a massive, continuous, north–south barrier of considerable elevation. This barrier is located on the periphery of the western lowland region. The western Andes contain, rather than divide as is the case of the Guiana Shield highlands, the hydrological basin of the adjacent lowland region.

Localized low-points in topography can also affect rainfall. Anomalously high annual precipitation has been recorded at a number of stations that share very few geographic attributes (Iquitos, Peru; San Carlos de Rio Negro, Venezuela; Macapa, Brazil), except that they are situated within or immediately adjacent to large, seasonally flooded sedimentary depressions (e.g. San Carlos and Casiquiere Rift in Fig. 2.9A). Large topographic depressions at the mouth of the Amazon, and along the central segment of the Amazon (Solimões) west of Manaus (Fig. 2.8) create vast evaporative moisture sources that may contribute to rainfall generation over the central and western Amazon through northwestward atmospheric conveyance.[10] Southeasterly trade winds travelling over the relatively dry Brazilian northeast become dry as they achieve equilibrium with the surface. When these dry winds encounter this depression, evaporation increases rapidly as the sensible heat of the air combines with incoming radiation. As the air adjusts to the wetter surface, this moisture is conveyed northwestward towards the ITCZ.

While the balance of ocean- and land-borne processes is clearly different for each location, local convection stimulated by enhanced surface to atmosphere water vapour flow (latent heat fluxes) may be enhancing precipitation otherwise predominantly delivered through the action of larger-scale mechanisms (Koster and Suarez, 1995).

VEGETATION COVER AND EVAPOTRANSPIRATIVE RECYCLING Salati and Margues (1984) estimated that 50% of the annual precipitation falling over central Amazonia is sourced from evapotranspiration. Eltahir and Bras (1994) more recently estimated a lower 25–35% of rainfall over the Amazon as sourced from regional evapotranspiration based on their model analysis of two different data sets. Evapotranspiration from vegetation cover and direct evaporation (latent heat fluxes) from the forest canopy and exposed water surfaces plays a crucial role in sustaining high regional rainfall levels in many parts of the tropics, but may be of particular importance over the relatively isolated western Guiana Shield, Sub-Andean Foredeep and Amazon Downwarp regions. Analyses carried out by Marques *et al.* (1979) show a westward gradient of increasing atmospheric water storage (precipitable water) over the Amazon Downwarp with little month-to-month variation. In part, this suggests that local land surface processes of vegetation cover and evapotranspirative recycling may play an equally or more important role in the maintenance of rainfall over the western Amazon and Guiana Shield than in the eastern half, where ocean-borne processes dominate (Ratisbona, 1976, p. 266). The strength of evapotranspirative recycling may also be acting to anchor the convergent arm of the zonal Atlantic Walker Circulation (Philander, 1990), reducing the amplitude of inter-annual variation it embraces relative to its Pacific counterpart. In this case, it may also be indirectly shaping rainfall variability over the eastern Guiana Shield region by modulating the range of zonal shifts in rainfall spurred by changing SSTA in the equatorial Atlantic.

SURFACE ALBEDO The amount of radiation reflected as a proportion of total incoming radiation is an important component of the global energy budget. It is known to have an important role in regulating energy fluxes that drive atmospheric convergence (divergence) and localized hydrostatic instability (stability). While the spatial variation in surface albedo is poorly resolved for the neotropics (Marengo and Nobre, 2001), differences associated with a change in surface cover are well known. Generally speaking, there is a steady increase in the albedo of surfaces covered by forest vegetation (0.13), pasture/grassland (0.18) and desert/exposed sandy soil (0.37), respectively (Robinson and Henderson-Seller, 1999). Simulated increases in surface albedo at low latitudes yielded a decrease in soil moisture and precipitation in one model (Lofgren, 1995). Marengo and Nobre noted, however, a wide range of disparate model results and emphasized the need for improved descriptions of land surface characteristics prior to establishing the climatological response to changes in attributes such as surface albedo. Nonetheless, seasonal variation in albedo across Amazonia is believed to be related principally to changes in soil moisture, rather than incoming solar radiation (Culf *et al.*, 1995). It has also been shown through field surveying that rapid secondary regrowth of vegetation after forest loss reduces albedo to background forest levels within a decade (Giambelluca *et al.*, 1997).

It stands to reason that regions with a higher proportion of sand-dominated soils and significant seasonal reductions in rainfall are likely to experience a greater increase in surface albedo after vegetation loss and more rapid development of atmospheric subsidence zones. Given the relative predominance of sandy soils, an established strong seasonality in rainfall over much of the central shield zone, relatively modest secondary biomass accumulation rates and a high susceptibility to periodic EN-driven drought, changes in surface albedo conditions are likely to have a relatively important effect on local rainfall formation in the Guiana Shield compared with lowland regions of western Amazonia or Central America that receive external moisture sourced from orographic, rather than convective, uplift along the upper eastern slopes of the Andes. River highstands at any given location typically lag peak seasonal rainfall by several months, attesting to the buffering capacity that surface flow can play on maintaining positive moisture balances. As a whole, sedimentary evidence suggests that the Amazon River system has experienced a delayed response to regional changes in climate (Latrubesse and Franzinelli, 2002). Conveyance of moisture from the ever-wet, eastern Andean slopes to lowland regions subject to relatively strong seasonal declines in rainfall has the invariable effect of buffering changes in surface albedo that would occur if this was a function of local rainfall alone.

Surface albedo changes in the Guiana Shield occur as a consequence of changes in vegetation cover and soil moisture levels. In the past, the climate of the Guiana Shield was believed to be quite different (see 'Prehistoric climates of the Guiana Shield', below) and the positive feedback effects of vegetation loss, soil moisture decline and increasing surface albedo may have further propelled some parts of the region into a period of much lower rainfall and more scattered forest cover (see 'Prehistoric climates of the Guiana Shield', below). At the same time, the effects of changing sea level throughout the Quaternary would alter surface run-off in the region, and some inland depressions may have experienced higher (lower) water tables of increased (decreased) duration than today (e.g. see 'Upland laterite soils', above), reducing (increasing) surface albedo effects.

LATENT VS SENSIBLE HEAT FLUXES AND SOIL TYPE EFFECTS Soil hydrological properties also influence regional rainfall by altering the relative contribution of sensible (H) and latent heat (LE) fluxes, depicted by the Bowen ratio (H/LE), to the overall process of energy transfer from surface to atmosphere. Water bodies, such as oceans and lakes, typically have very low Bowen ratios. Latent heat fluxes occur as water undergoes a phase change from liquid to gas (evaporation) to liquid (condensation) and these dominate the exchange over water bodies. Sensible heat fluxes (high to low air temperature changes) are minimal and temperature changes are modest. In extremely dry surfaces, such as deserts, the opposite is true with virtually all of the exchange occurring as sensible heat (and thus wide-ranging temperatures) since there is virtually no water available to drive evaporation. Regional surfaces that exhibit high Bowen ratios provide little energy to drive rainfall and attenuate temperature variations due to sensible heat fluxes. Considering surfaces with energy transfer processes dominated by sensible heat fluxes, the opposite is true. Generally speaking, the moisture-retention properties of surfaces are crucial in determining the Bowen ratio. For tropical forestlands, these properties are principally a function of rainfall intensity, topography, vegetation cover and soil type (Bonell and Balek, 1993).

This establishes a potentially important source of regional variation in the neotropics. Soil types dominated alternatively by clay-loam and sand are not equally distributed across the neotropics (see 'Soils and Soil Fertility', above). Clay-dominated soils will tend to retain a much higher surface water content than sand-dominated types and latent heat fluxes will play a more substantive role than sensible heat in defining the local energy budgets in areas where closed forest canopy is not in place. Closed forest canopy covers a wide range of soil types across the Guiana Shield and the greater neotropics. The latent heat flux associated with the forest canopy, when viewed as a surface, will always attenuate variation in the Bowen ratio relative to its underlying soil surface due to the added effects of transpiration and the ecophysiological adaptations (e.g. storage organs, tap roots) that work to avoid complete desiccation and death. Living plants, unlike soils, therefore, moderate rapid shifts from latent to sensible heat-dominated energy transfers that would otherwise occur if the underlying soils were exposed. When this canopy is removed, differences in soil properties, all else being equal, are more likely to spur greater spatial variation in precipitation and temperature responses to larger-scale forcing factors. The eastern Guiana Shield has the highest concentration of sand-dominated and laterite-capped soil surfaces in the wet neotropics. These soils, on average, experience faster dry-down rates than clay- and loam-dominated soil types (e.g. Jetten, 1994). The latter account for a larger proportion of the Amazon Downwarp and Sub-Andean Foredeep regions of western Amazonia (e.g. Sourdat, 1987) (Fig. 2.14).

SEA BREEZE Along the Atlantic coastline, a smaller land–ocean exchange process is believed to affect rainfall over the eastern Guiana Shield region when the low sensible heat release zone over the ocean comes in contact with the relatively higher absorption and release rates of the adjacent land surface. This means that the land heats up and cools down much faster than the ocean creating a temperature gradient perpendicular to the coastline (Robinson and Henderson-Seller, 1999). The consequent gradient in pressure forms a small-scale circulatory cell that drives a daily cycle of wind movement from relatively warm to cool zones. During the day, when the land is warmer than the adjacent water, the ascending branch of this cell occurs over land and an onshore surface wind predominates, called a sea breeze. The opposite condition, referred to as land breeze, is true at night when the slow heat loss rate of the ocean makes this zone warmer than the adjacent coastlands. In both cases, cumulus clouds formed along the ascending branch are conveyed towards the coastline and have been associated with enhanced precipitation in areas up to several hundred kilometres inland (Kousky, 1980). Kousky also suggests that seasonal variation in the contribution of sea and land breeze effects to precipitation along the coastline of northeastern Brazil reflects changes in SST and land heating. SSTAs are strongly linked to changes in wind stress and thermohaline current movements. This suggests that the contribution to annual rainfall made by the sea–land breeze cycle is sensitive to changes in global-scale circulatory patterns, such as ENSO and NAO, as well as changes to the amount of radiation received at the land surface that can be affected by external forcing factors (see 'Longer-term climate forcing factors', below).

Guiana Shield rainfall and temperature regulation in relation to other neotropical regions

Taking latitudinally averaged values for incoming solar and outgoing long-wave radiation emphasizes the spatial constancy of this important influence on temperature across the tropics *relative to regions of higher latitude*. But within the tropics there are important differences to be highlighted. The longer residence of the ITCZ north of the equator slightly reduces both ISR and OLR in this area due to the more persistent cloud cover compared to the austral portion of the equatorial trough. Across the Guiana Shield and Amazonia, there also appears to be a zonal increase in cloud cover persistence from east to west and a concomitant decline (of between 27 and 78 W/m^2) in annual average ISR received at the surface relative to eastern South America and the Caribbean (Lockwood, 1979). Satellite-based radar backscatter of rainfall illustrates this zonal gradient in Fig. 2.23.

Of course, annualized averages do not reflect spatial differences in temporal variation. As we have seen, Pacific ENSO, Atlantic meridional and zonal SST gradients, African easterly waves (and hurricanes), atmospheric Rossby waves and SACZ/SPCZ behaviour can all influence rainfall across the neotropics, but not uniformly. For example, the depth of precipitation response to Pacific ENSO and Atlantic SSTAs across eastern South America appears more pronounced than in western Amazonia (and the western tongue of the Guiana Shield). Western Amazonia can experience an anomalous increase or decline in rainfall during extreme ENSO phases, but these are modest by any measure when compared with the eastern Guiana Shield area (e.g. Fig. 2.6 in Marengo and Nobre, 2001). For example, Dai and Wigley's (2000) analysis of ENSO–precipitation relationships indicate an annual average decline (from the long-term mean) in rainfall during typical EN phases across western Amazonia of 0 to –100 mm per annum compared with a –50 to –300 mm decline over most of the Guiana Shield and northeastern Brazil. Considering the much higher average rainfall over most parts of western Amazonia only further emphasizes the contrast between these regional responses. Equatorial extension of mid-latitude atmospheric and oceanic systems, such as the Azores High AP (part of NAO) can have a pronounced effect on wind stress, latent heat flux and SST in the (sub-) tropical North Atlantic, and impact on rainfall in the Caribbean and Central America. The much greater southward extension of the ITCZ across eastern South America may contribute to a deeper equatorial extension of the Azores High across the eastern Guiana Shield. Land–ocean processes across the Guianas may buffer this area to some extent from this movement, leaving a high AP extension corridor to remain over the Savanna Trough, combining with reduced wind friction, higher albedo and more rapid moisture loss of these mixed open forest–grasslands. Anomalous phases of NAO-related AP changes register little or no change of rainfall in west-central or southern Amazonia. In contrast, unusual behaviour of the SACZ and SPCZ in the austral mid-latitudes or the austral arm of the Atlantic 'dipole' can strengthen or weaken precipitation in southern Amazonia and SE Brazil, but with relatively little impact on the Guiana Shield. Northward extension of cold air masses of polar origin can regularly bring seasonal drops in minimum temperatures over southern Amazonia, but much less frequently to the Guiana Shield.

In effect, mechanisms regulating rainfall across the neotropics overlap, but with impact epicentres occurring in different regions (after Marengo and Nobre, 2001). Variation in the influence of land surface processes, such as topography, act to create smaller-scale spatial anomalies in rainfall regimes regulated by larger-scale mechanisms. The upper Rio Negro–Rio Orinoco region receives some of the highest annual rainfall totals due to its position close to the meteorological equator, the influence of ITCZ behaviour and a high availability of precipitable water affected by the trough-like Casiquiare Rift. This rift receives tremendous surface run-off from both the eastern slopes of the Colombian Andes as well as the western ranges of the Guayana Highlands, culminating in extensive seasonal flooding as water flow is occluded along the Orinoco and Rio Negro River mainstems (see 'Casiquiare Rift', above). The coastal zone between Cayenne, French Guiana and northern Amapá also receives similarly high levels of rainfall due to geographic position. The upper Gran Sabana–Rio Branco–Rupununi region separating these two locations, despite falling in the same latitudinal band, receives much less rainfall. In part this has been attributed to a leeward location relative to the Guayana Highlands (Frost, 1968; Snow, 1976). But the fact that this region also rests along the watershed divide might suggest a more rapid seasonal decline in the soil moisture needed to maintain a strong internal precipitation cycle during periods when the ITCZ has migrated southward and high pressure from the north has filled in behind. Lowland areas of northwestern Amazonia within the ITCZ band and at the foothills of the Andes hold a mid-central position along a more severe watershed incline and the effects of seasonal rainfall decline are lagged by moisture moving downward along the main watershed axis. Closer to the mainstem of the Amazon/Solimões, the volume of water upstream attenuates the amplitude of discharge fluctuations (see 'River, Lake and Tidal Systems', below) and possibly the effects on water vapour and latent energy fluxes and thus precipitable water levels. For example, Marengo (1999) found that the river level at Manaus, Brazil and Iquitos, Peru appeared relatively insensitive to very strong EN/LN events due to the large size of the upstream area.

Taken as a whole, available information suggests that the three climatological regions of the Guiana Shield, as defined by Köppen, exhibit different rainfall patterns as a consequence of their *geographic position* in relation to the equator, oceans, mountains and sedimentary basins. This simple characteristic most cogently identifies those climate-regulating mechanisms that are largely responsible for observed differences in seasonal timing and, perhaps most importantly, the periodicity and amplitude of long-term rainfall and temperature variation. Changes in the fluctuating behaviour of these mechanisms, particularly those operating quasi-independently upon climate in the neotropics (e.g. human activity), are likely to result in new anisotropies in rainfall across the region that are not simply amplifying or attenuating existing response gradients. Examining patterns of climate indicators across the Guiana Shield and adjacent regions in relation to the modern climate-shaping mechanisms can help to establish to what extent former climate change has determined the spatial patchwork of varying composition and structure that forms modern tropical forests in the region.

Prehistoric climates of the Guiana Shield

Tropical forests in the Guiana Shield are arguably as much, if not more, the product of past climate conditions as they are contemporary ones. Yet the relative impact of prehistoric climatic fluctuations on modern-day forest distribution and composition is difficult to establish conclusively. Glacial and polar ice layers can trap air from past climatic periods in bubbles, providing information on the trace gas and isotopic composition of past atmospheres. Ocean floor sediment cores can also provide evidence of fluctuations in SSTs, salinity, current movement and biological productivity. The link between these and regional rainfall and temperature changes over the neotropics, however, is not a simple one. In the lowland tropics, these more precise measures of prehistoric atmospheres do not exist, so we do not have a direct indicator of varying rainfall and temperature across modern-day tropical forest regions. Instead, other proximate measures are typically used to frame regional prehistoric climate phases. Proximate measures of neotropical climate change taking place millions of years ago have relied heavily on analysis of geological features, such as loess deposits, geomorphological features (such as lateritic planation surfaces) and sedimentary rock type and distribution. Over the most recent 100 ka BP that define the Quaternary period, more precise measures have been developed using fossil organic remains that preserve well under certain lowland tropical conditions. These lake- and swamp-derived pollen and phytolith cores are being used to identify climatological shifts in the region. Palaeoecologists and archaeologists have also employed cultural transitions, measured by changes in archaeological, charcoal and phytolithic remnants, as a proximate measure of palaeoclimatic change in the neotropics over the last 10,000 years.

It is important to note, however, that the maximum temporal resolution of most of these records is 100 years (Table 12.1 in Robinson and Henderson-Seller, 1999), so the impact of rainfall fluctuations at the periodicities documented in studies of modern climatological phenomena, such as ENSO, may simply be integrated into the record along with other signals.

There are also issues related to the accuracy of radiocarbon dates assigned to different materials frequently used as indicators of past climatic change. Lake bottom sediments can often undergo varying levels of mixing and gyttja (pronounced 'yut-ya', Swed.) deposits can form from a variety of (non-atmospheric) carbon sources. Charcoal is considered the most reliable material for radiocarbon dating (Libby, 1955), but the assigned age will also reflect the source plant's pre-burn growth period, described as the 'in-built' or 'presample' age of the wood material. This can be considerable in the case of many modern Amazonian trees that have had their central heartwood radiocarbon dated from 150 to 1400 ^{14}C years BP (Zagt, 1997; Chambers *et al.*, 1998).[11] Stratigraphic variation of soil charcoal concentration is also subject to misinterpretation when the position in the soil profile is affected by factors other than climate change (e.g. post-fire downstream transport, biological activity, infiltration rates, stream meandering) and samples are not dated directly (Piperno *et al.*, 1990).

Further circumstantial evidence of climate change affecting the Guiana Shield could also be drawn from better-resolved proximate measures constructed from material in the mid-latitudes or high altitudes, such as tree rings, ice cores and lake sediments. This may assist in establishing important impacts of global-scale mechanisms (e.g. orbital forces), but the translation of impact from temperate to tropical latitudes again must rely in part on a correct interpretation of cascading interactions between regional atmospheric and oceanic circulations. Only the most pervasive of temperate–tropical climate connections would be reflected in the tropical records – changes in the internal mechanisms affecting rainfall (e.g. the balance between ocean- and evapotranspiration-derived contributions to Amazon precipitable water) would be more difficult to resolve. Ultimately, an acceptable reconstruction of prehistoric neotropical climate will require a compositing of the results from a growing number of study sites that, interpreted on their own, may variously reflect relative contributions of real changes in regional climatic conditions and the influences of local site attributes. A durable interpretation of the relationship between adjusting climate change mechanisms, hydrological and sedimentological responses to these adjustments, and the consequent deposition dynamics of fossil organic material has yet to be established. Sample sites not yielding significant material are typically disregarded rather than used to establish the impact of local site attributes on material deposition patterns. Modern pollen and phytolith spectra reflect depositional dynamics under current geomorphological and climate conditions – these may not be linearly related to prehistoric state conditions, particularly if fluctuating human occupation is considered as a factor over the last 10 ka BP. Climate change and human land-use signals are not necessarily in phase and could create different depositional outcomes.

This section aims to present a simple non-specialist's overview of the information used to reconstruct prehistoric climatic change in the Guiana Shield and neighbouring regions and to draw attention to the main reconstructions of palaeo-precipitation and temperature patterns across the region based on this information. The interpretation of palaeo-data fundamentally informs the approach adopted in reconstructing past climate change. From the stand-point of population and community dynamics, the impacts of former climate change on current forest composition and distribution within the Guiana Shield should be substantial given: (i) the established life expectancy of many tropical canopy trees; (ii) the geographical location and topography of the region in relation to the main climate-driving features of the global circulation; and (iii) the current geometry and position of the forest/savanna interface across the area (see Chapter 7). Given well-established concerns over the future impacts of a contracting neotropical forest cover on global climate, biological diversity levels and regional socio-economic well-being, a clear understanding of prehistoric 'cause and effect' pathways linking climate and tropical forests can substantively inform the range of likely responses to modern forest land-use.

PALAEO-CLIMATE PROXY DATA FOR THE GUIANA SHIELD Fewer than 60 sites have been sampled in the 'lowland' neotropics (Fig. 2.25) and, of these, material has been radiometrically dated in only a handful of cases (Colinvaux, 1996; Haberle, 1997; Piperno, 1997). Most data are palynological or sedimentary and collected from Andean highland locations (Fig. 2.25), although several phytolithic, isotopic and archaeological studies have also been carried out. Most published sample sites are located on the periphery or outside of the shield area and almost always in the large sedimentary basins (Berbice Depression, Takutu Graben, Sub-Andean Foredeep, Amazon Downwarp) or volcanic arcs (Central America) that range across the lowland neotropics (Fig. 2.25) (e.g. Fig. 2.1 in Romero, 1993). Few locations in the major areas of exposed Precambrian rock (Guayana Highlands, Tumucumaque Uplands, northern Brazilian Shield) have been examined and these derived almost exclusively from swamp or bog locations (terric Histosols) where stratified palynological profiles most typically develop (Rull, 1991). Less than 1% of the lowland neotropics is thought to be covered by Histosol soil types (Richter and Babbar, 1991) and mainly concentrated in sedimentary depressions of the Sub-Andean Foredeep (Schulman *et al.*, 1999), Berbice Depression (Fig. 2.11D) and other regional zones of downwarping (and inundation). Histic soil cover in areas dominated by Precambrian rock types is considerably reduced, although bog formations atop tepuis have proved useful for palynological studies of the Holocene (Schubert and Fritz, 1985; Rull, 1991). Lakes and swamps of an age needed to expand the record across the lowland neotropics are not common or evenly distributed. This relative rarity of pollen-preserving conditions presents difficulties in confidently discriminating the contribution of fluvial from aerial, local from remote, sources to the profiles and thus the spatial scale at which these profiles can be appropriately applied in reconstructing past vegetative responses to former climate conditions. Long-distance fluvial transport of pollen downslope to lowland sample sites is considered an insignificant transfer path for a variety rea-

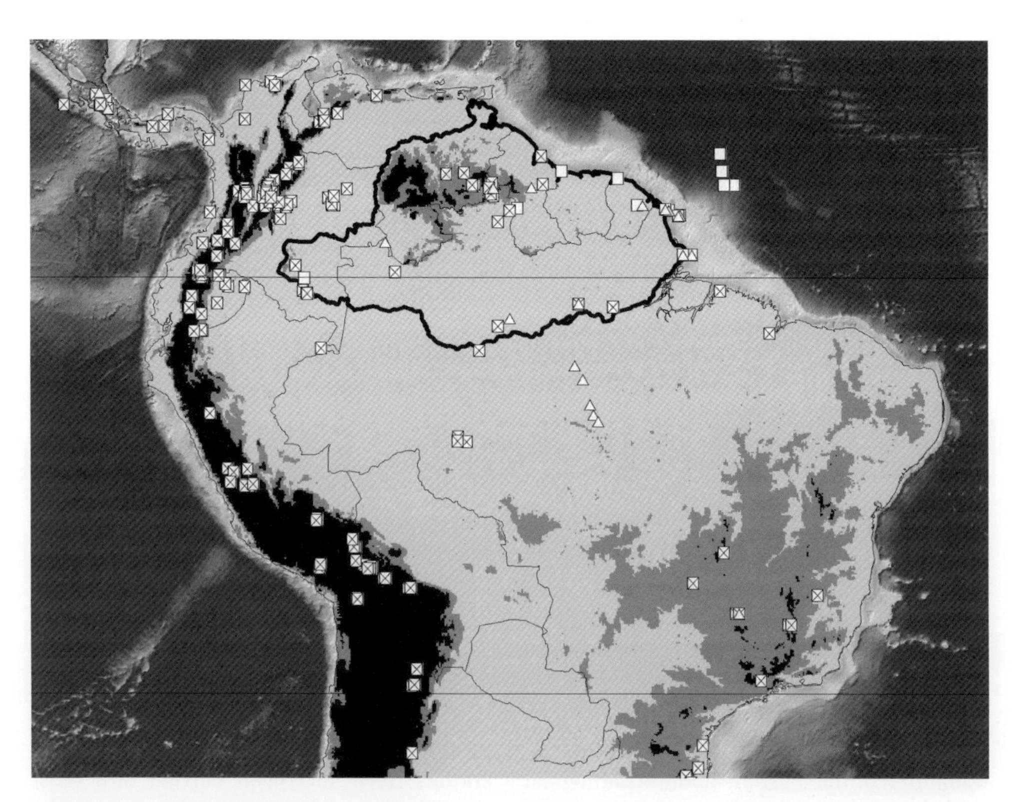

Fig. 2.25. Location of fossil pollen, phytolith and charcoal studies across the neotropics. Note disproportionate sampling of Andean highlands relative to Sub-Andean Foredeep, Amazon Depression and near absence of samples from Tumucumaque Upland region of Guiana Shield. Source: Latin American Pollen Database (crossed squares) and other non-LAPD listed pollen core locations (open squares) and additional charcoal study sites (triangles).

sons. Current understanding of modern fluvial systems across Amazonia repeatedly identifies the highland Andes as the source of the overwhelming majority of both particulate organic and total suspended solids sampled in stream and through sediment cores at the mouth of the Amazon (see 'River, Lake and Tidal Systems', below). It is difficult to reconcile current transport efficacy with a selective absence of pollen transport along similar distances of prehistoric river systems. Clearly, scaling up palaeoclimate reconstruction from a single sample site to the entire lowland neotropics assumes considerable spatial uniformity in the cause-and-effect of climate change. This spatial uniformity is not apparent in our current understanding of forces affecting the modern neotropical climate (see 'Modern climatic regions', above). Given the regional variation in influence of different modern climate forcing factors, past changes in rainfall most certainly varied by region rather than uniformly across the neotropics (Colinvaux, 1987, 1996).

Phytoliths, being mineral-based remnants of plant structures, are not constrained to local conditions that preserve organic materials (such as histic soils) and can cover a much wider range of substrate conditions (Piperno and Becker, 1996). To this end, they have greatly assisted in 'triangulating' results from pollen-based studies. Archaeological data have also been used to assess climate change, based on the assumption that early societies relocated and adjusted the style, quantity and composition of their utensil manufacturing and diet in response to changing climate and that this response is apparent from materials found at midden ('trash') and burial ('cemetery') sites (Meggers, 1994). Radiocarbon ages for buried charcoal have also been used to infer periods of relative aridity in the neotropics and/or former occupation by pre-Columbian inhabitants (Saldarriaga and West, 1986). Combinations of charcoal and phytolith data have greatly assisted in establishing plausible connections between palaeofire events and prehistoric human activities. Again, however, few neotropical sites have been examined and many soil types found throughout the region have not been studied. This presents another quandary in the scaling up of interpretation. Underlying soil types have asymmetric hydrological responses to fluctuations in rainfall and, therefore, susceptibilities to fire after an ignition event (see 'Soils and Soil Fertility', above; Hammond and ter Steege, 1998).

Apart from detailed technical concerns over site-specific influences, the main difficulty in understanding palaeoclimate from these types of data rests in the fact that they reflect a composite response to change, potentially occurring at many different spatial and temporal scales. Identifying the main signal associated with changes in the type of pollen, pottery and phytoliths encountered in a dated chronology is difficult. Not all signals necessarily reflect periodicities of the main climate forcing influences (e.g. anomalous social decision-making, war, disease and restoration, epidemic plant disease (Allison *et al.*, 1986), volcanic explosions (Behling, 2000), geoseismic impacts on soil movement and fluviatile dynamics (e.g. Frost, 1988)). Those that do should not necessarily be translated as a uniform impact on palaeoclimate throughout the neotropics.

Clearly the accrual of a substantial number of sampling sites is needed across the neotropics in order to recognize the spatial scale of anomalous rainfall responses to past climate change. Given the considerable zonal and meridional variation in rainfall seasonality that currently characterizes the Guiana Shield and adjacent areas, a more complete stratification of sample sites that embraces a wider range of landforms and soil types would increase the resolution of palaeoclimate change established by proxy for the region.

MAIN RECONSTRUCTIONS OF PAST NEOTROPICAL CLIMATE CHANGE Despite some lingering uncertainties, the interpretation of past climate based on the palynological, charcoal, phytolithic and archaeological record provides an important quantitative approach to estimating prehistoric fluctuations in temperature and precipitation across the neotropics. Compositing results from many different sites can assist in overcoming site-specific uncertainties and provides the basis for a cogent reconstruction of past climate. A detailed palaeoclimatological reconstruction for the Guiana Shield has yet to be made, although a number of scenarios have been generated based on best available information. The most well known of these is the CLIMAP project reconstruction of global climate during the Last Glacial Maximum (LGM) (18–25 ka BP), but the global scale of this early effort provides few details regarding the Guiana Shield region.

The distribution of extant plants and animals, geomorphological features, modern rainfall and temperature gradients, fossil pollen, phytoliths and charcoal have all been variously incorporated into hypotheses championing one or another aspect of climate change as the main force affecting neotropical forest cover. These data have historically been interpreted into two main scenarios of change. One is based on the view that rainfall was the predominant force for change across the neotropics during the Quaternary (Haffer, 1969; Vanzolini, 1973; Whitmore and Prance, 1987 and chapters therein), converting vast areas of forest to (wooded) savanna (Ab'Saber, 1977). The other contends that ambient temperature change largely explains variation in the palaeobotanical record during this period (van der Hammen, 1974; Liu and Colinvaux, 1985). Other climate change pathways have been proposed (for example, changes in [CO_2] or UV light (e.g. Flenley, 1998)), though these are not as easily testable using the type of proxy data collected in the region and discussed here. The strengths and weaknesses of each have been assessed by Colinvaux (1996) and Colinvaux and De Oliveira (2001). The weight of scientific opinion has oscillated between moisture and temperature over the past 35 years. In part, this is a consequence of a fragmented palaeo-botanical record that has failed to show strong spatial synchrony of vegetation responses to putative climatic fluctuations. This may reflect disparities due to local site factors, such as varying depositional responses to the same climate phases, the role of different plant successional trajectories, or varying importance of unconformable sedimentary records (often many periods are 'missing' from the core sample of sediments (e.g. see Ledru *et al.*, 1998)). A fuzzy signal may also be due to disparate methodological approaches (e.g. see Marchant *et al.*, 2001). Alternatively, inconsistent site interpretations of alternating wet and dry, cold and warm periods may reflect real regional differences in the way moisture and energy balances fluctuate in response to forcing factors (e.g. Hooghiemstra and van der Hammen, 1998). This would be consistent with the view that mechanisms affecting climate at any particular location tend to interact across a wide range of spatial and temporal scales to deliver anisotropic climate signals at the forest surface in a manner, but not magnitude, similar to present day systems.

The proximate record of climate change in the neotropics extends back to at least the mid-Cretaceous when the Guiana Shield was believed to have only started to separate from its West African counterpart, the Man and Benin Shields (see 'Internal planetary energetics', below) (Barron and Washington, 1982). Understanding the spatial variation of prehistoric climate across the Guiana Shield over the last 100,000 years, however, is arguably of greatest importance in establishing the effect of regional processes on past moisture and temperature variations that have shaped the composition and distribution of modern forests of the region.

HOLOCENE RAINFALL AND TEMPERATURE (0–10 KA BP) Anomalous periods of drought are believed to have affected the lowland neotropics over much of the Holocene (Absy *et al.*, 1991). Maximum achievable resolution of the palaeo-record, however, currently cannot characterize seasonal, annual or even decadal rainfall signals. An attenuation of wet season rainfall versus amplification of dry season drought would have different implications for rainforest composition and distribution, as it does today in distinguishing the vegetation of the Savanna Trough (upper Rupununi–Branco–Negro–Orinoco) from the eastern and western regions of the Guiana Shield. Evidence derived from fossil pollen suggests that existing forest and savanna distributions have been in place regionwide for at least 3000 years, but that they have undergone significant periods of heightened disturbance and local shifts in plant composition, perhaps in association with these intervening dry periods (Fig. 2.26A, C).[12] But only a handful of palynological studies have been conducted at locations currently covered by closed canopy lowland tropical forest (e.g. Colinvaux *et al.*, 1988; Piperno *et al.*, 1990; Ledru *et al.*, 1997).

DIFFICULTIES WITH FOSSIL-BASED ASSESSMENTS Most fossil pollen records originate from savanna, cerrado, montane or alpine sites and this imbalance in the meta-record makes a region-wide assessment of lowland climate change during the Holocene more difficult. The main problem is reconstructing lowland forestland moisture balances from pollen sample locations that are higher and/or drier than the modern lowland forest region. These records reflect importantly on the dynamics of tropical forest extension and contraction along the range perimeter, but are not as plausibly linked to the dynamics of the centre(s). Differences in the way that varying land processes, particularly vegetation–water relationships, shape local moisture regimes argues against simple spatial interpolations. The recycling of moisture originating from large, lowland evapotranspirative sources may be the single most important factor distinguishing wet from dry across regions affected equally by the seasonal trek of the ITCZ and functioning of the Hadley and Walker Circulations.

Savanna-based fossil pollen strata sampled in the llanos of Colombia (Behling and Hooghiemstra, 1998, 1999; Berrio *et al.*, 2002) and Rupununi Savanna of Guyana (Wijmstra and van der Hammen, 1966) are characterized principally by changes in the abundance of co-occurring species found today in these habitats. Contemporary climate over the Savanna Trough is noticeably drier than closed forest areas in the Guiana Shield and a shifting mosaic of palm swamp, gallery forest and wet/dry savannas driven by meandering stream courses and anthropogenic fire create a series of land units that constitute the lion's share of the modern savanna landscape (Eden, 1964). Transitions identified in the palynological record may simply reflect successional dynamics within this mosaic under fluctuating rainfall regimes not unlike those affecting change in modern vegetation communities (e.g. Berrio *et al.*, 2003).

Palynological transitions along the Atlantic coastline of the Guianas also appear to be largely a function of changing sea level (van der Hammen, 1963; Wijmstra, 1969; Tissot *et al.*, 1988) and thus oceanic conveyance of high-latitude climatic transitions into the tropics, rather than reflecting on local or regional shifts in rainfall.

Evidence from sites at relatively high elevations is close to or exceeds the temperature–moisture envelope that restricts many of the most common tropical forest taxa to the lowlands. These sites provide little information about changes in moisture availability independent of temperature. Water vapour traps thermal radiation, attenuating temperature variations. Orographic lifting consistently brings water vapour to central slopes of the Andes, but less predictably to higher elevations. This elevational cline in water vapour would plausibly alter the amplitude of temperature variation. If this is true, then vegetation response signatures indicating alpine transitional stages may not necessarily reflect on lowland sites downslope since orographic lift would not bring water vapour to all elevations in equal amounts. The shape of the temperature–moisture envelope would also not remain constant with descent, since up-slope moisture is transferred downslope, conveying an important source of latent heat to the lowlands and altering the seasonal variation in the Bowen ratio.[13] The importance of both forms of moisture conveyance, orographic lifting

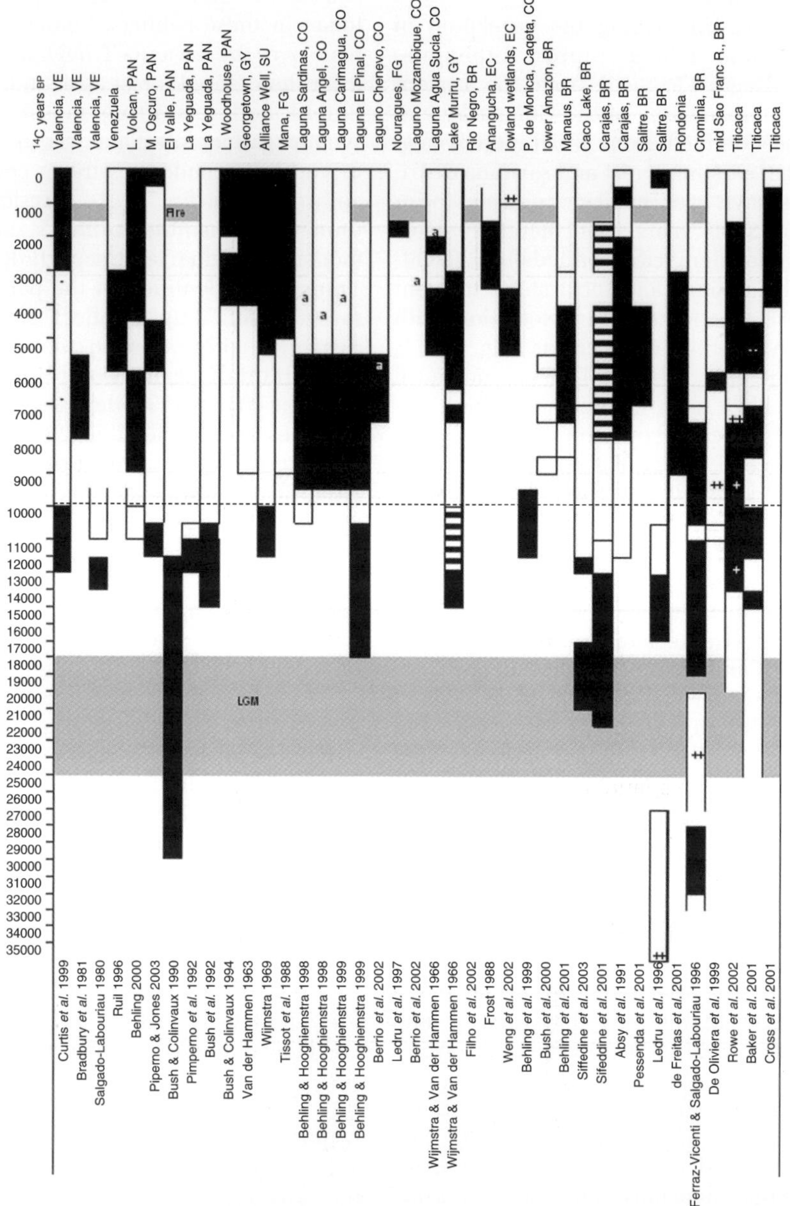

Fig. 2.26. Holocene/late Pleistocene distribution of (A) drier (solid column)/wetter (empty column) and (B) warmer (empty)/cooler (solid) climate phases based on interpretation of stratigraphic position of fossil pollen, phytoliths, geomorphological features and geochemical abundances in relation to stratigraphic position. Age of strata based on radiocarbon dates and projected sedimentation rates. Dashed line represents beginning of Holocene. Hatched bars indicate results showing high-frequency fluctuations between conditions. Grey bars indicate peak of charcoal radiocarbon dates from Fig. 2.26(C) and the expected span of the Last Glacial Maximum (LGM), a period of reduced sea level and lower global temperatures. (C) Distribution of ^{14}C dates for fossil charcoal samples collected at various lowland forest and savanna locations in the Guiana Shield and adjacent regions. Savanna is modern habitat at Gran Sabana and Salitre sites. Forest is modern habitat at all other sites. Other site features, soil sampling depth (D) and radiocarbon dating methods differ between sites. See Note reference in text for further clarification and site references. a = open aquatic sites; +/++ = peak phase.

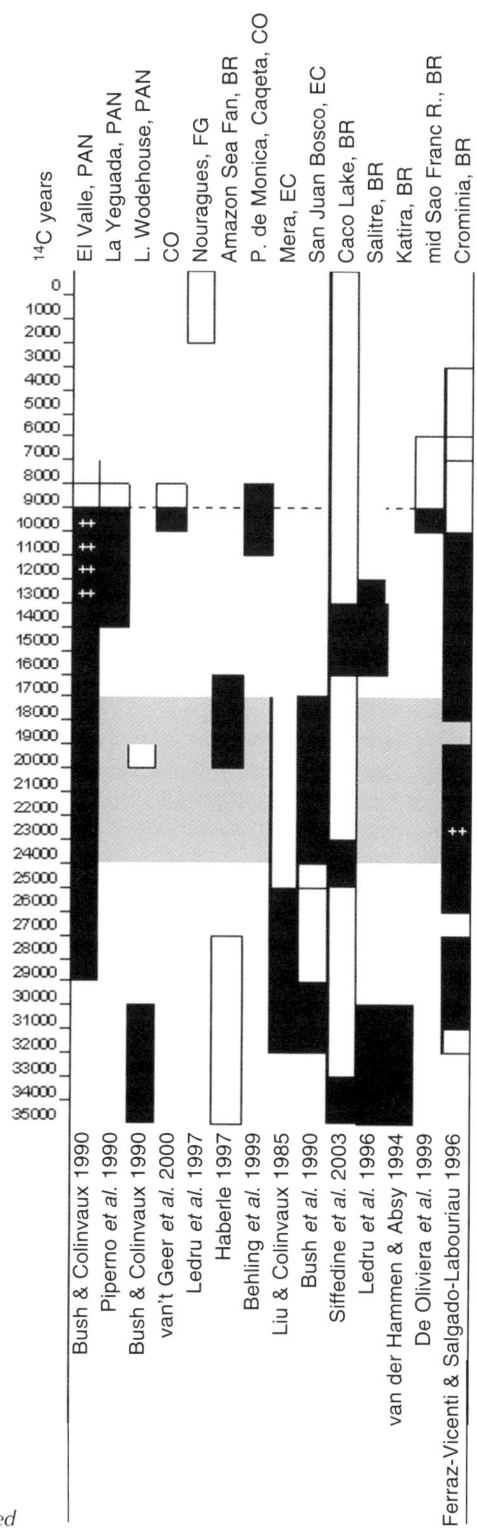

Fig. 2.26. *continued*

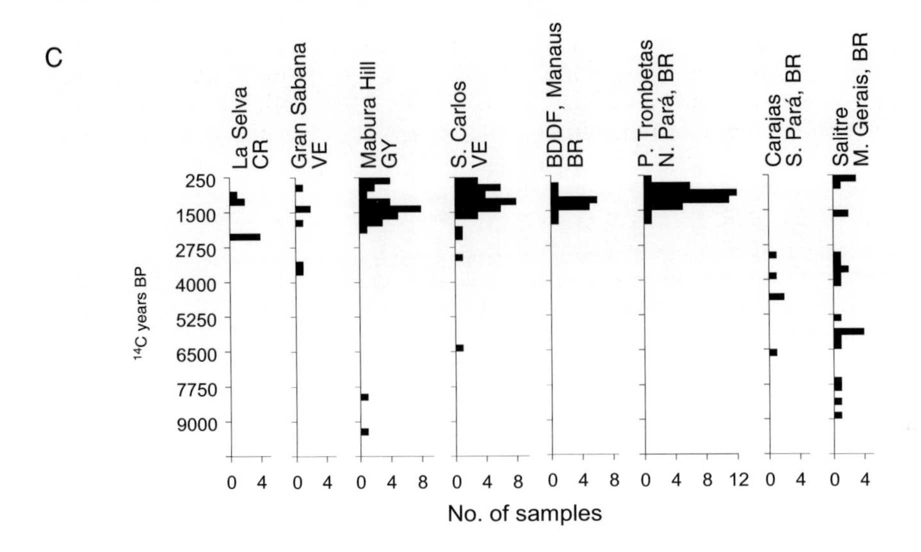

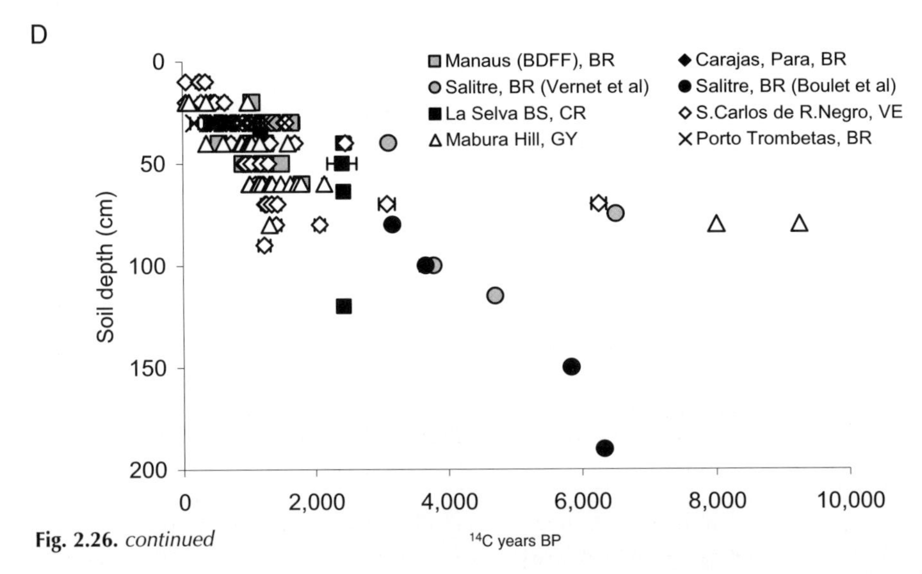

Fig. 2.26. *continued*

and downslope run-off, to temperature regulation suggest that a cooling of sub-alpine elevations may have only been possible if there was a concomitant contraction of moisture recycling between lowland and upland slopes in western Amazonia. In part, this could be produced through a sustained period of temperature reduction, but determining cause-and-effect in this relationship requires an assessment of extra-regional forcing factors that could alter the amount of energy or moisture available to the region. Discussion of temperature change in the neotropics independent of moisture is counterintuitive given the feedback effects between these two climatological parameters (e.g. Dai *et al.*, 1998).

Compositing the climate reconstructions for a large number of lowland fossil pollen profiles shows very little spatial uniformity in the duration or timing of relatively wet and dry periods assigned to the data over the last 10,000 years (Fig. 2.26A). Phytolith studies have been restricted largely to a few sites, mainly in Panama and along the Amazon mainstem. Adding these to the composite generally reinforces the view, however, that translating the palynological record into a prehistory of rainfall shows only modest spatial or temporal consistency over the Holocene (Piperno and Becker, 1996). It does, however, unequivocally show that the neotropical forest region has undergone significant change over the last 10,000 years.

Asynchrony in the fossil pollen, rock and isotopic abundance records over the Holocene suggests: (i) other forces have altered the pollen–vegetation–climate relationship; (ii) rainfall responses to climate change varied widely across the neotropics; or (iii) interpretations of pollen-type strata vary in their accuracy. All are likely to be true. Most records over the Holocene indicate that human activity has periodically altered the distribution of forest cover, particularly over the past 5 ka. At Lake Wodehouse, Panama, phytolithic evidence suggests that forests have only been recovering from prehistoric cultivation for 300 years, although modern rainfall seasonality is not expected to have varied significantly up to 1600 years earlier (Bush and Colinvaux, 1994; Piperno, 1994). Other natural catastrophes, such as flooding (Frost, 1988) or volcanic activity (Behling, 2000), can also disrupt the palynological/phytolithic record of response to prevailing climatic conditions. Bush and Colinvaux (1994) conclude that dry phases associated with the late Holocene record from Darien, Panama are most likely a reflection of local events. Ledru *et al.* (1996) identified several periods of aridity in the palynological record at the south-central Brazilian site at Salitre, but not at Serra Negra, only several hundred metres upslope from the lower peat bog location. Asynchrony in the record of geological responses to prehistoric climate change is also increasingly being viewed as a consequence of regional variation in the response to global climate forcing factors (Sifeddine *et al.*, 2003).

Local vegetation mosaics at the sample site may also act to filter out regional patterns of change. Berrio *et al.* (2002) opine that certain fossil pollen cores taken from lake locations in the Llanos Orientales of Colombia do not adequately reflect regional savanna elements when the sample site periphery is dominated by arboreal swamp vegetation.

Interpretation of the sample pollen strata may also introduce heterogeneity in the reconstruction of palaeoclimate. Bush (2002) warns that using fossil grass pollen as the main factor in reconstructing prehistoric precipitation change is potentially misleading, given long-ranging wind pollination of grasses and their ability to colonize both dry (savannas) and wet (floating grass mats and marshes) environments and strong associations with human-modified landscapes.

LOCAL AND REGIONAL DIFFERENCES These caveats further emphasize the role of localized effects on fossil pollen deposition and how they preclude wider extrapolation across Amazonia. The Guiana Shield, with its distinct geographical distribution along the thermal and meteorological equators, proximity to the western rim of the Atlantic Basin, and unique underlying geology, is unlikely to have been subject to the same climatic history as Central American, western Amazonian or northeastern Brazilian regions. Modern mechanisms with varying spatial centres of influence on neotropical rainfall further suggest that the prehistoric climate of the Guiana Shield may have varied 'out-of-phase' with adjacent regions. It is tempting to reconstruct the spatial variation of prehistoric rainfall by uniformly reducing or enhancing existing distributional patterns (e.g. Bush, 1994; Figs 3 and 4 in van der Hammen and Absy, 1994), but this logic in many ways is inconsistent with the geography of oceanic–land–atmosphere processes that govern the modern hydrological cycle. Forcing factors of tremendous amplitude would be needed to overshadow the dynamical adjustment processes that create spatial variation in climate conditions across the neotropics.

The few palynological cores that have

been examined within the Guiana Shield indicate that the area east of the Rio Negro/Rio Orinoco was subject to more pronounced changes in rainfall in comparison with western Amazonia over the past 10,000 years (Fig. 2.26A). The ubiquitous presence of charcoal across many parts of the eastern landscape further suggests that much of the existing forest/savanna mosaic in the Guiana Shield has been subject to drought-induced fire. Radiocarbon dates ranging from <100 to 10,000 ^{14}C years BP indicate that fire has maintained a presence over the entire Holocene (Fig. 2.26C). Few published, radiocarbon-dated charcoal records are available for western regions. Charcoal has been identified from soil profiles in some regions (e.g. Caqueta region, Colombia; Duivenvoorden and Lips, 1995) and as thick layers in terraces along the Colombian Solimões (D. Hammond, personal observation), so fire has occurred, perhaps catastrophically, in the upper Amazon Basin. The relatively infrequent occurrence of deposits, however, compared with the very high frequency of charcoal encountered in soil cores taken within (San Carlos de Rio Negro, Gran Sabana, Mabura Hill, Nouragues) or adjacent to (Manaus) the eastern Guiana Shield is noteworthy. This distinction may be due to fewer and smaller fires occurring under higher sustained rainfall, as it is today, in northwestern Amazonia, or reworking of charcoal deposits by the prominent regime of fluvial meandering that characterizes many parts of the western Amazonian lowlands (Salo *et al.*, 1986). The abundance of charcoal ages in the 1200–3000 years BP range may also be a testament to the rapid increase in anthropogenic lowland forest impacts during the late Holocene, although some locations in the Guianas appear to have been exposed to forest fires that pre-date evidence of human inhabitation (Tardy *et al.*, 2000; Hammond *et al.*, submitted).

PLEISTOCENE RAINFALL AND TEMPERATURE (10–100 KA BP) Global climate change during the Pleistocene is believed to have been dominated by a period of oscillating glacial–interglacial phases culminating in the Last Glacial Maximum (LGM), approximately 24–18 ^{14}C ka BP. Unlike the Holocene, palynological, geological and isotopic records for the Pleistocene generally suggest a uniform cooling phase across South America consistent with LGM onset. The composite record also indicates a period of drier climate extending from the LGM to the beginning of the Holocene (10 ka BP). A simultaneous decline in lowland moisture availability and temperature during the LGM would be consistent with a view that the lowland neotropics and upland slopes are bound by a temperature–moisture feedback cycle that can only be shifted when there is a change in the overall amount of energy or moisture available. Two external forcing factors have been identified as the most likely catalysts of declining moisture and temperature across the neotropics (see 'Longer-term climate forcing factors', below), but the spatial variation of impact remains poorly resolved. Most records extending through the LGM are derived from sediments sampled from alpine or savanna/cerrado locations on the periphery of the modern lowland forest region. The absence of LGM records from lowland neotropical locations is noticeable (Colinvaux, 1996, p. 359; Ledru *et al.*, 1998; Rull, 1999).

LAPSE RATES The Pleistocene lowland temperature depressions needed to explain altitudinal variation in the presence of modern alpine and montane pollen are typically estimated by assuming a linear relationship between temperature and elevation. The most common approach is to apply the mean lapse rate for tropical latitudes (currently figured at 6°C per km change in elevation) to the difference in elevation between the fossil pollen sample site and the modern elevational limits of the same indicator taxa.[14] By applying this rate to the distance, a change of temperature associated with downslope migration of indicator taxa is calculated.

Lapse rates, however, are far from constant across mountainous landscapes and have been shown to vary by elevation, aspect, latitude and proximity to ocean within the neotropics (Johnson, 1976; Walsh, 1996). The most appropriate application of a temperature–altitudinal relationship for these purposes is the environmental lapse rate (ELR) (Robinson and Henderson-Seller, 1999). ELRs vary continuously in space and time and are sensitive to simple changes in slope, aspect and other landscape features as well as diurnal and seasonal variations in cloud cover and incoming solar radiation. Local inflections in the temperature–altitude relationship are commonplace. An inversion can cause temperatures to increase with elevation over substantive areas. The flow of wind along mountain slopes can also set up local anomalies to assumed temperature gradients, as well as potentially influence the transport of wind-borne pollen. Energy losses on exposed upland ridges can occur more quickly than adjacent valleys as OLR from the valley and surrounding slopes is exchanged. This creates a downward flowing cold air mass, the katabatic wind. When the downward flow of this cold air is obstructed, it can settle at a lower elevation, creating a local negative temperature anomaly called a frost hollow. Colinvaux (1987) referred to this possibility as the 'katabatic wind hypothesis' and considered it as one explanation for the presence of putative montane or alpine pollen at elevations below their current range, but did not believe glaciers had descended far enough to sustain this explanation (Colinvaux, 1993). By return, however, an upward flowing mass, the anabatic wind, can also create strong flow of relatively warm air to higher elevations during the day. This shift in flow direction is a daily event with a wind force that is moderated by the level of cloud cover. This is important because temperature variation at tropical latitudes is greater between night and day than between seasons of the year. It complicates explanations of changing forest composition because the creeping descent of highland species would take on a highly fractal dimension when considering the action of local topography in creating katabatic/anabatic winds, frost hollows and inversion layers, among other anomalies impinging upon basic ELRs. Local mixing of lowland and highland taxa would in this instance vary considerably between locations, inflating the range of prehistoric temperature changes along the periphery of lowland neotropical forestlands. In fact, the consequence of these local effects on mixing may already be recorded in the many fossil pollen samples collected in the Andean and Panamanian highlands. Described as 'forest communities without modern analogue' (although it is worth noting that very similar mixes may still be found in the Guayana Highlands today), many strata derived from sediment cores contain pollen of both lowland and highland taxa in sufficient concentrations to suggest that prehistoric communities adjusted to temperature suppression during the LGM on a taxon-by-taxon basis. It is equally plausible that sediment strata are an integrated record of vegetation from both exposed (and colder) ridgelines and protected (and warmer) highland valleys.

GUIANA SHIELD VS. WESTERN AMAZONIA The cool aridity that is believed to have arrived in the neotropics during Pleistocene glacial advances is generally believed to have had the greatest impact on the Amazon periphery. There is some geological evidence to suggest that large areas of the eastern Guiana Shield were drier during these phases independent of sea-level change (Clapperton, 1993; Iriondo, 1997; Filho *et al.*, 2002), but pollen and phytolith data are generally scarce or absent from most of the region.

While mean adiabatic lapse rates would remain relatively constant across the latitudinal range embracing most of the neotropics (around 6–7°C/km elevation), ELRs at similar elevations are not likely to remain constant for slopes of the same aspect in the Andes and Guayana Highlands for several reasons related to variation in lapse rate control briefly mentioned above.

1. The general decline in tropical lowland elevation from the Andean foothills towards the Atlantic Ocean is in the range of 150–200 m. Applying the same simple lapse rate approach frequently used to estimate prehistoric temperature change from fossil pollen, the basic elevational difference between the Sub-Andean and Guianan lowlands would alone suggest that the Guiana Shield experienced temperatures 0.9–1.2°C higher than the western Amazonia lowlands during periods of global temperature depression.

2. The continuous line of the south-central Andes creates a formidable barrier to the south/northwestward-flowing trade winds, while the more highly weathered remnants of the Guayana Highlands consist of scattered tepuis and exposed intrusives intersected by low-lying valleys that have little orographic effect. Isolated tropical mountains and massifs also tend to collect and retain less incoming radiation than mountain ranges, such as the Andes. As a consequence, lower temperatures are sustained at lower elevations in isolated highland areas. This *Massenerhebung* effect (Grubb and Whitmore, 1966) represents a striking point of contrast in the way that temperature, and vegetation, would vary with elevation along the eastern slopes of the Guayana Highlands and Andes, respectively, in response to global cooling during the LGM. Considering this effect independent of other confounding variables, the highlands of the Guiana Shield would be cooler than the Andean slopes of western Amazonia at comparable elevations.

3. Trade winds have only travelled over forestland for a limited period before encountering the Guayana Highlands in comparison to the more substantial distances between the Atlantic and the eastern slopes of the Ecuadorian/Peruvian Andes. Latent heat fluxes are plausibly influenced by variation in western Atlantic SSTs in the case of the Guiana Shield as moisture is carried from ocean to mountain slopes. Fluxes of latent heat influencing ELRs along the Andean slopes are more likely drawn from the pool of lowland forest moisture. SSTs are strongly shaped not only by local ocean–atmosphere exchange, but also by oceanic conveyance. Thus, a source of latent heat potentially influencing the Guayana Highlands is not directly connected to highland rainfall and surface run-off. Latent heat source effects on temperature and moisture along the western Amazonian perimeter are more strongly dependent on precipitation recycling, linking latent heat sources created by lowland moisture pools with downslope run-off created by orographic rainfall through a cycle of sustained feedback. A change in the state condition of the lowland moisture pool (e.g. through deforestation) or run-off (e.g. through dam emplacement) could affect latent heat fluxes and the effect of moisture on ELRs. This suggests that the Guayana Highlands are more likely to experience attenuation of rainfall decline when lowland moisture pools or run-off are affected due to the compensating conveyance of ocean-borne latent heat. It also indicates that when oceanic latent heat fluxes decline, due to a lowering of SST in the western tropical Atlantic, intact lowland moisture recycling may attenuate, or delay, changes to moisture availability in the highland region. When there is both a decline in SSTs and a contraction of the terrestrial moisture pool, the Guayana Highland region would experience a more cataclysmic decline in moisture availability and lowering of temperature that would lead to substantial forest contraction. Given the much larger lowland moisture pool at the base of the Andes, a greater capacity to attenuate or delay forest decline in the face of diminishing moisture influx from extra-regional sources would appear plausible.

4. The role of local wind dynamics on temperature creep would also predictably differ between the two regions. The pediments, vertical walls and flat summits characterizing many of the highland formations in Guayana would not experience the same kind of daily wind flow adjustments that would be encountered in the alternating, ridge–valley relief (a trellis drainage pattern) that characterizes many of the eastern slopes of the Andes.

Whether local effects play a formative role in explaining elevational shifts in alpine and montane taxa, or simply represent minor spatial anomalies, will depend largely on the scale of impact discharged by longer-term external forcing factors that slowly shift the seasonal distribution and/or total amount of energy received at the surface (see below). Local and global processes deliver high and low frequency variations in temperature, respectively. The integration of these signals will ultimately define the conditions confronting inter-generational plant growth attributes (see Chapter 3), and the altitudinal migration and/or adaptation of cold and drought tolerant and intolerant species.

CRETACEOUS–PALAEOGENE TEMPERATURE AND MOISTURE Evaporites, laterite, bauxite, coal, kaolinite, Aeolian sandstone, carbonate and ironstone deposits are generally considered as strong indicators of prehistoric change in rainfall, evaporation, temperature and sea-level stand (e.g. Damuth and Kumar, 1975; Berrangé, 1977; Parrish *et al.*, 1982; Tardy, 1992; Gibbs and Barron, 1993; Ramón *et al.*, 2001; ODP, 2003) (see 'Sea-level change', below). Throughout the Cretaceous, the formation of ironstone, coal and bauxite deposits in the Guiana Shield suggests the region was warm and received, on average, rainfall only slightly less than modern levels. Around the boundary between Cretaceous and Tertiary periods (the K–T boundary), 70 million years BP, the geological, isotopic and palynological evidence indicates a more humid tropical climate for the Guiana Shield, one on par with modern conditions (Tardy *et al.*, 1990). Isotopic and palaeo-botanical evidence support the view that the amplification of this humid zone was most likely driven by a decrease in evaporation, rather than increase in rainfall. Evaporation would have dropped in concert with a decline in global temperature over the Maastrichtian.

Temperatures during the Eocene are believed to have been higher than at any other time during the Tertiary (Savin, 1977; Zachos *et al.*, 1994). During this epoch, extensive bauxitization occurred in the eastern Guiana Shield together with the formation of ironstone, kaolinite and coal deposits at various locations throughout northern South America (Prasad, 1983; Girard *et al.*, 2002). This high temperature–high humidity climate is believed to have dissipated during the Oligocene, about 30 million years ago, but was still capable of supporting further development of bauxite and laterite deposits in the region.

Longer-term climate forcing factors

Enveloping the wide range of oceanic, atmospheric and land-borne features that shape and influence the daily to decadal trek of rainfall and temperature change are processes operating at much longer timescales. These processes reflect on the internal energetics of our planet, the Sun and the astronomical relationship between these two bodies. Their planet-wide effects cut to the heart of the geophysical Earth.

But how would these ultimate forcing factors shape rainfall and temperature in the Guiana Shield? At low to very low frequencies, they change the basic amount and spatial distribution of insolation received in the region relative to the rest of the planet. Internal energetics of our planet fundamentally influence climate through two main mechanisms: volcanic emissions and crustal plate migration. Of these two, the influence of plate migration has the more substantial, but less punctuated, effect on climate in the Guiana Shield. Internal energetics of the Sun are also believed to affect change in the planet's climate, mainly through the frequency and magnitude of sunspots, flares and other phenomena that impinge upon our planet's magnetosphere, and amount and type of incoming radiation. Finally, cyclical variations in the astronomical relationship between Sun and Earth, called Milankovitch cycles, set up a series of slow-moving oscillations in the amount and distribution of energy reaching the planetary surface. No discussion of the engines driving variation in modern tropical forests should discount the impact of these subtle, but life-shaping, factors that establish the wider limits in which smaller processes, including the ascent of human predominance, ultimately operate.

Internal planetary energetics: plate tectonics

It is generally accepted, although not without uncertainties (Pratt, 2000), that the crustal surface of our planet is formed by a series of interconnecting, tectonic plates that are being continuously recycled along a circulating conveyor as destroyed surface crust moves downward through the mantle of underlying liquid rock and then rises again to harden into new crust. Recycling takes place mainly along the margins of each plate, subduction creating mantle from crust and sea-floor spreading zones creating crust from mantle. The rates of crustal subduction and sea-floor spreading vary between plates and between margins of the same plate. Differences in these rates create an unending, global shuffle of the continents, altering their geographical position, spawning volcanic chains, creating our tallest mountains and, through these, altering the effects of ocean and land surface on global climate. In the case of the oldest continental land masses, such as the Guiana Shield, the effect of plate tectonics spans billions of years. In relation to the evolution and development of modern tropical forests in this region, the action of plate dynamics since the early Cretaceous, approx. 135 million years ago, arguably embraces the most important period of influence (Raven and Axelrod, 1974; Romero, 1993).

EFFECTS ON GEOGRAPHICAL POSITION The latitudinal movement of the South American plate during the Cretaceous and into the Palaeogene coincided with an important period of forest evolutionary change in the Guiana Shield. Lowland forests dominated by conifers, ginkgos, cycads, ferns and early mammalian megafauna gave way to the modern-day forest flora and fauna. The most important impact of this movement on climate of the region was the change in geographic position of the Guiana Shield relative to the equator. It is well known from the fossil record and distribution of coal deposits that many regions now well-established in the temperate mid-latitudes once housed substantive, albeit primitive, tropical forests, as a consequence of plate movements carrying them equatorward. We also know that the modern distribution of land and ocean in relation to the equator plays a pivotal role in structuring the main mechanisms delivering neotropical rainfall and temperature, in particular the global system of interconnecting sea currents. Prior to the Cretaceous, the global oceanic circulation was characterized by circum-tropical surface current and a deep water system dominated by saline, rather than temperature, gradients. The break-up of the Gondwana supercontinent changed this by redistributing land mass towards the equator, shifting the main surface current feature from a circum-tropical to circum-polar position and a deep current system dominated by temperature rather than saline variation (Kennett and Barker, 1990). Growth of the Atlantic Ocean (through sea-floor spreading along the Mid-Atlantic Ridge), development of the northerly Caribbean and Cocos plates, initiation of the Andes and emergence of the Central American isthmus, created a ring of epic geological upheaval around the relatively inert Guiana Shield. The push-and-pull effect of activity along this ring and those of other plates is believed to have moved the Guiana Shield northwestward over the Cretaceous from a southern to northern hemispheric position, a latitudinal shift spanning approximately 13° (Dietz and Holden, 1970). By the late Palaeogene, 50 million years ago, the region is believed to have shifted to its present geographical position (Ziegler *et al.*, 1983). The Atlantic Ocean, however, had only grown to approximately half of its modern width in the northern tropics by the end of the Cretaceous (Sclater *et al.*, 1977). The changing distribution of land and ocean within the northern tropics would have created considerable adjustments in the way that coupled land–ocean–atmosphere processes affected climatic variation over the region, particularly influences of oceanic currents and SST migration on the formation and behaviour of Hadley and Walker circulations.

Given our basic understanding of how these processes affect modern-day climate in the shield region, a more southerly, landlocked position of the Guiana Shield prior to the Cretaceous break-up would foster a much drier, continental climate susceptible to strong polar air mass incursions during austral winters and a diminished influence of the ITCZ on seasonal rainfall as warm SSTs and the main equatorial trough remained largely north of the area. As the continent fractured, a narrow and shallow tropical Atlantic would have been much warmer and saltier than it is today (Leckie *et al.*, 2000). Mid-oceanic zones of sustained high pressure (anticyclones), which currently influence rainfall variation at multi-annual scales over the region through dipolar structuring of regional pressure systems, would have had more difficulty in forming and remaining positioned over a narrower Atlantic Ocean. The first of these, the south Tethyan subtropical high-pressure cell, probably did not develop until the Late Cretaceous as the southern Atlantic reached a width sufficient to accommodate the formation of a stable high-pressure centre (Parrish *et al.*, 1982). Subsequent establishment of the Caribbean Current through closure of the Central American Seaway and opening of the Pedro Channel along the Nicaraguan Rise during the Miocene represented the last crucial stage in the evolution of the western Atlantic/eastern Pacific basin geography and its influence on the global oceanic conveyor (Roth *et al.*, 2000). The separation of Atlantic and Pacific by the Panamanian Isthmus reorganized the thermohaline currents and thus would have changed the migratory responses of SSTs to the latitudinal swing of maximum insolation between northern and southern hemispheres each year.

EFFECTS OF VOLCANIC EMISSIONS Volcanic emissions are considered important climatic adjustors. Large eruption events inject considerable ash and aerosols into the stratosphere. The main effect of these large eruptions is an increase in the planetary albedo and therefore the amount of incoming solar radiation reflected back into space. The result is a cooling of the planet's surface for a period of several years following a major eruption (Angell and Korshover, 1985; Mass and Portman, 1989; Pyle, 1992). Cooling effects are small but consistent after major eruptions, registering a 0.1–0.2°C decrease in tropical temperatures up to 2 years after the event (Robock and Mao, 1995). An increase in planetary albedo occurs when the tremendous volume of sulphur dioxide emitted during blasts of very high magnitude is gradually transformed in the stratosphere into precipitating sulphuric acid aerosols. High concentrations of these aerosols also create conditions suitable for additional reactive chemistry that ultimately produce ozone-destroying molecules, and a consequent increase in UV light penetration.

Equatorial regions are the most important source of stratospheric volcanic aerosols. Most tropical eruptions do not create aerosol plumes that would lead to significant stratospheric aerosol loading, since the tropopause reaches its highest altitude over the tropics. However, the disproportionate concentration of major volcanic eruptions in the 0–30° region more than compensates for this higher plume height requirement (Halmer and Schminke, 2003). In the tropics, volcanic aerosols are quickly conveyed by the trade winds until they form a stratospheric belt of anomalously high aerosol concentrations above the equatorial regions with further dispersion into higher latitudes (Cadle *et al.*, 1976). Through this process of conveyance, source regions of eruptions registering high volcanic explosivity index (VEI) values in the tropics can rapidly impact the amount of insolation received by less active areas, such as the Guiana Shield. The eruption of Mount Pinatubo in the Philippines is an excellent example of a modern volcanic eruption with a very high VEI value that led to the formation of high concentrations of stratospheric aerosols during 1990–91 over the Guiana Shield (McCormick *et al.*, 1995). The injection of volcanic water vapour into the stratosphere may, however, counteract some of the negative radiative forcing effects caused by aerosol build-up (Joshi and Shine, 2003). Warming of the troposphere during EN periods has also been cited as a potential positive force that can mask the cooling effects of major volcanic eruptions (Robock and Mao, 1995). The effects of major eruptions on tropical rainfall have not been thoroughly assessed. Mass and Portman (1989) found no volcanic signal in their analysis of historical precipitation records. These minor cooling events may, however, decrease tropical evaporation, with consequent lagged effects on lowland moisture recycling mechanisms and run-off.

Lower magnitude eruptions often fail to inject gas or dust as high as the stratosphere and their resulting tropospheric emissions fail to remain airborne for long or travel as extensively, particularly across the lowland tropics where rainfall rates are very high. Eruptions in the active Andes–Central America–Caribbean region typically distribute aerosols and dust at tropospheric level in a northwestward direction (e.g. de Silva and Zielinski, 1998). Emissions from Caribbean volcanoes, therefore, can be deposited in the tropical forestlands of Central America. The absence of low-level tectonic activity southeastward of the Guiana Shield effectively eliminates a potent source of particulates and aerosols that can yield short-term impacts on local energy and nutrient budgets (Fig. 2.6A) (see Chapter 7).

Changes in the stratospheric influx of volcanic and anthropogenic aerosols, along with dust, alter the atmospheric optical depth (AOD). AOD is a good proxy indicator of changes in the amount of incoming solar radiation received at the surface as a consequence of periodic aerosol build-up.

Given the short-term effects of volcanic aerosol build-up on temperature, one can only conclude that stratospheric residence time of volcanic aerosols over the prehistoric Guiana Shield would have been most dependent on eruption frequency. Activation of the Greater Antilles, Panama–Costa Rica and Central American volcanic arc phases over a period of 45 million years during the late Cretaceous–early Tertiary would have brought more frequent volcanic eruptions and a plausible increase in volcanic aerosol control over short-term tropical temperature fluctuations during this period.

In the absence of long-term fluctuations in eruption frequencies, the most significant role of the relatively brief, but widespread, effect of volcanic eruptions on insolation levels would rest in its interaction with other climate forcing factors. The most notable of these interactions are attached to the modern relationship between volcanic eruptions and the ENSO. Based on an assessment of historical surface station temperatures and volcanic eruptions, Angell and Korshover (1985) and later Robock and Mao (1995) proposed that aerosol-induced cooling has been dampened by warming of the troposphere during EN phases, a conclusion similarly reached through modelling (Kirchner and Graf, 1995; Joshi and Shine, 2003). Given the strength of the historical EN–precipitation relationship in the Guiana Shield, it is most likely that large volcanic eruptions have combined in the past with other factors to modulate the consistency of this response. The massive eruption of El Chichon (Mexico, 17.4°N, 93.2°W) in 1982, for example, may assist in explaining the relatively modest precipitation response to the severe 1982–83 EN event recorded in Georgetown, Guyana compared with that recorded during 1997–98, an EN period without any major eruption events.[15]

Model results presented by Crowley (2000) suggest that, historically, volcanic emissions played a more prominent role in forcing temperature change prior to the onset of industrial greenhouse gas emissions around AD 1850. Between 22% and 49% of pre-industrial variation in a 1000-year temperature reconstruction could be explained by changes in ice-core derived, proximate measures ([SO_3]) of volcanic activity. At much longer timescales, Bryson (1988) proposed that the build-up of volcanic aerosols can modulate the effects of slow-changing planetary motion on prehistoric climate (see Milankovitch cycles, below). A 40,000-year reconstruction of atmospheric optical depth (AOD) estimates (Bryson, 2002) shows a strong increase during the LGM (Fig. 2.27), a period associated

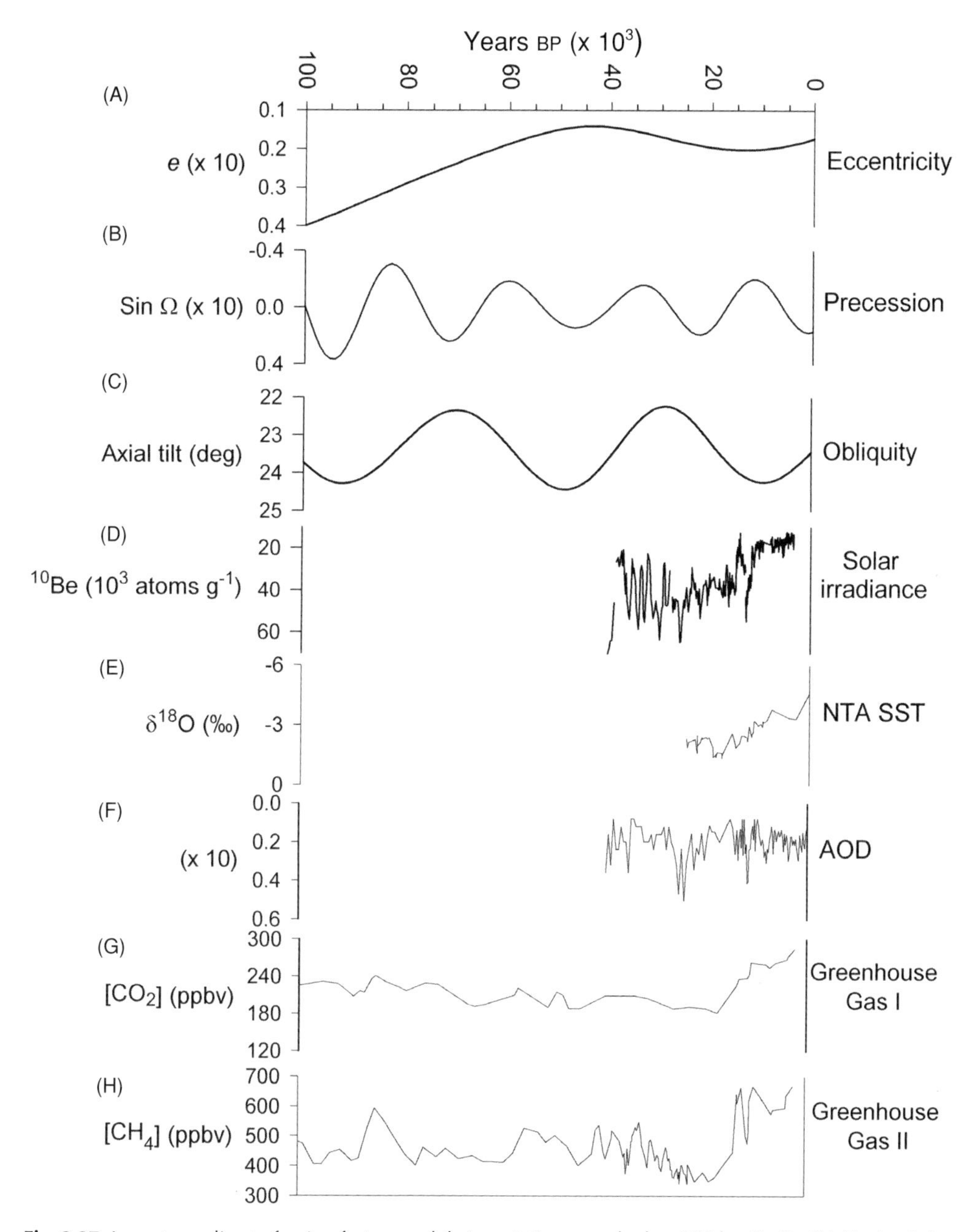

Fig. 2.27. Long-term climate forcing factors and their variation over the last 100 ka. (A, B, C) Milankovitch cycles (Berger and Loutre, 1991; Berger, 1992). (D) Solar irradiance inferred by variation in ^{10}Be concentrations from GISP-2 ice core (Finkel and Nishiizumi, 1997). (E) North tropical Atlantic SST variation based on Barbados coral d^{18}O proxy measure (Guilderson *et al.*, 1994, 2001a,b). (F) Variation in prehistoric aerosol optical depth (AOD) based on volcanicity index (Bryson, 1988, 2002). (G) Variation in atmospheric carbon dioxide (CO_2). (H) Methane (CH_4) concentrations based on Vostok ice cores samples (Petit *et al.*, 1999, 2001). Arrow direction indicates increasing temperature effects/responses. See references for details of data precision and assumptions.

with global cooling. Still other external forcing factors may interact with variation in volcanic aerosol build-up to exert a global-scale effect on climate change across the Guiana Shield. One of these forces, variation in solar irradiance, shows similar patterns of variation over historical and prehistoric time.

Internal solar energetics: solar cycles

In the early 1850s Rudolf Wolf, a Swiss astronomer, developed a tracking index of solar activity based on the number of sunspots observed either directly, or indirectly through the effects of varying solar winds on variation in the frequency of magnetic needle disturbances (Hoyt and Schatten, 1997). Sunspots had been recognized much earlier as a potential source of solar variability by other famous astronomers, but by 1868, it was Wolf who had reconstructed a time series of sunspots extending back to the year 1700. Building on the work of his predecessors, this series of observations, called the Wolf Sunspot Numbers (also later Zurich and Group Sunspot Numbers), revealed a repeating wave-like, 11-year cycle of variation between periods of maximum sunspot formation (Fig. 2.28).

Yet, up until the mid-20th century, most scientists still considered the amount of solar energy irradiating our planet as an unwavering constant, despite the record of Wolf Sunspot Numbers suggesting that solar activity was oscillatory. This was in part due to difficulties in detecting subtle shifts in the Sun's irradiance (Hoyt and Schatten, 1997).[16] Later experiments would provide further evidence in support of a variable solar output, but the pattern of activity established by Wolf and others was finally confirmed in the 1980s through a series of satellite-based observation missions operating over the two most recent solar cycles. The previously held total irradiance constant of 1367 W/m^2 has been observed to vary by as much as 5 W/m^2 since the 1978 mission launch (Willson and Hudson, 1991; Willson, 1997).

While confirmation of the solar cycle

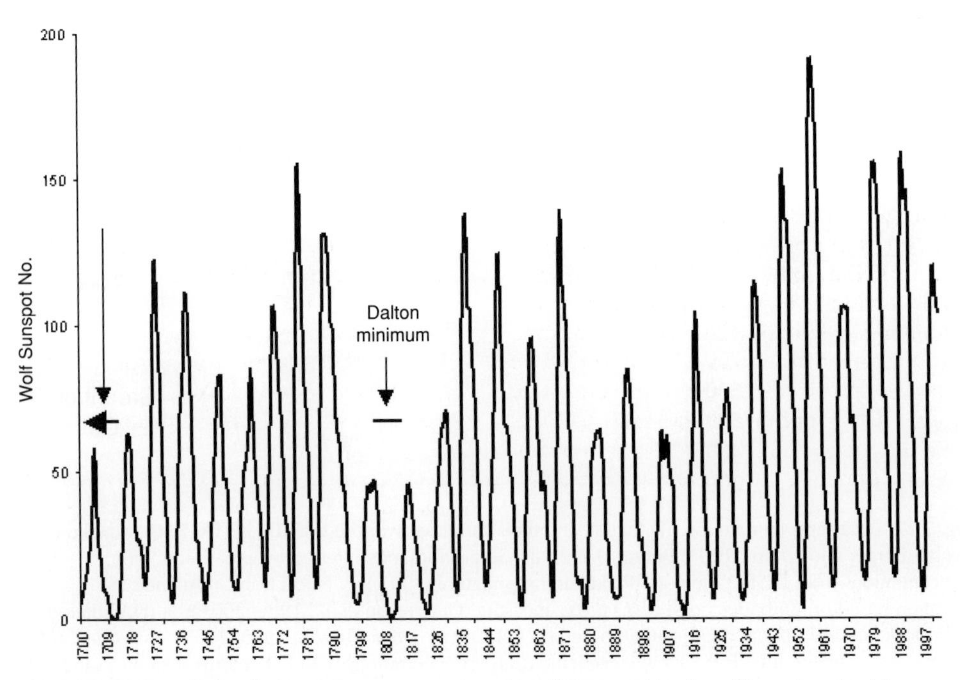

Fig. 2.28. Historical record of variation in mean annual Wolf Sunspot Numbers illustrating the 11-year cycle and local Maunder and Dalton minima (Waldmeier, 1961; NGDC, 2003).

through satellite-borne measurements has established sunspot activity as a potential climate forcing factor, documented variation is a mere 0.4% change in the amount of irradiant energy reaching the planetary atmosphere. This translates into a climate forcing effect that cannot directly account for historical changes in average temperature (Beer *et al.*, 2000). Calculated solar variation is also considered relatively small when compared to other forcing effects, such as measured increases in atmospheric CO_2 concentrations registered since 1850 (Fröhlich and Lean, 1998). A positive trend in surface temperature at scales greater than the 11-year solar cycle suggests that variability in the amount of solar irradiance at this timescale is not directly forcing climate change. It remains unclear, however, whether the amplitude and/or periodicity of the solar cycle measured most recently has remained the same over the past.

PALAEO-VARIATION IN SOLAR ACTIVITY Beyond the well-established 11-year Schwabe sunspot cycle, solar activity also fluctuates at lower frequencies as the amplitude of the 11-year cycle varies between cycles. The 22-year Hale,[17] 50–140-year Gleissberg and 170–260-year Suess cycles have received greatest attention among solar physicists and climatologists (Ogurtsov *et al.*, 2002; Tsiropoula, 2003). These medium-term fluctuations in solar output could exert a potent effect on tropical climate and, *inter alia*, modern tropical forest composition and distribution, because this timescale would embrace a physical landscape and taxonomic catalogue that is not substantively different from today. Since irradiance measurement series cover only the most recent solar cycles, proxy indicator patterns have been the main method used to examine medium-term variations in solar activity.

WOLF SUNSPOT NUMBERS – MULTI-DECADAL VARIATION The 300-year record of WSNs maintained by the Zurich Observatory strongly suggests considerable variation in solar cycle amplitude. The most important of these are identified by consistent low-amplitude Schwabe cycling over 3–6 cycles (30–70 years) or anomalously low sunspot minima. Historically, several such periods have been identified: (i) the Wolf grand minimum at approximately AD 1350; (ii) the Spörer Minimum (AD 1450–1540); (iii) the Maunder Minimum (AD 1645–1715) (Eddy, 1976); and (iv) the Dalton Minimum (AD 1795–1823) (Fig. 2.28).

COSMOGENIC ISOTOPES – MILLENNIAL VARIATION Varying concentrations of cosmogenic isotopes, such as beryllium-10 (^{10}Be) or carbon-14 (^{14}C), extracted from ice cores also point to considerable longer-term variation in solar irradiance (Beer *et al.*, 1993). Hoyt and Schatten (1997) urge caution, however, when using ^{10}Be to explore sun–climate relationships given its erratic correlation with historical WSN series. With caution in mind, and without considering collinearity with other climate forcing processes (see 'Obliquity' and 'Precession', below), a trend of increasing solar activity over the last 35,000 years can be tentatively postulated based on declining ^{10}Be concentrations in an ice core extracted from the Greenland ice sheet (Fig. 2.27D) (Finkel and Nishiizumi, 1997).[18] This trend embraces a number of important deviations that are synchronous with known periods of unusually high/low temperature throughout the Quaternary (Bond *et al.*, 2001; Shindell *et al.*, 2001). Perry and Hsu (2000) used harmonic progression to model variation of solar output over the last 90,000 years. They suggest that periodic variation within the last 90,000-year glacial cycle, as reconstructed by $^{16/18}$O ratios in deep sea sediment cores, fits well with the modelled variation in solar output based on a harmonic amplification of the basic 11-year sunspot cycle.

SOLAR EVOLUTION AND INCREASING LUMINOSITY At the largest timescale, it is well established that solar luminosity has increased by 30% since the birth of our planet (Hoyt and Schatten, 1997). Surface temperatures on Earth, on the other hand, have not increased in tandem with increasing luminosity. This illustrates the imperfect relationship between solar activity and surface temperature and emphasizes the important role of compensatory mechanisms (e.g. atmospheric water vapour) in regulating the amount of irradiance received at, and reflected from, the surface. The discrepancy between trends in solar luminosity and surface temperature indicate that solar activity cannot directly account for changes in planetary climate alone. While solar influence on climate is unequivocal, the impact may not be delivered directly through simple irradiation. Other forces are intervening to both amplify and attenuate variations in solar activity through a wide variety of backfeeding pathways. Thus the great change in luminosity that has occurred over the life of the Sun could be viewed as the founding force driving biophysical pools on Earth to undergo a long-term process of compensatory evolution. Compensatory evolution works towards energetic stability despite trending solar luminosity (i.e. the Gaia Hypothesis (Lovelock, 1979)). Tropical forests, through their role as a carbon pool, water store and albedo reducer, could be seen as both a product of and contributor to energetic adjustment in the face of trending change in solar irradiance over the last 40,000+ years.

PUTATIVE EFFECTS ON TEMPERATURE Without considering other compensatory mechanisms, average surface temperature should vary directly and in proportion to a change in solar activity. The analysis of Friis-Christensen and Lassen (1991) showed a strong correlation between variation in weighted Schwabe cycle length (9.7–11.8 years) and northern hemisphere land temperature anomalies over the last 140 years. Later, they extended this correlation back to AD 1500 (Lassen and Friis-Christensen, 1995).

This relationship does not necessarily reflect a direct causal linkage. The known level of solar variation (in W/m^2) is believed to be an order of magnitude smaller than that necessary to fully account for changes in surface temperature measured directly or through proxy indicator methods (Hoyt and Schatten, 1997; Rind, 2002). It could, however, reflect an initial forcing effect that is subsequently amplified or attenuated as a function of global climatic adjustment pathways. Recent research has suggested that the rather small variation in irradiance that accompanies movement through the solar cycle can trigger internal climatological adjustments that lead to more substantive shifts in temperature and rainfall via a wide range of intermediate components of the global climate system. Solar forcing of climate change via effects on stratospheric ozone concentrations in particular has stood out as one mechanistic explanation for oscillation of certain climate parameters (Haigh, 1996; Shindell *et al.*, 1999; Labitzke and Matthes, 2003). Dynamical models of the post-industrial warming period best fit observed temperature data when variation in solar activity, an increasing carbon dioxide concentration and major volcanic eruptions are incorporated (Hansen *et al.*, 1981; Schlesinger and Ramankutty, 1992; Thomson, 1997; Fröhlich and Lean, 1998; Meehl *et al.*, 2003; Stott *et al.*, 2003). Models examining the pre-industrial Holocene also indicate that solar forcing acts as one of several, if not the most, important factors forcing temperature change (Crowley, 2000). Crowley's energy balance climate model attributed 18–20% of reconstructed pre-industrial temperature changes to solar influences, based on variation in ^{10}Be data series, and 9–45% variation based on ^{14}C data.

Pegging palaeo-temperature change across the Guiana Shield as a consequence of solar variation is difficult. This is due in part to a dearth of data at observation frequencies sufficient to resolve a linkage between temperature variation in the region with known patterns of solar activity. Connecting neotropical change with better understood temperature changes at higher latitudes is also complicated due to complex conveyance dynamics within the Atlantic and eastern Pacific sectors (see sections on Atmosphere, Ocean and Land-borne components, above). An incomplete knowledge of energy transfer mechanisms linking low and high latitudes also cautions against assuming that there is a zonally synchronous response to lower-frequency variations in solar activity.

Rind and Overpeck (1993) constructed a GCM to examine the spatial variation in temperature change after prolonged solar irradiance dampening during periods such as the Maunder Minimum. Their results suggest that the Guiana Shield would have experienced a temperature decline of less than 0.5°C. It also suggests that southwestern Amazonia and parts of tropical West Africa may have experienced a more substantial temperature depression as a consequence of variation in solar activity. A more recent dynamic model constructed by Shindell *et al.* (2001) also shows a statistically significant relationship between projected decline in average temperature (–0.2 to –0.35°C) over the Guiana Shield and decline in sunspot activity during the 18th-century Maunder Minimum, but again the direct temperature response over the region appears modest in comparison to upper latitudes. Estimating longer-term temperature changes in the Guiana Shield due to solar variation alone is difficult. Cosmogenic isotope abundance can provide an estimate of varying solar activity at a millennial scale, but how and to what extent this leads to tropical temperature change remains poorly resolved.

PUTATIVE EFFECTS ON RAINFALL The impacts of varying solar activity on prehistoric rainfall in the Guiana Shield are most plausibly linked to changes in ITCZ behaviour over the region. The ITCZ is the main delivery mechanism for rainfall over most of the equatorial tropics. As part of the zonally driven Hadley Circulation, it is highly sensitive to oceanic and atmospheric conveyance mechanisms operating at higher latitudes.

Several studies have also concluded that the direct impact of solar forcing may vary by region, based on spatial variation in atmospheric scattering and absorption characteristics. Anti-cyclonic zones of high pressure that tend to reside over mid-ocean regions have been identified as important regional response zones. These stable areas of high pressure remain cloud-free and through this condition are most responsive to increases in solar irradiance. GCM results of Meehl *et al.* (2003) suggest that this maximum exposure to irradiance change enhances evaporation in these areas. This moisture is then advected towards the main tropical rainfall-producing zones, intensifying the Hadley and Walker circulations. In the case of the Guiana Shield, this would suggest that positive phases of the Schwabe cycle may bring enhanced rainfall over the eastern region most affected by subtropical oceanic moisture sources. By contrast, the authors suggest that this same pattern of solar variation would likely reduce rainfall over the Caribbean and Central America as intensification leads to further expansion of the subtropical high pressure zone along the descending arm of the Hadley Cell. Interestingly, Bond *et al.* (2001) concluded that a strong correlation between proxy measures of north Atlantic drift-ice abundance and periods of reduced solar irradiance was explained by heightened ozone production, consequent high-latitude cooling and a decrease in the intensity of the northern arm of the Hadley Circulation. They suggest that this dynamic adjustment process led to reduced precipitation in the (sub-)tropics during periods of reduced solar activity.

Taken alone, longer-term positive trends in solar irradiance would plausibly lead to considerable change in the magnitude of rainfall received in the region. But variation in atmospheric and oceanic responses combined with the effects of other forcing factors work to amplify and attenuate the magnitude of solar effects. While regional climatic effects of variation in solar output over decadal, centennial and millennial scales remain poorly understood, current research suggests that this external forcing factor has exerted a significant effect on past precipitation patterns over the Guiana Shield.

Solar–planetary relationships: Milankovitch Cycles

Another significant set of three external climate forcing factors, Milankovitch cycles, oscillate at ultra-low frequencies. These three external sources of variation, referred to as (i) equinoctial precession, (ii) obliquity, and (iii) orbital eccentricity, are linked to the way in which our planet moves in space relative to other planets and the Sun. Their impact is delivered through changes in both the amount and distribution of insolation received at the surface of our planet. Variations due to Milankovitch cycling occur very slowly. So slowly in fact that changes in precession, obliquity and eccentricity over the last 1000 years embrace only 0.7%, 5.2% and 0.6% of the total breadth of variation that has occurred since 5 million years BP, respectively (see Figs 2.27 and 2.28). The crucial climatic impact of these rhythms, however, rests with the fact that their low rates of change also sustain longstanding periods when the distribution and intensity of solar irradiance remains at relatively low and high levels.

OBLIQUITY Thermal seasonality that characterizes temperate and polar regions is largely created by a tilt of the spin axis linking the Earth's poles relative to the Sun. As our planet orbits the Sun, this axial tilt, or obliquity, leads to a pole-to-pole shift in the spatial distribution of maximum and minimum insolation. As a consequence, boreal and austral winters occur at opposite solstices and amplify toward the poles. Tropical latitudes are least affected, since this region of the planet remains well insolated regardless of orbital position.

Obliquity is a cyclical change in the angle of axial tilt. Earth's polar axis is currently tilted at 23.5° relative to the plane of the ecliptic, but this can vary between 22° and 24.5° (Fig. 2.27). A full shift between these limits occurs over a period of approximately 40,000 years. Since the tilt angle determines the latitudinal distribution of insolation across our planet over the course of each orbital pass, a change in this angle alters the energetic intensity of seasonality. It also would change the latitudinal limits of climate that could be considered tropical. As tilting becomes less severe, the latitudinal band of area that remains fully insolated throughout the year would expand. When the angle is increased, the tropical belt would narrow. Both contraction and expansion would be less than the change in axial tilt due to the effect of a decreasing incidence angle on insolation as latitude increases. As a result, the equatorial zone is not affected by changes in axial tilt. Obliquity will, however, force a change in the slope of the thermal gradient between poles and equator, an important anisotropy that drives transfer components of the global climate system. This intensifies seasonality at high latitudes.

PRECESSION Earth orbits the Sun along a slightly offset, elliptic path that controls the intensity of seasonal differences created through axial tilt. This orbital asymmetry is best described by two points along the path: one that positions Earth closest (perihelion) and another farthest (aphelion) from the Sun each year.

The orientation of the semi-major axis defined by these two points is not stable. Earth's orbital path is the product of competing gravitational forces exerted by the Sun and other planets within the solar system. The largest of these planets, Jupiter, exerts a particularly strong gravitational effect. This effect works to change the alignment of the perihelion as Jupiter moves through its orbit around the Sun at a much slower rate than Earth, changing the gravitational alignment between the Sun, Earth and Jupiter. This change in alignment 'drags' the perihelion around the orbital path, altering the relationship between seasonal timing (caused by axial tilt) and seasonal intensity (caused by the distance from the Sun).

At the same time, orientation of Earth's rotation axis is also moving by about 0.5° per century. This cyclical change, along a path best described as a 'conical wobble', creates further variation in the relationship between seasonal timing and intensity. This precession of the rotation axis combines with perturbations to the alignment of the orbital ellipse to govern the severity of seasons over the cycle. Forces creating precessional wobble are akin to an external object touching the side of a fast-spinning top. In this case, the external object is the gravitational torque exerted by the Sun and Moon on the rotating Earth.

It is not difficult to imagine that when the Earth reaches perihelion during the boreal winter, the decrease in distance between Sun and Earth will act to moderate the seasonal decline in insolation caused by axial tilt. When the perihelion shifts towards the solstice of boreal summer, seasonality intensifies as summers receive more insolation and winters (at the aphelion) receive less insolation. This is precisely the situation 11,000 years BP, when the perihelion was reached around June (Fig. 2.27). Currently, perihelion is reached in January and over the last millennium it has been aligned with boreal winter months, moderating temperature decline during this season. As the alignment of the rotation axis changes orientation through a full 'wobble' cycle, seasonal alignment of the perihelion occurs. A full cycle is believed to take approximately 22,000 years to complete, but can vary between 18 and 23 ka due to less predictable changes in orbital orientation.

ORBITAL ECCENTRICITY Earth's orbital path is not only subject to cyclical realignment of the semi-major axis, but also to variation in the ratio of major to minor axis length. In other words, the shape of the orbit continually oscillates between a perfect circle (eccentricity=0) and a parabola (eccentricity=1), but with a far more circular than parabolic shape to the ellipse (0–0.6) (Fig. 2.27). This pattern of variation taken alone alters the absolute distance between Earth and Sun. At lower eccentricity values, the orbital path is more circular and the difference in distance at perihelion and aphelion approaches zero. At higher eccentricity values, this difference increases. As a consequence, when orbital eccentricity is more elliptic, the distance between Earth and Sun is increased, and the amount of insolation is reduced (Fig. 2.29).

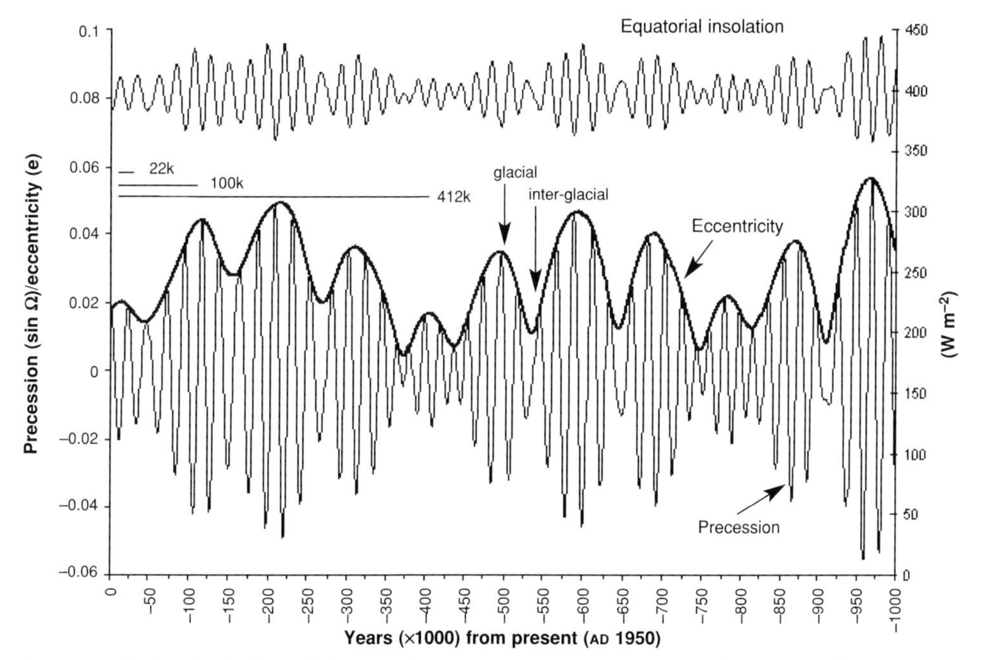

Fig. 2.29. Calculated Milankovitch forcing effects (precession, eccentricity) on the amount of incoming solar radiation received along the equator over the last 1 million years. Lengths of 22 ka, 100 ka and 412 ka represent glacial cycle periodicities. Note how peak glacial stadia associated with the 412 ka cycle are associated with higher amplitudinal swings in radiation and our current position (far left) leaving the LGM at a point of modest precessional forcing and relatively low solar flux variation in the tropics.

Orbital eccentricity is believed to account for a mere 2 W/m^2 ranging in received solar radiation over the course of a cycle. Two cycles have been identified for eccentricity variations, one at 100,000 and another at 412,000 years (Fig. 2.29). Earth's orbit currently is more circular than elliptic and has largely been so over the last 50,000 years, but with a relatively small increase in ellipticity around the time of the Last Glacial Maximum and onset of the current interglacial period (Figs 2.27 and 2.29).

ASTRONOMICAL THEORY OF CLIMATE CHANGE AND GLACIAL CYCLES Of the three Milankovitch cycles, only eccentricity variations affect changes in the total amount of solar radiation received. Obliquity and precession shape the distribution of incoming radiation, but not the total amount. Considered independently, each cycle creates relatively minor changes in the distribution and amount of insolation. However, taken together, all three cycles can produce severely contrasting scenarios. Precessional effects work to alter the timing of the seasons created by axial tilt, but wobbling of the rotation axis along a circular orbit would not alter the slope of the latitudinal insolation gradient, only the calendric timing of the shift between northern and southern hemispheres. The key rests with the addition of orbital eccentricity. As the orbit becomes more elliptic, the change in distance at perihelion/aphelion works with precessional and obliquity movements to amplify and attenuate effects on planetary climate. These effects have been shown to be most pronounced at high latitudes, but significant variation of insolation in the equatorial regions is also expected (Fig. 2.29).

The most significant change associated with high-latitude effects over the last 2.5 million years is believed to be a fluctuation in the amount of planetary water stored as ice. From 2.5 to 1.5 Ma BP, glacial maxima are believed to have occurred around every 41,000 years. At least ten ice ages are believed to have occurred over the last one million years, but with the frequency slowing to every 100,000 years, based on varying oxygen isotope ratios in sampled sediment cores (Hays *et al.*, 1976). These fluctuating periods of glacial expansion (stadia) and retreat (interstadia, interglacials) have been associated with precession, obliquity and eccentricity periodicities since the early 20th century (Milankovitch, 1941). Variation contained in a large number of sedimentary (e.g. Olsen and Kent, 1996; Scarparo Cunha and Koutsoukos, 2001), isotopic (e.g. Rial, 1999; Stirling *et al.*, 2001) and palynological records (e.g. Mayle *et al.*, 2000; Bush *et al.*, 2002) taken at both low and high latitudes appears consistent with Milankovitch cycle periodicities. The theory, despite its power to neatly explain variation in a wide array of biophysical processes linked to climate and glaciation, cannot account for all variation encountered over the composite proxy record of palaeoclimatic change (Imbrie and Imbrie, 1980). Nor has a 40,000-year range in glacial cycle duration over the Plio-Pleistocene been fully explained (Elkibbi and Rial, 2001), although Rial (1999) gives a compelling explanation based on frequency modulation of the 100 ka eccentricity cycle. In some instances, different cyclical beats may be shaping different aspects of environmental change during different climate phases (Muller and MacDonald, 1997), creating non-linear responses. Logically, the effects of varying internal and external forces are interacting within astronomical cycles to create a unique sequence, a prehistory, of conditions that trigger, amplify, attenuate and terminate climatic phases along a non-replicable path of planetary evolution.

IMPACTS ON GUIANA SHIELD PALAEO-CLIMATE The changing configuration of Milankovitch cycle phases is believed to have had its greatest impact at high latitudes, particularly in its regulatory effects on ambient temperature, polar ice formation and all that these influence. While past insolation changes may have had the weakest effect on the tropics, it also moved the slope of the latitudinal insolation gradient, altering the pace and extent of both oceanic conveyance and atmospheric advection processes. The considerable role that these transfer linkages between high and low latitudes play in shaping regional climates (see 'Remote forcing factors', above) suggests substantive consequences for low-latitude areas such as the Guiana Shield. At synoptic and larger scales, this is likely to have occurred through changes in the seasonal location and intensity of the ITCZ and SST anomalies. These mechanisms are inextricably bound to one another through a system of cascading feedback, so cannot be viewed as independent responses. However, they do exert different levels of impact upon regional climates across the neotropics.

The ITCZ is the principal mechanism that delivers strong seasonal rainfall across the Guiana Shield as it migrates with seasonal shifts in hemispheric insolation and SST belts. Anomalous shifts in ITCZ migratory behaviour, often driven by ENSO, currently bring anomalously high and low rainfall to the Guiana Shield at scales of weeks to years. At much greater timescales, shifts in the distribution of insolation due to Milankovitch cycling would have substantially altered the zonal limits of ITCZ ranging and the annual to multi-annual precipitation regimes across the region. Martin *et al.*'s (1997) comprehensive reconstruction of palaeoclimate over the neotropics indicates a +10° latitudinal extension of the ITCZ limit southward since the early Holocene/late Pleistocene as a consequence of astronomical forcing. Palynological evidence from the lowland tropics of Bolivia also suggests that tropical forests reached their modern-day extension into southwest Amazonia as late as 3000 BP, a consequence of an astronomically forced southward migration of the ITCZ (Mayle *et al.*, 2000). Variations in terrigenous titanium concentrations in the Cariaco Basin, Venezuela suggest that the ITCZ has been migrating further southward across South America over the last 10,000 years since the early Holocene 'Altithermal' (Haug *et al.*, 2001), a change that is consistent with variation in Milankovitch cycling (Fig. 2.27). Palaeolimnological data from two Amazonian lakes located in the western Guiana Shield (Amazonas, Brazil) show variations in potassium concentrations, an indicator of water level and leakage fluctuations, that are consistent with precessional-scale (19–22 ka cyclical) influences and a wetter LGM over north (western) Amazonia (Bush *et al.*, 2002) that would accompany a reduced poleward ranging of the ITCZ. A more northerly ranging of the ITCZ would also lead to more sustained polar incursions across southern Amazonia and more frequent seasonal cold mass surges across the equator. These would substantively reduce the annual temperatures across these areas, limiting growth and survival of most lowland tropical forest species.

These field data are consistent with GCM results that predict substantial changes in moisture and temperature across the neotropics. Reconstructions using the GENESIS model depict substantial regional responses to astronomical and other forcing effects during the LGM, 18,000–22,000 years ago (Clark *et al.*, 1999, Fig. 2 – LGM Anomaly). Climatic features over the Guiana Shield reconstructed as part of Clark *et al.*'s global assessment show a –2 (–4)°C to +4 (0)°C difference between modern and LGM boreal (austral) summer surface temperatures. In contrast, reconstructed surface temperatures over the adjacent western Amazonian (–8 (–8)°C to –2 (–2)°C) and Central American (–1 (–2)°C to –4 (–8)°C) regions during LGM boreal (austral) summers were noticeably cooler. Reconstructed net rainfall (precipitation–evaporation) rates indicate a much drier Guiana Shield relative to western Amazonia and Central America. Higher temperatures and lower rainfall over an LGM Guiana Shield are consistent with a more substantial decrease in cloud cover over the region relative to the other regions, a condition that parallels modern-day regional differences. LGM climate reconstructions using the CLIMBER-2 model show similar ranges and trends in temperature (–4°C to –6°C) and rain rate (–1 to 0.2 mm/day) changes across the neotropics but without sufficient spatial resolution to differentiate between regions (Ganopolski *et al.*, 1998, Fig. 6).

Polar cooling during peak glacial stages in the past is also believed to have been strengthened through enhanced ice albedo, prompting a further steepening of the equator-to-pole thermal gradient. Zonal steepening increases the strength of trade winds that convey energy and moisture towards the low pressure, equatorial trough (Trend-Staid and Prell, 2002). Trade winds also act to deepen the thermocline along the western rim of tropical oceans and promote upwelling of cold deep waters along equatorial zones (Bush and Philander, 1998). A substantial strengthening of these winds during season-intensifying Milankovitch phases would extend upwelling further across the basins and increase entrainment of colder subtropical waters. This process, reminiscent of strong La Niña ENSO phases, has repeated itself periodically at frequencies consistent with astronomical forcing effects. Prehistoric changes in wind gradients and seawater movement as a result of Milankovitch-driven insolation shifts are reflected in sea sediments recording variation in abundance of tropical organisms. Abundance of many tropical marine organisms is strongly responsive to changing seawater temperatures (e.g. algal blooms), and these have been shown to vary at frequencies consistent with astronomical forcing effects (McIntyre and Molfino, 1996; Perks *et al.*, 2002).

In the Atlantic, the net effect during glacial maxima would be a considerable decline in tropical and subtropical SSTs that normally drive both ITCZ and non-ITCZ precipitation across the eastern Guiana Shield, eastern Central America and the Lesser Antilles. A reduction in tropical SST would also have acted to further reduce temperature along the eastern rim of the Guiana Shield and Caribbean. Zonal transport of warm waters to higher latitudes would cease to moderate Milankovitch cycle insolation effects on temperature and thus act in concert with increases in ice sheet albedo to amplify temperature declines that are then conveyed back to the tropics (Bush and Philander, 1998; Ganopolski *et al.*, 1998; Clark *et al.*, 1999).

Prehistoric oceanic and atmospheric conditions affecting climate over the Guiana Shield were clearly very different. Results employing a wide range of climate change analogues and model assumptions show little uniform temperature depression over the prehistoric lowland neotropics as different land–ocean–atmosphere relationships altered responses to global-scale external forcing effects. Ultimately, however, it is the change in the amount, timing and spatial distribution of insolation regulated by astronomical forcing effects that governs the cascading adjustment chain and the way in which it has altered palaeoclimate across the Guiana Shield and adjacent regions of the neotropics.

Greenhouse gas concentrations

Of all the potential contributors to past and present climate change, greenhouse gases are perhaps the most widely acknowledged. Of these, the historic variation and forcing effects of carbon dioxide (CO_2), methane (CH_4) and nitrous oxide (N_2O) are arguably the best documented and most widely publicized, although other trace gases contribute (e.g. OH) and water vapour (i.e. cloud cover) is without doubt the most important in creating the greenhouse effect due to its major effects on deflecting ISR and trapping OLR.

Direct measurements of modern CO_2 concentrations in the atmosphere show that these have been increasing steadily and at a higher rate since 1950 and that this increase is largely attributed to anthropogenic emissions derived from agriculture, deforestation, fossil fuel burning and CO_2-producing industrial processes, such as cement manufacturing (Mackenzie and Mackenzie, 1995). Atmospheric CO_2 levels, however, also increase as a consequence of natural processes, such as changes in primary production (plants and phytoplankton), carbonate-rock formation and weathering, volcanic eruptions and changing SSTs, among others. These pre-industrial sources of variation alone are believed to have altered atmospheric concentrations of CO_2 substantively. Proximate measures of variation in CO_2 concentrations over the last

100 ka drawn from Antarctic (Vostok, Taylor Dome) ice core data show massive increases in both CO_2 (64%) and CH_4 (54%) over the 18 ka since the end of the LGM and prior to the onset of the Industrial Revolution (Fig. 2.27 – Greenhouse Gas I and II) (Petit *et al.*, 2001). The Vostok record extends back 430 ka, and several other periods with greenhouse gas concentrations as high or higher than those estimated for the modern pre-industrial period are evident around 125 ka, 325 ka and 415 ka BP. However, none of the ice core-derived measurements of prehistoric CO_2 or CH_4 equals or exceeds measured post-industrial levels.[19]

Unlike Milankovitch forcing that alters the meridional distribution and amount of insolation, the effect of changes in well-mixed greenhouse gases on the planetary energy balance is believed to have a relatively uniform spatial effect on temperature. In this case, mean annual air temperatures should be increasing across the Guiana Shield in line with global estimates. The IPCC calculates a global increase of 0.6±0.2°C in mean temperature over the last 100 years. This is consistent with the positive trend in Brazilian Amazon temperatures of +0.56°C per century calculated by Victoria *et al.* (1998) based on station records for the period 1913–1995 and spatially averaged temperature change for an area encompassing the Guiana Shield as a whole (Fig. 2.21).

Other external forcing factors

Past climatic changes over the Guiana Shield could also have been associated with other external forcing factors that are not traditionally incorporated in climate reconstruction efforts.

METEORIC IMPACTS One of these, meteoric impacts, has been invoked as a cause of the K–T extinction event. Meteoric impacts of sufficient magnitude would generate a period of cooling that is consistent with a phase of tremendous volcanic activity. Recent coring on the submerged eastern rim of the shield (the Demerara Rise) has uncovered 1–2 cm ejecta layers consistent with the massive K–T Chicxulub impact site off the coast of the Yucatán Peninsula (Fig. F14 in ODP, 2003).

WEAKENING OF THE MAGNETOSPHERE A reversal of geomagnetic polarity due to changes in the direction of flow within the Earth's molten internal core/inner mantle could also impinge substantially upon global and regional climate. Recent reconstructions suggest that the reversal process occurs as a process of field weakening (Glatzmaiers *et al.*, 1999). During a reversal, it is believed that the dipole circulation that typifies the modern magnetosphere slowly weakens and gives way to a period dominated by a polypolar field before re-establishing a reversed, dipole-dominated field (e.g. Li *et al.*, 2002). These openings are associated with a much higher amount of cosmic radiation at wavelengths that would fundamentally alter biological productivity, carbon storage, water cycling and sensible heat fluxes. How regional and global climate would adjust to this weak phase of the magnetic reversal is unclear. A chronology of this reversal process is well documented in the geological record, however, and a reversal of the magnetic pole is believed to have occurred several hundred times since the separation of South America from Africa and the expected beginning of angiosperm evolution (Kent and Gradstein, 1985). Its role in defining the palaeochronology of climate change cannot be discounted.

NON-VOLCANIC AEROSOLS Most of these are anthropogenic but, like volcanic emissions, have a relatively short atmospheric residence time and are deposited largely through dry (wind) and wet (rain/snow) precipitation. Their effects are varied. Deposited soot (black carbon) can reduce ice albedo and enhance glacial melting, while sulphates in the troposphere are believed to have a general cooling effect on regional climates. Anthropogenic sulphate concentration over the Guiana Shield is estimated to be one of the lowest globally and on par with levels more typical of austral high latitudes (Charlson *et al.*, 1991). Concentrations increase westward across

Amazonia, Central America and the Peruvian/Ecuadorian Andes based on Charlson *et al.*'s model, but remain low relative to north temperate centres of high concentrations. With little industry located eastward of the region, the main sources of allochthonous mineral aerosols are borne along African easterly wave fronts from the deserts of North Africa (Jones *et al.*, 2003).

Simple matter of rainfall?

The breadth and complexity of relationships ultimately responsible for characterizing rainfall over the Guiana Shield is difficult to embrace in even the most sophisticated climate models. Long-term variability in rainfall is strongly sensitive to spatial scale, and this sensitivity is at its greatest over the tropics (Giorgi, 2002), suggesting that there is a fractal-like scalar shaping the climate topology. Nonetheless, it is increasingly possible to chain together these eclectic relationships and draw several general conclusions about modern and prehistoric patterns of climate in the Guiana Shield and adjacent regions.

MODERN RAINFALL PATTERNS Firstly, it is clear that rainfall varies across the Guiana Shield in a manner that emphasizes the different role of oceanic and land-borne sources of moisture (a meridional gradient) and zonal variation in peak insolation (a zonal gradient). In the east, oceanic influences play an important role in a region highly sensitive to changes in coupled atmosphere–ocean processes, such as ENSO. Eastern shield forests are buffered from more severe rainfall seasonality by these maritime influences and a southward movement of the ITCZ over the western Atlantic that strongly lags the continental pattern. In the Savanna Trough, rainfall is most seasonal as the ITCZ reaches its maximum southward migration along this longitudinal swathe. Penetration of the Azores high pressure system behind this seasonal movement of the ITCZ southward interacts with land-borne feedbacks, such as increased albedo, southward-facing drainage bevel and reduced surface roughness, that amplify seasonal rainfall decline. The northern arm of this central region is interrupted by the Guayana highland archipelago that brings high local rainfall through orographic uplift. In the west, ITCZ ranging is more limited. Precipitation recycling occurs along a system characterized by the Andean orographic wall, an extensive forested landscape characterized by high surface roughness, high evapotranspiration, high west-to-east surface run-off and low albedo. Nonetheless, local topography, soils and land-use can substantively alter rainfall patterns associated with larger-scale circulatory features, creating local zones of anomalously high and low rainfall.

The variation in rainfall across the Guiana Shield straddles the crossroad of much larger meridional and zonal gradients driven by the seasonal swing of maximum insolation between hemispheres and a transition from marine- to terrestrial-driven systems. South and north of the region, rainfall becomes more seasonal as the range limits of the ITCZ are reached and the subsidence branches of the Hadley Circulation dominate rainfall. Westward, rainfall generally continues to increase as a consequence of higher rainfall rates and dampened seasonal decline. This increase is driven by a narrower seasonal ranging of the ITCZ and more substantial precipitation recycling fostered by sustained surface in-flow from the Andes. Both maintain high water tables, persistent cloud cover and steady evapotranspiration linked to an extensive forest-land system. Declining ENSO effects on western Amazonia rainfall further enhance inter-annual stability.

Temperature variations across the Guiana Shield and adjacent regions are largely shaped by daily and seasonal insolation cycling, topography and elevational isolation, proximity to the western Atlantic, increasing greenhouse gas concentrations and the interaction of these with the movement of the ITCZ, mainly via changes in cloud cover and equatorward incursion of cold polar fronts.

PAST RAINFALL PATTERNS The neotropics have not responded to global climate changes uniformly in the past and are unlikely to do so in the future. Sediment core data and GCM results do not support the view that palaeo-patterns of neotropical rainfall can be deduced simply by applying a uniform reduction in rainfall along modern isohyets or temperature along modern isotherms. This approach increasingly appears counterintuitive in light of evidence emphasizing a coupled land–ocean–atmosphere system dominated by short- and long-ranging transfer components.

Evidence also suggests that northwestern Amazonia experienced more moderate rainfall reductions during past climatic phases characterized by weakened Hadley and Walker Circulations and/or strongly altered ITCZ migratory behaviour. Most notable among these to occur in recent geological time is the Last Glacial Maximum. The Guiana Shield, in contrast, appears to be susceptible to more severe rainfall decline during these periods, possibly due to the greater influence of changing SSTs on regional climate combined with land-borne attributes (topography, soils) that move more rapidly to amplify ocean-derived changes to rainfall delivery. A substantial decline in sea level during the LGM would have also left current coastal regions of the Guiana Shield (the Guianas, Bolivar and Amapá states) much further inland (see 'Sea-level change', below). This may help to explain, from a mechanistic perspective, the discordance among palaeoclimate reconstructions derived from fossil pollen studies (Fig. 2.25), although local site attributes are also likely to prove a strong source of variation.

The early Holocene saw a substantial change in neotropical climate as ITCZ movement southward declined, sea level rose towards its current maximum and high temperatures and rainfall prevailed across much of the modern Guiana Shield. The influence of ENSO also began to increase, with pronounced inter-annual variation in rainfall and temperature across the shield area. Major volcanic eruptions also would have led to lower ISR levels and short-term anomalies in zonal movement of the ITCZ, cloud cover and rainfall.

MODERN VARIATION IN TEMPERATURE Variability in temperature is easier to define since this typically shows a much weaker sensitivity to spatial scale than rainfall (Giorgi, 2002). Given the very minor variation of insolation across the neotropics, other factors that subsequently alter available energy largely account for any broad spatial variation in daily maximum–minimum temperatures. The most significant of these appear to be: (i) proximity to warm oceanic currents; (ii) cloud cover persistence; and (iii) elevation.

The eastern Guiana Shield, Caribbean and Central American climate is buffered by the northward movement of the warm North Brazilian–Guiana Current and the long residence of warm waters forming the Western Hemisphere Warm Pool along the northern coast of South America (Fig. 2.22C). Cloud cover persists longer over western reaches of Amazonia and the Guiana Shield compared to eastern regions (Fig. 2.23) and this exerts a stronger modulation of diurnal temperature fluctuations throughout the year. Elevation is the strongest contributor to spatial variation in mean daily temperature across the Guiana Shield, with upper elevations in the Guayana Highlands showing the lowest daily minima. Polar incursions from southwestern South America, the *friagems*, can lead to short periods of much lower temperatures in western Amazonia, but these only rarely reach into the Guiana Shield area. Average temperatures across the Guiana Shield appear to be increasing over the last century in line with global warming predictions.

PAST VARIATION IN TEMPERATURE Prehistoric changes in neotropical temperatures, like those for rainfall, have also been proposed for the lowland forest regions. An increasing body of results, generated through both sediment cores and circulation models, reinforces the view derived from palynological studies that a temperature depression across the neotropics occurred during the LGM and temperatures have been generally increasing over the last 10,000 years. Several circulation models have produced results that indicate temperatures across the eastern Guiana Shield may have dropped less than in adjacent lowland regions of western Amazonia or Central America, probably due to declining tropical SSTs and strengthened high pressure systems that decrease cloud cover residence over the area. A wide range of field and model results suggests that precessional forcing of global insolation patterns is largely responsible for altering the coupled ocean–atmosphere influences on climate in the region at glacial timescales of 100–400 ka. The likelihood of a uniform variation of temperature across the Guiana Shield and the extended neotropics appears small when considering the influence of these coupled effects combined with local landscape attributes, such as topographic relief.

PRIMARY DETERMINANTS OF VARIATION Clearly many factors interact to bring about a change in rainfall and temperature over the Guiana Shield. However, several factors stand out among the most prominent. Most obvious, and perhaps least helpful, is geographic position. It encapsulates the integrated effect of land–ocean–atmosphere linkages over a wide range of time and spatial scales. The bulk of climatic research shows that the relative importance of different climate-governing mechanisms varies with geographic position and scale across the neotropics. Geographic position, therefore, is best viewed as a non-stationary indicator of any single climatic attribute, such as temperature or rainfall, since both larger-scale, high-energy features of the global climate, such as ENSO forcing, and smaller-scale, low-energy landscape attributes, such as topography, vary anisotropically, but not collinearly.

Changes in the ranging behaviour of the ITCZ are particularly powerful in linking many of the most significant forcing factors and the transfer linkages between land, sea and air that govern climate over the Guiana Shield. Changes in timing and extent of its latitudinal migration have been linked to variation in both drivers and conditions of modern, historical and prehistoric climate, such as ENSO, solar cycling, SSTs, cloud cover, NAO and Milankovitch cycling.

Evidence points to an interaction between Milankovitch cycling and topography as the most significant factor controlling tropical temperature fluctuations and the role these changes have played in forest landscape evolution across the neotropics, but factors (e.g. downstream mass transport of particulate carbon, edaphic and climatic tolerances of indicator taxa, local topographic effects on temperature) controlling the distribution of palaeo-pollen, an important proxy indicator of past climatic change, are still not adequately resolved.

Finally, tectonic forces altering the extent, geometry and latitudinal position of the South and Central American land masses have played a major part in the way land–atmosphere–ocean interactions shaped palaeoclimate across the ancient Guianan landscape – a legacy that continues to shape the distribution, structure and composition of forests across the region today.

River, Lake and Tidal Systems – A Tale of Two Basins

Guiana, Guayana, Guyana – the land of many waters. The region is awash, literally for much of the year, and water may ultimately prove to be one of the population's most valuable assets (see Chapter 8). River and tidal systems of the shield have traditionally delimited forest accessibility and with it the socio-economic landscape of this region. Only relatively recently have regional road and airfield networks expanded to such an extent that waterways have begun to decline from their historical role as backbone of the regional transportation network. Arguably, they continue to play a pre-eminent role in the way that people have accessed and continue to access the forest resources of the region (see Chapter 8). But compared with the year-round navigability of the Amazon, the seasonal impoundment of most major waterways in the shield historically has constrained travel to small craft, while shaping patterns of aquatic productivity.

The consequences of this subtle, but important, difference is reflected in the method and rate of (pre-)historic forest resource use and inhabitation (see Chapter 8).

Surface waters, however, also constitute an important pool in regional and global hydrological and biogeochemical cycles. They form a large part of the terrestrial compartment and link this directly with ocean and atmospheric components through run-off and evaporation. Water flow accounts for 85–95% of the material transported from land to ocean globally (Lerman, 1979, 1994). The export of Guiana Shield-borne terrigenous materials to the Atlantic occurs almost exclusively via water flow (see sections below, but also 'Nutrient Balance and Migration' in Chapter 7). Surface water distribution mainly reflects the interaction between rainfall and topography, though other contributing factors, such as soil and vegetation cover, also play significant roles in defining residence time and rate of movement. The distribution, persistence and depth of standing water in turn add a potent source of variation to the mix of factors shaping the tropical landscape. Changing river, lake and tidal stages influence regional soil formation and rock weathering (see 'Geology' and 'Soils' sections above), surface albedo (see 'Surface albedo', above), evaporation (Robinson and Henderson-Seller, 1999), decomposition, emissions of greenhouse gases (e.g. Quay *et al.*, 1989; Devol *et al.*, 1994; Artaxo, 2001) and vegetation types (e.g. Richards and Davis, 1933; Fanshawe, 1952; Schulz, 1960; Duivenvoorden and Lips, 1995; Huber, 1995b), just to name several of the more obvious responses.

The purpose of this section is to summarily describe the modern river, lake, wetland and tidal systems shaping the Guiana Shield. It reviews available information on a number of physical attributes that can be used to characterize the region in relation to adjacent neotropical areas. These include geomorphological controls and hydrological (watershed areas, discharge rates) and limnological characteristics (bottom, suspended and dissolved sediment loads, silica and carbon transport, clay fractions) of the region's main rivers. The location and size of major wetland and lake areas across the region are described. It also identifies a basic chronology and trends in sea-level change based on current scientific understanding and the coarse-scale spatial effects of eustatic adjustment on marine transgression and regression throughout the shield region.

Two major basins separating the Guiana Shield

The Precambrian geology of the shield area, although highly weathered, is not peneplanic and topography continues to shape the direction and rate of surface run-off (Gibbs and Barron, 1993). At the largest spatial scale, this run-off, flowing in opposite directions, is collected into two large basins divided by a band of higher elevation formed through the east–west extension of the Tumucumaque Uplands, Guayana Highlands and Chiribiquete Plateau from central Amapá to the base of the Colombian Andes (Fig. 2.30).

Waters shed southward flow into the Amazon River. This vast basin, the largest on our planet, drains an area of 7,050,000 km^2 (Sioli, 1984a). The part of the Amazon that drains all of the southern and western regions of the shield, referred to here as the North Amazon Basin (NAB) (North Peripheral Region, after Furch (1984)), is spread across five major river systems, the Caquetá (Japuerá), Negro, Uatuma/Jatapu, Trombetas/Mapuera and Jari along with 30 smaller watersheds (Fig. 2.30). Drainage into the mainstem waterways of these systems covers an estimated area of 1.3 million km^2, a little over 18% of the entire area of the Amazon Basin, but approximately 57% of the Guiana Shield.[20]

The remaining waters draining the Guiana Shield flow northward into the Atlantic Ocean and constitute the second major basin, referred to here as the Guiana (Guayana) Basin (GB). This network drains all of the eastern and northern shield regions, mainly through the Orinoco, Essequibo, Courentyne (Courentijn),

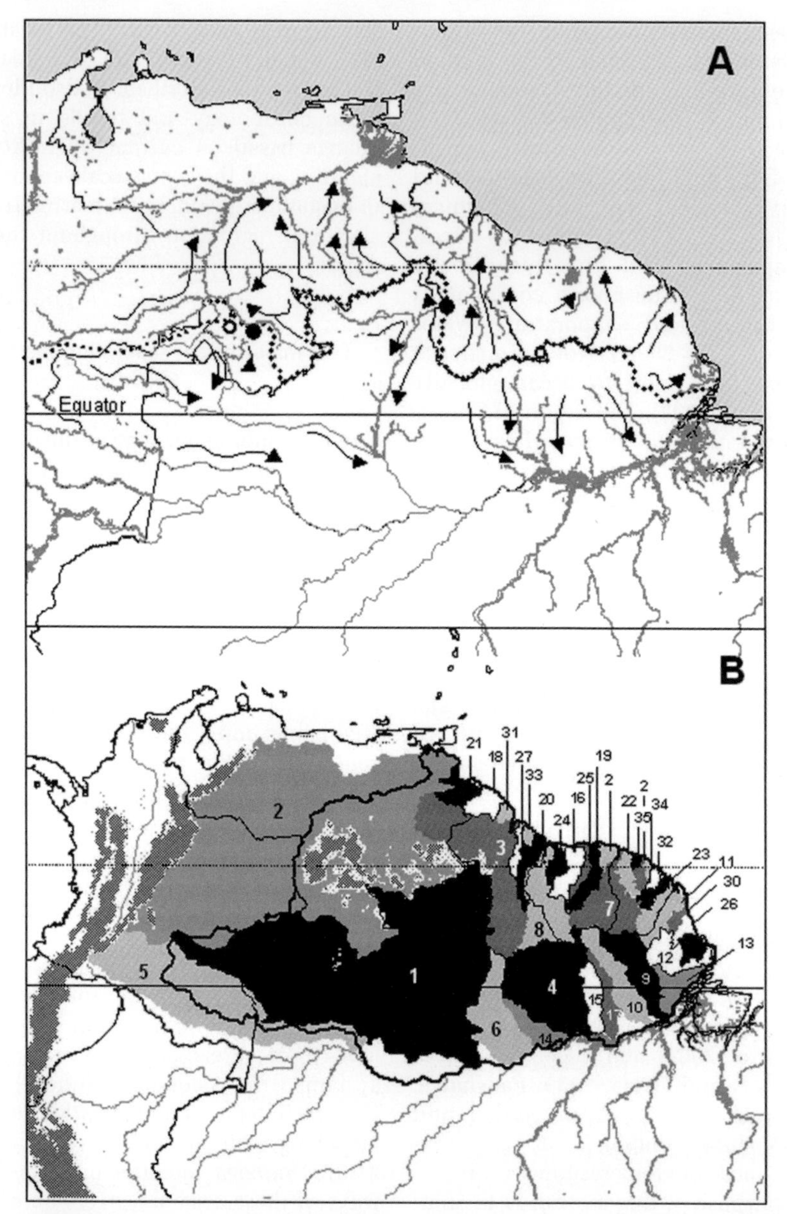

Fig. 2.30. (A) East to west divide (thick dotted line) separating two major basins draining the Guiana Shield. Arrows indicate major direction of flow across main watersheds. Basins are connected seasonally through Casiquiare Rift and Takutu Graben (solid circles) and possibly between upper tributaries of Rios Atabapo and Guainía (empty circle). Broken horizontal line is meteorological equator. (B) Major watersheds draining the Guiana Shield. Outflowing watersheds ranked by area they drain (within the shield region). (1) Negro, (2) Orinoco, (3) Essequibo, (4) Trombetas-Mapuera, (5) Caquetá/Japuerá, (6) Courentyne/Corantijn, (7) Marowijne/Maroni, (8) Jari, (9) Uatuma-Jatapu, (10) Oyapack/Oiapaque, (11) Paru de Este, (12) Maraca-Villa Nova, (13) Nhamunda, (14) Araguari, (15) Curua, (16) Waini-Barima, (17) Maicaru, (18) Berbice, (19) Suriname, (20) Amakura, (21) Koppename, (22) Mana, (23) Calcoene, (24) Approuague, (25) Nickerie, (26) Saramacca, (27) Demerara, (28) Commewijne, (29) Sinnamary, (30) Comté, (31) Cacipore, (32) Counamama, (33) Kourou, (34) Mahaica-Mahaicony-Abary, (35) Iracoubo. Only watersheds with an estimated area exceeding 1000 km^2 are delimited.

Maroni (Marowijne) and Oyapock (Oiapoque) Rivers (Fig. 2.30). Waters flowing across the Guiana (Guayana) Basin and ultimately into the Atlantic Ocean drain an estimated area of nearly 834,000 km^2 and 43% of the Guiana Shield. The NAB and GB are anchored by two parent rivers, the Rio Negro and Orinoco, respectively. These massive waterways receive run-off from vast areas of high rainfall, ranking them as two of the largest rivers in the world (Fig. 2.30A), dominating discharge from the Guiana Shield (Fig. 2.30B). The Orinoco and Caquetá/Japuerá, however, also receive considerable in-flow from tributaries draining the eastern slopes of the Andes and this can equal or exceed discharge from waterways originating in the shield region (Fig. 2.30B).

Major watersheds of the Guiana Shield

Watersheds fragment the two main basins into systems of semi-independent hydrological compartments. When organized around rivers leaving a basin, they delimit independent areas of run-off and the relative influence of differing regional physical attributes on water discharge, sediment transport and hydrochemistry (Degens *et al.*, 1990). Within watersheds, the transition from headwater perimeter to mainstem mouth embraces an important, spatially dependent hydrological gradient. Variation in both typically affects anisotropic changes in forest composition, structure and productivity (Richards, 1996), although mitigating variables, such as varying rainfall, soils, geology and population dynamics, may create patterns at intervening scales.

Thirty-four watersheds (>1000 km^2) collectively drain South America north of the Amazon and east of the Andes, including all of the Guiana Shield (Fig. 2.30B). Watersheds surrounding each mainstem river draining the region range in size from several hundred square kilometres to over 720,000 km^2 in the case of the Rio Negro, the shield's largest self-contained drainage area (Table 2.2). The median watershed size for the 34 main rivers flowing out of the Guiana Shield is estimated at 19,900 km^2. Measures of central tendency in this case are, however, misleading due to the bimodal distribution of watershed sizes created by a few, very large rivers draining the western reaches of the shield and a much larger number of relatively small watersheds aggregated within the Guiana Basin east of the Essequibo River (Fig. 2.30B).

Eight of the 30 largest watersheds belong to the national area of more than one country, including the three largest systems, the Negro, Orinoco and Essequibo (Table 2.2, Fig. 2.30B). The Rio Negro is principally a Brazilian river, flowing from the southern slopes of the Guayana Highlands and Chiribiquete Plateau, but also drains areas of southwest Guyana (east bank of Tacutu River) and Venezuela (Casiquiare Channel) due to localized fault zones (Casiquiare and Takutu) that have shifted the basin divide and led to a change in flow direction. This is most pronounced in the case of the Tacutu River (watershed area 9800 km^2). It drains a strip of southern Guyana as it flows northward before changing course 180° around the Serra do Tucano volcanics (see 'Prominent geological regions', above) and flowing southward into the Rio Branco less than 100 km westward of its headwaters.

The Orinoco, draining virtually the entire South American land area north of the Negro, Japuerá and Essequibo watersheds, is much larger than the Negro, but its headwaters are located both within and outside the Guiana Shield. In fact only about 40% of the estimated 1 million km^2 watershed area rests within the Guiana Shield, fed most notably by the Caroni (93,000 km^2), Caura (45,350) and Ventauri (41,760) tributaries – substantive rivers in their own right (Fig. 2.30B). The remaining part of the watershed flows from the headwaters of the Apure, Meta and Guaviare Rivers in the eastern Andes across the lowland sedimentary plains of Venezuela and Colombia (Figs 2.6B, 2.30B).

The Essequibo River watershed consists of two arms meeting in north-central Guyana. The northern arm flows eastward along two large rivers, from the Venezuelan Imataca region via the Cuyuni River (89,000

Table 2.2. The thirty largest watersheds exporting water from the Guiana Shield into the Amazon or Atlantic Ocean, their estimated surface areas and distribution among the six shield countries.

Ref.	River system	Colombia	Brazil	Venezuela	Guyana	Suriname	F. Guiana	Total (km^2)
1	Negro	78,600	589,300	40,500	12,300			720,700
2	Orinoco	18,200		359,000				377,200
3	Essequibo			39,000	118,500			157,500
4	Trombetas		136,400					136,400
5	Japuerá	73,700	17,800					91,500
6	Jatapu		74,700					74,700
7	Marowijne					39,100	30,900	70,000
8	Courentyne				27,900	40,700		68,600
9	Jari		54,600					54,600
10	Paru de Este		44,250					44,250
11	Oyapock		18,700				14,200	32,900
12	Araguari		31,500					31,500
13	Maraca Pucu		26,000					26,000
14	Nhamunda		25,500					25,500
15	Curua		24,400					24,400
16	Coppename					21,900		21,900
17	Maicaru		20,600					20,600
18	Waini-Barima			1,200	17,900			19,100
19	Suriname					17,200		17,200
20	Berbice				16,600			16,600
21	Amacuro-Aguirre			14,200				14,200
22	Mana						12,100	12,100
23	Approuague						10,250	10,250
24	Nickerie					10,100		10,100
25	Saramacca					9,150		9,150
26	Calcoene		9,100					9,100
27	Demerara				8,200			8,200
28	Commewijne					6,600		6,600
29	Sinnamary						6,600	6,600
30	Cacipore		5,500					5,500
	Misc. others*		125,000		11,100	2,650	11,200	149,950
	Total (km^2)	170,500	1,203,350	453,900	212,500	147,400	85,250	2,272,900

*Includes: Brazil: Urubu, Piorini, Parana, Cuncuaru, Amapá Grande and other smaller riverlets in Pará and Amapá; Guyana: Pomeroon, Mahaica, Mahaicony, Abary and others; Suriname: Coronie and other small coastal stream systems; French Guiana: Comté, Kourou, Counamama, Iracoubo, Organabo, Macouria, Montsinery, Mahury, Kaw and other small coastal creeks.

km^2) and the Guyanese North Pakaraimas through the Mazaruni River. The southern arm flows northward across lowland forest and savanna through the Kassikaityu, Kuyuwini, Kwitaro, Rewa (11,000 km^2), and Rupununi (22,000) Rivers from their sources in the Acarai, Kamoa, Amuku, Wassarai and Kanuku mountain ranges. While the Essequibo watershed constitutes 55% of Guyana's national land area, an estimated 25% of the total watershed area extends into Venezuela as part of the Cuyuni headwaters (Table 2.2).

The Trombetas system is the largest watershed contained within a single country of the shield region, Brazil. Covering most of Pará state north of the Amazon, the area surrounds a three-pronged system consisting of the Mapuera (26,500 km^2), Trombetas (73,400) and Paru de Oeste (36,500) watersheds, headwatered along the southern slopes of the Serra Acarai and Tumucumaque ranges along the Guyana–Suriname–Brazil border.

The Japuerá, or Caquetá as it is known in Colombia, finds its source west of the

Guiana Shield in the eastern slopes of the Colombian Andes. Its large watershed (242,000 km^2) is largely contained outside the Guiana Shield (62%) (Fig. 2.30B). Only the northern tributaries, most notably the Apaporis River, are considered here to drain from the Guiana Shield.

The Marowijne, or Maroni, River is a 70,000 km^2 watershed distributed almost equally between eastern Suriname and western French Guiana (Table 2.2). Flowing from the Tumucumaque Uplands via the Tapanahoni, Oelemort, Lawa, Waki, Tampoc (7650 km^2), Malani and Alitani Rivers, the Marowijne watershed is French Guiana's largest, accounting for an estimated 34% of French Guiana's land area and 25% of Suriname.

The third largest watershed in the Guiana Basin feeds the Courentyne (Courentijn) River mainstem (Fig. 2.30B) and is shared by Guyana (41% of area) and Suriname (59%) (Table 2.2). It originates in the Acarai range in Guyana and the Wilhelmina Mountains and Sipaliwini savannas of Suriname, draining along the New, Aramatau, Lucie and Sipaliwini Rivers.

The Oyapock (Oyapok, Oiapaque) watershed spreads westward from its mouth at the border between Amapá and French Guiana (Fig. 2.30B). Roughly 57% of the area falls within French Guiana along several major tributaries, most notably the Noussiri, Camopi (6000 km^2) and Yaloupi Rivers. In Amapá, it is fed principally by the Cricou, Anotaie, Iaué and Mutura Rivers.

The smallest of the watersheds shared between different countries is the Waini–Barima system spread across the northernmost regions of Guyana and Venezuela. Fed by Barama, Kaituma and Aruka Rivers in Guyana, the Barima swings northwestward along the Atlantic coast to drain along the southern bank of the Orinoco delta. The lower reaches of the Waini follow a parallel path but drain before reaching the border. Slightly more than 6% of the watershed area is located in Venezuela, but this importantly includes the Barima river mouth.

River directions and connections

The unique geomorphology of the Guiana Shield can exert a pronounced control on channel behaviour along many of the major waterways coursing through the region. The main features linking geomorphology to river dynamics in the shield area are a product of several interdependent processes, i.e. (i) repeated regional tilting and uplift; (ii) an accumulation of exposed faults, rifts and joints; (iii) accumulation of exposed dyke and sill lithologies; (iv) creation of an inverse topography through laterite formation in palaeochannels; (v) the prevalence of unconsolidated, reworked sediments in the major depositional basins; and (vi) retroflexion northwestward of coastal river mouths by strong wind-driven, ocean currents.

TECTONIC CONTROLS – UPLIFTING AND REGIONAL TILT The Precambrian landscape of the Guiana Shield is derived from a long series of intervening tectonic and eustatic processes that have repeatedly altered erosion bevel slope and aspect (Gibbs and Barron, 1993). Hydrography has changed as a consequence and many of the modern-day river courses once belonged to neighbouring watersheds or flowed in the opposite direction. There are numerous examples of stream capture as a consequence of regional tilting. These include the Rio Tacutu's upper Tertiary capture of the Rupununi River's west bank tributaries. Most notable among the waterways once flowing eastward is the Sawariwau River, whose headwaters are located near the seasonally inundated Dadanawa region (Berrangé, 1977). The upper Mazaruni's sudden change from a northward to eastward flow is also attributed to tilt-induced capture from its former downstream flow along the Urluowra River, a tributary of the Cuyuni River (Bracewell, 1950). This capture is indicative of a wider regional west to east downwarp across Guyana associated with Berbice Basin formation. Similarly, the headwaters of the Tapanahoni River in Suriname, a tributary of the Marowijne, were captured from Jai Creek as a consequence of eastward tilting (Haug, 1966; Leaflang *et al.*, 1978 cited in Gibbs and Barron, 1993). The capture of the Casiquiare Canal by the Rio Negro is believed to have been caused by a southeastern tilting along the Orinoco headwaters (Stern, 1970; Khobzi *et al.*, 1980). Gibbs and Barron (1993) also present several cases of stream capture that reflect more localized changes in erosion bevels. Of particular significance is the proposal that the current flow of the Essequibo River above Massara once formed the upper reaches of a much larger Berbice River watershed before a change in flow direction westward led to its capture by the lower Essequibo from Rattlesnake Creek.

STRUCTURAL CONTROLS – FAULTS AND JOINTS Regional tilting has occurred as both cause and consequence of fault, rift and joint formation. A series of tectonic events reactivated the igneous rock base and produced an extensive field of faults, rifts and joints across the region over a 2 billion year period (see Fig. 2.6A, B and 'Tectonic processes', above). Sea-level change as a consequence of climate variation has also led to further fracturing of the shield area as eustatic adjustments create alternating periods of downwarping and uplifting. Major fault zones have created wide, linear basins that constrain drainage direction and form the base of several regional erosion bevels. Most noticeable among the drainage effects of faulting and fracturing are abrupt and localized orthogonal channel swings and parallel stream spreading, often involving counter-flowing drainages. The counter-flowing upper Orinoco and Rio Negro drainages are constrained by the massive Casiquiare Rift (Fig. 2.9A, see 'Shield macro-features', above), a pattern also seen in the parallel channelling of the upper Rupununi and Tacutu Rivers in southwestern Guyana (Fig. 2.9B). The abrupt orthogonal shift in drainage alignment of the Berbice, Essequibo and Rupununi Rivers between Rappu Falls and Apoteri in central Guyana provides a more localized example of a major fault axis, in this instance the central section of the Takutu Graben (Fig. 2.6A, B), controlling parallel river flow direction. In other cases, direction along the entire length of mainstem flow is simply constrained by the fault major axis (Noordam, 1993). The southwestern flow of the Rio Branco follows the trough formed by the eastward extension of the Pisco-Jura Megafault and westward termination of the Takutu Graben (Fig. 2.6A, B). A major tributary of this river, the Uraricoera, has a flow constrained by the southeastern dip of the Aracaca–Urutamin horst-fault complex in north Roraima state (see Fig. 2.3 in Gibbs and Barron, 1993). Minor stress fractures, often formed as a consequence of eustatic adjustment, have exerted a less pronounced, but widespread, effect on localized river direction in the Guiana Shield (e.g. Suriname, Noordam (1993)). Localized, orthogonal swing in mainstem direction can be found along the courses of most major rivers draining the region, including parallel displacements along the middle Caroni and Paragua Rivers in Venezuelan Guayana and the upper Mazaruni and Essequibo Rivers in Guyana.

INVERTED RELIEF Water levels in palaeochannels over the Neogene fluctuated with alternating periods of relatively wet and dry climate (see 'Prehistoric climates of the Guiana Shield', above), creating ideal conditions for lateritic cementation of exposed basic rocks. Combining with other resistant parent rock structures, faults and fractures, the proliferation of laterite domes and hills across the landscape has also acted to limit local drainage direction and movement, contributing to the evolution of rectangular drainage patterns, a hydrographic feature commonly characterizing many parts of the shield region (e.g. central Suriname, Haug (1966)).

NATURAL IMPOUNDMENTS Almost every river draining the exposed crystalline basement of the Guiana Shield courses over exposed linear bands of intrusive rock that are more resistant than the surrounding material. The cataracts formed from the meeting of channel and typically an exposed dyke or sill are a unique feature of the central shield's waterways as they course across areas of relatively shallow sediment. Referred to as 'falls' in Guyana, 'saut' or 'soula' in French Guiana, 'sula' in Suriname (Zonneveld, 1952), 'cachicoera' in Brazil and 'raudal' or 'salto' in Venezuela, their impounding effect can push waters to open up a braided network of stationary, anastomosing channels that open and close with the seasonal change in river height. Dykes and sills can also act like faults or fractures in effectively extinguishing lateral channel migration when running parallel to the main erosion bevel, leading to stretches that appear unusually straight relative to other lengths of the river (e.g. Caqueta River, p. 34 in Duivenvoorden and Lips, 1995). Greenstone belts ranging across the exposed Precambrian rim of the eastern Guiana Shield have had a similar local-scale effect on drainage direction and sinuosity (Gibbs and Barron, 1993). As rivers move into the deeper sediments deposited along the edge of the shield, however, natural impoundments disappear altogether and water flow direction becomes increasingly controlled by the characteristics of the surrounding depositional environment (Sioli, 1984a; Goulding *et al.*, 1988), a feature that pervasively controls lowland dynamics across western Amazonia (Salo *et al.*, 1986; Räsänen *et al.*, 1987), albeit as a downstream response to more substantial elevational gradients and current velocities upstream.

SAND-DOMINATED SEDIMENTARY PLAINS The depositional environment of the Guiana Shield is characterized by an unusual abundance of heavily reworked quartz sand fields, a feature reflected in the particle-size distribution of upland soils, especially in the Rio Negro, Caquetá, Branco and Berbice basins (Duivenvoorden and Lips, 1995; van Kekem *et al.*, 1996; Dubroeucq *et al.*, 1999), Casiquiare rift valley (Table 4 in Saldarriaga, 1994) and coastal savannas of the Guianas (van der Eyk, 1957; Blancaneaux, 1973; Cooper, 1979). Clay- and silt-dominated soil types do occur (e.g. dystric Gleysols), but these are typically associated with stream channels and distributed as a consequence of river migration. In some instances, where past river capture has extinguished mainstem segments, these soils may appear at some distance from active floodplain areas (e.g. Kuruduni River in Guyana (van Kekem *et al.*, 1996)). Large substrate particle size and slow flow velocities combine to minimize river migration across the sand-dominated sedimentary plains. Instead, infiltration, the downward movement of water through the soil profile, dominates run-off across these areas (e.g. Jetten, 1994). This creates a drainage system characterized by a much lower stream density (e.g. Teunissen, 1993) than those dominated by surface run-off processes, such as those with headwaters in the Andes (e.g. Fig. 2.2 in Duivenvoorden and Lips, 1995).

DEFLECTED COASTAL RIVER MOUTHS Most rivers discharging into the Atlantic are characterized by downstream courses that rapidly move northwestward from their major channel axis. This sudden change in river course is caused by the northwestward movement of the Guiana Current that begins with entrainment of the Amazon freshwater river plume and continues through to the Orinoco. Larger rivers, particularly during peak discharge periods, experience less severe channel deflection of their lower reaches (e.g. Amazon, Essequibo, Courentyne, Maroni, Orinoco) compared to smaller waterways (e.g. Oyapock, Approuague, Mana, Commewijne, Nickerie, Pomeroon, Waini, Barima), although freshwater plumes, and sediment loads, are still diverted in line with the prevailing longshore currents.

Average water discharge rates

The surface export of freshwater from the Guiana Shield is substantial. Through approximately 47 medium to very large rivers (after Table 6.1 in Meybeck *et al.*, 1992) the region exports on average an estimated 2792 km^3 of freshwater each year. This amounts to 7–7.5% of the global total, estimated at between 37,300 and

40,000 km^3 per annum (Baumgartner and Reichel, 1975; Table 4 in Dai and Trenberth, 2002), and 25% of South America's total volume of freshwater, about 11,100 km^3 per annum (Degens *et al.*, 1990), discharged to the oceans (Table 2.3).[21] Northeastern South America is the world's largest regional contributor of freshwater to the global oceanic pool (Dai and Trenberth, 2002) and the two basins draining the

Table 2.3. Estimated average water discharge (Q) and specific discharge (*q*) rates for rivers exporting from the Guiana Shield region. See text and endnote no. 11 for details of base data sources, calculations and assumptions. Note: standard deviation is likely to be high for many river averages listed here due to inter-annual patterns of rainfall variation not fully embraced by record series.

Rank	Basin	Country[a]	River system	Q (km^3/year)	*q* (l/s/km^2)
1	NAB	Co-Ve-Br	Negro	1,400.0	59.2
2	GB	Ve	Orinoco (Guayana only)	565.0	35.8
3	NAB	Co-Br	Caquetá-Japuerá	168.3	22.0
4	GB	Gu (Ve)	Essequibo	154.0	32.3
5	NAB	Br	Trombetas	74.6	30.8
6	GB	Su-Fg	Marowijne-Maroni	56.3	27.1
7	GB	Gu-Su	Courentyne-Courentijn	49.6	23.1
8	NAB	Br	Jari	31.5	19.5
9	NAB	Br	Jatapu	30.5	42.1
10	GB	Fg-Br	Oyapock-Oiapoque	28.3	29.0
11	NAB	Br	Uatumá	21.1	32.9
12	NAB	Br	Paru de Oeste	17.3	15.1
13	GB	Su	Coppename	15.8	23.0
14	NAB	Br	Paru de Este	15.1	15.5
15	GB	Su	Suriname	13.4	25.8
16	GB	Fg	Approuague	11.7	36.1
17	GB	Gu	Berbice	11.0	21.1
18	GB	Br	Araguari	10.5	14.2
19	GB	Fg	Mana	9.7	25.4
20	GB	Fg	Sinnamary	9.1	44.2
21	NAB	Br	Nhamunda	7.3	21.7
22	GB	Su	Saramacca	7.1	25.0
23	GB	Fg	Comte	7.0	43.7
24	GB	Gu	Demerara	6.9	26.8
25	GB	Br	Caciporé	6.8	53.8
26	GB	Br	Calcoene	6.3	15.5
27	GB	Fg	Mahury	6.3	61.4
28	GB	Gu	Waini	6.3	36.7
29	GB	Su	Nickerie	5.6	17.6
30	NAB	Br	Padauari	4.4	20.6
31	NAB	Br	Curua	4.2	6.4
32	GB	Gu-Ve	Barima	4.0	8.8
33	GB	Su	Commewijne	3.8	18.2
34	NAB	Br	Maicaru	3.6	6.6
35	GB	Fg	Kourou	3.5	57.7
36	GB	Fg	Iracoubo	2.3	50.0
37	GB	Gu	Pomeroon	1.9	10.6
38	NAB	Br	Maraca Pucu	1.5	11.1
39	GB	Fg	Counamama	1.4	50.0
40	GB	Fg	Cayenne	1.3	58.3
	GB/NAB		Misc. others (7 rivers)	7.3	
			Guiana Shield	2,792	38.7

[a]Br, Brazil; Co, Colombia; Fg, French Guiana; Gu, Guyana; Su, Suriname; Ve, Venezuela.

Guiana Shield contribute significantly to this ranking.

NAB RIVERS Moving southward, 15 major rivers draining the Guiana Shield into the Amazon account for approximately one-third of the volume (5519 km^3 per annum (Richey *et al.*, 1990)) that is discharged, on average, from its mouth, give or take a 1% additional loss to evapotranspiration in-transit to the Atlantic as part of the Amazon mainstem flow. The Rio Negro dominates the distribution of surface water leaving the NAB, accounting for 79% of the total estimated discharge from the basin. This is equal to about 13% of the annual average discharge from the Amazon Basin (Molinier *et al.*, 1993). Two notable tributaries feed this flow. From the Chiribiquete region of Colombia, the Vaupés/Uaupés delivers an estimated 36 km^3 per annum (see Bodo (2001) and Notes for details) to the Negro south of its mainstem exit from Venezuelan Amazonas. The largest contributor, the Rio Branco, drains much of the southern slopes of the highland chain dividing the shield's two basins, before flowing through the Gran Sabana and into the lower Rio Negro. At this point, it delivers its water at an estimated average discharge rate of 377 km^3 per annum (Seyler and Boaventura, 2001), or 27% of the water exported each year to the Amazon via the Rio Negro. The Caquetá/Japuerá delivers approximately 6% of the water from the NAB to the Amazon mainstem, although this figure also includes waters draining from headwater tributaries in the Colombian Andes and inflates the contribution from shield-draining waterways downstream. Adequate discharge data are not available for the major rivers, Yarí, Mirití-Paraná and Apaporis, but given their watershed area and location within one of the major rainfall centres of the neotropics, their contribution (to water flow only) is likely to compensate for their modest elevational gradients.

GB RIVERS Moving northeastward, 32 major rivers flowing into the western tropical Atlantic account for a little more than 5% of the annual freshwater inflow to the Atlantic Ocean, estimated at 19,200 km^3 per annum (Dai and Trenberth, 2002). Over 56% of the surface water is estimated to leave the GB each year through the Orinoco, but the Orinoco receives only a little more than half of its estimated volume from shield-draining tributaries, most notably the Caura (110 km^3 per annum), Caroni (157 km^3 per annum) and Ventauri (63 km^3 per annum) (Weibezahn, 1990; Cressa *et al.*, 1993 and references therein). The remaining volume is discharged from western shield tributaries, such as the Inirida (95 km^3 per annum) and Atabapo Rivers that drain podzol/arenosol-dominated lowland regions of the Chiribiquete and Casiquiare regions and, most importantly, the eastern slopes of the Colombian and Venezuelan Andes, mainly through the Guaviare (230 km^3 per annum), Meta (157 km^3 per annum) and Apure (70 km^3 per annum) Rivers (Meade *et al.*, 1990, and references therein).

Together, the Orinoco and Rio Negro watershed make up over 70% of the estimated annual flow of surface water leaving the region (Table 2.3). Of the remainder, 445 km^3, about 40% of the total Orinoco flow, is generated on average each year from rivers draining northward through the Guianas and Amapá into the Atlantic. Most notable among the contributing rivers are the Essequibo, Courentyne/Courentijn, Marowijne/Maroni and Oyapock/Oiapoque (Table 2.3). Juxtaposed between the Orinoco and Amazon Rivers, this region appears insignificant in its annual export of freshwater. But the total volume is estimated as being over twice that discharged by the Danube (eastern Europe), one and three-quarters the volume of the Columbia (Pacific Northwest USA), similar to the Ganges (India) and nearly 75% of that flowing from the Mississippi (Dai and Trenberth, 2002), to place its composite contribution in a global context.

SPECIFIC DISCHARGE RATES More striking, at least perhaps to non-hydrologists, is the actual area drained to create these sizeable flow rates, referred to as the specific discharge rate.[21] In the Guiana-Amapá case, an area of 532 × 10^4 km^2 is drained (Σ A_{GB}–$A_{Orinoco}$ from Table 2.2, Σ Q_{GB}–$Q_{Orinoco}$ from Table 2.3), giving a specific discharge rate of 26.5 l/s/km^2. This is over three times that estimated for the Danube, 2.5 times that of the Columbia, nearly twice the specific discharge of the Ganges and over four times the volume of water estimated to leave the Mississippi for each square kilometre of land drained (Mouth Vol and DA – Table 2 in Dai and Trenberth, 2002).

Taken as a whole, the Guiana Shield averages a discharge rate of 2792 km^3 per annum from a drainage area that amounts to less than 2.3% of the global area draining to the oceans, achieving a specific discharge rate (*q*) of 38.5 l/s/km^2, one of the highest known for an area of this size. The Amazon Basin as whole is estimated to generate a slightly lower specific discharge rate, between 28 l/s/km^2 (Degens *et al.*, 1990) and 36 l/s/km^2 (calculated from Table 2 in Dai and Trenberth, 2002). The high value of *q* for the Rio Negro contributes significantly to estimations for both the Guiana Shield and Amazon Basin (Table 2.3). Other shield-draining tributaries of the Amazon also achieve high *q* values (Table 2.3) and compensate for the much lower values estimated for northward flowing Amazon tributaries, such as the Tapajos (26 l/s/km^2) and Xingu (19 l/s/km^2), that drain the drier eastern Brazilian Shield region (Fig. 2.16A). The contribution of shield-draining tributaries to the Orinoco's specific discharge is reflected in the larger *q* value calculated when considering the Guayanan portion alone (Table 2.3, 35.8 l/s/km^2) against that of the entire watershed (34.5 l/s/km^2). Again, the decline in annual rainfall recorded for lowland Venezuela/Colombia north of the Guiana Shield dampens the value of *q* despite several major rivers with headwaters sourced in the Andes. The presence of the Guayana Highlands along the high-rainfall-producing meteorological equator bolsters run-off and thus the specific discharge. The geographic distribution of the watershed in relation to the meteorological equator is further emphasized by the relatively high values of *q* estimated for much smaller rivers draining eastern French Guiana (e.g. Oyapock, Approuague, Sinnamary, Comté, Kourou and others in Table 2.3) and the decline in *q* moving westward along the Atlantic coast between the Maroni and Berbice Rivers. In part, this general trend appears to be more sensitive to the lithological make-up of the watershed area than changes in rainfall, with rivers largely draining deeper unconsolidated sands showing much lower specific discharge values. Infiltration rates can be substantive for arenosolic soils, especially where these are deep and do not rest above a relatively shallow sub-horizon of indurated material. This modulates the release rate into the surface river flow, especially in relation to lithological formations with shallow or high clay soils, or those with an exposed, indurated layer, such as laterite (see 'Main soils in the Guiana Shield', above). As a consequence, seasonal variation in the discharge rate from sand-dominated watersheds exhibits a relatively low amplitude compared to equivalent areas with less infiltration and groundwater flow capacities.

Seasonal variation in discharge

As is the case with rainfall, the average annual discharge of surface water gives a good indication of the spatial movement of water through and out of the Guiana Shield. It does not adequately describe, however, variation in surface flow, particularly in response to the behaviour of the mechanisms that deliver rainfall to the region. Annual averages mask the important seasonal variation component of the surface drainage flux and how this variation changes spatially. If rainfall is the seasonal pacesetter of terrestrial tropical life in the *terra firme* forests, the surface water response to alternating wet and dry seasons is certainly the pacesetter of life in floodplain forests.

Generally speaking, both discharge rate and stage (height) of the largest river mainstems peak one to three months after the peak in seasonal rainfall across the South American tropics. The latitudinal ranging of the ITCZ and the rainfall it delivers over the Guiana Shield and greater Amazon is clearly mirrored by the rise and fall of average monthly discharge rates (Fig. 2.31). Seasonal variation in discharge is not 'in-phase' or synchronous across the region, however (Figs 2.31, 2.32). With the onset of the 'dry' season over the Guiana Shield during the boreal autumn (September–November), and with the ITCZ recently arriving over the Brazilian Shield, discharge is at its lowest across both the Guiana Shield and southern Amazonia. Only the large Orinoco and its Andean-draining tributaries continue to flow at a rate above their seasonal minima (Fig. 2.31 – November). By February, rivers draining the western shield have diminished outflow, but eastern shield rivers in French Guiana and Amapá maintain higher rates as the ITCZ curls northward along the Atlantic seaboard of equatorial South America in tandem with the dampened poleward ranging of maximum SSTs. A rise in the discharge from rivers draining the Bolivian and Peruvian Andes and Brazilian Shield lags the southward movement of the ITCZ.

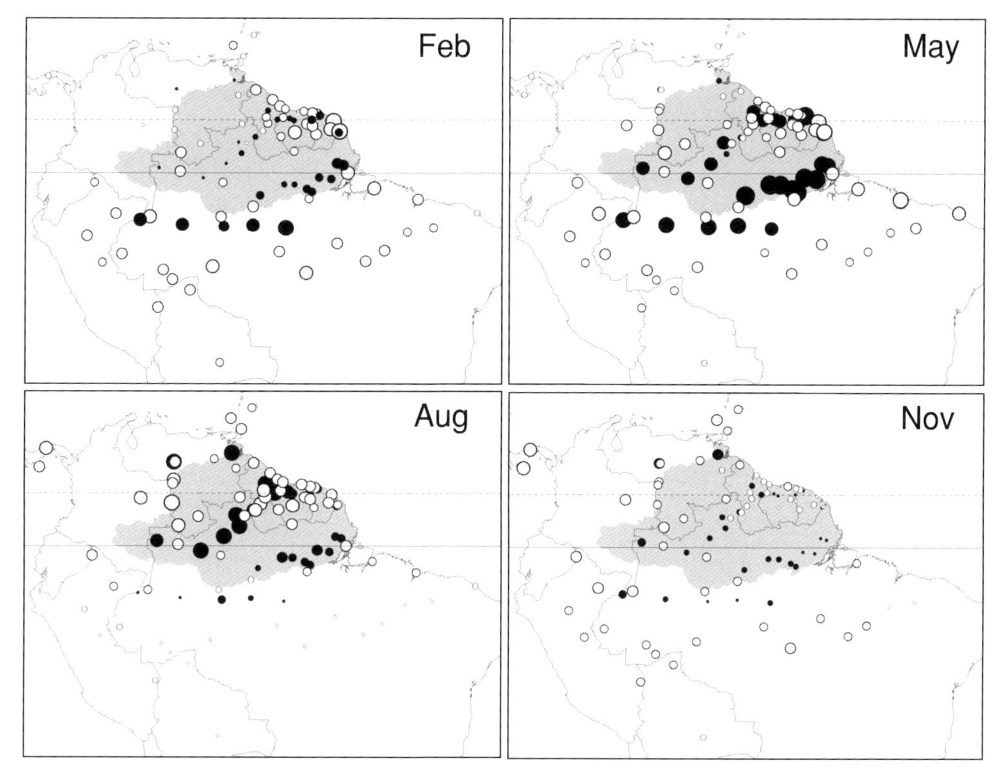

Fig. 2.31. Seasonal variation in river discharge rates (filled circles) and mean monthly rainfall (open circles) across the Guiana Shield and adjacent regions. Rainfall values are for the preceding month. February: period of minimum river discharge rates across most of shield region and maximum discharge from rivers south of Amazon reflecting southerly position of ITCZ . May: increase in discharge from major rivers across shield region as peak seasonal rainfall spreads across entire area and diminishes in the south with northward movement of ITCZ. August: wet season (June/July) nears end and shield rivers at maximum discharge as ITCZ returns southward from its northern range limit, while south Amazon rivers reach minima. November: discharge from shield rivers quickly declines as rainfall drops off as ITCZ moves southward, increasing rainfall in south Amazon. Hatched horizontal line is meteorological equator. Rainfall values range across seven intervals between 0–50 mm (smallest circle) and 500–600 mm (largest). Mean monthly discharge values are standard deviates (mean=0) for each station dataset and range across seven 0.5 intervals between –1.5 (smallest circle) and +2.0 (largest).

By April, the ITCZ has moved over the south-central Guiana Shield. Discharge from rivers draining the southeastern shield, viz. Amapá, North Pará and French Guiana, peak during this phase while rivers draining the south Amazon stay close to their seasonal highs (Fig. 2.31 – May, Fig. 2.32). Rivers across the north-northwestern Guiana Shield, including the Orinoco, continue to flow at rising, but still relatively low, rates. By the peak of the boreal summer in July, the high rainfall and river discharge rates have moved northward (Fig. 2.32). Discharge rates peak across the western and northern shield in July, while rates diminish across French Guiana, Amapá and along the southern rim of the region. Rainfall is at a minimum across southern Amazonia and discharge rates, as a consequence, approach their minima reached in October. At the same time, rainfall over southern Central America has just declined from its seasonal peak in July–August, generating a delayed peak in river discharge from September to November (Fig. 2.32).

Many factors other than latitudinal position (including stream order, sampling error and record length) can influence the timing of peak average discharge. These are reflected in the approximately four-month spread in peak averages recorded at the same latitude for 486 lowland stations from the ANEEL and UNESCO databases (Bodo, 2001) (Fig. 2.32). Nonetheless, within lowland systems of the neotropics, it is perhaps not too unreasonable to state that all rivers draining the Guiana Shield experience a peak discharge sometime between April and July, Central American rivers between May and November, and South Amazonian rivers between January and July (Fig. 2.32).

Inter-annual variation in discharge

In the Guiana Shield, surface discharge represents the major component of run-off. Run-off can be crudely equated to rainfall less atmospheric return through evapotranspiration. Logically, run-off should therefore be conditioned indirectly to the same

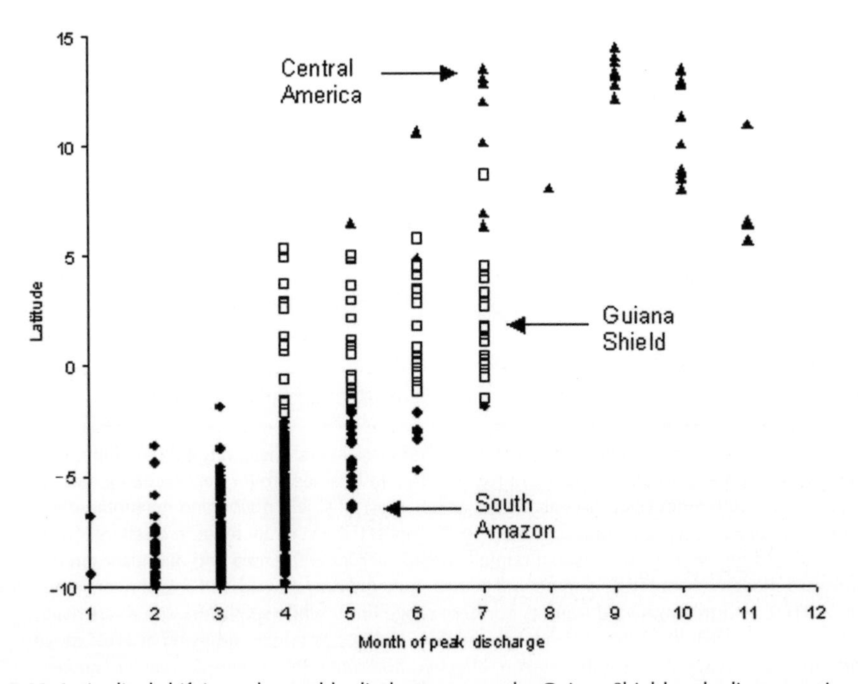

Fig. 2.32. Latitudinal shift in peak monthly discharge across the Guiana Shield and adjacent regions in the neotropics based on inter-annual averages. Note: Averages are not based on concomitant intervals for all rivers. Data sources: Bodo (2001), ANEEL, Hydrometeorological Service (1980, 1981, 1982, 1985, 2000).

factors that affect variation in these two dominant fluxes. Across the Guiana Shield, the two extreme phases of ENSO, El Niño and La Niña, are the most important climatological events affecting variation in modern rainfall and evapotranspiration (see 'Modern centres of high rainfall', above), and thus inter-annual variation in water flow rates and stage heights. This relationship has been well documented for the Amazon mainstem and several of its major tributaries (Molion and Moraes, 1987; Richey *et al.*, 1989; Marengo, 1995).

In the Guiana Shield, Lointier (1993) identified an 18–19-year interval between extreme low and high peak discharge rates of the Oyapock (Maripa) over the interval spanning 1953–1993. The variation from the long-term mean over this period clearly mirrors the incidence of strong EN and LN phases of ENSO, with discharge rates consistently declining by 12–50% from the long-term average during strong EN phase years and increasing by 12–80% during LN phases. Past annual discharge anomalies from other French Guiana rivers, such as the Approuague, Camopi, Mana, Maroni/Marowijne and Tampoc appear to similarly coincide with the onset of EN/LN conditions (Hiez and Dubreuil, 1964), as do neighbouring Surinamese and Guyanese rivers.

In fact, the spatial decline of a strong SOI–precipitation relationship north and south from its regional centre in the east-central Guiana Shield (see 'Historic variation in rainfall', above) should be reflected in the spatial variation in river discharge responses (Fig. 2.33).[22] Multi-decadal records of monthly river discharge within the region are extremely scarce, but several stations have sufficient data to evaluate responses to several of the more recent EN events. During the strong EN event of 1972–73, rivers draining the north and east of the shield (Essequibo, Rio Branco in Fig. 2.33) show a strong seasonal decline in monthly discharge compared to the decadal mean for each month over the period 1967–1977, registering only 20–40% of the average flow during the first quarter of the final EN year. The Orinoco also experienced a less drastic reduction in flow during the period at Ciudad Guayana (Puerto Angostura). In part, the relatively mild response of Orinoco discharge to EN-induced rainfall failure is due to modulation by its tributary connections with the Andes. Major contributors draining the Guayana Highlands appear to have experienced an ENSO response similar to the east-central shield (e.g. Caroni: Hastenrath *et al.*, 1999). Large rivers with strong connections to the Andes in south and western Amazonia showed a very weak or positive response to both the 1972–73 and very strong 1982–83 EN events (Caquetá/Japuerá, Jurua, Purus and Madeira Rivers in Fig. 2.33) (Sklar, 2000). Similar to 1972–73, shield-draining rivers experienced a sustained reduction in discharge from August (year 1) to April (year 2) as the EN event reached its 'mature' phase, noted by the large negative SOI score in Fig. 2.33 (right column, except Jari).

The anisotropic discharge response to EN and LN events across the lowland tropics of South America suggests that much of the Guiana Shield, without substantial fluvial connections to the Andes, would have experienced more severe and prolonged water-level fluctuations under extended palaeo-ENSO conditions, or other climatological phenomena that would have similarly forced a longer ITCZ residence north or south of the region (e.g. certain Milankovitch cycle phases). Based on current understanding of the climatological dynamics delivering rainfall to the region and effects of its surface geomorphology on run-off, the aquatic systems and prehistoric inhabitants of the Guiana Shield may have been disproportionately affected by a palaeoclimatic see-saw of flooding and impounded pooling not experienced in other neotropical regions.

Physico-chemical transport

A world-class, but ENSO-sensitive, surface discharge rate is not the only distinctive attribute of Guiana Shield waterways. Waters also erode and then transport partic-

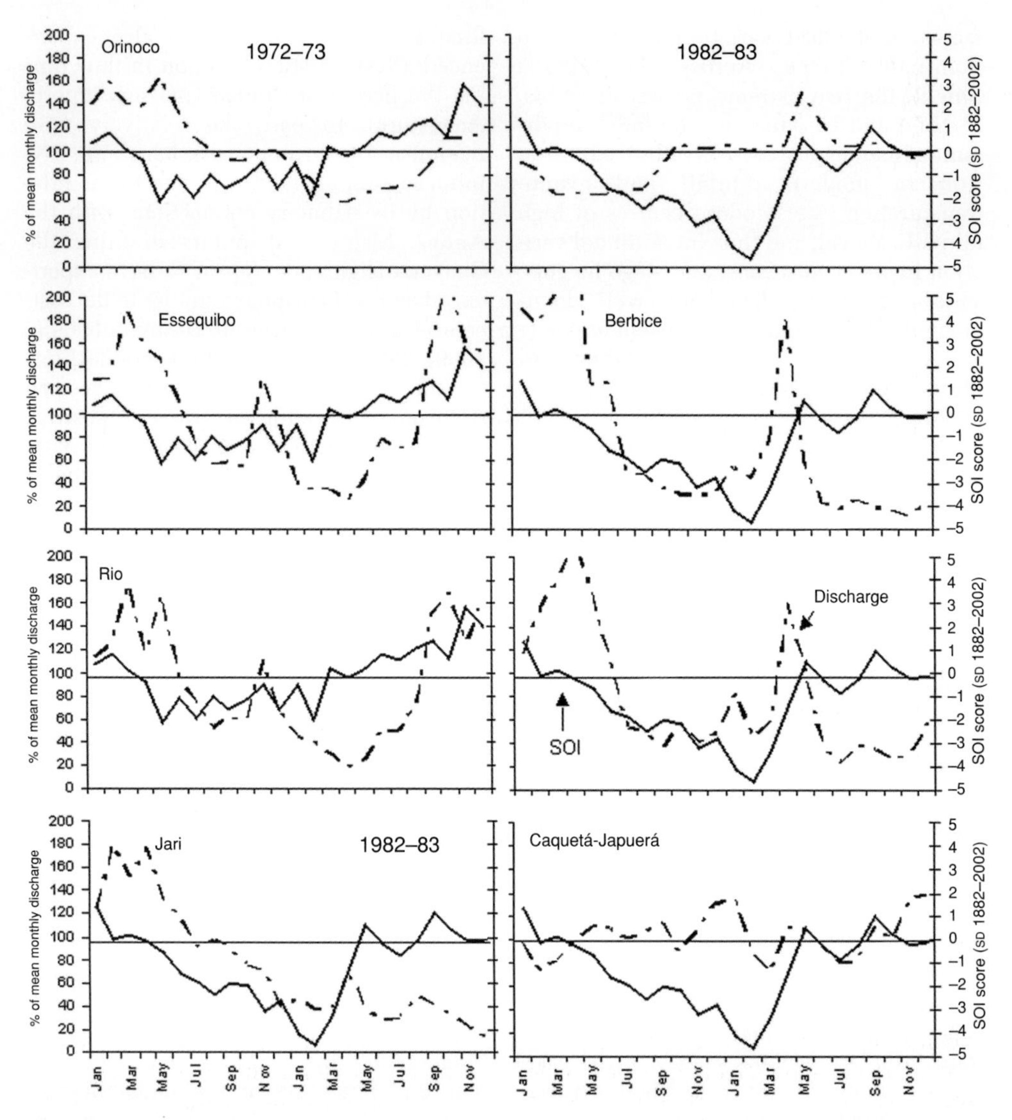

Fig. 2.33. River discharge responses to strong (1972–73) and very strong (1982–83) ENSO events across the Guiana Shield (top row) and southern Amazonia (bottom row). Monthly discharges during the ENSO events are expressed as a percentage of the long-term average for each month (=100). Data sources: see Notes for further details.

ulate and dissolved sediments that reflect the unique combination of lithological base, soil substrate and geographic position of their tributary headwaters (Sioli, 1975; Meybeck, 1994). Limnological conditions of freshwater systems also play an important role in the evolution and maintenance of aquatic life in the tropics (Lowe-McConnell, 1987; Goulding *et al.*, 1988) and represent the most important terrestrial mode of material flux on the planet (groundwater, wind and ice are other major pathways) (Lerman, 1994). Globally, tropical river systems account for an estimated three-quarters of the dissolved silica, nearly half of ions, over half of the organic carbon and one-third of the total suspended matter reaching oceans each year from only one-

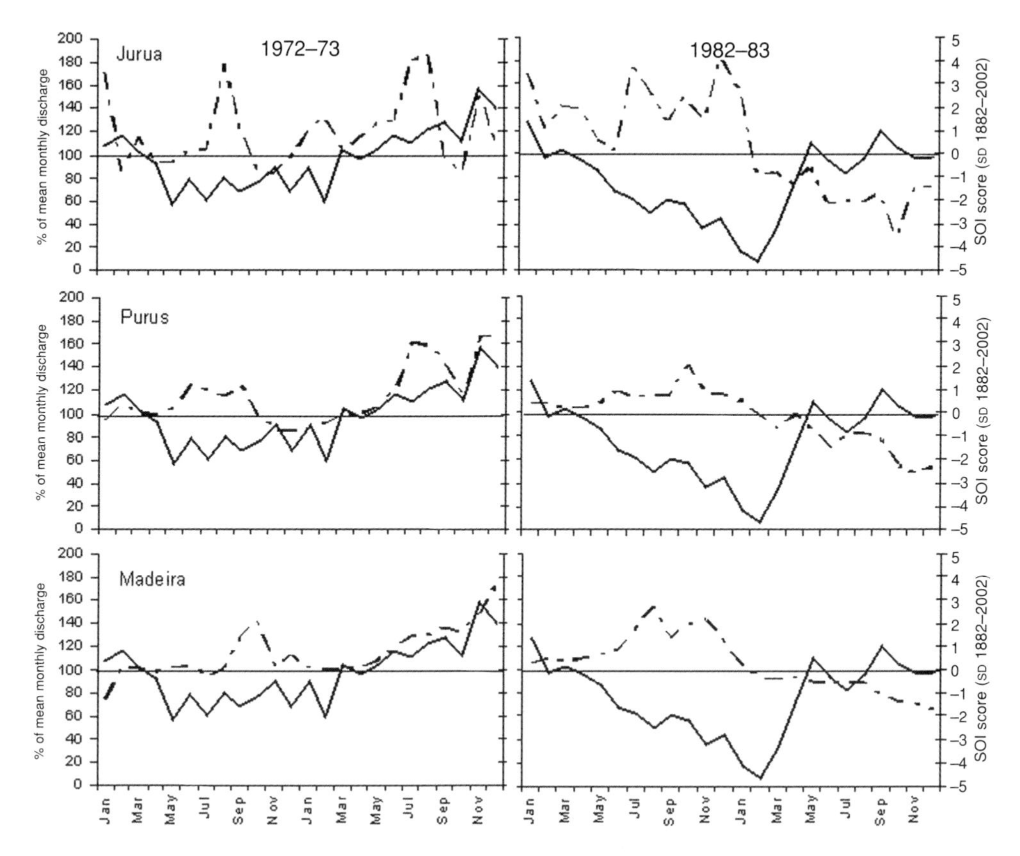

Fig. 2.33. *continued*

third of the total drainage area (Meybeck, 1982, 1988).

Up until very recently, the physico-chemical transport of materials was poorly studied across most of the Guiana Shield, consisting of a few, short-term measurement series at scattered locations (e.g. Edwards and Thornes, 1970). Studies along the Amazon and Orinoco mainstems were more substantial (Gibbs, 1967; Sioli, 1968; Stallard and Edmond, 1983; Furch, 1984, Irion, 1984), in part due to a productive Brazilian–German biogeochemical research programme in the 1960s and 1970s (for a review, see McClain and Elsenbeer, 2001), but still sparse relative to the spatial and temporal scale of fluvial processes attached to these two giant waterways. Two large international projects, CAMREX (Richey *et al.*, 1990) and SCOPE (Degens, 1982), combined with numerous other smaller studies undertaken over the last two decades or so, have further improved the general understanding of transport features and processes in the region. But a comprehensive network of long-term monitoring sites remains elusive and the combined spatial and temporal variation in the physico-chemical attributes of regional waterways is inadequately resolved. General trends in many of the key features, particularly as these relate to universal links between water chemistry, geomorphology and climate (change) can, however, be tentatively assembled for the Guiana Shield. It is important to note, however, that basic data for many common limnological parameters are not widely available for many significant rivers draining the region, particularly in Guyana, Suriname, North Pará, Roraima and Colombia. Most significant among these deficiencies is the lack of access to good data from Guyanese rivers, despite the lithological diversity of their drained watersheds and central position within the hydrography of the shield region.

PH – ACIDITY Waterways of the Guiana Shield register some of the lowest pH values in the neotropics, generally ranging from 4.5 to 7.0 along lower segments of the larger waterways (Fig. 2.34). Much smaller tributaries can show average pH readings of less than 4.0, but this depends primarily on the lithology of the drainage area. Waterways draining the north Amazon Basin are typically lower than those draining the Guiana Basin. This is principally a function of western shield tributaries of the Rio Negro watershed that have pH values consistently below 4.0 (Sioli, 1957; Furch, 1984; Seyler and Boaventura, 2001) and northern shield tributaries of the Orinoco River that have pH values consistently between 5.0 and 7.0 (Depetris and Paolini, 1990; Cressa *et al.*, 1993). Rivers draining the eastern shield area show a mixture of both low pH (Berbice, Maroni, Nickerie, Essequibo) and higher pH (Rio Branco, Trombetas) waters, but nearly all rivers can be classified as acidic.

All else being equal, tropical waters will typically show higher acidity values than temperate streams, but the role of watershed lithology can override the effects of tropical temperature on aquatic pH (hydrogen ion concentration). Waterways draining rock formations containing relatively high concentrations of acidity-buffering calcium tend to have a near-neutral to basic pH. The absence of significant calcium-bearing weathering surfaces in the lowlands of eastern Amazonia and the Guiana Shield generally prevents buffering that typifies most western Amazonian waterways sourced in the eastern slopes of the Andes (Fig. 2.34), but isolated exposure of carbonate sedimentaries in some depositional environments of the shield can lead to local exceptions (e.g. Takutu River: Berrangé, 1977).

A lowering of water pH can also occur with an increase in atmospheric carbon dioxide concentration, oxidation of nitrogen and sulphur emissions ('acid rain') (see 'Non-volcanic aerosols', above), volcanic emissions that form hydrochloric acid, sulphuric acid products of iron pyrite weathering, soil carbon dioxide release through oxidation of organic matter and the release of organic acids from plants and animals (Lerman, 1994). Of these, biological sources of hydrogen ions are the greatest contributors to the consistently low modern pH values of Guiana Shield waterways.

A biological provenance of hydrogen ion concentrations in surface waters of the Guiana Shield is in part facilitated by the quartz-rich lithologies that dominate the region and boost leachate transport through their relatively high infiltration capacities. High rates of necromass decomposition and microbial action in a highly acidic soil matrix delivers these leachates through infiltration and groundwater flow from the many sand-dominated substrates across the shield. These organic leachates are composed almost entirely of two carbon polymer groups, humic and fulvic acids (Stevenson, 1994) that act to reduce aquatic pH.

BEDLOAD TRANSPORT Most large materials are too heavy to be transported in suspension and these slide or jump (saltate) along the river beds with the current. Globally, bottom transport is estimated at 1.5×10^6 t/year, or around 10% of the total estimated sediment received by the world's oceans each year. Bedload sediments consist of materials with grain sizes typically larger than 60 μm, ranging from polycrystalline rock fragments (when spherically abraded, forming gravel) and grains (quartz sands) to simple minerals (such as feldspar, some quartz sands, heavy minerals), but the size distribution and material type also critically depend on various other factors, such as rock provenance, flow rates and pH (Nordin *et al.*, 1983; Irion, 1990; Guyot *et al.*, 1999).

Bedload transport along major rivers draining the Guiana Shield is nominal and consists almost exclusively of quartz sands intermixed with varying concentrations of heavier minerals (e.g. gold placers). Low

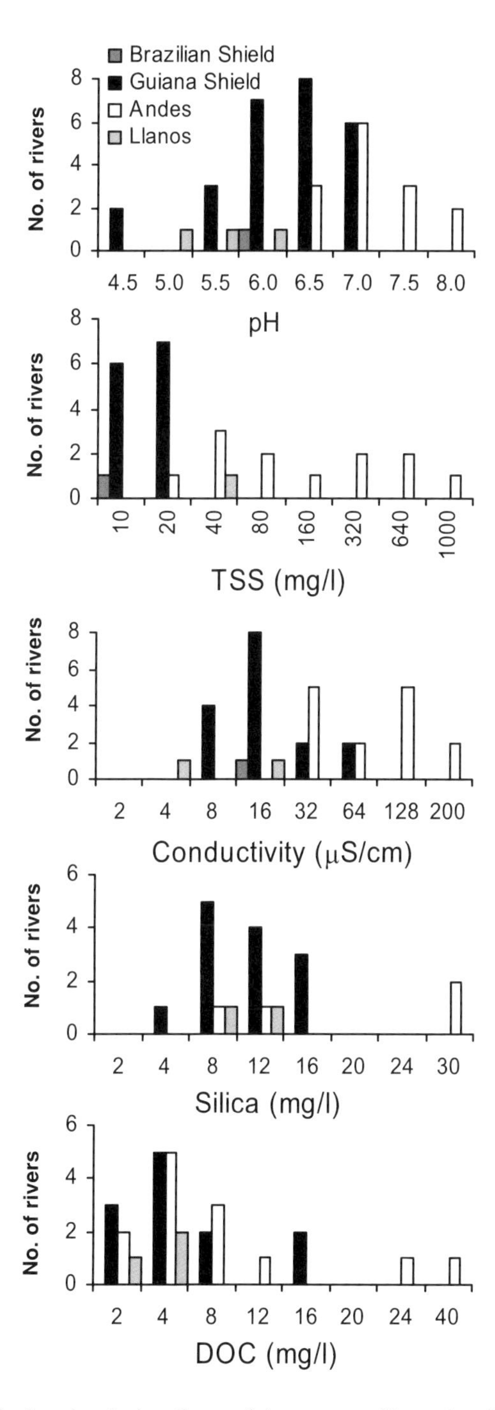

Fig. 2.34. Ranging of physico-chemical attributes of rivers grouped by region. Data sources: see Notes and main text for details.

flow velocities combined with dominance of crystalline basement lithologies and high acidities create conditions that lead to quartz dominance of bedload materials in most rivers draining the Guiana Shield. Potter (1978) ascribed the predominance of quartz to trailing edges of tectonic plates (i.e. shield areas) that typically show bedloads with a quartz:feldspar:rock fragment ratio of 7:1:2. In contrast, along collision rims of plates (e.g. the Andes) the ratio is typically 3:2:5. Rivers draining the Guiana Shield, such as the Rio Negro and Orinoco, show extreme quartz bedload dominance with average ratios of 9:1:1, some of the highest in the world (Young, 1976; Franzinelli and Potter, 1983; Franzinelli and Igreja, 2002).

SUSPENDED SEDIMENTS Bedload transport globally, however, accounts for only around 10% of the terrigenous sediment estimated to enter the oceans each year. In the Amazon mainstem, this figure declines to less than 2% of total sediment load (1–2% in Richey *et al.*, 1986 – but sampled 1982–84, a very strong EN event). Of far greater significance is the material exported in suspension, estimated at 13.5×10^9t each year, or 90% of the total global load (Milliman and Meade, 1983). Of this, 85% is believed to consist of minerals (Degens *et al.*, 1990). Minerals with a grain size less than 60 μm (fine suspended sediments (FSS)), typically silts and clays, form the lion's share of suspended matter, although much larger materials (coarse suspended sediments (CSS)) can also be transported in this manner, depending on current velocities and material densities (Meade, 1985).

Most rivers draining the Guiana Shield are consistently characterized by some of the lowest concentrations of total suspended solids (TSS) (collinear to turbidity (NTU)) known for neotropical waterways, averaging between 1 and 20 mg/l (Depetris and Paolini, 1990; Meade *et al.*, 1990; Weibezahn, 1990; Lointier, 1991; Cressa *et al.*, 1993 and references therein; Haripersad-Makhanlal and Ouboter, 1993; Vargas and Rangel, 1996b). In most instances, this represents a concentration range that is 1–3 orders of magnitude lower than averages recorded from rivers with Andean-sourced headwaters (10 to >1000 mg/l), though there is some overlap (Gibbs, 1967; Milliman and Meade, 1983; Sioli, 1984a and references therein; Depetris and Paolini, 1990; McClain *et al.*, 1995, Seyler and Boaventura, 2001) (Fig. 2.34). This asymmetry in sediment transport has been particularly well-resolved in the case of the Orinoco and its tributaries. Meade *et al.* (1990) estimate that tributaries draining the Guiana Shield make up no more than 5% of the Orinoco mainstem sediment load, despite contributing half of the water that it discharges, on average, each year. Only near the tide-influenced outlets of coastal rivers in the region are TSS concentrations likely to rise substantially as offshore fluid muds, upstream tidal flow and coastal agricultural run-off combine to hyper-concentrate suspended sediments along the lower reaches (e.g. Surinamese rivers (Amatali, 1993)).

Several rivers draining the central and eastern interior of the Guiana Shield, such as the Rio Branco or Rio Parú, are also known to register TSS concentrations as high as 100 mg/l. Irion (1984) notes that these rivers are actively channelling through weathering deposits rich in kaolinite. The TSSs exported through these rivers consist entirely of this kaolinite with some quartz, a quality that distinguishes them (and other shield waterways) from Andean-draining rivers carrying similar concentrations of suspended sediments (see 'Clay fractions', below). The Japuerá drains the western shield region and is also characterized by TSS at the upper end of the Guiana Shield range (28–30 mg/l), but receives much of this from its headwaters in the eastern slopes of the Colombian Andes. Based on other limnological attributes (Eden *et al.*, 1982), tributaries of the Japuerá exclusively draining the shield region are likely to have TSS concentrations similar to those feeding the Negro and Trombetas Rivers (5–15 mg/l) (Seyler and Boaventura, 2001). The wide distribution of TSS values for rivers sourced in the Andes depicted in

Fig. 2.34 reverses this logic. The Sub-Andean Trough acts as a giant sediment trap and TSS values upstream of this circum-Amazonian belt are in most instances defining the upper limits of TSS for these rivers. For example, at Angosto de Bala (12.3° S) along the upper reaches of the Beni River in Bolivia, Guyot *et al.* (1993) measured an average TSS concentration of 6348 mg/l (!) for March 1988, the seasonal peak in discharge (see Fig. 2.32), highest dilution, and therefore, annual minimum TSS concentration. Samples taken downstream of the Sub-Andean Trough characterize the lower limits of TSS values, reflecting both the loss of sediment to deposition in the trough and subsequent dilution by joining downstream waterways with little suspended load. Together, samples taken above and below the trough contribute to the wide range of TSS values gauged for Andean-sourced waterways.

A number of factors combine to create the low TSS environment of the Guiana Shield. Irion (1990) suggests TSS depends on four main factors: (i) tectonic uplifting (mountains); (ii) climate and vegetation cover; (iii) composition of bedrock material; and (iv) age and morphology of surfaces. Meybeck (1994) defines these factors more specifically as: (i) lithology; (ii) atmospheric aerosols; (iii) evapotranspiration; (iv) rainfall; (v) temperature; (vi) contact with rock; (vii) geomorphologic features; (viii) vegetation; (ix) past geological history; (x) lake retention; and (xi) tectonic and volcanic controls. Against all of these criteria, the modern state condition of the Guiana Shield places it at the lowest end of the predicted suspended load continuum. Since all rivers with large sediment loads invariably originate in mountains (Milliman and Syvitiski, 1994), this limits suspended solids production in the Guiana Shield to the scattered mountain peaks and chains of the Guayana Highland and Tumucumaque Upland regions. These highlands, however, do not typically reach elevations extending above the tree-line, limiting the area of rock and soil readily exposed to erosive forces in the absence of significant biological activity, a major contributor to dissolution rates (Eckhardt, 1985). Relatively high primary production across most of the shield's elevational gradient attests to the fact that virtually all of the landscape remains within a high temperature and rainfall envelope, except for the highest summits of the tepuis and peaks that form the Guayana Highlands.

Where highland formations are susceptible to physical weathering, they almost always show high silicate composition (e.g. Uatuma metavolcanics (Gibbs and Barron, 1993); Tumuc Humac inselbergs (Grimaldi and Riéra, 2001)) and/or, where they reside atop basic parent rock (e.g. Avanavero Suite), have surfaces protected by highly resistant lateritic duricrust formed from iron oxides, arguably the most stable minerals on the planet (Press and Siever, 1982). Both lithologies hinder physical weathering rates, even under tropical conditions, contributing to low TSS in the surface waters draining the Guiana Shield. Finally, the age of most geological formations in the shield and the virtual absence of significant new rock creation over the last 70+ million years (Cenozoic) has led to the selective extinction of highly weatherable lithologies across the region. Only the least weatherable crystalline basement formations and their unlithified sedimentary products remain, along with the last remnants of a once thick, Precambrian cover of mixed quartzic and arkosic sandstones, conglomerates and shales (Gibbs and Barron, 1993) that boost TSS in rivers draining parts of the Guayana Highlands, such as the Ireng and Cotingo tributaries of the Rio Branco (D. Hammond, personal observation). Isolated exposure of calcrete formations along the Takutu River in the south Rupununi, Guyana/Roraima, Brazil (Berrangé, 1977) also increase TSS contributions from this tributary.

CLAY FRACTIONS Most of the fine particulates that make up TSS in rivers are clay minerals. In the neotropics, most clay sediments in rivers can be classified as illite, smectite (montmorillonite), chlorite, kaolinite or gibbsite, but their relative contributions to TSS vary considerably between regions.

Illite (hydrous mica, clay mica) accounts on average for 45–60% of TSS clay fractions globally, making it the most common suspended clay mineral (Irion, 1990). It comprises 20–25% of the Amazon and Orinoco River mainstem suspended clay loads. This reflects both high contributions from Andean headwater tributaries (e.g. ~60% in Rio Beni, Bolivia (Guyot *et al.*, 1993)) and very low or no contributions from rivers draining shield areas (e.g. <1% in Rio Negro (Irion, 1984)). Illite is a weathering product of mountain granites, gneisses and slates (effectively cemented and compressed illites) and these surfaces are poorly represented in the Guiana Shield. X-ray crystallographic analyses of a wide range of shield soils show only traces of this mineral (see 'Soils and Soil Fertility', above).

Smectite (montmorillonite, bentonite) is the second most common component of river clay sediments globally, averaging 30–40% of suspended content (Irion, 1990). This group typifies western Amazonian rivers draining the Peruvian and Ecuadorian Andes, where it can account for more than three-quarters of the total clay fraction (e.g. >80% of Purús and Jurúa Rivers). The TSS clay mineralogy of the Beni, a major tributary of the Madeira, has also shown very low (4%) smectite contents during dry periods (Guyot *et al.*, 1993). It typically accounts, however, for 40–50% of the Amazon's clay load, but forms only a very small part of the (already low) suspended clays in the Orinoco (<10%), Negro (<1%) and other shield-draining rivers. Smectite originates from Tertiary (25–70 Ma BP), fine-grained sediments that cover large areas of the Sub-Andean Trough but are far less common in the sedimentary depressions of the shield (see 'Shield macro-features', above). Most sediments inhabiting the depressions are coarse-grained, quartzic materials (Gibbs and Barron, 1993) and this is reflected in the low smectite contributions to the TSS clay fraction discharged by rivers draining these sedimentary basins.

Chlorite typically makes up 15–25% of the global TSS clay fraction in rivers. On average, 20% of the Amazon clay fraction is chlorite, with some large tributaries draining the northwestern Andes (e.g. Huallaga, Napo) registering upwards of 50% content. Southern Amazon rivers (e.g. Madeira, Beni) linked to the Bolivian Andes register chlorite contents (~20%) more consistent with the Amazon mainstem (Martinelli *et al.*, 1989; Guyot *et al.*, 1993). In contrast, sampling of the Orinoco and Negro TSS load has yielded only trace quantities of chlorite in the clay fraction (Irion, 1984). Like illite, chlorite is a typical weathering product of alpine bedrock and opportunities for its production are far less common in the shield region in comparison with the western rim of Amazonia.

Kaolinite is a relatively scarce component of river TSS clay mineralogy at a global scale (<10% of average content), but underpins a fundamental contrast in the spatial occurrence of clay sediments throughout the river systems of the lowland neotropics. Unlike other minerals, kaolinite is dominant within the TSS clay fraction of rivers draining eastern shield regions. It accounts for >90% of the clay suspended in transport along the Rio Negro (Irion, 1984) and upwards of 40% of the Orinoco's clay content, but a much smaller fraction in the Amazon (~14%) and its western tributaries (e.g. Beni, ~13% (c.v. 0.39) (Guyot *et al.*, 1993)). In general, kaolinite dominance of the clay fraction occurs downstream along the highest stream orders in the Amazon (Seyler and Boaventura, 2001). Kaolinite is the (near) final chemical weathering product of silica-rich parent rock and also dominates the soil clay fraction across the Guiana Shield (see 'Soils and Soil Fertility', above) (e.g. Schulz, 1960; van Kekem *et al.*, 1996; Duivenvoorden and Lips, 1995; Franco and Dezzeo, 1994). This dominance declines in soils of the Central American tropical forestlands (e.g. Barro Colorado Island (Leigh, 1999, pp. 67–78)) and Western Amazonia, where smectites (montmorillonites) become more common.

SIGNIFICANCE OF GEOGRAPHIC VARIATION IN TSS CLAY FRACTION Clay minerals are formed as layered sheets of silica and alumina enclosed by oxygen or hydroxyl tetrahedrons. More importantly, the range of clays found in river suspended sediments and tropical forest soils exhibits different bonding properties that produce different affinities for both major and minor cations that are important for plant growth and determine standing levels of soil fertility. Strong hydrogen bonding between the silica and alumina sheets forming kaolinite limit the availability of negative charge sites. As a consequence, fewer cation-bearing water molecules are adsorbed. This creates a clay matrix that has limited hydration capacity and very low cation activity. Bonding between layers forming the crystalline structure of illite clays occurs as the formation of its base unit, a 2:1 complex of silica and alumina sheets, creates a net negative charge that is balanced by attracting potassium (or other) cations. These interlayer bonds are not as strong as the hydrogen bonds of kaolinite, and illite is capable of exchanging cations more readily, increasing fertility. Clays of the smectite group exhibit a 2:1 silicate layering similar to illite, but without silicon replacement by aluminium. Silica sheets stacks are thin and ephemeral, maintained only by very weak van der Waals' forces. Stacks disassociate easily as water and exchangeable ions attach freely to the negative charge sites between sheets. As a consequence, smectites generally have excellent hydration properties and support good cationic activity, both features that give soils with appreciable montmorillonite content a higher fertility in the tropics. The regional variation in abundance of FAO soil types with relatively higher CEC–pH/lower Al–Fe saturation (e.g. Vertisols, Ali-Acrisols, Lixi-Luvisols, transitional Nitisols) favours western Amazonia, Central America and to a lesser extent, the Amazon Downwarp (see Fig. 2.14).

Rivers draining the Guiana Shield carry a very small suspended sediment load containing a clay fraction dominated by kaolinite. This indicates that they play an insignificant role in the redistribution and replenishment of mineral nutrients from headwaters to lowlands, despite very substantial specific discharge rates. The transfer role of rivers headwatered in the Andes would seem altogether greater. Substantially greater TSS loads dominated by relatively more active smectite and illite groups are transferred from Tertiary sediments and highland bedrocks to the sedimentary plains of the Sub-Andean Trough, where most larger, non-clay minerals suspended in the discharge are deposited (Irion, 1984; Linna, 1993; Guyot *et al.*, 1999). Thus, western lowland Amazonia and the Amazon Downwarp are in receipt of an eternal external influx of exchangeable cations originating from an effectively limitless erosion surface. The subsequent insurgence of mainstem waters into the floodplain and 'spiralling' (after Richey *et al.*, 1990) lateral exchange of materials between these promotes deposition and long-term storage of Andean-derived clay fractions across the Sub-Andean and Amazon lowland floodplains (Meade *et al.*, 1985). Only substantial changes in climate (see 'Prehistoric climates of the Guiana Shield', above) would appear capable of modulating this transfer process and altering the balance between the modern ineffectiveness of river transport as a mode of nutrient redistribution across the Guiana Shield and the highly effective transfer function of rivers coursing the elevational gradient of the eastern Andean piedmont.

PARTICULATE CARBON Not all suspended matter in rivers originates from rock and sediment weathering. Approximately 2.25×10^9 t, or 15%, of the global sediment flux from rivers to oceans is believed to consist of organic matter (Degens *et al.*, 1990). Particulate carbon transported by rivers is typically classified as organic (POC) or inorganic (PIC). In the Amazon Basin, POC typically accounts on average for less than 2% of the TSS being delivered to the lower mainstem Amazon (Richey *et al.*, 1990; McClain *et al.*, 1995; Table 16.3 in Seyler and Boaventura, 2001). Rivers draining Andean carbonate rock formations can show high PIC concentrations at upper elevations. These dissolve rapidly with descent into warm, relatively acidic lowland reaches and PIC is practically absent from (neo)tropical river systems. For many rivers draining the Guiana Shield, the movement of POC as fine and coarse fractions of leaves, wood and other detrital matter can account on average for a much greater part of TSS (lower Caura: 30% (García, 1996); upper Rio Negro: 15–19%; Trombetas: 5%; Branco: 6% (Seyler and Boaventura, 2001); Orinoco: ~2–6% (Depetris and Paolini, 1990)). It is important to emphasize, however, that these higher contributions reflect on the very low concentrations of minerals, rather than unusually high POC, in Guiana Shield rivers (Meybeck, 1982; Fig. 15.14 in Devol and Hedges, 2001). Overall, POC appears to vary over the same magnitude in rivers draining both shield and downwarped landscapes, suggesting that coarse, large-scale geological features inadequately explain variation that may be better ascribed to smaller scale source (e.g. vegetation type and distribution, topography) and transport (flooding, impoundment) effects.

The flow of particulate carbon from land to ocean via rivers can serve an important storage function with the global carbon cycle. Contributions to both freshwater and estuarine biological productivity as well as storage in sea-shelf sediments represent non-trivial storage compartments. DeMaster and Aller (2001) estimate that 86% of the POC discharged at the Amazon mouth travels directly to the seabed and 31% ultimately remains buried, accounting for 70% of the total POC deposited in the Amazon shelf sediments (30% is of marine origin). Nearly 10% of the Amazon POC discharge is transported northwestward along the Guiana coastline to form a significant part of these shelf deposits.

DISSOLVED SEDIMENTS While the Precambrian age of the shield has largely exhausted modern suspended sediment production capacity, a long history of highly acidic conditions has catalysed chemical weathering of land materials and their movement as dissolved sediments. Dissolved sediments are typically grouped as: (i) cations and elements (major, trace and total (TZ^+)); (ii) silica (DSi); and (iii) organic (DOC) and inorganic (DIC) carbon.

CATION CONCENTRATIONS The concentration of major cations is the product of pH-dependent leaching and dissolution processes combined with the prevalence of clay types with varying cation carrying capacities (see above). Cation load depends on many of the same factors that determine the rate of suspended sediment accumulation in river mainstems and TZ^+ can vary widely between adjacent waterways draining small catchments with differing lithologies in the Andes (McClain *et al.*, 1995; Sobieraj *et al.*, 2002). Their concentration at any given time, like TSS, is also a function of dilution and therefore related to the seasonal variation in discharge rate (Meade, 1985).

The more active weathering associated with the Andes produces significantly higher average TZ^+ values in rivers draining into the western Amazonian lowlands compared to shield-draining waterways (Gibbs, 1967; Stallard and Edmond, 1983; Cressa *et al.*, 1993). Generally speaking, the latter rarely achieve average TZ^+ values greater than 200 μeq/l (Group 1 rivers after Stallard and Edmond (1983)). In a comparison of 50 major world rivers (watershed area $>10^5$ km^2), the TZ^+ value of Rio Negro waters was the lowest, by an order of magnitude (Meybeck, 1979). Cation sums for shield-draining rivers in the Guiana Basin exhibit similarly low concentrations (Cressa *et al.*, 1993; e.g. Maroni, in Négrel and Lachassagne, 2000; e.g. Caura, Caroni, Orinoco, in Colonnello, 2001). Another proximate measure of cation availability, the conductivity of water, shows a similar trend in relation to geographic origin of headwaters (Fig. 2.34). Conductivity values for shield-draining rivers generally range up to 64 μS/cm and Andean tributaries nearly three times this maximum (Fig. 2.34). Many rivers draining tropical forests in Panama and Costa Rica also show exceptionally high conductivity (e.g. 53–480 μS/cm in La Selva, CR; Sanford Jr *et al.*, 1994). Values for shield and sub-Andean regions, however, typically overlap, given their sensitivity to changes in discharge, dissolution and deposition rates affected by varying rainfall, temperature, pH and elevation.

DISSOLVED SILICATES (SILICIC ACID) Silica discharged from rivers constitutes more than 80±20% of the average annual influx to oceans globally, estimated at 169±56 megatonnes (Tréguer *et al.*, 1995). Nearly three-quarters of the riverborne material arrive via tropical waterways (Meybeck, 1994) and rivers draining northeastern South America are believed to account for around 20%, or 27.4 megatonnes (based on above estimate) of the global silica discharge to oceans each year. The Orinoco discharge probably accounts for approximately 10–15% of this amount (based on 1983–84 data from Paolini *et al.* (1987)) of which 50% (approx. 1.4 megatonnes) is thought to be sourced from the Guiana Shield (Edmond *et al.*, 1996). Rivers draining the Guianas probably contribute between 1 and 3 megatonnes (Lointier, 1991, 1993), but good data on dissolved silica fluxes are largely absent for most rivers draining Guyana and Suriname.

The super-abundance of silica-rich crystalline rocks in the shield limits overall weathering rates compared to other regions with more diverse, and recent, lithologies. But average concentrations of dissolved silicates (in the form of silicic acid, H_4SiO_4) in rivers draining the Guiana Shield appear comparable to other lowland rivers draining the Colombian and Venezuelan llanos and less than or equal to those originating in the Andes based on information available (Stallard and Edmond, 1983; Cressa *et al.*, 1993; Sobieraj *et al.*, 2002) (Fig. 2.34). Seasonal variations in dissolved silica, like TSS and TZ^+, show a negative correlation (or positive lag) with discharge rate (e.g. Lointier, 1993) and unbalanced sampling across the neotropics may belie regional uniformity given the latitudinal gradient in peak discharge (Figs 2.31, 2.32) and impacts of ENSO (Fig. 2.33). Silica dissolution far outstrips the chemical weathering of other cations in the Guiana Shield, however. This is reflected in its position as the major constituent of total dissolved solids transported in rivers draining the region (Edmond *et al.*, 1995). It forms a relatively minor component of the TDS concentration measured along rivers originating in the Andes, but principally due to the very high cationic, rather than excessively low silica, concentrations (Cressa *et al.*, 1993; Edmond *et al.*, 1996).

The absence of significant carbonate, modern volcanic or evaporite formations in the shield limits neutralization of high acidity through action with minerals in parent rock (Fölster, 1985). This creates optimal soil pH conditions for crystalline rock weathering and silica dissolution in the presence of an erosional surface dominated by this mineral. In the Andes, reduced biological activity and temperature combine with contributions from carbonate deposits to maintain higher pH levels (Fig. 2.34) and reduce dissolution potential of silica. Instead, physical erosion of the extensive surface bedrock exposed in the highlands combines to drive silica movement into rivers mainly as part of the suspended and bed loads, with chemical dissolution occurring downstream. This balance is reflected in the much higher silicate weathering rates assessed for the Andes in relation to lowland Amazonia. These higher rates support a decrease in soil profile thicknesses in the Andes, Sub-Andean Trough and Amazon Downwarp while soil profiles are deepening across the Guiana Shield (Mortatti and Probst, 2003).

Dissolved silica exported through rivers is thought to play an important role in marine biological uptake and storage of carbon dioxide (Wollast and Mackenzie, 1983). Silica is an essential base constituent in siliceous diatom and radiolarian development, both major contributors to offshore primary production along cold and warm water coastal margins, respectively (Milliman and Boyle, 1975). Silica recycling within the phytoplankton community is often minimal (DeMaster and Aller, 2001) and opal silicates released as zooplanktonic pellets are exported to coastal and abyssal floor deposits and/or re-enter solution and are re-exported to the surface via deep water upwelling currents. Changes in the amount of silica delivered to coastal oceanic waters via rivers can profoundly influence the balance of phytoplanktonic production (Egge and Aksnes, 1992; Turner *et al.*, 1998, and references therein), particularly where cold water upwelling influences are relatively weak in comparison to freshwater discharges, as is the case along the western rims of the tropical Atlantic and Pacific basins (e.g. Amazon shelf (DeMaster and Aller, 2001)). Water reservoirs and impoundments that efficiently trap silicates can have a particularly pronounced impact on river delivery rates (Milliman, 1997), and thus coastal productivity in these regions.

DISSOLVED CARBON The carbon loads of most (Table 16.3 in Seyler and Boaventura, 2001), but not all (e.g. Rio Beni, Table 7 in Guyot *et al.*, 1993), South American rivers are characterized by a very large dissolved carbon fraction (e.g. Paolini *et al.*, 1987). Blackwater rivers that typify many watersheds throughout the Guiana Shield (e.g. Rio Negro, Caroni, Caura, Essequibo, Demerara, Berbice, Nickerie) can exhibit high dissolved carbon concentrations, but the range of organic (DOC) fractions for various rivers appears consistent across lowland regions (Fig. 2.34), although the lowest DOC concentrations are found at higher elevation river segments draining the Andean piedmont (e.g. Fig. 4 in Guyot and Wasson, 1994), while POC fractions increase with increasing TSS concentrations (Meybeck, 1982).

Inorganic carbon (DIC) often accounts for an equally large or larger part of the dissolved load, mainly in the form bicarbonate, carbon dioxide and carbonate, although the latter is exclusive to the higher pH of whitewater Andean rivers (Gibbs, 1967; Stallard and Edmond, 1983). The DIC fraction in lowland rivers is principally bicarbonate, but the ratio of bicarbonate to carbon dioxide can vary along a spatial cline in biological activity from (mainstem) thalweg to floodplain (varzea), and therefore CO_2 production through (microbial) respiration (Devol and Hedges, 2001). Alkalinity (the sum of dissolved weak bases) is effectively a measure of this bicarbonate concentration. It is sensitive to pH and the relatively acidic waters of most rivers draining the Guiana Shield and eastern Amazonia express relatively low alkalinity values (e.g. Rio Caura (Lewis and Saunders, 1987; García, 1996), Caroni (Paolini, 1986), Sipaliwini, Tapanahoni (Haripersad-Makhanlal and Ouboter, 1993), Negro (Vegas-Vilarrúbia *et al.*, 1987)). Along the Amazon mainstem, DOC and DIC typically increase downstream as DOC-rich waters draining the shield regions combine with an increase in CO_2-producing varzea and floodplain area to heighten concentrations, particularly during the seasonal rise towards peak discharge (Devol and Hedges, 2001).

The organic fraction originates from both autochthonous and allochthonous sources. The autochthonous pathway centres on photosynthesis, while allochthonous carbon arrives through leaching of plant litter and soil organic matter. As a consequence, carbon arriving as leachates (refractory carbon) is typically of much higher molecular weight, having already been substantially broken down through decomposition and mineralized prior to arrival. Refractory carbon has typically undergone a diagenetic phase, combining with clay and other mineral particles through an adsorption process prior to arrival in the surface water system. In contrast, carbon biologically fixed in the water column is typically of low molecular weight, rich in nitrogen and available for mineralization (labile carbon). This distinction is important in defining the pathways each fraction takes through the carbon cycle as labile carbon contributes significantly to biological productivity, while refractory forms are generally shunted directly into riverine and coastal bottom sediments. Refractory DOC, in particular, forms a significant slow-turning compartment in the global pool of stored carbon. Analysis of isotopic carbon discrimination in materials sampled from the Amazon mainstem suggests that refractory DOC may reside in channel sediments as long as a century (but more typically decades) prior to final coastal discharge (Richey *et al.*, 1990; Devol and Hedges, 2001).

Dissolved carbon in rivers draining the Guiana Shield is principally organic, of high molecular weight and highly refrac-

tory. This reflects upon the diagenetic properties of the soil–DOC interactions. Humic substances (fulvic acids, humic acids, humin) produced through decomposition and varying in molecular weight and ionic charge are filtered through soils varying in clay content (Stevenson, 1994). Soils with little clay, such as Podzols or Arenosols, are heavily infiltrated and probably release both fulvic and humic acid forms of DOC. Humic acids tend to dominate the soil solution in these soil types underlying coastal savanna in French Guiana, although 30–50% remain immobile (Turenne, 1974). Forested ferrasolic soils with a higher clay fraction are more likely to immobilize soluble organic substances with an active charge through clay particle adsorption within the soil matrix, leaving refractory forms with neutral or negative charging (fulvic acids) to pass more quickly into the local streams. Fulvic acids are generally believed to contribute the largest carbon fraction to DOC of blackwater rivers draining the Guiana Shield due to the extensive forest cover, although humic acid contributions are greater in comparison to other Amazonian rivers due to the extensive savanna and podzolic scrubland areas drained in the Rio Negro, Berbice, Essequibo, Caroni, Orinoco and many smaller watersheds (Ertel *et al.*, 1986, cited in Richey *et al.*, 1990; Vegas-Vilarrúbia *et al.*, 1988). Thus, DOC carried by these rivers contributes little to in-channel and offshore productivity (e.g. bacteria (Farjalla *et al.*, 2002)). Their highly degraded state and low biological uptake leaves refractory DOC to form a substantial carbon store within river sediment beds. Organic matter produced through the turnover of living forest biomass is substantially degraded prior to leaching and transport into adjacent waterways.

DOC also readily binds to trace metals to form metallo-organic complexes through surface adsorption, although DOC may constitute only part of a river's complexing capacity (colloidal and particulate carbon also providing binding sites) (Degens *et al.*, 1990; Benaïm and Mounier, 1998). DOC's complexing capacity allows it to act as a transport vehicle for heavy metals, including mercury, arsenic and other materials generally capable of reaching toxic concentrations through bio-accumulation along the natural food chain as benthic feeders ingest carbon-complexed metals. Alternatively, they may be discharged into the oceans and assimilated into the coastal marine ecosystem.

Seasonally inundated floodplains

The seasonal or inter-annual inundation of riverine floodplains is a massive event that affects large areas of the Amazon Downwarp (Junk, 1993) and to a lesser extent, sedimentary depressions in the Guiana Shield (Fig. 2.35). The most significant of these is believed to account for, on average, nearly 3500 km^2 of the Rupununi/Rio Branco savannas, based on analysis of satellite-borne, radar reflectance signatures (Hamilton *et al.*, 2002). Other significant areas of floodplain forest in the region include the Casiquiare Rift–Rio Ventauri valleys of Venezuelan Amazonas (see Fig. 2.9A and Fig. 2.35, No. 4), the Takutu Rift extension across the Essequibo and upper Berbice Rivers in central Guyana (see Fig. 2.9B and Fig. 2.35, No. 3) and the middle Rio Negro (Fig. 2.35, No. 5).

While most floodplain centres across the Guiana Shield are inundated on a regular basis, the extent of inundation can vary by a factor of 4 or more in some locations (e.g. Rupununi maximum:average area flooded annually – 4.75) and generally more than other centres receiving surface water from the Andes (Apure-Meta – 3.1, Beni-Mamoré – 3.1, upper Amazon – 2.1) (Sippel *et al.*, 1998; Hamilton *et al.*, 2002). Given the relatively greater impact of ENSO events on rainfall and river discharge rate variation across the Guiana Shield, a higher variability in seasonal floodplain inundation is predictable.

Seasonally inundated floodplains in the tropics are considered important sources of methane, a potent greenhouse gas (Bartlett and Harris, 1993). Variation in the extent and duration of seasonal inundation can affect the rate of methane emission,

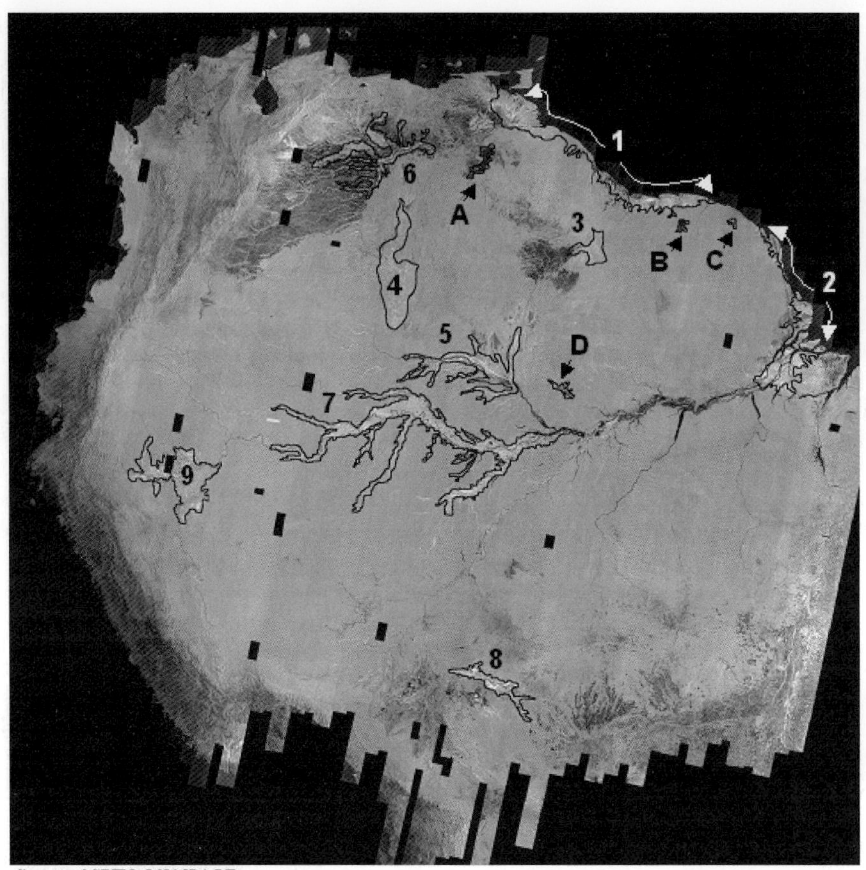

Fig. 2.35. General location of major inundated floodplains (numbers) and reservoirs (letters) across the Guiana Shield and adjacent regions of lowland Amazonia. Based on radar reflectance signatures (resolution=0.5 km) acquired by the JERS-1 satellite from February to July, 1996 (Chapman *et al.*, 2001). Only areas of flooded forest or open water are identified. Significant savanna areas (Rupununi, Sipaliwini) are also known to flood, but reflectances generated by water, grassland and other flat surfaces are not separable at the wavelength utilized by the JERS-1 SAR radiometer. Also note that sequential acquisition by the satellite spanned February–June along longitudinal swathes and all parts of the mosaic are not concordant with the latitudinal lag in peak discharge as presented in Fig. 2.32. Key: (1) Orinoco and coastal Guianan tidal swamps, (2) Amazon-Amapá tidal swamps, (3) Takutu Rift, (4) Casiquiare Rift, (5) Rio Negro, (6) Apuré-Meta, (7) Solimões mainstem, (8) Beni-Mamoré, (9) Pastaza-Marañón. Reservoirs: (A) Guri, (B) Brokopondo, (C) Petit Saut, (D) Balbina.

along with other factors. For example, an equivalent area of inundated forest along the Orinoco is believed to generate significantly less methane than along the Amazon due to differences in the duration and depth of coverage (Smith *et al.*, 2000). ENSO events are likely to reduce inundated areas along river systems. Without a buffering connection to the Andes, the contribution of these areas to total neotropical methane emissions is likely to be considerably less than western or central Amazonian flooded forests (e.g. Pastaza-Maranon) that are less responsive to changing ENSO phases, all else being equal.

Reservoirs

Permanently standing, or lentic, water bodies constitute a relatively minor fraction of the surface water area in the Guiana Shield, unlike lowland areas in the Amazon Downwarp and Sub-Andean Trough, where shallow tropical lakes are commonplace as a consequence of high-energy, fluvial dynamics (Sioli, 1984a; Räsänen *et al.*, 1987; Melack and Forsberg, 2001). Instead, many rivers draining the Guiana Shield consist of anastamosing channels created through constrained downstream transport of their quartz bedloads. In the case of the Rio Negro, the pattern of their migration across the mainstem can form a patchwork of sedimentary atolls each linked to faster-moving sub-channel waters through inlets and outlets that alter their water levels in line with changing river stage height (e.g. Goulding *et al.*, 1988, pp. 22–23). Long-standing inlets formed around geomorphological structures are also prevalent along many waterways of the region, but isolated, perennial lowland lakes are infrequent.

The largest and most significant bodies of year-around standing water are man-made. Large areas of former forest area have been permanently inundated to form reservoirs (*embalses*) along the lower reaches of several shield rivers. All reservoirs in the region have been developed for hydro-electric power generation and four of these account for the majority of area inundated and power produced (Fig. 2.35). These four include Guri Lake along the Caroni (Venezuela), Brokopondo Lake along the Suriname (Suriname), Petit Saut along the Sinnamary (French Guiana) and Cachoeira da Porteia along the Trombetas (Pará, Brazil) (Fig. 2.35). Combined, they cover an area of 6818 km^2 and have a combined hydro-electric generating capacity of 11,576 MW, although Guri, the second largest hydro-electric operation in South America, alone accounts for 86% of total power production.

Apart from the tremendous social issues attached to the creation of artificial lakes, the impoundment and regulation of natural river flow can have considerable effects on downstream hydrological and limnological conditions (WCD, 2000). Reservoirs can also become potent net greenhouse gas sources as DOC and POC is deposited and submerged forest remnants decompose under anoxic conditions, particularly when *terra firme*, rather than existing floodplain forests, are affected (e.g. Balbina, Tucurui dams, Brazil (Fearnside, 1995, 1997)). The initial pulse of methane and carbon dioxide emissions can be quite significant as original forest vegetation begins to decay (e.g. Petit Saut, FG (Galy-Lacaux *et al.*, 1997)). Impoundment of particulate and dissolved solids not only limits the functional economic life of reservoirs, but can lead to significant impacts at the coastal marine and estuarine interface. Restricted, modulated discharge that characterizes reservoir outflow extinguishes seasonal stage height fluctuations, constricts the size of the coastal freshwater plume and effectively eliminates the flow of dissolved solids (Colonnello, 2001), particularly silica, that play an important functional role in maintaining coastal water (phytoplankton) productivity and diversity (Turner *et al.*, 1998; DeMaster and Aller, 2001). Tidal saline wedges also penetrate further upstream as freshwater outflows diminish after dam construction, leading to changes in coastal vegetation (Colonnello and Medina, 1998; Echezuria *et al.*, 2002).

Coastal processes

ESTUARIES AND MUD-FLATS The northern coastline of the Guiana Shield is dominated by a series of prograding and degrading sections shaped by the influx, movement and deposition of clays and silts along the Guiana Shelf (that part of the South American craton current below sea-level). The movement of sediments discharged from the Amazon along the northwest-flowing Guiana Current accounts for the overwhelming majority of inputs and, as a consequence, control of Guianan coastline changes. Of the 11–13 $\times 10^8$ t of sediment estimated to leave the Amazon, on average, each year (Meade *et al.*, 1985), between 19% and 23% is thought to flow northwestward along the Guianan coast as suspended sediment and fluid muds (Wells and Coleman, 1981; Froidefond *et al.*, 1988; Muller-Karger *et al.*, 1988). These consist mainly of clays (Pujos *et al.*, 1989), similar in mineralogical composition to those found in the suspended sediment loads of western Amazonia and the Guiana Shield (see 'Clay fractions', above) (Parra and Pujos, 1998).

Fluid mud banks are large: 50–60 km long, 10–20 km wide and up to 5 m deep (Prost, 1987; Froidefond *et al.*, 1988). They are estimated to migrate northwestward along the French Guianan coastline at an average rate of 900 m/year, 1500 m/year along the Suriname coast and 1300 m/year off the Guyana coastline (Prost and Lointier, 1987). Rates vary in relation to local shoreline orientation relative to seasonal variation in wind stress, the main agent moving fluid mud banks along the coastline. Wind stress favouring northwestward movement along the coast is highest from June to October, when the ITCZ is moving northward, and thus southeasterly trade winds prevail across the shield latitudes. Slightly less fluid components of these mud banks, or 'slingmud', are active depositional sites that dampen wave action and grow land when they become attached to more solid foreshore extensions (Augustinus, 1980). Wave action is focused between the shifting mud banks, eroding the shoreline and creating a moving chain of alternating accretion and erosion zones.

Modern-day materials flowing northwestward along the coastline of the Guianas are almost entirely of Andean origin (Debrabant *et al.*, 1997; McDaniel *et al.*, 1997; DeMaster and Aller, 2001). In the past, putative dry periods (glacial maxima) and relatively low sea-level stages extended the terrestrial boundary seaward towards the edge of the craton. As a consequence, the supply of terrigenous sediments arriving from the Amazon is believed to have been mostly deposited in the deeper abyssal waters (north)eastward of the marine shelf with extinction of the longshore transport load. With nominal sediment being delivered to the palaeo-foreshore of the Guiana coastline from Amazonian sources, coastal waters would have been free of the modern-day fluid muds and high concentrations of suspended sediments. The remains of a barrier reef and presence of quartzic sands derived from shield-draining waterways (Pujos and Odin, 1986) attest to clear, productive coastal water conditions during the LGM that were more similar to present-day conditions in the south Caribbean Sea. Delivery of sediment from shield-draining rivers currently accounts for only a very small fraction of the coastal foreshore sediments (Parra and Pujos, 1998).

An overwhelming Andean provenance of sediment discharged at the mouth of the Amazon and the significant deflection northwestward of river discharge by the North Brazil/Guiana Current establishes an amazing linkage between land accretion along the Guiana Shield's eastern perimeter and physical weathering of the Andes. Combined with eustatic sea-level changes associated with glacial–interglacial periods, changes in the conditions affecting Andean erosion, lowland Amazonian transport and coastal wind-driven advection have governed change in shield coastline evolution and marine biogeochemical features throughout the Quaternary (Prost, 1987; Prost and Lointier; 1987, Pujos *et al.*, 1990).

Sea-level change

One of the most significant and obvious processes that has affected shield hydrology and surface area is sea-level change. Given the relative tectonic quiescence of the shield since the end of the Proterozoic (550 Ma BP) and absence of glacial ice, the cumulative impacts of a changing sea:land surface area in the region largely record adjustments to global seawater displacement and volume generated by events elsewhere, although localized subsidence (e.g. Berbice Basin) and uplifting continued to affect this balance (Gibbs and Barron, 1993). Thus, the range of sea-level change affecting the lowland shield region has been passively constrained by its geological stability and equatorial location. The highly weathered topography would have made it a significant contributor to global land loss during Mesozoic sea-level highstands. But rapid extinction of the crystalline basement

along the edge of the relatively narrow Guiana Shelf would have limited the plausible extension of land eastward during more recent Quaternary low stands.

The fluctuating change in the shield perimeter due to sea-level changes would have altered more than the shift in terrestrial land area. Both hydrology and regional climate, and thus forest cover and composition, of modern coastal and interior regions would have been profoundly affected as the shape and area of maritime influence (e.g. moisture, cation transport, wind stress) over terrestrial systems was modified (see 'Land-borne components', above). Adjusting erosion bevels would have supported alternating deposition and erosion of sediments and the consequent formation of environmentally and economically important deposits, such as peat (pegasse), white sands (quartzipsamments) and bauxite (Gibbs and Barron, 1993).

TIMELINE OF SEA-LEVEL CHANGE The geological, palaeo-oceanographic and marine palaeontological records are the most prominent indicators of past sea-level change. Geological approaches have been used extensively in the Guiana Shield (Gibbs and Barron, 1993), in the Amazon Downwarp (Irion, 1984), along the Orinoco, Guiana and Amazon Shelfs (Flood *et al.*, 1997; Shackleton *et al.*, 1997; ODP, 2003) and the Caribbean Sea (e.g. Jackson *et al.*, 1996; Leckie *et al.*, 2000), in part reflecting the important role that Phanerozoic sedimentary processes have had in concentrating economically important hydrocarbon, bauxite and carbonate (lime) deposits and in part a need to better understand the evolution of land–ocean dynamics in the region and its role in and response to global climate change.

Given the putative Cretaceous age of angiosperm lineages and break-up of Gondwana, the record of sea-level adjustments since the Cenomanian is of greatest direct relevance to modern forest evolution and development in the Guiana Shield. Sea-level is generally believed to have affected the peripheral areas of the shield region as a series of transgression and regression phases over the period following sea intrusion between the separating South American and African cratons, 100 Ma BP (Pitman *et al.*, 1993). The discrimination of these phases is based on proximate indicators of changing sea-level, such as terrestrial deposits of marine sediments, marine deposits of deltaic and lacustrine sediments, submerged and uplifted palaeo-reefs and mangrove sediments, weathering geomorphology and marine–terrestrial chemical signatures.

CRETACEOUS–PALAEOGENE The early Cenozoic witnessed a period of intense tectonic activity that shaped much of the modern-day neotropical realm. The Guiana Shield was in a unique transitional position during this period. Geological evidence indicates that it accounted for most of the terrestrial environment of northern South America, but had already divided from its counterpart region of modern-day West Africa and begun to develop an eastern shoreline environment. Stress fracturing and dolerite dyke emplacement across the region were commonplace as a result of mid-Atlantic rift formation (Choudhuri and Milner, 1971). Importantly, neither the western orogenic belts nor formation and eastward movement of the Caribbean Plate had initiated (Coates and Obando, 1996). Yet, palaeopalynological records indicate that the angiospermous flora was already rapidly developing and by the end of the Palaeogene, 24 Ma BP, many of the plant families that typify the modern forests of the region were well in place (Romero, 1993).[23] While virtually all of the modern-day petrographic features of the Guiana Shield were already in place, Atlantic seafloor spreading, the formation and migration of the Caribbean Plate, and subsequent tectonic uplift of the western and northern Andes over the last 100 million years, led to tremendous reworking of the South American coastline and surrounding Atlantic Basin. Subsidence and uplifting processes altered land–ocean relationships and regional compensatory changes in erosion bevels created a complex landscape that was not uniformly affected by climate-driven changes in sea-level. Glacial ice was apparently absent or insignificant in the Mesozoic world and average sea-level was considerably higher than today (Prentice and Matthews, 1988). As a consequence, the record of early Palaeogene sea-level change in relation to the shield area is occluded by subsequent tectonic events shaping the Andes, Central America and the Caribbean and the prior absence of glacial ice-driven processes. What evidence exists is closely tied to sedimentary-type classification along the shield margins (see 'Prominent geological regions', above) and is principally qualitative in nature (transgressive vs. regressive).

The production of a precise Cretaceous–Palaeogene sea-level curve for the Atlantic margin of the Guiana Shield is misled by the interaction between changing sea-levels due to external forcing and the substantial tectonic activity altering adjacent neotropical regions. This activity would have degraded, diachronously reorganized, or even ablated sedimentary layers recording sea-level high and low stands in the region. Instead, we have to rely on more precise characterizations of sea-level change from other, often extra-tropical, regions and assume that maximum high and low stand signatures dominate sea-level change around the Guiana Shield.

The available evidence, from both field sample analysis and theoretical assessments, shows surprising concordance in linking shield coastline to sea-level changes, albeit without the spatial or temporal precision attempted for Holocene estimates (see below). Expressed relative to modern sea level, the emerging coastline of the Guiana Shield is postulated to have experienced a significant transgression during the late-Cretaceous (Cenomanian–Turonian), classified as the Canje Formation (Naparima Hill Formation – Trinidad; La Luna Formation – Venezuela). The timing of this sedimentary formation correlates very well with a peak in production of glauconies, a common group of iron-rich, green clays precipitated directly on to marine bottom sediments from seawater, and favoured by deep water conditions that typify transgressive sea-level phases (Fig. 1 in Smith *et al.*, 1998). Vail *et al.* (1977) proposed a high stand phase of +350 m above modern sea-level based on coastal aggradation profiles two million years later during the Campanian that coincides with the putative age of the New Amsterdam Formation, although this technique carries significant source errors (e.g. see Christie-Black *et al.*, 1990).

The early Eocene benthic $\delta^{18}O$ record indicates that the planet was the warmest it had ever been (Zachos *et al.*, 1994; Bains *et al.*, 1999). This warming phase would have been associated with a lagging sea-level high stand of substantial magnitude. Another large increase in glaucony production during the late Eocene supports the view of a large transgressive phase directly following the Eocene Thermal Maximum. A chronology of successive transgression–regression phases affecting the eastern rim of a Palaeogene shield is consistent with palaeosolic sequences along the southern rim in northeastern Pará, Brazil belonging to the Piraruçu and Solimões formations (de Fátima Rossetti, 2001) and those constituting the Courentyne Group in Suriname (Krook, 1979). The Cretaceous–Palaeogene was characterized by a relatively young and isolated proto-Atlantic that increasingly became connected with much older basins as tectonic activity pulled the newly separated continents towards their present-day locations. The sedimentary record of sea-level change during this period is relatively scant (Rull, 1999), but available evidence suggests that direct impacts upon a tropical flora already in existence upon the Guiana Shield would have been substantial along a rim not yet loaded with the type and depth of sediment seen today. Uplifting events, such as that affecting the Guiana coast at the end of the Eocene (Gibbs and Barron, 1993), however, may have effectively attenuated the extent of sea ingress during these periods. The sedimentary record of the Rupununi end of the Takutu Graben region does not reflect a marine transgression since the early Jurassic (Berrangé, 1977).

NEOGENE Sea-level change during the neotropical Neogene (24 Ma BP to present) coincides with substantial tectonic uplift and isostatic adjustments across the South American continent, the formation of the Central American isthmus and expansion of the Caribbean island arc. In the early Neogene (Miocene), the northern rim of the Guiana Shield was transformed through its collision with the eastward-moving, southern edge of the Caribbean Plate and the emergence of the Andes across northern Colombia–Venezuela. Geosynclinal subsidence south of the emerging highlands initiated the formation of the northern arm of the Sub-Andean Trough (Fig. 2.6B). Sedimentary strata indicate that this was a large shallow marine or epicontinental inlet during the Miocene (Fig. 6 in Diaz de Gamero, 1996), a date that also is consistent with a registered glaucony peak, and putative sea high stand, from 19 to 15 Ma BP (Smith *et al.*, 1998). Marine deposits in western Amazonian regions of the Andean Trough date to the same period, suggesting that the rise in sea-level also covered much of the lowlands west of the shield (Hoorn, 1993; Räsänen *et al.*, 1995). Deposits along the southern shield rim record a similarly timed sea transgression, followed by a significant regressive phase by the end of the late Miocene (de Fátima Rossetti, 2001). This regression is consistent with interpretations of sedimentary formations along the Suriname coast (Krook, 1979), estimates of sea-level decline made by Vail *et al.* (1977) and a reduction in the area of the Venezuelan llanos affected by epicontinental marine conditions (Diaz de Gamero, 1996).

Constructing a chronology of sea-level change for the latter half of the Neogene (Plio-Pleistocene through Holocene) can draw upon a record of sedimentary evidence that has benefited from both their 'young' age and the timing of their emplacement along the geochronology. Recent deposits have not been exposed to the same breadth of diastrophic change as those emplaced during the late Mesozoic–early Cenozoic. The Andean orogeny had achieved much of its modern-day stature, the Central American Seaway had been closed by the Panamanian Isthmus and the Caribbean had moved into a sub-tropical Atlantic configured very similar to its modern-day size and shape by the middle Pliocene (3.5–3.1 Ma BP) (Coates and Obando, 1996). Over the Quaternary, changes in sea level attributable to eustatic and steric effects (see below) have become more important in driving fluctuations in the shield's coastline position relative to isostatic forces, although epeirogenic movements probably continue to play a formative role in defining variation among different locations within the region.[24]

The record of Plio-Pleistocene sea-level change adjacent to the eastern rim of the Guiana Shield suggests a degree of uniformity in sea-level response to eustatic and steric effects across the western tropical Atlantic. Sea-level heights derived from stacked Barbados coral $d^{18}O$ values[25] indicate high stands at 200 and 125 (^{230}Th) ka BP and alternating declines consonant with, but slightly higher than, eustatic adjustment over the late Quaternary series of glacial stadia–interstadia as calculated by Chappell and Shackleton (1986) for eastern Pacific deposits (Fig. 2.36A). The Barbados record indicates a high stand during these glacial periods above modern sea level (+5 m), as does the uplifted palaeoreef record of the Huon Peninsula, Papua New Guinea in the western Pacific (+6 m) (Bloom and Yonekura, 1990). Based on the profiles of SPECMAP $d^{18}O$ and Chappell and Shackleton's calculations, these are the only probable periods over the last 200 ka where sea level has met or exceeded Holocene levels in the region.

Since the LGM, 18–22 ka ago, sea rise has rapidly increased from a lowstand estimated worldwide at 120 m below present sea level (Fig. 2.36A). Holocene sea-level heights calculated from radiocarbon-dated sediments in the Lesser Antilles, Panama, Venezuela, Guyana, Suriname and Brazil concur with this rapid rise since the end of the LGM (Fig. 2.36B). A lowstand of this magnitude during the LGM would have increased the size of the eastern shield area to the edge of the continental shelf, adding

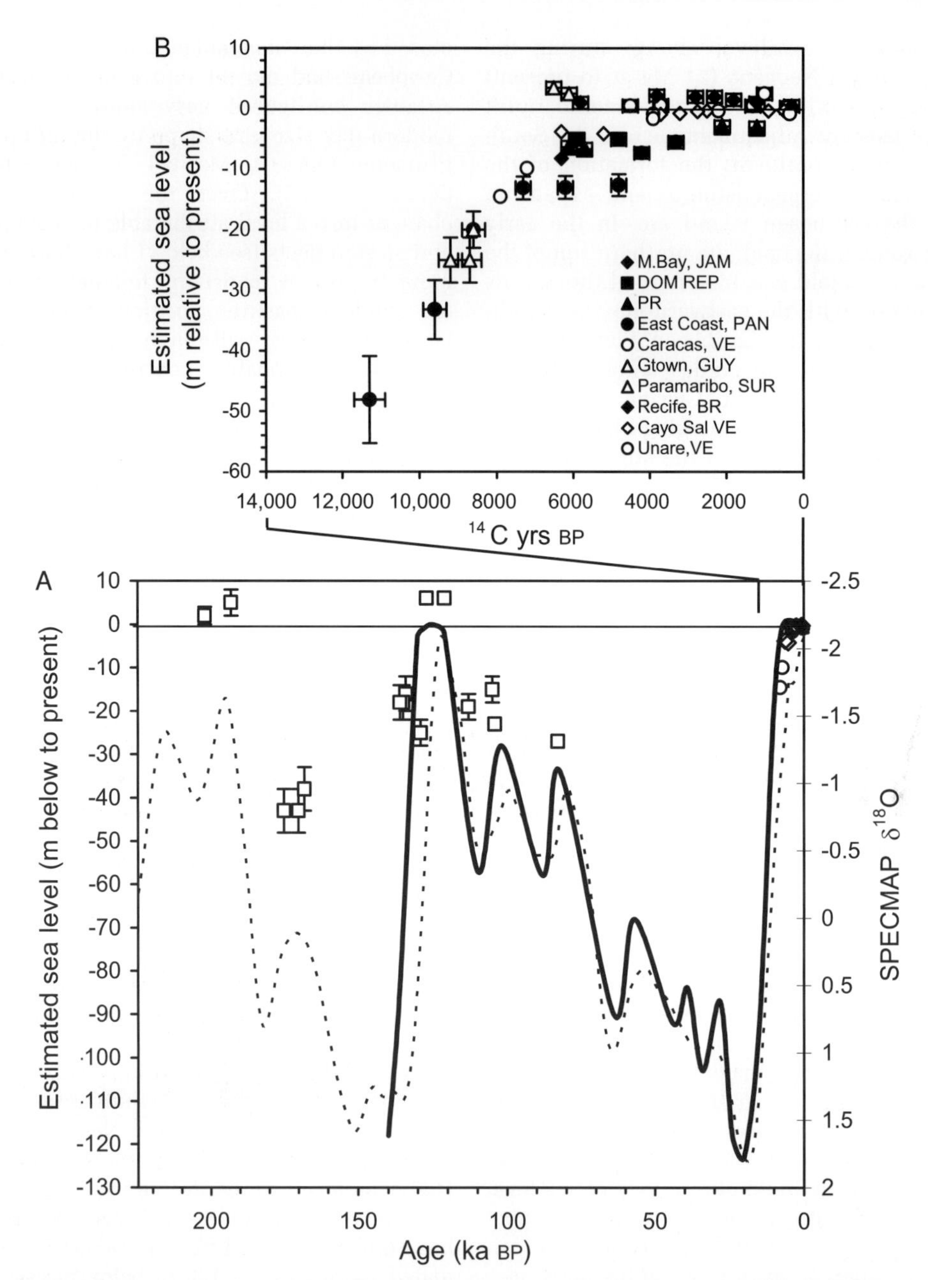

Fig. 2.36. Palaeo sea-level estimates. (A) Global estimates of ice-volume equivalent sea-level variation over last 140 ka (solid line) from Chappell and Shackleton (1986). SPECMAP $\delta^{18}O$ normalized values (dashed line) back to 225 ka BP included for reference through earlier isotopic glacial–interglacial stages (Imbrie *et al.*, 1984). Estimated sea levels for various tropical western Atlantic locations based on ^{230}Th–^{231}Pa ages of Barbados palaeo-reef coral deposits (open squares) (Edwards *et al.*, 1997; Gallup *et al.*, 2002) and ^{14}C ages of coastal Venezuelan corals and mangrove sediments at Cayo Sal (open circles) and Unare (open diamonds). Venezuelan sites also presented in (B) (without error bars) for cross-referencing with estimated Holocene sea-level changes at locations in the Caribbean, Panama and along the Atlantic coast of South America between Caracas, Venezuela and Recifé, Brazil. Data based on means (+/–SE) derived from Tushingham and Peltier (1993) for ^{14}C-dated estimates older than 0.3 ka BP.

approximately 30,000–40,000 km^2 to the Guiana Basin portion of the shield, created a terrestrial connection between Trinidad and the mainland and extended current coastal regions northeastward by an estimated 75 km (northwest Guyana coast) to 175 km (southern edge of Tobago) (Fig. 2.37). Patterns of alternating sediment deposition in the nearshore Amazon delta and offshore Amazon Fan are consistent with a withdrawal of the Atlantic beyond the edge of the shelf at the LGM (Damuth and Kumar, 1975; Milliman *et al.*, 1975) as are curves generated for the Atlantic coast of Venezuela (Rull, 1999) and various Caribbean islands (Pirazzoli, 1991). Radiometrically dated deposits along the Atlantic coastline from Panama to Recifé generate sea-level estimates conforming to the general view of a rapidly rising sea level along the coast of the Guiana Shield over the late Pleistocene–early Holocene in response to glacial retraction and oceanic warming (Fig. 2.36B).

Over the last 8000 years, however, the rate of global sea-level rise has generally declined, but along different regional trajectories (Clark and Lingle, 1979; Pirazzoli, 1991). An analysis of these trajectories along a latitudinal gradient by Bloom and Yonekura (1990), illustrates the substantially different sea-level change response experienced along the coast of Central America, the Caribbean and (by association) the Guiana Shield relative to other continental coastlines rimming the Atlantic Basin. Sites in the Caribbean and Panama show monotonic trends of decelerating sea-level rise, while those at higher latitudes show decelerating sea-level decline. Bloom and Yonekura suggest that this reflects the fact that most of the glacier meltwater had

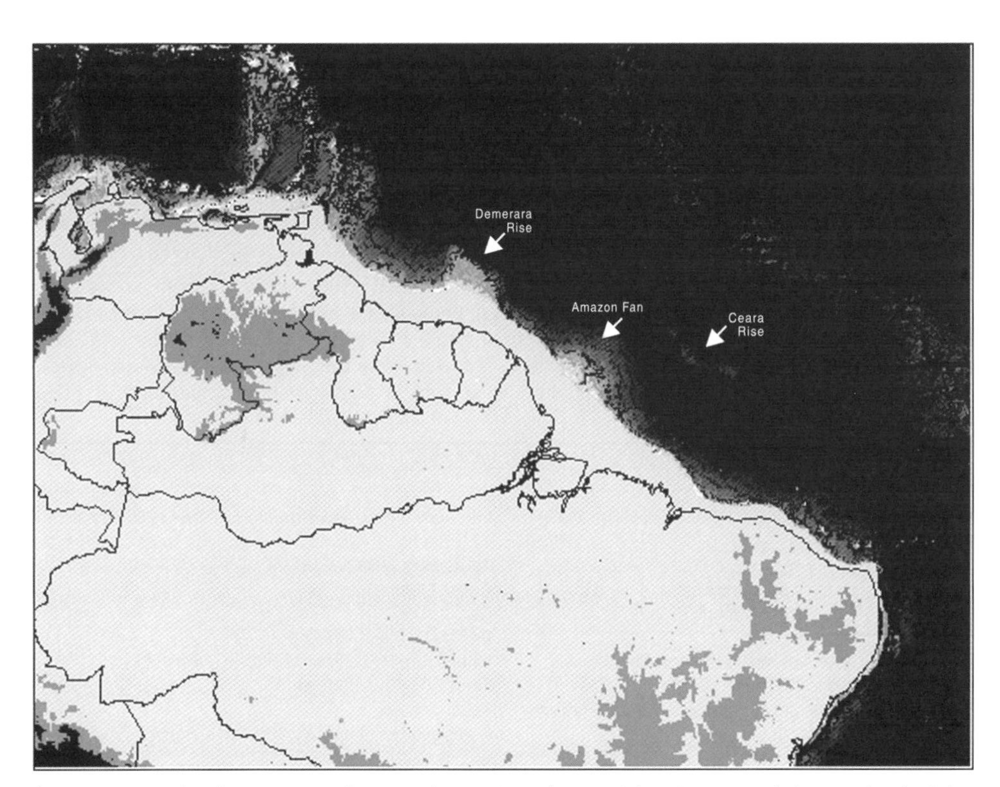

Fig. 2.37. Generalized impression of eastward extension of terrestrial environment during sea-level minimum of the LGM, 20 ka BP (based on data in Fig. 2.36), emphasizing land connections between Trinidad and Guiana Shield. Pleistocene shallow marine environments underpinning oceanographic studies of shield region palaeoclimate and terrigenous sedimentary dynamics of the Amazon Basin are identified.

returned to the oceans by 8 ka BP and that sea-level changes since then largely reflect regional variation in coastal isostatic adjustment (and also steric and epeirogenic effects). Responses among locations along the tropical Atlantic coast support this view. Sites south of Venezuela appear consistently at or above current levels, suggestive of a coastal region dominated by subsiding basins (e.g. Orinoco, Berbice and Amazon coastal regions). Estimates for sites north of Venezuela, in Panama and the Caribbean, show considerably lower sea levels, indicating that these areas have been uplifting (Fig. 2.36B). Williams (1998) similarly concluded from the distribution and stratigraphy of archaeological sites that the littoral zone between the Orinoco and Pomeroon Rivers along the Guiana coastline had experienced subsidence-driven sea incursion over the late Holocene, driving prehistoric Amerindian settlements inland.

MODERN AND FUTURE SEA-LEVEL CHANGE The notion of land- rather than ocean-driven sea-level change is important in realistically predicting future trends in sea-level change. The modern historical record of sea-level change in the western tropical Atlantic region continues to record this spatial variation in isostatic–epeirogenic adjustment processes with little concordance among regional time series (Aubrey *et al.*, 1988). The IPCC assessment of future sea-level change indicates a range of sea-level rise along the Guiana Shield coast of between 0 and 0.5 m over the next 100 years based on the results of nine different models incorporating thermal expansion and oceanic current shifts, but not isostatic or tectonic effects (Church *et al.*, 2001). Concordance among model predictions for the shield margin was one of the lowest of all tropical regions. The western and eastern equatorial Pacific and eastern tropical Atlantic displayed greater consistency among the model predictions.

Notes

[1] The average elevation for the Guiana Shield was calculated from 26,770 values contained in the raster-based elevational coverage GTOPO5 created by the USGS (EDC, 1996).

[2] A table comparing the main neotropical soil groups identified under the FAO/WBR, USDA and Brazilian soil classification systems based on property assignment in USDA (1996) and FAO (1998), and soil-type equivalents given in van Kekem *et al.* (1996, Appendices), Bernoux *et al.* (Table 11.1, 2001), Duivenvoorden and Lips (1995, Appendix II). The author was unable to locate an authoritative and comprehensive assessment of soil-type equivalents between the three main systems in use in the Guiana Shield, although this may exist and would facilitate more accurate soil assignment in region-wide comparisons of plant–animal–soil–water relationships.

FAO/WRB	USDA Soil Taxonomy	Sistema Brasileiro de Classificação de Solos
Acrisols	Oxisols, Ultisols (Paleudults/Hapluduits)	Latossolos, Podzolicos Vermehlo – Amarelos eutroficos
Andosols	Andisols	
Anthrosols		Terra preta (also Nitisol)
Arenosols	Entisols (Psamments)	Areias Quartzosas
Cambisols	Inceptisols (Tropepts)	Cambisolos
Chernozems	Chernozems	Brunizens Avermeihados
Ferraisols	Utisols (Kandiudults/Humults), Oxisols (Udoxs/Orthosx/Utoxs)	Latossolos, Podzolicos Vermehlo – Amarelo, Terras Roxas distroficas
Fluvisols	Inceptisols (Tropepts), Entisols (Fluvents)	Solus Aluviais
Gleysols	Inceptisols (Aquepts)	Cambissolos – Tropicais distroficos, Solos
Aluviais		
Histosols	Histosols (Saprists/Hemists	
Leptosols	Kandisols	
Luvisols	Kandisols	
Nitisols	Kandisols, Alfisols	Podzolicos Vermehlo – Amarelos plinticos, Terras Roxas estruturadas/eutroficos

Planosols		
Plinthosols	Ultisols (Plintaquults)	Podzois
Regosols		Solos Litolicos
Solonchaks	Aridisols	Solos salinos, Solonchaks
Solonetz	Aridisols	Solonetz-Solodizados
Umbrisols		
Vertisols	Vertisols	Vertissolos eutroico

[3] It is important to remember that station data reflect long-term rainfall at a single location. Rainfall received at each location reflects the influence exerted by the full gambit of factors operating at a wide range of scales, including any unfiltered instrumental and human bias. While data sources used here are considered reliable, the magnitude of rainfall anomalies and cross-correlations with SOI scores should be viewed in the context of other climatological analyses carried out on regional composites derived from these and scores from adjacent stations that are not presented here. In general, the precipitation patterns and their relationships with the SOI depicted here are consistent at larger scales with recent composite analyses (e.g. Ropelewski and Halpert, 1996).

Data depicting long-term variation in rainfall for seven of the nine stations included in Fig. 2.19 were primarily taken from records in the World Monthly Surface Station Climatology (NCDC). Published sources of data contributing to the WMSSC are available (Anonymous, 1927, 1934, 1947, 1959, 1961–2002, 1966, 1981–83). Ancillary data used to augment incomplete series available through the WMSSC were taken from the Global Historical Climatology Network (GHCN V2) (Vose *et al.*, 1992) and the IRL/LDEO Climate Data Library (ANEEL prcp sta and NOAA NCEP CDC EVE datasets) (http://www.ingrid.ldeo.columbia.edu). Data for Barro Colorado Island (Claro), Panama and La Selva, Costa Rica were provided by the Smithsonian Tropical Research Institute and the Organization for Tropical Studies (OTS) via anonymous ftp at http://www.stri.org/tesp/Metadata/details_bci_rain and http://www.ots.ac.cr/en/laselva/metereological, respectively. Readers are referred to Sanford *et al.* (1994), Rand and Rand (1999) and Leigh (1999) for further details and descriptions of rainfall collecting and hydrology at BCI and La Selva. Data from NOAA-NWSF (srh.noaa.gov) were used to augment the WMSSC series for San Juan, Puerto Rico. Data available for Georgetown, Guyana through the WMSSC were augmented by additional historical records provided by the Guyana Hydrometeorological Service (Hydromet) (Hydromet, 1971–1976, 1974–2002).

Normalized values of the Southern Oscillation Index (SOI) spanning the period 1882 to 2002 were taken directly via ftp from http://www.cpc.ncep.noaa.gov/data/indices For further information on the construction of these data series see Können *et al.* (1998), Allan *et al.* (1991) and Ropelewski and Jones (1987).

[4] Analysis of rainfall trends over the periods 1900–1946 and 1976–1998 for the region were not carried out as part of the IPCC assessment, probably due to poor quality or inaccessibility of rainfall station monitoring records. The period from 1940 to 1980 embraces the largest number of monitoring stations with the most complete records of monthly rainfall for the region.

[5] African easterly waves are believed to develop as temporary disturbances to the African easterly jet, a ribbon-like stream of rapidly moving air created by the low-level differences in atmospheric temperature created by the super-hot Sahara desert and much cooler coastal Guinea Current (Burpee, 1972). African easterly waves are considered the most important source of major tropical hurricanes (Landsea, 1993).

[6] A number of different indices are used to track changes in the Walker Circulation, particularly in an effort to identify the development and subsequent relaxation of ENSO conditions. Two main groups of indices are currently employed, one pressure-based and the other SST-based. Hanley *et al.* (2003) concluded that the SOI and the SST-based indices, Niño-3.4 and Niño-4, are the most sensitive indicators of EN events. SOI was considered less sensitive to LN than EN phases.

[7] This method identifies events by retaining only months that achieve both an SOI score and (Pacific) SST that fall within the 20th percentile of all monthly values over the period 1871–2002. El Niño events are delimited by periods where the SOI score falls below the lowest 20th percentile for two or more consecutive months separated by two or more months of weaker or positive index scores. La Niña events are identified by consecutive months where SOI scores fall within the uppermost 20th percentile. These periods are generally consistent with a strong to severe designation (Glantz, 1998).

[8] The analysis by Ropelewski and Halpert did not extend over the Sub-Andean Foredeep region in western Amazonia. Rainfall data from stations at Leticia (Fig. 2.22), Araracuara and Iquitos (Peru), do not show a strong below-average precipitation response during EN periods covered by the relatively short rainfall time series available for these locations. In part, this reflects the much higher monthly average rainfall. ENSO-related precipitation responses similar to those experienced in the eastern Guiana Shield have been identified for the northwestern Ecuadorian Andes (Vuille *et al.*, 2000).

[9] Latent heat flux in this case refers to the exchange of energy between the ocean and atmosphere as a consequence of evaporation from the ocean surface and subsequent condensation in the atmosphere. It represents an indirect transfer of heat occurring during a phase transition of water from liquid to vapour (evaporation) and vapour to liquid (condensation) or solid (ice). Feedback occurs through evaporation and precipitation amplifying (positive) or attenuating (negative) these heat fluxes.
[10] Differential SLP in the northwest and southeastern Amazon indicate major cross-equatorial northwesterly wind flow when the ITCZ is at its northernmost position and high seasonal evapotranspirative conditions predominate in the Amazon. Forward air mass trajectories from biomass burning in the southeastern Amazon region clearly trace this northwestward movement (Fig. 3.2 in Artaxo, 2001).
[11] It has been suggested that sapwood to heartwood movement of sap could lead to spurious radial chronologies (Worbes and Junk, 1989). If significant, then inward translocation of sap would lead to an underestimation of radiocarbon ages. Radiocarbon-dated heartwood from the base of these trees in this case would represent the minimum 'pre-set' age.
[12] Data sources. Fig. 2.26A: L. Volcan: Behling (2000); El Valle: Bush and Colinvaux (1990); L. Wodehouse: Bush and Colinvaux (1994); La Yeguada: Piperno *et al.* (1990); Anangucocha: Frost (1988); Mera-San Juan Bosco: Bush *et al.* (1990); River Lake 1, 2 and 3: Colinveaux *et al.* (1988); Pantano de Monica 1, 2 and 3: Behling *et al.* (1999); L. Agua Sucia, L. Chenevo, L. Mozambique: Berrio *et al.* (2002); L. Angel, L. Sardinas: Behling and Hooghiemstra (1998); L. Carimagua, L. El Pinal: Behling and Hooghiemstra (1999); Georgetown: Van der Hammen (1963); L. Moreiru (Muriru): Wijmstra and Van der Hammen (1966); Alliance Well: Wijmstra (1969); Mana: Tissot *et al.* (1988); Nouragues: Ledru *et al* (1997); Salitre: Ledru *et al.* (1996); Crominia: Salgado-Labouriau *et al.* (1997). Fig. 2.25B: La Selva: Horn and Sanford (1992), Kennedy and Horn (1997); Gran Sabana: Fölster (1992); Mabura Hill: Hammond, ter Steege and van der Borg (unpublished manuscript); San Carlos de Rio Negro: Saldarriaga and West (1986); BDFF/Manaus: Piperno and Becker (1996); Porto Trombetas: Francis and Knowles (2001); Carajas: Soubies (19790); Salitre: Vernet *et al.* (1994), Boulet *et al.* (1995).
[13] The Bowen ratio (H/LE) is the ratio of sensible to latent heat at any point in time. It links water availability (through latent heat flux) with temperature (through sensible heat flux). Maritime climates at mid-latitudes are strongly affected by seasonal declines in sensible heat flux, but experience less severe temperature drops than expected given their latitude due to compensating latent heat flux from the adjacent ocean to atmosphere. Similarly, a shift in the sensible heat flux due to changes in global temperatures would be compensated by an increase in the latent heat flux, particularly where there is abundant soil moisture, such as in flooded forest regions, to drive this process.
[14] This would take the form: $-6\,\frac{(E_m - E_s)}{1000}$

where E_m is the minimum modern elevational limit for indicator taxon I and E_s is the elevation of the sampled fossil pollen for indicator taxon I.

Indicator taxa used in assessing Pleistocene temperature changes include many putative highland or high-latitudinal species (Colinvaux, 1996): South America – Alpine: *Alnus* (Betulaceae), *Araucaria, Humiria* (Humiriaceae), *Podocarpus* (Podocarpaceae), *Weinmannia* (Cunoniaceae); Montane: Melastomataceae, *Hedyosmum* (Chloranthaceae), *Rapanea* (Myrsinaceae), *Ilex* (Aquifoliaceae). Central America – *Quercus* (Fagaceae), *Ilex, Magnolia* (Magnoliaceae), *Gunnera* (Haloragidaceae), *Symplocos* (Symplocaceae). Only *Araucaria* unequivocally qualifies as a mid-to-high latitude inhabitant. Gentry (1993, p. 263) assigns the modern *Alnus acuminata* to montane second-growth forests in northwest South America and indicates that *Podocarpus* can be found at scattered lowland locations on psamment soils. *Ilex inundata* is a common species found throughout the lowland tropical forests of the Amazon Downwarp, Sub-Andean Trough and on sandy soils or seasonally inundated floodplains (Gentry, 1993; Ribeiro *et al.*, 1999). Several species of *Ilex* are also commonly encountered in the low-lying hills of the Iwokrama Forest (Clarke *et al.*, 2001), Central Guyana and on dry sand savannas and savanna forest throughout Guyana, Suriname and French Guiana (van Roosmalen, 1985; Boggan *et al.*, 1998). Species of *Humiria* are also common components of lowland plant communities on podzol and arenosol (white sand) soils in the Guianas and Venezuela Guayana (Richards, 1952; Boggan *et al.*, 1998; ter Steege, 2000), as well as western Amazonia (Gentry, 1993). Both *Ilex* (*divaricata*) and *Humiria* typify bana-like shrublands found on white sand dune formations in the Sipapo, Atabapo and Guainia lowlands (50–200 m asl) of Venezuelan Guayana (Huber, 1995b). Gentry (1993) assigns *Rapanea* (aka *Myrsine*) to mid-elevational cloud forests 'in rather exposed situations' (i.e. drier). Julian Steyermark (1966, 1967) found a collection of taxa at the 1400–2200 m asl range of the Guayanan tepui region that shows a mix of lowland and highland genera very similar to that identified from fossil pollen cores in the Andes. The highland genera *Podocarpus, Weinmannia* and *Magnolia* were found mixing with common lowland genera (but not species!) such as *Schefflera, Caryocar, Moronobea, Dimorphandra* and *Byrsonima.*

Taxa assigned an upland distribution in fossil pollen analyses appear as modern components of relatively dry lowland habitats. Are these taxa indicative of Pleistocene aridity or cooling? Virtually all of the taxa indicative of cooling, except *Araucaria* and *Hedyosmum*, can be associated with relatively dry lowland (*Ilex*, *Humiria*, *Podocarpus*), or upland (*Rapanea*) habitats, suggesting the aridity hypothesis cannot be independently discounted.

[15] This refers to the absence of tropical eruptions of sufficient magnitude to create a sulfate spike in both polar ice caps. Several much smaller, mainly phreatic (steam-driven) eruptions took place in the tropics during the 1997–98 El Niño event, but none of these injected aerosol-producing chemicals into the stratosphere in amounts capable of significantly altering insolation levels.

[16] Solar irradiance is a measure of the portion of total solar energy output, or luminosity, that radiates towards Earth at a distance of one astronomical unit ($= 1.5 \times 10^8$ km). It is used in preference to the previously used term 'solar constant', since this has now been shown to be not strictly correct.

[17] Hoyt and Schatten (1997) note, however, that the Hale cycle is based on a doubling of the Schwabe cycle to explain 20–25 year oscillation in certain meteorological phenomena, rather than a distinct periodicity within the sunspot record. Hoyt and Schatten suggest that these may be better explained by the 18.6 year Saros lunar tidal cycle (Camuffo, 1999).

[18] Cosmogenic isotope concentrations deposited on Earth are believed to be regulated by the strength of the solar magnetic field. During periods of high solar activity, this field is strengthened, reducing Earth's exposure to cosmic rays and the consequent production of ^{10}Be, ^{14}C and other isotopes. Solar activity and isotope concentrations are thereby considered inversely related (Bard *et al.*, 1997). Lower concentrations would indicate higher solar activity and a warming of the planetary atmosphere (Hoyt and Schatten, 1997, p. 179).

[19] The measurement intervals for ice core-derived concentrations and using gas autoanalysers differ considerably. For example, the interval between consecutive samples taken from the 400 ka Vostok ice core series of gas measurements ranges between 43 and 5966 years, while modern concentrations at Mauna Loa, Hawaii have been directly sampled daily over a mere 45 years. The possibility of prehistoric spikes of post-industrial magnitude occurring during one or more of the long intervals between ice core measurements cannot be fully discounted due to the large discrepancy between direct and proxy sampling resolutions. However, for this to have occurred, a CO_2-generating 'event' or 'phase' of sufficient magnitude, driven by a plausible (group of) mechanism(s) capable of naturally raising concentrations to modern levels would need to be identified (e.g. carbonate weathering, volcanic emissions).

[20] The fractional contribution of these rivers to the Amazon and to the Guiana Shield differ by the area estimated to be covered by the upper Japurá region not located within the delimited shield region.

[21] Discharge (Q) is the downstream rate of water flux expressed in cubic metres or cubic kilometres per second. Specific discharge (*q*) is this rate (Q) divided by the size of the draining watershed (A) (see Table 2.3) and is typically expressed in litres per second per square kilometre (Lerman, 1994). It offers a way of comparing flow rates of rivers with strongly contrasting surface drainage areas.

Discharge rates for rivers draining the Guiana Shield are based on a variety of published sources, often covering different time intervals with varying degrees of overlap. Varying monitor intervals can potentially create significant differences between the stated and the true long-term average discharges if the data intervals used to calculate averages differ in their width and inclusion of anomalies related to strong to severe ENSO events. Where flow data covered less than 10 years but included severe ENSO events (1982–83, 1990–91, 1997–98), these years were not included in the calculation of the average annual discharge. Brazilian station data are drawn from station records published by ANEEL and made available through hidroweb.aneel.gov.br(vazões) Station data from other countries were sourced from the following: Venezuela – UNESCO ds552.0 v1.3 via dss.ucar.edu/datasets/ds552.0 (Bodo, 2001), Table 1 in Weibezahn (1990), Table 2 in Cressa *et al.* (1993) and cross-checked for consistency, given available river data, with figures given in Yanez and Ramirez (1988), Edmond *et al.* (1995, 1996) and Vargas and Rangel (1996b); Guyana – UNESCO ds552.0, Hydrometeorological Service (1981, 1982, 1985, 2000) and Dai and Trenberth (2002); Suriname – UNESCO ds552.0, Amatali (1993), Dai and Trenberth (2002); French Guiana – UNESCO ds552.0, Hiez and Dubreuil (1964), Lointier (1991) and Dai and Trenberth (2002).

[22] River discharge data taken from UCAR compilation UNESCO ds552.0 v1.3 via dss.ucar.edu/datasets/ds552.0 (Bodo, 2001) and ANEEL via hidroweb.aneel.gov.br, except for Berbice River, 1982–83, taken from Hydrometeorological Service (2000). SOI values are standard deviates from the long-term average for the 200-year period spanning 1882 to 2002. The discharge for each month of the EN years evaluated was calculated as a percentage of a decadal average for that month. For the 1972–73 and 1982–83 events, the averages were based on the periods 1967–1977 and 1978–1988, respectively.

[23] It is important to note, however, that virtually all of the 65 fossil study sites examined by Romero (1993)

were located on the periphery of the South American craton. About one-third of these are located along the eastern shield regions (mainly the Brazilian Shield) and thus could be considered exclusively characteristic of Palaeogene vegetation on the passive margin of a Precambrian landscape. While the northwestern margin of the craton had not yet fully evolved in response to the emerging Andean orogeny, volcanic arc belts associated with the southern perimeter of the emerging Caribbean Plate are believed to have been active since the Cenomanian or earlier (Pitman *et al.*, 1993).

[24] *Eustatic movements* refer to the global change in sea level, typically as a result of glacial expansion and contraction. *Steric effects* refer here to changes in sea level caused by changes in seawater temperature. When seawater increases (decreases) in temperature, it expands (contracts), raising (lowering) sea level. Isostatic movements are generated by long-term rise or subsidence of crust, typically in response to mountain-building and weathering, which can influence coastline changes relative to other effects. Epeirogenic movements are vertical shifts in land elevation believed to be caused by mantle convection currents or thermal effects below the lithosphere (Harrison, 1990).

[25] $\delta^{18}O$ refers to the deviation in the ratio of the relatively rare (and heavy) oxygen isotope (^{18}O) and the relatively more common (and lighter) isotope (^{16}O) within a foraminifera (coral) sample from that of standard mean ocean water (SMOW). Algebraically, $\delta=([^{18}O/^{16}O_S]/[^{18}O/^{16}O_{SMOW}])-1$. Glacial ice is isotopically light (and therefore the remaining seawater is isotopically heavy) and low (high) sea stands are characterized by larger (smaller) $\delta^{18}O$ values.

References

Ab'Saber, A.N. (1977) Espacos ocupados pela expansão dos climas secos na América do Sul, por ocasião dos períodos glaciais quaternários. *Paleoclimas Universidad São Paulo, Instituto de Geografia* 3, 1–19.

Absy, M.L., Cleef, A., Fournier, M., Martin, L., Servant, M., Sifeddine, A., Ferreira da Silva, M., Soubiés, F., Suguio, K., Turcq, B. and van der Hammen, T. (1991) Mise en évidence de quatre phases d'ouverture de la forêt dense dans le sud-est de l'Amazonie au cours des 60000 dernières années. Première comparaison avec d'autres régions tropicales. *Comptes Rendu de l'Academie des Sciences, Paris* 312, Série II, 673–678.

Allan, R.J., Nicholls N., Jones, P.D. and Butterworth, I.J. (1991) A further extension of the Tahiti-Darwin SOI, early SOI results and Darwin pressure. *Journal of Climate* 4, 743–749.

Allison, T.D., Moeller, R.E. and Davis, M.B. (1986) Pollen in laminated sediments provides evidence for a mid-Holocene forest pathogen outbreak. *Ecology* 67, 1101–1105.

Amatali, M. (1993) Climate and surface water hydrology. In: Ouboter, P.E. (ed.) *The Freshwater Ecosystems of Suriname.* Kluwer Academic, Dordrecht, The Netherlands, pp. 29–52.

Angell, J.K. and Korshover, J. (1985) Surface temperature changes following the six major volcanic episodes between 1780 and 1980. *Journal of Applied Meteorology* 24, 937–951.

Anonymous (1927) *World Weather Records – 1920.* Smithsonian Institution, Washington, DC.

Anonymous (1934) *World Weather Records – 1921 to 1930.* Smithsonian Institution, Washington, DC.

Anonymous (1947) *World Weather Records – 1931 to 1940.* Smithsonian Institution, Washington, DC.

Anonymous (1959) *World Weather Records – 1941 to 1950.* US Department of Commerce, Washington, DC.

Anonymous (1961–2002) *Monthly Climatic Data for the World – 1961 to Present.* US Department of Commerce, Asheville, North Carolina.

Anonymous (1966) *World Weather Records – 1951 to 1960* US Department of Commerce, Washington, DC.

Anonymous (1981–1983) *World Weather Records – 1961 to 1970.* US Department of Commerce, Washington, DC.

Artaxo, P. (2001) The atmospheric component of biogeochemical cycles in the Amazon Basin. In: Richey, J.E. (ed.) *The Biogeochemistry of the Amazon Basin.* Oxford University Press, Oxford, UK, pp. 42–52.

Ashton, P.S. (1977) A contribution of rain forest research to evolutionary theory. *Annals of the Missouri Botanical Garden* 64, 694–705.

Aubert, G. (1965) Classification des sols. *Cahiers ORSTOM, Série Pédologie* 3, 269–288.

Aubrey, D.G., Emery, K.O. and Uchupi, E. (1988) Changing coastal levels of South America and the Caribbean region from tide-gauge records. *Tectonophysics* 154, 269–284.

Augustinus, P.G.E.F. (1980) Actual development of the chenier coast of Suriname. *Sedimentary Geology* 26, 91–113.

Aymard, G., Cuello, N. and Schargel, R. (1998) Floristic composition, structure, and diversity in the moist forest communities along the Casiquiare channel, Amazonas state, Venezuela. In: Comiskey, J.A. (ed.) *Forest Biodiversity in North, Central and South America, and the Caribbean*. Parthenon, Paris, pp. 495–506.

Baillie, I.C. (1989) Soil characteristics and classification in relation to the mineral nutrition of tropical wooded ecosystems. In: Proctor, J. (ed.) *Mineral Nutrients in Tropical Forest and Savanna Ecosystems*. Blackwell Scientific, Oxford, UK, pp. 15–26.

Baillie, I.C. (1996) Soils of the humid tropics. In Richards, P.W. (ed.) *The Tropical Rain Forest*. Cambridge University Press, Cambridge, UK, pp. 256–286.

Bains, S., Corfield, R.M. and Norris, R.D. (1999) Mechanisms of climate warming at the end of the Paleocene. *Science* 285, 724–727.

Baize, D. and Girard, M.C. (1995) *Référentiel Pédologique*. INRA, Paris.

Baker, P.A., Seltzer, G.O., Fritz, S.C., Dunbar, R.B., Grove, M.J., Tapia, P.M., Cross, S.L., Rowe, H.D. and Broda, J.P. (2001) The history of South American tropical precipitation for the past 25,000 years. *Science* 291, 640–643.

Balée, W. (1989) The culture of Amazonian forests. *Advances in Economic Botany* 7, 1–21.

Bard, E., Raisbeck, G.M., Yiou, F. and Jouzel, J. (1997) Solar modulation of cosmogenic nuclide production over the last millennium: comparison between ^{14}C and ^{10}Be records. *Earth and Planetary Science Letters* 150, 453–462.

Barron, E.J. and Washington, W.M. (1982) Cretaceous climate: a comparison of atmospheric simulations with the geologic record. *Palaeogeography, Palaeoclimatology, Palaeoecology* 40, 103–133.

Bartlett, K.B. and Harris, R.C. (1993) Review and assessment of methane emissions from wetlands. *Chemosphere* 26, 261–320.

Bates, H.W. (1879) *The Naturalist on the River Amazons*, 5th edn. John Murray, London.

Baumgartner, A. and Reichel, E. (1975) *The World Water Balance*. Elsevier, New York.

Beer, J., Joos, F., Lukasczyk, C., Mende, W., Rodriguez, J., Siegenthaler, U. and Stellmacher, R. (1993) ^{10}Be as an indicator of solar variability and climate. In: Nesme-Ribes, E. (ed.) *The Solar Engine and its Influence on Terrestrial Atmosphere and Climate*. Springer, London, pp. 221–234.

Beer, J., Mende, W. and Stellmacher, R. (2000) The role of the sun in climate forcing. *Quaternary Science Reviews* 19, 403–415.

Behling, H. (2000) A 2860 yr high-resolution pollen and charcoal record from the Cordillera de Talamanca in Panama: a history of human and volcanic forest disturbance. *The Holocene* 10, 387–393.

Behling, H. and Hooghiemstra, H. (1998) Late Quaternary palaeoecology and palaeoclimatology from pollen records of the savannas of the Llanos Orientales in Colombia. *Palaeogeography, Palaeoclimatology, Palaeoecology* 139, 251–267.

Behling, H. and Hooghiemstra, H. (1999) Environmental history of the Colombian savannas of the Llanos Orientales since the Last Glacial Maximum from lake records El Piñal and Carimagua. *Journal of Paleolimnology* 21, 461–476.

Behling, H., Carlos Berrio, J. and Hooghiemstra, H. (1999) Late Quaternary pollen records from the middle Caquéta river basin in central Colombian Amazon. *Palaeogeography, Palaeoclimatology, Palaeoecology* 145, 193–213.

Behling, H., Keim, G., Irion, G., Junk, W.J. and Nunes de Mello, J. (2001) Holocene environmental changes in the central Amazon Basin inferred from Lago Calado (Brazil). *Palaeogeography, Palaeoclimatology, Palaeoecology* 173, 87–101.

Benaïm, J.-V. and Mounier, S. (1998) Metal transport by organic carbon in the Amazon Basin. *Croatica Chemica Acta* 71, 405–419.

Berger, A. (1992) *Orbital Variations and Insolation Database*. IGBP Pages/World Data Center – A for Paleoclimatology Data Contribution Series No. 92-007, NOAA/NGDC Paleoclimatology Program, Boulder, Colorado.

Berger, A. and Loutre, M.F. (1991) Insolation values for the climate of the last 10 million of years. *Quaternary Sciences Review* 10, 297–317.

Berrangé, J.P. (1977) *The Geology of Southern Guyana, South America*. HMSO, London.

Berrio, J.C., Hooghiemstra, H., Behling, H., Botero, P. and van der Borg, K. (2002) Late-Quaternary savanna history of the Colombian Llanos Orientales from Lagunas Chenevo and Mozambique: a transect synthesis. *The Holocene* 12, 35–48.

Berrio, J.C., Arbeláez, M.V., Duivenvoorden, J., Cleef, A. and Hooghiemstra, H. (2003) Pollen representation and successional vegetation change on the sandstone plateau of Araracuara, Colombian Amazonia. *Review of Palaeobotany and Palynology* 126, 163–181.

Beurlen, K. (1970) *Geologie von Brasilien.* Gebrüder Borntraeger, Berlin.
Bjerknes, J. (1966) A possible response of the atmospheric Hadley circulation to equatorial anomalies of ocean temperature. *Tellus* 18, 820–829.
Bjerknes, J. (1969) Atmospheric teleconnections from the equatorial Pacific. *Monthly Weather Review* 97, 163–172.
Blancaneaux, P. (1973) Podzols et sols ferrallitiques dans le Nord-Ouest de la Guyane Francaise. *Cahiers ORSTOM, Série Pédologique* 11, 121–154.
Blancaneaux, P. and Pouyllau, M. (1977) Les relations géomorpho-pédologiques e la retombée nord-occidentale du massif guyanais (Vénézuela). *Cahiers ORSTOM, Série Pédologique* 15, 437–448.
Bleakley, D. (1957) Observations on the geomorphology and recent geological history of the coastal plain of British Guiana. *Bulletin of the British Guiana Geological Survey* 30, 1–46.
Bleakley, D. (1964) Bauxites and laterites of British Guiana. *Bulletin of the British Guiana Geological Survey* 34, 156.
Bleakley, D. and Khan, E.J.A. (1963) Observations on the white sand areas of the Berbice Formation, British Guiana. *Journal of Soil Science* 14, 44–51.
Bloom, A.L. and Yonekura, N. (1990) Graphic analysis of dislocated Quaternary shorelines. In: Revelle, R.R. (ed.) *Sea-Level Change.* National Academy Press, Washington, DC, pp. 104–115.
Bodo, B.A. (2001) *Monthly Discharge Data for World Rivers (excluding former Soviet Union).* University of Toronto, Toronto.
Boggan, J., Funk, V., Kelloff, C., Hoff, M., Cremers, G. and Feuillet, C. (1998) *Checklist of the Plants of the Guianas (Guyana, Suriname, French Guiana),* 2nd edn. BDGP – NMNH Smithsonian Institution, Washington, DC.
Bond, G., Kromer, B., Beer, J., Muscheler, R., Evans, M.N., Showers, W., Hoffmann, S., Lotti-Bond, R., Hajdas, I. and Bonani, G. (2001) Persistent solar influence on North Atlantic climate during the Holocene. *Science* 294, 2130–2136.
Bonell, M. and Balek, J. (1993) Recent scientific developments and research needs in hydrological processes of the humid tropics. In: Gladwell, J.S. (ed.) *Hydrology and Water Management in the Humid Tropics.* Cambridge University Press, Cambridge, UK, pp. 167–272.
Boulet, R., Pessenda, L.C.R., Telles, E.C.C. and Melfi, A.J. (1995) Une évaluation de la vitesse de l'accumulation superficielle de matiére par la faune du sol á partir de la datation des charbons et de l'humine du sol. Exemple des latosols des versants du lac Campestre, Salitre, Minas Gerais, Brésil. *Comptes Rendu de l'Academie des Sciences, Paris* 320, 287–294.
Bracewell, S. (1950) The search for minerals in British Guiana. *Timehri* 29, 40–43.
Bradbury, J.P., Leyden, B.W., Salgado-Labouriau, M.L., Lewis, W.M., Schubert, C., Binford, M.W., Frey, D.G., Whitehead, D.R. and Weibezahn, F.H. (1981) Late Quaternary environmental history of lake Valencia, Venezuela. *Science* 214, 1199–1305.
Briceño, E., Balbás, L. and Blanco, J.A. (1997) Bosques ribereños del bajo rio Caura: vegetacion, suelos y fauna. In: Rosales, J. (ed.) *Ecología de la Cuenca del Río Caura, Venezuela 2.* FIBV/UNEG, Caracas, pp. 259–289.
Bridger, C.S. (1984) A summary of Colombian cratonic cover and its significance to Amazonian geology. In: DNPM (ed.) *2nd Symposium Amazonica.* DNPM, Manaus, Brazil, pp. 20–21.
Broecker, W.S. (1991) The great ocean conveyor. *Oceanography* 4, 84–89.
Bryson, R.A. (1988) Late Quaternary volcanic modulation of Milankovitch climate forcing. *Theoretical and Applied Climatology* 39, 115–125.
Bryson, R.A. (2002) *Volcanic Eruptions and Aerosol Optical Depth Data.* NOAA/NGDC Paleoclimatology Program, Boulder, Colorado.
Burnham, C.P. (1989) Pedological processes and nutrient supply from parent material in tropical soils. In: Proctor, J. (ed.) *Mineral Nutrients in Tropical Forest and Savanna Ecosystems.* Blackwell Scientific, Oxford, UK, pp. 27–41.
Burpee, R.W. (1972) The origin and structure of easterly waves in the lower troposphere of North Africa. *Journal of Atmospheric Science* 29, 77–90.
Bush, A.B.n.G. and Philander, S.G.H. (1998) The role of ocean–atmosphere interactions in tropical cooling during the last glacial maximum. *Science* 279, 1341–1344.
Bush, M.B. (1994) Amazonian speciation: a necessarily complex model. *Journal of Biogeography* 21, 5–17.
Bush, M.B. (2002) On the interpretation of fossil Poaceae pollen in the lowland humid neotropics. *Palaeogeography, Palaeoclimatology, Palaeoecology* 177, 5–17.
Bush, M.B. and Colinvaux, P.A. (1990) A pollen record of a complete glacial cycle from lowland Panama. *Journal of Vegetation Science* 1, 105–118.

Bush, M.B. and Colinvaux, P.A. (1994) Tropical forest disturbance: paleoecological records from Darien, Panama. *Ecology* 75, 1761–1768.

Bush, M.B., Colinvaux, P.A., Wiemann, M.C., Piperno, D.R. and Liu, K.-b. (1990) Late Pleistocene temperature depression and vegetation change in Ecuadorian Amazonia. *Quaternary Research* 34, 330–345.

Bush, M. B., Piperno, D.R., Colinvaux, P.A., De Oliveira, P.E., Krissek, L., Miller, M.C. and Rowe, W. (1992) A 14,300 year paleoecological profile of a lowland tropical lake in Panama. *Ecological Monographs* 62, 251–275.

Bush, M.B., Miller, M.C., De Oliveira, P.E. and Colinvaux, P.A. (2000) Two histories of environmental change and human disturbance in eastern lowland Amazonia. *The Holocene* 10, 543–553.

Bush, M.B., Miller, M.C., De Oliveira, P.E. and Colinvaux, P.A. (2002) Orbital forcing signal in sediments of two Amazonian lakes. *Journal of Paleolimnology* 27, 341–352.

Cadle, R.D., Kiang, C.S. and Louis, J F. (1976) The global scale dispersion of the eruption clouds from major volcanic eruptions. *Journal of Geophysical Research* 81, 3125–3132.

Campos, C.W. and Bacoccoli, G. (1973) Os altos sincronos e a psequisa de petróleo no Brasil. In *27th Congreso Brasiliero do Geologia*, CBG, Aracajú, Brazil, pp. 373–415.

Camuffo, D. (1999) Lunar influences on climate. *Earth, Moon and Planets* 85/86, 99–113.

Carton, J.A., Cao, X., Giese, B.S. and Da Silva, A.M. (1996) Decadal and interannual SST variability in the tropical Atlantic Ocean. *Journal of Physical Oceanography* 26, 1165–1175.

Case, G.O. (1943) Coastal sediments. *Timehri* 29, 92–106.

Chambers, J.Q., Higuchi, N. and Schimel, J.P. (1998) Ancient trees in Amazonia. *Nature* 391, 135–136.

Chang, P., Ji, L. and Li, H. (1997) A decadal climate variation in the tropical Atlantic ocean from thermodynamic air–sea interactions. *Nature* 385, 516–518.

Chapman, B., Rosenqvist, A. and Wong, A. (2001) *JERS-1 SAR Global Rain Forest Mapping Project.* Vol. AM-1, *South America (1995–1996).* CD-ROM. National Space Development Agency of Japan, Earth Observation Research Center; National Aeronautics and Space Administration, Jet Propulsion Laboratory; European Commission Joint Research Centre; Earth Remote Sensing Data Analysis Center of Japan; Remote Sensing Technology Center of Japan; and Alaska SAR Facility. Available from ORNL Distributed Active Archive Center, Oak Ridge, TN (http://www.daac.ornl.gov).

Chappell, J. and Shackleton, N.J. (1986) Oxygen isotopes and sea level. *Nature* 324, 137–140.

Charlson, R.J., Langer, J., Rodhe, H., Leovy, C.B. and Warren, S.G. (1991) Perturbation of the northern hemisphere radiative balance by backscattering from anthropogenic sulfate aerosols. *Tellus* 43AB, 152–163.

Chelton, D.B. and Schlax, M.G. (1996) Global observations of oceanic Rossby waves. *Science* 272, 234–238.

Choubert, B. (1957) *Essai sur la Morphologie de la Guyane Francaise – Ses Relations avec l'Histoire Géologique.* Service Carte Geologique du France, Paris.

Choudhuri, A. and Milner, M. (1971) Basic magmatism in Guiana and continental drift. *Nature* 232, 154–155.

Christie-Black, N., Mountain, G.S. and Miller, K.G. (1990) Seismic stratigraphic record of sea-level change. In: Revelle, R.R. (ed.) *Sea-Level Change.* National Academy Press, Washington, DC, pp. 116–140.

Church, J.A., Gregory, J.M., Huybrechts, P., Kuhn, M., Lambeck, K., Nhuan, M.T., Qin, D. and Woodworth, P.L. (2001) Changes in sea level. In: Johnson, C.A. (ed.) *Climate Change 2001: the Scientific Basis.* Contribution of the working group I to the third assessment report of the intergovernmental panel on climate change. Cambridge University Press, Cambridge, UK.

Clapperton, C.M. (1993) *Quaternary Geology and Geomorphology of South America.* Elsevier, London.

Clark, D.A. (1994) Plant demography. In: McDade, L.A., Bower, K.S., Hespenheide, H.S. and Hartshorn, G.S. (eds) *La Selva: Ecology and Natural History of a Neotropical Rain Forest.* University of Chicago Press, London, pp. 90–105.

Clark, J.A. and Lingle, C.S. (1979) Predicted relative sea-level changes (18 ka BP to present) caused by late glacial retreat of the Antarctic ice sheet. *Quaternary Research* 11, 279–298.

Clark, P.U., Alley, R.B. and Pollard, D. (1999) Northern hemisphere ice-sheet influences on global climate change. *Science* 286, 1104–1111.

Clarke, H.D., Funk, V.A. and Hollowell, T. (2001) *Plant Diversity of the Iwokrama Forest, Guyana.* Botanical Research Institute of Texas, Fort Worth, Texas.

Clinebell, R.R., Phillips, O.L., Gentry, A.H., Stark, N. and Zuuring, H. (1995) Prediction of neotropical tree and liana species richness from soil and climatic data. *Biodiversity and Conservation* 4, 56–90.

Coates, A.G. and Obando, J.A. (1996) The geological evolution of the Central American isthmus. In: Coates, A.G. (ed.) *Evolution and Environment in Tropical America.* Chicago University Press, London, pp. 21–56.

Colinvaux, P.A. (1987) Amazon diversity in light of the paleoecological record. *Quaternary Science Review* 6, 93–114.

Colinvaux, P.A. (1993) Pleistocene biogeography and diversity in tropical forests of South America. In: Goldblatt, P. (ed.) *Biological Relationships Between Africa and South America.* Yale University Press, New Haven, Connecticut, pp. 473–499.

Colinvaux, P.A. (1996) Quaternary environmental history and forest diversity in the neotropics. In: Coates, A.G. (ed.) *Evolution and Environment in Tropical America.* University of Chicago Press, London, pp. 359–406.

Colinvaux, P.A. and De Oliveira, P.E. (2001) Amazon plant diversity and climate through the Cenozoic. *Palaeogeography, Palaeoclimatology, Palaeoecology* 166, 51–63.

Colinvaux, P.A., Frost, M., Frost, I., Liu, K.-B. and Steinitz-Kannan, M. (1988) Three pollen diagrams of forest disturbance in the western Amazon basin. *Review of Palaeobotany and Palynology* 55, 73–81.

Colonnello, G. (2001) Physico-chemical comparison of the Mánamo and Macareo rivers in the Orinoco delta after the 1965 Mánamo dam construction. *Interciencia* 26, 136–143.

Colonnello, G. and Medina, E. (1998) Vegetation changes induced by dam construction in a tropical estuary: the case of the Mánamo river, Orinoco Delta (Venezuela). *Plant Ecology* 139, 145–154.

Coomes, D.A. and Grubb, P.J. (1996) Amazonian caatinga and related communities at La Esmeralda, Venezuela: forest structure, physiognomy and floristics, and control by soil factors. *Vegetatio* 122, 167–191.

Cooper, A. (1979) Muri and white sand savannah in Guyana, Suriname and French Guiana. In: Specht, R.L. (ed.) *Heathlands and Related Shrublands.* Elsevier, Amsterdam, pp. 471–481.

Cordani, U.G. and de Brito Neves, B.B. (1982) The geologic evolution of South America during the Archaean and Early Proterozoic. *Revista Brasiliera Geociencias* 12, 78–88.

Covey, D.L. and Hastenrath, S. (1978) The Pacific El Niño phenomenon and the Atlantic circulation. *Monthly Weather Review* 106, 1280–1287.

Crawford, F.D., Szelewski, C.E. and Alvey, G.D. (1984) Geology and exploration in the Takutu graben of Guyana and Brazil. *Journal of Petroleum Geology* 8, 5–36.

Cressa, C., Vasquez, E., Zoppi, E., Rincon, J.E.and Lopez, C. (1993) Aspectos generales de la limnologia en Venezuela. *Interciencia* 18, 237–248.

Critchfield, H.J. (1983) *General Climatology,* 4th edn. Prentice-Hall, New York.

Cross, S.L., Baker, P.A., Seltzer, G.O., Fritz, S.C. and Dunbar, R.B. (2001) Late Quaternary climate and hydrology of tropical South America inferred from an isotopic and chemical model of Lake Titicaca, Bolivia and Peru. *Quaternary Research* 56, 1–9.

Crowley, T.J. (2001) Causes of climate change over the past 1000 years. *Science* 289, 270–289.

Cuevas, E. (2001) Soil versus biological controls on nutrient cycling in terra firme forests. In: Richey, J.E. (ed.) *The Biogeochemistry of the Amazon Basin.* Oxford University Press, Oxford, UK, pp. 53–67.

Culf, A.D., Fisch, G. and Hodnett, M.G. (1995) The albedo of Amazonian forest and ranch land. *Journal of Climate* 8, 1544–1554.

Curtis, J.H., Brenner, M. and Hodell, D.A. (1999) Climate change in the Lake Valencia basin, Venezuela, 12600 yr BP to present. *The Holocene* 9, 609–619.

CVG-TECMIN (1991) *Proyecto Inventario de los Recursos Naturales de la Región Guayana.* CVG-TECMIN, Ciudad Bolivar, Venezuela.

Dai, A. and Trenberth, K.E. (2002) Estimates of freshwater discharge from continents: latitudinal and seasonal variations. *Journal of Hydrometeorology* 3, 660–687.

Dai, A. and Wigley, T.M.L. (2001) Global patterns of ENSO-induced precipitation. *Geophysical Research Letters* 27, 1283–1286.

Dai, A., Trenberth, K.E. and Karl, T.R. (1998) Effects of clouds, soil moisture, precipitation, and water vapor on diurnal temperature range. *Journal of Climate* 12, 2451–2473.

Damuth, J.E. and Kumar, N. (1975) Amazon cone: morphology, sediments, age and growth pattern. *Bulletin of the Geological Society of America* 86, 863–878.

Daniel, J.R.K. (1984) *Geomorphology of Guyana: An Integrated Study of Natural Environments.* Release Books, Georgetown.

Data+ and ESRI (1996) *ArcAtlas: Our Earth.* ESRI, Redlands, California.

de Fátima Rossetti, D. (2001) Late Cenozoic sedimentary evolution in northeastern Pará, Brazil, within the context of sea level changes. *Journal of South American Earth Sciences* 14, 77–89.

de Freitas, H.A., Pessenda, L.C.R., Aravena, R., Gouveia, S.E.M., de Souza Ribeiro, A. and Boulet, R. (2001) Late Quaternary vegetation dynamics in the southern Amazon Basin inferred from carbon isotopes in soil organic matter. *Quaternary Research* 55, 39–46.

de Oliveira, P.E., Barreto, A.M.F. and Suguio, K. (1999) Late Pleistocene/Holocene climatic and vegetational history of the Brazilian caatinga: the fossil dunes of the middle Sao Francisco River. *Palaeogeography, Palaeoclimatology, Palaeoecology* 152, 319–337.

de Silva, S.L. and Zielinski, G.A. (1998) Global influence of the AD 1600 eruption of Huaynaputina, Peru. *Nature* 393, 455–458.

Debrabant, P., Lopez, M. and Chamley, H. (1997) Clay mineral distribution and significance in Quaternary sediments of the Amazon Fan. In: Peterson, L.C. (ed.) *Proceedings of the Ocean Drilling Program, Scientific Results.* ODP, College Station, Texas, pp. 177–192.

Degens, E.T. (1982) *Transport of Carbon and Minerals in Major World Rivers, Part 1.* SCOPE/UNEP, Hamburg.

Degens, E.T., Kempe, S. and Richey, J. (1990) Summary: biogeochemistry of major world rivers. In: Degens, E.T., Kempe, S. and Richey, J. (eds) *Biogeochemistry of the Major World Rivers.* John Wiley & Sons, Chichester, UK, pp. 323–332.

Delecluse, P., Servain, J., Levy, C., Arpe, K. and Bengtsson, L. (1994) On the connection between the 1984 Atlantic warm event and the 1982–83 ENSO. *Tellus* 46A, 448–464.

DeMaster, D.J. and Aller, R.C. (2001) Biogeochemical processes on the Amazon shelf: changes in dissolved and particulate fluxes during river/ocean mixing. In: Richey, J.E. (ed.) *The Biogeochemistry of the Amazon Basin.* Oxford University Press, Oxford, UK, pp. 328–357.

Depetris, P. and Paolini, J. (1990) Biochemical aspects of South American rivers: the Paraná and Orinoco. In: Degenes, E.T., Kempe, S. and Richey, J. (eds) *Biogeochemistry of the Major World Rivers.* John Wiley & Sons, Chichester, UK, pp. 105–125.

Déqué, M. and Servain, J. (1989) Teleconnections between tropical Atlantic sea surface temperatures and midlatitude 50 kPa heights during 1964–1986. *Journal of Climate* 2, 929–945.

Devol, A.H. and Hedges, J.I. (2001) Organic matter and nutrients in the mainstem Amazon River. In: Richey, J.E. (ed.) *The Biogeochemistry of the Amazon Basin.* Oxford University Press, Oxford, UK, pp. 275–306.

Devol, A.H., Richey, J., Forsberg, B.R. and Martinelli, L.A. (1994) Environmental methane in the Amazon river floodplain. In: Mitsch, W.J. (ed.) *Global Wetlands.* Elsevier, London, pp. 151–165.

Diaz de Gamero, M.L. (1996) The changing course of the Orinoco River during the Neogene: a review. *Paleogeography, Palaeoclimatology, Palaeoecology* 123, 385–402.

Dickinson, R.E. (1978) Rossby waves – long-period oscillations of oceans and atmospheres. *Annual Review of Fluid Mechanics* 10, 185–195.

Dietz, R.S. and Holden, J.C. (1970) The breakup of Pangaea. *Scientific American* 223, 30–41.

Dommenget, D. and Latif, M. (2001) Interannual to decadal variability in the tropical Atlantic. *Journal of Climate* 13, 777–792.

Drexel, J.F., Preiss, W.V. and Parker, A.J. (eds) (1993) *The Geology of South Australia 1. The Precambrian.* South Australia State Print, South Australia.

Dubroeucq, D. and Volkaff, B. (1998) From oxisols to spodosols and histosols: evolution of the soil mantles in the Rio Negro basin (Amazonia). *Catena* 32, 245–280.

Dubroeucq, D., Volkaff, B. and Faure, P. (1999) Les couvertures pédologiques á Podzols du Bassin du Haut Rio Negro (Amazonie). *Étude et Gestion des Sols* 6, 131–153.

Duivenvoorden, J.F. and Lips, J.M. (1995) *A Land-Ecological Study of Soils, Vegetation and Plant Diversity in Colombian Amazonia.* Tropenbos Foundation, Wageningen, The Netherlands.

Duivenvoorden, J.F. and Lips, J.M. (1998) Mesoscale patterns of tree species diversity in Colombian Amazonia. In: Comiskey, J.A. (ed.) *Forest Biodiversity in North, Central and South America, and the Caribbean.* Parthenon, Paris, pp. 535–550.

Echezuria, H., Cordova, J., Gonzalez, M., Gonzalez, V., Mendez, J. and Yanes, C. (2002) Assessment of environmental changes in the Orinoco River delta. *Regional Environmental Change* 3, 20–35.

Eckhardt, F.E.W. (1985) Solubilization, transport, and deposition of mineral cations by microorganisms – efficient rock weathering agents. In: Drever, J.I. (ed.) *The Chemistry of Weathering.* D. Reidel, Dordrecht, The Netherlands, pp. 161–174.

EDC (1996) *GTOPO30 (a Digital Elevation Model).* USGS EROS Data Center, Washington, DC.

Eddy, J.A. (1976) The Maunder minimum. *Science* 192, 1189–1192.

Eden, M.J. (1964) *The Savanna Ecosystem – Northern Rupununi, British Guiana.* Department of Geography, McGill University, Montreal.

Eden, M.J., McGregor, D.F.M. and Morelo, J.A. (1982) Geomorphology of the middle Caquetá basin of eastern Colombia. *Zeitschrift für Geomorphologie N.F.* 26, 343–364.

Eden, M.J., Bray, W., Herrera, L. and McEwan, C. (1984) Terra preta soils and their archaeological context in the Caquetá basin of southeast Colombia. *American Antiquity* 49, 125–140.
Edmond, J.M., Palmer, M.R., Measures, C.I., Grant, B. and Stallard, R.F. (1995) The fluvial geochemistry and denudation rate of the Guayana Shield in the Venezuela, Colombia, and Brazil. *Geochemica et Cosmochimica Acta* 59, 3301–3325.
Edmond, J.M., Palmer, M.R., Measures, C.I., Brown, E.T. and Huh, Y. (1996) Fluvial geochemistry of the eastern slope of the northeastern Andes and its foredeep in the drainage of the Orinoco in Colombia and Venezuela. *Geochemica et Cosmochimica Acta* 60, 2949–2974.
Edwards, A.M.C. and Thornes, J.B. (1970) Observations on the dissolved solids of the Casiquiare and upper Orinoco, April–June 1968. *Amazoniana* 2, 245–256.
Edwards, M.H. (1986) Digital image processing of local and global bathymetric data. Masters thesis. Washington University, St Louis, Missouri.
Edwards, R.L., Chang, H., Murrell, M.T. and Goldstein, S.J. (1997) Protactinium-231 dating of carbonates by thermal ionization mass spectrometry: implications for Quaternary climate change. *Science* 276, 782–786.
Egge, J.K. and Aksnes, D.L. (1992) Silicate as a regulating nutrient in phytoplankton competition. *Marine Ecol. Prog. Ser.* 83, 281–289.
Eisma, B.R. and van der Marel, H.W. (1971) Marine mud along the Guiana coast and their origin from the Amazon Basin. *Contributions to Mineralogy and Petrology* 31, 321–334.
Elkibbi, M. and Rial, J.A. (2001) An outsider's review of the astronomical theory of the climate: is the eccentricity-driven insolation the main driver of the ice ages? *Earth-Science Reviews* 56, 161–177.
Eltahir, E.A.B. and Bras, R.L. (1994) Precipitation recycling in the Amazon basin. *Quarterly Journal of the Royal Meteorological Society* 120, 861–880.
EMBRAPA-CNPS (1999) *Sistema Brasileiro de Classificacao de Solos.* Embrapa Produção da Informação, Rio de Janeiro.
Enfield, D.B. (1996) Relationships of inter-American rainfall to tropical Atlantic and Pacific SST variability. *Geophysical Research Letters* 23, 3505–3508.
Enfield, D.B. and Alfaro, E.J. (1999) The dependence of Caribbean rainfall on the interaction of the tropical Atlantic and Pacific Oceans. *Journal of Climate* 12, 2093–2103.
Enfield, D.B. and Mayer, D.A. (1997) Tropical Atlantic SST variability and its relation to El Niño-Southern Oscillation. *Journal of Geophysical Research* 102, 929–945.
Ertel, J., Hedges, J.I., Richey, J., Devol, A.H. and dos Santos, U. (1986) Dissolved humic substances of the Amazon river system. *Limnology and Oceanography* 31, 739–754.
Estrada, J. and Fuertes, J. (1993) Estudios en la Guayana colombiana IV. Notas sobres la vegetación y la flora de la sierra de Chiribiquete. *Revista de la Academia Colombiana de Ciencias Exactas, Físicas y Naturales* 18, 483–497.
Fanshawe, D.B. (1952) *The Vegetation of British Guiana: A Preliminary Review.* Imperial Forestry Institute, Oxford, UK.
FAO (1965) *A report to accompany a general soil map of British Guiana.* FAO, Rome.
FAO (1988) *Soil Map of the World.* Revised Legend. FAO, Rome.
FAO (1998) *World Reference Base for Soil Resources.* FAO, Rome.
FAO-UNESCO (1974) *Soil Map of the World 1:5000000.* Vol. 1. Legend. UNESCO, Paris.
Farjalla, V.F., Esteves, F.A., Bozelli, R.L. and Roland, F. (2002) Nutrient limitation of bacterial production in clear water Amazonian ecosystems. *Hydrobiologia* 489, 197–205.
Fearnside, P.M. (1995) Hydroelectric dams in the Brazilian Amazon as sources of 'greenhouse gasses'. *Environmental Conservation* 22, 7–19.
Fearnside, P.M. (1997) Greenhouse gas emissions from Amazonian hydroelectric reservoirs: the example of Brazil's Tucurui dam as compared to fossil fuel alternatives. *Environmental Conservation* 24, 64–75.
Ferraz-Vicentini, K.R. and Salgado-Labouriau, M.L. (1996) Palynological analysis of a palm swamp in Central Brazil. *Journal of South American Earth Sciences* 9, 207–219.
Figueroa, N. and Nobre, C.A. (1990) Precipitation distribution over central and western tropical South America. *Climanálise* 5, 36–42.
Filho, A.C., Schwartz, D., Tatumi, S.H. and T. Rosique (2002) Amazonian paleodunes provide evidence for drier climate phases during the late Pleistocene-Holocene. *Quaternary Research* 58, 205–209.
Finkel, R.C. and Nishiizumi, K. (1997) Beryllium 10 concentrations in the Greenland Ice Sheet Project 2 ice core from 3–40 ka. *Journal of Geophysical Research* 102, 26699–26706.

Fittkau, E.J. (1971) Ökologische Gliederung des Amazonas-Gebietes auf geochemischer Grundlage. *Münstersche Forschungen zur Geologie und Paläontologie* 20, 35–50.

Flenley, J.R. (1998) Tropical forests under the climates of the last 30,000 years. *Climatic Change* 39, 177–197.

Flood, R.D., Piper, D.J.W., Klaus, A. and Peterson, L.C. (eds) (1997) *Amazon Fan.* ODP, College Station, Texas.

Folland, C.K., Karl, T.R., Christy, J.R., Clarke, R.A., Gruza, G.V., Jouzel, J., Mann, M.E., Oerlemans, J., Salinger, M.J. and Wang, S.-W. (2001) Observed climate variability and change. In: Johnson, C.A. (ed.) *Climate Change 2001: The Scientific Basis.* Contribution of the working group I to the third assessment report of the intergovernmental panel on climate change. Cambridge University Press, Cambridge, UK, pp. 101–181.

Fölster, H. (1985) Proton consumption rates in Holocene and present-day weathering of acid forest soils. In: Drever, J.I. (ed.) *The Chemistry of Weathering.* D. Reidel, Dordrecht, The Netherlands, pp. 197–210.

Fölster, H. (1992) Holocene autochthonous forest degradation in southeast Venezuela. In: Goldammer, J.G. (ed.) *Tropical Forests in Transition.* Birkhäuser, Basel, pp. 25–44.

Francis, J.K. and Knowles, O.H. (2001) Age of A2 horizon charcoal and forest structure near Porto Trombetas, Pará, Brazil. *Biotropica* 33, 385–392.

Franco, W. and Dezzeo, N. (1994) Soil and soil water regime in the terra firme-caatinga forest complex near San Carlos de Rio Negro, state of Amazonas, Venezuela. *Interciencia* 19, 305–316.

Franzinelli, E. and Igreja, H. (2002) Modern sedimentation in the lower Negro river, Amazonas State, Brazil. *Geomorphology* 44, 259–271.

Franzinelli, E. and Potter, P.E. (1983) Petrology, chemistry and texture of modern river sands, Amazon river system. *Journal of Geology* 91, 23–29.

Friis-Christensen, E. and Lassen, K. (1991) Length of the solar cycle: an indicator of solar activity closely associated with climate. *Science* 254, 698–700.

Fröhlich, C. and Lean, J. (1998) The Sun's total irradiance: cycles, trends and related climate change uncertainties since 1976. *Geophysical Research Letters* 25, 4377–4380.

Froidefond, J.M., Pujos, M. and Andre, X. (1988) Migration of mud banks and changing coastline in French Guiana. *Marine Geology* 84, 19–30.

Frost, D.B. (1968) *The Climate of the Rupununi Savannas.* McGill University, Montreal.

Frost, I. (1988) A Holocene sedimentary record from Anañgucocha in the Ecuadorian Amazon. *Ecology* 69, 66–73.

Fu, R., Dickinson, R.E., Chen, M. and Wang, H. (2001) How do tropical sea surface temperatures influence the seasonal distribution of precipitation in the equatorial Amazon. *Journal of Climate* 14, 4003–4026.

Fuentes, J. and Madero, A.J. (1996) Suelos. In: Huber, O. (ed.) *Ecología de la Cuenca del Río Caura, Venezuela. I. Caracterización General.* UNEG/FIBV, Caracas, pp. 44–47.

Furch, K. (1984) Water chemistry of the Amazon basin: the distribution of chemical elements among freshwaters. In: Sioli, H. (ed.) *The Amazon. Limnology and Landscape Ecology of a Mighty Tropical River and its Basin.* Dr. W. Junk, Dordrecht, The Netherlands, pp. 167–195.

Gallup, C.D., Cheng, H., Taylor, F.W. and Edwards, R.L. (2002) Direct determination of the timing of sea level change during termination II. *Science* 295, 310–313.

Galvis, J., Huguett, A. and Ruge, P. (1979) Geologia de la Amazonia Colombiana. *Boletín Geologica Ingeominas* 22, 3–86.

Galy-Lacaux, C., Delmas, R., Jambert, C., Dumestre, J.-F., Labroue, L. and Richard, S. (1997) Gaseous emissions and oxygen consumption in hydroelectric dams. A case study in French Guiana. *Global Biogeochemical Cycles* 11, 471–483.

Ganopolski, A., Rahmstorf, S., Petoukhov, V. and Claussen, M. (1998) Simulation of modern and glacial climates with a coupled global model of intermediate complexity. *Nature* 391, 351–356.

Gansser, A. (1954) Observations on the Guiana Shield (South America). *Ecologae Geologicae Helvetiae* 47, 77–112.

Gansser, A. (1974) The Roraima problem (South America). *Verhandlungen Naturforschenden Gesellschaft Basel* 84, 80–100.

García, S. (1996) Limnologia. In: Rosales, J. and Huber, O. (eds) *Ecología de la Cuenca del Río Caura, Venezuela. I. Caracterización General.* Refolit, Caracas, pp. 54–59.

Gavaud, M., Blancaneaux, P., Dubroeucq, D. and Pouyllau, M. (1986) Les paysages pédologiques de l'Amazonie vénézuélienne. *Cahiers ORSTOM, Série Pédologique* 22, 265–284.

Gentry, A.H. (1990) Floristic similarities and differences between southern Central America and upper and central Amazonia. In Gentry, A.H. (ed.) *Four Neotropical Forests.* Yale University Press, New Haven, Connecticut, pp. 141–157.

Gentry, A.H. (1993) *A Field Guide to the Families and Genera of Woody Plants of Northwest South America.* Conservation International, Washington, DC.

Giambelluca, T.W., Hölscher, D., Bastos, T.X., Frazão, R.R., Nullet, M.A. and Ziegler, A.D. (1997) Observations of albedo and radiation balance over postforest land surfaces in the Eastern Amazon Basin. *Journal of Climate* 10, 919–928.

Giannini, A., Cane, M.A. and Kushnir, Y. (2001) Interdecadal changes in the ENSO teleconnection to the Caribbean region and the North Atlantic Oscillation. *Journal of Climate* 14, 2867–2879.

Gibbs, A.K. and Barron, C.N. (1993) *The Geology of the Guiana Shield.* Oxford University Press, Oxford, UK.

Gibbs, R.J. (1967) The geochemistry of the Amazon river system. Part 1. The factors that control the salinity and the composition and concentration of the suspended solids. *Geological Society of America Bulletin* 78, 1203–1232.

Gill, A.E. (1982) *Atmosphere–Ocean Dynamics.* Academic Press, New York.

Giorgi, F. (2002) Dependence of the surface climate interannual variability on spatial scale. *Geophysical Research Letters* 29, 2101–2106.

Girard, J.-P., Freyssinet, P. and Morillon, A.-C. (2002) Oxygen isotope study of Cayenne duricrust paleosurfaces: implications for past climate and laterization processes over French Guiana. *Chemical Geology* 191, 329–343.

Glantz, M.H. (1998) *Currents of Change: Impacts of El Nino and La Nina on Climate and Society.* Cambridge University Press, Cambridge, UK.

Glatzmaiers, G.A., Coe, R.S., Hongre, L. and Roberts, P.H. (1999) The role of the Earth's mantle in controlling the frequency of geomagnetic reversals. *Nature* 401, 885–890.

Goldblatt, P. (1993) Biological relationships between Africa and South America: an overview. In: Goldblatt, P. (ed.) *Biological Relationships Between Africa and South America.* Yale University Press, New Haven, Connecticut.

Goodwin, A. (1996) *Principles of Precambrian Geology.* Academic Press, New York.

Goulding, M., Carvalho, M.L. and Ferreira, E.G. (1988) *Rio Negro: Rich Life in Poor Water.* SPB Academic, The Hague, The Netherlands.

Graham, N.E. and Barnett, T.P. (1987) Sea surface temperature, surface wind divergence and convection over tropical oceans. *Science* 238, 657–659.

Grimaldi, M. and Riéra, B. (2001) Geography and climate. In: Théry, M. (ed.) *Nouragues – Dynamics and Plant–Animal Interactions in a Neotropical Rain Forest.* Kluwer Academic, Dordrecht, The Netherlands, pp. 9–18.

Groeneweg, W. and Bosma, W. (1969) Review of the stratigraphy of Suriname. In: *8th Guiana Geological Conference*, Georgetown, Guyana, pp. 1–29.

Grubb, P.J. and Whitmore, T.C. (1966) A comparison of montane and lowland forest in Ecuador. II. The climate and its effects on the distribution and physiognomy of the forests. *Journal of Ecology* 54, 303–333.

Guilderson, T.P., Fairbanks, R.G. and Rubenstone, J.L. (1994) Tropical temperature variations since 20,000 years ago: modulating interhemispheric climate change. *Science* 263, 663–665.

Guilderson, T.P., Fairbanks, R.G. and Rubenstone, J.L. (2001a) *Barbados Coral Oxygen Isotope Data.* #2001-010. World Data Center for Paleoclimatology, Boulder, Colorado.

Guilderson, T.P., Fairbanks, R.G. and Rubenstone, J.L. (2001b) Tropical Atlantic coral oxygen isotopes: glacial–interglacial sea surface temperatures and climatic change. *Marine Geology* 172, 75–89.

Guyot, J.L. and Wasson, J.G. (1994) Regional pattern of riverine dissolved organic carbon in the Amazon drainage basin in Bolivia. *Limnology and Oceanography* 39, 452–458.

Guyot, J.L., Jouanneau, J.M., Quintanilla, J. and Wasson, J.G. (1993) Les flux de matières dissoutes et particulaires exportés des Andes par le Rio Béni (Amazonie Bolivienne), en période de crue. *Geodinamica Acta (Paris)* 6, 233–241.

Guyot, J.L., Jouanneau, J.M. and Wasson, J.G. (1999) Characterisation of river bed and suspended sediments in the Rio Madeira drainage basin (Bolivian Amazonia). *Journal of South American Earth Sciences* 4, 401–410.

Haberle, S. (1997) Upper quaternary vegetation and climate history of the Amazon Basin: correlating marine and terrestrial pollen records. In: Peterson, L.C. (ed.) *Proceedings of the Ocean Drilling Program, Scientific Results.* ODP, College Station, Texas, pp. 381–396.

Haffer, J. (1969) Speciation in Amazonian forest birds. *Science* 165, 131–137.

Haigh, J.D. (1996) The impact of solar variability on climate. *Science* 272, 981–984.

Häkkinen, S. and Mo, K.C. (2002) The low-frequency variability of the tropical Atlantic Ocean. *Journal of Climate* 15, 237–250.
Hall, J.B. and Swaine, M.D. (1976) Classification and ecology of closed canopy forest in Ghana. *Journal of Ecology* 64, 913–951.
Halmer, M.M. and Schminke, H.-J. (2003) The impact of moderate-scale explosive eruptions on statospheric gas injections. *Bulletin of Volcanology* 65, 433–440.
Hameed, S., Meinster, A. and Sperber, K. R. (1993) Teleconnections of the Southern Oscillation in the tropical Atlantic sector in the OSU coupled upper ocean–atmosphere GCM. *Journal of Climate* 6, 487–498.
Hamilton, S.K., Sippel, S.J. and Melack, J.M. (2002) Comparison of inundation patterns among major South American floodplains. *Journal of Geophysical Research* 107 (D20), 8038.
Hammond, D.S. and ter Steege, H. (1998) Propensity for fire in the Guianan rainforests. *Conservation Biology* 12, 944–947.
Hanley, D.E., Bourassa, M.A., O'Brien, J.J., Smith, S.R. and Spade, E.R. (2003) A quantitative evaluation of the ENSO indices. *Journal of Climate* 16, 1249–1258.
Hansen, J., Johnson, D., Lacis, A., Lebedeff, S., Lee, P., Rind, D. and Russell, G. (1981) Climate impact of increasing atmospheric carbon dioxide. *Science* 213, 957–966.
Hardy, F. and Follet-Smith, R.R. (1931) Studies in tropical soils. 2. Some characteristic igneous rock soil profiles in British Guiana, South America. *Journal of Agricultural Science* 21, 739–761.
Haripersad-Makhanlal, A. and Ouboter, P.E. (1993) Limnology: physico-chemical parameters and phytoplankton composition. In: Ouboter, P.E. (ed.) *The Freshwater Ecosystems of Suriname.* Kluwer Academic, Dordrecht, The Netherlands, pp. 53–75.
Harrison, C.G.A. (1990) Long-term eustasy and epeirogeny in continents. In: Revelle, R.R. (ed.) *Sea-Level Change.* National Academy Press, Washington, DC, pp. 141–158.
Harrison, D.E. and Vecchi, G.A. (1999) On the termination of El Niño. *Geophysical Research Letters* 26, 1593–1596.
Harrison, J.B. and Reid, K.D. (1911) Formation of laterite from a practically quartz-free diabase. *Geological Magazine* 8, 120, 355, 477.
Hastenrath, S. (1984a) Interannual variability and annual cycle: mechanisms of circulation and climate in the tropical Atlantic sector. *Monthly Weather Review* 112, 1097–1107.
Hastenrath, S. (1984b) Predictability of Northeast Brazil drought. *Nature* 307, 531–533.
Hastenrath, S. and Heller, L. (1977) Dynamics of climate hazards in northeast Brazil. *Quarterly Journal of the Royal Meteorological Society* 103, 77–92.
Hastenrath, S., Greischar, L., Colón, E. and Gil, A. (1999) Forecasting the anomalous discharge of the Caroni river, Venezuela. *Journal of Climate* 12, 2673–2678.
Haug, G.H., Hughen, K.A., Sigman, D.M., Peterson, L.C. and Rohl, U. (2001) Southward migration of the Intertropical Convergence Zone through the Holocene. *Science* 293, 1304–1308.
Haug, G.M.W. (1966) *Verslag van het veldwerk in het stroomgebeid van de Paloemeu.* GMD Informatie 17/66, Geologie Mijnbouw Dienst Suriname, Paramaribo.
Hawkes, M.D. and Wall, J.R.D. (1993) *The Commonwealth and Government of Guyana Iwokrama Rain Forest Programme: Phase 1 Site Resource Survey.* Natural Resources Institute, Chatham, UK.
Haxby, W.F. (1983) Digital images of combined oceanic and continental data sets and their use in tectonic studies. *EOS Transactions of the American Physical Union* 64, 995–1004.
Hays, J.D., Imbrie, J. and Shackleton, N.J. (1976) Variations in the earth's orbit: pacemaker of the ice ages. *Science* 194, 1121–1132.
Heyligers, P.C. (1963) Vegetation and soil of a white-sand savanna in Suriname. *Verhandelingen der Koninklijke Nederlandse Akademie van Wetenschappen, Afd. Natuurkunde* 54, 1–148.
Hiez, G. and Dubreuil, P. (1964) *Les Régimes Hydrologiqes en Guyane Francaise.* ORSTOM, Paris.
Hoag, R.E. (1987) Characterization of soils on floodplain of tributaries flowing into the Amazon River in Peru. PhD thesis. North Carolina State University, Raleigh, North Carolina.
Hoffmann, P.F. (1989) Precambrian geology and tectonic history of North America. In: Bally, A.W. and Palmer, A.R. (eds) *The Geology of North America – an Overview.* The Geological Society of America, Boulder, Colorado, pp. 447–512.
Hole, F.D. (1961) A classification of pedoturbations and some other processes and factors of soil formation in relation to isotropism and anisotropism. *Soil Science* 91, 375–377.
Hooghiemstra, H. and van der Hammen, T. (1998) Neogene and Quaternary development of the neotropical rain forest: the forest refugia hypothesis, and a literature review. *Earth-Science Reviews* 44, 147–183.

Hoorn, C. (1993) Marine incursion and the influence of Andean tectonics on the Miocene depositional history of northwestern Amazonia: results of a palynostratigraphic study. *Palaeogeography, Palaeoclimatology, Palaeoecology* 105, 267–309.
Horn, S.P. and Sanford, R.L. (1992) Holocene fires in Costa Rica. *Biotropica* 24, 354–361.
Houghton, R.W. and Tourre, Y.M. (1992) Characteristics of low-frequency sea surface temperature fluctuations in the tropical Atlantic. *Journal of Climate* 5, 765–772.
Hoyt, D.V. and Schatten, K.H. (1997) *The Role of the Sun in Climate Change.* Oxford University Press, Oxford, UK.
Huang, B. and Shukla, J. (1997) Characteristics of the interannual and decadal variability in a general circulation model of the tropical Atlantic Ocean. *Journal of Physical Oceanography* 27, 1693–1712.
Huang, J.-P., Higuchi, K. and Shabbar, A. (1998) The relationship between the North Atlantic Oscillation and the El Niño-Southern Oscillation. *Geophysical Research Letters* 25, 2707–2716.
Huber, O. (1995a) Geographical and physical features. In: Berry, P.E., Holst, B.K. and Yatskievych, K. (eds) *Flora of the Venezuelan Guayana.* Timber Press, Portland, Oregon, pp. 1–62.
Huber, O. (1995b) Vegetation. In: Berry, P.E., Holst, B.K. and Yatskievych, K. (eds) *Flora of the Venezuelan Guayana.* Timber Press, Portland, Oregon, pp. 97–160.
Huston, M.A. (1980) Soil nutrients and tree species richness in Costa Rican forests. *Journal of Biogeography* 7, 147–157.
Huston, M.A. (1994) *Biological Diversity: the Coexistence of Species on Changing Landscapes.* Cambridge University Press, Cambridge, UK.
Hydromet (1971–1976) *Annual Climatological Data Summary.* Guyana Hydrometeorological Service (Ministry of Public Works and Transport), Georgetown.
Hydromet (1974–2002) *Monthly Weather Bulletin.* Guyana Hydrometeorological Service (Ministry of Agriculture), Georgetown.
Hydrometeorological Service (1980) *Annual Surface Water Data – 1969.* Hydrometeorological Service, Georgetown, Guyana.
Hydrometeorological Service (1981) *Annual Surface Water Data – 1970.* Hydrometeorological Service, Georgetown, Guyana.
Hydrometeorological Service (1982) *Annual Surface Water Data – 1971.* Hydrometeorological Service, Georgetown, Guyana.
Hydrometeorological Service (1985) *Annual Surface Water Data – 1972.* Hydrometeorological Service, Georgetown, Guyana.
Hydrometeorological Service (2001) Berbice river basin, surface water data (1975–1995) Hydrometeorological Service, Georgetown, Guyana.
Ichiye, T. and Peterson, J. (1963) The anomalous rainfall of the 1957–58 winter in the equatorial central Pacific arid area. *Journal of the Meteorological Society of Japan* 41, 172–182.
IGAC (1993) *Aspectos Ambientales para el Ordenamiento Territorial del Occidente del Departamento del Caquetá.* Tomo I, Capítulos I–III. IGAC, Bogotá.
Imbrie, J. and Imbrie, J.Z. (1980) Modeling the climate response to orbital variations. *Science* 202, 943–953.
Imbrie, J., Hays, J.D., Martinson, D.G., McIntyre, A., Mix, A.C., Moreley, J.J., Pisias, N.G., Prell, W.L. and Shackleton, N.J. (1984) The orbital theory of Pleistocene climate: support from a revised chronology of the marine 18O record. In: Berger, A. (ed.) *Milankovitch and Climate.* D. Reidel, Dordrecht, The Netherlands, pp. 269–306.
Irion, G. (1976) Quaternary sediments of the upper Amazon lowlands of Brazil. *Biogeographica* 7, 163–167.
Irion, G. (1984) Sedimentation and sediments of the Amazonian rivers and evolution of the Amazonian landscape since Pliocene times. In: Sioli, H. (ed.) *The Amazon. Limnology and Landscape Ecology of a Mighty Tropical River and its Basin.* Dr. W. Junk, Dordrecht, The Netherlands, pp. 201–213.
Irion, G. (1990) Minerals in rivers. In: Degenes, E.T., Kempe, S. and Richey, J. (eds) *Biogeochemistry of the Major World Rivers.* John Wiley & Sons, Chichester, UK, pp. 265–281.
Iriondo, M.H. (1997) Models of deposition of loess and loessoids in the upper Quaternary of South America. *Journal of South American Earth Sciences* 10, 71–79.
Jackson, J.B.C., Budd, A.F. and Coates, A.G. (1996) *Evolution and Environment in Tropical America.* Chicago University Press, Chicago, Illinois.
Jenny, H. (1980) *The Soil Resource: Origin and Behaviour.* Springer, New York.
Jetten, V.G. (1994) *Modelling the Effects of Logging on the Water Balance of a Tropical Rain Forest. A Study in Guyana.* Tropenbos Foundation, Wageningen, The Netherlands.

Johnson, A.M. (1976) The climate of Peru, Bolivia and Ecuador. In: Schwerdtfeger, W. (ed.) *Climates of Central and South America.* Elsevier, Oxford, UK, pp. 147–218.
Jones, C., Mahowald, N. and Luo, C. (2003) The role of easterly waves on African desert dust transport. *Journal of Climate* 16, 3617–3628.
Jones, P.D., Parker, D.E., Osborn, T.J. and Briffa, K.R. (1999) Global and hemispheric temperature anomalies – land and marine instrument records. In: Carbon Dioxide Information Analysis Center (ed.) *Trends: A Compendium of Data on Global Change.* Carbon Dioxide Information Analysis Center, Oak Ridge National Laboratory, US Department of Energy, Oak Ridge, Tennessee.
Jordan, C.J. and Herrera, R. (1981) Tropical rainforests: are nutrients really critical? *American Naturalist* 117, 167–180.
Joshi, M.M. and Shine, K.P. (2003) A GCM study of volcanic eruptions as a cause of increased stratospheric water vapour. *Journal of Climate* 16, 3525–3534.
Junk, W.J. (1993) Wetlands of tropical South America. In: Whigham, D.F. (ed.) *Wetlands of the World: Inventory, Ecology and Management.* Kluwer Academic, Dordrecht, The Netherlands, pp. 679–739.
Kalliokoski, J. (1965) Geology of north-central Guyana Shield, Venezuela. *Geological Society of America Bulletin* 76, 1027–1050.
Kennedy, L.M. and Horn, S.P. (1997) Prehistoric maize cultivation at the La Selva biological station, Costa Rica. *Biotropica* 29, 368–370.
Kennett, J.P. and Barker, P.F. (1990) Latest Cretaceous to Cenozoic climate and oceanographic developments in the Weddell Sea, Antarctica: an ocean-drilling perspective. In: Kennet, J.P. (ed.) *Proceedings of the Ocean Drilling Program, Scientific Results.* ODP, College Station, Texas, pp. 937–960.
Kent, D.V. and Gradstein, F.M. (1985) A Cretaceous and Jurassic geochronology. *Bulletin of the Geological Society of America* 96, 1419–1427.
Kessler, W.S. and McPhaden, M.J. (1995) Forcing of intraseasonal Kelvin waves in the equatorial Pacific. *Journal of Geophysical Research* 100, 10613–10631.
Khan, Z., Paul, S. and Cummings, D. (1980) *Mabura Hill – Upper Demerara Forestry Project: Soils Investigation.* Report No. 1. NARI, Ministry of Agriculture, Georgetown, Guyana.
Khobzi, J., Kroonenberg, S.B., Favre, P. and Weeda, A. (1980) Aspectos geomorfologicos de la Amazonia y Orinoquia Colombianas. *Revista CIAF* 5, 97–126.
Kidson, J.W. (1975) Tropical eigenvector analysis and the Southern Oscillation. *Monthly Weather Review* 103, 187–196.
Kiladis, G.N. and Diaz, H.F. (1986) An analysis of the 1877–78 ENSO episode and comparison with the 1982–83. *Monthly Weather Review* 114, 1035–1047.
Kiladis, G.N. and Diaz, H.F. (1989) Global climatic anomalies associated with extremes in the Southern Oscillation. *Journal of Climate* 2, 1069–1090.
Kirchner, I. and Graf, H.-F. (1995) Volcanoes and El Nino: signal separation in northern hemisphere winter. *Climate Dynamics* 11, 341–358.
Klammer, G. (1984) The relief of the extra-Andean Amazon basin. In: Sioli, H. (ed.) *The Amazon: Limnology and Landscape Ecology of a Mighty Tropical River and its Basin.* Dr. W. Junk, Dordrecht, The Netherlands, pp. 47–83.
Klinge, H. (1968) *Report on Tropical Podzols.* FAO, Rome.
Kloosterman, J.B. (1973) Vulcoes gigantes do tipo anelar no Escudo das Guianas. *Miner. Metal.* 36, 52–58.
Können, G.P., Jones, P.D., Kaltofen, M.H. and Allan, R.J. (1998) Pre-1866 extensions of the Southern Oscillation Index using early Indonesian and Tahitian meteorological readings. *Journal of Climate* 11, 2325–2339.
Köppen, W. (1918) Klassifikation der Klimate nach temperatur, niederschlag und jahreslauf. Peterm. *Geographische Mitteilungen* 64, 193–203.
Koster, R.D. and Suarez, M.J. (1995) Relative contributions of land and ocean processes to precipitation variability. *Journal of Geophysical Research* 100, 13775–13790.
Kouadio, Y.K., Ochou, D.A. and Servain, J. (2003) Tropical Atlantic and rainfall variability in Cote d'Ivoire. *Geophysical Research Letters* 30, 8005–8012.
Kousky, V.E. (1980) Diurnal rainfall variation in northeast Brazil. *Monthly Weather Review* 108, 488–498.
Kousky, V.E., Kagano, M.T. and Cavalcant, I.F. (1984) A review of the Southern Oscillation: oceanic–atmospheric circulation changes and related rainfall anomalies. *Tellus* 36, 490–504.
Krook, L. (1969) The origin of bauxite in the coastal plains of Suriname and Guyana. *Geologie en Mijnbouw Dienst van Suriname* 20.
Krook, L. (1979) Sediment petrographical studies in Northern Suriname. PhD thesis, Vrije Universiteit,

Amsterdam.
Krook, L. and de Roever, E.W.F. (1975) Some aspects of bauxite formation in the Bakhuis Mountains, western Suriname. In: *Decima Conferencia Geologia Interguianas,* Belem, Brazil, pp. 686–695.
Labitzke, K. and Matthes, K. (2003) Eleven-year solar cycle variations in the atmosphere: observations, mechanisms and models. *The Holocene* 13, 311–317.
Landsea, C.W. (1993) A climatology of intense (or major) Atlantic hurricanes. *Monthly Weather Review* 121, 1703–1713.
Lassen, K. and Friis-Christensen, E. (1995) Variability of the solar cycle length during the past five centuries and the apparent association with terrestrial climate. *Journal of Atmospheric and Terrestrial Physics* 57, 835–845.
Latrubesse, E.M. and Franzinelli, E. (2002) The Holocene alluvial plain of the middle Amazon River, Brazil. *Geomorphology* 44, 241–257.
Latrubesse, E.M. and Nelson, B.W. (2001) Evidence for late Quaternary aeolian activity in the Roraima-Guyana region. *CATENA* 43, 63–80.
Leckie, R.M., Sigurdsson, H., Acton, G.D. and Draper, G. (eds) (2001) *Caribbean Ocean History and the Cretaceous/Tertiary Boundary Event.* ODP, College Station, Texas.
Ledru, M.-P., Soares Braga, P.I., Soubiés, F., Fournier, M., Martin, L., Suguio, K. and Turcq, B. (1996) The last 50,000 years in the neotropics (Southern Brazil): evolution of vegetation and climate. *Palaeogeography, Palaeoclimatology, Palaeoecology* 123, 239–257.
Ledru, M.-P., Blanc, P., Charles-Dominique, P., Fournier, M., Martin, L., Riera, B. and Tardy, C. (1997) Reconstituion palynologique de la forêt guyanaise au cours des 3000 dernières années. *Comptes Rendu de l'Academie des Sciences, Paris* 324, 469–476.
Ledru, M.-P., Bertaux, J., Sifeddine, A. and Suguio, K. (1998) Absence of last glacial maximum records in lowland tropical forests. *Quaternary Research* 49, 233–237.
Leeflang, E.C., Kolander, J.H. and Kroonenberg, S.B. (1976) *Suriname in Geografisch Perspektief: een Geografieleerboek voor het Midelbaar Onderwijs.* Bolivar Editions, Paramaribo, Suriname.
Leenheer, J.A. and Santos, U.M. (1980) Consideracoes sobre os processos de sedimentacao na água preta ácida do Rio Negro (Amazonia Central). *Acta Amazonica* 10, 343–355.
Leigh, E.G.J. (1999) *Tropical Forest Ecology – a View from Barro Colorado Island.* Oxford University Press, Oxford, UK.
Lerman, A. (1979) *Geochemical Processes – Water and Sediment Environment.* John Wiley & Sons, London.
Lerman, A. (1994) Surficial weathering fluxes and their geochemical controls. In: BESR, CGER and NRC (eds) *Material Fluxes on the Surface of the Earth.* National Academy Press, Washington, DC, pp. 28–45.
Lescure, J.-P. and Boulet, R. (1985) Relationships between soil and vegetation in a tropical rain forest in French Guiana. *Biotropica* 17, 155–164.
Leveque, A. (1961) *Memoire Explicatif de la Carte des Sols de Terres Basses de Guyane Francaise.* Institut Francais d'Amerique Tropicale, Paris.
Lewis, W.M.J. and Saunders, J.F. (1987) Major element chemistry, weathering and element yields for the Caura river drainage, Venezuela. *Biogeochemistry* 4, 159–181.
Li, J., Sato, T. and Kageyama, A. (2002) Repeated and sudden reversals of the dipole field generated by a spherical dynamo action. *Science* 295, 1887–1890.
Libby, W.F. (1955) *Radiocarbon Dating.* University of Chicago Press, Chicago, Illinois.
Liebmann, B. and Marengo, J.A. (2001) Interannual variability of the rainy season and rainfall in the Brazilian Amazon basin. *Journal of Climate* 14, 4308–4318.
Linna, A. (1993) Factores que contribuyen a las caracteristicas del sedimento superficial en la selva baja de la Amazonia Peruana. In: Kalliola, R.J., Puhakka, M. and Danjoy, W. (eds) *Amazonia Peruana – vegetacion húmeda tropical en el llano subandino.* PAUT/ONERN, Jyväskylä, Finland, pp. 87–97.
Liu, K.-B. and Colinvaux, P.A. (1985) Forest changes in the Amazon basin during the last glacial maximum. *Nature* 318, 556–557.
Lockwood, J.G. (1979) *The Causes of Climate.* Edward Arnold, London.
Lofgren, B.M. (1995) Sensitivity of land–ocean circulations, precipitation, and soil moisture to perturbed land surface albedo. *Journal of Climate* 8, 2521–2542.
Lointier, M. (1991) *Hydrologie en Guyane et Qualité Physico-Chimique des Eaux.* Journées Departementales de l'eau en Guyane ORSTOM, Cayenne, French Guiana.
Lointier, M. (1993) Variations saisonnières et flux de quelques éléments majeurs dans trois rivières de Guyane Française. In: Boulègue, J. and Olivry, J.-C. (eds) *Grands Bassin Fluviaux Périatlantiques: Congo, Niger, Amazone.* PEGI/INSU/CNRS/ORSTOM, Paris, pp. 391–410.

Lovelock, J.E. (1979) *Gaia: a New Look at Life on Earth.* Oxford University Press, Oxford, UK.
Lowe-McConnell, R.H. (1987) *Ecological Studies in Tropical Fish Communities.* Cambridge University Press, Cambridge, UK.
Mackenzie, F.T. and Mackenzie, J.A. (1995) *Our Changing Planet.* Prentice Hall, New York.
Madden, R.A. and Julian, P.R. (1972. Description of global scale circulation cells in the tropics with a 40–50 day period. *Journal of Atmospheric Science* 29, 1109–1123.
Malmgren, B.A., Winter, A. and Chen, D. (1998) El Niño–Southern Oscillation and North Atlantic Oscillation control of climate in Puerto Rico. *Journal of Climate* 11, 2713–2718.
Marchant, R., Behling, H., Berrio, J.C., Cleef, A., Duivenvoorden, J., Hooghiemstra, H., Kuhry, P., Melief, B., van Geel, B., van der Hammen, T., van Reenen, G. and Wille, M. (2001) Mid- to late-Holocene pollen-based biome reconstructions for Colombia. *Quaternary Science Review* 20, 1289–1308.
Marengo, J. (1995) Variations and change in South American streamflow. *Climatic Change* 31, 99–117.
Marengo, J.A. (1999) Factores medio-ambientales del area de estudio: climatologia en la zona de Iquitos. Parte III. In: Flores, S. (ed.) *Geoecologia y Desarrollo de la Zona de Iquitos, Peru.* University of Turku, Turku, Finland, pp. 35–55.
Marengo, J.A. and Hastenrath, S. (1993) Case studies of extreme climatic events in the Amazon Basin. *Journal of Climate* 6, 617–627.
Marengo, J.A. and Nobre, C.A. (2001) General characteristics and variability of climate in the Amazon Basin and its links to the global climate system. In: Richey, J.E. (ed.) *The Biogeochemistry of the Amazon Basin.* Oxford University Press, Oxford, UK, pp. 17–41.
Marengo, J.A., Nobre, C.A. and Culf, A.D. (1997) Climatic impacts of 'Friagens' in forested and deforested areas of the Amazon basin. *Journal of Applied Meteorology* 36, 1553–1566.
Mario, P. and Pujos, M. (1998) Origin of late Holocene fine-grained sediments on the French Guiana shelf. *Continental Shelf Research* 18, 1613–1629.
MARNR (1985) *Atlas de la Vegetación de Venezuela.* MARNR, Caracas.
Marques, J., Santos, J.M. and Salati, E. (1979) O armazenamento atmosférico de vapor de água sobre a região Amazonica. *Acta Amazonica* 9, 715–721.
Martin, L., Bertaux, J., Corrége, T., Ledru, M.-P., Mourguiart, P., Sifeddine, A., Soubiés, F., Wirmann, D., Suguio, K. and Turcq, B. (1997) Astronomical forcing of contrasting rainfall changes in tropical South America between 12,400 and 8800 cal yr BP. *Quaternary Research* 47, 117–122.
Martinelli, L.A., Victoria, R.L., Devol, A.H., Richey, J. and Forberg, B.R. (1989) Suspended sediment load in the Amazon basin: an overview. *Geojournal* 19, 381–389.
Mass, C.F. and Portman, D.A. (1989) Major volcanic eruptions and climate: a critical evaluation. *Journal of Climate* 2, 566–593.
Mayle, F.E., Burbridge, R. and Killeen, T.J. (2001) Millennial-scale dynamics of southern Amazonian rain forests. *Science* 290, 2291–2294.
McClain, M. and Elsenbeer, H. (2001) Terrestrial inputs to Amazon streams and internal biogeochemical processing. In: McClain, M.E., Victoria, R.L. and Richey, J.E. (eds) *The Biogeochemistry of the Amazon Basin.* Oxford University Press, Oxford, UK, pp. 183–208.
McClain, M., Richey, J. and Victoria, R.L. (1995) Andean contributions to the biogeochemistry of the Amazon river system. *Bulletin de l'Institut français d'études andines* 24, 1–13.
McConnell, R.B. (1958) *Provisional Stratigraphical Table for British Guiana.* GSBG, Georgetown.
McConnell, R.B. (1959) Fossils in the North Savannas and their significance in the search for oil in British Guiana. *Timehri* 38, 64–85.
McConnell, R.B. (1968) Planation surfaces in Guyana. *Geographical Journal* 134, 506–520.
McConnell, R.B. (1969) The succession of erosion bevels in Guyana. *Records of the Geological Survey of Guyana* 6, 1–17.
McConnell, R.B. and Williams, E. (1969) Distribution and provisional correlation of the Precambrian of the Guiana Shield. In: *8th Guiana Geology Conference.* Geological Survey Dept, Georgetown, Guyana, pp. 1–22.
McCormick, M.P., Thomason, L.W. and Trepte, C.R. (1995) Atmospheric effects of the Mt. Pinatubo eruption. *Nature* 373, 399–404.
McDaniel, D.K., McLennan, S.M. and Hanson, G.N. (1997) Provenance of Amazon fan muds: constraints from Nd and Pb isotopes. In: Peterson, L.C. (ed.) *Proceedings of the Ocean Drilling Program, Scientific Results.* ODP, College Station, Texas, pp. 169–176.
McIntyre, A. and Molfino, B. (1996) Forcing of Atlantic equatorial and subpolar millennial cycles by precession. *Science* 274, 1867–1870.

Meade, R.H. (1985) *Suspended Sediment in the Amazon River and its Tributaries in Brazil During 1982–1984.* USGS Open File Report 85-333. USGS, Denver, Colorado.

Meade, R.H., Dunne, T., Richey, J., de Santos, U.M.and Salati, E. (1985) Storage and remobilization of sediment in the lower Amazon river of Brazil. *Science* 228, 488–490.

Meade, R.H., Weibezahn, F.H., Lewis, W.M.J. and Hernández, D.P. (1990) Suspended sediment budget for the Orinoco River. In: Weibezahn, F.H., Alvarez, H. and Lewis, W.M.J. (eds) *El Rio Orinoco Como Ecosistema.* Impresos Rubel CA, Caracas, pp. 55–79.

Meehl, G.A., Washington, W.M., Wigley, T.M.L., Arblaster, J.M. and Dai, A. (2003) Solar and greenhouse gas forcing and climate response in the twentieth century. *Journal of Climate* 16, 426–444.

Meggers, B.J. (1994) Archaeological evidence for the impact of mega-Niño events on Amazonia during the past two millennia. *Climatic Change* 28, 321–338.

Mehta, V.M. (1998) Variability of the tropical ocean surface temperatures at decadal-multidecadal timescales. Part I: the Atlantic Ocean. *Journal of Climate* 11, 2351–2375.

Mehta, V.M. and Delworth, T. (1995) Decadal variability of the tropical Atlantic Ocean surface temperature in shipboard measurements and in a global ocean–atmosphere model. *Journal of Climate* 8, 172–190.

Meisner, B.N. and Arkin, P.A. (1987) Spatial and annual variations in the diurnal cycle of large-scale tropical convective cloudiness and precipitation. *Monthly Weather Review* 115, 2009–2032.

Melack, J.M. and Forsberg, B.R. (2001) Biogeochemistry of the Amazon floodplain lakes and associated wetlands. In: McClain, M.E., Victoria, R.L. and Richey, J.E. (eds) *The Biogeochemistry of the Amazon Basin.* Oxford University Press, Oxford, UK, pp. 235–274.

Mélice, J.-L. and Servain, J. (2003) The tropical Atlantic meridional SST gradient index and its relationships with the SOI, NAO and Southern Ocean. *Climate Dynamics* 20, 447–464.

Mendoza, V. (1977) Evolucion tectonica del Escudo de Guayana. *Boletín de Geologia (Publicacion Especial)* 7, 2237–2270.

Mendoza, V. (1980) Abstract – Petrotectonic provinces of the Amazonas territory, Guiana Shield, Venezuela. In: *26th International Geological Congress.* IGC, Paris, p. 99.

Mestas-Nuñez, A.M. and Enfield, D.B. (2001) Eastern Equatorial Pacific SST variability: ENSO and non-ENSO components and their climatic associations. *Journal of Climate* 14, 391–402.

Meybeck, M. (1979) Concentration des eaux fluviales en éléments majeurs et apports en solution aux océans. *Revue de Geologie Dynamique et de Geographie Physique* 21, 215–246.

Meybeck, M. (1982) Carbon, nitrogen and phosphorus transport by world rivers. *American Journal of Science* 282, 401–450.

Meybeck, M. (1988) How to establish and use world budgets of river material. In: Lerman, A. and Meybeck, M. (eds) *Physical and Chemical Weathering in Geochemical Cycles.* Kluwer Academic, Dordrecht, The Netherlands, pp. 247–272.

Meybeck, M. (1994) Origin and variable composition of present day riverborne material. In: BESR, CGER and NRC (eds) *Material Fluxes on the Surface of the Earth.* National Academy Press, Washington, DC, pp. 61–73.

Meybeck, M., Friedrich, G., Thomas, R. and Chapman, D. (1992) Rivers. In: Chapman, D. (ed.) *Water Quality Assessments – A Guide to the Use of Biota, Sediments and Water in Environmental Monitoring.* Chapman & Hall, London, pp. 241–320.

Milankovitch, M.M. (1941) *Canon of Insolation and the Ice-Age Problem.* Koniglich Serbische Akademie, Belgrade.

Milliman, J.D. (1997) Blessed dams or dammed dams. *Nature (London)* 386, 325–327.

Milliman, J.D. and Boyle, E. (1975) Biological uptake of dissolved silica in the Amazon River estuary. *Science* 189, 995–997.

Milliman, J.D. and Meade, R.H. (1983) World-wide delivery of river sediment to the oceans. *Journal of Geology* 91, 1–21.

Milliman, J.D. and Syvitiski, J.P.M. (1994) Geomorphic/tectonic control of sediment discharge to the ocean: the importance of small mountainous rivers. In: BESR, CGER and NRC (eds) *Material Fluxes on the Surface of the Earth.* National Academy Press, Washington, DC, pp. 74–85.

Milliman, J.D., Summerhayes, C.P. and Barretto, H.T. (1975) Quaternary sedimentation on the Amazon continental margin: a model. *Bulletin of the Geological Society of America* 86, 610–614.

Mo, K., Bell, G.D. and Thiawn, W.M. (2001) Impact of sea surface temperature anomalies on the Atlantic tropical storm activity and West African rainfall. *Journal of the Atmospheric Sciences* 58, 3477–3496.

Mo, K.C. and Häkkinen, S. (2001) Interannual variability in the tropical Atlantic and linkages to the Pacific. *Journal of Climate* 14, 2740–2762.

Molinier, M., Guyot, J.L., de Oliveira, E., Guimarães, V. and Chaves, A. (1993) Hydrologie du bassin de l'Amazone. In: Boulègue, J. and Olivry, J.-C. (eds) *Grands Bassin Fluviaux Périatlantiques: Congo, Niger, Amazone.* PEGI/INSU/CNRS/ORSTOM, Paris, pp. 335–344.

Molion, L.C.B. and Moraes, J.C. (1987) Oscilação Sul e descarga de rios na America do Sul tropical. *Revista Brasiliera Eng.* 5, 53–63.

Montgomery, C.W. (1979) Uranium–lead geochronology of the Archaen Imataca Series, Venezuelan Guayana Shield. *Contributions to Mineralogy and Petrology* 69, 167–176.

Moran, E. (1982) *Developing the Amazon.* University of Indiana Press, Bloomington, Indiana.

Moran, E.F. (1995) Rich and poor ecosystems of Amazonia: an approach to management. In: Uitto, J.I. (ed.) *The Fragile Tropics of Latin America.* United Nations University Press, Tokyo, pp. 45–67.

Morellato, L.P.C. and Haddad, C.F.B. (2001) Introduction: the Brazilian Atlantic forest. *Biotropica* 32, 786–792.

Mortatti, J. and Probst, J.-L. (2003) Silicate rock weathering and atmospheric/soil CO_2 uptake in the Amazon basin estimated from river water geochemistry: seasonal and spatial variations. *Chemical Geology* 197, 177–196.

Muller, R.A. and MacDonald, G.J. (1997) Glacial cycles and astronomical forcing. *Science* 277, 215–218.

Muller-Karger, F., McClain, M. and Richardson, P. (1988) The dispersal of the Amazon's water. *Nature (London)* 333, 56–59.

Naqvi, S.M. and Rogers, J.J.W. (1987) *Precambrian Geology of India.* Clarendon Press, New York.

Négrel, P. and Lachassagne, P. (2001) Geochemistry of the Maroni river (French Guiana) during the low water stage: implications for water–rock interaction and groundwater characteristics. *Journal of Hydrology* 237, 212–233.

NGDC (2003) Solar Data Services-Sunspot Numbers, http://www.ngdc.noaa.gov/stp/SOLAR/SSN/ssn

Nobre, C., Zebiak, S.E. and Kirtman, B.P. (2003) Local and remote sources of tropical Atlantic variability as inferred from the results of a hybrid ocean–atmosphere coupled model. *Geophysical Research Letters* 30, 8008.

Nobre, P. and Shukla, J. (1996) Variations of sea surface temperature, wind stress, and rainfall over the tropical Atlantic and South America. *Journal of Climate* 9, 2464–2479.

Noordam, D. (1993) The geographical outline. In: Ouboter, P.E. (ed.) *The Freshwater Ecosystems of Suriname.* Kluwer Academic, Dordrecht, The Netherlands, pp. 13–28.

Nordin, C.F.J., Meade, R.H., Cranston, C.C. and Curtis, W.F. (1983) *Data from Sediment Studies of the Rio Orinoco, Venezuela, August 15–25 1982.* USGS Open File Report 83-679. USGS, Denver, Colorado.

Nota, D.J.G. (1969) Geomorphology and sediments of Western Suriname Shelf: a preliminary note. *Geologie en Mijnbouw* 48, 185–188.

ODP (ed.) (2003) *Demerara Rise: Equatorial Cretaceous and Paleogene Paleoceanographic Transect, Western Atlantic.* ODP, College Station, Texas.

Ogden, J. (1966) Ordination studies on a small area of tropical rain forest. MSc thesis, University of Wales, Bangor, UK.

Ogurtsov, M.G., Nagovistyn, Y.A., Kocharov, G.E. and Jungner, H. (2002) Long-period cycles of the Sun's activity recorded in direct solar data and proxies. *Solar Physics* 211, 371–394.

Olsen, P.E. and Kent, D.V. (1996) Milankovitch climate forcing in the tropics of Pangaea during the Late Triassic. *Palaeogeography, Palaeoclimatology, Palaeoecology* 122, 1–26.

Olson, D.M., Dinerstein, E., Wikramanayake, E.D., Burgess, N.D., Powell, G.V.N., Underwood, E.C., D'Amico, J.A., Itoua, I., Strand, H.E., Morrison, J.C., Loucks, C.J., Allnutt, T.F., Ricketts, T.H., Kura, Y., Lamoreux, J.F., Wettengel, W.W., Hedao, P. and Kassem, K.R. (2001) Terrestrial ecoregions of the world: a new map of life on Earth. *Bioscience* 51, 933–938.

Osher, L.J. and Buol, S.W. (1998) Relationship of soil properties to parent material and landscape position in eastern Madre de Dios, Peru. *Geoderma* 83, 143–166.

Paegle, J.N. and Mo, K.C. (2002) Linkages between summer rainfall variability over South America and sea surface temperature anomalies. *Journal of Climate* 15, 1389–1407.

Paolini, J. (1986) Transporte de carbono y minerales en el río Caroni. *Interciencia* 11, 295–297.

Paolini, J., Hevia, R. and Herrera, R. (1987) Transport of carbon and minerals in the Orinoco and Caroni rivers during the years 1983–84. In: Degens, E.T., Kempe, S. and Weibin, G. (eds) *Transport of Carbon and Minerals in Major World Rivers. Part 4.* SCOPE/UNEP, Hamburg, pp. 325–338.

Parra, M. and Pujos, M. (1998) Origin of late Holocene fine-grained sediments on the French Guiana shelf. *Continental Shelf Research* 18, 1613–1629.

Parrish, J.T., Ziegler, A.M. and Scotese, C.R. (1982) Rainfall patterns and the distribution of coals and evaporites in the Mesozoic and Cenozoic. *Palaeogeography, Palaeoclimatology, Palaeoecology* 40, 67–101.

Peña, O. (1996) Hidrografia. In: Huber, O. (ed.) *Ecología de la Cuenca del Río Caura, Venezuela. I. Caracterización General.* FUNDACITE, Caracas, pp. 29–33.

Perks, H.M., Charles, C.D. and Keeling, R.F. (2002) Precessionally forced productivity variations across the equatorial Pacific. *Paleoceanography* 17, 1037–1043.

Perry, C.A. and Hsu, K J. (2001) Geophysical, archaeological, and historical evidence support a solar-output model for climate change. *PNAS* 97, 12433–12438.

Pessenda, L.C.R., Boulet, R., Aravena, R., Rosolen, V., Gouveia, S.E.M., Ribeiro, A.S. and Lamotte, M. (2001) Origin and dynamics of soil organic matter and vegetation changes during the Holocene in a forest-savanna transition zone, Brazilian Amazon region. *The Holocene* 11, 250–254.

Petit, J.R., Jouzel, J., Raynaud, D., Barkov, N.I., Barnola, J.M., Basile, I., Bender, M., Chappellaz, J., Davis, J., Delaygue, G., Delmotte, M., Kotlyakov, V.M., Legrand, M., Lipenkov, V., Lorius, C., Pépin, L., Ritz, C., Saltzman, E. and Stievenard, M. (1999) Climate and atmospheric history of the past 420,000 years from the Vostok Ice Core, Antarctica. *Nature* 399, 429–436.

Petit, J R., Jouzel, J., Raynaud, D., Barkov, N.I., Barnola, J.M., Basile, I., Bender, M., Chappellaz, J., Davis, J., Delaygue, G., Delmotte, M., Kotlyakov, V.M., Legrand, M., Lipenkov, V., Lorius, C., Pépin, L., Ritz, C., Saltzman, E. and Stievenard, M. (2001) *Vostok Ice Core Data for 420,000 Years.* Paleoclimatology Data Contribution Series No. 2001-76, NOAA/NGDC Paleoclimatology Program, Boulder, Colorado.

Philander, S.G.H. (1983) El Niño Southern Oscillation phenomena. *Nature* 302, 295–301.

Philander, S.G.H. (1990) *El Nino, La Nina, and the Southern Oscillation.* Academic Press, London.

Philander, S.G.H., Gu, D., Lambert, G., Li, T., Halpern, D., Lau, N.-C. and Pacanowski, R.C. (1996) Why the ITCZ is mostly north of the equator. *Journal of Climate* 9, 2958–2972.

Pindell, J.L. and Dewey, J.F. (1982) Permo-Triassic reconstruction of western Pangaea and the evolution of the Gulf of Mexico/Caribbean region. *Tectonics* 1, 1179–1211.

Piperno, D.R. (1994) Phytolith and charcoal evidence for prehistoric slash-and-burn agriculture in the Darien rain forest of Panama. *The Holocene* 4, 321–325.

Piperno, D.R. (1997) Phytoliths and microscopic charcoal from leg 155: a vegetational and fire history of the Amazon basin during the last 75 k.y. In: Peterson, L.C. (ed.) *Proceedings of the Ocean Drilling Program, Scientific Results.* ODP, College Station, Texas, pp. 411–418.

Piperno, D.R. and Becker, P. (1996) Vegetational history of a site in the central Amazon basin derived from phytolith and charcoal records from natural soils. *Quaternary Research* 45, 202–209.

Piperno, D.R. and Jones, J.G. (2003) Paleoecological and archaeological implications of a Late Pleistocene/Early Holocene record of vegetation and climate from the Pacific coastal plain of Panama. *Quaternary Research* 59, 79–87.

Piperno, D.R., Bush, M.B. and Colinvaux, P.A. (1990) Paleoenvironments and human occupation in late-glacial Panama. *Quaternary Research* 33, 108–116.

Pirazzoli, P.A. (1991) *World Atlas of Holocene Sea-Level Change.* Elsevier, Amsterdam.

Pitman, W.C.I., Cande, S., LaBrecque, J.L. and Pindell, J.L. (1993) Fragmentation of Gondwana: the separation of Africa from South America. In: Goldblatt, P. (ed.) *Biological Relationships Between Africa and South America.* Yale University Press, London, pp. 15–34.

Poldevaart (1955) Chemistry of the Earth's crust. *Geological Society of America Special Paper* 62, 119–144.

Potter, P.E. (1978) Petrology and chemistry of modern big river sands. *Journal of Geology* 86, 423–449.

Prado, H.D. (1996) *Manual de Classificacão de Solos do Brasil.* FUNEP, São Paulo.

Prasad, G. (1983) A review of the early Tertiary bauxite event in South America, Africa and India. *Journal of African Earth Sciences* 1, 305–313.

Pratt, D. (2001) Plate tectonics: a paradigm under threat. *Journal of Scientific Exploration* 14, 307–352.

Prentice, I.C. and Matthews, R.K. (1988) Cenozoic ice-volume history: development of a composite oxygen isotope record. *Geology* 16, 963–966.

Press, F. and Siever, R. (1982) *Earth,* 3rd edn. W.H. Freeman and Co., San Francisco, California.

Priem, H.N.A., Boelrijk, N.A.I.M., Hebeda, E.H., Verschure, R.H. and Verdurmen, E.A.T. (1968) Isotopic age determinations on Suriname rocks, 3. Proterozoic and Permo-Triassic basalt magmatism in the Guiana Shield. *Geologie en Mijnbouw* 47, 17–20.

Priem, H.N.A., Andriessen, P.A.M., Boelrijk, N.A.I.M., de Boorder, H., Hedeba, E.H., Huguett, A., Verdurmen, E.A.T. and Verschure, R.H. (1982) Geochronology of the Precambrian in the Amazonas region of southeastern Colombia (western Guiana Shield). *Geologie en Mijnbouw* 61, 229–242.

PRORADAM (1979) *La Amazonia Colombiana y sus Recursos.* IGAC/CIAF/MINDEFENSA, Bogotá.

Prost, M. (1987) Shoreline changes in French Guiana. In: Rabassa, J. (ed.) *Quaternary of South America and Antarctic peninsula.* A.A. Balkema, Rotterdam, The Netherlands, pp. 291–299.

Prost, M. and Lointier, M. (1987) Sedimentology and stratigraphy of the Holocene formations of the French Guiana coastal plain. In: *IGCP Project 201 Mérida (Venezuela) Meeting.* ORSTOM, Mérida, Venezuela, pp. 55–83.

Pujos, M. and Froidefond, J.M. (1995) Water masses and suspended matter circulation on the French Guiana continental shelf. *Continental Shelf Research* 15, 1157–1171.

Pujos, M. and Odin, G.S. (1986) La sédimentation au Quaternaire terminal sur la plate-forme continentale de la Guyane française. *Oceanologica Acta* 9, 363–382.

Pujos, M., Bobier, C., Chagnaud, M., Fourcassies, C., Froidefond, J.M., Gouleau, D., Guillaume, P., Jouanneau, J.M., Parra, M., Pons, J.C., Pujos, A., DeRessequier, A. and Viguier, C. (1989) *Les caractères de la sédimentation fine sur le littoral de la Guyane Française (région de Cayenne): nature, origine et périodicité de l'envasement.* Rapport CORDET 87001/88310.

Pujos, M., Bouysse, P. and Pons, J.C. (1990) Sources and distribution of heavy minerals in Late Quaternary sediments of the French Guiana continental shelf. *Continental Shelf Research* 10, 59–79.

Putzer, H. (1984) The geological evolution of the Amazon basin and its mineral resources. In: Sioli, H. (ed.) *The Amazon: Limnology and Landscape Ecology of a Mighty Tropical River and its Basin.* Dr. W. Junk, Dordrecht, The Netherlands, pp. 15–46.

Pyle, D.M. (1992) On the 'climate effectiveness' of volcanic eruptions. *Quaternary Research* 37, 125–129.

Quay, P., King, D., Wilbur, D., Wofsy, S. and Richey, J. (1989) $^{13}C/^{12}C$ of atmospheric CO_2 in the Amazon Basin: forest and river sources. *Journal of Geophysical Research* 94, 18327–18336.

Rabinowitz, P.D. and LaBrecque, J.L. (1979) The isostatic gravity anomaly: key to the evolution of the ocean–continent boundary at passive continental margins. *Earth and Planetary Science Letters* 35, 145–150.

RADAMBRASIL (1972) *Levantamento de Recursos Naturais.* Ministério das Minas e Energia/IBGE, Rio de Janeiro.

RADAMBRASIL (1973–1978) *Levantamento de Recursos Naturais.* MME, Rio de Janeiro.

Ramón, J.C., Dzou, L.I., Hughes, W.B. and Holba, A.G. (2001) Evolution of the Cretaceous organic facies in Colombia: implications for oil composition. *Journal of South American Earth Sciences* 14, 31–50.

Ramos, B. (1997) Los suelos en las cumbres de las Sierras de Maigualida y uasadi-jidi, Guayana Venezolana. In: Rosales, J. (ed.) *Ecología de la Cuenca del Río Caura, Venezuela.* Refolit C.A., Caracas, pp. 423–440.

Ramos, B. and Blanco, J.A. (1997) Los suelos del medio Rio Caura, Sector Rio Chanaro – Rio Marik, Estado Bolivar, Venezuela. In: Rosales, J. (ed.) *Ecología de la Cuenca del Río Caura, Venezuela 2.* FIBV/UNEG, Caracas, p. 473.

Rand, A.S. and Rand, W.M. (1982) Variation in rainfall on Barro Colorado Island. In: Windsor, D.M. (ed.) *The Ecology of a Tropical Forest Seasonal Rhythms and Long-term Changes.* Smithsonian Institution Press, Washington, DC, pp. 47–60.

Räsänen, M. (1993) La geohistoria y geologia de la Amazonia Peruana. In: Danjoy, W. (ed.) *Amazonia Peruana. Vegetacion Humeda Tropical en el Llano Subandino.* PAUT/ONERN, Jyväskylä, Finland, pp. 43–65.

Räsänen, M., Linna, A.M., Santos, J.C.R. and Negri, F.R. (1995) Late Miocene tidal deposits in the Amazonian foreland basin. *Science* 269, 386–390.

Räsänen, M.E., Salo, J.S. and Kalliola, R.J. (1987) Fluvial perturbance in the western Amazon basin: regulation by long-term sub-Andean tectonics. *Science* 238, 1398–1401.

Rasmusson, E.M. and Wallace, J.M. (1983) Meteorological aspects of the El Niño/Southern Oscillation. *Science* 222, 1195–1202.

Ratisbona, L.R. (1976) The climate of Brazil. In: Schwerdtfeger, W. (ed.) *Climates of Central and South America.* Elsevier, Oxford, UK, pp. 219–289.

Raven, P.H. and Axelrod, D.I. (1974) Angiosperm biogeography and past continental movements. *Annals of the Missouri Botanical Garden* 61, 539–673.

Rial, J.A. (1999) Pacemaking the ice ages by frequency modulation of Earth's orbital eccentricity. *Science* 285, 564–568.

Ribeiro, J E.L.d.S., Hopkins, M.J.G., Vicentini, A., Sothers, C.A., Costa, M.A.d.S., de Brito, J.M., de Souza, M.A.D., Martins, L.H.P., Lohmann, L.G., Assuncao, P.A.C.L., Pereira, E.d.C., da Silva, C.F., Mesquita, M.R. and Procopio, L.C. (1999) *Flora da Reserva Ducke.* INPA-DfID, Manaus.

Richards, P.W. (1952) *The Tropical Rain Forest.* Cambridge University Press, Oxford, UK.

Richards, P.W. (1996) *The Tropical Rain Forest: An Ecological Study.* Cambridge University Press, Cambridge, UK.

Richards, P.W. and Davis, T.A. (1933) The vegetation of Moraballi Creek, British Guiana: an ecological study of a limited area of tropical rain forest. *Journal of Ecology* 21, 350–384.

Richardson, P.L. and Walsh, D.W. (1986) Mapping climatological seasonal variations of surface currents in the tropical Atlantic using ship drifts. *Journal of Geophysical Research* 91, 10537–10550.

Richey, J., Meade, R.H., Salati, E., Devol, A.H., Nordin, C.F. and dos Santos, U. (1986) Water discharge and suspended sediment concentrations in the Amazon River: 1982–1984. *Water Resources Research* 22, 756–764.

Richey, J., Nobre, C.A. and Deser, C. (1989) Amazon river discharge and climate variability. *Science* 246, 101–103.

Richey, J., Victoria, R.L., Salati, E. and Forsberg, B.R. (1990) The biogeochemistry of a major river system: the Amazon case study. In: Degenes, E.T., Kempe, S. and Richey, J. (eds) *Biogeochemistry of the Major World Rivers.* John Wiley & Sons, Chichester, UK, pp. 57–74.

Richter, D.D. and Babbar, L.I. (1991) Soil diversity in the tropics. *Advances in Ecological Research* 21, 369–389.

Riehl, H. (1977) Venezuelan rain systems and the general circulation of the summer tropics. I. Rain systems. *Monthly Weather Review* 105, 1402–1420.

Rind, D. (2002) The sun's role in climate variations. *Science* 296, 673–677.

Rind, D. and Overpeck, J.T. (1993) Hypothesized causes of decade-to-century scale climate variability-climate model results. *Quaternary Science Review* 12, 357–374.

Robertson, A.W., Mechoso, C.R. and Kim, Y.-J. (2001) The influence of Atlantic sea surface temperature anomalies on the North Atlantic Oscillation. *Journal of Climate* 13, 122–138.

Robinson, P.J. and Henderson-Seller, A. (1999) *Contemporary Climatology,* 2nd edn. Longman, London.

Robock, A. and Mao, J. (1995) The volcanic signal in surface temperature observations. *Journal of Climate* 8, 1086–1103.

Rogers, J.C. (1984) The association between the North Atlantic Oscillation and the Southern Oscillation in the northern hemisphere. *Monthly Weather Review* 112, 1999–2015.

Rogers, J.C. (1988) Precipitation variability over the Caribbean and tropical Americas associated with the Southern Oscillation. *Journal of Climate* 1, 172–182.

Romero, E.J. (1993) South American paleofloras. In: Goldblatt, P. (ed.) *Biological Relationships between Africa and South America.* Yale University Press, New Haven, Connecticut pp. 62–85.

Roosevelt, A.C. (1991) *Moundbuilders of the Amazon: Geophysical Archaeology on Marajó Island, Brazil.* Academic Press, London.

Ropelewski, C.F. and Halpert, M.S. (1987) Global and regional scale precipitation patterns associated with the El Nino/Southern Oscillation. *Monthly Weather Review* 115, 1606–1626.

Ropelewski, C.F. and Halpert, M.S. (1989) Precipitation patterns associated with the high index phase of the Southern Oscillation. *Journal of Climate* 3, 268–284.

Ropelewski, C.F. and Halpert, M.S. (1996) Quantifying Southern Oscillation–precipitation relationships. *Journal of Climate* 9, 1043–1059.

Ropelewski, C.F. and Jones, P.D. (1987) An extension of the Tahiti–Darwin Southern Oscillation Index. *Monthly Weather Review* 115, 2161–2165.

Roth, J.M., Droxler, A.W. and Kameo, K. (2001) The Caribbean carbonate crash at the middle to late Miocene transition: linkage to the establishment of the modern global ocean conveyor. In: Draper, G. (ed.) *Proceedings of the Ocean Drilling Program, Scientific Results.* ODP, College Station, Texas, pp. 249–273.

Rowe, W., Dunbar, R.B., Mucciarone, D.A., Seltzer, G.O., Baker, P.A. and Fritz, S.C. (2002) Insolation, moisture balance and climate change on the South American altiplano since the last glacial maximum. *Climatic Change* 52, 175–199.

Ruiz-Barradas, A., Carton, J.A. and Nigam, S. (2001) Structure of interannual-to-decadal climate variability in the tropical Atlantic sector. *Journal of Climate* 13, 3285–3297.

Rull, V. (1991) *Contribución a la Paleoecología de Pantepui la Gran Sabana (Guayana Venezolana): Clima, Biogeografía, y Ecología.* Fundacite-Guayana/UNEG, Caracas.

Rull, V. (1996) Late Pleistocene and Holocene climates of Venezuela. *Quaternary International* 31, 85–94.

Rull, V. (1999) Palaeoclimatology and sea-level history in Venezuela: new data, land–sea correlations, and proposals for future studies in the framework of the IGBP-PAGES project. *Interciencia* 24, 92–101.

Ryther, J.H., Menzel, D.W. and Corwin, N. (1967) Influence of the Amazon River outflow on the ecology of the western tropical Atlantic I: hydrology and nutrient chemistry. *Journal of Marine Research* 25, 69–82.

Salati, E. and Marques, J. (1984) Climatology of the Amazon region. In: Sioli, H. (ed.) *The Amazon, Limnology and Landscape Ecology of a Mighty Tropical River and its Basin.* Dr. W. Junk, Dordrecht, The Netherlands, pp. 85–126.
Saldarriaga, J.G. (1994) *Recuperación de la Selva de 'Tierra Firme' en el Alto Río Negro Amazonia Colombiana–Venezolana.* Tropenbos Colombia Programme, Bogotá.
Saldarriaga, J.G. and West, D.C. (1986) Holocene fires in the northern Amazon basin. *Quaternary Research* 26, 358–366.
Salgado-Labouriau, M.L. (1980) A pollen diagram of the Pleistocene-Holocene boundary of Lake Valencia, Venezuela. *Review of Palaeobotany and Palynology* 30, 297–312.
Salgado-Labouriau, M.L., Casseti, V., Ferraz-Vicentini, K.R., Martin, L., Soubiés, F., Suguio, K. and Turcq, B. (1997) Late Quaternary vegetational and climatic changes in cerrado and palm swamp from Central Brazil. *Palaeogeography, Palaeoclimatology, Palaeoecology* 128, 215–226.
Salo, J., Kalliola, R., Hakkinen, L., Makinen, Y., Niemela, P., Puhakka, M. and Coley, P.D. (1986) River dynamics and the diversity of Amazon lowland forest. *Nature* 322, 254–258.
Sanchez, P.A. (1976) *Properties and Management of Soils in the Tropics.* John Wiley & Sons, New York.
Sanchez, P.A. (1981) Soil management in the Oxisol savannahs and Ultisol jungles of tropical South America. In: Greenland, D.J. (ed.) *Characterisation of Soils.* Oxford University Press, Oxford, UK, pp. 214–253.
Sanford Jr, R.L., Paaby, P., Luvall, J.C. and Phillips, E. (1994) Climate, geomorphology and aquatic systems. In: McDade, L.A., Bawa, K.S., Hespenheide, H.A. and Hartshorn, G.S. (eds) *La Selva: Ecology and Natural History of a Neotropical Rain Forest.* Chicago University Press, Chicago, Illinois, pp. 19–33.
Saravanan, R. and Chang, P. (1999) Oceanic mixed layer feedback and tropical Atlantic variability. *Geophysical Research Letters* 26, 3629–3632.
Savin, S.M. (1977) The history of the Earth's surface temperature during the past 100 million years. *Annual Review of Earth and Planetary Sciences* 5, 319–356.
Scarparo Cunha, A.A. and Koutsoukos, A.M. (2001) Orbital cyclicity in a Turonian sequence of the Cotinguiba Formation, Sergipe Basin, NE Brazil. *Cretaceous Research* 22, 529–548.
Schlesinger, M.E. and Ramankutty, N. (1992) Implications for global warming of intercycle solar irradiance variations. *Nature* 360, 330–333.
Schopf, P.S. and Suarez, M.J. (1987) Vacillations in a coupled ocean–atmosphere model. *Journal of Atmospheric Science* 45, 549–566.
Schubert, C. and Fritz, S.C. (1985) Radiocarbon ages of peat, Guayana highlands (Venezuela). *Naturwissenschaften* 72, 427–429.
Schulman, L., Ruokolainen, K. and Tuomisto, H. (1999) Parameters for global ecosystem models. *Nature* 399, 535–536.
Schulz, J.P. (1960) Ecological studies on rain forest in Northern Suriname. *Verhandelingen der Koninklijke Nederlandse Akademie van Wetenschappen, Afd. Natuurkunde* 163, 1–250.
Schwabe, G.H. (1969) Towards an ecological characterization of the South American continent. In: Fittkau, E.J. (ed.) *Biogeography and Ecology in South America.* Dr. W. Junk, The Hague, The Netherlands.
Sclater, J.G., Hellinger, S. and Tapscott, C. (1977) The paleobathymetry of the Atlantic Ocean from the Jurassic to the present. *Journal of Geology* 85, 509–552.
Seyler, P.T. and Boaventura, G.R. (2001) Trace elements in the mainstem Amazon River. In: McClain, M.E., Victoria, R.L. and Richey, J.E. (eds) *The Biogeochemistry of the Amazon Basin.* Oxford University Press, Oxford, UK, pp. 307–327.
Shackleton, N.J., Curry, W.B., Richter, C. and Bralower, T.J. (eds) (1997) Ceara Rise. ODP Vol. 154 in *Proceedings of the Ocean Drilling Program, Scientific Results.* College Station, Texas.
Shindell, D., Rind, D., Balachandran, N., Lean, J. and Lonergan, P. (1999) Solar cycle variability, ozone, and climate. *Science* 284, 305–308.
Shindell, D.T., Schmidt, G.A., Mann, M.E., Rind, D. and Waple, A. (2001) Solar forcing of regional climate change during the Maunder minimum. *Science* 294, 2149–2152.
Sifeddine, A., Martin, L., Turcq, B., Volkmer-Ribeiro, C., Soubiés, F., Cordeiro, R.C. and Suguio, K. (2001) Variations of the Amazonian rainforest environment: a sedimentological record covering 30,000 years. *Palaeogeography, Palaeoclimatology, Palaeoecology* 168, 221–235.
Sifeddine, A., Albuquerque, A.L.S., Ledru, M.-P., Turcq, B., Knoppers, B., Martin, L., Zamboni de Mello, W., Passenau, H., Dominquez, J.M L., Cordeiro, R.C., Abrao, J.J. and da Silva Pinto Bittencourt, A.C. (2003) A 21000 cal years paleoclimatic record from Cacó Lake, northern Brazil: evidence from sedimentary and pollen analyses. *Palaeogeography, Palaeoclimatology, Palaeoecology* 189, 25–34.

Sinha, N.K.P. (1968) *Geomorphic Evolution of the Northern Rupununi Basin, Guyana.* McGill University, Montreal.
Sioli, H. (1957) Valores de pH de águas amazônicas. *Boletin Museu Paraense E.Goeldi (Geólogia)* 1, 1–37.
Sioli, H. (1968) Hydrochemistry and geology in the Brazilian Amazon region. *Amazoniana* 1, 267–277.
Sioli, H. (1975) Tropical rivers as expressions of their terrestrial environments. In: Golley, F.B. and Medina, E. (eds) *Tropical Ecosystems: Trends in Terrestrial and Aquatic Research.* Springer, New York, pp. 275–287.
Sioli, H. (1984a) The Amazon and its main affluents: hydrography, morphology of the river courses, and river types. In: Sioli, H. (ed.) *The Amazon: Limnology and Landscape Ecology of a Mighty Tropical River and its Basin.* Dr. W. Junk, Dordrecht, The Netherlands, pp. 127–195.
Sioli, H. (ed.) (1984b) *The Amazon: Limnology and Landscape Ecology of a Mighty Tropical River and its Basin.* Dr. W. Junk, Dordrecht, The Netherlands.
Sippel, S.J., Hamilton, S.K., Melack, J.M. and Novo, E.M. (1998) Passive microwave observations of inundation area and area/stage relation in the Amazon River. *International Journal of Remote Sensing* 19, 3055–3074.
Sklar, L. (2000) *Report on Hydrological and Geochemical Processes in Large Scale River Basins.* World Commission on Dams, Manaus, Brazil.
Smith, L.K., Lewis, W.M.J., Chanton, J.P., Cronin, G. and Hamilton, S.K. (2001) Methane emissions from the Orinoco River floodplain, Venezuela. *Biogeochemistry* 51, 113–140.
Smith, P.E., Evenson, N.M., York, D. and Odin, G.S. (1998) Single-grain ^{40}Ar–^{39}Ar ages of glauconies: implications for the geologic time scale and global sea level variations. *Science* 279, 1517–1519.
Snow, J.W. (1976) The climate of northern South America. In: Schwerdtfeger, W. (ed.) *Climates of Central and South America.* Elsevier, Oxford, UK, pp. 295–478.
Sobieraj, J.A., Elsenbeer, H. and McClain, M. (2002) The cation and silica chemistry of a Subandean river basin in western Amazonia. *Hydrological Processes* 16, 1353–1372.
Sollins, P., Sancho, F.M., Mata, C.R. and Sanford Jr, R.L. (1994) Soils and soil process research. In: McDade, L.A., Bawa, K.S., Hespenheide, H.A. and Hartshorn, G.S. (eds) *La Selva: Ecology and Natural History of a Neotropical Rain Forest.* Chicago University Press, Chicago, Illinois, pp. 34–53.
Sombroek, W. (1999) Annual rainfall and strength of dry season strength in the Amazon region and their environmental consequences. *Ambio* 30, 388–396.
Sombroek, W.G. (1990) *Amazon Landforms and Soils in Relation to Biological Diversity in Priority Areas for Conservation in Amazonia.* Conservation International/INPA, Manaus, Brazil.
Sombroek, W.G. (1991) Amazon landforms and soils in relation to biological diversity. In: ISRIC (ed.) *Annual Report 1990.* ISRIC, Wageningen, The Netherlands, pp. 7–25.
Soubiés, F. (1979) Existence d'une phase sèche en Amazonie Brésilienne datée par la présence de charbons dans les sols (6.000–3.000 ans B.P.). *Cahiers ORSTOM, Série Géologie* 11, 133–148.
Sourdat, M. (1987) Reconnaissances pédologiques en Amazonie péruvianne. Problèmes de pédogenése et de mise en valeur. *Cahiers ORSTOM, Série Pédologie* 23, 95–109.
Stallard, R.F. (1988) Weathering and erosion in the humid tropics. In: Lerman, A. and Meybeck, M. (eds) *Physical and Chemical Weathering in Geochemical Cycles.* Kluwer, Dordrecht, The Netherlands, pp. 225–246.
Stallard, R.F. and Edmond, J.M. (1983) Geochemistry of the Amazon: 2. The influence of geology and weathering environment on the dissolved load. *Journal of Geophysical Research* 88, 9671–9688.
Stallard, R.F. and Edmond, J.M. (1987) Geochemistry of the Amazon 3. Weathering chemistry and limits to dissolved inputs. *Journal of Geophysical Research* 92, 8293–8302.
Stark, J., Hill, H., Rutherford, G.K. and Jones, T.A. (1959) *Soil and Land Use Survey No. 5. British Guiana. Part II – the Bartica Triangle.* ICTA Regional Research Centre, Port-of-Spain, Trinidad.
Stern, K.M. (1970) Der Casiquiare-Kanal, einst und jetzt. *Amazoniana* 2, 401–416.
Sternberg, H.O.R. (1995) Waters and wetlands of Brazilian Amazonia: an uncertain future. In: Uitto, J.I. (ed.) *The Fragile Tropics of Latin America – Sustainable Management of Changing Environments.* United Nations University Press, Tokyo, pp. 113–179.
Stevenson, F.J. (1994) *Humus Chemistry – Genesis, Composition, Reactions.* John Wiley & Sons, New York.
Steyermark, J. (1966) Contribuciones a la flora de Venezuela, parte 5. *Acta Botanica Venezuelica* 1, 9–256.
Steyermark, J. (1967) Flora del Auyán-tepui. *Acta Botanica Venezuelica* 2, 5–370.
Stirling, C.H., Esat T.M., Lambeck, K., McCulloch, M.T., Blake, S.G., Lee, D.-C. and Halliday, A.N. (2001) Orbital forcing of the marine isotope stage 9 interglacial. *Science* 291, 290–293.
Stott, P.A., Jones, G.S. and Mitchell, J.F.B. (2003) Do models underestimate the solar contribution to recent climate change? *Journal of Climate* 16, 4079–4093.

Sutton, R.T., Jewson, S.P. and Rowell, D.P. (2001) The elements of climate variability in the tropical Atlantic region. *Journal of Climate* 13, 3261–3284.
Szatmari, P. (1983) Amazon rift and Pisco-Juruá fault: their relation to the separation of North America from Gondwana. *Geology* 11, 300–304.
Tanaka, M., Tsuchiya, A. and Nishizawa, T. (1995) Distribution and interannual variability of rainfall in Brazil. In: Uitto, J.I. (ed.) *The Fragile Tropics of Latin America*. United Nations University Press, Tokyo, pp. 94–109.
Tardy, C., Kobilsek, B., Roquin, C. and Paquet, H. (1990) Influence of peri-Atlantic climates and paleoclimates on the distribution and mineralogical composition of bauxites and ferricretes. In: *Geochemistry of the Earth's Surface and of Mineral Formation*, 2nd International Symposium. CNRS/ORSTOM, Provence, France, pp. 179–182.
Tardy, C., Vernet, J.-L., Servant, M., Fournier, M., Leprun, J.-C., Pessenda, C.L., Sifeddine, A., Solari, M.-E., Soubies, F., Turcq, B. and Wengler, L. (2001) Feux, sols et écosystèmes forestiers tropicaux. In: Servant-Vildary, S. (ed.) *Dynamique à Long Terme des Ecosystèmes Forestiers Intertropicaux*. UNESCO, Paris, pp. 343–348.
Tardy, Y. (1992) Geochemistry and evolution of lateritic landscapes. In: Chesworth, W. (ed.) *Weathering, Soils and Paleosols*. Elsevier, London, pp. 408–443.
Tardy, Y. (1997) *Petrology of Laterites and Tropical Soils*. Ashgate, London.
Tassinari, C.C.G. (1984) *Evolução Geotectonica da Provincia Rio Negro-Juruena na Regiao Amazonica*. Universidade de São Paulo, São Paulo.
Teixeira, W., Tassinari, C.C.G., Cordani, U.G. and Kawashita, K. (1989) A review of the geochronology of the Amazonian Craton: tectonic implications. *Precambrian Research* 42, 213–227.
ter Steege, H. (2001) *Plant Diversity in Guyana*. Tropenbos Foundation, Wageningen, The Netherlands.
ter Steege, H. and Hammond, D.S. (2001) Character convergence, diversity, and disturbances in tropical rain forest in Guyana. *Ecology* 82, 3197–3212.
ter Steege, H., Jetten, V.G., Polak, A.M. and Werger, M.J.A. (1993) Tropical rain forest types and soil factors in a watershed area in Guyana. *Journal of Vegetation Science* 4, 705–716.
Terborgh, J. (1992) *Diversity and the Tropical Rain Forest*. Scientific American Library, New York.
Teunissen, P.A. (1993) Vegetation and vegetation succession of the freshwater wetlands. In: Ouboter, P.E. (ed.) *The Freshwater Ecosystems of Suriname*. Kluwer Academic, Dordrecht, The Netherlands, pp. 77–98.
Thomson, D.J. (1997) Dependence of global temperatures on atmospheric CO_2 and solar irradiance. *PNAS* 94, 8370–8377.
Tissot, C., Djuwansah, M.R.and Marius, C. (1988) Evolution de la mangrove en Guyane au cours de l'Holocene: étude palynologique. *Inst. fr. Pondichéry, trav. sec. sci. tech.* 25, 125–137.
Tomczak, M. and Godfrey, S.J. (1994) *Regional Oceanography: An Introduction*. Pergamon Press, New York.
Tréguer, P., Nelson, D.M., van Bennekom, A.J., DeMaster, D.J., Leynaert, A. and Quéguiner, B. (1995) The silica balance in the world ocean: a reestimate. *Science* 268, 375–379.
Trend-Staid, M. and Prell, W.L. (2002) Sea surface temperature at the Last Glacial Maximum: a reconstruction using the modern analog technique. *Paleoceanography* 17, 1065–1169.
Tsiropoula, G. (2003) Signatures of solar activity variability in meteorological parameters. *Journal of Atmospheric and Solar-Terrestrial Physics* 65, 469–482.
Turenne, J.F. (1974) Molecular weights of humic acids in podzol and ferrallitic soils of the savannas of French Guyana and their evolution related to soil moisture. *Tropical Agriculture (Trinidad)* 51, 134–144.
Turner, R.E., Qureshi, N., Rabalais, N.N., Dortch, Q., Justic, D., Shaw, R.F. and Cope, J. (1998) Fluctuating silicate:nitrate ratios and coastal plankton food webs. *Proceedings of the National Academy of Sciences USA* 95, 13048–13051.
Tushingham, A.M. and Peltier, W.R. (1993) *Relative Sea Level Database*. IGBP PAGES/World data center – A for paleoclimatology data contribution series 93-016. NOAA/NGDC Paleoclimatology Program, Boulder, Colorado.
USGS and T. M. Corporacion Venezolana de Guayana (1993) *Geology and Mineral Resource Assessment of the Venezuelan Guayana Shield*. US GPO, Washington, DC.
UNEP, ISSS, ISRIC and FAO (1995) *Global and National Soils and Terrain Digital Databases (SOTER)*, 2nd edn. LWWD-FAO, Rome.
USDA (1996) *Keys to Soil Taxonomy*, 7th edn. USDA, Washington, DC.
Uvo, C.B., Repelli, C.A., Zebiak, S.E. and Kushnir, Y. (1998) The relationships between tropical Pacific and Atlantic SST and northeast Brazil monthly precipitation. *Journal of Climate* 11, 551–562.
Vail, P.R., Mitchum, M.R. and Thompson, S. (1977) Seismic stratigraphy and global changes in sea level, part

4: global cycles of relative changes of sea level. In: Payton, C.E. (ed.) *Seismic Stratigraphy – Applications to Hydrocarbon Exploration.* AAPG, Tulsa, Oklahoma, pp. 49–212.

van Dam, O. (2001) *Forest Filled with Gaps: Effects of Gap Size on Water and Nutrient Cycling in Tropical Rain Forest. A Study in Guyana.* Tropenbos Guyana Programme, Wageningen, The Netherlands.

van der Eyk, J.J. (1957) *Reconnaissance Soil Survey in Northern Suriname.* WAU, Wageningen, The Netherlands.

van der Hammen, T. (1963) A palynological study of the Quaternary of British Guiana. *Leidse Geologische Mededelingen* 29, 125–180.

van der Hammen, T. (1974) The Pleistocene changes of vegetation and climate in tropical South America. *Journal of Biogeography* 1, 3–26.

van der Hammen, T. and Absy, M.L. (1994) Amazonia during the last glacial. *Palaeogeography, Palaeoclimatology, Palaeoecology* 109, 247–261.

van der Hammen, T. and Wymstra, T.A. (1964) A palynological study on the Tertiary and Upper Cretaceous of the British Guiana. *Leidse Geologische Mededelingen* 30, 183–241.

van Kekem, A.J., Pulles, J.H.M. and Khan, Z. (1996) *Soils of the Rainforest in Central Guyana.* Tropenbos-Guyana, Georgetown.

van Loon, H. and Rogers, J.C. (1978) The seesaw in winter temperatures between Greenland and Northern Europe. Part I: general description. *Monthly Weather Review* 106, 296–310.

van Roosmalen, M.G.M. (1985) *Fruits of the Guianan Flora.* Institute of Systematic Botany, Utrecht, The Netherlands.

Vanzolini, P.E. (1973) Paleoclimates, relief and species multiplication in equatorial forests. In: Duckworth, W.D. (ed.) *Tropical Forest Ecosystems in African and South America: a Comparative Review.* Smithsonian Institution Press, Washington, DC, pp. 255–258.

Vargas, H. and Rangel, J. (1996a) Clima: comportamiento de las variables. In: Huber, O. (ed.) *Ecología de la Cuenca del Río Caura, Venezuela. I. Caracterización General.* UNEG/FIBV, Caracas, pp. 34–39.

Vargas, H. and Rangel, J. (1996b) Hidrologia y sedimentos. In: Rosales, J. and Huber, O. (eds) *Ecología de la Cuenca del Río Caura, Venezuela. I. Caracterización General.* Refolit, Caracas, pp. 48–53.

Vegas-Vilarrúbia, T., Paolini, J. and García-Miragaya, J. (1988) Differentiation of some Venezuelan black water rivers based upon physico-chemical properties of their humic substances. *Biogeochemistry* 6, 59–77.

Vegas-Vilarrúbia, T., Paolini, J. and Herrera, R. (1987) A physico-chemical survey of blackwater rivers from the Orinoco and the Amazon basins of Venezuela. *Archiv Hydrobiologie* 111, 491–506.

Verhofstad, J. (1971) Geology of the Wilhelmina Mountains, with special reference to the occurrence of Precambrian ash-flow tuffs. *Amsterdam Geologie Mijnbouw Dienst Sur. Med.* 21, 9–97.

Vernet, J.-L., Wengler, L., Solari, M.-E., Ceccantini, G., Fournier, M., Ledru, M.-P. and Soubiés, F. (1994) Feux, climats et végétations au Brésil central durant l'Holocène: les données d'un profil de sol à charbons de bois. *Comptes Rendus de l'Académie de Science, Paris* 319, 1391–1397.

Victoria, R.L., Martinelli, L.A., Moraes, J.M., Ballester, M.V., Krusche, A.V., Pellegrino, G., Almeida, R.M.B. and Richey, J.E. (1998) Surface air temperature variations in the Amazon region and its borders during this century. *Journal of Climate* 11, 1105–1110.

Vitousek, P.M. (1984) Litterfall, nutrient cycling, and nutrient limitation in tropical forests. *Ecology* 65, 285–298.

Vitousek, P.M. and Denslow, J.S. (1987) Differences in extractable phosphorus among soils of the La Selva Biological Station, Costa Rica. *Biotropica* 19, 167–170.

Vizy, E.K. and Cook, K.H. (2001) Mechanisms by which Gulf of Guinea and Eastern North Atlantic sea surface temperature anomalies can influence African rainfall. *Journal of Climate* 14, 795–821.

Volkaff, B. (1984) Organisations régionales de la couverture pédologique du Brésil. Chronologie des différenciations. *Cahiers ORSTOM, Série Pédologique* 21, 225–236.

Vose, R.S., Schmoyer, R.L., Steurer, P.M., Peterson, T.C., Heim, R., Karl, T.R. and Eischeid, J. (1992) *The Global Historical Climatology Network: Long-term Monthly Temperature, Precipitation, Sea Level Pressure, and Station Pressure Data.* ORNL/CDIAC-53, NDP-041, Oak Ridge National Laboratory, Oak Ridge, Tennessee.

Vuille, M., Bradley, R.S. and Keimig, F. (2001) Climate variability in the Andes of Ecuador and its relation to tropical Pacific and Atlantic sea surface temperature anomalies. *Journal of Climate* 13, 2520–2535.

Waldmeier, M. (1961) *The Sunspot – Activity in the Years 1610–1960.* Schulthess and Co. AG, Zurich.

Walker, G.T. (1923) Correlation in seasonal variations of weather. VIII. A preliminary study of world weather. *Memoirs of the Indian Meteorological Dept* 24, 75–131.

Walker, G.T. (1924) Correlation in seasonal variations of weather. IX. A further study of world weather. *Memoirs of the Indian Meteorological Dept* 24, 275–332.

Wallace, A.R. (1878) *Tropical Nature and Other Essays.* Macmillan, London.

Walsh, R.P.D. (1996) Climate. In: Richards, P.W. (ed.) *The Tropical Rain Forest.* Cambridge University Press, Cambridge, pp. 159–205.

Wang, C. (2002) Atlantic climate variability and its associated atmospheric circulation cells. *Journal of Climate* 15, 1516–1536.

Wang, C. and Enfield, D.B. (2003) A further study of the tropical western hemisphere warm pool. *Journal of Climate* 16, 1476–1493.

Waylen, P.R., Caviedes, C.N. and Quesada, M.E. (1996) Interannual variability of monthly precipitation in Costa Rica. *Journal of Climate* 9, 2606–2613.

WCD (2001) Ecosystems and large dams: environmental performance. In: *Dams and Development: A New Framework for Decision-making.* Earthscan, London, pp. 73–96.

Weibezahn, F.H. (1990) Hidroquimica y solidos suspendidos en el alto y medio Orinoco. In: Weibezahn, F.H., Alvarez, H. and Lewis, W.M.J. (eds) *El Rio Orinoco como Ecosistema.* Impresos Rubel CA, Caracas, pp. 151–210.

Wells, J.T. and Coleman, J.M. (1981) Physical processes and fine-grained sediment dynamics, coast of Suriname, South America. *Journal of Sedimentary Petrology* 51, 1053–1068.

Weng, C., Bush, M.B. and Athens, J.S. (2002) Holocene climate change and hydrarch succession in lowland Amazonian Ecuador. *Review of Palaeobotany and Palynology* 120, 73–90.

Westin, F.C. (1962) *Report to the Government of Venezuela on the Major Soils of Venezuela.* FAO, Rome.

Whitmore, T.C. and Prance, G.T. (1987) *Biogeography and Quaternary History in Tropical America.* Oxford University Press, Oxford, UK.

Wijmstra, T.A. (1969) Palynology of the Alliance Well. *Geologie en Mijnbouw* 48, 125–133.

Wijmstra, T.A. and van der Hammen, T. (1966) Palynological data on the history of tropical savannas in northern South America. *Leidse Geologische Mededelingen* 38, 71–90.

Williams, D. (1998) The archaic colonization of the Western Guiana Littoral and its aftermath. *Archaeology and Anthropology* 12, 22–41.

Williams, E., Cannon, R.T. and McConnell, R.B. (1967) *The Folded Precambrian of Northern Guyana Related to the Geology of the Guiana Shield.* GSG, Georgetown, Guyana.

Willson, R.C. (1997) Total solar irradiance trend during solar cycles 21 and 22. *Science* 277, 163–165.

Willson, R.C. and Hudson, H.S. (1991) The Sun's luminosity over a complete solar cycle. *Nature* 351, 42–44.

Wollast, R. and Mackenzie, F.T. (1983) Global cycle of silica. In: Aston, S.R. (ed.) *Silicon Geochemistry and Biogeochemistry.* Academic Press, New York, pp. 39–76.

Wood, W.I. and McCann, J.M. (1999) The anthropogenic origin and persistence of Amazonian dark earths. *Yearbook of the Conference of Latin American Geographers* 25, 7–14.

Woodruff, S.D., Slutz, R.J., Jenne, R.L. and Steurer, P.M. (1987) A comprehensive ocean-atmosphere data set. *Bulletin of the American Meteorological Society* 68, 1239–1250.

Woodruff, S.D., Diaz, H.F., Elms, J.D. and Worley, S.J. (1998) COADS Release 2 data and metadata enhancements for improvements of marine surface flux fields. *Physics and Chemistry of the Earth* 23, 517–527.

Worbes, M. and Junk, W.J. (1989) Dating tropical trees by means of ^{14}C from bomb tests. *Ecology* 70, 503–507.

Xie, S.P., Tanimoto, Y., Noguchi, H. and Matsuno, T. (1999) How and why climate variability differs between the tropical Atlantic and Pacific. *Geophysical Research Letters* 26, 1609–1612.

Yanez, C. and Ramírez, A. (1988) Estudio geoquímico de grandes ríos venezolanos. *Memorias de la Sociedad de Ciencias Naturales La Salle* 48, 41–58.

Yavitt, J.B. (2001) Nutrient dynamics of soil derived from different parent material on Barro Colorado Island, Panama. *Biotropica* 32, 198–207.

Yavitt, J.B. and Wieder, R.K. (1988) Nitrogen, phosphorus and sulfur properties of some forest soils on Barro Colorado Island, Panama. *Biotropica* 20, 2–10.

Young, S.W. (1976) Petrographic textures of detrital polygonal quartz as an aid to interpreting crystalline source rocks. *Journal of Sedimentary Petrology* 34, 595–603.

Zachos, J.C., Stott, L.D. and Lohmann, K.C. (1994) Evolution of early Cenozoic marine temperatures. *Palaeoceanography* 9, 353–387.

Zagt, R.J. (1997) *Tree Demography in the Tropical Rain Forest of Guyana.* Tropenbos-Guyana Programme, Georgetown.

Zarin, D.J. (1999) Spatial heterogeneity and temporal variability of some Amazonian floodplain soils. In:

Henderson, A. (ed.) *Várzea: Diversity, Development, and Conservation of Amazonia's Whitewater Floodplains.* NYBG, New York, pp. 313–321.
Zebiak, S.E. (1993) Air–sea interaction in the equatorial Atlantic region. *Journal of Climate* 6, 1567–1586.
Ziegler, A.M., Scotese, C.R. and Barrett, S.F. (1983) Mesozoic and Cenozoic paleogeographic maps. In: Sundermann, J. (ed.) *Tidal Friction and the Earth's Rotation II.* Springer, London, pp. 240–252.
Zonneveld, J.I.S. (1952) Watervallen in Suriname. *Tijdschrift Koninklijke Nederlandse Aardkundig Genootschap* 69, 499–507.

3 Ecophysiological Patterns in Guianan Forest Plants

Thijs L. Pons[1], Eustace E. Alexander[2], Nico C. Houter[1], Simmoné A. Rose[2] and Toon Rijkers[3]

[1]*Department of Plant Ecophysiology, Utrecht University, Utrecht, The Netherlands;* [2]*Planning and Research Development Division, Guyana Forestry Commission, Kingston, Georgetown, Guyana;* [3]*Department of Forest Ecology and Forest Management, Wageningen University, Wageningen, The Netherlands*

Introduction

Ecophysiology seeks to understand the physiological traits that explain the distribution of plants in space and time. Starting from ecological observations about habitat characteristics and population dynamics, physiological mechanisms are investigated that are at the basis of these phenomena. Agronomical questions, including those from forestry, can also be the starting point for these types of studies. Most ecophysiological studies operate at the level of the whole plant and its interaction with the (a)biotic environment. The tremendous progress in physiology and environmental sciences has greatly stimulated the development of ecophysiology into a field of its own right (Lambers *et al.*, 1998). Plant traits are often divided between physiological and morphological or structural, which is also done in this chapter. However, one should realize that structure is a result of development and our understanding about the physiological control of development has increased substantially, which makes it a rather artificial dichotomy.

Ecophysiological research on tropical forests started more recently, mainly in the neotropics and later in other regions. Technical advances in field instrumentation that have brought the laboratory to the forest and up into the canopy have principally made this possible. Much of this work has been reviewed recently by Mulkey *et al.* (1996) and Lüttge (1997). A number of problems that make it generally difficult to extrapolate ecophysiological research to the community and ecosystem level are exacerbated by the vast species diversity of large, long-lived plants present in tropical forests. The size and longevity of the individuals also pose technical and conceptual problems, and make essential experimental approaches that are commonly applied elsewhere difficult. Nevertheless, much progress has been made in understanding the ecophysiological underpinnings of tropical forest plant functioning during the past two decades. We now have a much better understanding of the physiological traits of tropical forest plants that are particularly relevant in examining plant distribution and succession. However, these conclusions inevitably are based on assumptions about the adaptive value of plant traits. This is a potential pitfall, because critical tests of these assumptions are lacking. Since these tests are difficult to perform, if not impossible, we must find other tools.

Critical reasoning supported by mechanistic modelling is one approach. Correlating physiological traits with distribution and population biological parameters is another.

An important tool in ecophysiological work is the investigation of the response of plants to a range of growth conditions. In tropical forest studies, the light climate has often been used as an environmental variable. Plants adjust to these growth conditions. However, for this adjustment to be called acclimation, it should result in a better performance than without the adjustment. Acclimation potential for a particular environmental factor is an important trait, which determines the range of conditions where the plant can occur. Tropical forest tree species are often divided in two functional groups with respect to their light requirements for regeneration (Swaine and Whitmore, 1988). Shade-intolerant pioneers that require large disturbances are included in one group and late successional shade-tolerant species belong to the second group. Prototypes of the two groups occupy extremes of a continuum. They are characterized by contrasting trait sets (Swaine and Whitmore, 1988). Other species occupy positions along the continuum between the two groups. Many recent ecophysiological studies carried out in the Guianas were set up to identify the position of species in the successional continuum and identify the traits associated with that position. A complicating factor with forest trees is that their environment changes with development. A late successional tree that can establish as a seedling in the understorey or small canopy gap must be able to tolerate shade as a juvenile, but must be capable of efficiently exploiting full daylight once it reaches adult stature. This may place conflicting demands on the plant and may require ontogenetic changes in traits. Adaptations to the light climate are considered to be of overriding importance for establishment and succession after disturbance. However, nutrient and water availability may play an important role as well, particularly on nutrient-poor soils (Coomes and Grubb, 2000). Variation in forest composition on different substrates must at least to some extent be associated with the species water and nutrient relations and tolerance to toxic elements where relevant. However, less progress is made with the unravelling of the physiological mechanisms of adaptation of forest trees to these edaphic factors as compared to the light climate.

This chapter reviews the currently available knowledge on traits relevant for resource acquisition and stress tolerance in forest plants of the Guianas. Photosynthesis is discussed with particular attention to acclimation of plants to light climate changes and some data about ontogenetic changes associated with this. Growth performance based on photosynthetic carbon gain and morphological adjustments in different gap environments is treated next. We then discuss water relations which may be particularly relevant in very dry years and when floods occur. Acquisition of nutrient elements is reviewed, with particular attention to the importance of nitrogen fixation among the abundant leguminous species, and the possible role of toxic concentrations of aluminium. Most of these studies are from French Guiana and Guyana, since hardly any ecophysiological studies have been carried out in Suriname. Data from the Guianas are supplemented with data from other neotropical forests where relevant and they are discussed in that wider perspective.

Photosynthetic Performance

Photosynthetic pathways

Plant growth in terms of biomass accumulation occurs as the result of carbon gain achieved through the photosynthetic process. Of the three main types of photosynthetic metabolism that can be distinguished, tropical forest plants mostly have the C_3 pathway. The C_4 type typically is restricted to herbaceous weeds that can invade forest clearings. Crassulacean acid metabolism (CAM) is most prominent in (hemi-)epiphytes, and can be found in the families Bromeliaceae, Orchidaceae, Cactaceae, Gesneriaceae, Piperaceae and Clusiaceae. Several genera of these families that have members capable of CAM occur in the Guianas (Table 3.1). However, their photosynthetic performance has not been studied there. Although CAM is generally interpreted as an adaptation to dry (micro) habitats, the pathway is also found in humid forest and even in understorey plants (Medina, 1996). However, CAM species are most frequent in drier forest, and the pathway is expressed most prominently in facultative CAM plants during the dry season (Winter and Smith, 1996).

A special case is the genus *Clusia*. All of the 20 species that have been investigated are probably capable of CAM at least to some extent (Lüttge, 1999), but there is substantial variation in the degree of CAM performance. Five of the species that have been studied occur in the Guianas. Of these only *C. rosea* assimilates a substantial proportion of its CO_2 through the CAM pathway as judged from its high $\delta^{13}C$ ratio, nocturnal CO_2 uptake and acid accumulation. *Clusia nemerosa* (only based on $\delta^{13}C$) and *C. minor* show this to a lesser extent, whereas CAM activity was hardly detectable in *C. parviflora* and *C. grandiflora*. Further discussion will focus on the C_3 photosynthetic pathway only, since a considerable body of research in the Guianas has dealt with these plants.

Light response of photosynthesis

The response of photosynthesis per unit leaf area to photon flux density (PFD) has a general form for which two examples are shown in Fig. 3.1. One of these is for *Goupia glabra*, a shade-avoiding species sampled in a gap, and another is *Eschweilera sagotiana*, a shade-tolerant plant sampled from the understorey of closed forest. The two response curves are typical of leaves from open and closed canopy sites, respectively. Light response curves are characterized by the rate of dark respiration, the slope of the light-limited part of the response (apparent quantum yield), the light-saturated rate (A_{max}) and the transition from the light-limited to the light-saturated rate. A_{max} largely determines the form of the curve, since dark respiration typically covaries with it (Pons *et al.*, 1989). The quantum yield of photosynthetic CO_2 assimilation is rather similar among healthy leaves of C_3 plants on the basis of absorbed light. Since little variation exists between leaves of tropical forest plants in light absorption (Poorter *et al.*, 1995), apparent quantum yield based on incident PFD was also found to be rather constant between species and light conditions in the forest (Valladares *et al.*, 1997; Rijkers, 2000). Decrease in quantum yield below normal values of around 0.05 mol CO_2 per mol photons is found in senescing

Table 3.1. Families and genera that occur in the Guianas for which epiphytic CAM species have been reported. Taxa are derived from the enumeration of Medina (1996) in combination with the checklist of plants for the Guianas (Boggan *et al.*, 1997).

Family	Genera with CAM		
Bromeliaceae	*Aechmea*	*Billbergia*	*Quesnelia*
	Araeococcus	*Hohenbergia*	*Tillandsia*
Orchidaceae	*Bulbophyllum*	*Cattleya*	*Oncidium*
	Campylocentrum	*Epidendrum*	
Cactaceae	*Epiphyllum*	*Hylocereus*	*Rhipsalis*
Clusiaceae	*Clusia*	*Oedematopus*	
Gesneriaceae	*Codonanthe*		
Piperaceae	*Peperomia*		

leaves or is the result of stresses that cause a decrease in stomatal conductance, chlorophyll content and thus light absorption, or photoinhibition (Kao and Forseth, 1992). The transition from the light-limited part of the response curve to the light-saturated rate is generally more acute for shade-grown leaves and more gradual for leaves from more exposed sites (Valladares *et al.*, 1997), as was true for the measured tree saplings also (Fig. 3.1).

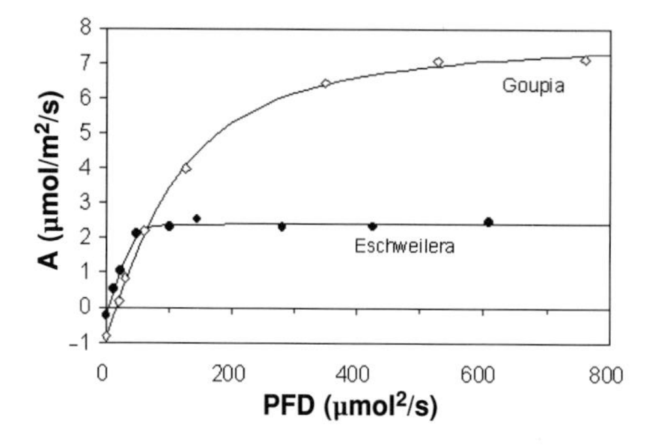

Fig. 3.1. Light-response curves of photosynthesis for leaves of saplings of two species in the rainforest around Mabura Hill, Central Guyana. Measurements were carried out on detached leaves at around 28°C. *Eschweilera sagotiana* was sampled in the understorey under a closed forest canopy; *Goupia glabra* was sampled in a logging gap. Data points were fitted to the non-rectangular hyperbola as described in Lambers *et al.* (1998) (T.L. Pons, unpublished data).

A substantial amount of data on A_{max} of forest tree species is available for the Guianas. However, in the following discussion, data will also be used for species of interest that occur in the Guianas but that were measured elsewhere. Low values of A_{max} on a leaf area basis between 1 and 4 µmol/m²/s are found in epiphytes (Zotz and Winter, 1993) and in plants growing in the shaded forest understorey (Figs 3.1 and 3.2) (Raaimakers *et al.*, 1995; Rijkers *et al.*, 2000a,b). Most saplings and young individual late successional trees that are growing in rather exposed conditions have A_{max} values of 5–10 µmol/m²/s (Figs 3.1 and 3.2; Huc *et al.*, 1994; Raaimakers *et al.*, 1995; Reich *et al.*, 1995; Rijkers *et al.*, 2000a,b). However, for some species, such as *Chlorocardium rodiei*, lower values are reported (Fig. 3.2). Not many data are available for mature trees in the Guianas. Those that are available suggest that values for young trees are indicative for mature ones. In the study by Rijkers *et al.* (2000a), A_{max} values of young trees were similar to those of mature trees in *Dicorynia guianensis* and *Vouacapoua americana*, though seedlings and saplings had lower values. However, the low A_{max} measured for a mature individual of the canopy tree *Qualea rosea* (Roy and Salager, 1992) was also found for saplings of the same species that were grown in exposed conditions by Barigah *et al.* (1998). Several light-demanding species have higher rates of A_{max}. Values of up to

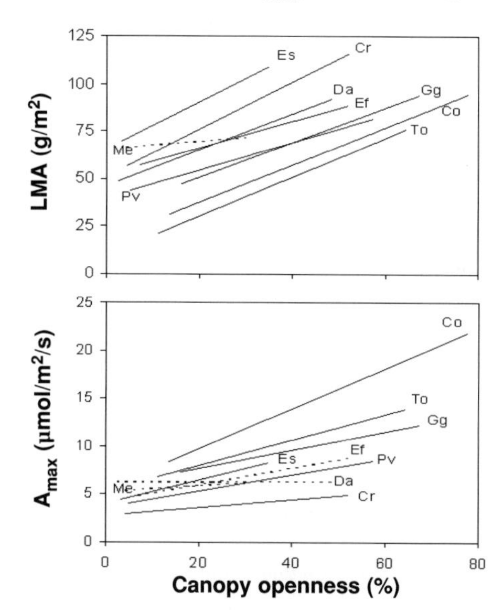

Fig. 3.2. Relationships of leaf mass per unit area (LMA) and light-saturated rate of photosynthesis per unit area (A_{max}) with canopy openness for nine species in the forest of the Mabura area in Central Guyana. Continuous lines represent significant correlations; dotted lines refer to non-significant correlations. The gradient in canopy openness resulted from logging activity. A_{max} was measured in situ on leaves of saplings (Raaimakers *et al.*, 1995). *Cecropia obtusa* (Co), *Tapirira obtusa* (To), *Goupia glabra* (Gg), *Eschweilera sagotiana* (Es), *Peltogyne venosa* (Pv), *Eperua falcata* (Ef), *Dycimbe altsonii* (Da), *Mora excelsa* (Me), *Chlorocardium rodiei* (Cr).

15 μmol/m^2/s have been reported for *Piper* spp. (Chazdon and Field, 1987), *Vismia* spp. (Reich *et al.*, 1995), *Jacaranda copaia* (Huc *et al.*, 1994), *Tapirira marchandii*, *Goupia glabra* (Raaimakers *et al.*, 1995), *Virola surinamensis* and *Diplotropis purpurea* (Bonal *et al.*, 2000b). Even higher values of up to 30 μmol/m^2/s have been found for fast-growing early successional species such as *Ceiba pentandra* and *Ficus insipida* (Zotz and Winter, 1996), and *Cecropia obtusa* (Raaimakers *et al.*, 1995).

High irradiances incident on leaves are encountered frequently in a tropical environment when there is no substantial shading by neighbours. Hence, PFD frequently will be above the saturation point where photosynthesis operates at A_{max}. A positive association of A_{max} with daily photosynthetic carbon gain in exposed upper canopy leaves can thus be expected. However, it is surprising that there is a close linear relationship (Zotz and Winter, 1993), since leaves operate only part of the day at light saturation and stress effects during midday reduce photosynthetic rates below their potential. Nevertheless, the relationship with A_{max} established by Zotz and Winter (1993) creates the possibility to estimate daily carbon gain of leaves on the basis of measurement of A_{max} only.

Environmental effects on photosynthetic parameters

The light-saturated rate of photosynthesis (A_{max}) not only varies between species, but is also under strong environmental control. For instance, A_{max} typically increases with an increase in nutrient (Pons *et al.*, 1994) and light availability during growth (Björkman, 1981). For instance, A_{max} of *Goupia glabra*, a species that has been studied extensively, varied substantially within and between studies from the Guianas and elsewhere. A_{max} measured in Guyana (Raaimakers *et al.*, 1995) was higher than A_{max} reported for French Guiana (Huc *et al.*, 1994; Rijkers *et al.*, 2000a) and from Venezuela (Reich *et al.*, 1995). A_{max} typically varies proportionally with leaf N,

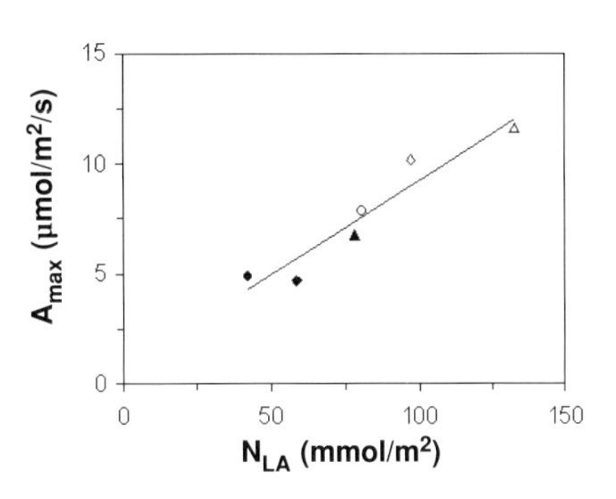

Fig. 3.3. Relationship of light-saturated rate of photosynthesis (A_{max}) with leaf nitrogen (N_{LA}), both per unit leaf area, for *Goupia glabra* in two different studies combined. Raaimakers *et al.* (1995) categories of canopy openness (<25%, ○; 25–50%, ◊; >50%, Δ) and Rijkers *et al.* (2000a), three tree sizes (1 m, •; 5 m, ♦;15 m, ▲).

because a substantial fraction of it is involved in the photosynthetic apparatus (Field and Mooney, 1986; Pons *et al.*, 1989). This relationship remains consistent across observations made on *Goupia glabra* in Guyana and French Guiana (Fig. 3.3). Hence, the higher photosynthetic rates measured in Guyana are the result of higher leaf N contents per unit leaf area. This in turn may be the result of a higher light and/or higher N availability for the plants.

Raaimakers *et al.* (1995) measured A_{max} of saplings growing in selectively logged forest in central Guyana where large variation in light availability existed. A_{max} was positively associated with canopy openness in most species (Fig. 3.2). This was particularly true for the three early successional species involved in the study (*Cecropia obtusa*, *Tapirira obtusa* and *Goupia glabra*), which responded strongly to increased canopy openness, whereas this was less apparent for the late successional species (only significant in three out of the six species studied). Within species, individuals growing in low light had thin leaves, as evident from the low leaf mass per unit area (LMA) (Fig. 3.2), particularly in the early successional species. The late successional species, *Eschweilera* and *Chlorocardium*, had thick leaves, particularly when grown in high light.

Although A_{max} per unit leaf area decreases with decreasing light availability (Fig. 3.2), the amount of chlorophyll per unit leaf area often remains rather constant across light conditions in many species (Rijkers *et al.*, 2000a). Consequently, A_{max} per unit chlorophyll also decreases. This aspect of acclimation to the light environment involves complicated structural changes in chloroplasts (Anderson *et al.*, 1995). It fine tunes the balance between the investment in photon absorption and photosynthetic capacity to the available light. High investment in photon absorption at the expense of photosynthetic capacity improves photosynthetic efficiency in shaded environments, whereas the reverse applies to high irradiance conditions (Evans and Poorter, 2001). Species vary in their plasticity with respect to A_{max} per unit chlorophyll (Murchie and Horton, 1997). Houter (unpublished results) investigated this variation in photosynthetic capacity per unit chlorophyll in four species growing in gaps of different sizes. Photosynthetic capacity was estimated by means of chlorophyll fluorescence measured at saturating PFD. The light-saturated rate of photosynthetic electron transport (ETR_{max}) was calculated from these data, which scales with A_{max}. The shade-tolerant species *Oxandra asbeckii* and *Chlorocardium rodiei* had low ETR_{max} and high chlorophyll concentrations when growing under a closed canopy (Fig. 3.4). This resulted in a low ETR_{max} per unit chlorophyll (ETR_{max}/chl) in these species. In contrast, the light-demanding species *Goupia glabra* and *Cecropia obtusa* had higher ETR_{max} and somewhat lower chlorophyll concentrations in shade, leading to a higher ETR_{max}/chl. When growing in large gaps, all species had a substantially higher ETR_{max}/chl. Only *Oxandra*, a small tree that remains below the upper canopy when mature, did not reach the high values found for the other three species that normally encounter full daylight conditions at least when mature (Fig. 3.4). The large variation in ETR_{max}/chl found for *Chlorocardium* under different light conditions contrasts strongly with the small variation in A_{max}/area in this species (Fig. 3.2). This species is shade tolerant in early developmental stages, but is exposed to full daylight when mature. Greater photosynthetic efficiency in high light of *Chlorocardium* leaves is apparently not achieved by increasing capacity per unit area but by lowering chlorophyll content. The parameter photosynthetic capacity per unit chlorophyll thus better reveals the potential of photosynthetic acclimation of this species to the large variation in light conditions to which it is normally exposed during its life cycle than capacity per unit area.

Rijkers *et al.* (2000a) investigated four tree species for the response of their leaf

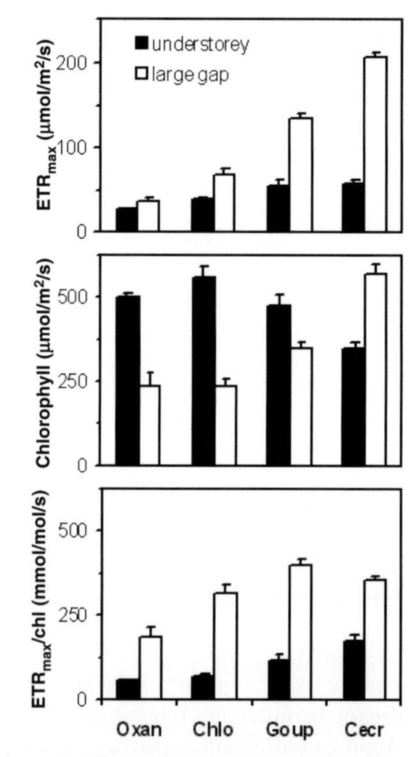

Fig. 3.4. Photosynthetic capacity (ETR_{max}) and chlorophyll both per unit leaf area, and the ratio of both parameters (ETR_{max}/chl) for four species in the forest of the Mabura Hill area in Central Guyana. As a measure of photosynthetic capacity, the rate of electron transport was measured by means of chlorophyll fluorescence at light saturation ($\Delta F/F_m'$ * PFD * 0.425). The species are: *Oxandra asbeckii* (Oxan), *Chlorocardium rodiei* (Chlo), *Goupia glabra* (Goup) and *Cecropia obtusa* (Cecr) (N.C. Houter, unpublished data).

characteristics to light availability in individuals of different sizes in Nouragues, French Guiana. Individuals in natural gaps and under a closed canopy were measured. The range in light availability was less than in the study of Raaimakers *et al.* (1995), resulting in a smaller effect of canopy openness on A_{max} and leaf mass per unit area (LMA) (Rijkers *et al.*, 2000a). However, most remarkable was the association of these parameters with tree size (Fig. 3.5). The larger individuals had higher A_{max} and LMA, which was independent of light availability. A_{max} measured on sunlit leaves of large mature trees was not higher than the value for trees of around 20 m, although LMA was substantially higher (Rijkers *et al.*, 2000a). Such an ontogenetic change in leaf traits has earlier been found for temperate trees (Niinemets, 1997), and appears to be present in tropical trees as well.

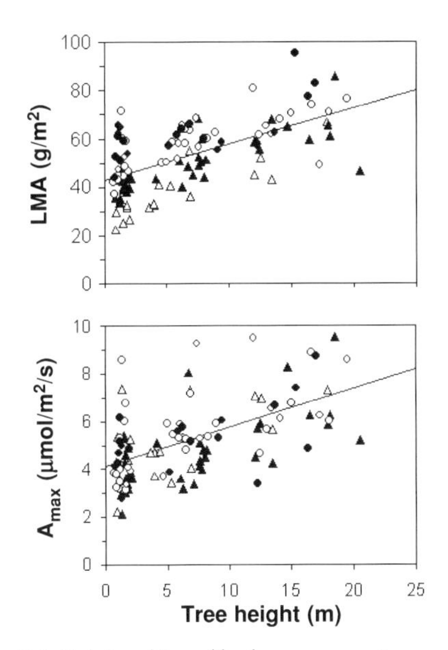

Fig. 3.5. Relationships of leaf mass per unit area (LMA) and light-saturated rate of photosynthesis per unit leaf area (A_{max}) with tree size of four species in the forest of Nouragues, French Guiana (Rijkers *et al.*, 2000a). Species are: *Duguetia surinamensis* (•), *Vouacapoua americana* (○), *Dicorynia guianesis* (▲) and *Goupia glabra* (Δ). Individual trees were selected in a way that tree height and canopy openness varied independently to a considerable extent. Ancova with species as main factor and tree height and canopy openness as covariables showed significant independent effects of the three factors on LMA (r^2=0.77) and A_{max} (r^2=0.55).

Photosynthetic rates are constrained by the low availability of light under a closed forest canopy. This causes a long payback time of resources invested in a leaf. The low LMA that plants have when growing in shade (Fig. 3.2) indicates a low investment of carbon compounds and thus low construction costs per unit area. Hence, this adaptive trait reduces the inherently long payback times in shaded understorey environments. Rijkers (2000) calculated payback times on the basis of the temporal distribution of PFD over a cloudless day and the photosynthesis–light response curves that were rather constant over the life span of the leaves. Results are a minimum estimate, since aspects like cloudiness and midday stress effects (see below) at the carbon gain side and carbon costs of supporting structures were not taken into account. Payback times of *Dicorynia guianensis* and *Vouacapoua americana* were indeed substantially longer (approx. 20–40 days) for a low light understorey environment compared to about 8 days for leaves in gaps. *Dicorynia* showed no variation in leaf life span with light environment, but *Vouacapoua* had longer leaf life spans when growing in shade (Rijkers, 2000), similar to several other species in Guyana (Rose, 2000). Presumably, longer leaf life span in shade is the result of selection for this trait in an environment with inherently long payback times. The predominantly long leaf life spans in the forests of the Guiana Shield and elsewhere in the neotropics (Bongers and Popma, 1990; Reich *et al.*, 1995; Rijkers, 2000; Rose, 2000) are associated with selection for shade-tolerant genotypes in these forests.

Dynamics of photosynthesis

The data on photosynthetic performance described so far are based on steady-state responses. However, photon flux densities in particular are highly dynamic. On clear days, the diffuse understorey light environment is interrupted by occasional short

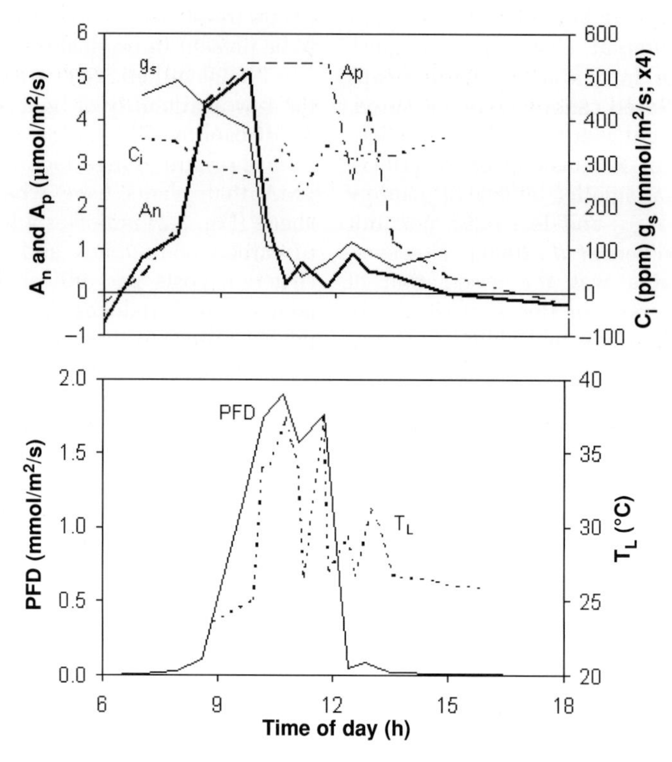

Fig. 3.6. Course over a day of gas exchange parameters of a *Goupia glabra* leaf (upper panel) and environmental parameters (lower panel). Measurements were carried out on a sapling in a large gap (c. 3200 m^2) on a bright day in the dry season (19 November 1999) in the forest in the Mabura Hill area in Central Guyana. Shown are net photosynthesis (A_n), the potential rate of net photosynthesis (A_p) calculated on the basis of the light response curve of that leaf and the course of the photon flux density (PFD), stomatal conductance (g_s), the intercellular CO_2 concentration (C_i) and the leaf temperature (T_L) measured when the leaf was not in the cuvette (N.C. Houter, unpublished data).

periods of high light when direct sunlight penetrates through small holes in the canopy (Chazdon, 1988). In a gap, diffuse light is alternated by longer periods of direct light (Fig. 3.6). The photosynthetic apparatus in a leaf requires some time after exposure to high light before the full steady-state rate is achieved, referred to as photosynthetic induction (Pearcy, 1990).

The photosynthetic response to a sudden increase in PFD was investigated in saplings of three species growing in the shaded understorey and in gaps in Nouragues, French Guiana (Rijkers *et al.*, 2000b). Times until 90% of full induction were rather short, at 7–11 minutes. Increase in chloroplast activity was the main limitation in *Dicorynia guianensis* and *Vouacapoua americana*, whereas stomatal limitations were more important in *Pourouma bicolor*. Plants growing in the shaded understorey maintained a high induction state when switched from high to low light, enabling them to respond rapidly to a sequence of short-lived light flecks. In contrast, *Dicorynia* saplings growing in gaps showed a rapid loss of photosynthetic induction, principally limited by chloroplast rather than stomatal performance. These experiments were carried out in the morning. However, studies done on trees and shrubs at other neotropical locations indicate that induction times tend to be longer and loss of induction faster in the afternoon and during the dry season (Poorter and Oberbauer, 1993; Allen and Pearcy, 2000).

Stress response of photosynthesis

Light-saturated rates of photosynthesis in forest canopies are not typically maintained during the whole period of the day when PFD exceeds the light-saturation point. Often at midday hours photosynthetic rates decline while PFD remains high (Huc and Guehl, 1989; Roy and Salager, 1992; Huc *et al.*, 1994; Bonal *et al.*, 2000b). This is referred to as midday depression of photosynthesis. The photosynthetic rate may or may not increase again later in the afternoon. This causes a decrease in daily carbon gain to a level below the maximum potential accrued when calculated simply on the course of PFD over the day. Young plantation trees of the shade-intolerant species *Jacaranda copaia*, *Diplotropis purpurea* and *Virola surinamensis* showed a more pronounced midday depression in comparison to the more shade-tolerant *Eperua falcata* (Huc and Guehl, 1989; Huc *et al.*, 1994; Bonal *et al.*, 2000b).

The higher light availability in gaps, compared to understorey locations, potentially increases productivity by stimulating photosynthetic activity. However, the effects of midday depression may substantially reduce the daily carbon gain below its potential. Figure 3.6 depicts a leaf of *Goupia glabra* that showed reduced photosynthetic activity after exposure of the leaf to direct sunlight. No substantial recovery of photosynthetic activity occurred when the plant was back in diffuse light conditions later in the afternoon. Such daily courses of photosynthetic activity were measured for six tree species growing in gaps in central Guyana. Potential rates were based on the photosynthesis–PFD relationship that was derived from morning observations and the course of PFD over the day as shown in Fig. 3.6. The reduction of daily carbon gain relative to its potential was calculated. *Cecropia obtusa* and *Mora gongrijpii* showed little relative reduction in carbon gain due to midday depression. However, daily carbon gain was around 60% of its potential in *Goupia glabra*, *Catostemma fragrans*, *Chlorocardium rodiei* and *Hymenaea courbaril*. Stomatal conductance (g_s) declined parallel with photosynthetic activity during midday depression of photosynthesis, indicating that at least part of the midday depression is due to stomatal closure. However, in several cases, the intercellular CO_2 partial pressure (C_i) did not decrease much, which would be expected if stomatal opening were the only limiting factor (Jones, 1985). Hence, other limiting factors may be involved that must be associated with the functioning of mesophyll cells. Supra-optimal leaf temperatures in combination with the high PFD can downregulate the efficiency of photosystem II (PSII) (Mulkey and Pearcy, 1992). However, the ratio F_v/F_m, a measure of efficiency of PSII, did not decrease to any great extent during midday hours, indicating that photoinhibition did not play an important role in the decline of photosynthetic activity. Another possible factor is that photorespiration inevitably increases in response to the high midday temperatures. This reduces net photosynthetic rates in the afternoon if it is not compensated by an increase in gross photosynthesis. Experiments with *Eperua grandiflora* indicate that this is indeed important at a leaf temperature of 38°C (Pons and Welschen, 2003). Another reason for an apparent constancy of C_i with decreasing g_s could be inhomogeneity of stomatal closure (Eckstein *et al.*, 1996). However, that would be an artefact caused by assumptions used with the calculations of C_i.

When plants in the understorey that are only exposed to direct sunlight during short duration sun flecks are suddenly exposed to full sunlight after a tree fall event, they typically show signs of photoinhibition. The high PFD in combination with the high leaf temperatures is probably responsible for this effect on these non-acclimated leaves (Mulkey and Pearcy, 1992; Lovelock *et al.*, 1994). In a forest in central Guyana, quantum yield of PSII in understorey saplings around three weeks after creation of gaps of different sizes was measured by means of chlorophyll fluorescence (F_v/F_m) (Houter and Pons, 2005). Results for the response of two species are shown in Fig. 3.7. They showed a decline of F_v/F_m with increasing gap size and values substantially below the normal 0.8 value for healthy leaves in the larger gaps. The predawn values are only slightly higher than the midday measurements, which indicates that most of the photoinhibition could be attributed to long-lasting damage to PSII rather than to a temporary photoprotective down-regulation in quantum yield. The shade-tolerant subcanopy species *Oxandra asbeckii* suffered strongly from photoinhibition in gaps larger than 100 m^2, whereas F_v/F_m of the canopy species *Catostemma fragrans* reached low values only in the largest gaps of 1600 m^2. The data suggest that the sensitivity to photoinhibition is

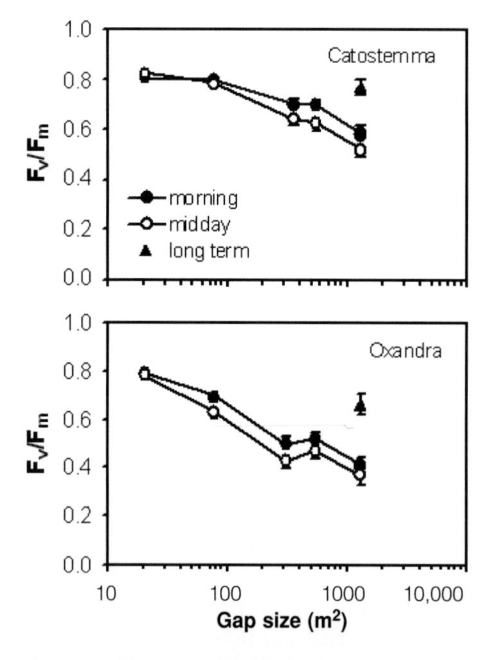

Fig. 3.7. Quantum yield of photosystem II measured after dark adjustment (F_v/F_m) of leaves of saplings three weeks after the creation of openings of different sizes in the forest canopy (means of three similar sized gaps ± SE). Measurements were carried out early in the morning and during midday hours in the forest in the Mabura Hill area in central Guyana. Morning values for plants that had been exposed for half a year to conditions in a large gap are also included. A closed forest canopy is arbitrarily put at a gap size of 25 m^2 (Houter and Pons, 2005). Species are: *Catostemma fragrans* and *Oxandra asbeckii.*

related to a plant's shade tolerance, which was also found for rainforest plants in New Guinea (Lovelock *et al.*, 1994).

Gas exchange processes at the canopy and ecosystem level

The increase of photosynthetic capacity in seedlings and saplings that occurs with an increase in light availability in treefall gaps is also found to occur along the vertical profile of single trees. In an Amazonian rainforest, Carswell *et al.* (2000) found that the photosynthetic capacity of leaves decreased with distance from the canopy surface and that this was accompanied by a decrease in nitrogen content. Such a parallel distribution of photosynthetic capacity with light intensity improves resource use efficiency for photosynthesis at the whole plant and community level (Pons *et al.*, 1989).

Exchange processes between forest and atmosphere were investigated in French Guiana by using carbon and oxygen stable isotope composition of air, plants and litter (Buchmann *et al.*, 1997). During daytime, CO_2 concentrations in the upper canopy regions were about 12 ppm lower than in the bulk air above the canopy due to photosynthetic CO_2 uptake processes (Table 3.2). Conversely, measured CO_2 values at ground surface locations were well above the normal atmospheric values due to CO_2 generation from soil-based microbial respiration. Night-time values were all higher than free atmosphere, with a strong gradient to the soil surface where CO_2 concentration reached 562 ppm. Plant-derived organic matter has substantially less ^{13}C compared to CO_2 in the atmosphere (approx. 20‰) due to discrimination against ^{13}C during the photosynthesis process (Farquhar *et al.*, 1982). The CO_2 originating from respiration had a similar isotopic carbon composition as plants and litter, resulting in a decline in $\delta^{13}C$ during the night and towards the soil surface due to CO_2 originating from soil and plants mixing with that from the atmosphere (Table 3.2). CO_2 from the atmosphere apparently dominates the composition of the air in the canopy during daytime,

Table 3.2. Gradients in CO_2 and ^{13}C concentrations in the dry season in a forest in French Guiana (Paracou). The stable isotope ^{13}C is expressed as a ratio relative to the PDB standard ($\delta^{13}C$). $\Delta^{13}C$ is the discrimination against ^{13}C during the assimilation of leaf carbon. $\delta^{13}C$ of soil is the mean of soil organic matter and litter (Buchmann *et al.*, 1997).

	CO_2 concentration (ppm)		$\delta^{13}C$ of CO_2 in air (‰)		$\delta^{13}C$ leaf, soil (‰)	$\Delta^{13}C$ (‰)
	Day	Night	Day	Night		
Troposphere	355		−7.7			
28 m	343	434	−7.9	−11.5	−28.6	21.7
13 m	342	465	−7.9	−11.5	−33.2	26.5
2 m	350	499	−8.0	−11.8	−33.6	26.1
Near surface	444	562	−9.7	−16.5		
Soil			−26.5		−29.4	

because the $\delta^{13}C$ values were not much different from free atmospheric values. During the night, decreased air turbulence leads to a localized build-up of plant- and soil-derived CO_2 in the air in the forest (Buchmann *et al.*, 1997).

Leaves sampled from different canopy positions showed decreasing $\delta^{13}C$ with decreasing height (Table 3.2). When the values are corrected for the different isotopic composition of CO_2 in the air at the height of the leaf, then it appears that ^{13}C discrimination ($\Delta^{13}C$) is higher in the mid-canopy and understorey compared to the upper canopy regions. This means that upper canopy leaves operate at a lower intercellular CO_2 concentration (C_i) compared to mid-canopy and understorey leaves (Farquhar *et al.*, 1982). Hence, leaves in the lower regions of the forest canopy have their stomata more widely open relative to their low photosynthetic rates in the shade as compared to upper canopy leaves (Dolman *et al.*, 1991).

Gas exchange processes at the ecosystem level have also been studied in Amazonian rainforest in Brazil using stable isotope and Eddy covariance techniques which permit conclusions on net carbon exchange between forest and atmosphere (Grace *et al.*, 1996; Lloyd *et al.*, 1996). The concept of a climax forest in most of Amazonia would imply no net positive or negative exchange of carbon of these forests with the atmosphere. However, these studies challenge this concept and suggest that Amazonian forests are net sinks for CO_2. The rising atmospheric CO_2 concentration drives this carbon sequestering, which is periodically reversed in El Niño years when these forests are sources to the atmosphere (Tian *et al.*, 1998). Model calculations and further evidence indicate that tropical forests in general act as carbon sinks (Malhi and Grace, 2000), which would thus also apply to the predominantly primary forests of the Guiana Shield. However, these conclusions are based on data from a limited number of sites. Measurements in the Guianan forests are needed to quantify their contribution to the global carbon cycle.

Relative Growth Rate and its Components

The growth rate of plants in terms of biomass increase is not only determined by the rate of photosynthetic carbon gain of their leaves, but also by the amount of photosynthetically active tissue. Hence, photosynthetic characteristics of leaves as discussed above are just one category of traits that determine growth performance. One of these is the light-saturated rate of photosynthesis, which is a good predictor of daily carbon gain at the leaf level (Zotz and Winter, 1996). To obtain the daily net carbon gain of a whole plant that is equivalent to biomass increase, the daily release of

CO_2 through respiration must be subtracted from daily photosynthesis. Respiration is a substantial part of daily photosynthesis: about 40% in high light and 75% in low light conditions has been measured for tropical tree seedlings (Lehto and Grace, 1994). The other major component that influences growth rate is distribution of biomass over plant organs, particularly photosynthetic leaf area. The leaf area ratio (LAR; leaf area per unit plant biomass), the parameter used to describe this biomass distribution, was found to explain most of the variation in relative growth rate (RGR; increase in biomass per unit biomass present and time) in herbaceous species under growth chamber conditions, whereas the net assimilation rate (NAR; increase in biomass per unit leaf area and time) was of minor importance (Poorter and van der Werf, 1998). This contrasts with growth analysis carried out with tropical tree seedlings under conditions that better resemble natural habitat conditions. In a review of 43 studies, Veneklaas and Poorter (1998) showed that slow-growing climax species that are shade tolerant in early development generally have a low NAR in high light conditions, compared to shade-intolerant pioneer species. This high NAR in pioneers is largely due to a higher light-saturated rate of photosynthesis which more than compensates for the respiratory losses of CO_2 that are also somewhat higher (Veneklaas and Poorter, 1998). The high photosynthetic capacity of pioneers enables them to exploit high light conditions more efficiently than climax species, which generally have lower capacities (Fig. 3.2). Hence, photosynthetic characteristics are important determinants of growth rate in high light conditions for tree saplings in the forest. The morphological parameter LAR is generally a more important determinant of growth rate under low light conditions. The LAR is typically higher in pioneer species compared to climax species, which results in a paradoxically higher relative growth rate (RGR) of shade-intolerant trees in low light (Veneklaas and Poorter, 1998). The most important determinant of LAR is the specific leaf area (SLA; leaf area per unit leaf dry mass), which is often high in pioneers, although they may also be characterized by a somewhat higher leaf mass fraction (LMF; leaf mass per unit plant mass).

In many studies where the effect of light availability on growth has been analysed, plants are grown under shade devices that do not fully simulate variation in the complex light climate as found in the forest. One important point is that the spectral component of light is absent in most of these studies. The low red:far-red ratio found under a leaf canopy usually influences allocation and morphological parameters and thus growth rates (Corré, 1983), although this may be less in shade-tolerant tropical rainforest saplings (Kitajima, 1994). Furthermore, the daily course of photon flux density (PFD) in canopy gaps is characterized by periods of high PFD in direct sunlight with periods of low PFD in diffuse light (Fig. 3.6). This contrasts with a more homogeneous distribution of PFD over a day under experimental shade screen conditions. Photosynthetic capacity is probably more important for photosynthetic performance and thus NAR and RGR in a gap light environment compared to the more homogeneous reduction in light under shade screens, although this has not been explicitly investigated. The large variation in PFD over the day in canopy gaps is also the reason why daily carbon gain is expected to be lower in a gap compared to the more equal light distribution under shade screen (Pons *et al.*, 1994).

Growth in relation to canopy openness

Experiments were conducted in Guyana where seedlings were planted in gaps of different sizes and growth was monitored by means of destructive harvesting (Boot, 1993, 1996; Rose, 2000). Since seedlings did not survive in all conditions, particularly at the lower light availabilities, these experiments provide some information on shade tolerance in early development. Small-seeded species such as *Cecropia obtusa*, *Goupia glabra* and *Laetia procera* showed reduced survival under a closed canopy and in a small gap of about 50 m^2, with increasing survival times in the order of the species mentioned. Several other species were included in the experiments, but all showed substantial survival in deep shade of up to a year or more (Boot, 1996; Rose, 2000).

Relative growth rates generally increased with increasing gap size, but levelled off above a gap size of around 800 m^2 (Fig. 3.8). There were clear differences between species in maximum RGR achieved in the largest gaps. Small-seeded species had much higher RGR_{max} than large-seeded ones, particularly the very large-seeded *Mora gongrijpii* (Fig. 3.8), *M. excelsa* and *Chlorocardium rodiei* (ter Steege, 1993, 1994b; ter Steege *et al.*, 1994), which had very low RGR_{max}. This resulted in a close correlation between inherent RGR_{max} and seed mass (Rose, 2000; Rose and Poorter, 2003). Some general trends associated with the RGR_{max}–seed mass relationship could be distinguished. The higher RGR of the small-seeded species was partly caused by their high LAR, which was in turn the result of a high SLA. The LMF, the other component of the LAR, was only slightly higher in the fast-growing species than in the slow-growing ones (Fig. 3.8). The NAR, the other component of the RGR, was also substantially higher in the fast-growing species. As argued above, the driving force of NAR is the daily assimilation, which is to a large extent determined by A_{max} (Zotz and Winter, 1996). The high A_{max} of the small-seeded pioneer species thus contributes to their high growth rate through its effect on NAR. The large plasticity in SLA expressed by *Cecropia* and *Goupia* compared to the other species was particularly impressive, both showing very high SLA when growing in gaps of 200 m^2 (Fig. 3.8).

Although clear trends could be distinguished in RGR_{max} and its components, there were also important exceptions. *Carapa guianensis* had a higher RGR_{max} compared to other species with similar seed mass. This large-seeded, but not very shade-tolerant species (Favrichon, 1994) had a rather high NAR, probably based on a high A_{max} (Huc *et al.*, 1994). The high RGR of *Cecropia obtusa* was not based on a high SLA, but its thick leaves have a high A_{max} leading to a high NAR. The reverse was true for *Goupia glabra*, which had a NAR similar to several larger-seeded climax species, but had thin leaves (high SLA) resulting in a high LAR that led to the high RGR (Fig. 3.8). Another exception is the shade-tolerant *Duguetia neglecta* that had a high SLA compared to other shade-tolerant species (Boot, 1993, 1996). Its RGR was low due to a low NAR, but whether that is associated with an even lower A_{max} than the other shade-tolerants is unknown.

Plants with small seeds that have a high RGR_{max} start their development with very little initial biomass but increase this rapidly after germination. This contrasts with plants with large seeds that have a low RGR. They start growth with a relatively high biomass, but that increases slowly after germination. At a certain moment the accumulated biomass of both plant types will equal each other (Boot, 1996). These moments of equal plant mass have been calculated assuming constant RGR (Rose and Poorter, 2003). It would take *Cecropia obtusa* 174 days to attain the same plant mass as *Mora gongrijpii*, which has, on average, a seed mass that is six orders of magnitude greater than *Cecropia*. The slower-growing *Goupia glabra* and *Hymenaea courbaril* would take 264 and 786 days respectively to attain mass equal to *Mora*. Hence, if the residual vegetation is not too dense after a large disturbance, the small-seeded pioneers can exploit their inherently higher growth rate to compensate for the small initial plant size and gain advantage over the slow-growing climax species.

However, RGR typically declines during plant growth (Veneklaas and Poorter, 1998). The low RGR of larger plants can be due to a larger investment in stem mass, thus causing a low LMF and LAR, and self-shading may be more prevalent in their crowns leading to a lower NAR. Since large seeds give rise to large seedlings, there is thus a fair chance that at least part of the

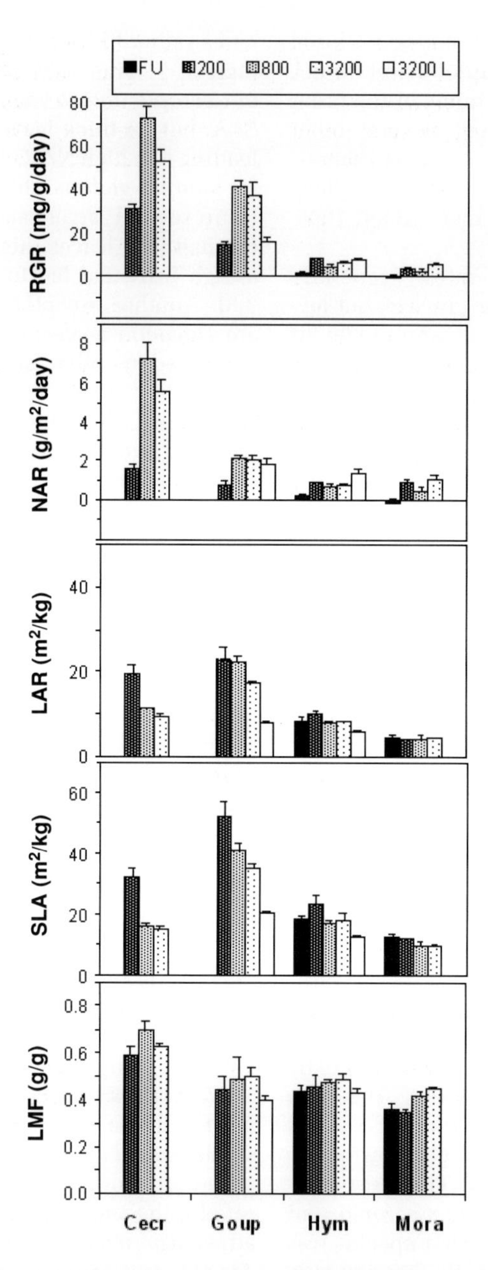

Fig. 3.8. Relative growth rate (RGR) and its components net assimilation rate (NAR), leaf area ratio (LAR), specific leaf area (SLA) and leaf mass fraction (LMF) of four species differing by five orders of magnitude in seed size. Species order from low to high seed mass is: *Cecropia obtusa, Goupia glabra, Hymenaea courbaril* and *Mora gongrijpii.* Growth parameters of the seedlings were determined for a period of up to five months after seed reserves were exhausted. The experiment was carried out in the forest understorey under a closed canopy (FU), and in gaps of exponentially increasing sizes as indicated in the legend in m^2. In the largest gap, *Goupia* and *Hymenaea* were allowed to grow until similar or larger plant size was reached than *Mora* (3200 L). *Cecropia* and *Goupia* did not survive in sufficient numbers in FU (after Rose, 2000).

association of RGR_{max} with seed mass is caused by this decreasing RGR with plant size. Furthermore, the assumptions of the model calculations referred to above may not be valid. To investigate if RGR does indeed decrease with plant size in the species under investigation, *Goupia* and *Hymenaea* were allowed to grow in the largest gap until they reached a similar or larger size than *Mora*, the species with the highest seed mass (Fig. 3.8). The predicted decline in RGR was indeed found for *Goupia* and was the result of a decrease in LAR, but not NAR, which even increased. The *Hymenaea* plants that were allowed to grow further also showed a stimulation of NAR compared to the smaller plants which compensated fully for the decrease in LAR, resulting in a similar RGR across a plant mass range of two orders of magnitude. Although RGR of *Goupia* decreased with plant size, it was still substantially higher than that of the larger-seeded climax species *Hymenaea* and *Mora*. Hence, there is indeed an inherent difference in RGR between small-seeded pioneers and large-seeded shade-tolerants, independent of plant size. Furthermore, although the calculations based on stable RGR may underestimate the time it takes to reach equal plant mass, the principle that small-seeded species can outcompete large-seeded slow growers in larger gaps based on their high RGR remains valid.

The high biomass that small-seeded pioneer species such as *Jacaranda copaia* can accrue after a relatively brief period of growth at high light availability, compared to lower growth rates of large-seeded, shade-tolerant species, has also been demonstrated in an experiment in French Guiana (Barigah *et al.*, 1998). The hemi-shade-tolerant species with larger seeds, *Platonia insignis* and two *Carapa* spp. (*C. guianensis* and *C. procera*) had similar plant masses at the end of the experimental growth period compared to *Jacaranda*. This indicates that the latter species had also a higher RGR, which enables it to surpass larger-seeded slower-growing species in high light conditions.

Fast growth of small-seeded pioneers in large gaps was also demonstrated in an experiment where eight species were planted in gaps and allowed to grow for three years (Rose, 2000). The initially small *Cecropia obtusa*, *Goupia glabra* and *Sclerolobium guianense* dominated the canopy after one year in the largest experimental gaps (1600 m^2 and 3200 m^2). Surprisingly, the largest-seeded species used in the experiment, *Catostemma fragrans* and *Chlorocardium rodiei*, were not the tallest in the small gap (50 m^2). That position was achieved by *Pentaclethra macroloba*, which also showed substantial height growth in the largest gaps. The remarkable combination of tolerance of deep shade and fast growth in gaps was also demonstrated for this species in Costa Rica (Fetcher *et al.*, 1994).

Seedling establishment and seed size

The results of the experiments with Guianan tree species described above fit quite well with general ideas about the role of seed size in establishment (Leishman *et al.*, 2000). One of the advantages of a large seed is associated with survival in shade. If seed reserves are not immediately allocated to the slow-growing seedling, then the reserves in the cotyledons can sustain the seedling for a prolonged period until conditions become more favourable. This is particularly evident in the very large-seeded *Chlorocardium rodiei*, which keeps its reserves for up to a year after germination (Boot, 1993; ter Steege *et al.*, 1994). In the process, plant mass, including cotyledons, gradually decreases due to a negative carbon balance in the shaded understorey environment, but survival is high. Another advantage of a large seed is associated with the competitive advantage of the large initial size. *Carapa* spp. combine these two traits. Large seeds always come with the disadvantage of the trade-off with small numbers (Smith and Fretwell, 1974). Large numbers of seeds are essential for species that exploit the unpredictable window for establishment after an occasional heavy disturbance. The small seed size that comes with larger numbers improves dispersal

(Hammond *et al.*, 1996) and survival in the soil seed bank is often positively associated with small seed size (Leishman *et al.*, 2000). The likelihood of being in the right time and place for establishment is thus maximized. A high RGR is, however, important to compensate for the low initial plant mass once favourable conditions for establishment have been encountered.

Water Relations

The diffusion of CO_2 into the leaf during the photosynthetic process is inevitably accompanied by water loss through the opened stomata. This transpirational water loss is more than two orders of magnitude larger than photosynthetic carbon gain expressed on a molar basis. For large trees, water loss to the atmosphere can amount to several hundred kilograms per day (Wullschleger *et al.*, 1998). According to model calculations, severe water stress is not likely to develop in forests on most soils in the moderate dry seasons in the Guianas (Jetten, 1994). However, measurements showed that limitations in growth performance can be caused by low water availability and uptake and transport capacity of trees (Bonal *et al.*, 2000a,b). Furthermore, in exceptionally dry seasons that occur during El Niño events, growth and survival of young trees may be severely affected by drought (ter Steege, 1994a; Condit *et al.*, 1996).

Transpiration

A common method used to study water loss by trees estimates sap flow rates in stems using a heat dissipation technique. Measurements using this technique have been carried out on plantation-grown trees and in the natural forest in French Guiana (Granier *et al.*, 1992, 1996; Bonal *et al.*, 2000b). The data were scaled up to estimate whole tree transpiration rates by taking sap wood area per tree into account, and to the stand level by using data on basal sap wood area for the stand. Young plantation-grown *Simarouba amara* and *Goupia glabra* had high sap flux densities of approx. 4 $kg/dm^2/h$. Similar values (3–4 $kg/dm^2/h$) were also measured for large canopy trees in the forest, but for *Eperua falcata* and *Dicorynia guianensis* only. Other forest trees had lower rates, with the lowest values for *Vouacapoua americana* and *Carapa procera* (1.0–1.5 $kg/dm^2/h$). Transpiration of the whole tree stand did not increase linearly with vapour pressure deficit. Hence, calculated canopy conductance for water vapour decreased with increasing VPD as a result of stomatal closure at midday. This effect was most pronounced for the young plantation-grown trees (Granier *et al.*, 1992; Bonal *et al.*, 2000b), but also clear for the mature forest trees (Granier *et al.*, 1996).

Decrease in stomatal conductance in the afternoon is a widespread phenomenon in tropical trees, as mentioned above (Fig. 3.6). Stomata are highly sensitive to the difference in vapour pressure between air inside and outside of the leaf (ΔW) that increases as a result of increasing leaf temperature in the course of the day. The sensitivity of the stomata to ΔW is illustrated for *Eperua grandiflora* measured under controlled conditions (Fig. 3.9). Partial stomatal closure kept transpiration rather

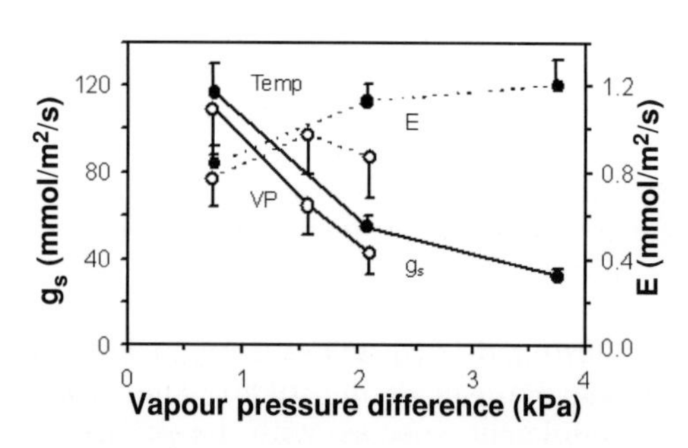

Fig. 3.9. The response of stomatal conductance for water vapour (g_s: ––) and transpiration rate (E: - - -) of *Eperua grandiflora* on the difference in vapour pressure between inside and outside the leaf (vapour pressure difference, ΔW). Plants were grown in a greenhouse and ΔW was varied in two ways. Vapour pressure was kept constant and leaf temperature was increased from 28°C to 33°C and 38°C (•), and leaf temperature was kept constant at 28°C and vapour pressure was decreased (○) (Pons and Welschen, 2003).

constant across a wide range of ΔW, as was also found in growth room experiments for *Eperua falcata*, *Diplotropis purpurea* and *Virola michelii* (Bonal, 2000). No difference in the ΔW-response of the stomata was evident between these species. However, the closure response may be so strong in other species that transpiration rates even decline with increasing ΔW, as found for the shrub *Piper auritum* (Tinoco-Ojanguren and Pearcy, 1993). The ΔW-response is an important reason for the midday decrease in stomatal opening. It limits water loss, and if intercellular CO_2 concentration decreases with partial stomatal closure, it also enhances water-use efficiency (WUE) of the photosynthetic process.

Midday water potentials (Ψ_w) measured in plantation-grown trees varied between –1 and –2 MPa in the study of Huc *et al.* (1994). The two shade-intolerant pioneer species that were included in the study, *Jacaranda copaia* and *Goupia glabra*, had high midday Ψ_w, and the two late successional species, *Dicorynia guianensis* and *Eperua falcata*, had low midday Ψ_w (more negative). *Carapa guianensis* had values in between these extremes. No decrease in pre-dawn or midday Ψ_w was evident in the course of the dry season, when the measurements were performed. This indicates that the trees had access to moist soil layers. The higher midday Ψ_w in the two pioneer species, in combination with similar stomatal conductance, suggests a lower resistance for water transport and/or uptake in these species compared to the late successional species (Huc *et al.*, 1994).

Midday water potentials were more contrasting between shade-tolerant and -intolerant species in another study carried out on plantation-grown trees in French Guiana (Bonal *et al.*, 2000b). *Eperua falcata* again had a low midday Ψ_w of –2 MPa, which was stable across wet and dry seasons (Fig. 3.10). *Diplotropis pupurea* had a midday Ψ_w of –1.6 MPa in the wet season, but surprisingly, a substantially higher Ψ_w in the dry season, when a lower value would be expected. The other shade-intolerant species included in the study, *Virola surinamensis*, which generally occurs on wet soils, exhibited strong, isohydric behaviour. Midday water potential was never lower than –0.3 MPa, which was close to the predawn Ψ_w of –0.2 MPa (Fig. 3.10). These water potentials are unusually high for a tree and suggest a very low resistance for uptake and transport of water. Stomatal and canopy conductance of *Diplotropis* and *Virola* were high during the wet season (Bonal *et al.*, 2000b). Conductances decreased substantially in *Diplotropis* in the dry season, which apparently limited transpiration to such an extent that Ψ_w increased. Stomata almost completely closed in the dry season in *Virola*, which is a condition for the very high Ψ_w. These seasonal changes in water relations contrast with those of *Eperua*, which showed only a limited decrease in conductances in the dry season. This is consistent with the view that *Eperua* can access water in deeper soil layers, based on similar $d^{18}O$ in soil and xylem water and a relatively deep rooting depth (Bonal *et al.*, 2000a). Pot experiments carried out under controlled growth room conditions (Bonal, 2000) confirmed that *Virola* (in this case *Virola michelii*) maintained high Ψ_w and *Eperua* a low Ψ_w across a wide range of soil water contents, and that *Diplotropis* increased its Ψ_w in response to moderately dry soil conditions.

Flooding and drought tolerance in *Mora* spp.

Mora gongrijpii is found more frequently in upland well-drained conditions and may experience occasional droughts in dry years, whereas *M. excelsa* mostly occurs along creeks and rivers and experiences occasional floods. Ter Steege (1993, 1994a) carried out several experiments in Central Guyana in order to establish the species specific traits that are responsible for this habitat segregation. The study concentrated on early growth stages because these were thought to be most vulnerable to environmental stress, particularly drought. Midday water potentials of saplings of the two shade-tolerant *Mora* spp. growing in the forest understorey were high in the dry sea-

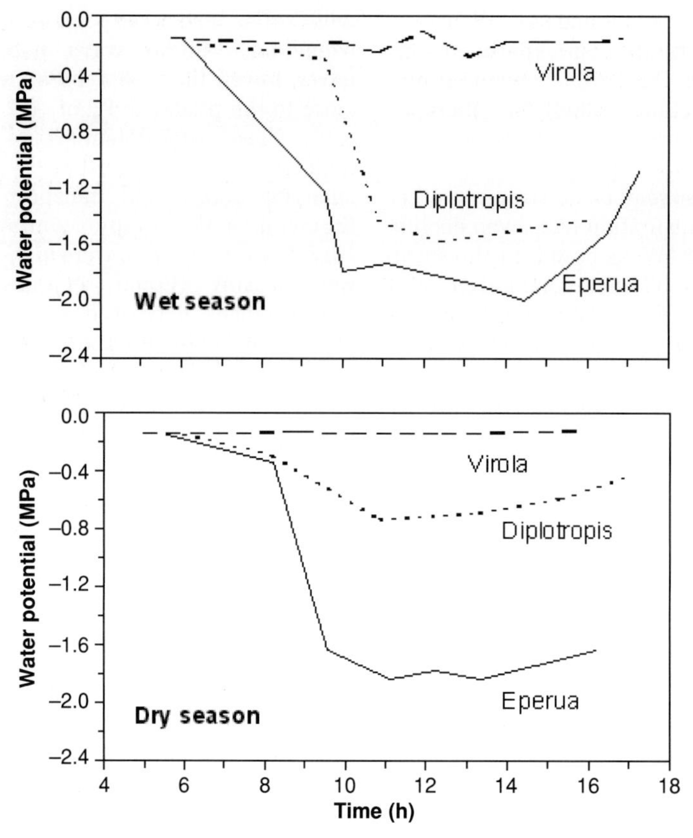

Fig. 3.10. Course of leaf water potential (Ψ_w) of three species over a day in the wet season and in the dry season in the forest at Paracou, French Guiana. Species are *Virola surinamensis, Diplotropis purpurea* and *Eperua falcata*. Predawn Ψ_w was the same for the three species in both seasons. Midday Ψ_w was low in both seasons for *Eperua*, low for *Diplotropis* in the wet season but high in the dry season, and *Virola* exhibited very high Ψ_w that hardly decreased below predawn values (Bonal *et al.*, 2000b).

son (about –0.2 MPa) compared to values measured on trees growing in more exposed conditions (Alexander, 1991; Huc *et al.*, 1994; Bonal *et al.*, 2000b). Apparently, the low evaporative demand in the moist understorey keeps water potential high. These high Ψ_w values were far from Ψ_w at turgor loss (–1.6 MPa for *M. gongrijpii*, –1.8 MPa for *M. excelsa*), which was similar to turgor loss potentials of saplings of three other species growing in more exposed sites (Alexander, 1991). Experiments showed that *M. excelsa* seeds and seedlings did indeed tolerate floods better than those of *M. gongrijpii*, but the differences that might point to a better drought tolerance of *M. gongrijpii* were only small. Hence, the reason for the exclusion of *M. gongrijpii* from occasionally flooded areas is more straightforward than the exclusion of *M. excelsa* from upland sites.

Intrinsic water-use efficiency

Stable carbon isotope ratios ($\delta^{13}C$) have been used to derive intrinsic water-use efficiencies of photosynthesis in C_3 plants. The balance between demand for CO_2 by the chloroplasts inside the leaf relative to the conductance of supply of CO_2 through the stomata determines intercellular CO_2 concentration (C_i) in leaves. The difference in CO_2 concentration inside and outside the leaf (C_a–C_i) corresponds to the ratio of CO_2 assimilation over stomatal conductance for water vapour (A/g_s), also commonly referred to as the intrinsic water-use efficiency (WUE). A low C_i resulting from relatively closed stomata thus indicates a high WUE of the gas exchange process. Since discrimination against the heavier carbon isotope ^{13}C during photosynthetic CO_2 assimilation depends on C_i, WUE can be derived from carbon isotope ratios of plant dry matter (Farquhar *et al.*, 1989). This relationship, however, is complicated by the fact that fractionation of stable isotopes also occurs during further transport and metabolism of assimilates (Leavitt and Long, 1982; Schmidt and Gleixner, 1998). And, particularly relevant in trees, structural organic compounds are deposited in different plant parts at different times when conditions for CO_2 assimilation may be different (Pate and Arthur, 1998). Nevertheless, $\delta^{13}C$ ratios in plant carbon are widely used to estimate intrinsic water-use efficiency.

Substantial variation in $\delta^{13}C$ values of 21 tree species in a natural forest and a nearby plantation was found in a study by Guehl *et al.* (1998). The $\delta^{13}C$ of sunlit leaves ranged from –26.7‰ to –31.4‰, a difference of about 5‰, which would represent a range in C_i of 80 ppm. Groups with different levels of shade tolerance exhibited differences in isotopic signature. Species classified as shade hemi-tolerant, those that can establish under a closed canopy but need canopy openings for their growth to canopy height, had the highest $\delta^{13}C$. Although there was overlap between the groups, heliophilic and true shade-tolerant trees tended to have lower (more negative) $\delta^{13}C$ values. A more recent study on a much larger number of species (102) was carried out on three sites in natural forest in French Guiana by Bonal *et al.* (2000c). They found an even larger variation in $\delta^{13}C$ of sunlit leaves with some extremely low (negative) $\delta^{13}C$ values (range –34.8‰ to –27.5‰). This range is representative of a threefold difference in WUE. This larger data set confirmed the conclusion of the earlier study that species classified as shade hemi-tolerant had the highest $\delta^{13}C$, and that heliophilic and true shade-tolerant species had 1–2‰ lower values. A negative correlation was found between $\delta^{13}C$ and midday water potential, which suggests a negative relationship of $\delta^{13}C$ with hydraulic conductance. Furthermore, evergreen trees had substantially lower $\delta^{13}C$ than deciduous ones. The full meaning of these results can not yet be fully appreciated in ecophysiological terms until the causes behind the correlations have been further investigated. However, it is clear that the substantial variation in carbon isotope composition between species in a forest contains important information on plant functioning with respect to carbon acquisition and water relations. It clearly shows that these functional aspects differ considerably between co-occurring species.

Relationships between $\delta^{13}C$ and physiological parameters have been investigated in a comparison of plantation grown trees in French Guiana. Leaf $\delta^{13}C$ values were also less negative (–27.3‰) for two late successional species compared to three early successionals (–29.9‰) (Huc *et al.*, 1994), which is consistent with the larger range of species in the studies mentioned above. However, the difference in C_i as concluded from the $\delta^{13}C$ data could not be confirmed by gas exchange measurements. In the more recent study on three species of these plantation trees by Bonal *et al.* (2000b), similar differences in $\delta^{13}C$ were found between species differing in shade tolerance, as in the other studies. The shade-intolerant *Virola surinamensis* and *Diplotropis purpurea* had low $\delta^{13}C$ (–29.9‰ and –30.9‰, respectively) and the shade hemi-tolerant *Eperua falcata* had a higher value (–28.6‰). C_i measured by gas exchange was more consistent with $\delta^{13}C$ values in this study. High C_i and thus low WUE prevailed during most of the day in *Virola* and *Diplotropis* in the wet season, when these perform most of their CO_2 assimilation. *Eperua* maintained a lower C_i in both wet and dry seasons.

In several studies in French Guiana, both leaf and wood $\delta^{13}C$ was determined (Huc *et al.*, 1994; Guehl *et al.*, 1998; Bonal, 2000). Leaf carbon originates largely from the time when the leaf was formed with an unknown fraction of more recently assimilated carbon. Large variation in $\delta^{13}C$ between leaves of a tree exists, depending, for example, on canopy position (Buchmann *et al.*, 1997) and time of leaf development (Pate and Arthur, 1998). In contrast, stem wood carbon is laid down more continuously and supposedly originates from all leaves of the tree in proportion to their photosynthetic activity. Wood may thus provide a more long-term average for ^{13}C discrimination during the assimilation process of the whole tree. Cellulose of sunlit leaves had a consistently more negative $\delta^{13}C$ value compared to wood cellulose, with substantial interspecific variation in the difference between the two compartments (Huc *et al.*, 1994; Guehl *et al.*, 1998; Bonal, 2000). The causes of these differences are not well understood, apart from the above-described differences in timing of carbon deposition, but that probably does not explain it fully.

Wood $\delta^{13}C$ was used by H. ter Steege (unpublished results) to compare dominant tree species along a gradient in elevation above a creek on soils differing in drainage and water-holding capacity. Trees on the wet Fluviosol soils along the creek had lower $\delta^{13}C$ values than those growing on the well-drained white sand soils on the top of the ridge (Table 3.3). The other two soils that had intermediate positions on the slope, Leptosol and ferralic Arenosol (lateritic clay and brown sand), had trees with intermediate $\delta^{13}C$ values. Only *Pentaclethra macroloba* was sampled along the creek and higher up the ridge, which showed a lower $\delta^{13}C$ at the lower elevation which has a more year-round high water availability (Table 3.3). Apparently, *Pentaclethra* growing at the lower elevation operated at a lower WUE compared to the higher site. A low WUE and thus a high C_i, however, means a higher rate of photosynthesis with the same size of the photosynthetic apparatus and thus a high nutrient use efficiency (Fredeen *et al.*, 1991). The systematic variation of wood $\delta^{13}C$ with soil drainage contrasts with the study of Bonal *et al.* (2000c), where no difference in leaf $\delta^{13}C$ was found between trees sampled on soils of different drainage characteristics. The studies are difficult to compare since wood (Table 3.3) and leaves (Bonal *et al.*, 2000c) were used. However, a comparison

Table 3.3. Carbon stable isotope ratios ($\delta^{13}C$ in ‰) of wood samples taken from common tree species along an altitudinal gradient from a creek border up to a ridge top south of Mabura Hill in Central Guyana. Different soil types occurred along the gradient. Samples were taken from the cambium until a depth of 5 cm, which represents probably at least 20 years of radial increment. Means (SE) of five replications per species soil type combination are presented. The effect of elevation is statistically significant (H. ter Steege, unpublished results).

Soil type	Dystric Fluvisol	Dystric Leptosol	Ferralic Arenosol	Albic Arenosol
Height above creek (m)	5	25	25	60
Species				
Eperua rubiginosa	–28.1 (0.8)	–	–	–
Mora excelsa	–30.0 (0.5)	–	–	–
Pentaclethra macroloba	–27.7 (0.3)	–	–26.0 (0.6)	–
Chlorocardium rodiei	–	–28.3 (0.6)	–28.4 (0.2)	–
Mora gongrijpii	–	–27.8 (0.3)	–28.1 (0.3)	–
Vouacapoua macropetala	–	–26.9 (0.5)	–	–
Dicymbe altsonii	–	–	–	–27.5 (0.1)
Eperua falcata	–	–	–	–25.9 (0.3)
Eperua grandiflora	–	–	–	–26.5 (0.5)
Mean per soil type	–28.5 (0.4)	–27.6 (0.3)	–27.5 (0.4)	–26.6 (0.3)

suggests that some tropical rainforest tree species may respond to water availability in terms of their intrinsic WUE and others may not.

Nutrient Relations

Most of the soils of the Guiana Shield are old and highly leached. They are characterized as poor in available mineral nutrients (Poels, 1987; van Kekem *et al.*, 1996). Nitrogen may have a higher availability for plant growth in the large areas of relatively undisturbed forest, as this has predominantly an organic origin. Most plants have symbiosis with mycorrhiza and the forest trees in the Guianas are no exception. Some inventory work on these fungi has been done, but no ecophysiological work is available for our area. Furthermore, most soils are acidic and many of these may have an excess of potentially toxic elements such as aluminium. Not much work has been done on nutrient relations of forest plants in the Guianas. With the limited information available, the question addressed is what is known about how trees cope with this low nutrient availability and possible high concentrations of toxic elements. Differences between species with respect to modes of nutrient acquisition and use are also examined.

Nutrient concentrations in leaves

In the framework of several studies, leaves have been analysed for their nutrient concentrations. Data on the elements nitrogen and phosphorus are most widely available and they are summarized in Table 3.4 for

Table 3.4. Concentrations of nitrogen and phosphorus in dry matter of leaves and P/N ratios measured in various studies in the Guianas. Number of species sampled (n) and means (SD) are provided. Species are grouped in categories as indicated by the authors, such as successional status and diameter at breast height (dbh).

	Number of species (n)	N (mg/g) Mean (SD)	P (mg/g) Mean (SD)	N/P (g/g) Mean (SD)
Kabo, Suriname[a]				
Trees > 5 cm dbh		16.8	0.85	19.8
Trees < 5 cm dbh and other small plants		15.2	0.78	19.5
Palms		9.4	0.68	13.8
Lianas		17.3	0.93	18.6
Mabura Hill, Guyana[b]				
Late successionals on white sands				
Saplings	6	17.1 (4.1)	0.77 (0.26)	23.0 (4.9)
Mature trees	6	16.1 (2.3)	0.64 (0.15)	26.3 (8.6)
Late successionals on brown sands				
Saplings	6	13.9 (2.7)	0.60 (0.07)	23.6 (5.7)
Mature trees	6	16.6 (1.0)	0.59 (0.13)	29.6 (7.0)
Early successionals (both soils types)				
Saplings	3	18.9 (4.5)	0.70 (0.18)	27.2 (4.6)
Mature trees	3	19.5 (2.7)	0.64 (0.14)	30.9 (4.6)
Paracou, French Guiana[c]				
Mature trees				
Early successional and heliophilic sp.	5	18.2 (5.5)	0.60 (0.09)	30.9 (9.8)
Late successional and shade (hemi-)tolerant sp.	3	14.4 (3.7)	0.58 (0.09)	28.1 (3.9)
Nouragues, French Guiana[d]				
Late successional sp.	3	25.2 (1.3)	1.18 (0.12)	21.7 (3.3)

[a]Poels (1987), mixed samples from 12 plots; [b]Raaimakers (1995); [c]Guehl *et al.* (1998); [d]Rijkers *et al.* (2000a), supplemented with unpublished results.

different areas. The range of N concentrations of dried leaf material is between 9 and 25 mg/g, with most values around 17 mg/g. Phosphorus concentrations are low and range from 0.6 to 1.2 mg/g, with most values around 0.7 mg/g. The studies in Mabura Hill, Guyana (Raaimakers, 1995) and Paracou, French Guiana (Guehl *et al.*, 1998) show similar leaf N and P concentrations, the study at Kabo, Suriname (Poels, 1987) slightly higher ones, but the study of Rijkers *et al.* (2000b) in Nouragues, French Guiana, indicates substantially higher values of both elements (Table 3.4). This suggests a higher nutrient availability in this forest compared to the other ones, at least as far as these elements are concerned. However, the limited number of data does not allow a firm conclusion. Several early successional species such as *Cecropia obtusa*, *Jacaranda copaia* and *Sclerolobium melinonii* have higher values compared to late successional species for leaf N and P in the studies of Raaimakers (1995), Guehl *et al.* (1998) or for N only (Roggy *et al.*, 1999b), but the averages for early and late successionals are not much different (Table 3.4). Saplings and mature trees also have similar N and P concentrations. High N concentrations are found in some species that have symbiotic nitrogen fixation such as *Sclerolobium*, *Swartzia* and *Inga* spp., but not in all (Roggy *et al.*, 1999b; Perreijn, 2002).

P concentrations in the rainforest tree leaves in the Guianas are among the lowest found in tropical forests (Lathwell and Grove, 1986; Vitousek and Sanford, 1986) and are similar to other *terra firme* sites in South America. The values are substantially lower than for tropical rainforests on more fertile soils (Vitousek and Sanford, 1986) and those that are considered satisfactory for agricultural fruit trees (Bennett, 1993). Although concentrations of nutrient elements cannot be used to draw firm conclusions on their limitation for growth, the data suggest that the availability of phosphate may limit tree productivity in these forests. The N/P ratios are highest in the Mabura Hill and Paracou samples (Table 3.4), further indicating that not N but P is the limiting nutrient for growth there. Nitrogen can accumulate in undisturbed ecosystems as a result of biotic and abiotic inputs (Jordan, 1985), whereas phosphorus cannot accumulate on strongly leached low phosphorus soils, or is sequestered in soils high in aluminium (Al) and iron (Fe) oxides.

Brouwer (1996) and Poels (1987) also analysed concentrations of other macronutrient elements in leaves: K, Ca and Mg. The values for these elements that they report for Guyana and Suriname, respectively, are rather low. Hence, these elements are also potentially growth limiting. Extremely low values were found for calcium in late successional trees on Oxisols in the Rio Negro area in Venezuela (about 0.5 mg/g; Reich *et al.*, 1995). There were some indications that these low calcium concentrations do indeed limit photosynthetic rates. Since similar soil types also occur in the Guianas, such a calcium limitation may be expected there as well. Specific adaptations to low availability of some mineral nutrients such as P and Ca may be expected on the highly leached soils of the Guianas, which needs further investigation.

Effects of phosphorus availability

Since phosphorus was identified as a possible limiting nutrient for plant growth in the forests of the Guianas, experiments were carried out to investigate this hypothesis further (Raaimakers, 1995; Raaimakers and Lambers, 1996). Two species, *Tapirira obtusa*, a fast-growing pioneer, and *Lecythis corrugata*, a slow-growing late successional, both mainly occurring on white sands in Guyana, were grown on a range of P addition rates in pots. Increasing addition of P did not stimulate the growth of *Lecythis*, although the plants did absorb the nutrient as evident from the increasing P concentrations (Fig. 3.11). Phosphorus from the seed reserves was apparently still sufficient to sustain growth during the 6-month period of the experiment. The results suggest that the inherent growth rate of *Lecythis* is so low that the extra phosphorus

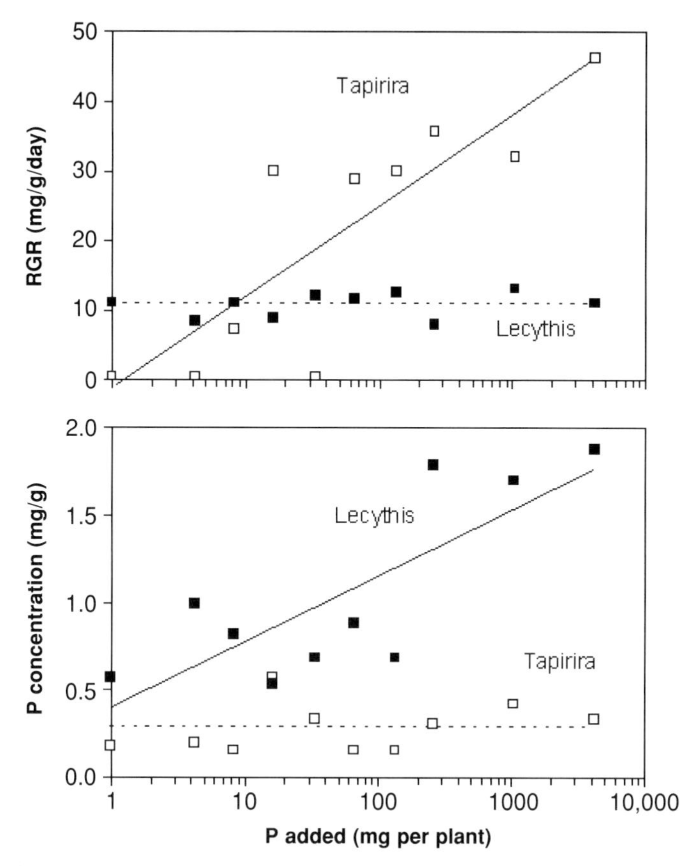

Fig. 3.11. Relationship of relative growth rate (RGR) and concentration of phosphorus in plant dry mass (P concentration) with P added to the plants during growth (P added). Two species were investigated, the early successional *Tapirira obtusa* (□) and the late successional *Lecytis corrugata* (■). Plants were grown in pots with sand in a greenhouse. Continuous lines represent significant correlations; dotted lines refer to non-significant correlations (after Raaimakers and Lambers, 1996).

cannot be used to increase growth. The reverse situation was found for *Tapirira*. Relative growth rate of this species increased with increasing P addition, but P concentrations remained very low, indicating that growth of this inherently faster-growing species is more easily limited by P availability. Since P concentrations remained low independent of P availability, it further suggests that the low P concentrations found in leaves (Table 3.4) may not be an indication of actual P limitation, but rather a reflection of the species adaptation to the low P environment.

Nutrient addition experiments have also been carried out with seedlings in the forest. A similar result as in the greenhouse experiment was found (Raaimakers, 1995). The growth of *Dicymbe altsonii* was stimulated by P addition and concentrations increased only slightly. However, the growth of *Chlorocardium rodiei* was not stimulated, but P concentrations increased, again indicating that the plant had taken up more P, but this was stored mostly in roots and stems, and not utilized for increased growth.

A pot experiment carried out in Guyana with the early successional *Cecropia obtusa* and *Goupia glabra* and the late successional *Chamaecrista adiantifolia* essentially confirmed the results described above with respect to growth stimulation and P concentration in early and late successional species (Raaimakers, 1995). The roots of the pioneers were much thinner than those of the climax species. However, the rates of P absorption per unit root length were not different between species. Supposedly, the thin, fast-growing roots of the pioneers are effective for exploring the soil for nutrients in a gap environment, thus enabling fast growth. However, this probably occurs at the expense of a high root turnover.

Efficiency of nutrient use

Nutrient elements are used for growth and maintenance processes in the plant. The rates of these processes per unit of a particular element present in the plant can vary substantially between species and growth conditions. An important process that drives growth is photosynthesis. The light-saturated rate of photosynthesis (A_{max}) per unit N present in the leaf, often referred to as photosynthetic N use efficiency (PNUE), was high in the three early successional tree species measured by Raaimakers *et al.* (1995), although still low compared to many herbaceous plants (Pons *et al.*, 1994). The six late successional species had about half the PNUE of the early successionals, and similar PNUE to other evergreen species (Field and Mooney, 1986). These PNUE calculations are based on instantaneous rates of photosynthesis. However, when the carbon acquisition over the whole lifetime of a leaf is considered, N use efficiencies in the two species groups are more similar, because leaves of late successionals may live more than twice as long as those of early successionals (see below). At the prevailing low tissue concentrations, phosphorus may limit photosynthetic capacity. That this might be the case is suggested by the closer association of A_{max} with P compared to the relationship with N, and that A_{max} per unit P was high (Raaimakers *et al.*, 1995).

During the senescence process and before abscission, a plant typically resorbs compounds from a leaf, thus increasing the efficiency of the utilization of nutrients. Nitrogen and phosphorus are among the nutrient elements that are resorbed, and the percentage resorption of both elements is around 50% across all plant species (Aerts, 1996). Resorption of N and P has been investigated by Raaimakers (1995) in Guyana. Although P concentrations are very low, as mentioned above, P resorption appeared to be 57% on average for the four species studied, with no difference between early and late successional species. However, there was no or hardly any resorption of N, which is exceptional, in comparison with other studies on nutrient-poor soils elsewhere in the Amazon area (Medina, 1984; Reich *et al.*, 1995). High resorption efficiency in forests on Oxisols and Ultisols is also concluded from the very low P concentrations in leaf litter (Vitousek, 1984; Vitousek and Sanford, 1986).

Nutrient use efficiency by plants is determined by residence time in the plant (Berendse and Aerts, 1987). One factor influencing that parameter is the resorption of nutrients from senescing tissue as mentioned above; another factor is the life span of plant tissue. Calculations show that the variation in leaf life span contributes more to nutrient conservation than variation in resorption efficiency (Aerts, 1995). Hence, plants growing on the predominantly nutrient-poor soils in the Guianas would benefit from long life spans of their leaves and other tissues with respect to conservation of nutrients. Evergreen tropical forest trees are characterized by long leaf life spans. However, a large range in leaf life spans can be found in tropical forests (Reich *et al.*, 1991; Veneklaas and Poorter, 1998). Data have recently become available from studies in the Guianas. Leaf life spans appeared to be about 3 years for *Dicorynia guianensis* and for larger trees of *Vouacapoua americana*. This increased to 7 years for saplings of *V. americana* growing in the shaded understorey (Rijkers, 2000). Saplings of various species growing in gaps in Guyana had shorter median leaf life spans (Rose, 2000). Values for the late successional species

Catostemma fragrans, *Chlorocardium rodiei*, *Ormosia coccinea* and *Pentaclethra macroloba* ranged between 1 and 2 years. Two early successionals, *Cecropia obtusa* and *Goupia glabra*, had leaf life spans of less than half a year. However, two other early successional species, *Laetia procera* and *Sclerolobium guianensis* had similar life spans as the late successionals. Hence, there are examples among the selected species that illustrate the commonly held view that early successionals have high leaf turnover rates, but important exceptions exist. The adaptive value of a high leaf turnover rate in pioneers is supposed to be associated with fast growth and competition for light in early succession (Ackerly and Bazzaz, 1995). This is apparently not always the overriding factor accounting for variation in leaf turnover rate in Guianan trees. The factors influencing leaf longevity deserve further attention.

Symbiotic nitrogen fixation

Most nitrogen in ecosystems originates from an organic source, i.e. biological nitrogen fixation. This process can occur in free-living bacteria in soil or in the phyllosphere, as Ruinen (1965) showed in Suriname, and in legumes with their enhanced capability to form symbiosis with *Rhizobium*-type bacteria or other bacterial symbioses. Legumes are very abundant in the forests of the Guianas (see Chapter 1), but not all of them are active in N_2 fixation. This raises two questions. First, which of the leguminous species are active in N_2 fixation? Secondly, what is the importance of N_2 fixation in the nitrogen cycle of forests in the Guiana Shield? Although nitrogen is not likely to be an important growth-limiting element in primary forest, as argued before, losses to the ecosystem, such as by leaching of nitrate following disturbance (Poels, 1987; Brouwer, 1996), and denitrification, should be compensated by N inputs mainly from N_2 fixation.

Leguminous species have been investigated for the presence of root nodules in a large survey comprising 172 species in Brazil (De Souza Moreira *et al.*, 1992). The survey included some species that also occur in the Guianas, and several other species are closely related. Nodulation in the natural habitat and/or in nursery-grown plants was found in 47% of the species examined. Some species that were nodulating in the nursery were not found to do so in the natural habitat, particularly on nutrient-poor upland soils. Hence, species may have the potential of symbiotic N_2 fixation, but may not do so in the forest, which may be due to deficiency of mineral elements such as phosphorus (De Souza Moreira *et al.*, 1992). A high availability of soil-borne nitrogen relative to other nutrient elements may also limit the occurrence of symbiosis and/or suppress the nitrogen-reducing activity in the nodules (Marschner, 1995).

In a survey in French Guiana (Roggy and Prevost, 1999), 67% of the 62 leguminous taxa that were investigated appeared to be nodulated, with the lowest frequency in the Caesalpiniacae (50%) that are the most abundant in the forest. A combination of surveys from Guyana and French Guiana (Table 3.5) reveals that there are a number of important, sometimes locally dominant, genera that lack rhyzobia symbiosis, such as *Dicorynia*, *Dicymbe*, *Eperua*, *Hymenaea*, *Mora*, *Peltogyne* and *Vouacapoua*, although an occasional nodule-like structure may be found, such as in *Dicorynia* (Béreau and Garbaye, 1994). Species in both the Mimosaceae (*Parkia* sp.) and in the Papilionaceae (*Dipteryx* sp.) also showed no root nodules in Guyana (Norris, 1969; Perreijn, 2002) and French Guiana (Guehl *et al.*, 1998; Roggy and Prevost, 1999). Important Guianan tree genera that have been shown to nodulate include: *Cassia*, *Chamaecrista*, *Sclerolobium* and *Swartzia* of the Caesalpininiaceae, *Abarema*, *Inga* and *Pentaclethra* of the Mimosaceae, and *Dioclea*, *Hymenolobium* and *Ormosia* of the Fabaceae (Table 3.5).

Pot experiments were carried out in Guyana to investigate the N_2-fixation capacity of several nodulating species (Perreijn, 2002). Local forest soil was enriched with ^{15}N and uptake of total N and ^{15}N was measured in potential N_2-fixers

Table 3.5. Nodulation of legumes and nitrogen-fixing activity of selected taxa. Nodulation was derived from four studies in Guyana (Gu) and French Guiana (FG). Presence (+) or absence (–) of root nodules was recorded (blank=not sampled); Nor: Norris (1969); Per: Perreijn *et al.* (2002); Gue: Guehl *et al.* (1998), Rog: Roggy *et al.* (1999a). Percentages in the study of Roggy *et al.* (1999a) refer to the proportion of species that showed nodulation. The percentage of leaf N derived from atmospheric N_2 (%Ndfa) was calculated from ^{15}N stable isotope ratios by Perreijn *et al.* (2002) and Roggy *et al.* (1999b). The two assumptions used for fractionation associated with N_2-fixation used in the last reference were averaged. Values are presented for sample sizes >5 only. Only taxa that (are supposed to) occur in the three Guianas or are considered important otherwise are presented. Nomenclature after Bogan *et al.* (1997).

	Nodulation score				% Ndfa	
	Nor Gu	Per Gu	Gue FG	Rog FG	Perreijn[a] Gu (n)	Roggy[b] FG (n)
Caesalpiniaceae						
Bauhinia kuntheana	–					
Cassia cowanii	+					
Chamaecrista spp.	+	+			23 (2)	
Dialum guianense				–		
Dicorynia guianensis			–	–		
Eperua spp.	–	–	–	–		
Hymenea courbaril		–				
Mora spp.	–	–				
Peltogyne venosa	–	–	–	–		
Recordoxylon speciosum			+			
Sclerolobium spp.	+	+	+	+	33 (1)	
Senna latifolia	–					
Senna quinquangulare				+		
Swartzia spp.	+	+		+	57 (3)	75 (11)
Tachigali paniculata				+		
Vouacapoua americana				–		41 (11)
Family				50%	**41** (6)	**40** (57)
Mimosaceae						
Abarema spp.		+		+		
Balizia pedicellaris		+		+		
Enterolobium schomburgkii				+		
Inga spp.	+	+		83%	0 (1)	52 (30)
Mimosa myriadena	–					
Parkia spp.		–	–	–		
Pentacletra macroloba	+	+			58 (1)	
Pseudopiptadenia suaveolens				–		
Stryphnodendron polystachium				+		
Zygia racemosa				-		
Family				71%	**17** (4)	**44** (36)
Papilionaceae						
Dioclea spp.	+	+		+		
Diplotropis purpurea	–	+	+			71 (1)
Dipterix spp.	–			–		
Hymenolobium spp.	+			+		
Mucuna urens	+					
Ormosia spp.	+	+		+	85 (2)	
Poecilanthe hostmanii				+		
Pterocarpus officinalis		+				
Family				77%	**66** (7)	**50** (19)

[a]Family and genera means over species, n=number of species. [b]Means over individuals, n=number of samples.

and in non-nodulating reference species. These experiments showed clearly that when supplied with all other mineral nutrients except N, conditions that stimulate N_2 fixation, *Pentaclethra macroloba* and *Sclerolobium guianensis* derived around 90% of their N from atmospheric N_2 (%Ndfa). *Chamaecrista adiantifolia* had a lower N_2-fixation activity, which is consistent with its lower %Ndfa calculated from natural abundance of ^{15}N (Table 3.5). N_2 fixation decreased when no extra P was supplied and also when extra ammonium was added. Hence, on soils that are known to have low P availability, this element is probably limiting for N_2 fixation. Furthermore, a high availability of soil N relative to other nutrient elements, which may occur in undisturbed forest, can down-regulate N_2 fixation. The low light conditions in the understorey may also limit nodulation and thus N_2-fixation activity of tree seedlings, as found for *Pentaclethra macroloba* in Costa Rica (Fetcher *et al.*, 1994). Nevertheless, evidence from natural abundance of ^{15}N indicates that *Swartzia* sp. can have N_2 fixation in the forest understorey (Shearer and Kohl, 1986). However, further information is required on the effect of light availability on N_2-fixation activity.

Stable nitrogen isotope natural abundance studies

The difference in the ratio of $^{15}N/^{14}N$ between a plant or soil sample and atmospheric N_2 ($\delta^{15}N$) can be used to identify nitrogen sources for plant growth, provided that the isotopic signatures are sufficiently different. Guehl *et al.* (1998) investigated the natural variation in $\delta^{15}N$ in the forest in Paracou, French Guiana. Among the seven leguminous species, there were three that had nodules (Table 3.5) and these also had high leaf N concentrations (>2%). Two of the nodulating species had $\delta^{15}N$ values close to zero, which is indicative of active N_2 fixation. The other nodulating species had a $\delta^{15}N$ in the range of the non-nodulating species, which might be due to low N_2-fixing activity. Interpretation, however, is complicated by the fact that some non-nodulating legumes and non-leguminous species also had low $\delta^{15}N$ ratios. Thus, further diversification in N-source in combination with different isotopic signatures of these sources may be involved, as well as species-specific isotopic discrimination during acquisition and metabolism of N-compounds, and several other possible reasons (as discussed by Guehl *et al.* (1998) and Roggy *et al.* (1999b)).

Roggy *et al.* (1999b) included more nodulating leguminous species in their investigation and found large variation in $\delta^{15}N$ and leaf N concentration, ranging between −1‰ and 7‰ for $\delta^{15}N$ and between 1.0% and 3.2% for N concentration. The results indicate that active N_2-fixers are characterized by a low $\delta^{15}N$ and a high N concentration, non-N_2-fixing legumes by a high $\delta^{15}N$ and a low N concentration, and non-leguminous pioneers by a low $\delta^{15}N$ and a low N concentration. They conclude that $\delta^{15}N$ can be used for distinguishing functional groups with respect to N acquisition and metabolism when used in combination with other traits such as nodulation, N concentration and taxonomy. Pioneer *Piper* spp. preferentially use nitrate as a nitrogen source, whereas shade-tolerant species of the same genus use ammonium (Fredeen *et al.*, 1991; Fredeen and Field, 1992). Nitrate is found in higher concentrations in soils in gaps compared to the understorey of undisturbed forest where ammonium prevails (Brouwer, 1996). The different isotopic signatures of non-nodulating pioneers and late successionals as found by Roggy *et al.* (1999b) is probably related to this uptake of different forms of soil N, because the $\delta^{15}N$ of pioneers resembled that of nitrate and the $\delta^{15}N$ of late successionals was close to that of ammonium (J.C. Roggy, unpublished data).

Natural abundance of ^{15}N potentially can be used to quantify actual symbiotic N_2-fixation activity in the forest (Shearer and Kohl, 1986). Calculations are based on a two-source model, hence a condition for successful use of the method is that soil N and air N have a sufficiently different $\delta^{15}N$. Reference species are selected that do not fix N_2, but otherwise share similar soil-bound N sources and use a similar mode of N acquisition. Their $\delta^{15}N$ value represents soil N as a source. Plants that have N_2 as their only source of N have $\delta^{15}N$ values between 0‰ and –2‰. The exact values can be determined experimentally, but these are difficult to obtain for tropical forest trees. Hence, assumptions are made about ^{15}N fractionation during N_2 fixation. Although the method has its limitations, it is the only method available for estimating symbiotic N_2 fixation in the forest.

The method has been used at Piste de St Elie in French Guiana (Roggy *et al.*, 1999a) and in the forest south of Mabura Hill in Guyana (Perreijn, 2002). Samples from forests on white sand soils at both sites in the Guiana Shield area appeared to have similar isotopic signatures in nodulating legumes and in non-fixing reference species, which made it impossible to estimate N_2-fixation rates. However, on loamy sands and lateritic clays (ferallic Arenosols and Ferralsols) the difference was large enough to allow the calculation of the percentage of N in plants that is derived from atmospheric N_2 fixation (%Ndfa). The study from French Guiana was at the community level and thus had few replicates for most species, which did not allow conclusions at the species level. The Guyana study was designed to draw conclusions at the species level. For that purpose, fewer species were sampled more frequently. Still, comparisons can be made between the two sites for specific taxa. At the family level, the two studies had similar results for the Caesalpiniaceae, but the Guyana study had lower values for the Mimosaceae and higher %Ndfa for the Papilionaceae compared to the study in French Guiana (Table 3.5). Not surprisingly, the nodulating Papilionaceae were most active with a mean of 58%Ndfa. Caesalpiniaceae (41%) and Mimosaceae (31%) had lower %Ndfa. At the genus level, *Swartzia*, which is sometimes included in the Papilionaceae, has several actively N_2-fixing members. *Inga* also has several active N_2-fixing members, but also non-nodulating and nodulating species that did not register any significant N_2 fixation. *Ormosia* had the highest calculated N_2-fixation rates (Table 3.5). Indication for active N_2 fixation from $\delta^{15}N$ ratios was also found for tree species such as *Pentaclethra macroloba*, *Diplotropis purpurea*, *Pterocarpus officinalis*, *Clathrotropis* spp. and *Sclerolobium* spp., and in the liana *Dioclea elliptica* (Perreijn, 2002). Interestingly, the low $\delta^{15}N$ of the non-nodulating legume *Vouacapoua americana* suggests that it can fix N_2 (Roggy *et al.*, 1999a), but this phenomenon requires independent confirmation.

Data on the fraction of nitrogen derived from N_2 fixation, together with estimates of growth rates and biomass partitioning between N_2-fixers and non-fixers, were used by Roggy *et al.* (1999a) to calculate symbiotic N_2-fixation rates at the community level. Roggy *et al.* (1999a) conclude that in the Piste de St Elie forest of French Guiana, 5.5% of total leaf N was derived from N_2 fixation. The annual rate of symbiotic N_2 fixation was calculated at 7 kg/ha/year. N_2 fixation by free-living microorganisms in the phyllosphere or in the soil cannot be estimated in this way. Bentley and Carpenter (1984) estimated that between 10% and 25% of leaf N may be derived from phyllospheric N_2 fixation in a Costa Rican forest, but large uncertainties remain. Hence, although no basis for a good quantitative estimate exists, free-living microorganisms may contribute a substantial amount of fixed N. These first estimates of symbiotic N_2 fixation can be compared with other N inputs and N losses due to leaching, volatilization and denitrification in order to further improve N budgets at the ecosystem level. Estimates at the species level can also be used to improve forest management practices. For example, individuals of species capable of N_2 fixation should be spared in forest operations in order to keep the population of N_2 fixers large enough to compensate for N that is inevitably lost with exploitation.

Aluminium toxicity on acid soils

Several soil types found in the Guiana Shield have a high content of aluminium.

At the low pH found in these soils, Al^{3+} builds up to such high concentrations that it becomes toxic to many plants (Marshner, 1995). A drop in pH below 4.5 with accompanied increase in Al^{3+} in the soil solution was found after gap creation on brown sandy ferralic Arenosols in Guyana (Brouwer, 1996). White sandy albic Arenosols almost completely lack aluminium and no build up of Al^{3+} can be expected (van Kekem *et al.*, 1996). Hence, plants growing on Al-containing brown loamy sands and lateritic clays (ferralic Arenosols and Ferralsols) are exposed to potentially toxic levels of Al^{3+}. The difference in Al content between soil types could be one of the causes of the floristic differences between the forests on these soils. Aluminium tolerance should be an important trait of plants establishing on Al-containing soils, particularly after disturbance because of the high Al^{3+}-concentrations that predominate during these early successional periods.

The hypothesis that aluminium is an important environmental variable explaining edaphic effects on forest species composition in central Guyana was investigated (Alexander and ter Steege, unpublished results). Non-shade-tolerant pioneer species growing in gaps in the forest canopy on Al-soils appeared to accumulate substantial quantities of Al in their leaves (Table 3.6). This was not found in pioneers growing on non-Al-containing white sand. Climax species sampled in gaps did not accumulate Al in their leaves and Al concentrations were not different between these plants when growing on Al-containing Ferralsols and Leptosols or non-Al-containing albic Arenosols (Table 3.6). The degree of Al accumulation in leaf tissue was substantially different between pioneer species. *Palicourea guianensis* and *Miconia* spp. were strong accumulators and had Al concentrations up to 1% of dry mass. *Goupia glabra* and *Cecropia* spp. (*C. obtusa* and *C. sciadophylla*) were in an intermediate group with Al concentrations of >1‰. The non-accumulators were the climax species *Mora gongrijpii*, *Chlorocardium rodiei* and *Vouacapoua macropetala*; they had Al concentrations <0.4‰. No increase in Al concentration in leaves of Al-accumulating species was evident with increasing gap size. Such an increase was expected, since earlier research had indicated that free Al^{3+} increases with gap size (Brouwer, 1996). One aspect of Al tolerance is that the toxic element is sequestered in ageing tissue that is abscised later (Marshner, 1995). Accumulation of Al in leaf tissue on Al-containing soils is thus considered as an expression of Al tolerance. The high tissue

Table 3.6. Aluminium concentrations (means ± SE; mg/g) in leaves of pioneer and climax species sampled in gaps in the forest on different soil types in central Guiana (Mabura Hill area) (Alexander and ter Steege, unpublished results).

Species	Dystric Leptosol	Ferralic Arenosol	Albic Arenosol
Pioneers			
Cecropia spp.	1.87 (0.10)	1.22 (0.20)	0.28 (0.04)
Goupia glabra	1.09 (0.18)	2.53 (0.21)	0.49 (0.03)
Miconia spp.	5.44 (0.41)	8.07 (0.70)	–
Palicourea spp.	11.22 (0.87)	0.77 (0.05)	–
Mean	5.86 (0.63)	4.12 (0.32)	0.38 (0.04)
Climax			
Chlorocardium rodiei	0.14 (0.01)	0.51 (0.04)	–
Mora gongrijpii	0.26 (0.07)	0.33 (0.04)	–
Voucapoua macropetala	0.32 (0.03)	0.36 (0.04)	–
Mean	0.26 (0.03)	0.40 (0.02)	–

concentrations of pioneers on Al-containing soils indicate that Al tolerance exists in these species. These plants lose much of the accumulated toxic element due to their high leaf turnover. Al tolerance should also be present in climax species that show healthy growth in gaps where high soil concentrations of free Al^{3+} occur, although they do not accumulate the element.

Aluminium tolerance can also be based on real tolerance of exposed tissue or avoidance of exposure by exclusion of the toxic element (Marshner, 1995). Root growth is one good indicator of tolerance. The question if Al tolerance differs between species growing on Al-free and Al-containing soils was investigated experimentally (Alexander and ter Steege, unpublished results). A total of 20 species were grown from seed and transferred to aerated nutrient solution containing different concentrations of Al^{3+} (0, 2.5 and 5 g/m^3). Most species showed inhibition of root growth, root tip necrosis and other signs of toxicity with increasing Al^{3+} concentration in the nutrient solution. However, no clear pattern emerged of smaller effects and thus larger tolerance in species that are restricted to Al-containing soils. This was expected to be particularly prominent in pioneer species on these soils. The conclusion of the study is that species occurring in the area are tolerant of the Al that they encounter in their environment on Al-containing soils. However, no indications were found that variation in Al tolerance as a plant trait is an important factor for partitioning of species between soil types.

Concluding Remarks

As illustrated above, substantial progress has been made over the past decade in ecophysiological research in the Guianas. A broad range of plant parameters has been investigated and conclusions drawn on physiological characteristics of species. This has added to our broader knowledge of forests in the Guiana Shield. The association of these physiological traits with functional groups was the subject of several investigations. Functional groups are mostly defined on life history, demographic, distribution and/or morphological criteria. The most frequently used division was between early and late successional species or between shade-avoidance and shade-tolerance in early development; groups that are largely identical in the dense tropical rainforest, but stress different criteria. Not surprisingly, similar associations of physiological traits with these functional groups have been found in the Guianas, as they have been found elsewhere in the tropics (Swaine and Whitmore, 1988; Ackerly, 1996). That refers to typical representatives, but several species appeared to share traits associated with both groups, which supports the idea of a continuum between early and late successional species.

The physiological traits that have been found to be associated with each of the two functional groups are summarized in Table 3.7. The qualification of some of the traits is sometimes based on a limited number of species, as indicated. Hence, the generalizations presented in Table 3.7 are to some extent tentative. Typical members of the early successional shade-intolerant group are species with small seeds and low-density wood. It was found that they have a short leaf life span, probably also a high root turnover, thin roots, a high nutrient absorption rate per unit root mass, a high fraction of plant biomass that is allocated to leaves and/or a high specific leaf mass, a high plasticity in specific leaf mass, a high photosynthetic capacity per unit leaf area, per unit leaf mass and per unit nutrient elements in leaves, and a low shade acclimation potential of their chloroplasts. Furthermore, several representatives had a low water-use efficiency of photosynthesis, a high midday water potential, a low resistance for water uptake and transport, a low $\delta^{15}N$ in leaves and there are indications that they use nitrate as the principal source of N. These traits allow fast growth in the highly competitive environment after disturbance when light and nutrient availability are high. This contrasts with typical late successional canopy tree species that have large seeds and high-density wood and are shade-tolerant in early growth stages. This functional group appeared to be more diverse, but the following generalizations may be made on the basis of the studies in the Guianas. Late successional shade-tolerants were found to be characterized by a long leaf life span, probably also a low turnover of roots, thick roots, a low nutrient absorption rate per unit root mass, a low specific leaf area, a low plasticity for this parameter, a low photosynthetic capacity per unit leaf area, leaf mass and nutrient elements in leaves, a high plasticity for acclimation of chloroplasts to light availability, a high $\delta^{15}N$ in leaves and ammonium is probably the most important source of N. With respect to water relations, a high water-use efficiency of photosynthesis, a low midday water potential and a high resistance for water uptake and transport were found in several representatives, but the low $\delta^{13}C$ values of particularly the most shade-tolerant species suggests that this is not typical for all late successionals. Several of these traits can be identified as important for the survival of populations of the species in the highly competitive environment of the climax forest, where limited amounts of nutrients become continuously available through cycling and light is only abundant at the top of the canopy.

However, some species combine traits that are identified here as specific for one functional group. Examples are *Pentaclethra macroloba*, which does not have a high photosynthetic capacity (Fetcher *et al.*, 1994), has a long leaf life span and good growth in shade, but combines that with good growth in large gaps (Rose, 2000). *Sclerolobium guianense* also has a long leaf life span and a low plasticity in leaf mass

Table 3.7. Summary of ecophysiological differences between early and late successional trees in the rainforest of the Guianas. They are identified on the basis of studies carried out in the Guianas as discussed in this chapter and are supported by data from elsewhere. Some statements are speculative when based on limited evidence as indicated. The groups are not as homogeneous as suggested by this table, since many species share traits of both groups as discussed in the text.

	Early successional	Late successional
General characteristics		
Shade tolerance[a]	low	high
Seed mass	low	high
Wood density	low	high
Ecophysiological traits		
Leaf life span	short	long
Root turnover[b]	high	low
Specific root length[b]	high	low
Nutrient absorption per root mass[b]	high	low
Leaf mass fraction (in early development)	high	low
Specific leaf area	high	low
Plasticity in specific leaf area	high	low
Photosynthetic capacity (A_{max})		
Per unit leaf area, mass, P and N	high	low
Plasticity in A_{max} per unit leaf area	high	low
Plasticity in A_{max} per unit chlorophyll[b]	low	high
Photosynthetic water-use efficiency[b]	low	variable
Resistance for water transport[b]	low	high
Midday water potential[b]	high	low
Principal nitrogen source[b]	nitrate	ammonium

[a]Shade tolerance as evident from distribution in gap and closed forest understorey habitats.
[b]Limited evidence for a generalization.

per unit area, but combines that with fast growth in large gaps (Rose, 2000). *Carapa guianensis* has large seeds, a trait that is typical for late successional species, but also has a relatively high photosynthetic capacity (Huc *et al.*, 1994) and a relatively high growth rate in large gaps, but shows good survival in small gaps (Rose, 2000). Such species apparently occupy intermediate positions not only with respect to their physiological traits, but also with respect to their requirements for establishment from seed. Possible association of such traits with the further subdivision in functional groups, as suggested by Swaine and Whitmore (1988) and Favrichon (1994) or based on other criteria than used by these authors, needs to be investigated in more detail. Physiological information may also be used as criteria for distinguishing functional groups, such as the capacity for N_2 fixation and intrinsic water-use efficiency. However, further research is needed to point out the functional significance of particular physiological traits and how they contribute to niche differentiation in tropical rainforest plants.

Most of the ecophysiological studies either dealt with young growth stages or with mature trees; only a few compared different growth stages. It appeared that physiological traits do not remain constant throughout development. These shifts with growth stage probably reflect changing demands, both internally and with the changing environment when trees grow larger. An example is the decrease in shade tolerance with tree size as a result of the inevitable decrease in leaf area ratio (Givnish, 1988). The ontogenetic pattern of character development in tropical forest plants, the species-specific nature of these patterns and what is the functional significance of the changes are largely unknown. Studies that address this issue would need to integrate architectural with physiological data through model-based approaches. Such approaches have been used elsewhere (Ackerly, 1996; Pearcy and Yang, 1996) and need to be further developed in order to increase our understanding of whole tree processes in the complex environment of the tropical rainforest.

References

Ackerly, D.D. (1996) Canopy structure and dynamics: integration of growth processes in tropical pioneer trees. In: Mulkey, S.S., Chazdon, R.L. and Smith, A.P. (eds) *Tropical Forest Plant Ecophysiology.* Chapman & Hall, New York, pp. 619–658.

Ackerly, D.D. and Bazzaz, F.A. (1995) Leaf dynamics, self-shading and carbon gain in seedlings of a tropical pioneer tree. *Oecologia* 101, 289–298.

Aerts, R. (1995) The advantages of being evergreen. *Trends in Ecology and Evolution* 10, 402–407.

Aerts, R. (1996) Nutrient resorption from senescing leaves of perennials: are there general patterns? *Journal of Ecology* 84, 597–608.

Alexander, D.Y. (1991) Comportement hydrique au cours de la saison sèche et place dans la succession de trois arbres Guyanais: *Trema micrantha, Goupia glabra* et *Eperua grandiflora. Annales des Sciences Forestières* 48, 101–112.

Allen, M.T. and Pearcy, R.W. (2000) Stomatal behavior and photosynthetic performance under dynamic light regimes in a seasonally dry tropical rain forest. *Oecologia* 122, 470–478.

Anderson, J.M., Chow, W.S. and Park, Y.I. (1995) The grand design of photosynthesis: acclimation of the photosynthetic apparatus to environmental cues. *Photosynthesis Research* 46, 129–139.

Barigah, S.T., Imbert, P. and Huc, R. (1998) Croissance et assimilation nette foliaire de jeune plants de dix arbres de la forêt Guyanaise, cultivés a cinq niveaux d'éclairement. *Annales des Sciences Forestières* 55, 681–706.

Bennett, W.F. (1993) *Nutrient Deficiencies and Toxicities in Crop Plants.* APS Press, The American Phytopathological Society, St Paul, Minnesota.

Bently, B.L. and Carpenter, E.J. (1984) The direct transfer of newly fixed nitrogen from free-living epiphyllous micro-organisms to their host plant. *Oecologia* 63, 52–56.

Béreau, M. and Garbaye, J. (1994) First observations on the root morphology symbioses of 21 major tree species in the primary tropical rain forest of French Guyana. *Annales des Sciences Forestières* 51, 407–416.

Berendse, F. and Aerts, R. (1987) Nitrogen-use-efficiency: a biological meaningful definition? *Functional Ecology* 1, 293–296.

Björkman, O. (1981) Responses to different quantum flux densities. In: Lange O.L., Nobel, P.S., Osmond, C.B. and Ziegler, H. (eds) *Physiological Plant Ecology. I. Responses to the Physical Environment.* Springer, Berlin, pp. 57–107.

Boggan, J., Funk, V., Kelloff, C., Hoff, M., Cremers, G. and Feuillet, C. (1997) *Checklist of the Plants of the Guianas (Guyana, Surinam, French Guiana).* Smithsonian Institution, Washington, DC.

Bonal, D. (2000) Variabilité interspecifique de l'efficience d'utilisation de l'eau en forêt tropicale humide guyanaise. PhD thesis, University of Nancy, France.

Bonal, D., Atger, C., Barigah, T.S., Ferhi, A., Guehl, J.M. and Ferry, B. (2000a) Water acquisition patterns of two wet tropical canopy tree species of French Guiana as inferred from (H_2O)-^{18}O extraction profiles. *Annals of Forest Science* 57, 717–724.

Bonal, D., Barigah, T.S., Granier, A. and Guehl, J.M. (2000b) Late-stage canopy tree species with extremely low delta C^{13} and high stomatal sensitivity to seasonal soil drought in the tropical rain forest of French Guiana. *Plant Cell and Environment* 23, 445–459.

Bonal, D., Sabatier, D., Montpied, P., Tremeaux, D. and Guehl, J.M. (2000c) Interspecific variability of delta ^{13}C among trees in rainforests of French Guiana: functional groups and canopy integration. *Oecologia* 124, 454–468.

Bongers, F. and Popma, J. (1990) Leaf dynamics of seedlings of rain forest species in relation to canopy gaps. *Oecologia* 82, 122–127.

Boot, R. (1993) *Growth and Survival of Tropical Rain Forest Tree Seedlings in Forest Understorey and Gap Openings; Implications for Forest Management.* Tropenbos Documents 6. Tropenbos Foundation, Wageningen, The Netherlands.

Boot, R. (1996) The significance of seedling size and growth rate of tropical rainforest tree seedlings for regeneration in canopy openings. In: Swaine, M.D. (ed.) *The Ecology of Tropical Forest Tree Seedlings.* Parthenon, Paris, pp. 267–284.

Brouwer, L. (1996) *Nutrient Cycling in Pristine and Logged Tropical Rain Forest.* Tropenbos-Guyana Series 1. Tropenbos-Guyana programme, Georgetown, Guyana.

Buchmann, N., Guehl, J.M., Barigah, T.S. and Ehleringer, J.R. (1997) Interseasonal comparison of CO_2 concentrations, isotopic composition, and carbon dynamics in an Amazonian rain forest (French Guiana). *Oecologia* 110, 120–131.

Carswell, F.E., Meir, P., Wandelli, E.V., Bonates, L.C.M., Kruijt, B., Barbosa, E.M., Nobre, A.D., Grace, J. and Jarvis, P.G. (2000) Photosynthetic capacity in a central Amazonian rain forest. *Tree Physiology* 20, 179–186.

Chazdon, R.L. (1988) Sunflecks and their importance to forest understorey plants. *Advances in Ecological Research* 18, 1–63.

Chazdon, R.L. and Field, C.B. (1987) Determinants of photosynthetic capacity in six rainforest *Piper* species. *Oecologia* 73, 222–230.

Condit, R., Hubbell, S.P. and Foster, R.B. (1996) Changes in tree species abundance in a neotropical forest: impact of climate change. *Journal of Tropical Ecology* 12, 231–256.

Coomes, D.A. and Grubb, P.J. (2000) Impacts of root competition in forests and woodlands: a theoretical framework and review of experiments. *Ecological Monographs* 70, 171–207.

Corré, W.J. (1983) Growth and morphogenesis of sun and shade plants. II. The influence of light quality. *Acta Botanica Neerlandica* 32, 185–202.

De Souza Moreira, F.M., Da Silva, M.F. and De Faria, S.M. (1992) Occurrence of nodulation in legume species in the Amazon region of Brazil. *The New Phytologist* 121, 563–570.

Dolman, A.J., Gash, J.H.C., Roberts, J. and Shuttleworth, W.J. (1991) Stomatal and surface conductance of tropical rainforest. *Agricultural and Forest Meteorology* 54, 303–318.

Eckstein, J., Beyschlag, W., Mott, K.A. and Ryel, R.J. (1996) Changes in photon flux can induce stomatal patchiness. *Plant Cell and Environment* 19, 1066–1074.

Evans, J.R. and Poorter, H. (2001) Photosynthetic acclimation of plants to growth irradiance: the relative importance of specific leaf area and nitrogen partitioning in maximizing carbon gain. *Plant Cell and Environment* 24, 755–767.

Farquhar, G.D., O'Leary, M.H. and Berry, J.A. (1982) On the relationship between carbon isotope discrimination and the intercellular carbon dioxide concentration in leaves. *Australian Journal of Plant Physiology* 9, 121–137.

Farquhar, G.D., Ehleringer, J.R. and Hubick, K.T. (1989) Carbon isotope discrimination and photosynthesis. *Annual Review of Plant Physiology* 40, 503–537.

Favrichon, V. (1994) Classification des espèces arborées en groupes fonctionelles en vue de la réalisation d'un modèle de dynamique de peuplement en forêt Guyanese. *Revue de Ecologie – la Terre et la Vie* 49, 379–403.

Fetcher, N., Oberbauer, S.F. and Chazdon, R. (1994) Physiological ecology of plants. In: McDade, L.A., Bawa, K.S., Hespenheide, H.A. and Hartshorn, G.S. (eds) *La Selva Ecology and Natural History of a Neotropical Rain Forest.* University of Chicago Press, Chicago, Illinois, pp. 128–141.

Field, C. and Mooney, H.A. (1986) The photosynthesis–nitrogen relationship in wild plants. In: Givnish, T.J. (ed.) *On the Economy of Plant Form and Function.* Cambridge University Press, Cambridge, UK, pp. 25–55.

Fredeen, A.L. and Field, C.B. (1992) Ammonium and nitrate uptake in gap, generalist and understory species of the genus *Piper. Oecologia* 92, 207–214.

Fredeen, A.L., Griffin, K. and Field, C.B. (1991) Effects of light quantity and quality and soil nitrogen status on nitrate reductase activity in rainforest species of the genus *Piper. Oecologia* 86, 441–446.

Givnish, T.J. (1988) Adaptation to sun and shade: a whole plant perspective. *Australian Journal of Plant Physiology* 15, 63–92.

Grace, J., Malhi, Y., Lloyd, J., McIntyre, J., Miranda, A.C., Meir, P. and Miranda, H.S. (1996) The use of eddy covariance to infer the net carbon dioxide uptake of Brazilian rain forest. *Global Change Biology* 2, 209–217.

Granier, A., Huc, R. and Colin, F. (1992) Transpiration and stomatal conductance of two rain forest species growing in plantations (*Simarouba amara* and *Goupia glabra*) in French Guyana. *Annales des Sciences Forestieres* 49, 17–24.

Granier, A., Huc, R. and Barigah, S.T. (1996) Transpiration of natural rain forest and its dependence on climatic factors. *Agricultural and Forest Meteorology* 78, 19–29.

Guehl, J.M., Domenach, A.M., Bereau, M., Barigah, T.S., Casabianca, H., Ferhi, A. and Garbaye, J. (1998) Functional diversity in an Amazonian rain forest of French Guyana: a dual isotope approach ($d^{15}N$ and $d^{13}C$). *Oecologia* 116, 316–330.

Hammond, D.S., Gourlet-Fleury, S., van de Hout, P., Ter Steege, H. and Brown, V.K. (1996) A compilation of known Guianan timber trees and the significance of their dispersal mode, seed size and taxonomic affinity to tropical rain forest management. *Forest Ecology and Management* 83, 99–116.

Houter, N.C. and Pons, T.L. (2005) Gap size effects on photoinhibition in understorey saplings in tropical rainforest. *Plant Ecology* (in press).

Huc, R. and Guehl, J.M. (1989) Environmental control of CO_2 assimilation rate and leaf conductance in two species of tropical rain forest of French Guiana (*Jacaranda copaia* D. Don and *Eperua falcata* Aubl.). *Annales des Sciences Forestières* 46 suppl., 443–447.

Huc, R., Ferhi, A. and Guehl, J.M. (1994) Pioneer and late stage tropical rainforest tree species (French Guiana) growing under common conditions differ in leaf gas exchange regulation, carbon isotope discrimination and water potential. *Oecologia* 99, 297–305.

Jetten, V.G. (1994) *Modelling the Effects of Logging on the Water Balance of a Tropical Rain Forest.* Tropenbos Series 6. Tropenbos Foundation, Wageningen, The Netherlands.

Jones, H.G. (1985) Partitioning stomatal and non-stomatal limitations to photosynthesis. *Plant Cell and Environment* 8, 95–104.

Jordan, C.F. (1985) *Nutrient Cycling in Tropical Forest Ecosystems.* John Wiley & Sons, Chichester, UK.

Kao, W.Y. and Forseth, I.N. (1992) Diurnal leaf movement, chlorophyll fluorescence and carbon assimilation in soybean grown under different nitrogen and water availabilities. *Plant Cell and Environment* 15, 703–710.

Kitajima, K. (1994) Relative importance of photosynthetic traits and allocation patterns as correlates of seedling shade tolerance of 13 tropical trees. *Oecologia* 98, 419–428.

Lambers, H., Chapin III, F.S. and Pons, T.L. (1998) *Plant Physiological Ecology.* Springer, New York.

Lathwell, D.J. and Grove, T.L. (1986) Soil–plant relationships in the tropics. *Annual Review of Ecology and Systematics* 17, 1–16.

Leavitt, S.W. and Long, A. (1982) Evidence for $^{13}C/^{12}C$ fractionation between tree leaves and wood. *Nature* 298, 742–745.

Lehto, T. and Grace, J. (1994) Carbon balance of tropical tree seedlings: a comparison of two species. *New Phytologist* 127, 455–463.

Leishman, M.R., Wright, I.J., Moles, A.T. and Westoby, M. (2000) The evolutionary ecology of seed size. In: Fenner, M. (ed.) *Seeds: The Ecology of Regeneration in Plant Communities.* CAB International, Wallingford, UK, pp. 31–57.

Lloyd, J., Kruijt, B., Hollinger, D.Y., Grace, J., Francey, R.J., Wong, S.C., Kelliher, M., Miranda, A.C., Farquhar, G.D., Gash, J.H.C., Vygodskaya, N.N., Wright, I.R., Miranda, H.S. and Schultze E.D. (1996) Vegetation effects on the isotopic composition of atmospheric CO_2 at local and regional scales: theoretical aspects and a comparison between rain forest in Amazonia and a Boreal forest in Siberia. *Australian Journal of Plant Physiology* 23, 371–399.

Lovelock, C.E., Jebb, M. and Osmond, C.B. (1994) Photoinhibition and recovery in tropical plant species: response to disturbance. *Oecologia* 97, 297–307.

Lüttge, U. (1997) *Physiological Ecology of Tropical Plants.* Springer, Berlin.

Lüttge, U. (1999) One morphotype, three physiotypes: sympatric species of *Clusia* with obligate C_3 photosynthesis, obligate CAM and C_3-CAM intermediate behaviour. *Plant Biology* 1, 138–148.

Malhi, Y. and Grace, J. (2000) Tropical forests and atmospheric carbon dioxide. *Trends in Ecology and Evolution* 15, 332–337.

Marschner, H. (1995) *Mineral Nutrition of Higher Plants.* Academic Press, London.

Medina, E. (1984) Nutrient balance and physiological processes at the leaf level. In: Medina, E., Mooney, H.A. and Vazquez-Yanes, C. (eds) *Physiological Ecology of Plants in the Wet Tropics.* Junk Publishers, The Hague, pp. 139–154.

Medina, E. (1996) CAM and C_4 plants in the humid tropics. In: Mulkey, S.S., Chazdon, R.L. and Smith, A.P. (eds) *Tropical Forest Plant Ecophysiology.* Chapman and Hall, New York, pp. 56–88.

Mulkey, S.S. and Pearcy, R.W. (1992) Interactions between acclimation and photoinhibition of photosynthesis of a tropical forest understorey herb, *Alocasia macrorrhiza,* during simulated canopy gap formation. *Functional Ecology* 6, 719–729.

Mulkey, S.S., Chazdon, R.L. and Smith, A.P. (1996) *Tropical Forest Plant Ecophysiology.* Chapman and Hall, New York.

Murchie, E.H. and Horton, P. (1997) Acclimation of photosynthesis to irradiance and spectral quality in British plant-species – chlorophyll content, photosynthetic capacity and habitat preference. *Plant Cell and Environment* 20, 438–448.

Niinemets, U. (1997) Distribution patterns of foliar carbon and nitrogen as affected by tree dimensions and relative light conditions in the canopy of *Picea abies. Trees – Structure and Function* 11, 144–154.

Norris, D.O. (1969) Observations on the nodulation status of rainforest leguminous species in Amazonia and Guyana. *Tropical Agriculture, Trinidad* 46, 145–151.

Pate, J. and Arthur, D. (1998) Delta ^{13}C analysis of phloem sap carbon: novel means of evaluating seasonal water stress and interpreting carbon isotope signatures of foliage and trunk wood of *Eucalyptus globulus. Oecologia* 117, 301–311.

Pearcy, R.W. (1990) Sunflecks and photosynthesis in plant canopies. *Annual Review of Plant Physiology* 41, 421–453.

Pearcy, R.W. and Yang, W. (1996) A three dimensional crown architecture model for assessment of light capture and carbon gain by understorey plants. *Oecologia* 108, 1–12.

Perreijn, K. (2002) *Symbiotic Nitrogen Fixation by Leguminous Trees in Tropical Rain Forest in Guyana.* Tropenbos-Guyana Series 11. Tropenbos-Guyana programme, Georgetown, Guyana.

Poels, R.L.H. (1987) Soils, water and nutrients in a forest ecosystem in Suriname. PhD thesis, Wageningen University, Wageningen, The Netherlands.

Pons, T.L. and Welschen, R.A.M. (2003) Midday depression of net photosynthesis in the tropical rainforest tree *Eperua grandiflora*: contributions of stomatal and internal conductances, respiration and *Rubisco* functioning. *Tree Physiology* 23, 937–947.

Pons, T.L., Schieving, F., Hirose, T. and Werger, M.J.A. (1989) Optimization of leaf nitrogen allocation for canopy photosynthesis in *Lysimachia vulgaris.* In: Lambers, H., Cambridge, M.L., Konings, H. and Pons, T.L. (eds) *Causes and Consequences of Variation in Growth Rate and Productivity of Higher Plants.* SPB Academic Publishing, The Hague, pp. 175–186.

Pons, T.L., Van der Werf, A. and Lambers, H. (1994) Photosynthetic nitrogen use efficiency of inherently slow- and fast-growing species: possible explanations for observed differences. In: Roy, J. and Garnier, E. (eds) *A Whole Plant Perspective on Carbon–Nitrogen Interactions.* SPB Academic Publishing, The Hague, pp. 61–77.

Poorter, L. and Oberbauer, S.F. (1993) Photosynthetic induction responses of two rain forest tree species in relation to light environment. *Oecologia* 96, 193–199.

Poorter, H. and Van der Werf, A. (1998) Is inherent variation in RGR determined by LAR at low irradiance and by NAR at high irradiance? A review of herbaceous species. In: Lambers, H., Poorter, H. and van Vuuren, M.M.I. (eds) *Inherent Variation in Plant Growth: Physiological Mechanisms and Ecological Consequences.* Backhuys Publishers, Leiden, Belgium, pp. 309–336.

Poorter, L., Oberbauer, S.F. and Clark, D.B. (1995) Leaf optical properties along a vertical gradient in a tropical rain forest canopy in Costa Rica. *American Journal of Botany* 82, 1257–1263.

Raaimakers, D. (1995) *Growth of Tropical Rain Forest Trees as Dependent on Phosphorus Supply*. Tropenbos Series 11. Tropenbos Foundation, Wageningen.

Raaimakers, D. and Lambers, H. (1996) Response to phosphorus supply of tropical tree seedlings: a comparison between a pioneer species *Tapirira obtusa* and a climax species *Lecythis corrugata*. *New Phytologist* 132, 97–102.

Raaimakers, D., Boot, R.G.A., Dijkstra, P., Pot, S. and Pons, T.L. (1995) Photosynthetic rates in relation to leaf phosphorus content in pioneer versus climax tropical rainforest trees. *Oecologia* 102, 120–125.

Reich, P.B., Uhl, C., Walters, M.B. and Ellsworth, D.S. (1991) Leaf life span as a determinant of leaf structure and function among 23 amazonian tree species. *Oecologia* 86, 16–24.

Reich, P.B., Ellsworth, D.S. and Uhl, C. (1995) Leaf carbon and nutrient assimilation and conservation in species of differing successional status in an oligotrophic Amazonian forest. *Functional Ecology* 9, 65–76.

Rijkers, T. (2000) Leaf function in tropical rain forest canopy trees. PhD thesis, Wageningen University, Wageningen, The Netherlands.

Rijkers, T., Pons, T.L. and Bongers, F. (2000a) The effect of tree height and light availability on photosynthetic leaf traits of 4 neotropical species differing in shade tolerance. *Functional Ecology* 14, 77–86.

Rijkers, T., de Vries, P.J., Pons, T.L. and Bongers, F. (2000b) Photosynthetic induction in saplings of three shade-tolerant tree species: comparing understorey and gap habitats in French Guiana rain forest. *Oecologia* 125, 331–340.

Roggy, J.C. and Prévost, M.F. (1999) Nitrogen fixing legumes and silvigenesis in a rain forest in French Guiana: a taxonomic and ecological approach. *The New Phytologist* 144, 283–294.

Roggy, J.C., Prévost, M.F., Garbaye, J. and Domenach, A.M. (1999a) Nitrogen cycling in the tropical rain forest of French Guiana: comparison of two sites with contrasting soil types using $d^{15}N$. *Journal of Tropical Ecology* 15, 1–22.

Roggy, J.C., Prévost, M.F., Gourbiere, F., Casabianca, H., Garbaye, J. and Domenach, A.M. (1999b) Leaf natural ^{15}N abundance and total N concentration as potential indicators of plant N nutrition in legumes and pioneer species in a rain forest of French Guiana. *Oecologia* 120, 171–182.

Rose, S.A. (2000) *Seeds, Seedlings and Gaps – Size Matters*. Tropenbos-Guyana Series 9. Tropenbos-Guyana programme, Georgetown, Guyana.

Rose, S.A. and Poorter, L. (2003) The importance of seed mass for early regeneration in tropical forest: a review. In: ter Steege, H. (ed.) *Long-term Changes in Tropical Tree Diversity: Studies from the Guiana Shield, Africa, Borneo and Melanesia*. Tropenbos Series Vol. 2. Tropenbos International, Wageningen, The Netherlands, pp. 19–35.

Roy, J. and Salager, J.L. (1992) Midday depression of net CO_2 exchange of leaves of an emergent forest tree in French Guiana. *Journal of Tropical Ecology* 8, 499–504.

Ruinen, J. (1965) The phyllosphere. III. Nitrogen fixation in the phyllosphere. *Plant and Soil* 22, 375–394.

Schmidt, H.L. and Gleixner, G. (1998) Carbon isotope effects on key reactions in plant metabolism and ^{13}C-patterns in natural compounds. In: Griffiths H. (ed.) *Stable Isotopes, Integration of Biological, Ecological and Geochemical Processes*. Bios Scientific Publishers, Oxford, UK, pp. 13–25.

Shearer, G. and Kohl, D.H. (1986) N_2-fixation in field settings: estimations based on natural ^{15}N abundance. *Australian Journal of Plant Physiology* 13, 699–756.

Smith, C.C. and Fretwell, S.D. (1974) The optimal balance between size and number of offspring. *American Naturalist* 108, 499–506

Swaine, M.D. and Whitmore, T.C. (1988) On the definition of ecological species groups in tropical rain forests. *Vegetatio* 75, 81–86.

ter Steege, H. (1993) *Patterns in Tropical Rain Forest in Guyana*. Tropenbos Series 3. Tropenbos Foundation, Wageningen, The Netherlands.

ter Steege, H. (1994a) Flooding and drought tolerance in seeds and seedlings of two *Mora* species segregated along a soil hydrological gradient in the tropical rain forest in Guyana. *Oecologia* 100, 356–367.

ter Steege, H. (1994b) Seedling growth of *Mora gonggrijpii*, a large seeded climax species, under different soil and light conditions. *Vegetatio* 112, 161–170.

ter Steege, H., Bokdam, C., Boland, M., Dobbelsteen, J. and Verburg, I. (1994) The effects of man made gaps on germination, early survival and morphology of *Chlorocardium rodiei* seedlings in Guyana. *Journal of Tropical Ecology* 10, 245–260.

Tian, H.Q., Melillo, J.M., Kicklighter, D.W., McGuire, A.D., Helfrich, J.V.K., Moore, B. and Vorosmarty, C.J.

(1998) Effect of interannual climate variability on carbon storage in Amazonian ecosystems. *Nature* 396, 664–667.

Tinoco-Ojanguren, C. and Pearcy, R.W. (1993) Stomatal dynamics and its importance to carbon gain in two rainforest *Piper* species. I. VPD effects on the transient stomatal response to lightflecks. *Oecologia* 94, 388–394.

Valladares, F., Allen, M.T. and Pearcy, R.W. (1997) Photosynthetic responses to dynamic light under field conditions in 6 tropical rain forest shrubs occurring along a light gradient. *Oecologia* 111, 505–514.

van Kekem, A.J., Pulles, J.H.M. and Khan, Z. (1996) *Soils of the Rainforest in Central Guyana.* Tropenbos-Guyana Series 2. Tropenbos-Guyana programme, Georgetown, Guyana.

Veneklaas, E.J. and Poorter, L. (1998) Growth and carbon partitioning of tropical tree seedlings in contrasting light environments. In: Lambers, H., Poorter, H. and van Vuuren, M.M.I. (eds) *Inherent Variation in Plant Growth: Physiological Mechanisms and Ecological Consequences.* Backhuys Publishers, Leiden, The Netherlands, pp. 337–361.

Vitousek, P.M. (1984) Litterfall, nutrient cycling and nutrient limitation in tropical forests. *Ecology* 65, 285–298.

Vitousek, P.M. and Sanford, R.L. (1986) Nutrient cycling in moist tropical forest. *Annual Review of Ecology and Systematics* 17, 137–167.

Winter, K. and Smith, J.A.C. (1996) *Crassulacean Acid Metabolism: Biochemistry, Ecophysiology and Evolution.* Springer, Berlin.

Wullschleger, S.D., Meinzer, F.C. and Vertessy, R.A. (1998) A review of whole plant water use studies in trees. *Tree Physiology* 18, 499–512.

Zotz, G. and Winter, K. (1993) Annual carbon balance and nitrogen-use efficiency in tropical C_3 and CAM epiphytes. *The New Phytologist* 126, 481–492.

Zotz, G. and Winter, K. (1996) Diel patterns of CO_2 exchange in rainforest canopy plants. In: Mulkey, S.S., Chazdon, R.L. and Smith, A.P. (eds) *Tropical Forest Plant Ecophysiology.* Chapman and Hall, New York, pp. 89–113.

4 Rainforest Vertebrates and Food Plant Diversity in the Guiana Shield

Pierre-Michel Forget[1] and David S. Hammond[2]

[1]*Département Ecologie et Gestion de la Biodiversité, Museum National d'Histoire Naturelle, Brunoy, France;* [2]*Iwokrama International Centre for Rain Forest Conservation and Development, Georgetown, Guyana. Currently: NWFS Consulting, Beaverton, Oregon, USA*

Introduction

In the tropics, many animals are a fundamental component of plant reproductive biology. Animals underpin plant fecundity as a pollinator while feeding on nectar, increase their spatial distribution and the probability of survival of their offspring as a seed disperser, and regulate the regeneration of their sympatric competitors as a seed and seedling predator (Janzen, 1970, 1971; Howe and Smallwood, 1982; Wheelwright and Orians, 1982). Conversely, plants produce the leaves, roots, sap, flowers, fruit, seeds, wood and bark that provide vital resources to animal reproduction and survival, including carnivorous and insectivorous species that rely indirectly upon plants via their primary consumers. Apart from their fundamental role as food, plants also provide animals with shelter and support when foraging or resting. Together, the myriad evolutionary forces which combine to define relationships between plants and animals play a significant role in shaping the diversity of tropical forests (Terborgh, 1988, 1992). For the last 30 years, the bulk of studies dealing with plant–animal relationships in the tropical forests of the Guianas has mirrored studies at other tropical research sites, focusing on frugivory as a mutualistic interaction (Fleming, 1991). These studies have illustrated the wide variety of feeding habits and diets among flying, arboreal and terrestrial vertebrates (reviewed by Howe, 1986).

Howe and Westley (1988) later separated the ecological relationships of plants and animals into two types, viz. herbivory and mutualism, though in reality both overlap. For instance, large neotropical scatter-hoarding rodents act as herbivores when they consume seeds, but dispersers when the cached seeds are forgotten, and this deposition enhances germination and seedling establishment (Forget, 1990, 1994). In contrast to Howe and Westley's (1988) definition, mutualistic associations between plants and animals are here considered as being included in the process of herbivory. Herbivory is broadly defined here as the consumption of plant material such as leaves, fruit, seed, flowers and bark, and may clearly be advantageous to both animals and plants under certain conditions (Boucher *et al.*, 1982; see also Boucher, 1985; Lewis, 1985). However, where it is clear that most animals behave as herbivores at one time or another during their life, it is not certain if any mutual benefit is derived from this action. For example, the ultimate consequences of fruit ingestion and seed dispersal are often uncertain (Howe and Smallwood, 1982) and thus characterizing plant–frugivore relationships as mutualistic may often prove premature.

Though the majority of Guianan vertebrates are classified as fruit-eating animals (Dubost, 1987), their diets are not restricted to fruit year-round, as a period of low availability associated with a seasonal decline in rainfall forces animals to shift to alternative resources such as leaves, flowers, insects and small vertebrates (Primates: van Roosmalen, 1985; Guillotin *et al.*, 1994; Simmen and Sabatier, 1996. Marsupials: Julien-Laferrière and Atramentowicz, 1990. Ground-dwelling vertebrates: Guillotin, 1982a, 1982b; Henry, 1994a. Birds: Erard *et al.*, 1991).

In the next section, we introduce the reader to the Guianan vertebrate community, but with an emphasis on mammals and their specialized diets in relation to the forest types they are most likely to inhabit. In subsequent sections we review the current knowledge of the diversity of plants and plant parts that are incorporated in the diet of vertebrate groups known to actively feed on plants during some period in their lifetime. We then show how many Guianan vertebrate herbivores play a significant ecological role in pollination, seed dispersal and seed–seedling predation. After analysing the diversity of plant families that are included in the diets of Guianan animal species or groups, a special interest is devoted to the palm family, a keystone plant group in the Guianan forests. The last section aims at underlining the importance of herbivores to the maintenance of plant diversity, and the conservation of managed tropical rainforest in the Guiana Shield.

General Characteristics of the Vertebrate Fauna in the Guianas

Taxonomic richness[1]

The faunal diversity of the Guianan region is generally considered to be relatively low when compared to other neotropical sites, particularly those mega-diverse sites along the eastern foot slopes of the Andes in western Amazonia (Voss and Emmons, 1996; Kay *et al.*, 1997). While work by taxonomists and ecologists in the Guianas generally supports this notion, several vertebrate groups have recently been more thoroughly examined and appear to rival, if not exceed, counts made from other sites. Several of these groups are important plant-feeders and will be discussed more thoroughly later in the chapter. Table 4.1 shows the number of vertebrate species, by class, known to occur in each of the three Guianan countries, the result of a long record of zoological investigations. Each country presently houses over 1000 vertebrate species. These numbers are certain to increase. Several individuals sighted in the field are still awaiting identification, including several small mammals from Suriname (Charles-Dominique, 1993a). Many arboreal vertebrates have yet to be sighted at all due to the traditional difficulties in accessing large areas of the forest canopy (Voss and

Table 4.1. Number of described vertebrates known to occur in each of the three Guianan countries.

Class	Guyana	Suriname	French Guiana	Guianas
Mammals	221	180	191	234
Birds	752	672	699	812
Fish, skates and eels[a]	420	318	334	≅440
Amphibians	81	94	72	118
Reptiles[a]	140	143	162	206
Total	1426	1365	1416	≅1765

[a]Freshwater only.

Table 4.2. Number of described mammals known to occur in each of the three Guianan countries.

Order	Guyana	Suriname	French Guiana
Marsupialia	15	11	12
Xenarthra	10	10	10
Chiroptera	126	105	109[a]
Primates	8	8	8
Carnivores	16	15	15
Perissodactyla	1	1	1
Artiodactyla	5	5	5
Rodentia	40	25	31
Total	221	180	191

[a]Including a new species and name revisions in Simmon and Voss (1998) and Voss *et al.* (2001).

Emmons, 1996). For instance, a newly discovered nocturnal, arboreal rodent (*Isothrix surinamensis*, Echimyidae) was captured during the flooding of the Petit-Saut Dam when tree crowns were easily surveyed (Vié *et al.*, 1996).

Based on a considerable catalogue of information,[2] it is possible to present a list of all mammals (excluding Pinnipedia, Sirenia and introduced species) that are known to occur in Guianan rainforests (Appendix 4.1). Overall, the number of species occurring per order in each of the three countries is similar (Table 4.2).

With more than 140 species identified to date, bats stand out as being a particularly important mammalian order in the Guianan Shield (Table 4.3), accounting for 55–60% of all known species in the neotropics. The bat communities in Guianan forests are as diverse (Brosset and Charles-Dominique, 1990; Smith and Kerry, 1996; Simmons and Voss, 1998; Simmons *et al.*, 2000; Lim *et al.*, 1999; Lim and Engstrom, 2001a,b; see also Kalko and Handley, 2001; Bernard and Fenton, 2002) as those described from surveys carried out at other neotropical sites (Simmons and Voss, 1998). Over 86 species have been recorded to date from several, small collection centres in the Iwokrama Rain Forest area in central Guyana (Smith and Kerry, 1996; Lim and Engstrom, 2000, 2001a,b). Using mist nets as well as a systematic search for roosts, a bat inventory by Simmons and Voss (1998) in the mature forest at Paracou, French Guiana, recorded at least 78 species (62% in Phyllostomidae), among them a new species, *Micronycteris brosseti*. The bat community at Paracou is estimated to be composed of about 85–95 species (Simmons and Voss, 1998), and 100–106 species at Iwokrama (B. Lim, personal communication; see also Lim and Engstrom, 2001b), based on statistical extrapolation (rarefaction). Forty-seven per cent of bats at Paracou have a pantropical distribution, most of them being known from habitats other than the mature rainforest (Simmons and Voss, 1998). For comparison, using the same techniques, an inventory on Barro Colorado Island, Panama, recorded 66 species (Kalko *et al.*, 1996), and 70 species (expected 80 species) on an adjacent mainland (Pipeline road) site (E. Kalko, personal communication). Using new acoustic identification, Kalko (1997) identified two new species for Barro Colorado Island (BCI), and a similar result is expected for Paracou, French Guiana (see Simmon and Voss, 1998). In comparison, up to 65 species have been recorded as occurring at La Selva, Costa Rica, using the same capture techniques (Timm, 1994).

In comparing vertebrate communities of the three main Guianan countries there is a need to emphasize natural barriers which restrict a species to a particular region, such as the Demerara, Essequibo, Berbice, Courantyne, Taranchany, Maroni and Oyapock Rivers (e.g. Hoogmoed, 1979), forest–savanna boundaries in southern

Table 4.3. Number of mammalian species recorded and expected (in brackets) at six Guianan rain forest sites, with area of site and no. of years in which collections or sightings have been made.

Order	Imataca (Venezuela) (3.2×10^6 ha) 3 years	Iwokrama (Guyana) (3.6×10^5 ha) 4 years	Kartabo (Guyana) (4.3×10^3 ha) 5 years	Bronwsberg Nature Park (Suriname) (8.4×10^3 ha) 2 years	Nouragues-Arataye (French Guiana) (1×10^3 ha) 17 years	Paracou (French Guiana) (1×10^3 ha) 10 years
Marsupialia	9	7 (10)	7 (8)	6	11	12
Xenarthra	9	9	9	6	9	9
Chiroptera	78	83 (106)	12 (75)[b]	57	69	78 (85-95)
Primates	3	5 (7)	6	8	7	6[a]
Carnivora	16	13 (14)	13	8	13	10
Perissodactyla	1	1	1	1	1	1
Artiodactyla	5	4	4	4	4	4
Rodentia	22	15 (27)	20 (22)	22	24	21
Total	143	137 (178)	72 (128)	112	138	141 (148-158)

[a]Black spider monkey *Ateles paniscus* and brown capuchin *Cebus apella* are nearly extinct (R. Voss, personal communication; P.-M. Forget, personal observation).
[b]Disproportionately low sampling effort was expended on bats relative to other mammals.

Guyana and Suriname or even changes in forest habitat types, rather than artificial boundaries. For example, Eisenberg's (1989) distribution maps often suggest geopolitical limits to the geographic ranges of mammals. Obviously, the real extent and diversity of each forest habitat, especially in regard to those plant resources utilized by vertebrates, can strongly influence patterns of regional vertebrate diversity in the Guianas. Voss and Emmons (1996) analysed the diversity of the mammalian fauna recorded at Arataye, French Guiana, and Kartabo, Guyana (Table 4.3), and compared them to other Amazonian sites. They showed that sites in similar biomes have comparable mammalian assemblages despite being in different countries (Appendix 4.1). However, the mammalian fauna in the Guianas is composed of far fewer species than like assemblages found in other regions of lowland forest in Amazonia (Voss and Emmons, 1996). The number of mammal species, notably primates, in Venezuelan Guayana (Ochoa, 1995, 2000), the Iwokrama Forest (Engstrom and Lim, personal communication; see Lim *et al.*, 2002), the Brownsberg Nature Park (Fitzgerald *et al.*, 2002; Fitzgerald, 2003) and the Nouragues reserve (Bongers *et al.*, 2001; Feer and Charles-Dominique, 2001) are also generally fewer than elsewhere in the neotropics, though a number of specific groups are more speciose in the Guianas than at other neotropical sites (Table 4.4). With an average rainfall for the region of more than 2000 mm per year (see Chapter 2), the lower diversity of primates recorded in the Guiana Shield is more likely related to lower productivity than to the relatively low tree diversity found in some parts of the shield area compared to other Amazonian sites, where higher productivity may be related to soil nutrient richness and available energy in the ecosystem (Kay *et al.*, 1997; Lehman, 2000).

Ecological characteristics

Up until the 1970s there was a strong emphasis on mammal identification, classification, geographical distributions and reproductive/mating behaviour in the Guianas, with a few anecdotal observations of arboreal mammal feeding behaviour and foraging preferences (Beebe, 1917; Tate, 1939; Muckenhirn *et al.*, 1976; Husson, 1978). Since the 1980s, however, substan-

Table 4.4. Diet distribution among 232 (out of 234) mammalian species for which diet is known (see Appendix 4.2).

Order	Carnivores	Herbivores	Insectivores
Artiodactyla	0	5	0
Carnivores	10	6	0
Chiroptera	8	46	85
Marsupiala	2	4	9
Perissodactyla	0	1	0
Primates	0	8	0
Rodentia	0	36	2
Xenarthra	0	2	8
Total	20 (8.6%)	108 (46.5%)	104 (44.8%)

tial effort has been directed towards characterizing which plants are eaten by which animals, the effect this has on plant regeneration, distribution and diversity and how fluctuations in the availability of plant resources influence the direct and indirect benefits received by both plant and animals.

According to our review of mammalian species for which diet is known (N=232 out of 234), nearly 46% of animals are herbivores, including nectarivores and folivores (Table 4.4). Based on the type of food items consumed, vertebrates eating plant parts to meet at least part of their daily nutritional requirement may be crudely categorized as being: (i) completely or (ii) partially frugivorous (including those that eat seeds, i.e. granivores), the latter category consisting of feeding habits which are mainly: (i) frugivorous–insectivorous, (ii) frugivorous–carnivorous, (iii) frugivorous–folivorous or (iv) frugivorous–nectarivorous (Dubost, 1987; see also Robinson and Redford, 1986).

Overall, 44% of the mammals found in the Guiana Shield principally feed on plants. This rises to nearly 66% when considering mammals other than bats. Nearly 29% of bats found in the Guianas have been observed to feed on plants. Among these primary consumers, one may note that frugivores are well-represented, accounting for 38% of mammalian species (N=79 of 209) which, added to the species of frugivorous birds (e.g. 72 out of 575 species in French Guiana (Charles-Dominique, 1986)) compose a guild of animals capable of significantly influencing forest regeneration (see Charles-Dominique, 1995). It is important to note, however, that not all herbivorous vertebrates have a restricted diet in Guianan forests. For instance, although large rodents such as the red-rumped agouti (*Dasyprocta leporina*) and the red acouchy (*Myoprocta exilis*) may be classified as frugivores, they may also eat leaves, flowers, insects and other animal matter (e.g. Henry, 1994a, 1996).

The same statement holds true for birds (Erard *et al.*, 1991), bats (Emmons and Feer, 1990), monkeys (van Roosmalen, 1985; Guillotin *et al.*, 1994; Julliot and Sabatier, 1993; Pack *et al.*, 1998) and large terrestrial ungulates (Henry, 1994a), of which few are exclusively frugivorous. Dubost (1987) suggested that terrestrial species are more likely to be folivorous, feeding during both day and night, while arboreal species tend to be diurnal and frugivorous, and flying species (mainly bats) are typically nocturnal, insectivorous or insectivorous–carnivorous. Feeding throughout both day and night is thought to be necessary for strict folivores to ingest a large volume of food of relatively poor nutritional value. On the other hand, feeding purely on nutrient-rich fruits and seeds does not really occur (e.g. Guillotin *et al.*, 1994). Even the most frugivorous species in the Guiana Shield, the black spider monkey (*Ateles paniscus paniscus*), consumes leaves and flowers in the dry season (van Roosmalen, 1985). Frugivores are better viewed as vertebrates whose diets consist mainly of seeds and fruits, in terms of the total volume and energy consumed per year, but are not restricted to these items.

Arboreal vertebrates are unique among forest macrofauna in that they have direct (indirectly through insects) access to the massive primary productive capacity of the tropical forest canopy. Floral parts, nectar, pollen, ripe and unripe fruit and seeds, young foliage, bark, sap and a multitude of insects are available to arboreal vertebrates. These same resources are, in principle, never or only secondarily available to sympatric vertebrate species restricted to the forest floor, though on occasion a plant's defences may prevent initial feeding by arboreal vertebrates. Particularly important among canopy-inhabiting animals are the mammals, which of all forest vertebrates are responsible for consuming the overwhelming majority of plant biomass. The community of arboreal, non-volant mammals inhabiting Guianan forests consists of 21 species, 19 of these feeding on plants wholly or in part. This group represents nearly 10% of all mammal species expected to occur in the Guiana Shield. An additional four plant-eating mammals could be considered scansorial, foraging consistently on the forest floor and in the trees, while another three species are primarily terrestrial, but may occasionally forage above the forest floor for fruit, flowers, foliage and bark.

Forest habitat and mammalian diet diversity

Guianan forest tree diversity is highly dependent on soil type, drainage class and topographical features, which ultimately influence the landscape-scale spatial distribution of species and the overall representation of different plant families in the community (Lescure and Boulet, 1985; ter Steege *et al.*, 1993; Sabatier *et al.*, 1997; ter Steege, 1998; ter Steege and Zondervan, 2000). At a small spatial scale in the Guiana Shield, occurrence of soil with good drainage, often on slopes, hilltops and ridges, alternates with conditions where soil drainage is delayed, blocked (downslope) or absent, leading to hydromorphic soils alongside creeks and rivers (Lescure and Boulet, 1985). Small-scale, local soil conditions change abruptly. A distance of several hundred metres can often cover a mosaic of forest types consisting of highly mixed and diversified assemblages through to forests with strong monodominance. This patchwork of soil conditions thus strongly influences the distribution of food plants at a smaller scale. High tree-species richness (approx. >100–150 species per ha) (Sabatier *et al.*, 1997) is often encountered at the mid-point along the soil drainage gradient on well-drained, sandy and lateritic Ferrasols. Richness declines in both directions from this mid-point (Richards, 1996; ter Steege *et al.*, 1993). Excessively well-drained, pure sand Psamments are characterized by several dominating species, such as occur in Wallaba forest (*Eperua grandiflora*, *E. falcata*). Soils that are coincident with strong monodominance in the canopy, such as *Mora excelsa* (Caesalpiniaceae) or *Mauritia flexuosa*, tend to occur on poorly drained floodplain (Fluvisolic) and swamp (Histisolic) soils (Rankin, 1978; Richards, 1996) (Table 4.5).

An important finding of Dubost (1987) is that upland forest at a site in northern French Guiana had a richer faunal community than that close to the edge of a river. The abundance of several large-seeded gravity, water or rodent-dispersed species throughout the Guianas (*Alexa imperatricis*, *Dicymbe* spp., *Elisabetha coccinea*, *Eperua* spp., *Mora* spp., *Vouacapoua* spp., Caesalpiniaceae; *Carapa* spp. Meliaceae; *Catostemma commune*, Bombacaceae; *Chlorocardium rodiei*, Lauraceae), and the trend for these taxa to dominate the stand at the local scale in many areas (Davis and Richards, 1933, 1934; Hammond and Brown, 1995; Richards, 1996; see Chapter 7, this volume) lowers the likelihood that species which produce fleshy fruits will establish in the stand, due to infrequent visitation by arboreal or flying frugivores and the strong competitive edge local dominants have achieved. Consequently, the few vertebrates which spend most, if not all, of their time in habitats adjoining rivers, lakes

Table 4.5. Description of the three tree stands found in the Guianan forests with dominant mammalian guild and diet.

	Forest type		
	Monodominant stand[a]	Ecotone[b]	Mixed forest[c]
Habitat type	Flooded forest nearby river, lakes, permanent creek	Floodplain and footslope	Upper slope and ridge
Dominant plant families dbh > 30 cm[d]	Caesalpiniaceae Fabaceae Mimosaceae	Caesalpiniaceae (30.7%) Lecythidaceae (19%) Chrysobalanaceae (12.1%) Sapotaceae (10.8%)	Burseraceae (15.8%) Lecythidaceae (12%) Moraceae (8.8%) Sapotaceae (8.4%)
Dominant plant genera or species[d]	*Mora, Brownea, Inga Pentaclethra macroloba, Euterpe, Eperua*	*Eperua, Eschweilera, Licania, Chlorocardium rodiei, Dicymbe altsonii*	*Protium, Tetragastris, Eschweilera, Couratari, Ficus, Brosimum, Pouteria, Micropholis*
Dominant vertebrate guild[e]	Terrestrial, nocturnal		Arboreal, diurnal
Dominant diet[e]	Folivores, Insectivores–Carnivores		Frugivores

[a]Rankin (1978), see also Richards (1996)
[b]Lescure and Boulet (1985), Sabatier *et al.* (1997)
[c]Rankin (1978), after Lescure (1981), Sabatier and Prévost (1990), Sabatier *et al.* (1997)
[d]Rankin (1978), Lescure (1981), Maury-Lechon and Poncy (1986), ter Steege *et al.* (1993)
[e]After Dubost (1987).

and permanent creeks are predominantly terrestrial, folivorous and insectivorous–carnivorous (Dubost, 1987). Arboreal frugivores are more likely to reside outside of swamp or riparian habitats which, in Guianan forests, are represented by upland sandy plateaus and laterite-covered, doleritic domes and ridges. Indeed, upland high forest is especially well-stocked with nutritionally superior resources for animal such as the spider monkey/toucan-dispersed *Virola* spp. (Myristicaceae) (Sabatier, 1983; van Roosmalen, 1985; Forget, 1991a; Forget and Sabatier, 1997) which preferentially grow on upper slopes where soils are often less than 1.5 m deep (Sabatier *et al.*, 1997). In contrast to Dubost's findings in French Guiana, however, data from Peres (1994b) on 13 primate species suggest that several highly frugivorous species (e.g. *Ateles paniscus chamek*) utilized only seasonally inundated igapó and/or creekside forest, despite these habitats only covering a minor part of the study area. It was suggested that igapó forest was rarely under seasonal water stress and thus was able to maintain a relatively steady production of fruit relative to upland *terra firme* habitat. Interestingly, igapó forests do not occur along most lower sections of Guianan rivers, the riverside vegetation flooding sporadically but rarely affording the same consistency in fruitfall as suggested by Peres (1994b) for his central Amazonian site. This pattern is not entirely consistent across the Guiana Shield, however. Nearer to their southern headwaters south of the 4th parallel, localized riverside forests in the Guiana Shield are more consistent with the Amazon (see Chapter 7). These forests are less structured, occur alongside narrower river channels, and are more typical of stands responding to stronger meandering action and dominated by species with fleshy fruits, such as *Inga* (Leguminosae), *Astrocaryum* (Palmae), *Virola*, *Mouriri* and *Bellucia* (Melastomataceae).

Home range size of herbivorous animals is strongly influenced by the diversity of local *in situ* plant resources (fruit, flower, nectar, seeds, resins) combined with the quantity and type of seeds imported to an area by primary dispersal agents. Spider monkeys and howler monkeys consume leaves and/or flowers of *Eperua falcata* (Caesalpiniaceae) in the dry season, but the bulk of their diet is made of myristicaceous and sapotaceous tree species, respectively (van Roosmalen, 1985; Julliot, 1993), two families that are most often found in high forest on well-drained soils. Acouchies have small home ranges (about 1 ha) and prefer forest near unflooded riverbanks, whereas agoutis have larger home ranges, including forest at both inundated and well-drained sites (Dubost, 1988). Though they consume many species with seeds that simply fall and accumulate on the ground (Meliaceae, Caesalpiniaceae, Chrysobalanaceae, Lauraceae; Forget, 1990, 1996; Hammond *et al.*, 1992), both rodent species also rely upon an influx of seeds of many other plant families that are not found within their home ranges, but are dropped in transit or beneath perches by bats (e.g. Anacardiaceae, Caryocaraceae, Chrysobalanaceae, Lecythidaceae), birds (e.g. Burseraceae) or monkeys (e.g. Sapotaceae).

Bats as Herbivores

Diversity and dietary distribution

More than 137 bat species have been catalogued for the entire Guianan Shield region (see Appendix 4.2). The majority of these are insectivores (61.3%), while plant-eating bats account for nearly 32% (N=35) of the total. Fifty-four percent (76 out of 137) of the Guianan bat species belong to the family Phyllostomidae, which includes the fruit (N=37) and nectar-eating specialists (N=13) in the Guianas (Table 4.6, see also Quelch, 1892; Greenhall, 1959; Husson, 1962; Brosset and Charles-Dominique, 1990; Cosson, 1994; Smith and Kerry, 1996; Brosset *et al.*, 1996; Kalko, 1997; Simmons and Voss, 1998; Simmons *et al.*, 2000; Kalko and Handley, 2001; Lim and Engstrom, 2001a,b; Bernard and Fenton, 2002). Of the 109 bat species recorded so far from French Guiana (see Appendix 4.2), 75 and 61 of these were found in primary and secondary forests, respectively (Brosset *et al.*, 1996; see also Brosset *et al.*, 1995), most of these feeding on insects or fruit (Charles-Dominique, 1993a). For instance, in the mature forest of Paracou, at least 74 bat species were captured, including 17 frugivores and five nectarivores (Simmons and Voss, 1998).

Little and large fruit-eating bats

The abundance of fruit-eating bat species is directly attributable to the occurrence of fruit trees on which they prefer to feed, this frequency being a consequence of the level of disturbance to the forest stand. For example, Brosset *et al.* (1996) found that the population growth achieved by some frugivorous bats (e.g. the short-tailed fruit

Table 4.6. Number of bat species according to their principal dietary item in the Guianas. Some species consume more than one dietary item.

Dietary item	Guyana[a]	Suriname[b]	French Guiana[c]	All Guianas[d]
Fruit	37	26	23	31
Nectar	11	14	7	13
Insects	60	66	53	84
Vertebrates	1	6	4	6
Blood	2	1	2	2

[a]After Greenhall (1959), Emmons and Feer (1990), Smith and Kerry (1996)
[b]After Husson (1962), Genoways and Williams (1979)
[c]After Charles-Dominique (1994), Brosset *et al.* (1996)
[d]This study (Appendix 4.2).

bat, *Carollia perspicillata*; see Cosson, 1994) in secondary forests was linked to the abundance of fruit from early secondary recolonizers (*Cecropia obtusa*, *Ficus* spp., *Piper* spp., *Solanum* spp., *Vismia* spp.) which, due to their early establishment and high fecundity, are consumed and dispersed in large numbers even further by bats (De Foresta *et al.*, 1984; Charles-Dominique, 1986; see also Marinho-Filho, 1991). In contrast, the populations of other large fruit-eating bats (e.g. *Artibeus jamaicensis* and *A. lituratus*) appear to decline in largely degraded areas, possibly due to a decline in the mature forest tree species whose fruits they principally feed upon. Conversely, populations of *A. obscurus*, a common large fruit-eating bat with broad feeding preferences and typically abundant in primary forest but not in secondary stands (Brosset *et al.*, 1996), showed a favourable short-term response to forest fragmentation resulting from the building of the Petit Saut Dam in French Guiana (Cosson *et al.*, 1999; Granjon *et al.*, 1997). Following fragmentation, the abundance and diversity of the bat communities was lower on smaller islands compared to larger islands (40 ha) and the mainland peninsula (>1500 ha) (Cosson *et al.*, 1999; Granjon *et al.*, 1997; Pons and Cosson, 2002).

As an example of diet specialization in frugivorous bats, Cockle (1997) found that fruits of (hemi-)epiphytes in the Cyclanthaceae (*Evodianthus funifer*, *Asplundia* spp., *Thoracocarpus bissectus*) and Araceae (*Philodendron* spp., *Rhodospatha* spp.) accounted for 57% of the diet of the small fruit-bat *Rhinophylla pumilio* at a mature forest in French Guiana. Seeds of these families were also found in the faeces of the short-tailed fruit bats, *Carollia brevicaudata* and *C. perspicillata*, and the yellow-shouldered fruit bats, *Sturnira tildae* and *S. lilium*. Fruits from species in the Solanaceae (*Solanum* spp.), Moraceae (*Ficus* spp.), Piperaceae (*Piper* spp.) and Bignonaceae (*Schlegelia* spp.) also figured prominently in the diet of *R. pumilio* during the course of the study (Cockle, 1997). In total, 47 species in 15 plant families were identified from faeces beneath feeding roosts as being consumed and dispersed by *R. pumilio*.

Though underestimated, the protein-rich foliage resource might be also essential for frugivorous bats such as the large fruit-bats *A. jamaicensis* and *A. lituratus*, which, in forests of Puerto Rico at least, meet their micro-nutrient and vitamin requirements through leaves, an abundant and easy-to-capture resource in comparison to insects (Kunz and Diaz, 1995).

Arboreal Herbivorous Mammals

Diurnal foragers

Flowers

Reproductive parts are a staple resource for many diurnal, arboreal frugivores during the dry season, especially primates. Julliot and Sabatier (1993) estimated that the trees *Micropholis cayennensis* (Sapotaceae), *Eperua falcata* (Caesalpiniaceae) and the liana *Odontadenia* sp. (Apocynaceae) constituted more than 50% of flower consumption by red howler monkeys, *Alouatta seniculus*, in French Guiana. Nearly 13% of the diet of howler monkeys is composed of flowers, based on feeding records (Julliot, 1994). Flowers are also part of capuchin and spider monkey diets in French Guiana and Suriname, though to a much lesser extent than in the case of howler monkeys (van Roosmalen, 1985; Guillotin *et al.*, 1994; Simmen and Sabatier, 1996; Simmen *et al.*, 2001). Oliveira *et al.* (1985) observed that a troop of Guianan sakis, *Pithecia pithecia*, ate only flowers from species in the Bignoniaceae, Leguminosae and Passifloraceae during the dry season (14% of records) at a site near Manaus, Brazil. Tamarins (*Saguinus* spp.), capuchin monkeys (*Cebus* spp.) and white-faced saki all turned to *Symphonia* and *Mabea* cf. *eximia* nectar during the July–October dry season trough in fruit production at a site on the Rio Urucu in central Amazonia (Peres, 1994b).

Fruit and seeds

A large part of zoological research has been directed towards primates, arguably being the most influential, non-volant, vertebrate consumers of plant materials in Guianan tropical rainforests. Guillotin *et al.* (1994) compared the diversity of fruit and seeds contained in the stomach contents of a random sample of black spider monkeys, red howler monkeys and brown capuchin monkeys in French Guiana. They concluded that competition for fruits is of little importance since each species concentrated on fruits of different sizes. The larger species (spider monkey and howler monkey) focused primarily on the top end of the fruit size spectrum, while relatively small individuals of brown capuchins *Cebus apella* ate smaller fruits. During the dry season, howler monkeys and capuchins shifted to foliage and insects, respectively, whereas spider monkeys maintained a frugivorous diet (see also van Roosmalen, 1985). Based on diversity of plant species consumed, Guillotin *et al.* (1994) found that spider monkeys and howler monkeys fed on canopy, emergent (>25 m) and subcanopy trees (15–25 m) more often than in understorey treelets (<15 m), whereas brown capuchins did not show any preferences between strata, though they were found to prospect more frequently in epiphytes and lianas. The diets of the three monkeys overlap by 35–59%, while the proportion of fruit in the diet is directly related to the total number of species identified from stomach contents (42, 34 and 63 species, respectively). The higher the number of fruit species eaten by a given monkey, the lower the percentage of those shared with other monkey species (Guillotin *et al.*, 1994).

The diversity of plant species consumed by monkeys depends on body size, habitat type and location. The relative occurrence of each plant item consumed is then dependent on the method used to describe the diet. For instance, based on stomach content analysis, Guillotin *et al.* (1994) quoted only four species of Sapotaceae consumed by howler monkeys at Arataye river area, whereas, after observing monkeys for more than 2 years, Julliot (1996a,b) listed 21 species at the Nouragues station only 8 km away (see also Julliot, 1994). At Nouragues, the family Sapotaceae (27 species) is a major source of fruit for howler monkeys (59% of feeding records), spider monkeys (35%) and capuchins (28%). Other species, mainly Moraceae (e.g. *Bagassa guianensis*), Myristicaceae (*Virola* spp.) and Mimosaceae (*Inga* spp.) fruit from February through April when fruit production is highest (Simmen and Sabatier, 1996). The diversity of fruiting species consumed by the largest monkeys is similar, though when considering these together with the number of species whose leaves were consumed (at least 19 species), the diet of howler monkeys is much more varied (67 items) than that of spider monkeys (23 items) (Simmen and Sabatier, 1996; see also Guillotin *et al.*, 1994). When restricted to highly disturbed, flooded forest remnants with leafless trees, an isolated troop of howler monkeys survived for some time by feeding exclusively on leaves of Araceae (*Philodendron* spp.) and Cyclanthaceae (*Ludovia lancifolia*) (de Thoisy and Richard-Hansen, 1997).

A fourth important Guianan monkey, the golden-handed tamarind (*Saguinus midas midas*) is insectivorous–frugivorous all year round, with important seasonal variation in dietary composition. Insects (64% of diet) are predominately taken during the period of high fruit production (Pack *et al.*, 1998). A total of 31 mainly small-seeded species, belonging to 25 plant families, were consumed by tamarins during the course of Pack *et al.*'s (1998) study. Tamarins have a small home range of about 31–48 ha, with an average density of 19.4 individuals/km^2 (Day, 1997) and capture insects (e.g. Acrididae: K.S. Pack, O. Henry, personal communication) on leaves and lianas at between 10 and 20 m height within low-statured forest types. In contrast to larger monkeys, tamarins prefer edge and disturbed habitat types (Mittermeir and van Roosmalen, 1981; Kessler, 1995), such as forests dominated by lianas or *Dimorphandra* spp. on well-drained, nutrient-poor soils which are susceptible to frequent treefall or fire events (see Chapters 2 and 7). This preference for disturbed habitats is reflected by the low diversity of plant families eaten (Pack *et al.*, 1998).

Troops of brown bearded sakis (*Chiropotes satanus*) in Suriname depend almost entirely upon the immature seeds and ripe fruits of 53 tree species (van Roosmalen *et al.*, 1988), other plant parts (flower, leaves) accounting for only a marginal portion of their diet. Van Roosmalen *et al.* (1988) observed that fruit comprises 96% (of which 69% are seeds) of the diet of brown bearded saki. According to Kinzey and Norconk (1993), fruit and seeds (52–91% of total) account for 91–99% of the annual dietary intake of brown bearded saki, with substantial variation in the monthly totals occurring as a result of seasonal and habitat differences in resource availability.

Details of the frugivorous diet of Guianan sakis are fragmentary (e.g. Oliveira *et al.*, 1985; Setz, 1993, 1996; for details of a closely related species, *P. albicans*, from northeast Brazil, see also Peres, 1993a). Mittermeier and van Roosmalen (1981) report that the diet of Guianan sakis is composed of 93% fruit (of which 36% are seeds), 6.7% flowers, and no leaves at a site in Suriname. Setz (1992, 1993, 1996) showed that Guianan sakis rely upon 189 species in 51 families found within a 10-ha fragment in north-central Amazonia (see Bierregaard *et al.*, 1992), which represents the longest list of plants consumed by Guianan sakis to date. More recently, Vié (personal communication) studied Guianan sakis at two sites in French Guiana, one covered by primary forest (Vié, 1997, 1998; Vié and Richard-Hansen, 1997; see also Vié *et al.*, 2001; Richard-Hansen *et al.*, 1999), and the other on a 30–40 ha island near to the Saint Eugene research station situated within the fragmented Petit Saut Dam area (see Granjon *et al.*, 1997; Claessens *et al.*, 2001). Based on more than 1500 visual feeding counts on the island, Vié (personal communication) found that the diet of Guianan sakis was primarily composed of fruits and seeds (78%), with some alternate consumption of buds, stems and leaves (10%) and flowers (1%). This is consistent with the 79% figure for fruit consumption (of which 22% are immature seeds) in a conspecific troop in Brazil (Oliveira *et al.*, 1985). Although ongoing, an incomplete identification of fruits and seeds in Vié's study covers a total of 33 plant families, of which two were not present in Setz's (1993) inventory. Apart from intensive seed predation by saki monkeys, capuchins (brown capuchin and wedge-capped capuchin, *C. olivaceus*) also act as seed predators during the Guianan dry season, consuming immature fruit and seeds of palms in particular and discarding the fruits of many other species on the ground below the parent tree (e.g. Zhang, 1994).

Foliage

In the canopy, the folivorous pale-throated three-toed sloth (*Bradypus tridactylus*) and the southern two-toed sloth (*Cloelopus didactylus*) may reach high densities in some Guianan forests (see Charles-Dominique *et al.*, 1981), though the impact of high densities upon the amount of leaf matter consumed (see Foley *et al.*, 1995) is as yet unknown. Red howler monkeys also consume a large amount of young (3% of observational feeding units) and mature (54%) leaves, especially in the dry season when fruit resources are scarce at Nouragues field station (Julliot and Sabatier, 1993). By studying the stomach contents of howler monkeys, Guillotin *et al.* (1994) found similar mean percentages (53%) of leaves ingested, with a maximum in June–September (73%) within the Arataye area, 8 km from Nouragues. A total of 96 plant species were recorded as being consumed by howler monkeys during a 2-year period at Nouragues (Julliot, 1993). The Leguminosae (Caesalpiniaceae, Fabaceae and Mimosaceae), Burseraceae, Chrysobalanaceae and Araceae are particularly important in the leaf diet of howler monkeys, accounting for 71% of all species (N=29 species) consumed by monkeys in the wet season (Simmen and Sabatier, 1996; see also Julliot and Sabatier, 1993). Leaves eaten by monkeys are poor in lipids, have low concentrations of soluble sugar and have a higher protein content than fruits – young leaves being richer than mature leaves (Simmen and Sabatier, 1996).

Gum

While *Ficus* fruit may serve as a staple key resource in the dry season (Terborgh, 1986), Peres (2000) suggested that other plant items such as gum of *Parkia* spp. pods (Mimosaceae) play the same role for several arboreal, flying and terrestrial vertebrates in Amazonian and Guianan rainforests. For instance, among arboreal diurnal foragers, Peres reports that seeds and gums of *Parkia* pods were one of the most important dry-season food items for brown capuchin monkeys in Amazonian forests (Peres, 1994a, 2000). In Suriname, golden-handed tamarins and spider monkeys also consume gum of *Parkia pendula* and *P. nitida* (van Roosmalen, 1985, personal communication, in Peres, 2000).

Nocturnal foragers

Nocturnal, arboreal mammals of the Guianan forests have also received much attention. Charles-Dominique and his research team analysed resource use and niche partitioning in three rodents, five marsupials and one carnivore in an 80-year-old secondary forest (Cabassou) near an urban centre in French Guiana. Arboreal rodents typically eat immature seeds, whereas marsupials and the kinkajou *Potosi flatus* are principally nectar and pulp-eating, disseminating the seeds of many short-lived (e.g. *Decrepit* spp., *Ficus* spp., *Vismia* spp., Melastomataceae) and longer-living (*Inga* spp., *Virola* spp., Mimosaceae, Lauraceae) species in the process (Charles-Dominique *et al.*, 1981). A total of 31 plant species in 19 families (of 127 species and 55 families; Charles-Dominique *et al.*, 1981) accounts for the majority of resources utilized by the bare-tailed woolly opossum (*Caluromys philander*), the common opossum (*Didelphis marsupialis*) and the common grey four-eyed opossum (*Philander opossum*) (Atramentowicz, 1988). When considering the feeding records of the same animal species in the mature forest of Piste de Saint Elie in addition to those from Cabassou, the number of plant species increases to 44 species in 21 families, of which only 19 species, all with seeds less than 1 cm in length (largest were *Passiflora*), are dispersed after ingestion and defecation (Atramentowicz, 1988). The diet of the bare-tailed woolly opossum is composed mainly of fruit pulp (75%) and arthropods (25%) during periods of heavy fruitfall, but includes nectar and tree exudates (gum) when fruit become less available (see also Julien-Laferrière and Atramentowicz, 1990). The diets of the mainly terrestrial common opossum and common grey four-eyed opossum are more varied, fruits accounting for only 50% of the diet, amphibians and small mammals making up the difference (Atramentowicz, 1982, 1988). At Piste de Saint-Elie, bare-tailed woolly opossum feed upon 23 plant species in 13 families (Julien-Laferrière, 1989, 1999b; for information on foraging behaviour, see also Julien-Laferrière, 1990, 1995, 1997, 1999a). Up to 30 plant species in 21 families were consumed over the course of a year by kinkajou at Piste de Saint-Elie (Julien-Laferrière, 1989, 1999a). Combining feeding records collected at Nouragues station (dry season, Julien-Laferrière, 1993; wet season, Julien-Laferrière, 1999b, 2001) and at Cabassou (Charles-Dominique *et al.*, 1981), it is possible to state that at least 61 species in 34 families are included in the diet of the kinkajou, many of them being dispersed via defecation (Julien-Laferrière, 1989, 1999a,b).

Ground-dwelling Mammals

Ungulates

Red brocket deer (*Mazama americana*), grey brocket deer (*M. gouazoubira*), white-tailed deer (*Odocoileus virginianus*), collared peccaries (*Tayassu tajacu*), white-lipped peccaries (*T. pecari*) and Brazilian tapir (*Tapirus terrestris*) are thought to feed at least in part on seeds or fruit in the Guianas. Here, we document their diet after reviewing the literature from the Guianas and countries outside the limit of the Guianan Shield extending throughout Amazonia where the same species occur. We also briefly consider data on relative species from Central America when little information is available for Amazonia in general.

Deer

Deer in Suriname consume fruit and seeds primarily during the months encompassing the maximum and minimum periods of rainfall, vegetative plant parts being preferred during the intervening transitional periods (Branan *et al.*, 1985; Gayot, 2000), when new leaf production is likely to be highest (see Aide, 1988, 1993). An increase in the consumption of reproductive plant parts from October to December is synchronous with the birth of red brocket offspring during these months (Branan and Marchinton, 1986). Red brocket deer (57 species in 36 families; N=57 stomachs) had a richer diet than either grey brocket deer (15 species in 12 families; N=5 stomachs) or white-tailed deer (14 species in 10 families; N=13 stomachs) (Branan *et al.*, 1985). Stomach content analyses of plant consumption by deer in French Guiana (Gayot, 2000; Gayot *et al.*, 2004) showed that red brocket deer (N=28 stomachs) and grey brocket deer (N=34) share a total of 30 families (of N=38 and 40 plant families identified so far, respectively). *Symphonia* appears to be an important resource for red brocket deer, while both *Bellucia grossularioides* (Melastomataceae) and *Swartzia benthamiana* (Caesalpiniaceae) are important components of the grey brocket deer diet in Suriname. In French Guiana, four families are dominant so far in both deer's stomachs, including Myristicaceae (seeds of *Virola surinamenis*, *V. michelii*, *V. kwatae*), Caesalpiniaceae (seeds of *Dicorynia guianensis*, flowers of *Eperua falcata*), Lecythidaceae (flowers and seeds of *Eschweilera* spp.) and Moraceae (fruits of *Ficus* spp.). Seeds and fruits from other important neotropical plant families such as Palmae and Leguminosae in general were also commonly observed in stomach contents of these species in Suriname (Branan *et al.*, 1985) and French Guiana (*Oenocarpus bacaba*; Gayot, 2000) as well as in Amazonian Peru (except *Odocoileus*, which is absent) (e.g. Bodmer, 1990b). Stalling (1984) lists a total of 23 plant families from which leaves, fruits, twigs and flowers are consumed by grey brocket deer in Paraguay. Because the number of identifiable plant species from stomach contents is often limited, our current knowledge of dietary preferences and composition in deer remains fragmentary. However, field observations suggest that most deer intensively browse the young seedlings of many species following the decline in fruitfall and increase in germination which takes place from May to July each year. At Mabura Hill, red brocket deer repeatedly browsed over 100 seedlings of the large-seeded sapotaceae, *Pouteria speciosa*, beneath a single parent tree, often returning more than ten times to consume the compensatory growth (Hammond, unpublished results), which is a typical response of large-seeded species to initial loss of above-ground seedling biomass (Harms and Dalling, 1997; Hammond and Brown, 1998).

Tapir

Recently, inspection of the tapir's dietary patterns at Nouragues, French Guiana and on Maracá Island, Roraima, Brazil, has shown that Brazilian tapir may ingest and disperse a large number of small- to large-seeded species while foraging in *terra firme* forest (Fragoso, 1997; Fragoso and Huffman, 2000; Henry *et al.*, 2000). For instance, in French Guiana, 81% of tapir stomachs (N=27) contained intact seeds, a majority of them being less than 1 cm in diameter, but occasionally swallowing seeds up to 2.5 cm in diameter, in this case belonging to the liana *Elephantomene eburnea* (Menispermaceae) (F. Feer, personal communication). The most abundant species found were *Spondias mombin* (Anacardiaceae), *Bagassa guianensis* (Moraceae) and the liana *Landolfia guianensis* (Apocynaceae). In contrast, the dominant species in faeces of tapirs at Maracá Island, Roraima, Brazil are *Mauritia flexuosa* and *Maximiliana maripa* (*Attalea maripa*) (Arecaceae), *Cassia moschata*, and *Enterolobium schomburgkii* (Fabaceae), *Spondias mombin* and *Anacardium giganteum* (Anacardiaceae). Contrasting dietary regimes are likely a consequence of difference of habitat dominance between sites (Fragoso and Huffman, 2000; see also Fragoso, 1998a, 1998b). Salas and Fuller (1996) and Salas (1996) documented habitat use and the diet of tapir in Venezuelan Guayana (see description of this forest in Rollet, 1969), and recorded that it ate fruit of 33 species of trees, palms and lianas belonging to 12 plant families and browsed at least 88 different species of plants, the most common being *Cecropia sciadophylla* (Cecropiaceae). In contrast, the large faecal deposits (up to 10 kg fresh weight, N=8) of tapir in central Guyana are composed almost totally of leaf with no discernible seed remnants, though a number of small seeds (e.g. *Solanum*) were enmeshed within the fibrous mass (D. Hammond, unpublished data) of the few samples observed. The predisposition of tapirs to defecate in water (D. Hammond, personal observation; Janzen, 1981b, 1983b; Fragoso, 1997) would not make for effective dispersal of seeds, unless they are adapted to a swamp or riparian environment (e.g. *Catostemma commune*, Palmae, Rapateaceae). While sampling effort varied tremendously, the differences in seed abundance in fresh dung recorded from different sites in the Guianas suggests that diet and dietary choice are being affected by seasonal and/or spatial variation in fruit and seed resources and tapir feeding behaviour.

At a neotropical location outside of the Guiana Shield, Rodrigues *et al.* (1993) found numerous seeds of the palm *Euterpe edulis*, the liana *Maytenus* sp. (Celastraceae) and the tree *Virola oleifera* (Myristicaceae), in addition to twigs and leaves of the giant bamboo, *Guadea augustifolia*, in two dung samples taken from the Atlantic forest in Brazil. Bodmer (1990a) analysed gut (stomach and caecal) content and faecal samples of tapir in Amazonian Peru and found that tapirs consumed a large quantity of fruit (33% of dry weight), principally from the families Arecaceae (*Mauritua flexuosa*, *Jessenia* sp., *Scheelea* sp.), Sapotaceae, Araceae, Chrysobalanaceae, Leguminosae and Menispermaceae. By comparison, Foerster (personal communication; see Foerster and Vaughan, 2002) tallied a total of 126 for the plant families ingested by a Baird's tapir during a 10-month period in Corcovado National Park, Costa Rica (see Matola and Todd, 1997). Of this total, 35% were vines, 34% trees, 16% shrubs and 15% herbs (C. Foerster, personal communication). This is of similar magnitude to the 94 plant species (45 families) listed by Terwillinger (1978) as being eaten by Baird's tapir on Barro Colorado Island, Panama. Araceae (*Monstera* sp. vine), Lauraceae (*Persea* sp. fruit) and Rubiaceae (*Psychotria* sp. bark) accounted for 40% of diet in Costa Rica. In both Central American forests, occurrence of fruits in the diet is small, accounting for 19% of diet according to Foerster (1998, personal communication) and only three fruit species in Terwillinger (1978), which contrast with a high frugivorous diet of tapir in Amazonian Peru (Bodmer, 1990a) and Guianas (F. Feer, personal communication). Such differences are possibly related to the nature of the habitats foraged by animals, i.e. essentially secondary vegetation in Central America (Terwillinger, 1978; Foerster and Vaughan, 2002) (despite the clumped population of *Attalea* palms) versus primary tall rainforest in the Guianas.

Peccaries

Information on the diet, feeding behaviour and habits of peccaries is rare, perhaps because these species are threatened with local extinction in many parts of their former geographic range (see Bodmer, 1995; Peres, 1996; Beck, 2005). Virtually no information of this kind exists for peccaries in the Guiana Shield, though most accounts from local people and researchers suggest that their populations are largely intact across much of the Guianan interior. A recent study by Judas and Henry (1999) shows that seasonal variation in home range size is correlated with fruit availability: the size of the core area was smaller during the low-fruit season (June–September) than during the high-fruit season (February–May). In addition, Judas (1999) was able to test casual fruits and seeds encountered by a captive collared peccary that was kept on a lead along forest trails: up to 17 plant families and species of fruits and seeds were eaten by the peccary. Leaves of at least five plant families (Araceae, Heliconiaceae, Marantaceae, Melastomataceae and Sapotaceae) were also consumed; the ratio leaves:fruit depended on fruit seasonality, the greatest amount of fruit and seeds being harvested during the high fruiting season. Reproductive plant parts represent a major part (>60%) of stomach contents of collared and white-lipped peccaries in Amazonian Peru and the Guianas (Bodmer, 1991; Henry, 1994a, 1996; Feer *et al.*, 2001a). Data from other neotropical countries are informative here because many of the same plant species or closely related plant species occur in the Guianas. Fragoso (1999) list some species eaten by white-lipped and collared peccaries in Amazonian Brazil. *Attalea (Maximiliana) maripa* and *Mauritia flexuosa*, two palms abundant in *terra firme* forests and palm wetlands, are consumed by both peccary species either in the wet or the dry season. Other plant items eaten include pulp and/or seeds of Caesalpiniaceae, Mimosaceae and Sapotaceae. Based on the analyses of faecal samples collected over an 11-month period, Altrichter and colleagues (Altrichter *et al.*, 2000, 2001; see also Altrichter and Almeida, 2002) similarly found a high proportion of fruit (62% of contents) being consumed frequently (95% of samples) by white-lipped peccaries in Corcovado National Park, Costa Rica; at least 30 different species were identified, mainly belonging to the Moraceae, Arecaceae, Chrysobalanaceae, Myristicaceae and Anacardiaceae; palm fruits were also the most common dietary constituent for peccaries in western Amazonian forests (Bodmer, 1990b; Kiltie, 1980, 1981; see also Kiltie, 1982). Vegetative parts of the families Araceae, Cyclanthaceae, Heliconiaceae and Sapindaceae and flowers of Araceae, Arecaceae and Clusiaceae were also ingested by white-lipped peccaries (Altrichter *et al.*, 2000; Beck, 2005).

Fruit-eating rodents

Most caviomorphs in the Guianas are relatively large-bodied, dedicated seed-eaters, while myomorphs are smaller and tend to be more omnivorous (Eisenberg, 1989 and references therein). In Trinidad, Rankin (1978) found that seeds of a Guianan tree, *Clathrotropis brachypetala* (Fabaceae) (hereafter *Clathrotropis*), contained significant quantities of the known toxicants, anagyrine and cytisine. Oral administration of *Clathrotropis* extract was fatal to mice and Rankin (1978) suggested that, despite its high average protein (15.8%), lipid (20.4%) and moisture (82.2%) content, the secondary chemicals render *Clathrotropis* an undesirable resource. She found that attack on *Clathrotropis* was the lowest of five species studied, though attack was monitored for only less than a month, and it is clear that caviomorph rodents may turn to large, persistent seeds of species, like *Clathrotropis*, during periods of resource scarcity (Hammond *et al.*, 1999). Interestingly, seeds of the congener, *C. macrocarpa*, are heavily attacked in Guyana (Hammond *et al.*, 1992) and consumed or scatterhoarded by agouti individuals in 20 × 20 m paddocks, despite the availability of poorly defended resources, such as coconut (M. Swagemakers and J. van Essen, personal communication). Uncooked beans (*Phaseolus vulgaris*, Fabaceae) contain high concentrations of toxic phytohemagglutinins (lectins) (Janzen *et al.*, 1976a) and are fatal to the myomorph rat *Liomys* in Costa Rica (Janzen, 1981a), yet are readily consumed by the caviomorph agouti in Guyana (D. Hammond *et al.*, unpublished data). In French Guiana, however, the common rice rat *Oryzomys capito* (Muridae) and spiny rat *Proechimys* spp. (Echimyidae) both consumed the seeds of most species offered while in captivity (Guillotin, 1982a; see also Adler, 1995).

Information on the diet of free-ranging red-rumped agoutis is vague and anecdotal (e.g. for congener *D. punctata*, Smythe *et al.*, 1982; Sabatier, 1983; van Roosmalen, 1985). Using different sources and observations in Panama and French Guiana (Smythe, 1970; Sabatier, 1983, P.-M. Forget, personal observation), however, one may establish a preliminary list of 37 plant families as the basis of the agouti diet in the Guianas (see Appendix 4.2). This is also true for the paca (*Cuniculus paca*), which feeds upon 21 plant families in Central America (Smythe *et al.*, 1982; Beck-King *et al.*, 1999).

Similar to granivorous birds, many terrestrial rodents are wholly dependent upon the type and quantity of fruitfall. In many cases, these rodents profit from the dispersal of seeds by flying and non-volant, arboreal mammals, especially where these resources may not be available due to absence of parent trees within their limited home ranges (Hammond *et al.*, 1999). The home ranges of pacas and agoutis are unusually small in comparison to their respective body sizes (1.0–5.5 kg and 6–14 kg) and the immigration of resources through frugivore dispersal may play a crucial role if the seed production within their home ranges is insufficient to meet base metabolic requirements throughout the year (e.g. Smythe *et al.*, 1982; Beck-King *et al.*, 1999). If fat reserves (Smythe, 1970) and hoarded seeds (Smythe *et al.*, 1982) are insufficient to meet demands during periods of resource scarcity in pacas and agoutis, respectively, then they may shift to alternate resources such as insects (spiny rats, agoutis: Guillotin, 1982a,b; Henry, 1994a), lizards (agoutis: R. Thomas, personal communication) and/or seedlings (pacas, agoutis: Smythe *et al.*, 1982) when seeds and fruit are scarce.

Little is known of the three squirrel species that are expected to occur in Guianan forests, and a large number of arboreal and terrestrial rats and mice (mainly Muridae) have yet to be studied in any great detail. Information concerning the feeding ecology of the long-haired, prehensile-tailed porcupine, *Coendou prehensilis* (Erethizontidae) is scant (but see Husson, 1978). Most of these rodents are relatively small in size and the cumulative impact of these groups upon plant regeneration remains unclear.

Other plant-eating terrestrial mammals

Several neotropical carnivores, such as the tayra (*Eira barbara*) and the South American coati (*Nasua nasua*) are also known to consume fruits (Emmons and Feer, 1990; Feer *et al.*, 2001b), most of the available information being from Barro Colorado Island, Panama. On Barro Colorado Island, both the tayra and Central American coati (*N. narica*) consume *Dipteryx panamensis* fruits in the dry season (Bonaccorso *et al.*, 1980; P.-M. Forget, personal observation) and *Spondias mombin* in the wet season (Anacardiaceae) (Gompper, 1995). Important resources of coatis include palm fruit (*Attalea butyracea*, formerly *Scheelea zonensis*), *Ficus insipida* (Moraceae) and *Tetragastris panamensis* (Burseraceae). Overall, coatis will consume fruits from 54 species in 27 plant families on BCI (Gompper, 1994). Though there are eight to ten non-feline species of carnivores found commonly throughout Guianan tropical forests, next to nothing is known about their ecology, dietary habits and the extent to which they may influence forest dynamics (see Terborgh, 1988).

Birds

Snow (1981) described tropical frugivorous birds as: (i) endozoochorous seed-dispersers; (ii) seed-predators; or (iii) scatter-hoarding dispersers (e.g. jays: Vander Wall, 1990), though the last type is not known to occur in the neotropics. In this section, only the first group will be discussed, the second being developed in the section on seed predation.

According to Snow (1981), the most important resources for avian fruit specialists are found in the Annonaceae, Burseraceae, Lauraceae, Melastomataceae, Myristicaceae and Rubiaceae. Studies of frugivorous birds in the Guianas have documented the dietary breadth of many common terrestrial and canopy frugivores. Erard *et al.* (1991) and Théry *et al.* (1992) analysed the stomach contents of 76 common trumpeters (*Psophia crepitans*, hereafter *Psophia*, Psophiidae) and 43 marail guans (*Penelope marail*, hereafter *Penelope*, Cracidae), respectively. Despite comparable body weights (1071 vs. 990 g, respectively), the diet of trumpeters is more diverse than that of marail guans: 55 species in 21 families versus 24 species in 17 families. Interestingly, dietary overlap between the two species in the dry season, i.e. from August to November when fruitfall is at its lowest, is only important with regard to the use of two important resources, *Eugenia coffeifolia* (Myrtaceae) and *Euterpe oleracaea* (Palmae) (Erard and Théry, 1994). Trumpeters are also able to subsist on invertebrates during this season (Théry *et al.*, 1992). The largely ground-dwelling trumpeters find most of their food on the forest floor, moving into the forest canopy only to sleep or to hide from predators. In contrast, marail guans actively feed in the canopy. The contrast in forage zone preference and the concomitant differences in resource availabilities between forest canopy and floor might explain, in part, why trumpeters exploit more fruit species than marail guans and the apparent partitioning of habitat resources (Théry *et al.*, 1992).

Erard *et al.* (1989) observed the cock-of-the-rock (*Rupicola rupicola*) while it was nesting, and identified fruit species that were taken to young. They completed their list by observing females during foraging bouts. As a result, 65 species in 31 families were identified. Species in the Annonaceae (*Pseudoxandra cuspidata*), Burseraceae (*Protium apiculatum*) and Lauraceae (*Ocotea* spp.) accounted for more than half (52.9%) of all feeding observations. Supplementary information on the biology and diet of the related Peruvian cock-of-the-rock (*R. peruviana*) are available in Benalcázar and Silva de Benalcázar (1984), completing the list of plant families eaten by these birds.

Another important bird group in the Guianas, the Pipridae, is mainly composed of specialized frugivores (Snow, 1981; Théry, 1990b; see also Théry, 1997). Piprids are often the most abundant (or most easily netted) birds in the neotropics, accounting for the largest number of individuals caught at a variety of sites in Central and South America (Karr *et al.*, 1990). In French Guiana, Théry (1990b) showed that a community of seven piprids exploits a total of 20 and 22 species of Melastomataceae (*Miconia*, *Clidemia*, *Henrietella*, *Loreya mespiloides*) and Rubiaceae (*Psychotria*), respectively, two families that are typical of the early phase of regrowth following natural and anthropogenic disturbance events (see also Poulin *et al.*, 1999). Other plants commonly consumed included species in the Flacourtiaceae (*Laetia procera*), Goupiaceae (*Goupia glabra*), Moraceae (*Ficus guianensis*) and Ulmaceae (*Trema guianensis*), all being typically long-lived, early colonizers after disturbance in the Guianas. The most complete information on piprid diet was collected by Snow (1962a,b) in Trinidad for the black and white manakin *Manacus manacus* and the golden-headed manakin *Pipra erythrocephala*, two species which are also present in the Guianas (Théry, 1990b, 1992). Direct observations of feeding birds in addition to the collection of regurgitated seeds from below nests, from display grounds and from trapped birds, allowed Snow (1962a) to establish a list of 73 species belonging to 27 plant families. Fruits from the Rubiaceae and Melastomataceae were the most representative families, with 17 and 15 species, respectively, while other families were represented by 1–4 species (14 families with only one species). In addition, at least 43 species in 18 plant families were consumed by the golden-headed manakin (Snow, 1962b), 15 of them also being eaten by black and white manakins (Snow, 1962a). In Trinidad, Snow (1964) found that 19 species of light-demanding *Miconia* (Melastomataceae) formed a large part of the diet of several manakin species, also known to occur in Guianan forests, these birds efficiently dispersing seeds into gaps and disturbed areas. In comparison, the red-capped manakin (*Pipra mentalise*) and the golden-collared manakin (*Manacus vitellinus*), studied near Barro Colorado Island, Panama were recorded as consuming 62 species in 23 families, most of these being light-demanding epiphytes (e.g. *Anthurium* spp.) and short (e.g. *Palicourea elliptica*) or long-lived (e.g. syn. *Schefflera morototoni*) pioneers (Worthington, 1982). Together, data from Trinidad and Panama give 40 plant families that may be expected to provide food for piprids (Appendix 4.2).

Members of the Cotingidae are among the most frugivorous bird species in the neotropics (Erard *et al.*, 1989) and constitute the most speciose frugivorous bird group in the tropics. The relatively large body size and gape width of many cotingids means that they are able to feed on many of the largest fruits which other smaller, sympatric frugivores, such as manakins, must forego. A particularly important component of the large avian frugivore diet is fruits from *Ocotea* and *Nectandra* (Lauraceae), probably due to the combination of a large diameter (Wheelwright, 1985) and relatively high fat and protein content of their pericarp (e.g. Snow, 1981). Lauraceae constituted 74% of the seeds and fruits collected below the nest of a bearded bellbird (*Procnias averano*) in Trinidad during the fledging period (May–June); lauraceous fruits, such as *Ocotea oblonga* and *Cinnamomum elongatum* are known to have a relatively high nutritional value (Snow, 1970). Snow (1971) noted that the purple-throated fruit-crow (*Querula purpurata*) consumed eight different kinds of fruits in the wet season (January–April) in Guyana, 90% of the feeding records consisting of only four species (*Didymopanax* (=*Schefflera*) *morototoni*, Araliaceae; *Guarea trichiloides*, Meliaceae; *Hirtella* sp., Chrysobalanaceae; an unidentified Lauraceae). Fruit of *Cecropia* sp. (Moraceae) were also consumed and the largest seed ingested was a lauraceous species 2.7 × 1.5 cm. Snow (1972) listed 37 species of seeds found beneath the perches of the lek-forming capuchinbird (*Perissocephalus tricolour*) in Guyana and Trinidad, of which 12 species were found in the Lauraceae (lek: an assembled group or assembly area for communal courtship display). Many of the largest, fleshy fruits in these genera can be found in the Guianas (e.g. in Guyana: *Nectandra grandis*, 12.2 g fresh weight; *Endilicheria chalisea*, 14.3 g; *Ocotea* sp. nov., 9.8 g) and the unusually high number of cotingas (10 species), including large numbers of several of the largest-bodied species (3 of 11 species >250 g body weight), co-occurring in the Kanuku Mountains was duly noted as one of the most prominent features of the area by Parker *et al.* (1993). Cotinga communities at Mabura Hill (ter Steege *et al.*, 1996), Iwokrama (Ridgely and Agro, 1999) and Nouragues (Thiollay, 1994) also appear to be relatively rich, and are dominated by the screaming piha, *Lipaugus vociferans* (Cotingidae). The bearded bellbird (*Procnias averano*) is one of the largest cotingids and most abundant components of the Brownsberg avifauna in Suriname (Reichart, 1991). Trail and Donahue (1991) observed the Guianan red cotinga (*Phoenicircus carnifex*) in the early wet season (December–March) at Brownsberg and report casual data on their diet which includes Myrtaceae (*Eugenia* sp.), Clusiaceae (*Clusia* sp.), Melastomataceae (*Miconia* sp.) and an unidentified Lauraceae. Though many cotingids enjoy a pan-Amazonian distribution, several species, including the screaming piha, appear to be particularly abundant at Guianan sites compared to other neotropical regions. As part of Conservation International's Rapid Assessment Programme, team-member Ted Parker identified 11 cotingids and only six of the smaller-bodied piprids during the Kanuku Mountains survey, whereas lowland forest site surveys in southwestern Amazonia yielded seven cotingids and 10 piprids at Tambopata, Peru and five and seven, respectively, at Alto Madidi, Bolivia (Parker *et al.*, 1991, 1993; Foster *et al.*, 1994).

Fish

The vegetarian diet of Amazonian fish is well-documented and examples of icthyochory abound. Several species of Euphorbiaceae and *Gnetum venosum* (Gnetaceae) are regularly transported by fish during the high-water season in varzea forest (Kubitski, 1985). In the Guianas, however, little published information exists regarding fish dietary selection and seed dispersal in particular. Beebe (1925) found plant parts in the stomachs of 35% of the 77 species of fish examined, but only one of these, *Doras granulosus* (syn. *Pterodoras granulosus*), was found to have been feeding on seeds at the time. One-third of the species feeding on plant parts were exclusively vegetarian, mainly in the family Serrasalmidae (subfamily Myleinae). These species feed mainly on Podostemaceae which grow in the shallow waters along the waters edge. The true Brazilian pacu, *Myleus rhomboidalis*, feeds on the hard seeds of the palms *Euterpe oleracea* (49–69% of items) and *Mauritia flexuosa*, other large-seeded species such as *Dipteryx* sp. (Fabaceae), *Carapa guianensis* (Meliaceae), *Virola surinamensis*, *Macrolobium* sp. (Caesalpiniaceae), *Psidium* sp. (Myrtaceae) and *Inga* spp. (Mimosaceae) during the wet season and turns to the aquatic plants, *Mourera fluviatilis* and *Apinagia richardiana* (Podostemaceae), when the waters recede (Boujard *et al.*, 1990; Planquette *et al.*, 1996). However, only the small seeds of the common climber, *Passiflora laurifolia*, were found wholly intact in stomachs from individuals caught in the Approuague and Sinnamary Rivers in French Guiana (Boujard *et al.*, 1990). In contrast, the diet of the congeneric, *Myleus ternetzi*, appears to consist mainly of allochthonous leaves, from such riverine species as *Inga meissneriana* and *Cydista aequinoctialis* (Bignoniaceae) (Boujard *et al.*, 1990).

More substantial data on the diet of herbivorous species inhabiting the rivers of the Guianas can be gleaned from studies of the Rio Negro aquatic community in southern Venezuelan Guayana and Brazil (Goulding *et al.*, 1988). Along the course of this enormous blackwater river, there are 79 species of fish which feed on the fruits and seeds of riverine plants, nearly 28% of these also feeding on the leaves and flowers of woody plants. Herbaceous, mainly aquatic, vegetation also figures prominently in the community dietary intake, nearly 50 species regularly feeding on the roots, stems and leaves of plants in the families Podostemaceae, Araceae, Lentibulariaceae, Pontederiaceae, Salviniaceae and Poaceae. The berries of *Montrichardia arborescens* (muku-muku), a giant, semi-woody aroid typical of most Guianan waterways are regularly consumed by Rio Negro fish.

Fish in the Rio Negro consume the seeds and fruits of 94 genera in 32 families, though only a fraction of these appear to be defecated or regurgitated intact (Table 4.7).

Table 4.7. Fish food and the potential for dispersal in the Rio Negro based on whether seed material found in stomach contents was intact or had been masticated. Adapted from Goulding *et al.* (1988). Note how the seeds of several small-seeded families (*) always remain intact while many of the typically large-seeded families are mainly broken down (‡).

Family	No. of genera	Masticated	Intact
Annonaceae	2	+	+
Arecaceae‡	3	+	+
Bignoniaceae	6	+	
Chrysobalanaceae	3		+
Euphorbiaceae	6	+	+
Lauraceae	1	+	+
Lecythidaceae‡	1	+	
Leguminosae‡	16	+	
Malphigiaceae	4	+	
Melastomataceae*	3		+
Moraceae*	3		+
Myristicaceae	1		+
Polygonaceae	2	+	+
Quiinaceae	1	+	+
Rubiaceae*	11	+	+
Simaroubaceae	1	+	+

Foremost among these food items are palm fruits, which represent an important dietary item for the catfish *Phractocephalus* spp. (Pimelodidae) and *Megaladoras* spp. (Doradidae) as well as numerous large-bodied characins and serrasalmids, most notably in the genera *Hemigrammus*, *Metynnis*, *Moenkhausia* and *Myleus*. The role of fish in the dispersal of palms along rivers is exemplified by the success of the aggressive colonizer *Astrocaryum jauary* along newly formed point-slope bar deposits (Piedade, 1985). Other important woody plant families include the Chrysobalanaceae (*Couepia*, *Licania*, *Hirtella*), Leguminosae and Myrtaceae (*Eugenia*, *Psidium*, *Myrcia*, *Myrciaria*). Plant matter is a relatively important contributor to the energetic needs of the Rio Negro fish community, but this is seasonally dependent. Species which feed on terrestrial plant parts, typically during the wet season, turn to aquatic vegetation when the river level lowers (Goulding *et al.*, 1988).

Lizards and Turtles

Very few reptiles and no amphibians have been recorded as feeding on fruits or other vegetable matter in the Guianas, which is consistent for South American herpetofaunas in general (Duellman, 1990). Beebe (1925) found that of the 30 lizard species inhabiting Kartabo, Guyana, only one, *Iguana iguana*, was strictly vegetarian and only half the diet of another iguanid, *Polychrus marmoratus*, was found to consist of berries and leaves. The diet of the more common Guianan macroteiid, *Cnemidophorus lemniscatus lemniscatus*, consisted of only 2% fruit and 1% flower material. Several other species, notably basaliscs, *Basaliscus* spp., are known to feed on fruit and other plant material but are rare, introduced or absent from Guianan forests. Of the ten species of turtle known to inhabit Guianan forestlands, the most frugivorous is the common, mainly terrestrial, red-footed tortoise *Geochelone denticulata* (Testudinidae), which feeds on leaves, fruits, seeds (up to 1.5 × 1.0 cm in size) and insects on the forest floor, and its congeneric sympatric, *G. carbonaria*, which is more frequently found along savanna edges. Faeces of the Amazon mud turtle, *Kinosternon scorpioides* (Kinosternidae), were found to contain fragments of palm fruits in French Guiana and *Phrynops nasutus* (Chelidae) was observed to consume the fruits of the common aroid climber, *Philodendron dolimoesense* (Métrailler and Le Gratiet, 1996), but few systematic studies of reptilian dietary patterns have been undertaken in the Guianas. The giant Amazon river turtle, *Podocnemis expansa* (Pelomedusidae), which is thought to occur in Guyana and whose herbivorous relative *P. unifilis* is found throughout the Guianas, has a diet consisting largely of fruit (80–90%) and leaves and stems (4%) (Ojasti, 1967). Typically Guianan taxa found in the diet of *P. expansa* include *Campsiandra comosa* (Caesalpiniaceae), a common riverine species in central Guyana, *Bactris* sp., *Astrocaryum* spp., *Micropholis* sp. (Sapotaceae), *Symphonia globulifera*, *Simaba guianensis* and *Hevea* spp. (Euphorbiaceae) (Best, 1984).

Plant–Vertebrate Relationships: Dispersal vs. Predation

Plant pollination

By bats

Bats that feed on flower nectar (Glossophaginae) are represented by at least seven species in the Guianas (Emmons and Feer, 1990; Brosset *et al.*, 1996; Smith and Kerry, 1996; Simmons and Voss, 1998) (Appendix 4.1). Several species of long-tongued bats (e.g. *Glossophaga soricina*) are specific to edge habitats and are rarely found in primary forest, unless this is dissected by roads linking its interior to more open habitats such as littoral savannas (Brosset *et al.*, 1996). There are almost no studies on flower–bat interactions in the Guianas, and most information on neotropical glossophaginids was reviewed by

Fig. 4.1. Numerous elongated stamens in the bat-polinated riverine tree, *Pachira aquatica* (Sterculiaceae). Photo P.-M. Forget.

Heithaus (1982). Any information on the glossophaginid bats inhabiting the Guianas is currently based on the few studies of these species carried out in other neotropical countries, mainly in Central America (see also Heithaus *et al.*, 1974, 1975).

Typical bat-pollinated flowers have stamens, pistil and nectar that are fully exposed (shaving brush or penicillate effect), often positioned away from foliage, open at night, often without colour, have a stale, fermenting odour, and produce copious pollen (Heithaus, 1982). In the Guianas, many Bombacaceae, such as *Pachira aquatica* (Fig. 4.1), and Caesalpiniaceae and Caryocaraceae are likely to be pollinated by bats. The flowers of *Eperua falcata* (Fig. 4.2), which hang at the end of a long pedicel, are often visited by *Anoura* spp. and *Lichonycteris* spp. (P. Charles-Dominique, personal communication) as are the light yellow, capitulate inflorescences of the emergent tree, *Parkia pendula* (Mimosaceae), visited by *G. soricina* at

Fig. 4.2 Pendulate flower position in the bat-pollinated *Eperua falcata* (Caesalpiniaceae). Photo P.-M. Forget.

Mabura Hill, Guyana (D. Hammond, personal observation; for a focused look at *Parkia* spp., see also Hopkins and Hopkins, 1983).

By other mammals

Apart from bats, at least 13 species of non-flying, frugivorous mammals may consume flowers and pollen as alternative resources to fruits (e.g. Prance, 1980; Janson *et al.*, 1981; van Roosmalen, 1985; Julien-Laferrière and Atramentowicz, 1990). Because visiting animals do not damage flowers, Janson *et al.* (1981) postulated that they might act as pollinators. Plants that were visited by non-flying mammals in the dry season in Peru include species in the family Bombacaceae (*Ceiba pentandra*, *Quararibea cordata*, *Ochroma pyramidale*), Combretaceae (*Combretum fruticosum*) and Fabaceae (*Erythrina* spp.). The night monkey *Aotus trivirgatus*, marsupials (*Caluromysiops irrupta*, *Didelphis marsupialis*, *Caluromys lanatus*), and procyonids (kinkajou and olingo, *Bassaricyon alleni*) were observed foraging at flowers of *Q. cordata* (Janson *et al.*, 1981). Several monkeys, including common squirrel monkeys (*Saimiri sciureus*), brown capuchins, *Cebus apella* (hereafter *Cebus*) and spider monkeys (*A. paniscus*) exploited the infructescence of *C. pentandra*, their heads covered with pale yellow pollen as a result (Janson *et al.*, 1981). The authors also casually observed other Amazonian monkeys feeding on the flowers of species in the Anacardiaceae, Bignoniaceae, Clusiaceae and Moraceae. The flowers and nectar of *Symphonia globulifera* (hereafter *Symphonia*) (Clusiaceae) are a preferred food item for many non-flying mammals throughout the neotropics (e.g. French Guiana: Julien-Laferrière and Atramentowicz, 1990; Brazil: Peres, 1994b).

In French Guiana, *Caluromys philander* L., *Philander opposum* L., *Didelphis marsupialis* L. (hereafter *Caluromys*, *Philander*, *Didelphis*), *Marmosa cinerea* and *Potos* were also observed feeding on nectar, especially when fruit was scarce in the September–November dry season (Charles-Dominique *et al.*, 1981; Julien-Laferrière and Atramentowicz, 1990). For instance, *Caluromys* visits flowers of *Inga thibaudiana*, *I. ingoides* (Mimosaceae), *Hymenaea courbaril* (Caesalpiniaceae), *Eperua falcata* (Caesalpiniaceae), *Symphonia*, and many other unidentified plant species whose anthers were found among the stomach contents (Charles-Dominique *et al.*, 1981; Julien-Laferrière and Atramentowicz, 1990). A closely related species, *C. lanatus*, forages at flowers of *Quararibea cordata* (Bombacaceae) in Peru and Ecuador (Janson *et al.*, 1981), and Gribel (1988) observed that flowers of *Pseudobombax tomentosum* (Bombacaceae) were also visited by *C. lanatus* in central Brazil.

By birds

Bittrich and Amaral (1996) studied the pollination biology of *Symphonia* in Manaus, Brazil. The flowers of this typically Guianan tree were mainly pollinated by hummingbirds, the blue-chinned sapphire, *Chlorestes notatus*, and the fork-tailed woodnymph, *Thalurania furcata*. Other nectar-feeding birds included flocks of the icterid *Cacicus cela*. In the Guianas, pollination by birds so far is poorly known. In a recent study Gill and colleagues (1998) observed birds visiting *Symphonia globulifera* flowers in central French Guiana: these mostly belong to the family Thraupidae, i.e. the black-faced dacnis, blue dacnis, green honeycreeper, purple honeycreeper and red-legged honeycreeper. Other casual visitors feeding on floral nectar included three tanagers, two Parulidae, the bananaquit and the waved woodpecker, with hummingbirds only infrequently and briefly visiting flowers. Given the accessibility of the axillary inflorescences to birds, the flower structure and colour (staminal tube, red, glued pollen) and the diurnal availability of sugar-poor nectar, Gill *et al.* (1998) were prompted to suggest that Thraupidae are the most important pollinators of *Symphonia globulifera*, whereas hummingbirds are merely incidental visitors or even nectar robbers.

Plant dispersal

By marsupials

High densities of *Caluromys*, especially in secondary forests (see Atramentowicz, 1986), suggests they may have an important role in disseminating small-seeded plant species (Atramentowicz, 1988). In fact, marsupials will drop large seeds that they cannot ingest but may disperse small-seeded species of pioneer plants (*Cecropia*, *Ficus*, Melastomataceae) ingested when foraging in the lower stratum of edge habitat (Charles Dominique *et al.*, 1981).

By bats and birds

The combined outcome of variation in resource and habitat preferences of bats directly influences plant recolonization after disturbance (De Foresta, 1983). Bats often use holes in the canopy to forage in the understorey where they may roost or feed on items they collected above (Charles-Dominique, 1986, 1993b). Since defecation occurs soon after feeding and typically when an individual is flying, the repeated actions of many bats produces an even seed rain across trunks, branches, logs and ground within gaps (De Foresta *et al.*, 1984; Charles-Dominique, 1986). Studying the seed rain at night, De Foresta *et al.* (1984) showed that bats were directly responsible for the dispersal of pioneer species such as *Solanum* spp., *Piper* spp. and *Cecropia obtusa* (see Prévost, 1983) in open areas more than 50 m from the forest edge as well as to the centre of a small gap within the forest, mainly because they defecate while flying (see also Charles-Dominique, 1986). Large fruit-eating bats, *Artibeus jamaicensis* and *A. lituratus*, two of the larger-bodied species inhabiting Guianan forests, consume mainly small-seeded fruits, ingesting the minute seeds that are eventually dispersed after passing through the digestive tract, but also consume larger-seeded species whose stony seeds are released beneath perches after the pulp has been scraped off (Charles-Dominique, 1986; see also Janzen *et al.*, 1976b) (Fig. 4.3).

Fleming and Heithaus (1981) observed that the density of small-seeded species declines with distance from a fruit tree in a Costa Rican dry forest. Seeds of a wide range of species were deposited beneath

(a)

(b)

Fig. 4.3. (a.) Pericarp of large-seeded fruit being scraped by *Platyrrhinus helleri* (Sternodermatinae), later to be dispersed to forest floor. Photo courtesy of P. Charles-Dominique. (b.) Clumped deposit of *Licania heteromorpha* (Chrysobalanceae) beneath feeding roost of fruit-eating bat. These are secondarily dispersed by seed-eating rodents, germinate to form high density seedling aggregates and/or are attacked by insect predators. Photo D. Hammond.

roosts, often in clumps. In French Guiana, with a home range of around 8–13 ha, the little fruit bats *Rhinophylla pumilio* disperse seeds at distances ranging from 100 to 300 m. The seed rain generated by this species is mainly composed of *Evodianthus funifer* seeds, which are mainly dispersed in humid valleys as opposed to adjacent, drier ridges (Cockle, 1997).

Dispersal of small seeds by birds is also important when remnant trees allow them to perch in gaps or clearings during the defecation process (Théry, 1990a). Consequently, both bats and birds (see above) contribute to the seed rain and the content and density of the seed bank, which are characterized by bat/bird-dispersed species such as *Cecropia* spp. and *Solanum* spp. in secondary forests and degraded areas (Holthuijzen and Boerboom, 1982; De Foresta and Prévost, 1986), and Araceae, Cyclanthaceae and Melastomataceae in mature forests (Cockle, 1997; Krijger *et al.*, 1997). The contribution of small-seeded bat/bird dispersal to regeneration is normally only important after the canopy is opened by either single or multiple treefalls (Riéra, 1985) or large, anthropogenic disturbances, such as clear-cut logging operations (De Foresta, 1983), or along forest access roads (D.S. Hammond, unpublished data). However, for some small-seeded, climbing species dispersed by bats, a closed canopy forest with a wide range of tree trunk sizes available for support provides the best conditions for establishment and reproduction (Cockle, 1997).

Though the relationships between frugivorous bats and small-seeded species are well described from both plant (De Foresta, 1983; De Foresta *et al.*, 1984; see also Fleming and Heithaus, 1981; Fleming *et al.*, 1985; Cockle, 1997) and animal (Charles-Dominique, 1991, 1995; Cockle, 1997; Charles-Dominique and Cockle, 2001; see also Fleming *et al.*, 1972) perspectives, scant literature is devoted to the role of large fruit-bats on the recruitment of large-seeded species which are typical of Guianan rainforests (Hammond and Brown, 1995). This group includes *Andira* spp. and *Dipteryx odorata* (Fabaceae), *Bocoa prouacensis*, *Swartzia* spp. (Caesalpiniaceae), *Eschweilera* spp. (Lecythidaceae), *Symphonia globulifera* (Guttiferae), *Couepia* spp., *Parinari* spp. and *Licania* spp. (Chrysobalanaceae) and *Caryocar glabrum* (Caryocaraceae). De Foresta *et al.* (1984) describe how a large fruit-eating bat *A. lituratus* can disperse the seed of *S. globulifera* up to 200 m into secondary regrowth. Such seed dispersal by bats is most likely responsible for the recuitment of *S. globulifera* as well as that of other large-seeded, bat-dispersed species in the families Chrysobalanaceae (*Licania* spp.) and Lecythidaceae (*Lecythis* spp. and *Eschweilera micrantha*) in areas of forest clearcut nearly 15 years ago (Larpin, 1989).

Greenhall (1965) observed that the large spear-nosed bat, *Phyllostomus hastatus*, fed upon the fleshy aril attached to the large seeds of the common Guianan sapucaia nut tree, *Lecythis zabucajo*. He also noted that piles of seeds developed below the roosts of the large spear-nosed bat *P. hastatus*, but not below those of the large fruit-eating bats *A. jamaicensis* or *A. lituratus*, two other bats large enough to carry the nut. Given the high diversity of *Lecythis* spp. and other bat-adapted fruit species in Guianan rainforests, and given that these may be either eaten or secondarily dispersed by scatterhoarding rodents (P.-M. Forget, personal observation), as is the case with *Dipteryx panamensis* in Panama (see Forget, 1993), the ultimate effect of bat seed dispersal on the recruitment of many large-seeded species in the Guianas is potentially of great consequence, yet still largely unknown.

By monkeys

Despite their less diverse diet, spider monkeys act as the main seed dispersers in Guianan forests, in terms of both the number and the diversity of seeds ingested, and the total forest area over which seeds are dispersed (van Roosmalen, 1985; Guillotin *et al.*, 1994; Simmen and Sabatier, 1996; see review in Simmen *et al.*, 2001). At Voltzberg, Suriname, van Roosmalen (1985) recorded that 81%, 6% and 14% of 171 species (in 58 families) were dispersed, dropped beneath trees and predated by black spider monkeys, respectively. With a home range of 225 ha, encompassing mostly high rainforest habitat (>90% of sightings), spider monkeys can carry seeds internally for more than 1 km and then deposit these (72%) beneath both conspecific and allospecific trees (>25 m height) while travelling through the canopy (van Roosmalen, 1985; see also Norconk and Kinzey, 1994; Forget and Sabatier, 1997). In comparison, Julliot (1996b) observed that howler monkeys in Nouragues were capable of carrying seeds internally over a distance of 550 m, but more typically dispersed seeds at a distance of 255 m from the parent tree. Of the 96 species consumed by howler monkeys at Nouragues, 94% were dispersed via defecation, often encased in a foliage-rich dung, and deposited beneath sleeping areas (60% of species) (Julliot, 1994; see also Julliot, 1997). The concentrated deposition of seeds beneath trees by both howler monkeys and spider monkeys (Julliot *et al.*, 2001) often creates localized patches of high seed density which are then subsequently preyed upon or scatterhoarded in large numbers by various ground-dwelling vertebrates (P.-M. Forget, personal observation; F. Feer, personal communication). The largest seeds dispersed internally by either spider monkeys or howler monkeys in Suriname and French Guiana are *Platonia insignis* (4 × 2 cm, Clusiaceae; van Roosmalen, 1985), *Pouteria laevigata* (4 × 2.5 cm, 9 g fresh weight, Sapotaceae; Julliot, 1996a; F. Feer, personal communication) and *Virola kwatae* (up to 4 × 2 cm, 7 g fresh weight, Myristicaceae; Sabatier, 1983; Forget and Sabatier, 1997).

In comparison, brown capuchins eat fruit pulp from 135 species (in 42 families), dispersing the smaller seeds of 113 species and discarding the larger-seeded species (14 families) beneath the parent trees (Zhang, 1994). The combination of a typical home range of 355 ha and daily forays averaging 2.3 km means that capuchins disperse seeds of many small-seeded species over considerable distances in high forest. Seeds are often deposited below canopy trees, but unlike their larger primate cousins, capuchins will forage in the lower strata, especially during intermediate periods of fruit availability in the upper strata (Zhang, 1994).

Given their preference for disturbed habitats, their limited gut volume and their relatively small body size, the smaller *Saguinus* are most likely dispersers of small seeds of light-demanding ruderal (e.g. Convolvulaceae, Curcurbitaceae, Melastomataceae) and long-lived pioneer (e.g. *Goupia glabra*, Celastraceae) species in the Guianas. The most frequently (annual occurrence greater than 10%, N=69 seeds) observed species in stomach contents (N=40) were *Pourouma* sp. (Cecropiaceae) in October–January (dry to early wet season), and an unidentified Gesneriaceae and *Inga* sp. (Mimosaceae) in February–September (wet to early dry season) (Pack *et al.*, 1998).

There is no clear evidence that passage through the digestive tract increases germination success of seeds ingested and defecated by spider monkeys, howler monkeys or capuchins (Julliot, 1994; Zhang, 1995; P.-M. Forget, personal observation). Moreover, seed dispersal effectiveness can differ between monkey species based on the number of dispersed seeds which actually establish (Zhang and Wang, 1995b). However, it is difficult to predict the fate of seeds dispersed in monkey faeces since the patterns of deposition may vary according to season, the distance between fruiting trees which are visited during a foray, the diversity of plant parts consumed and the activity levels of these arboreal frugivores at the time of consumption.

By carnivores

Kinkajou disperse seeds of a much broader size range (Julien-Laferrière, 1999a). Given the breadth of their diet, the importance of kinkajou as a seed disperser is probably underestimated due to their highly nocturnal habit. In forests where populations of diurnal primates have been severely reduced through hunting, the relatively high abundance of kinkajou in these stands (Charles-Dominique *et al.*, 1981 vs. Rylands and Keuroghlian., 1988, and references therein; see also Kays and Gittleman, 1995) may compensate to some degree for the loss of these important seed-dispersers, though the need to conserve primates in the Guianas is clear (Mittermeier, 1987). Moreover, kinkajou appear to be much more efficient dispersers of large seeds than marsupials, mainly because they restrict their feeding to the upper strata of the forest and thus often defecate seeds at heights exceeding 20 m (D. Julien-Laferrière personal communication, P.-M. Forget, personal observation).

By ungulates

In Guianan forests, only a few plant species such as *Spondias mombin* (Anacardiaceae) are adapted to regurgitative dispersal by deer and peccaries, and gut dispersal by tapir (Kiltie, 1982; Janzen, 1986; Feer *et al.*, 2001b; Beck, 2005). While the role of deer and peccaries as seed dispersers is weak, given their ability to destroy most of the seeds they ingest (only those tiny seeds of Moraceae, Cecropiaceae and Passifloraceae may pass the filter; P.-M. Forget, D.S. Hammond, personal observations), the tapir is a potential seed disperser for many plant species (Henry *et al.*, 2000). Given the range of seed sizes ingested, and its ability to defecate intact seeds in large faecal deposits, the tapir could be considered as the last of the late, great neotropical megaherbivores whose dietary make-up and role as wide-ranging disperser of both small- and large-seeded (Janzen and Martin, 1982) (e.g. *Attalea butyracea*, *Maximilliana maripa*) species has been inherited by no other indigenous, neotropical forest animal (Silvius, 2002; Fragoso and Huffman, 2000). Escape from insect parasitism of tapir-dispersed seeds is often, however, dependent on subsequent scatterhoarding by either caviomorph (Hallwachs, 1986; Forget *et al.*, 1994; Fragoso, 1997; Silvius and Fragoso, 2002; see also Quiroga-Castro and Roldan, 2001) or myomorph (Janzen, 1986) rodents at a much smaller scale.

Interestingly, Fragoso (1997) suggests that the distribution of a common palm, *M. maripa* (syn. *Attalea maripa*, *A. regia*), on Maracá Island is largely due to tapir dispersal. However, *M. maripa* is more commonly associated with anthropogenic disturbance (as is *A. butyracea* in Central America), particularly fire-fed forest clearance (Balée, 1988; Kahn and De Granville, 1992). The widespread presence of charcoal on Maracá and record of former human inhabitation (Thompson *et al.*, 1992) suggests that the influence of historical events upon adult *M. maripa* distribution cannot be discounted. The widespread burning of the island reserve during the El Niño-induced drought in early 1998 is testimony to the frequency with which fire can affect some relatively wet tropical forest areas. Nonetheless, the survivorship of the small fraction (<1%) of *M. maripa* seeds taken by tapirs is clearly enhanced by their long-distance dispersal and faecal deposition, and is likely to be intertwined with subsequent short-distance seed dispersal by caviomorph rodents as is the case with *A. butyracea* in Central America (Forget *et al.*, 1994). Note that *M. maripa* occurs at low density in Guianan forests, and was not observed in tapir stomachs taken in French Guiana (F. Feer and O. Henry, personal communication). Tapirs at Maracá defecated seeds of 38 tree species (Fragoso, 1997), but no details were given of the vegetative plant parts consumed.

By rodents: predation or dispersal?

The fact that some animals consume seeds does not necessarily imply that they are always predators (Forget *et al.*, 2005). Hoarding behaviour (Morris, 1962; Vander Wall, 1990), for example, is often intertwined with seed predation. Squirrels, mice, spiny rats, agoutis and acouchies will hoard seeds in many single, scattered caches in the axiles of tree branches, below ground, in burrows or dens, and on the soil surface. Seeds may have been dispersed previously by bats (*Dipteryx* spp.: see Forget, 1993) or monkeys, birds and kinkajous (*Virola* spp., Myristicaceae: see Forget and Milleron, 1991). Only those seeds buried several centimetres deep (e.g. *Vouacapoua americana*, Caesalpiniaceae: Forget, 1990; *Moronobea coccinea*: Forget, 1991a; *Carapa procera*, Meliaceae: Forget, 1996) or covered by litter (*Astrocaryum paramaca*, Arecaceae: Forget, 1991b) would be placed in the most favourable microsites for recruitment (Forget, 1991a, 1994, 1997a), since this increases the chance that they are not recovered later or found by other rodents and peccaries (Kiltie, 1981). Finally, although they may ultimately consume many cached seeds later on, many may be left to establish under the favourable microsite conditions created through the burial and the local redistribution of seeds (Feer *et al.*, 2001b; Feer, 1999).

Another understated aspect of seed–granivore interactions is the notion of sub-lethal attack. Seeds of many Guianan trees and lianas are capable of germination and seedling establishment after withstanding substantial loss in seed mass (Hammond and Brown, 1998). In many instances, the embryo in these seeds is positioned so that the most common feeding methods employed by vertebrates leave the embryo intact (Fig. 4.4). In some cases, partial consumption of seed mass stimulates germination (e.g. *Clathrotropis brachypetala*), similar to the effect that burial during caching can have on other species.

Whether a rodent acts as a disperser or

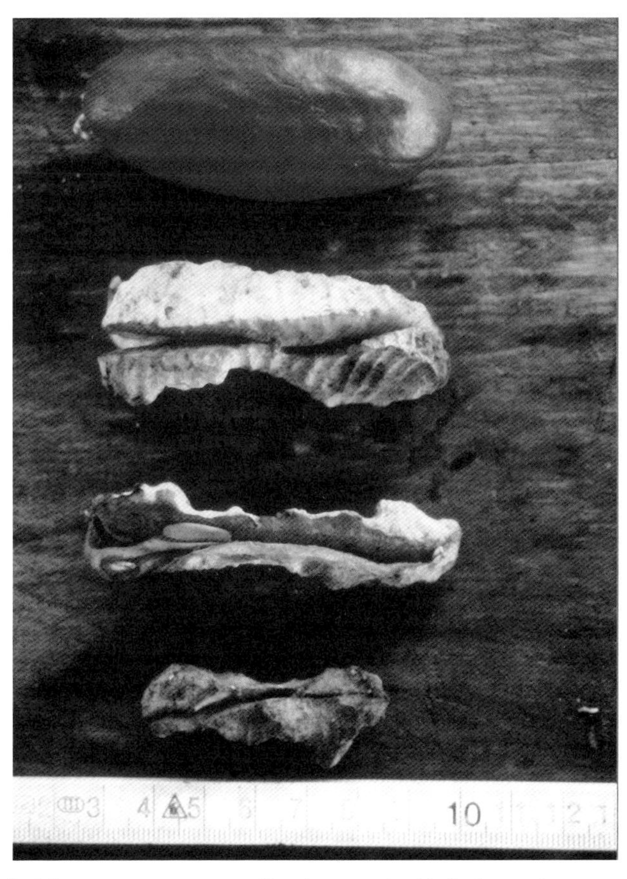

Fig. 4.4. Intact seed of *Catostemma commune* (Bombacaceae) with fleshy seed coat and subject to various levels of seed mass loss as a consequence of rodent feeding. Note intact and differentiating embryo at left end of major axis and its persistence despite considerable loss of surrounding storage tissue. Photo D. Hammond.

predator with respect to any given resource depends largely on phylogeny (Forget *et al.*, 1998) and the characteristics which define the resource base, i.e. (i) how much fruit is available over the home range; (ii) how large is the seed reward; (iii) how toxic are the seeds, especially ungerminated seeds compared to those which have germinated; (iv) what is the risk of predation during foraging; and (v) how important are food reserves to offspring survival and maternal care.

Once seeds have been released from the parent through natural dehiscence or through the feeding activities of arboreal and flying frugivores, they fall into the foraging realm of terrestrial granivores, among which rodents such as spiny rats, acouchies and agoutis are potential secondary seed dispersers. The potential role of rodents in the regeneration and recruitment of tropical plants is enormous, given their relatively high densities and reproductive rates. Since these animals have relatively small home ranges, the probability of seeds being transported long distances is low, although distances of up to 30–50 m are possible. In general, seeds will be taken and cached short distances (<10–20 m) from the site of initial deposition. Because most of the large seeds (those that offer ample energetic reward and yet can still be handled by rodents) are in the first instance transported by frugivores such as toucans and large monkeys that avoid gaps, it is likely that such seeds will be secondarily dispersed in the understorey, along forest edges, and to a lesser degree in the centre of a forest clearing if this offers some protection to rodents from their own predators through dense, low-lying logs and branches (e.g. Schupp and Frost, 1989). Through a complex net of seed consumption, caching and dropping, rodents can influence the spatial distribution of regeneration and, ultimately, the hierarchy in which juvenile, woody plants must compete at any particular microsite. This process is particularly important for large-seeded species.

Often there are few alternative agents capable of effectively redistributing these seeds which are, for the most part, deposited beneath the parent tree. Escape (after Howe and Smallwood, 1982) from density-dependent attack by insects via dispersal away from adult trees is one possible means in which rodents might positively influence a species success in an area. Alternatively, dispersal away from trees by arboreal frugivores may decrease the likelihood of seeds being encountered by largely predaceous rodents. However, most tropical rodents appear not to attack seeds with regard to distance from parent trees (Hammond and Brown, 1998), since they tend to forage for a wide range of seeds and fruits within a limited, but plastic, home range (e.g. Smythe, 1978). Thus, the putative advantage of dispersal by arboreal frugivores may be offset by the distance-independent foraging behaviour of seed-eating rodents (e.g. Janzen, 1982; Forget, 1993).

Due to the high relative abundance of large-seeded woody plants (Hammond and Brown, 1995) with fruits adapted for synzoochorous dispersal (van Roosmalen, 1985), seed- and fruit-eating rodents must rank among the most important vertebrate herbivores in Guianan forests. For the offspring of many canopy tree species, it appears that removal by rodents is the most effective means of avoiding an early death from their density/distance-dependent invertebrate or fungal predators (e.g. *Chlorocardium rodiei* in Guyana (Hammond *et al.*, 1999), *Carapa guianensis* in Trinidad (Rankin, 1978) and Guyana (Hammond *et al.*, 1992)) and colonizing suitable upland sites on ridges and hillsides (Hammond *et al.*, 1992; Forget, 1992a). Thus, the way in which those factors which shape resource availability (Hammond and Brown, 1998) and use (Janzen, 1969) by seed foragers interplay is paramount in determining the outcome of rodent attack on seeds (e.g. Schupp, 1988; Forget *et al.*, 1994; Forget, 1994). For example, there are thought to be distinct differences in the capacity of caviomorph and myomorph rodents to cope with the relatively large concentrations of toxic secondary chemicals in seeds (Rankin, 1978), possibly due to contrasting detoxification pathways (Williams, 1972; Freeland and Janzen, 1974).

By birds

Birds often systematically carry fruit away from the source tree. Toucans, for instance, regularly fly on a round-trip schedule with a few large fruits or seeds taken at each passage, and then rapidly regurgitate these at a short distance (e.g. <40 m) from the focal fruiting trees (Howe and Primack, 1975; Howe *et al.*, 1985). Smaller birds, such as manakins, which intensively visit fruiting trees, disperse seeds along a gradient of microhabitats extending from mature forest to the largest clearings while foraging for resources or displaying at lekking sites (Théry, 1990a,b,c, 1992; Krijger *et al.*, 1997). Defecation can be rapid (around 12–15 min; Worthington, 1989) and takes place from isolated trees in open vegetation, along gap edges, roads and riverbanks, and other sites far from the parent source (Théry, 1990b). Many of the lek-forming Pipridae feed rapidly during display bouts, and will disperse seeds directly into well-lit microhabitats which enhance colour display and courtship behaviour (Théry, 1987a, 1990b,c; Théry and Vehrencamp, 1995; Endler and Théry, 1996). Some light-demanding Melastomataceae species are preferentially disseminated by piprids which display at forest edges, in gaps or in high closed forest (Krijger *et al.*, 1997; see also Poulin *et al.*, 1999).

The cock-of-the-rock (*R. rupicola*) illustrates another example of directional (after Howe and Smallwood, 1982) seed dispersal by birds. The cock-of-the-rock forms leks, feeds upon a wide variety of fruits (1–3 cm in length) collected from small shrubs and trees up to 40 m into the canopy (Schuchmann, 1984; Erard *et al.*, 1989) and deposits the seeds on to the arena (Théry and Larpin, 1993). The arena itself is often located at the centre of a dense thicket of saplings, shrubs and treelets which may have grown from long-term seed deposition by the cock-of-the-rock (Théry and Larpin, 1993).

A faithfully frugivorous diet can have a tremendous long-term effect on a species' behaviour. The elaborate mating rituals, especially lekking and communal behaviour, displayed by manakins and other common frugivorous birds in the Guianas (e.g. cotingids) are thought to be largely a product of their specialization on fruits and nectar, which are more rapidly located and consumed than other resources. This efficient capture of resources has made the paternal role in offspring care redundant, leading to the display behaviour characteristic of these groups (Snow, 1963). In turn, the evolution of such displays in manakins would solidify the consummate nature of their frugivory, enhancing the co-evolutionary relationship between consumer and producer.

Seed predation and seedling recruitment limitation

Pre-dispersal predation

BY MONKEYS While most of the Guianan monkeys (see 'Arboreal Herbivorous Mammals', above) do not damage the seeds contained within the fruits they feed on, the brown bearded saki and the Guianan saki are known as true canopy seed-predators, spending 95–99% of their feeding time eating fruit whose seeds they digest (van Roosmalen *et al.*, 1988; Kinzey and Norconk, 1993; Norconk *et al.*, 1997). The average crushing resistance value of seeds eaten by *Pithecia* is intermediate in hardness between those eaten by *Chiropotes* and those swallowed intact by spider monkeys (Kinzey and Norconk, 1993). Seeds masticated by brown bearded saki are soft compared with seeds masticated by the Guianan saki, and especially soft compared with those swallowed by spider monkeys (Kinzey and Norconk, 1993). A closely related species, the white-nosed bearded saki (*C. albinasus*) was observed eating the swollen receptacle of *Anacardium excelsum* (Anacardiaceae), the mesocarp of the babaçu palm *Orbyginia martiana* (Ferrari, 1995) and the fruit pulp of *Astrocaryum vulgare* (Arecaceae) in Brazil (Ayres, 1981, 1989). Apart from intensive seed predation by saki monkeys, capuchins also act as seed predators during the Guianan dry season, consuming immature fruit and seeds of palms in particular and discarding the fruits of many other species on the ground below the parent tree (e.g. Zhang, 1994).

BY BIRDS Snow (1981) also identified two main groups of seed-eating birds: (i) those that consume seeds adapted to dispersal by other frugivores; and (ii) those that feed upon seeds adapted for abiotic, often wind, dispersal. In the Guianas, as in most neotropical forests, the first group would include such frugivorous–granivorous birds as guans and curassows, which pick up fresh fruit and seeds (here below referred to as ripe fruit) on the ground after they are dropped by arboreal consumers or dehisce. The second group consists mainly of parrots and macaws, whose strong, powerful beaks allow them to exploit the soft, green seeds which are contained within the indurate pericarp of unripe fruit. Parrots manipulate fruit with their feet, beak and tongue. They open hardened fruits and seeds and then sculpt the softer seed mass into a bolus of a size and shape that is easily swallowed (Hopkins and Hopkins, 1983; Norconk *et al.*, 1997). As has been generally observed elsewhere in the neotropics (e.g. Higgins, 1979), the large confamilial group of parrots, parakeets and macaws (Psittacidae) destroy vast quantities of green fruit and seeds in the Guianas every year. Often these are from wind-dispersed tree species such as *Dicorynia guianensis* (Caesalpiniacaeae), *Couratari guianensis* (Lecythidaceae) and *Qualea rosea* (Voshysiaceae), and bat-dispersed species such as *Swartzia remiger*, *S. schomburgkii*, *S. leiocalycina* (Caesalpiniacaeae) (Forget, 1988; Loubry, 1993; D. Hammond, personal observation). Another important food source for parrots is the long, black legumes of trees in the genus *Parkia*, which are locally known as macaw-bean soup tree ('faveira arra tucupi') in Brazil (Hopkins and Hopkins, 1983). Several species of *Parkia* have been recorded as being regularly attacked by parrots and macaws, such as *Pionus fuscus* at *P. panurensis*, and *Ara araurana*, *A. macao*, *Deropytus acciptrinus* in Manaus, Brazil; similar damage was recorded from herbarium specimens of *P. multijuga*, *P. pendula* and *P. nitida* from Guyana (Hopkins and Hopkins, 1983). Norconk *et al.* (1997) report that *Ara chloroptera* fed upon seeds of Anacardiaceae (*Spondias mombin*), Burseraceae (*Protium tenuifolium*), Euphorbiaceae (*Sapium glandulosum*), Fabaceae (*Centrolobium paraense*, *Pterocarpus acapulcensis*) and Sapotaceae (*Chrysophyllum lucentifolium*) in nearby Guri Lake, Venezuela. Rankin (1978) noted that immature pods of *Mora excelsa* were attacked by parrots in Trinidad. Such destruction, unfortunately, is rarely quantified, despite its potential regulatory role in seedling establishment and recruitment. Forget (1988) estimated that up to 63% of the crop of green fruit of a single *Swartzia remiger* tree was attacked by the large parrot *Amazona farinosa*. As a consequence, many intact and partly damaged seeds were dropped beneath the parent tree, but this waste (after Howe, 1980) had little effect on early recruitment, since very few seeds were subsequently attacked by insects or rodents (Forget, 1988). At another tree with no apparent parrot activity, most fruits matured, the pods opened and the lack of regeneration beneath the tree crown suggests that bats may have effectively dispersed seeds in the absence of pre-dispersal attack by parrots.

Post-dispersal predation

BY BIRDS Erard *et al.* (1991) analysed the stomach contents of 69 crested curassows (*Crax alector*, Cracidae) and 17 great tinamous (*Tinamus major*, Tinamidae) and identified the debris of fruits and seeds as belonging to 80 (in 28 families) and 38 species (in 18 families), respectively. Despite the fact that both birds restrict their foraging to the forest floor, there is a weak overlap of their diet (Erard and Théry, 1994). Erard *et al.* (1991) also remarked that Meliaceae (*Guarea* spp.) and Myristicaceae (*Virola* spp.) accounted for more than 50% of the diet (in terms of g dry weight) in crested curassows and great tinamous, respectively. No ingested seeds with a length >1 mm pass through the digestive tract of these birds intact, due to the efficient grinding action in their gizzards (Erard and Sabatier, 1989; P.-M. Forget, personal observation). Therefore, in contrast to other large terrestrial birds (see previous section, e.g. trumpeters), crested curassows and great tinamous can be considered true seed-predators.

BY GROUND-DWELLING MAMMALS Large rodents, peccaries, deer and tapir are important ground-dwelling, vertebrate seed/seedling predators in the Guianas. Vegetative plant parts can make up nearly 10% of the diet (dry weight of stomach content) of large, frugivorous rodents such as agouti during the peak (February–March) and trough (October–November) of fruit production in French Guiana (Henry, 1994a, 1999). Plant parts consumed during periods of fruit scarcity often consist of germinating seeds and young seedlings, while older, established seedlings are often consumed during the dry season (P.-M. Forget, personal observation) when fruitfall is generally greater. The number of Guianan plant species whose stems and leaves are consumed by large rodents has not been directly quantified, but it probably includes species whose seeds are scatterhoarded as well, such as *Carapa procera* (Meliaceae) (Forget, 1996; Jansen *et al.*, 2002, 2004, 2005), Chrysobalanaceae (Jansen and Forget, 2001) and various palm species (Sist, 1989a,b; Forget, 1991b), for instance. In contrast to the seasonal use of vegetative parts by large terrestrial rodents, leaves and fibres constitute a negligible fraction of the diet of smaller, frugivorous–insectivorous rodents such as spiny rats (*Proechimys* spp.), and rat mice (*Oryzomys*), regardless of season (Henry, 1994a, 1996).

Seeds and seedlings are also eaten by peccaries (Kiltie, 1981) and deer (see Bodmer, 1991). In February–June, for instance, collared peccaries are often and regularly observed feeding on fallen seeds at large-seeded trees such as *Carapa procera* (Meliaceae) in French Guiana (see Forget, 1996), and if seeds are not taken and cached further away by rodents, there is very little chance to find some intact ones (Jansen and Forget, 2001; Jansen *et al.*, 2002). Similarly, Ayres (personal communication in Hopkins and Hopkins, 1983) noted that white-lipped peccaries were observed feeding on *Parkia multijuga* (Mimosaceae) seeds, which are adapted to scatterhoarding by rodents (Hopkins and Hopkins, 1983). A general rule is that both soft and hard seeds are thoroughly masticated (e.g. Kiltie, 1982), or swallowed and digested in the stomach where detoxification takes place, in deer, for instance (Feer *et al.*, 2001b). Deer and peccaries may thus be defined as true predators. Only those seeds hard enough to resist being cracked and ground down might be dispersed either via regurgitation or via defecation (Janzen, 1983a, Feer *et al.*, 2001b; Beck, 2005).

Henry (1994a) found that consumption of plant parts (leaves, fibre and wood) by collared peccaries was high in the wet season (up to 45% of diet in December–January) and lower in the dry season (June throughout September) when animals shift to insect larvae and seeds, many of which are those cached by rodents in previous months (see also Kiltie, 1981). Leaves, stems, bark and roots accounted for over 50% of rumen contents in red brocket deer in Suriname during August–September (Branan *et al.*, 1985), though rumens were not sampled during the peak fruitfall period from January to April. Interestingly, fungal fruiting bodies (probably Polyporaceae, Aphyllophorales) accounted for 10–50% of contents in any given month of the survey. The importance of vegetative plant parts in the diet of red brocket deer and tapir in Guianas is not yet fully analysed, but it is crucial for their reproduction in light of other studies at various Amazonian sites (see Bodmer, 1990a,b; Rodrigues *et al.*, 1993). Extrapolation of ungulate resource use in forests of western Amazonia to those in the Guianas, however, requires some caution, since Guianan forests are never nearly as inundated as those in floodplain forests of the Peruvian Amazon, and are generally characterized by higher family richness (see Bodmer, 1990b).

Food Plant and Vertebrate Dietary Diversity: a Synthesis

After reviewing the diet of 23 animal or species groups (six birds, eight arboreal and nine terrestrial mammals), with body weights ranging from several grams (averaging 15 g in four piprid species) to 300 kg, one may rank plant families by their relative contribution to the diet of the faunal community (Appendix 4.2) and discuss such diversity in light of dietary choice. Except for species-rich families such as Orchidaceae, Piperaceae, Bromeliaceae, Cyperaceae and Poaceae, which contribute little to dietary diversity, the 32 families contributing to the diet of at least 50% of the studied vertebrate species (i.e. fed upon by at least 12 species) are among the richest families with the number of species fed upon ranging from 4 to 69 (Table 4.8). The most frequent families occurring in the diet of (mainly French) Guianan vertebrates are Moraceae (100% of vertebrate species consuming at least one species from the family), Burseraceae, Annonaceae, Arecaceae (see previous subsection), Rubiaceae, Myrtaceae and Sapotaceae. Most of these are among the 20 most-abundant families with diameter at breast height >10 cm at the Nouragues forest in French Guiana (Poncy *et al.*, 1998; Belbenoit *et al.*, 2001), but several families which are abundant in French Guiana, notably the Moraceae, Burseraceae and Arecaceae, tend to decline significantly as one sweeps northwestward towards Roraima and Venezuelan Guayana (see Fig. 7.8, Chapter 7) and are substantially less important in northern and central Guyana than more easterly or southerly sites in the Guianas (Ek, 1997; ter Steege and Hammond, 2001; ter Steege, 1998).

The larger the body size of an animal, the greater the movement across its home range (Swihart *et al.*, 1988) and the higher the diversity of plants it may encounter. Also, regardless of feeding niche, i.e. frugivore–omnivore/herbivore/granivore/browser, there is a significant relationship between body mass and population density (Robinson and Redford, 1986). When plotting body weight against the percentage of all known Guianan forest plant families included in the diet, we only found a significantly positive relationship for arboreal mammals (r^2=0.653, P=0.015, N=8 species) (Fig. 4.5).

One may remark that, for most arboreal vertebrates, the information obtained from visual records of feeding in the wild, especially when whole troops of monkeys were followed for several years, or from stomach content analyses is likely to be incomplete. Stomach content analysis alone does not

Table 4.8. Rank occurrence of plant families represented in the diet of 23 herbivorous vertebrate species or groups and the number of plant species in these families found at Nouragues (this study, Poncy *et al.* 1998; Belbenoit *et al.*, 2001).

Family	In diet	In forest
Moraceae	23	29
Burseraceae	21	24
Annonaceae	20	26
Rubiaceae	20	69
Myrtaceae	19	37
Arecaceae	18	23
Lauraceae	18	37
Sapotaceae	18	64
Clusiaceae	17	20
Anacardiaceae	16	9
Euphorbiaceae	16	27
Mimosaceae	16	65
Myristicaceae	16	8
Sapindaceae	16	25
Chrysobalanaceae	15	37
Caesalpiniaceae	14	30
Cucurbitaceae	14	6
Fabaceae	14	23
Melastomataceae	14	46
Meliaceae	14	19
Convolvulaceae	13	6
Flacourtiaceae	13	9
Bombacaceae	12	9
Humiriaceae	12	5
Lecythidaceae	12	29
Menispermaceae	12	6
Polygalaceae	12	6
Sterculiaceae	12	7
Apocynaceae	11	23
Araceae	11	57
Bignoniaceae	11	14
Boraginaceae	11	6

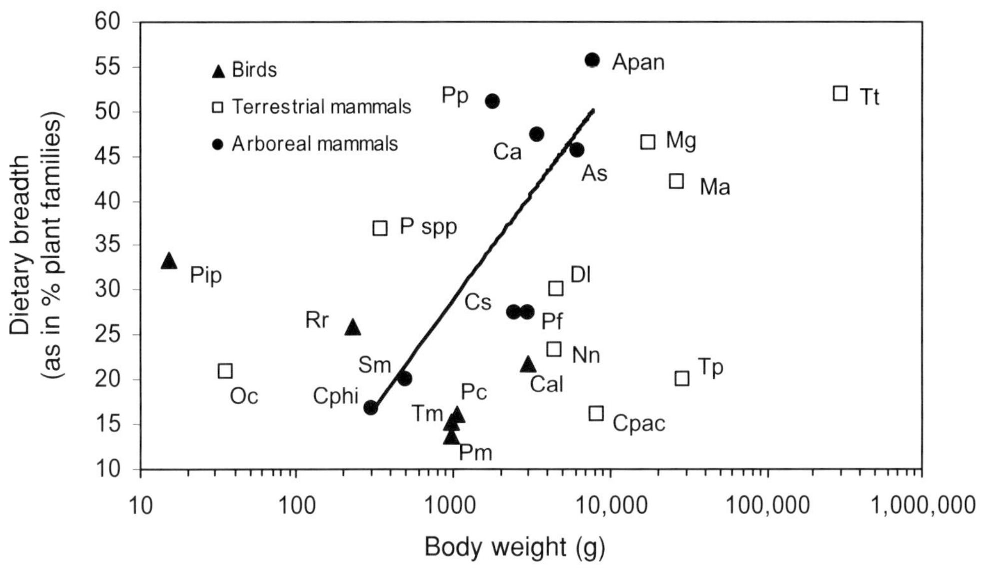

Fig. 4.5. Relationship between body mass and dietary breadth within three Guianan bird and mammal species or groups. Dietary breadth is the known number of plant families consumed for each vertebrate species at Nouragues as a percentage of the overall number of plant families present at the site. Species name abbreviations: Birds – Pip: Piprid group; Rr: *Rupicola rupicola*; Tm: *Tinamus major*; Pm: *Penelope marail*; Pc: *Psophia crepitans*; Cal: *Crax alector*. Arboreal mammals – Cp: *Caluromys philander*; Sm: *Saguinus midas*; Pp: *Pithecia pithecia*; Pf: *Potos flavus*; Cs: *Chiropotes satanas*; Cap: *Cebus apella*; As: *Alouatta seniculus*; Apan: *Ateles paniscus*. Terrestrial mammals – Oc: *Oryzomys capito*; P spp.: *Proechimys cuvieri* and *P. guianensis*; Nn: *Nasua narica*; Dl: *Dasyprocta leporina*; Apac: *Agouti paca*; Mg: *Mazama gouazoubira*; Ma: *Mazama americana*; Tp: *Tayassu pecari*; Tt: *Tapirus terrestris*. See text for data sources. Regression line is for arboreal mammals: r^2=0.61, P=0. 018, N=8 species.

always allow a correct description of the diet diversity because it merely represents a single moment in feeding history, delimited by the time between ingestion and assimilation/defecation and, thus, wholly dependent on gut retention times. For these reasons, it is more than likely that the small samples for some frugivorous species (e.g. large birds) has underestimated the range of plant families consumed. In the piprid group, on the contrary, our expectation of the diet diversity is high, with at least 32% of plant families likely to be eaten by these small birds, weighing 11–17 g on average. Note that many of these families have small fruits and, except Rubiaceae and Melastomataceae, are represented by a relatively small number of species in the Guianas. There is a trend though, not significant, among terrestrial mammals for an increased body size to be associated with a more diverse diet, especially after incorporating the large number of grass species consumed by deer and tapir. However, information remains incomplete for many less apparent, small-bodied rodents and bats and, in particular, for the much larger-bodied paca and peccaries.

Palms: an Outstanding Plant Group in the Guianas

Despite the predominance of the fig family in the diet of all studied vertebrates (see previous section) and unlike many New and Old World tropical forests, figs (*Ficus* spp.) are relatively rare in most undisturbed mature Guianan forests. There is very little evidence to suggest that this genus acts as a keystone (after Terborgh, 1986) plant resource where it does occur and may not fulfil a pivotal role in sustaining vertebrate frugivores in eastern Amazonia (e.g. Peres, 1993b, 1994b) as is suggested for sites in tropical lowland Peru (Terborgh, 1986) or parts of SE Asia (Leighton and Leighton, 1983). However, in comparison, palms are one of the most successful woody plant families in the Guianan forest ecosystems (e.g. Sist, 1989b,c), albeit never reaching the pervasiveness, in terms of density or diversity, of their confamilials in western Amazonia or Central America (Kahn and De Granville, 1992).

Much of the success of palms could be attributed to their many relationships with animals, foremost among these being seed dispersal (Zona and Henderson, 1989). A broad spectrum of animals, invertebrate and vertebrate, rely upon the pulp and seeds of palms for their survival and reproduction throughout the year. In our survey at least 18 animal species or groups of birds, primates and ground-dwelling mammals are the predominant consumers and dispersers of palm fruit and seeds, though other arboreal and flying mammals occasionally feed on palms as well.

Fruit from the palm *Jessenia bataua* are intensively exploited by brown capuchins in Nouragues forests (Zhang, 1994) due to the extended period of fruiting in the dry season (September–November), when overall fruit availability is low (Sist, 1989c; Sist and Puig, 1987; Zhang and Wang, 1995a; see also Peres, 1994a). Several species of palms in central Guyana are known to fruit mainly during the trough in production among sympatric woody dicots (June–December) (ter Steege *et al.*, 1996), or fruit continuously throughout the year (D. Hammond, unpublished data; e.g. *Astrocaryum aculeatum, Bactris humilis, Euterpe oleracea*). *Oenocarpus oligocarpa* produces fruit mainly between peaks in community-wide production at Piste de Saint Elie (Sabatier, 1985). In secondary forests of French Guiana, fruits of *Attalea regia* and *Astrocaryum vulgare* are produced during several months in the wet season and are two important food resources for marsupials, kinkajou and rodents, this last group of animals being largely responsible for seed dispersal (Charles-Dominique *et al.*, 1981). *Attalea maripa* fruit is also an important resource for tapirs, rodents and other mammals when they ripen during the lull in overall fruitfall on Maracá Island, Brazil (Fragoso, 1997; Silvius, 2002). Sist and Puig (1987) noted that parrots consume large fruit of *Jessenia*, dropping many seeds beneath the parent stem and adjacent trees used as perches. Their role as seed dispersers is therefore somewhat weak in comparison to that of toucans (e.g. *Rhamphastos tucanus*) and other large frugivorous birds, and spider monkeys (van Roosmalen, 1985), which are capable of dispersing seeds much further than parrots. Because *J. bataua* seedlings require a period of growth in shade before reaching adulthood, it is unlikely that aggregated seeds and seedlings beneath parents will find favourable conditions in the immediate vicinity of parents which often remain standing after large disturbance events (De Granville, 1978).

Other palms, such as species of *Astrocaryum* and *Bactris*, are often understorey specialists, and exclusively dispersed by scatterhoarding rodents, acouchis, agoutis and squirrels (van Roosmalen, 1985; Sist, 1989a,b; Forget, 1991b). It is probable that seed dispersal efficiency (i.e. the ratio between hoarding rate and predation rate) may vary according to the degree of habitat specificity of either rodent (see Dubost, 1988) or palm (Sist, 1989c) as well as the overall diversity of resources available at the time of fruit maturation in palms ('frequency-dependent selection', after Greenwood, 1985).

Palm fruit is also likely to be a major resource for ungulates in Guianan forests. However, compared to the predominance of *Mauritia flexuosa* fruits in the diet of ungulates in varzea forests in central and western Amazonia (Bodmer, 1990b, 1991), it is unlikely that this resource would represent such an important resource for ungulates in Guianan forests where well-drained soils predominate and *M. flexuosa* is poorly represented (P.-M. Forget, personal communication). Localized depressions and perched catchments along dolerite dykes in central Guyana are often exclusively inhabited by *M. flexuosa* and *J. bataua* and the importance of these concentrations of monotypic palm resources to ungulates and other forest fauna requires investigation. At the Mabura Hill Forest Reserve, the incidence of tapir tracks and faeces is two to four times higher in palm swamps during the dry season compared to other riverine and upland forest types (D. Hammond, unpublished data). Conversely, the role of rodents and ungulates as either predators or dispersers of palm seeds is poorly understood in Guianan forests, despite the advances made in this regard at other neotropical sites (e.g. Smythe, 1989; Forget *et al.*, 1994). Many aquatic vertebrates, most notably river turtles and fish, inhabiting Guianan blackwater rivers and streams also utilize palm fruits extensively as an important source of energy in an otherwise relatively oligotrophic habitat.

Once dispersed by vertebrates, palm seeds may eventually be damaged by insects if they have not already infested fruit prior to dispersal. Smythe (1989) showed that buried *Astrocaryum standleyanum* seeds are protected from bruchid infestation when pulp is scraped off by rodents before scatterhoarding them. Forget *et al.* (1994) suggested that removal (predation and hoarding) of *A. butyracea* seeds by rodents may limit the bruchid population by reducing the number of oviposition sites.

Palms not only provide food but they also serve as roosts and dens. Indeed, with a high diversity and density of juveniles with large leaves in the understorey (Sist, 1989c), palms in the genera *Astrocaryum*, *Bactris*, *Geonoma* and *Scheelea* provide shelter for many bats (e.g. *Artibeus* spp.), which either perch at night during feeding sessions or during the day while sleeping in groups (Brosset and Charles-Dominique, 1990; Charles-Dominique, 1993c; see also Foster and Timm, 1976). Spiny leaves of the palm *Astrocaryum sciophyllum*, an abundant palm in the Guianan forest understorey (Sist, 1989b), are also used by *Campylopterus largipennis* (Trochilidae) to nest (Théry, 1987b) as well as by other hummingbirds in French Guiana (Tostain *et al.*, 1992). Litter cones which form at the base of *A. paramaca* palms (De Granville, 1977) are inhabited by a diverse community of amphibians and reptiles (Gasc, 1986; Kahn and De Granville, 1992). Low-lying palm cones are also often used by caviomorph rodents as deposition sites during scatterhoarding (Forget, 1990), which they or other terrestrial seed-eaters (e.g. peccaries; Kiltie, 1981) may use to relocate cached seeds.

Forest Management and Implications for Conservation

In the Guianas, mixed forests (after Richards, 1996) growing on both poorly and well-drained soils (ter Steege *et al.*, 1993; Sabatier *et al.*, 1997) offer a diversity of fruit to local fauna (see above, 'Forest habitat and mammalian diet diversity') and thus are most likely to maintain a diverse assortment of animal–plant relationships. In contrast, inundated areas on hydromorphic soils, where flooded forests are dominated by one or several tree species with seeds dispersed by gravity, explosion or water, often provide little reward to vertebrate and invertebrate herbivores (see Chapter 5). In comparison to other neotropical forests, the Guianas also host a very large proportion of trees with seeds weighing more than 1 g in average, even when the very largest seeds (often with unassisted mode of dispersal) are excluded from the calculation (Hammond and Brown, 1995). Most plant families which together constitute the major portion of vertebrate diets in the Guianas also contain commercial timber trees (Hammond *et al.*, 1996), the remaining dietary items coming from understorey trees in the families Annonaceae, Myrtaceae, Rubiaceae, Melastomataceae, and other plants such as palms, epiphytes (Araceae) and lianas (e.g. Convolvulaceae, Curcurbitaceae, Gnetaceae, Malpighiaceae, Passifloraceae).

According to Hammond *et al.* (1996), a vast majority of the medium- to large-seeded plants are primarily dispersed by mammals (44%) and birds (38%), 20% of the remaining species also being secondarily dispersed by other vertebrates, such as caviomorph rodents. Comparing dispersal modes of harvested and unharvested timber species, Hammond *et al.* (1996) stressed the importance of animals for recruitment of Guianan timber species. Of 172 potential timber species, 132 (76.7%) are currently harvested in at least one of the three Guianan countries, almost 50% and 21% of them being transported by mammals and birds, respectively (Hammond *et al.*, 1996; see also Hammond *et al.*, 1992). Animal–plant relationships may therefore contribute to the coexistence of the mosaic of forest patches through seed dispersal and seed predation processes, two key mechanisms of tree regeneration, and contribute to maintenance of diversity in Guianan forests (see Janzen, 1974; Connell and Lowman, 1989; Hart *et al.*, 1989).

Forest management and timber extraction which does not account for the support services provided by vertebrate dispersal agents may reduce the breadth of large mammal dietary items by commercially extinguishing those species which yield large, edible fruits or seeds but are relatively rare (e.g. *Pouteria* spp., *Ocotea* spp.) in comparison to the dominant species, a number of which are not harvested at all (e.g. *Dicymbe altsonii* (Caesalpiniaceae) in Guyana). Combined with a reduction of population sizes of larger vertebrates due to the increased wildlife harvesting which accompanies logging, the formative role played by the relationship between large-bodied vertebrate and large-seeded trees in shaping Guianan forests will diminish. The change in the mosaic of microhabitat structure which accompanies logging will favour light-demanding species and the frugivorous bats or birds which disperse their small seeds (Johns, 1997). Recruitment is likely to decline for those trees losing their specialist seed dispersers (see Hammond *et al.*, 1992).

The impact of logging necessarily varies between central Guyana, Suriname and French Guiana, depending on the level of harvesting imposed, being highest in the former region and lowest in the latter (Hammond *et al.*, 1996). In light of the great diversity of plant families upon which vertebrates depend and they, in turn, service through pollination and seed dispersal, the maintenance of forest habitat diversity, from monodominant stands to species-rich mixed forest, is paramount if sustained recruitment of these species is to be achieved in Guianan and other neotropical rainforests. Forest management models need to embrace the future cost of dwindling plant family diversity and increased structural fragmentation. The design of forest operational procedures should attempt to offset these costs through well-coordinated preharvest planning and the adoption of low-impact harvesting routines. This would decrease future losses due to inadequate seed dispersal by allowing animals to forage effectively across their entire home range, which often includes all types of forest habitats, i.e. from hilltops to creek, river and lake edges (see, for instance, Bodmer, 1990b), with the consequent scattering of seeds enhancing seedling survival and growth and, ultimately, contributing to the maintenance of large-seeded tree species diversity in the Guianas. Meanwhile, an increase in the density of light-demanding, small-seeded species, most often dispersed by small vertebrates and wind in Guianan forests (Sabatier, 1983, 1985; Hammond and Brown, 1995) is unavoidable, and may lead to a lower diversity and density of large frugivores in logged areas (Hammond *et al.*, 1992), preventing a rapid return to preharvest conditions, given the slow pace of recruitment in vertebrate-dispersed, large-seeded species (e.g. Forget, 1991a, 1994, 1997b). With a reduced volume of seeds being distributed at larger scales, regional forests may evolve at smaller scales alone, producing a patchwork of polarized forest processes where some areas are dominated by small-seeded, vertebrate-dispersed species (see Charles-Dominique *et al.*, 1981) and others by poor-mixed forests, dominated by large-seeded species, such as *Eperua* spp., with life-history traits adapted to regenerate in a world without vertebrates (Forget, 1989, 1992a; see also Connell and Lowman, 1989). Data available on the composition of forests over a 15 million ha area in Guyana suggest that forests undergo a form of spatial character convergence, with smaller-seeded wind, bird and primate-dispersed species being typical of more diverse forest tree communities in the south, while rodent and unassisted dispersal of large seeds is typical of relatively species-poor forests which are subject to fewer catastrophic disturbances in the central region (ter Steege and Hammond, 2001). The growing body of information on seed dispersal and predation and the influence of varying plant attributes upon this process in Guianan forests suggests that vertebrates provide an important support service to forest management (Hammond, 1998), restocking areas previously depleted and regulating the mix of competing species after the canopy has been opened.

Acknowledgements

The authors would like to acknowledge all persons who took some time correcting the text, sharing unpublished material and communicating published information not available to us. So many thanks to Marianna Altricher, Mark Angstrom, Harald Beck, Pierre Charles-Dominique, Gérard Dubost, François Feer, Charles Foerster, Matt Gompper, Cécile Hansen, Olivier Henry, M. Hoogmoed, Didier Julien-Laferrière, Burton Lim, Elisabeth Kalko, Andy Plumptre, Nancy Simmons, Peter Sherman, Marc Théry, Jean-Christophe Vié and Robert Voss. Our appreciation to the Zoological Library at the Wildlife Conservation Society, Bronx Zoo for access to unpublished archival material and reprints.

Notes

[1] Throughout text nomenclature follows Emmons and Feer (1990).

[2] Sources: *Mammals*: Beebe (1919), Tate (1939), Husson (1978), Eisenberg (1989), Ochoa (1995), Voss and Emmons (1996), Brosset *et al.* (1996), ter Steege *et al.* (1996), Simmon and Voss (1998), Voss *et al.* (2001), P. Charles-Dominique (personal communication), R. Voss (personal communication), B. Lim (personal communication), M. Engstrom (personal communication), Lim and Engstrom (2001a,b), Engstrom and Lim (2002). *Birds*: Beebe (1925), Haverschmidt (1968), Meyer de Schauensee and Phelps (1978), Parker *et al.* (1993), Thiollay (1994), ter Steege *et al.* (1996). *Fish, skates and eels*: Eigenmann (1912), Puyo (1949), Boesman (1952), Lowe-McConnell (1964), Géry (1972), Planquette *et al.* (1996), Ouboter and Mol (1993). *Reptiles and Amphibians*: Parker (1935), Beebe (1946), Gasc (1976, 1981), Hoogmoed (1973), Hoogmoed and Lescure (1975), Hoogmoed (1979), Gasc and Rodrigues (1980), Abuys (1982a,b,c,d, 1983a,b,c,d, 1984a,b), Chippaux (1986), Métrailler and Le Gratiet (1996).

References

Abuys, A. (1982a) De slangen van Suriname, deel I: De wormslangen. *Litteratura Serpentium* 2, 64–82.

Abuys, A. (1982b) De slangen van Suriname, deel II: De families Anilidae en Boidae. *Litteratura Serpentium* 2, 112–133.

Abuys, A. (1982c) De slangen van Suriname, deel III: De families Colubridae. *Litteratura Serpentium* 2, 226–245.

Abuys, A. (1982d) De slangen van Suriname, deel IV: Onderfamilie Colubrinae. *Litteratura Serpentium* 2, 274–287.

Abuys, A. (1983a) The snakes of Surinam, part V: Subfamily Colubrinae (genera *Geophis, Leptophis, Mastigodryas* and *Spilotes*). *Litteratura Serpentium* 3, 22–31.

Abuys, A. (1983b) The snakes of Surinam, part VI: Subfamily Xenodontinae (general data and genus *Atractus*). *Litteratura Serpentium* 3, 66–81.

Abuys, A. (1983c) The snakes of Surinam, part VII: Subfamily Xenodontinae (genera *Clelia and Dipsas*). *Litteratura Serpentium* 3, 111–120.

Abuys, A. (1983d) The snakes of Surinam, part VIII: Subfamily Xenodontinae (genera *Erythrolamprus, Helicops and Hydrodynastes*). *Litteratura Serpentium* 3, 203–212.

Abuys, A. (1984a) The snakes of Surinam, part IX: Subfamily Xenodontinae (genera *Hdrops, Imantodes and Leimadophis*). *Litteratura Serpentium* 4, 63–74.

Abuys, A. (1984b) The snakes of Surinam, part X: Subfamily Xenodontinae (genera *Leptodeira, Liophis and Lygophis*). *Litteratura Serpentium* 4, 160–172.

Adler, G.H. (1995) Fruit and seed exploitation by central American spiny rats, *Proechimys semispinosus*. *Studies on Neotropical Fauna and Environment* 30, 237–244.

Aide, T.M. (1988) Herbivory as a selective agent on the timing of leaf production in a tropical understory community. *Nature* 336, 574–575.

Aide, T.M. (1993) Patterns of leaf development and herbivory in a tropical understory community. *Ecology* 74, 455–466.

Altrichter, M. and Almeida, R. (2002) Exploitation of white-lipped peccaries *Tayassu pecari* (Artiodactyla: Tayassuidae) on the Osa Peninsula, Costa Rica. *Oryx* 36, 126–132.

Altrichter, M., Saenz, J.C., Carrillo, E. and Fuller, T.K. (2000) Dieta estacional del *Tayassu pecari* (Artiodactyla: Tayassuidae) en el Parque Nacional Corcovado, Costa Rica. *Revista Biologia Tropicale* 48, 689–702.

Altrichter, M., Carrillo, E., Saenz, J. and Fuller, T.K. (2001) White-lipped peccary (*Tayassu pecari*, Artiodactyla: Tayassuidae) diet and fruit availability in a Costa Rican rain forest. *Revista de Biologia Tropical* 49, 1183–1192.

Atramentowicz, M. (1982) Influence du milieu sur l'activité locomotrice et le reproduction de *Caluromys philander* (L.). *Revue d'Ecologie (Terre et Vie)* 36, 373–395.

Atramentowicz, M. (1986) Dynamique de population chez trois marsupiaux Didelphidés de Guyane. *Biotropica* 18, 136–149.

Atramentowicz, M. (1988) La frugivorie opportuniste de trois marsupiaux Didelphidés de Guyane. *Revue d'Ecologie (Terre et Vie)* 43, 47–57.

Ayres, J.M. (1981) Observacoes sobre a ecologia e o comportamento dos cuxiús (*Chiropotes albinasus* e *Chiropotes satanas*, Cebidae, Primates). MSc thesis, University of Amazonas, Manaus, Brazil.

Ayres, J.M. (1989) Comparative feeding ecology of the uakari and bearded saki, *Cacajao* and *Chiropotes*. *Journal of Human Evolution* 18, 697–716.

Balée, W. (1988) Indigenous adaptation to Amazonian palm forests. *Principes* 32, 47–54.

Beck, H. (2005) Seed predation and dispersal by peccaries throughout the Neotropics and its consequences: a review and synthesis. In: Forget, P.-M., Lambert, J.E., Hulem, P.E. and Vander Wall, S.B. (eds) *Seed fate: predation, dispersal and seedling establishment.* CABI Publishing, Wallingford, UK, pp. 77–115.

Beck-King, H., von Helversen, O. and Beck-King, R. (1999) Home range, population density, and food resources of *Agouti paca* (Rodentia: Agoutidae) in Costa Rica: a study using alternative methods. *Biotropica* 31, 675–685.

Beebe, W. (1917) *Tropical Wild Life in British Guiana.* New York Zoological Society, New York.

Beebe, W. (1919) The higher vertebrates of British Guiana with special reference to the fauna of the Bartica district, no. 7. List of Amphibia, Reptilia and Mammalia. *Zoologica* 2, 205–238.

Beebe, W. (1925) Studies of a tropical jungle. One quarter of a square mile of jungle at Kartabo, British Guiana. *Zoologica* 6, 1–193.

Beebe, W. (1946) Field notes on the snakes of Karatabo, British Guiana, and Caripito, Venezuela. *Zoologica* 31, 11–52.

Belbenoit, P., Poncy, O., Sabatier, D., Prévost, M.-F., Riéra, B., Blane, P., Larpin, D. and Sarthou, C. (2001) Appendix 1, Floristic checklist of the Nouragues area. In: Bongers, F., Charles-Dominique, P., Forget, P.-M. and Thery, M. (eds) *Nouragues: dynamics and plant-animal interactions in a neotropical rainforest.* Kluwer Academic Publisher, Dordrecht, The Netherlands, pp. 301–341.

Benalcázar, C.E. and Silva de Benalcázar, F. (1984) Historia natural del gallo de roca andino. *Cespedesia* 47–48, 59–92.

Bernard, E. and Fenton, M.B. (2002) Species diversity of bats (Mammalia: Chiroptera) in forest fragments, primary forests, and savannas in central Amazonia, Brazil. *Canadian Journal of Zoology – Revue Canadienne de Zoologie* 80, 1124–1140.

Best, R.C. (1984) The aquatic mammals and reptiles of the Amazon. In: Sioli, H. (ed.) *The Amazon.*

Limnology and Landscape Ecology of a Mighty Tropical River and its Basin. Junk Publishers, Dordrecht, The Netherlands, pp. 371–412.

Bierregaard, Jr, R.O., Lovejoy, T.E., Kapos, V., dos Santos, A.A. and Hutchings, R.W. (1992) The biological dynamics of tropical rainforest fragments. *Bioscience* 42, 859–866.

Bierregaard, Jr, R.O., Gascon, C., Lovejoy, T.E. and Mesquita, R. (2001) *Lessons from Amazonia: the Ecology and Conservation of a Fragmented Forest.* Yale University Press, New Haven, Connecticut.

Bittrich, V. and Amaral, C.E. (1996) Pollination biology of the tree *Symphonia globulifera* (Clusiaceae). *Plant Systematics and Evolution* 200, 101–110.

Bodmer, R.E. (1990a) Fruit patch size and frugivory in the lowland tapir (*Tapirus terrestris*). *Journal of Zoology of London* 222, 121–128.

Bodmer, R.E. (1990b) Responses of ungulates to seasonal inundations in the Amazon floodplain. *Journal of Tropical Ecology* 6, 191–201.

Bodmer, R.E. (1991) Strategies of seed dispersal by ungulates and seed predation in Amazonian ungulates. *Biotropica* 23, 255–261.

Bodmer, R.E. (1995) Susceptibility of mammals to overhunting in Amazonia. In: Bissonette, J.A. and Krausman, P.R. (eds) *Integrating People and Wildlife for a Sustainable Future.* The Wildlife Society, Bethesda, Maryland, pp. 292–295.

Boesman, M. (1952) A preliminary list of Surinam fishes not included in Eigenmann's enumeration of 1912. *Zoologica Medelingenen Leiden* 31, 179–200.

Bonaccorso, F.J., Glanz, W.E. and Sandford, C.M. (1980) Feeding assemblages of mammals at fruiting *Dipteryx panamensis* (Papilionaceae) trees in Panama: seed predation, dispersal, and parasitism. *Revista Biologia Tropical* 28, 61–72.

Bongers, F., Charles-Dominique, P., Forget, P.-M. and Théry, M. (2001) *Nouragues: Dynamics and Plant–Animal Interactions in a Neotropical Rainforest.* Kluwer Academic, Dordrecht, The Netherlands.

Boucher, D.H. (ed.) (1985) *The Biology of Mutualism.* Oxford University Press, Oxford, UK.

Boucher, D.H., James, S. and Keeler, K.H. (1982) The ecology of mutualism. *Annual Review Ecology Systematics* 13, 315–347.

Boujard, T., Sabatier, D., Rojas-Beltran, R., Prévost, M.-F. and Renno, J.-F. (1990) The food habits of three allochthonous feeding characoids in French Guiana. *Revue d'Ecologie* (*Terre et Vie*) 45, 247–258.

Branan, W.V. and Marchinton, R.L. (1986) Reproductive ecology of white-tailed and red brocket deer in Suriname. In: Wemmer, C. (ed.) *Biology and Management of the Cervidae.* Smithsonian Institution Press, Washington, DC, pp. 972–976.

Branan, W.V., Werkhoven, M.C.M. and Marchinton, R.L. (1985) Food habits of brocket and white-tailed deer in Suriname. *Journal of Wildlife Management* 49, 972–976.

Brosset, A. and Charles-Dominique, P. (1990) The bats from French Guiana: a taxonomic, faunistic and ecological approach. *Mammalia* 54, 509–560.

Brosset, A., Cosson, J.-F., Gaucher, P. and Masson, D. (1995) Les chiroptères d'un marécage côtier de Guyane; composition du peuplement. *Mammalia* 59, 527–535.

Brosset, A., Charles-Dominique, P., Cockle, A., Cosson, J.-F. and Masson, D. (1996) Bat communities and deforestation in French Guiana. *Canadian Journal of Zoology* 74, 1974–1982.

Charles-Dominique, P. (1986) Inter-relations between frugivorous vertebrates and pioneer plants, *Cecropia*, birds and bats in French Guyana. In: Estrada, A. and Fleming, T.H. (eds) *Frugivores and Seed Dispersal.* Dr. W. Junk Publishers, The Hague, pp. 119–135.

Charles-Dominique, P. (1991) Feeding strategy and activity budget of the frugivorous bat *Carollia perpicillata* (Chiroptera: Phyllostomidae) in French Guiana. *Journal of Tropical Ecology* 7, 243–256.

Charles-Dominique, P. (1993a) Ecologie des mammifères forestiers de Guyane française: origines biogéographiques et organisation du peuplement. In: *Gestion de l'Ecosystème Forestier et Aménagement de l'Espace Régional. Nature Guyanaise* (supplement), 145–152.

Charles-Dominique, P. (1993b) Speciation and coevolution: an interpretation of frugivory phenomena. *Vegetatio* 107/18, 75–84.

Charles-Dominique, P. (1993c) Tent-use by the bat *Rhinophylla pumilio* (Phyllostomidae: Carolliinae) in French Guiana. *Biotropica* 25, 111–116.

Charles-Dominique, P. (1995) Interactions plantes–animaux frugivores, conséquences sur la dissémination des graines et la régénération forestière. *Revue d'Ecologie (Terre Vie)* 50, 223–237.

Charles-Dominique, P. and Cockle, A. (2001) Frugivory and seed dispersal by bats. In: Bongers, F., Charles-Dominique, P., Forget, P.-M. and Théry, M. (eds) *Nouragues: Dynamics and Plant–Animal Interactions in a Neotropical Rainforest.* Kluwer Academic, Dordrecht, The Netherlands, pp. 207–215.

Charles-Dominique, P., Atramentowicz, M., Charles-Dominique, M., Gérard, H., Hladik, A., Hladik, C.-M. and Prévost, M.-F. (1981) Les mammifères frugivores arboricoles nocturnes d'une forêt guyanaise: inter-relations plantes–animaux. *Revue d'Ecologie (Terre Vie)* 35, 341–435.

Chippaux, J.-P. (1986) *Les Serpents de la Guyane française.* Faune Tropicale XXVII. ORSTOM, Paris.

Claessens, O., Granjon, L., De Massary, J.-C. and Ringuet, S. (2002) La station de recherche de Saint-Eugène: situation, environment et présentation générale. In: Forget, P.-M. (ed.) *Fragmentation of tropical rainforest: Petit Saut dam, Sinnamary river, French Guiana. Revue d'Ecologie (Terre et Vie)*, Supplement 8, Paris, pp. 21–37.

Cockle, A. (1997) Modalités de dissémination et d'établissement de lianes de sous-bois (Cylanthaceae et Philodendron) en forêt guyanaise. Thèse de doctorat de l'Université Paris VI, Paris.

Connell, J.H. and Lowman, M.D. (1989) Low-diversity tropical rain forests: some possible mechanisms for their existence. *American Naturalist* 134, 88–119.

Cosson, J.-F. (1994) Dynamique de population et dispersion de la Chauve-Souris frugivore *Carollia perspicillata* en Guyane Française. Thèse de doctorat de l'Université Paris XI, Villetaneuse.

Davis, T.A.W. and Richards, P.W. (1933) The vegetation of Moraballi creek, British Guiana: an ecological study of a limited area of tropical forest. Parts I. *Journal of Ecology* 21, 350–384.

Davis, T.A.W. and Richards, P.W. (1934) The vegetation of Moraballi creek, British Guiana: an ecological study of a limited area of tropical forest. Part II. *Journal of Ecology* 22, 106–155.

Day, R. (1997) Behavioural ecology of the tamarin, *Saguinus midas midas,* in a Guianense primate community. PhD thesis, The Queen's University of Belfast, UK.

De Foresta, H. (1983) Hétérogénéité de la végétation pionnière en forêt tropicale humide: exemple d'une coupe papetière en forêt guyanaise. *Acta Oecologica, Oecologia Applicata* 4, 221–235.

De Foresta, H. and Prévost, M.-F. (1986) Végétation pionnière et graines du sol en forêt guyanaise. *Biotropica* 18, 279–286.

De Foresta, H., Charles-Dominique, P., Erard, C. and Prévost, M.-F. (1984) Zoochorie et premiers stades de la régénération naturelle après coupe en forêt guyanaise. *Revue d'Ecologie (Terre et Vie)* 39, 369–400.

De Granville, J.J. (1977) Notes biologiques sur quelques palmiers guyanais. *Cahier ORSTOM, Série Biologie* vol. XII, 4, 347–353.

de Thoisy, B. and Richard-Hansen, C. (1997) Diet and social behaviour in a red howler monkey (*Alouatta seniculus*) troop in a highly degraded rainforest. *Folia Primatologica* 68, 357–361.

Dubost, G. (1987) Une analyse écologique de deux faunes de mammifères forestiers tropicaux. *Mammalia* 51, 415–436.

Dubost, G. (1988) Ecology and social life of the red acouchy, *Myoprocta exilis*; comparisons with the orange-rumped agouti, *Dasyprocta leporina. Journal of Zoology of London* 214, 107–123.

Duellman, W.E. (1990) Herpetofaunas in neotropical rainforests: comparative composition, history and resource use. In: Gentry, A.H. (ed.) *Four Neotropical Forests.* Yale University Press, New Haven, Connecticut, pp. 455–505.

Eigenmann, C.H. (1912) The fresh water fish of British Guiana, including a study of the ecological grouping of species and relation of the fauna of the plateau to that of the lowlands. *Memoirs of the Carnegie Museum* 5, 1–578.

Eisenberg, J.F. (1989) *Mammals of the Neotropics. The Northern Neotropics;* Vol. 1. University of Chicago Press, Chicago, Illinois.

Ek, R. (1997) *Botanical Diversity in the Tropical Rain Forest of Guyana.* Tropenbos-Guyana Series 4. Tropenbos-Guyana, Georgetown.

Emmons, L.H. and Feer, F. (1990) *Neotropical Rainforest Mammals: a Field Guide.* University of Chicago Press, Chicago, Illinois.

Endler, J.A. and Théry, M. (1996) Interactions effects of lek placement, display behavior, ambient light, and color patterns in three neotropical forest-dwelling birds. *The American Naturalist* 148, 421–452.

Engstrom, M.D. and Lim, B.K. (2002) Mamíferos de Guyana. In: Ceballos, G. and Simonetti, J.A. (eds) *Diversidad y conservación de los mamiferos neotropicales.* Comisión Nacional para el Conocimiento y Uso de la Biodiversidad, Mexico, pp. 329–375.

Erard, C. and Sabatier, D. (1989) Rôle des oiseaux frugivores terrestres dans la régénération forestière en Guyane française. *Proceedings of International Ornithological Congress* 19, 803–815.

Erard, C. and Théry, M. (1994) Frugivorie et ornithochorie en forêt guyanaise: l'exemple des grands oiseaux terrestres et de la pénélope marail. *Alauda* 62, 27–31.

Erard, C., Théry, M. and Sabatier, D. (1989) Régime alimentaire de *Rupicola rupicola* (Cotingidae) en Guyane française: relations avec le frugivorie et la zoochorie. *Revue d'Ecologie (Terre et Vie)* 44, 47–74.

Erard, C., Théry, M. and Sabatier, D. (1991) Régime alimentaire de *Tinamus major* (Tinamidae), *Crax alector* (Cracidae) et *Psophia crepitans* (Psophidae), en forêt guyanaise. *Gibier Faune Sauvage* 8, 183–210.

Feer, F. and Charles-Dominique, P. (2001) Mammals of the Nouragues and lower Arataye areas. In: Bongers, F., Charles-Dominique, P., Forget, P.-M. and Théry, M. (eds) *Nouragues: Dynamics and Plant–Animal Interactions in a Neotropical Rainforest.* Kluwer Academic, Dordrecht, The Netherlands, pp. 351–355.

Feer, F., Henry, O., Forget, P.-M. and Gayot, M. (2001a) Frugivory and seed dispersal by terrestrial mammals. In: Bongers, F., Charles-Dominique, P., Forget, P.-M. and Théry, M. (eds) *Nouragues: Dynamics and Plant–Animal Interactions in a Neotropical Rainforest.* Kluwer Academic, Dordrecht, The Netherlands, pp. 207–215.

Feer, F., Julliot, C., Simmen, B., Forget, P.M., Bayart, F. and Chauvet, S. (2001b) Recruitment, a multi-stage process with unpredictable result: the case of a Sapotaceae in French Guianan forest. *Revue d'Ecologie Terre et Vie* 56, 119–145.

Ferrari, S.F. (1995) Observations on *Chiropotes albinasus* from the Rio dos Marmelos, Amazonas, Brazil. *Primates* 36, 289–293.

Fitzgerald, K.A. (2003) Utilizing ecological indicators to assist in the management of Brownsberg Nature Park, Suriname, South America. MSc thesis, Washington State University.

Fitzgerald, K.A., de Dijn, B.P.E. and Mitro, S. (2002) *Brownsberg Nature Park Ecological Research and Monitoring Program: 2001–2006.* STINASU, Paramaribo, Suriname.

Fleming, T.H. (1991) Fruiting plant–frugivore mutualism: the evolutionary theater and the ecological play. In: Price, P.W., Lewinsohn, T.M., Fernandes, G.W. and Benson, W.W. (eds) *Plant–Animal Interactions: Evolutionary Ecology in Tropical and Temperate Regions.* John Wiley & Sons, New York, pp. 119–144.

Fleming, T.H. and Heithaus, E.R. (1981) Frugivorous bats, seed shadows, and the structure of tropical forests. *Biotropica* (supplement *Reproductive Botany*), 45–53.

Fleming, T.H., Hooper, E.T. and Wilson, D.E. (1972) Three Central American bat communities: structure, reproductive cycles, and movement patterns. *Ecology* 53, 555–569.

Fleming, T.H., Williams, C.F., Bonaccorso, F.J. and Herbst, L.H. (1985) Phenology, seed dispersal, and colonization in *Muntingia calabura*, a neotropical pioneer tree. *American Journal of Botany* 72, 383–391.

Foerster, C.R. and Vaughan, C. (2002) Home range, habitat use, and activity of Baird's tapir in Costa Rica. *Biotropica* 34, 423–437.

Foley, W.J., Engelhardt, W.V. and Charles-Dominique, P. (1995) The passage of digesta, particle size, and in vitro fermentation rate in three-toed sloth *Bradypus tridactylus* (Edentata: Bradypodidae). *Journal of Zoology of London* 236, 681–696.

Forget, P.-M. (1988) Dissémination et régénération naturelle de huit espèces d'arbres en forêt Guyanaise. Thèse de l'Université Pierre et Marie Curie, Paris VI.

Forget, P.-M. (1989) La régénération naturelle d'une espèce autochore de la forêt guyanaise: *Eperua falcata* Aublet (Caesalpiniaceae). *Biotropica* 21, 115–125.

Forget, P.-M. (1990) Seed-dispersal of *Vouacapoua americana* (Caesalpiniaceae) by caviomorph rodents in French Guiana. *Journal of Tropical Ecology* 6, 459–468.

Forget, P.-M. (1991a) Comparative recruitment pattern of two non-pioneer tree species in French Guiana. *Oecologia* 85, 434–439.

Forget, P.-M. (1991b) Scatterhoarding of *Astrocaryum paramaca* by *Proechimys* in French Guiana: comparison with *Myoprocta exilis. Tropical Ecology* 32, 155–167.

Forget, P.-M. (1992a) Regeneration ecology of *Eperua grandiflora* (Caesalpiniaceae), a large-seeded tree in French Guiana. *Biotropica* 24, 146–156.

Forget, P.-M. (1992b) Seed removal and seed fate in *Gustavia superba* (Lecythidaceae). *Biotropica* 24, 408–414.

Forget, P.-M. (1993) Post-dispersal predation and scatterhoarding of *Dipteryx panamensis* (Papilionaceae) seeds by rodents in Panama. *Oecologia* 94, 255–261.

Forget, P.-M. (1994) Regeneration pattern of *Vouacapoua americana* (Caesalpiniaceae), a rodent-dispersed tree species in French Guiana. *Biotropica* 26, 420–426.

Forget, P.-M. (1996) Removal of seeds of *Carapa procera* (Meliaceae) by rodents and their fate in rainforest in French Guiana. *Journal of Tropical Ecology* 12, 751–761.

Forget, P.-M. (1997a) Effect of microhabitat on seed fate and seedling performance in two rodent-dispersed tree species in rainforest in French Guiana. *Journal of Ecology* 85, 693–703.

Forget, P.-M. (1997b) Ten-year seedling dynamics in *Vouacapoua americana* in French Guiana: a hypothesis. *Biotropica* 29, 124–126.

Forget, P.-M. and Milleron, T. (1991) Evidence for secondary seed dispersal in Panama. *Oecologia* 87, 596–599.

Forget, P.-M. and Sabatier, D. (1997) Dynamics of a seedling shadow in a frugivore-dispersed tree species in French Guiana. *Journal of Tropical Ecology* 13, 767–773.

Forget, P.-M., Munoz, E. and Leigh Jr, E.G. (1994) Predation on seeds falling late in the fruiting season of *Scheelea* palm on Barro Colorado Island, Panama. *Biotropica* 26, 408–419.

Forget, P.-M., Milleron, T. and Feer, F. (1998) Patterns in post-dispersal seed removal by neotropical rodents and seed fate in relation with seed size. In: Newbery, D.M., Prins, H.T.T. and Brown, N.D. (eds) *Dynamics of Tropical Communities.* Blackwell Scientific, Cambridge, UK, pp. 25–50.

Forget, P.-M., Lambert, J., Hulme, P. and Vander Wall, S.B. (2005) *Seed fate: predation, dispersal and seeding establishment.* CAB International, Wallingford, UK.

Foster, M.S. and Timm, R.M. (1976) Tent-making by *Artibeus jamaicensis* (Chiroptera Phyllostomatidae) with comments on plants used by bats for tents. *Biotropica* 8, 265–269.

Foster, R.A., Parker III, T.A., Gentry, A.H., Emmons, L.H., Chicchón, A., Schulenberg, T., Rodríguez, L., Lamas, G., Ortega, H., Icochea, J., Wust, W., Romo, M., Castillo, J.A., Phillips, O., Reynel, C., Kratter, A., Donahue, P.K. and Barkley, L.J. (1994) *The Tambopata-Candamo Reserved Zone of Southeastern Perú: A Biological* Evodianthus funifer *Assessment.* RAP Working Papers 6. Conservation International, Washington, DC.

Fragoso, J.M.V. (1997) Tapir-generated seed shadows: scale-dependent patchiness in the Amazon rain forest. *Journal of Ecology* 85, 519–529.

Fragoso, J.M.V. (1998a) Home range and movement patterns of white-lipped peccary (*Tayassu pecari*) herds in the Northern Brazilian Amazon. *Biotropica* 30, 458–469.

Fragoso, J.M.V. (1998b) White-lipped peccaries and palms on the ilha de Maraca. In: Milliken, W. and Ratter, J.A. (eds) *Maraca: the Biodiversity and Environment of an Amazonian Rainforest.* John Wiley, Chichester, UK, pp. 151–163.

Fragoso, J.M.V. (1999) Perception of scale and resource partitioning by peccaries: behavioral causes and ecological implications. *Journal of Mammalogy* 80, 993–1003.

Fragoso, J.M.V. and Huffman, J.M. (2000) Seed-dispersal and seedling recruitment patterns by the last neotropical megafaunal element in Amazonia, the tapir. *Journal of Tropical Ecology* 16, 369–385.

Freeland, W.J. and Janzen, D.H. (1974) Strategies in herbivory by mammals: the role of plant secondary compounds. *American Naturalist* 108, 269–289.

Gasc, J.-P. (1976) Contribution á la connaissance des Squamates (Reptilia) de la Guyane française. Nouvelles localités pour les Sauriens. *Comptes Rendus de la Société Biogéographique* 454, 17–36.

Gasc, J.-P. (1981) Quelques nouvelles données sur le répartition et l'écologie des sauriens en Guyane française. *Revue d'Ecologie (Terre et Vie)* 35, 273–325.

Gasc, J.-P. (1986) Le peuplement herpétologique d'*Astrocaryum paramaca* (Areacaceae), un palmier important dans la structure de la forêt en Guyane française. *Mémoires du Muséum National d'Histoire naturelle, Nouvelle Série, Série A, Zoologie* 132, 97–107.

Gasc, J.-P. and Rodrigues, M.T. (1980) Liste préliminaire des serpents de Guyane française. *Bulletin du Museum National Histoire Naturelle*, Serie 4, 2, 559–598.

Gayot, M. (2000) Stratégie alimentaire de deux espèces de Cervidés de Guyane française (*Mazama* spp.) et comparaison avec des Bovidés du Gabon (*Cephalophus* pp.). DEA thesis, Université Pierre et Marie Curie, Paris.

Gayot, M., Henry, O., Dubost, G. and Sabatier, D. (2004) Comparative diet of the two forest cervids of the genus *Mazama* in French Guiana. *Journal of Tropical Ecology* 20, 31–43.

Genoways, H.H. and Williams, S.L. (1979) Records of bats (Mammalia: Chiroptera) from Suriname. *Annals of the Carnegie Museum* 48, 323–335.

Géry, J. (1972) Poissons characoïdes des Guyanes. I. Généralités. II. Famille des Serrasalmidae. *Zoologica Verhandelening* 122, 1–260.

Gill Jr, G.E., Fowler, R.T. and Mori, S.A. (1998) Pollination biology of *Symphonia globulifera* (Clusiaceae) in central French Guiana. *Biotropica* 30, 139–144.

Gompper, M.E. (1994) The importance of ecology, behavior, and genetics in the maintenance of coati (*Nasua narica*) social structure. PhD thesis, University of Tennessee, Knoxville.

Gompper, M.E. (1995) *Nasua narica. Mammalian Species* 487, 1–10.

Goulding, M., Leal Carvalho, M. and Ferreira, E.G. (1988) *Rio Negro, Rich Life in Poor Water.* SPB Academic, The Hague, The Netherlands.

Greenhall, A.M. (1959) Bats of Guiana. *Journal of the British Guiana Museum* 22, 55–57.

Greenhall, A.M. (1965) Sapucaia nut dispersal by greater spear-nosed bats in Trinidad. *Caribbean Journal of Science* 5, 167–171.

Greenwood, J.J.D. (1985) Frequency-dependent selection by seed-predators. *Oikos* 44, 195–210.

Gribel, R. (1988) Visits of *Caluromys lanatus* (Didelphidae) to flowers of *Pseudobombax tomentosum* (Bombacaceae): a probable case of pollination by marsupials in central Brazil. *Biotropica* 20, 344–347.

Guillotin, M. (1982a) Activités et régimes alimentaires: leurs interactions chez *Proechimys cuvieri* et *Oryzomys capito velutinus* (Rodentia) en forêt guyanaise. *Revue d'Ecologie (Terre et Vie)* 36, 337–371.

Guillotin, M. (1982b) Place de *Proechimys cuvieri* (Rodentia, Echimidae) dans les peuplements micromammaliens terrestres de la forêt guyanaise. *Mammalia* 46, 299–318.

Guillotin, M., Dubost, G. and Sabatier, D. (1994) Food choice and food composition among three major primate species of French Guiana. *Journal of Zoology* 233, 551–579.

Hallwachs, W. (1986) Agoutis *Dasyprocta punctata*: the inheritors of guapinol *Hymenaea courbaril* (Leguminosae). In: Estrada, A. and Fleming, T.H. (eds) *Frugivores and Seed Dispersal.* Dr. W. Junk Publishers, The Hague, pp. 119–135.

Hammond, D.S. (1998) Biodiversity conservation as a sustainable forest management tool. In: Zagt, R.J. (ed.) *Research in Tropical Rain Forests: Its Challenges for the Future.* Tropenbos Foundation, Wageningen, The Netherlands, pp. 231–247.

Hammond, D.S. and Brown, V.K. (1995) Seed size of woody plants in relation to disturbance, dispersal, soil type in wet neotropical forests. *Ecology* 76, 2544–2561.

Hammond, D.S. and Brown, V.K. (1998) Disturbance, phenology and life-history characteristics: factors influencing distance/density-dependent attack on tropical seeds and seedlings. In: Newbery, D.M., Prins, H.T.T. and Brown, N.D. (eds) *Dynamics of Tropical Communities.* Blackwell Scientific, Cambridge, UK, pp. 51–78.

Hammond, D.S., Schouten, A., van Tienen, L., Weijerman, M. and Brown, V.K. (1992) The importance of being a forest animal: implications for Guyana's timber trees. *Proceedings of the 6th Annual NARI/CARDI Review Conference,* Georgetown, pp. 144–151.

Hammond, D.S., Gourlet-Fleury, S., van der Hout, P., ter Steege, H. and Brown, V.K. (1996) A compilation of known Guianan timber trees and the significance of their dispersal mode, seed size and taxonomic affinity to tropical rain forest management. *Forest Ecology and Management* 83, 99–116.

Hammond, D.S., Brown, V.K. and Zagt, R.J. (1999) Spatial and temporal patterns of seed attack and germination in a large-seeded neotropical tree species. *Oecologia* 119, 208–218.

Harms, K.E. and Dalling, J.W. (1997) Damage and herbivory tolerance through resprouting as an advantage of large seed size in tropical trees and lianas. *Journal of Tropical Ecology* 13, 617–621.

Hart, T.B., Hart, J.A. and Murphy, P.G. (1989) Monodominant and species-rich forests of the humid tropics: causes for their co-occurrence. *American Naturalist* 133, 613–633.

Haverschmidt, F. (1968) *Birds of Surinam.* Oliver and Boyd, London.

Heithaus, E.R. (1982) Coevolution between bats and plants. In: Kuntz, T.H. (ed.) *Ecology of Bats.* Plenum Press, London, pp. 327–367.

Heithaus, E.R., Opler, P.A. and Baker, H.G. (1974) Bat activity and pollination of *Bauhinia pauletia*: plant–pollinator coevolution. *Ecology* 55, 412–419.

Heithaus, E.R., Fleming, T.H. and Opler, P.A. (1975) Foraging pattern and resources utilization in seven species of bats in a seasonal tropical forest. *Ecology* 56, 841–854.

Henry, O. (1994a) Caractéristiques et variations saisonnières de la reproduction de quatre mammifères forestiers terrestres de Guyane française: *Oryzomys capito velutinus* (Rodentia, Cricetidae), *Proechimys cuvieri* (Rodentia, Echimyidae), *Dasyprocta leporina* (Rodentia, Dasyproctidae) and *Tayassu tajacu* (Artiodactyle, Tayassuidae). Influence de l'age, des facteurs environnementaux et de l'alimentation. Thèse de doctorat de l'Université Paris VII, Paris.

Henry, O. (1994b) Saisons de reproduction chez trois Rongeurs et un Artiodactyle en Guyane française, en fonction des facteurs du milieu et de l'alimentation. *Mammalia* 58, 183–200.

Henry, O. (1996) The influence of sex and reproductive state on diet preference in four terrestrial mammals of the French Guianan rainforest. *Canadian Journal of Zoology* 75, 929–935.

Henry, O. (1999) Dietary choice of the orange-rumped agouti (*Dasyprocta leporina*) in French Guiana. *Journal of Tropical Ecology* 13, 291–300.

Henry, O., Feer, F. and Sabatier, D. (2000) Diet of the lowland tapir (*Tapirus terrestris* L.) in French Guiana. *Biotropica* 32, 364–368.

Higgins, M.L. (1979) Intensity of seed predation on *Brosimum utile* by mealy parrots (*Amazona farinosa*). *Biotropica* 11, 80.

Holthuijzen, A.M.A. and Boerboom, J.H.A. (1982) The *Cecropia* seedbank in the Surinam lowland rain forest. *Biotropica* 14, 62–68.

Hoogmoed, M.S. (1973) *Notes on the Herpetofauna of Surinam, IV. The Lizards and Amphisbaenians of Surinam.* Biogeographica No. 4. Junk, The Hague, The Netherlands.
Hoogmoed, M.S. (1979) The herpetofauna of the Guianan region. In: Duellman, W.E. (ed.) *The South American Herpetofauna: Its Origin, Evolution and Dispersal.* Monographs of the Museum of Natural History University of Kansas 7, 241–279.
Hoogmoed, M.S. and Lescure, J. (1975) An annotated checklist of the lizards of French Guiana, mainly based on two recent collections. *Zoologischen Mededlingen* 49, 141–171.
Hopkins, H.C. and Hopkins, M.J.G. (1983) Fruit and seed biology of the neotropical species of *Parkia.* In: Sutton, S.L., Whitmore, T.C. and Chadwick, A.C. (eds) *Tropical Rain Forest Ecology and Management.* Blackwell, Oxford, UK, pp. 197–209.
Howe, H.F. (1980) Monkey dispersal and waste of a neotropical fruit. *Ecology* 61, 944–959.
Howe, H.F. (1986) Seed dispersal by fruit-eating birds and mammals. In: Murray, D.R. (ed.) *Seed Dispersal.* Academic Press Australia, Sydney, pp. 123–189.
Howe, H.F. and Primack, R.B. (1975) Differential seed dispersal by birds of the tree *Casearia nitida* (Flacourtiaceae). *Biotropica* 7, 278–263.
Howe, H.F. and Smallwood, J.H. (1982) Ecology of seed dispersal. *Annual Review of Ecology and Systematics* 13, 201–228.
Howe, H.F. and Westley, L.C. (1988) *Ecological Relationships of Plants and Animals.* Oxford University Press, Oxford, UK.
Howe, H.F., Schupp, E.W. and Westley, L.C. (1985) Early consequences of seed dispersal for a neotropical tree (*Virola surinamensis*). *Ecology* 66, 781–791.
Husson, A.M. (1962) *The Bats of Suriname.* Zoologische Verhandelingen No. 58. E.J. Brill, Leiden, The Netherlands.
Husson, A.M. (1978) *The Mammals of Suriname.* Zoöl. Monogr. Rijksmuseum Nat. Historie No. 2. E.J. Brill, Leiden, The Netherlands.
Jansen, P.A. and Forget, P.-M. (2001) Scatterhoarding by rodents and tree regeneration in French Guiana. In: Bongers, F., Charles-Dominique, P., Forget, P.-M. and Théry, M. (eds) *Nouragues: Dynamics and Plant–Animal Interactions in a Neotropical Rainforest.* Kluwer Academic Publishers, Dordrecht, The Netherlands, pp. 275–288.
Jansen, P.A., Bartholomeus, M., Bongers, F., Elzinga, J.A., Den Ouden, J. and van Wieren, S.E. (2002) The role of seed size in dispersal by a scatter-hoarding rodent. In: Levey, D.J., Silva, W.R. and Galetti, M. (eds) *Seed Dispersal and Frugivory: Ecology, Evolution and Conservation.* CAB International, Wallingford, UK, pp. 209–225.
Jansen, P.A., Bongers, F. and Hemerik, L. (2004) Seed mass and mast seeding enchance dispersal by a neotropical scatter-hoarding rodents. *Ecological Monographs* 74, 569–589.
Jansen, P. A. and den Ouden, J. (2005) Observing seed removal: remote video monitoring of seed selection, predation and dispersal. In: Forget, P.-M., Lambert, J.E., Hulme, P.E. and Vander Wall, B. (eds) *Seed fate: predation, dispersal and seedling establishment.* CAB International, Wallingford, UK, pp. 363–378.
Janson, C.H., Terborgh, J.W. and Emmons, L.H. (1981) Nonflying mammals as pollinating agents in the Amazonian forests. *Biotropica* 13 (supplement), 1–6.
Janzen, D.H. (1969) Seed-eaters versus seed size, number, toxicity and dispersal. *Evolution* 23, 1–27.
Janzen, D.H. (1970) Herbivores and the number of species in tropical forests. *American Naturalist* 104, 501–528.
Janzen, D.H. (1971) Seed predation by animals. *Annual Review of Ecology and Systematics* 2, 465–492.
Janzen, D.H. (1974) Tropical black water rivers, animals, and mast fruiting by the Dipterocarpaceae. *Biotropica* 6, 69–103.
Janzen, D.H. (1981a) Lectins and plant–herbivore interactions. *Recent Advances in Phytochemistry* 15, 241–258.
Janzen, D.H. (1981b) Digestive seed predation by a Costa Rican Baird's tapir. *Biotropica* 13, 59–63.
Janzen, D.H. (1982) Removal of seeds from horse dung by tropical rodents: influence of habitat and amount of dung. *Ecology* 63, 1887–1900.
Janzen, D.H. (1983a) Dispersal of seeds by vertebrates guts. In: Futuyma, D.J. and Slatkin, M. (eds) *Coevolution.* Sinauer Associates, Sunderland, pp. 232–262.
Janzen, D.H. (1983b) *Tapirus bairdii.* In: Janzen, D.H. (ed.) *Costa Rican Natural History.* University of Chicago Press, Chicago, Illinois, pp. 496–497.
Janzen, D.H. (1986) Mice, big mammals, and seeds: it matters who defecate what where. In: Estrada, A. and Fleming, T.H. (eds) *Frugivores and Seed Dispersal.* Dr. W. Junk Publishers, The Hague, pp. 251–271.

Janzen, D.H. and Martin, P.S. (1982) Neotropical anachronisms: the fruits the gomphotheres ate. *Science* 215, 19–27.

Janzen, D.H., Juster, H.B. and Liener, I.E. (1976a) Insecticidal action of the phytohemagglutinin in black beans on a bruchid beetle. *Science* 192, 795–796.

Janzen, D.H., Miller, G.A., Hackforth-Jones, J., Pond, C.M., Hooper, K. and Janos, D.P. (1976b) Two Costa Rican bat-generated seed shadow of *Andira inermis* (Leguminosae). *Ecology* 57, 1068–1075.

Johns, A.G. (1997) *Timber Production and Biodiversity Conservation in the Tropical Rain Forests.* Cambridge University Press, Cambridge, UK.

Judas, J. (1999) Ecologie du Pécari à collier (*Tayassu tajacu*) en forêt tropicale humide de Guyane française. Thèse de doctorat de l'Université F. Rabelais, Tours, France.

Judas, J. and Henry, O. (1999) Seasonal variation of home range of collared peccary in tropical rain forest of French Guiana. *Journal of Wildlife Management* 63, 546–552.

Julien-Laferrière, D. (1989) Utilisation de l'espace et des ressources alimentaires chez *Caluromys philander* (Marsupiala, Didelphidae) en Guyane Française. Comparaison avec *Potos flavus* (Eutheria, Procyonidae). Thèse de doctorat de l'Université Paris XIII, Villetaneuse.

Julien-Laferrière, D. (1990) Actographie chez *Caluromys philander* (Marsupiala, Didelphidae) en Guyane française. *Compte Rendus de l'Académie des Sciences Paris, Série III,* 311, 25–30.

Julien-Laferrière, D. (1993) Radio-tracking observations on ranging and foraging patterns by kinkajous (*Potos flavus*) in French Guiana. *Journal of Tropical Ecology* 9, 19–32.

Julien-Laferrière, D. (1995) Use of space by the woolly opossum *Caluromys philander* (Marsupiala, Didelphidae) in French Guiana. *Canadian Journal of Zoology* 73, 1280–1289.

Julien-Laferrière, D. (1997) The influence of moonlight on activity of woolly opossum (*Caluromys philander*). *Journal of Mammalogy* 78, 251–255.

Julien-Laferrière, D. (1999a) Foraging strategies and food partitioning in the neotropical frugivorous mammals *Caluromys philander* and *Potos flavus. Journal of Zoology, London* 247, 71–80.

Julien-Laferrière, D. (1999b) Fruit consumption, seed dispersal and seed fate in the vine *Strychnos erichsonii* in a French Guianan forest. *Revue d' Ecologie (Terre et Vie)* 54, 315–326.

Julien-Laferrière, D. (2001) Frugivory and seed dispersal by Kinkajous. In: Bongers, F., Charles-Dominique, P., Forget, P.-M. and Théry, M. (eds) *Nouragues: Dynamics and Plant–Animal Interactions in a Neotropical Rainforest.* Kluwer Academic, Dordrecht, The Netherlands, pp. 217–225.

Julien-Laferrière, D. and Atramentowicz, M. (1990) Feeding and reproduction of three Didelphid Marsupials in two neotropical forests (French Guiana). *Biotropica* 22, 404–415.

Julliot, C. (1993) Diet of the red howler monkeys (*Alouatta seniculus*) in French Guiana. *International Journal of Primatology* 14, 527–550.

Julliot, C. (1994) Frugivory and seed dispersal by red howler monkeys: evolutionary aspect. *Revue d'Ecologie (Terre et Vie)* 49, 331–341.

Julliot, C. (1996a) Fruit choice by red howler monkeys (*Alouatta seniculus*) in a tropical rain forest. *American Journal of Primatology* 40, 261–282.

Julliot, C. (1996b) Seed dispersal by red howler monkeys (*Alouatta seniculus*) in the tropical rain forest of French Guiana. *International Journal of Primatology* 17, 239–258.

Julliot, C. (1997) Impact of seed dispersal by red howler monkeys (*Alouatta seniculus*) on the seedling population in the understorey of tropical rainforest. *Journal of Ecology* 85, 431–440.

Julliot, C. and Sabatier, D. (1993) Diet of the red howler monkey (*Alouatta seniculus*) in French Guiana. *International Journal of Primatology* 14, 527–550.

Julliot, C., Simmen, B. and Zhang, S.Y. (2001) Frugivory and seed dispersal by monkeys at the Nouragues station in French Guiana. In: Bongers, F., Charles-Dominique, P., Forget, P.-M. and Théry, M. (eds) *Nouragues: Dynamics and Plant–Animal Interactions in a Neotropical Rainforest.* Kluwer Academic, Dordrecht, The Netherlands, pp. 197–215.

Kahn, F. and De Granville, J.J. (1992) *Palms in Forest Ecosystems of Amazonia.* Ecological Studies No. 95. Springer, Berlin.

Kalko, E.K.V. (1997) Diversity in tropical bats. In: Ulrich, H. (ed.) *Tropical Biodiversity and Systematics.* Proceedings of the international symposium on biodiversity and systematics in tropical ecosystems. Zoologisches Forschunginstitut und Museum Alexander Koenig, Bonn, pp. 13–43.

Kalko, E.K.V. and Handley, C.O. (2001) Neotropical bats in the canopy: diversity, community structure, and implications for conservation. *Plant Ecology* 153, 319–333.

Kalko, E.K.V., Handley Jr, C.O. and Handley, D. (1996) Organization, diversity and long-term dynamics of a neotropical bat community. In: Cody, M.L. and Smallwood, J.A. (eds) *Long-term Studies of Vertebrate Communities*. Academic Press, New York, pp. 503–533.

Karr, J.R., Robinson, S.K., Blake, J.G. and Bierregaard Jr, R.O. (1990) Birds of four neotropical forests. In: Gentry, A.H. (ed.) *Four Neotropical Forests*. Yale University Press, London, pp. 237–269.

Kay, R.F., Madden, R.H., van Schaik, C. and Higdon, D. (1997) Primate species richness is determined by plant productivity: implications for conservation. *Proceedings of the National Academy of Sciences, USA* 94, 13023–13027.

Kays, R.W. and Gittleman, J.L. (1995) Home range size and social behavior of kinkajous (*Potos flavus*) in the Republic of Panama. *Biotropica* 27, 530–534.

Kessler, P. (1995) Preliminary field study of the red-handed tamarin, *Saguinus midas*, in French Guiana. *Neotropical Primates* 3, 184–185.

Kiltie, R.A. (1980) Stomach content of rain forest peccaries (*Tayassu tajacu* and *T. pecari*). *Biotropica* 13, 234–236.

Kiltie, R.A. (1981) Distribution of palm fruit on a rain forest floor: why white-lipped peccaries forage near objects. *Biotropica* 6, 69–103.

Kiltie, R.A. (1982) Bite force as a basis for niche differentiation between rain forest peccaries (*Tayassu tajacu* and *T. pecari*). *Biotropica* 14, 188–195.

Kinzey, W.G. and Norconk, M.A. (1993) Physical and chemical properties of fruit and seeds eaten by *Pithecia* and *Chiropotes* in Surinam and Venezuela. *International Journal of Primatology* 14, 207–228.

Krijger, C.L., Opdam, M., Théry, M. and Bongers, F. (1997) Courtship behaviour of manakins and seed bank composition in a French Guianan rain forest. *Journal of Tropical Ecology* 13, 631–636.

Kubitski, K. (1985) Ichthyochory in *Gnetum venosum*. *Anais da Academia Brasiliera de Ciéncia* 57, 513–516.

Kunz, T.H. and Diaz, C.A. (1995) Folivory in fruit-eating bats, with new evidence from *Artibeus jamaicensis* (Chiroptera: Phyllostomatinae). *Biotropica* 27, 106–120.

Larpin, D. (1989) Evolution floristique et structurale d'un recru forestier en Guyane française. *Revue d'Ecologie (Terre et Vie)* 44, 209–224.

Lehman, S.M. (2000) Primate community structure in Guyana: a biogeographic analysis. *International Journal of Primatology* 21, 333–351.

Leighton, M. and Leighton, D.R. (1983) Vertebrate responses to fruiting seasonality within a Bornean rain forest. In: Sutton, S.L., Whitmore, T.C. and Chadwick, A.C. (eds) *Tropical Rain Forest: Ecology and Management*. Blackwell Scientific, London, pp. 181–196.

Lescure, J.-P. (1981) La végétation et la flore dans la région de la Piste de Saint-Ellie. *Bulletin d'ECEREX* 3, 4–23.

Lescure, J.-P. and Boulet, R. (1985) Relationships between soil and vegetation in a tropical rain forest in French Guiana. *Biotropica* 17, 155–164.

Lewis, D.H. (1985) Symbiosis and mutualism: crisp concepts and soggy semantics. In: Boucher, D.H. (ed.) *The Biology of Mutualism*. Oxford University Press, Oxford, UK, pp. 29–39.

Lim, B.K. and Engstrom, H.D. (2000) Preliminary survey of bats from the upper Mazaruni of Guyana. *Chiroptera Neotropical* 6, 119–123.

Lim, B.K. and Engstrom, M.D. (2001a) Bat community structure at Iwokrama Forest, Guyana. *Journal of Tropical Ecology* 17, 647–665.

Lim, B.K. and Engstrom, M.D. (2001b) Species diversity of bats (Mammalia: Chiroptera) in Iwokrama Forest, Guyana, and the Guianan subregion: implications for conservation. *Biodiversity and Conservation* 10, 613–657.

Lim, B.K., Engstrom, H.D., Timm, R.M., Anderson, R.P. and Watson, L.C. (1999) First records of 10 bat species in Guyana and comments on diversity of bats in Iwokrama forest. *Acta Chiropterologica* 1, 179–190.

Lim, B.K., Peterson, A.T. and Engstrom, M.D. (2002) Robustness of ecological niche modeling algorithms for mammals in Guyana. *Biodiversity and Conservation* 11, 1237–1246.

Loubry, D. (1993) Les paradoxes de l'angélique (*Dicorynia guianensis* Amshoff): dissémination et parasitisme des graines avant dispersion chez un arbre anémochore de forêt guyanaise. *Revue d'Ecologie (Terre et Vie)* 48, 353–363.

Lowe-McConnell, R.H. (1964) The fishes of the Rupununi savanna district of British Guiana, Pt. 1. Groupings of fish species and effects of the seasonal cycles on the fish. *Journal of the Linnean Society* (*Zoology*) 45, 103–144.

Marinho-Filho, J.S. (1991) The coexistence of two frugivorous bats species and the phenology of their food plants in Brazil. *Journal of Tropical Ecology* 7, 59–67.

Maury-Lechon, G. and Poncy, O. (1986) Dynamique forestière sur 6 hectares de forêt dense humide de Guyane française, à partir des données de quelques espèces de forêt primaire et de cicatrisation. *Mémoires du Muséum national d'Histoire naturelle, Nouvelle Série, Série A, Zoologie* 132, 211–242.

Matola, S. and Todd, S. (1997) Tapir specialist group. Current field work. *Species* 29, 52.

Métrailler, S. and Le Gratiet, G. (1996) *Tortues Continentales de Guyane Française.* PMS, Bramois, Switzerland.

Meyer de Shauensee, R. and Phelps Jr, W.H. (1978) *A Guide to the Birds of Venezuela.* Princeton University Press, London.

Mittermeier, R.A. (1987) Hunting and its effects on wild primate populations in Suriname. In: Robinson, J.G. and Redford, K.H. (eds) *Neotropical Wildlife Use and Conservation.* University of Chicago Press, Chicago, Illinois, pp. 93–110.

Mittermeier, R.A. and van Roosmalen, M.G.M. (1981) Preliminary observations on habitat utilisation and diet in eight Suriname monkeys. *Folia Primatologica* 36, 1–39.

Morris, D.O. (1962) The behavior of the green acouchi (*Myoprocta pratti*) with special reference to scatter-hoarding. *Proceedings of the Royal Zoological Society* 139, 701–732.

Muckenhirn, N.A., Mortensen, B., Vessey, S., Fraser, C.E. and Singh, B. (1976) *Report on a Primate Survey in Guyana.* Pan-American Health Organization, Washington, DC.

Norconk, M.A. and Kinzey, W.G. (1994) Challenge for neotropical frugivory: travel patterns of spider monkeys and bearded sakis. *American Journal of Primatology* 34, 171–183.

Norconk, M.A., Wertis, C. and Kinzey, W.G. (1997) Seed predation by monkeys and macaws in Eastern Venezuela: preliminary findings. *Primates* 38, 177–184.

Ochoa, J.G. (1995) Los mamiferos de la region de Imataca, Venezuela. *Acta Cientifica Venezolana* 46, 274–287.

Ochoa, J.G. (2000) Effects of logging on small-mammal diversity in the lowland forests of the Venezuelan Guyana region. *Biotropica* 32, 146–164.

Ojasti, J. (1967) Consideraciones sobre la ecologia y conservación de la tortuga *Podocnemis expansa* (Chelonia: Pelomedusidae). *Atlas Simposios de Biota Amazônica* 7, 201–206.

Oliveira, J.M.S., Lima, M.G., Bonvincino, C., Ayres, J.M. and Fleagle, J.G. (1985) Preliminary note on the ecology and behavior of the Guianan saki (*Pithecia pithecia,* Linnaeus 1766; Cebidae, Primate). *Acta Amazonica* 15, 249–263.

Ouboter, P. and Mol, J.H.A. (1993) The fish fauna of Suriname. In: Ouboter, P. (ed.) *The Freshwater Ecosystems of Suriname.* Kluwer Academic, Dordrecht, The Netherlands, pp. 133–154.

Pack, K.S., Henry, O. and Sabatier, D. (1999) The insectivorous-frugivorous diet of the golden-handed tamarin (*Saguinus midas midas*) in French Guiana. *Folia Primatologica* 70, 1–7.

Parker, H.W. (1935) The frogs, lizards, and snakes of British Guiana. *Proceedings Zoological Society of London* 2, 505–530.

Parker III, T.A., Foster, R.B., Emmons, L.H., Gentry, A.H., Beck, S., Estenssoro, S. and Hinojosa, F. (1991) *A Biological Assessment of the Alto Madidi Region.* RAP Working Papers 1. Conservation International, Washington, DC.

Parker III, T.A., Foster, R.B., Emmons, L.H., Freed, P., Forsyth, A.B., Hoffman, B. and Gill, B.D. (1993) *A Biological Assessment of the Kanuku Mountain Region of Southwestern Guyana.* RAP Working Papers 5. Conservation International, Washington, DC.

Peres, C.A. (1993a) Notes on the ecology of buffy saki monkeys (*Pithecia albicans,* Gray 1860): a canopy seed-predator. *American Journal of Primatology* 31, 129–140.

Peres, C.A. (1993b) Structure and spatial organization of an Amazonian terra firme forest primate community. *Journal of Tropical Ecology* 9, 259–276.

Peres, C.A. (1994a) Composition, density, and fruiting phenology of arborescent palms in an Amazonian terra firm forest. *Biotropica* 26, 285–294.

Peres, C.A. (1994b) Primate responses to phenological changes in an Amazonian terra firme forest. *Biotropica* 26, 98–112.

Peres, C.A. (1996) Population status of while-lipped *Tayassu pecari* and collared peccaries *T. tajacu* in hunted and unhunted Amazonian forests. *Biological Conservation* 77, 115–123.

Peres, C.A. (2000) Identifying keystone plant resources in tropical forests: the case of gums from *Parkia* pods. *Journal of Tropical Ecology* 16, 287–317.

Piedade, M.T.F. (1985) Ecologia e biologia reproductiva de *Astrocaryum jauaru* Mart. (Palmae) como exemplo de populacão adaptada as areas inundáveis do Rio Negro (*Igapós*). MSc thesis, INPA.

Planquette, P., Keith, P. and Le Bail, P.-Y. (1996) *Atlas des Poissons d'Eau Douce de Guyane.* Tome 1. MNHN, Paris.

Poncy, O., Riéra, B., Larpin, D., Belbenoit, P., Jullien, M., Hoff, M. and Charles-Dominique, P. (1998) The permanent field research station 'Les Nouragues' in the tropical forest of French Guiana: current projects and preliminary results on tree diversity, structure, and dynamics. In: Dallmeier, F. and Comiskey, J.A. (eds) *Forest Diversity in North, Central and South America and the Caribbean: Research and Monitoring. Measuring and Monitoring Tropical Forest Diversity: a Network of Biodiversity Plots.* Man and the Biosphere Series 21. Pergamon Press, Washington, DC, pp. 385–410.

Pons, J.M. and Cosson, J.F. (2002) Use of forest fragments by animalivorous bats in French Guiana. *Revue d'Ecologie (Terre et Vie),* 117–130.

Poulin, G., Wright, S.J., Lefebvre, G. and Calderon, O. (1999) Interspecific synchrony and asynchrony in the fruiting phenologies of congeneric bird-dispersed plants in Panama. *Journal of Tropical Ecology* 15, 213–227.

Prance, G. (1980) Pollination of *Combretum* by monkeys. *Biotropica* 12, 239.

Prévost, M.F. (1983) Les fruits et les graines des espèces végétales pionnières de Guyane française. *Revue d'Ecologie (Terrre et Vie)* 38, 122–145.

Puyo, J. (1949) *Poissons de la Guyane française. Faune de l'empire Français XII.* Office de la Recherche Scientifique d'Outre-mer, Paris.

Quelch, J.J. (1892) The bats of British Guiana. *Timehri* 6, 90–109.

Quiroga-Castro, V.D. and Roldan, A.I. (2001) The fate of *Attalea phalerata* (Palmae) seeds dispersed to a tapir latrine. *Biotropica* 33, 472–477.

Rankin, J.M. (1978) The influence of seed predation and plant competition on tree species abundances in two adjacent tropical rain forest communities in Trinidad, West Indies. PhD thesis, University of Michigan.

Reichart, H.A. (1991) *Brownsberg Nature Park Management Plan 1991–1995.* WWF and Stichting Natuurbehoud Suriname, Paramaribo, Suriname.

Richard-Hansen, C., Vié, J.-C., Vidal, N. and Kéravec, J. (1999) Body measurements of 40 species of mammals from French Guiana. *Journal of Zoology* 247, 419–428.

Richards, P.W. (1996) *The Tropical Rain Forest,* 2nd edn. Cambridge University Press, Cambridge, UK.

Ridgely, R. and Agro, D. (1999) *Birds of the Iwokrama Forest.* Iwokrama International Centre, Georgetown, Guyana.

Riéra, B. (1985) Importance des buttes de déracinement dans la régénération forestière en Guyane française. *Revue d'Ecologie (Terre et Vie)* 40, 1–9.

Robinson, J.G. and Redford, K.H. (1986) Body size, diet, and population density of neotropical forest mammals. *American Naturalist* 128, 665–680.

Rodrigues, M., Olmos, F. and Galetti, M. (1993) Seed dispersal by tapir in Southeastern Brazil. *Mammalia* 57, 460–462.

Rollet, B. (1969) La régénération naturelle en forêt dense humide sempervirente de plaine de la guyane vénézuelienne. *Bois et Forêts des Tropiques* 124, 19–38.

Rylands, A.B. and Keuroghlian, A. (1988) Primate populations in continuous forest and forest fragments in Central Amazonia. *Acta Amazonica* 18, 29–1307.

Sabatier, D. (1983) Fructification and dissémination en Forêt Guyanaise. L'exemple de quelques espèces ligneuses. Thèse de doctorat de l'Université USTL, Montpellier, France.

Sabatier, D. (1985) Saisonnalité et déterminisme du pic de fructification en forêt guyanaise. *Revue d'Ecologie (Terre et Vie)* 40, 289–320.

Sabatier, D. and Prévost, M.-F. (1990) Quelques données sur la composition floristique et la diversité des peuplements forestiers de Guyane française. *Bois et Forêts des Tropiques* 219, 31–55.

Sabatier, D., Grimaldi, M., Prévost, M.-F., Guillaume, J., Godron, M., Dosso, M. and Curmi, P. (1997) The influence of soil cover organization on the floristic and structural heterogeneity of a Guianan rain forest. *Plant Ecology* 131, 81–108.

Salas, L.A. (1996) Habitat use by lowland tapirs (*Tapirus terrestris* L.) in the Tabaro River valley, southern Venezuela. *Canadian Journal of Zoology* 74, 1452–1458.

Salas, L.A. and Fuller, T.K. (1996) Diet of the lowland tapir (*Tapirus terrestris* L.) in the Tabaro River valley, southern Venezuela. *Canadian Journal of Zoology* 74, 1444–1451.

Schuchmann, K.L. (1984) Zur ernährung des Cayenne Felsenhahnes (*Rupicola rupicola,* Cotingidae). *Journal für Ornithologie* 125, 239–241.

Schupp, E.W. (1988) Factors affecting post-dispersal seed survival in a tropical forest. *Oecologia* 76, 526–530.
Schupp, E.W. and Frost, E.J. (1989) Differential predation of *Welfia georgii* seeds in treefall gaps and the forest understorey. *Biotropica* 21, 200–203.
Setz, E. (1992) Comportamento de alimentação de *Pithecia pithecia* (Cebidae, Primatas) em um fragmento florestal. *Primatologia no Brasil* 3, 327–330.
Setz, E.Z.F. (1993) Ecologia alimentar de um grupo de paravacus (*Pithecia pithecia chrysocephala*) em um fragmento florestal na amazonia central. PhD thesis, University of Campinas, Brazil.
Setz, E.Z.F. (1996) Foraging ecology of golden-faced sakis in a forest fragment in central Amazon. *XVIth Congress of the International Primatological Society and the XIXth Conference of the American Society of Primatologists,* 11–16 August, University of Wisconsin-Madison, p. 157.
Silvius, K.M. (2002) Spatio-temporal patterns of palm endocarp use by three Amazonian forest mammals: granivory or 'grubivory'? *Journal of Tropical Ecology* 18, 707–723.
Silvius, K.M. and Fragoso, J.M.V. (2002) Pulp handling by vertebrate seed dispersers increases palm seed predation by bruchid beetles in the northern Amazon. *Journal of Ecology* 90, 1024–1032.
Simmen, B. and Sabatier, D. (1996) Diet of some French Guianan primates: food composition and food choice. *International Journal of Primatology* 17, 661–693.
Simmen, B., Julliot, C., Pagès, E. and Bayart, F. (2001) The primate community: population densities and feeding ecology. In: Bongers, F., Charles-Dominique, P., Forget, P.-M. and Théry, M. (eds) *Nouragues: Dynamics and Plant–Animal Interactions in a Neotropical Rainforest.* Kluwer Academic, Dordrecht, The Netherlands, pp. 89–101.
Simmons, N.B. and Voss, R.S. (1998) Mammals of Paracou, French Guiana: a neotropical rainforest fauna, part 1. Bats. *Bulletin of the American Museum of Natural History* 237, 1–219.
Simmons, N.B., Voss, R.S. and Peckham, H.C. (2000) The bat fauna of the Saul region, French Guiana. *Acta Chiropterologica* 2, 23–36.
Sist, P. (1989a) Demography of *Astrocaryum sciophilum,* an understorey palm of French Guiana. *Principes* 33, 142–151.
Sist, P. (1989b) Stratégies de régénération de quelques palmiers en forêt guyanaise. Thèse de doctorat de l'Université Paris VI, Paris.
Sist, P. (1989c) Peuplement et phénologie des palmiers en forêt guyanaise (Piste de St Elie). *Revue d'Ecologie (Terre et Vie)* 44, 113–151.
Sist, P. and Puig, H. (1987) Régénération, dynamique des populations et dissémination d'un palmier de Guyane française: *Jessenia bataua* (Mart.) Burret subsp. *Oligocarpa* (Griseb. and H. Wendl.) Balick. *Bulletin du Muséum national d'Histoire naturelle, 4ème Série, 9, section B, Adansonia* 3, 317–336.
Smith, P.G. and Kerry, S.M. (1996) The Iwokrama Rain Forest Programme for sustainable development: how much of Guyana's bat (Chiroptera) diversity does it encompass? *Biodiversity and Conservation* 5, 921–942.
Smythe, N. (1970) Relationships between fruiting seasons and seed dispersal methods in a neotropical forest. *The American Naturalist* 104, 25–35.
Smythe, N. (1978) The natural history of the Central American agouti (*Dasyprocta punctata*) *Smithsonian Contribution to Zoology* 257, 1–52.
Smythe, N. (1989) Seed survival in the palm *Astrocaryum standleyanum*: evidence for dependence upon its seed dispersers. *Biotropica* 21, 50–56.
Smythe, N., Glanz, W.E. and Leigh Jr, E.G. (1982) Population regulation in some terrestrial frugivores. In: Leigh Jr, E.G., Rand, A.S. and Windsor, D.M. (eds) *The Ecology of a Tropical Forest.* Smithsonian Institution Press, Washington, DC, pp. 227–238.
Snow, B.K. (1970) A field study of the beared bellbird in Trinidad. *Ibis* 112, 299–329.
Snow, B.K. (1972) A field study of the calfird *Perissocephalus tricolor. Ibis* 114, 139–162.
Snow, D.W. (1962a) A field study of the black and white manakin, *Manacus manacus* in Trinidad. *Zoologica: New York Zoological Society* 47, 65–63.
Snow, D.W. (1962b) A field study of the golden-headed manakin, *Pipra erythrocephala,* in Trinidad, W. I. *Zoologica: New York Zoological Society* 47, 183–198.
Snow, D.W. (1963) The evolution of manakin displays. In: *Proceedings of the XIIIth Ornithological Congress.* Syracuse University, Ithaca, New York, pp. 553–561.
Snow, D.W. (1964) A possible selection factor in the evolution of fruiting seasons in tropical forest. *Oikos* 15, 274–281.
Snow, D.W. (1971) Observations on the purple-throated fruit-crow in Guyana. *The Living Bird* 10, 5–17.

Snow, D.W. (1981) Tropical frugivous birds and their food plant: a world survey. *Biotropica* 13, 1–14.

Stalling, J.R. (1984) Notes on the feeding habits of *Mazama gouazoubira* in the Chaco boreal of Paraguay. *Biotropica* 16, 155–157.

Swihart, R.K., Slade, N.A. and Bergstrom, B.J. (1988) Relating body size to the rate of home range use in mammals. *Ecology* 69, 393–399.

Tate, G.H.H. (1939) The mammals of the Guiana region. *Bulletin of the American Museum of Natural History* 76, 151–229.

ter Steege, H. (1998) The use of forest inventory data for a National Protected Area Strategy in Guyana. Biodiversity and conservation. *Biodiversity and Conservation* 7, 161–170.

ter Steege, H. and Hammond, D.S. (2001) Character convergence, diversity, and disturbance in tropical rain forest in Guyana. *Ecology* 82, 3197–3212.

ter Steege, H. and Zondervan, G. (2000) A preliminary analysis of large-scale forest inventory data of the Guiana Shield. In: ter Steege, H. (ed.) *Plant Diversity in Guyana*. Tropenbos Foundation, Wageningen, The Netherlands, pp. 35–54.

ter Steege, H., Jetten, V., Polak, M. and Werger, M. (1993) Tropical rain forest types and soils of a watershed in Guyana, South America. *Journal of Vegetation Science* 4, 705–716.

ter Steege, H., Boot, R.G.A., Brouwer, L.C., Caesar, J.C., Ek, R.C., Hammond, D.S., Haripersaud, P.P., van der Hout, P., Jetten, V.G., van Kekem, A.J., Kellman, M.A., Khan, Z., Polak, A.M., Pons, T.L., Pulles, J., Raaimakers, D., Rose, S.A., van der Sanden, J.J. and Zagt, R. (1996) *Ecology and Logging in a Tropical Rain Forest in Guyana*. Tropenbos Series 14. Tropenbos Foundation, Wageningen, The Netherlands.

Terborgh, J.W. (1986) Keystone plant resources in the tropical forest. In: Soule, M.E. (ed.) *Conservation Biology II*. Sinauer, Sunderland, pp. 330–344.

Terborgh, J.W. (1988) The big things that run the world – a sequel to E. O. Wilson. *Conservation Biology* 2, 402–403.

Terborgh, J.W. (1992) Maintenance of diversity in tropical forests. *Biotropica* (supplement) 24, 283–292.

Terwillinger, V.J. (1978) Natural history of Baird's tapir on Barro Colorado Island, Panama Canal Zone. *Biotropica* 10, 211–220.

Théry, M. (1987a) Influence des caractéristiques lumineuses sur la localisation des sites traditionnels, parade et baignade des manakins (Passeriformes, Pipridae). *Compte Rendus de l'Académie des Sciences, Paris, Série III* (1), 304, 19–24.

Théry, M. (1987b) Nidification de *Campylopterus largipennis* (Trochilidae) en Guyane française. *L'oiseau et R.F.O.* 57, 141–144.

Théry, M. (1990a) Display repertoire and social organization of the white-fronted and white-throated manakins. *The Wilson Bulletin* 102, 123–130.

Théry, M. (1990b) Ecologie et comportement des oiseaux Pipridae en Guyane: leks, frugivorie et dissémination des graines. Thèse de doctorat de l'Université Paris VI, Paris.

Théry, M. (1990c) Influence de la lumière sur le choix de l'habitat et le comportement sexuel des Pipridae (Aves: Passeriformes) en Guyane française. *Revue d'Ecologie (Terre et Vie)* 45, 215–236.

Théry, M. (1992) The evolution of leks through female choice: differential clustering and space utilization in six sympatric manakins. *Behavioral Ecology and Sociobiology* 30, 227–237.

Théry, M. (1997) Wing-shape variation in relation to ecology and sexual selection in five sympatric lekking manakins (Passeriformes: Pipridae). *Ecotropica* 3, 9–19.

Théry, M. and Larpin, D. (1993) Seed dispersal and vegetation dynamics at a cock-of-the-rock's lek in the tropical forest of French Guiana. *Journal of Tropical Ecology* 9, 109–116.

Théry, M. and Vehrencamp, S.L. (1995) Light patterns as cues for mate choice in the lekking white-throated manakin (*Corapipo gutturalis*). *The Auk* 112, 133–145.

Théry, M., Erard, C. and Sabatier, D. (1992) Les fruits dans le régime alimentaire de *Penelope marail* (Aves, Cracidae) en Guyane française: frugivorie stricte et sélective. *Revue d'Ecologie (Terre et Vie)* 47, 383–401.

Thiollay, J.-M. (1994) Structure, density and rarity in an Amazonian rainforest bird community. *Journal of Tropical Ecology* 10, 449–481.

Thompson, J., Proctor, J., Viana, V., Milliken, W., Ratter, J.A. and Scott, D.A. (1992) Ecological studies on a lowland evergreen rain forest on Maracá Island, Roraima, Brazil. I. Physical environment, forest structure and leaf chemistry. *Journal of Ecology* 80, 689–703.

Timm, R.M. (1994) The mammal fauna. In: McDade, L.A., Bawa, K.S., Hespenheide, H.A. and Hartshorn, G.S. (eds) *La Selva: Ecology and Natural History of a Neotropical Rain Forest*. The University of Chicago Press, Chicago, Illinois, pp. 229–237.

Tostain, O., Dujardin, J.-L., Erard, C. and Thiollay, J.-M. (1992) *Oiseaux de Guyane. Biologie, Ecologie, Protection et Répartition.* Sociétés d'études ornithologiques, MNHN, Brunoy, France.
Trail, P.W. and Donahue, P. (1991) Notes on the behavior and ecology of the red-cotingas (Cotingidae: *Phoenicircus*). *The Willson Bulletin* 103, 539–551.
Vander Wall, S.B. (1990) *Food Hoarding in Animals.* Chicago University Press, Chicago, Illinois.
van Roosmalen, M.G.M. (1985) Habitat preferences, diet, feeding strategy and social organization of the black spider monkey (*Ateles paniscus* Linnaeus 1758) in Surinam. *Acta Amazonica* (supplement) 15, 1–238.
van Roosmalen, M.G.M., Mittermeier, R.A. and Fleagle, J.G. (1988) Diet of the northern bearded saki (*Chiropotes satanas chiropotes*): a neotropical seed predator. *American Journal of Primatology* 14, 11–36.
Vié, J.-C. (1997) Ecology and behavior of white-faced sakis (*Pithecia pithecia*) in French Guiana. Preliminary results. *XVIth Congress of the International Primatological Society and the XIXth Conference of the American Society of Primatologists.* 11–16 August, University of Wisconsin-Madison.
Vié, J.-C. (1998) Les effets d'une perturbation majeure de l'habitat sur deux espèces de primates en Guyane française: translocation de singes hurleurs roux (*Alouatta seniculus*) et translocation et insularisation de sakis à face pâle (*Pithecia pithecia*). Thèse de l'Université de Montpellier II, Montpellier, France.
Vié, J.-C. and Richard-Hansen, C. (1997) Primate translocation in French Guiana. A preliminary report. *Neotropical Primates* 5, 1–3.
Vié, J.-C., Volobouev, V., Patton, J.L. and Granjon, L. (1996) A new species of *Isothrix* (Rodentia: echimyidae) from French Guiana. *Mammalia* 60, 393–406.
Vié, J.-C., Richard-Hansen, C. and Fournie-Chambrillon, C. (2001) Abundance, use of space, and activity patterns of white-faced sakis (*Pithecia pithecia*) in French Guiana. *American Journal of Primatology* 55, 203–221.
Voss, R.S. and Emmons, L.H. (1996) Mammalian diversity in neotropical lowland rainforests: a preliminary assessment. *Bulletin of the American Museum of Natural History* 230, 1–230.
Voss, R.S., Lunde, D.P. and Simmons, N.B. (2001) Mammals of Paracou, French Guiana: a neotropical rainforest fauna part 2. Nonvolant species. *Bulletin of the American Museum of Natural History* 263, 1–236.
Wheelwright, N.T. (1985) Fruit size, gape width, and the diets of fruit-eating birds. *Ecology* 66, 808–818.
Wheelwright, N.T. and Orians, G.H. (1982) Seed dispersal by animals: contrasts with pollen dispersal, problem with terminology, and constraint on coevolution. *American Naturalist* 119, 402–413.
Williams, R.T. (1972) Species variations in drug biotransformations. In: La Du, B.N., Mandel, H.G. and Way, E.L. (eds) *Fundamentals of Drug Metabolism and Drug Distribution.* Williams and Wilkins, Baltimore, Maryland, pp. 187–205.
Worthington, A.H. (1982) Population sizes and breeding rhythms of two species of manakins in relation to food supply. In: Leigh Jr, E.G., Rand, A.S. and Windsor, D.M. (eds) *The Ecology of a Tropical Forest.* Smithsonian Institution Press, Washington, DC, pp. 213–225.
Worthington, A.H. (1989) Adaptations for avian frugivory: assimilation efficiency and gut transit time of *Manacus vitellinus* and *Pipra mentalis*. *Oecologia* 80, 381–389.
Zhang, S.-Y. (1994) Utilisation de l'espace, stratégies alimentaires et rôle dans la dissémination des graines du singe capucin (*Cebus apella*) en Guyane Française. Thèse de doctorat de l'Université Paris VI, Paris.
Zhang, S.-Y. (1995) Activity and ranging patterns in relation to fruit utilization by brown capuchins (*Cebus apella*) in French Guiana. *International Journal of Primatology* 16, 489–507.
Zhang, S.-Y. and Wang, L.-X. (1995a) Comparison of three fruit census methods in French Guiana. *Journal of Tropical Ecology* 11, 281–294.
Zhang, S.-Y. and Wang, L.-X. (1995b) Fruit consumption and seed dispersal of *Ziziphus cinnamomum* (Rhamnaceae) by two sympatric primates (*Cebus apella* and *Ateles paniscus*) in French Guiana. *Biotropica* 27, 397–401.
Zona, S. and Henderson, A. (1989) A review of animal-mediated seed dispersal of palms. *Selbyana* 11, 6–21.

Appendix 4.1. Plant families found to be part of the diet of 22 herbivorous vertebrates in Guiana Shield forests.

	Birds						Arboreal mammals								Terrestrial mammals								
Species	Pip	Rr	Tm	Pm	Pc	Ca	Cp	Sm	Pp	Pf	Cs	Ca	As	Ap	Oc	P spp	Nn	Dl	Cpac	Mg	Ma	Tp	Tt
Body mass	15	230	975	990	1071	2985	300	500	1809	2490	2986	3445	6185	7775	35	350	4400	5000	8227	17350	26100	28550	300000
Acanthaceae																				+	+		+
Anacardiaceae							x	x	+	x	x	x	x	x		+	+	+	+	x	x	+	x
Annonaceae	+	x	x	x	x	x	x	x	x	x	x	x	x	x	x	x	+			x	x		x
Apocynaceae								x	x			x	x	x	x	x		x		x	x		x
Aquifoliaceae	+																		+	+	+	+	+
Araceae	+	x							x		x	x	x	x		+	+			x	x		+
Araliaceae	+	x				x		x				x	x							x	x	+	
Arecaceae		x	x	x	x	x			x		x	x		x	x	x	+	x	+	x	x	+	x
Asclepiadaceae																							+
Asteraceae	+																						
Bignonaceae							x		x	x	x	x	x	x		x				+	x		+
Bombacaceae			x						+			x	x	x	x	x	+	+	+	x	x	+	
Bonnetiaceae									+														
Boraginaceae	+		x						x	x	x	x	x	x		x		+		+			x
Bromeliaceae								x				x				+	+			+			+
Burseraceae	+	x	x	x	x	x	x	x	+	x	x	x	x	x	x	x	+	x	+		x		+
Cactaceae									+					x						+			
Caesalpiniaceae					x		x		x	x	x	x	x	x	x	x		x		x	x	+	+
Campanulaceae																					+		
Canellaceae													x	x									
Capparidaceae	+								x					x									
Caricaceae								x				x	x	x		+	+			+	x		
Caryocaraceae												x	x					x		x	x		
Cecropiaceae		x		x		x		x	+	x		x	x	x				+		x	x		+
Celastraceae	x					x		x	x		x	x	x	x	x		+						
Chloranthaceae			x			x																	
Chrysobalanaceae				x			x		x	x	x		x	x	x	x	+	x	+	x		+	x
Clusiaceae		x	x				x	x	x	x	x	x		x	x	x	+	+		x	x	+	x
Combretaceae									+				x	x	x	x				+			+
Commelinaceae																				+			

Connaraceae	+			x					x									+					
Convolvulaceae							x	x	+	x		x	x	x				+	+	x	x	+	+
Cucurbitaceae		x	x		x	x		x	+	x	x	x	x	x						x	x		x
Cyclanthaceae	+							x				x	x	x		+					x	+	+
Cyperaceae																					+		+
Dilleniaceae	+	x							x			x	x			+	+	x			x		+
Dioscoreaceae														x									
Ebenaceae								x	+			x		x				x		x	x		
Elaeocarpaceae												x			x	x							
Ericaceae									+							+							
Euphorbiaceae	+		x		x	x	x		x	x	x	x	x	x				x	+	x	x		+
Eythroxylaceae	+								x														
Fabaceae					x	x			+			x	x	x		x	+	x	+	x	c	+	+
Flacourtiaceae	+	x			x		x		x	x	x		x	x	x	x	+	+					
Gesneriaceae								x						x									
Gnetaceae		x					x		x	x			x	x	x			x		x			
Hippocrateaceae				x					x			x	x	x				x		x	x		x
Hugoniaceae																				x	x	x	x
Humiriaceae			x	x	x	x	x			x		x		x	x	x		x		x			x
Hymenophyllaceae																					+		
Icacinaceae						x						x	x	x				x		x	x		
Lacistemaceae									+														+
Lauraceae	+	x	x	x	x	x	x		+	x	x	x	x	x		x	+	+	+	x	x		+
Lecythidaceae									x		x	x	x	x	x	x	+	+		x	x		+
Liliaceae		x																					
Linaceae	+	x					x		+	x		x											
Loganiaceae							x			x	x	x	x	x		+		x		x	x		x
Loranthaceae											x		x	x									
Malpighiaceae		x			x				+		x	x	x			+	+				+		x
Malvaceae																							+
Maranthaceae	+					x																	+
Marcgraviaceae		x			x			x	+	x		x	x	x									
Melastomataceae	x	x			x			x	x	x		x	x	x			+	x	+		+		x
Meliaceae	+	x	x		x	x	x		x		x	x		x	x	x		x		x	x		
Mendonciaceae								x			x	x	x	x						x			

continued

Appendix 4.1. *continued*

	Birds						Arboreal mammals								Terrestrial mammals								
Species	Pip	Rr	Tm	Pm	Pc	Ca	Cp	Sm	Pp	Pf	Cs	Ca	As	Ap	Oc	P spp	Nn	Dl	Cpac	Mg	Ma	Tp	Tt
Body mass	15	230	975	990	1071	2985	300	500	1809	2490	2986	3445	6185	7775	35	350	4400	5000	8227	17350	26100	28550	300000
Mennispermaceae			x	x		x			+	x		x	x	x				x			+	+	x
Mimosaceae			x	x		x		x	+	x	x	x	x	x	x	x		+	+	x	x	+	x
Monimiaceae		x							+	x													
Moraceae	x	x	x	x	x	x	x	x	x	x	x	x	x	x	x	x	+	x	+	x	x	+	x
Musaceae	+	x															+		+			+	+
Myristicaceae		x	x			x	x		x	x			x	x		x	+	x	+	x	x	+	+
Myrsinaceae	+				x											+							
Myrtaceae	+		x	x	x	x	x	x	x	x	x	x	x	x	x	x	+	x		x	+	+	+
Nyctaginaceae	+					x			+			x	x	x			+			+			
Ochnaceae	+			x		x							x										
Olacaceae	+								x		x	x	x	x	x	x				x	+		+
Opiliaceae									x														
Orchidaceae														x									
Oxalidaceae																			+				+
Passifloraceae	+	x							x	x	x	x		x		+			+	x	x		+
Phytolaccaceae	+																						
Piperaceae	+													x							+		x
Poaceae	+					x																+	+
Polygalaceae	+	x						x	x	x	x	x	x	x		x				x	x	+	
Polygonaceae		x	x						+			x	x	x			+			+			
Polypodiaceae																							+
Portulacaceae																				+	+		
Quiinaceae									x			x	x	x									x
Rhamnaceae				x		x				x		x	x	x						+			+
Rhizophoraceae																+							+
Rosaceae															x	x							
Rubiaceae	x	x	x	x	x	x		x	x	x	x	x	x	x	x	x	+	x	+	x	x		x
Rutaceae									x		x	x		x									
Sapindaceae	+	x			x	x			+		x	x	x	x	x	x	+	+		x	x	+	+

Sapotaceae		x			x	x		x	x	x	x	x	x	x	x	+	x	+	x	x	+	x
Simaroubaceae	+	x						x	x			x	x		+				+			
Smilacaceae		x																	x			+
Solanaceae	+	x						+			x	x		x	x				x	x		+
Sterculiaceae						x			x	x	x	x	x		+	+	+	+	x			+
Styracaceae											x		x									
Symplocaceae		x																				
Tiliaceae	+								x	x	x	x	x		x	+	+	+				
Ulmaceae	x						x	+					x						+			+
Urticaceae	+																				+	+
Verbenaceae	+							x				x	x							x		+
Violaceae								x					x	x	x							+
Viscaceae											x											
Vitaceae			x				x													+	+	+
Vochysiaceae								+		x			x									x
Zingiberaceae	+			x																x	+	
Zygophyllaceae																			+			

Species name abbreviations: Birds – Pip: Piprids; Rr: *Rupicola rupicola*; Tm: *Tinamou major*, Pm: *Penelope marail*; Pc: *Psophia crepitans*; Ca: *Crax alector*. Arboreal mammals - Cp: *Caluromys philander*, Sm: *Saguinus midas*; Pp: *Pithecia pithecia*; Pf: *Potos flavus*; Cs: *Chiropotes satanas*; Ca: *Cebus apella*; As: *Alouatta seniculus*; Ap: *Ateles paniscus*. Terrestrial mammals - Oc: *Oryzomys capito*; P spp: *Proechimys cuvieri* and *P. guianensis*; Nn: *Nasua narica*; Dl: *Dasyprocta leporina*; Cpac: *Cuniculus paca*; Mg: *Mazama gouazoubira*; Ma: *Mazama americana*; Tp: *Tayassu pecari*; Tt: *Tapirus terrestris*.
Body mass is according to Erard *et al.* (1989), Erard *et al.* (1991) and Théry *et al.* (1992) for birds, and Robinson and Redford (1986) for all mammals but Nn (Gompper, 1994), Oc, P spp, Dl (Henry, 1994a), and Tt (Emmons and Feer, 1990) after taking the average between male and female or the midpoint for range of values.

Symbols: x, data from Guianas; +, data from outside the Guianan region.

References: Birds – Pip : Thery (1990b), Worthington (1982); Rr: Erard *et al.* (1989); Tm, Pm and Ca: Erard *et al.* (1991); Pm Théry *et al.* (1992). Arboreal mammals – Cp: Julien-Laferrière (1989); Sm: Pack *et al.* (1997); Pp: J.C. Vié (pers. comm.), Kinzey and Norconk 1993; Pp: Zetz (1993), Oliveira *et al.* (1985), Peres (1993); Pf: Julien-Laferrière (1989), (1993), (pers. comm.); Cs: Roosmalen *et al.* (1988), Kinzey and Norconk (1993), Norconk *et al.* (1997); Ca: Shu-Yi (1994), Guillotin *et al.* (1994), Simmen and Sabatier (1996); As: Julliot (1992), (1994); Ap: Roosmalen (1980), Guillotin *et al.* (1994), Simmen and Sabatier (1996). Terrestrial mammals – Oc: Guillotin (1982); P spp.: Guillotin (1982), Adler (1995); Nn: Gompper (1994); Dl: D. Sabatier (1983), P.-M. Forget, (personal observation), Smythe *et al.* (1982); Ap: Beck *et al.* (1998); Mg: Gayot et al. (2004), Bodmer (1990b); Ma: Gayot *et al.* (2004), Stallings (1984); Tp: F. Feer (personal communication), Bodmer (1990a), M. Altrichter, (personal communication), Tt: Salas and Fuller (1996), Bodmer (1990a), Terwillinger (1978).

Appendix 4.2. List of mammalian species present in the three Guianas (G: Guyana; S: Suriname; FG: French Guiana) characterizing their activity (N: nocturnal; D: diurnal), their habit (A: arboreal; Aq: aquatic; B: arboreal/terrestrial; F: flying; T: terrestrial) and their diet (Fr: fruit; S: seed; L: leaves/shoots; Po: pollen/nectar; E: exudates; Fu: fungi; Pu: pulp; I: insects; V: vertebrates; Bl: blood) after Emmons and Feer (1990), Eisenberg (1990), Voss and Emmons (1996) and Brosset *et al.* (1996).

Order	Family	Genus	Species	G	S	FG	Active Period	Habit	Diet										
									Fr	S	L	Po	E	Fu	Pu	In	V	Bl	Fish
Marsupiala	Didelphidae	*Caluromys*	*philander*	x	x	x	N	A	x							x			
		Caluromys	*lanatus*	x			N	A											
		Chironectes	*minimus*	x	x	x	N	Aq								x			x
		Didelphis	*marsupialis*	x	x	x	N	B	x			x				x	x		
		Didelphis	*albiventris*	x		x	N	B											
		Gracilianus	*emiliae*	x	x	x	N	A	x							x			
		Lutreolina	*crassicaudata*	x	x		N	A								x	x		x
		Marmosa	*tyleriana*	x			N	A	x							x			
		Marmosa	*lepida*	x	x	x	N	A	x							x			
		Marmosa	*murina*	x	x	x	N	A	x							x	x		
		Marmosops	*parvidens*	x	x	x	N	A	x							x			
		Metachirus	*nudicaudatus*	x	x	x	N	T	x							x			
		Micoureus	*demerarae*	x		x	N	A	x							x	x		
		Monodelphis	*brevicaudata*	x	x	x	D	T								x			
		Philander	*opposum*	x	x	x	N	B	x							x	x		
Xenarthra	Myrmecophagidae	*Myrmecophaga*	*tridactyla*	x	x	x	B	T								x			
		Tamandua	*tetradactyla*	x	x	x	B	B								x			
		Cyclopes	*didactylus*	x	x	x	N	A								x			
		Bradypus	*tridactylus*	x	x	x	B	A			x								
	Choloepidae	*Choloepus*	*didactylus*	x	x	x	N	A	x.		x								
	Dasypodidae	*Dasypus*	*novemcinctus*	x	x	x	B	T	x					x		x	x		
		Dasypus	*kappleri*	x	x	x	N	T								x			
		Cabassous	*unicinctus*	x	x	x	N	T								x			
		Priodontes	*maximus*	x	x	x	N	T								x			
		Euphractus	*sexcinctus*	x	x	x	D	T			x					x	x		
Chiroptera	Emballonuridae	*Centronycteris*	*maximiliani*	x	x	x	N	F								x			
		Cormura	*brevirostris*	x	x	x	N	F								x			
		Cyttarops	*alecto*	x	x	x	N	F								x			
		Diclidurus	*albus*	x	x	x	N	F								x			
		Diclidurus	*ingens*	x			N	F								x			

	Diclidurus	isabellus	x			N	F			x	
	Diclidurus	*scutatus*	x	x	x	N	F			x	
	Peropteryx	*leucoptera*	x	x	x	N	F			x	
	Peropteryx	*kappleri*	x	x	x	N	F			x	
	Peropteryx	*macrotis*	x	x	x	N	F			x	
	Peropteryx	*trinitatis*			x	N	F			x	
	Rhynchonycteris	*naso*	x	x	x	N	F			x	
	Saccopteryx	*bilineata*	x	x	x	N	F			x	
	Saccopteryx	*canescens*	x	x	x	N	F			x	
	Saccopteryx	*gymnura*	x	x	x	N	F			x	
	Saccopteryx	*leptura*	x	x	x	N	F			x	
Noctilionidae	*Noctilio*	*albiventris*	x	x	x	N	F			x	
	Noctilio	*leporinus*	x	x	x	N	F			x	x
Mormoopidae	*Pteronotus*	*parnellii*	x	x	x	N	F			x	
	Pteronotus	*gymnonotus*	x		x	N	F			x	
	Pteronotus	*personatus*	x	x	x	N	F			x	
	Natalus	*tumidirostris*		x	x	N	F			x	
Phyllostomidae	*Ametrida*	*centurio*	x	x	x	N	F	x			
	Anoura	*amplus*	x			N	F		x		
	Anoura	*caudifera*	x	x	x	N	F	x	x	x	
	Anoura	*geoffroyi*	x	x	x	N	F	x	x	x	
	Anoura	*latidens*	x						x		
	Artibeus	*cinereus*	x	x	x	N	F	x	x	x	
	Artibeus	*concolor*	x	x	x	N	F	x	x	x	
	Artibeus	*glaucus*	x					x	x	x	
	Artibeus	*gnomus*	x	x	x	N	F	x	x	x	
	Artibeus	*hartii*	x			N	F	x	x	x	
	Artibeus	*jamaicensis*	x	x	x	N	F	x	x	x	
	Artibeus	*lituratus*	x	x	x	N	F	x	x	x	
	Artibeus	*obcurus*			x	N	F	x	x		
	Artibeus	*planirostris*	x		x	N	F	x	x		
	Carollia	*brevicauda*	x	x	x	N	F	x	x	x	
	Carollia	*perspicillata*	x	x	x	N	F	x	x	x	
	Chiroderma	*trinitatum*	x	x	x	N	F	x			
	Chiroderma	*villosum*	x	x	x	N	F	x			
	Choeroniscus	*godmani*	x	x		N	F		x		
	Choeroniscus	*intermedius*	x			N	F		x		

continued

Appendix 4.2. *continued*

Order	Family	Genus	Species	G	S	FG	Active Period	Habit	Diet										
									Fr	S	L	Po	E	Fu	Pu	In	V	Bl	Fish
		Choeroniscus	*minor*	x	x	x	N	F				x							
		Chrotopterus	*auritus*	x	x	x	N	F								x	x		
		Desmodus	*rotundus*	x	x	x	N	F										x	
		Diaemus	*youngi*	x		x	N	F										x	
		Glossophaga	*commissarisi?*	x					x			x							
		Glossophaga	*longirostris*	x			N	F	x			x				x			
		Glossophaga	*soricina*	x	x	x	N	F	x			x				x			
		Glyphonycteris	*daviesi*	x	x	x	N	F	x							x			
		Glyphonycteris	*sylvestris*	x	x	x	N	F	x							x			
		Lampronycteris	*brachyotis*	x	x	x	N	F	x							x			
		Lichonycteris	*obscura*	x	x	x	N	F				x							
		Lionycteris	*spurrelli*	x	x	x	N	F				x							
		Lonchophylla	*thomasi*	x	x	x	N	F	x			x				x			
		Lonchorhina	*aurita*	x	x		N	F	x							x			
		Lonchorhina	*fernandezi*			x	N	F											
		Macrophyllum	*macrophyllum*	x	x	x	N	F								x			
		Ectophylla	*macconnelli*	x	x	x	N	F	x										
		Micronycteris	*brachyotis*	x	x	x	N	F	x							x			
		Micronycteris	*brossetti*	x		x	N	F	x							x			
		Micronycteris	*hirsuta*	x		x	N	F	x							x			
		Micronycteris	*homezi*	x		x	N	F	x							x			
		Micronycteris	*megalotis*	x	x	x	N	F	x							x			
		Micronycteris	*microtis*	x		x	N	F	x										
		Micronycteris	*minuta*	x	x	x	N	F	x							x			
		Micronycteris	*schmidtorum*			x	N	F	x										
		Mimon	*bennettii*	x	x	x	N	F								x	x		
		Mimon	*crenulatum*	x	x	x	N	F								x	x		
		Phylloderma	*stenops*	x	x	x	N	F	x							x			
		Phyllostomus	*discolor*	x	x	x	N	F	x			x				x			
		Phyllostomus	*elongatus*	x	x	x	N	F	x			x				x			
		Phyllostomus	*hastatus*	x	x	x	N	F	x			x				x			
		Phyllostomus	*latifolius*	x	x	x	N	F	x			x				x			
		Platyrrhinus	*aurarius*	x	x		N	F	x			x				x			

	Platyrrhinus	*brachycephalus*	x	x	x	N	F	x	x	x	
	Platyrrhinus	*helleri*	x	x	x	N	F	x	x	x	
	Platyrrhinus	*lineatus*	X	x	x	N	F	x			
	Pygoderma	*bilabiatum*		x		N	F	x			
	Rhinophylla	*pumilio*	x	x	x	N	F	x			
	Sphaeronycteris	*toxophyllum*	x			N	F				
	Sturnira	*lilium*	x	x	x	N	F	x	x		
	Sturnira	*ludovici*	x			N	F	x	x		
	Sturnira	*tildae*	x	x	x	N	F	x	x		
	Tonatia	*brasiliense*	x	x	x	N	F	x		x	
	Tonatia	*carrikeri*	x	x	x	N	F	x		x	
	Tonatia	*saurophila*	x	x	x	N	F			x	
	Tonatia	*schulzi*	x	x	x	N	F	x		x	
	Tonatia	*silvicola*	x	x	x	N	F	x		x	
	Trachyops	*cirrhosus*	x	x	x	N	F			x	x
	Trinycteris	*nicefori*	x	x	x	N	F	x		x	
	Uroderma	*bilobatum*	x	x	x	N	F	x	x	x	
	Uroderma	*magnirostrum*	x			N	F	x	x	x	
	Vampyressa	*bidens*	x	x	x	N	F	x			
	Vampyressa	*brocki*	x	x	x	N	F	x			
	Vampyressa	*melissa*	x		x	N	F	x			
	Vampyressa	*pusilla*	x		x	N	F	x			
	Vampyrodes	*caraccioli*	x	x	x	N	F	x			
	Vampyrum	*spectrum*	x	x	x	N	F	x		x	x
Furipteridae	*Furipterus*	*horrens*	x	x	x	N	F			x	
Thyropteridae	*Thyroptera*	*tricolor*	x	x	x	N	F			x	
Vespertilionidae	*Eptesicus*	*brasiliensis*	x	x		N	F			x	
	Eptesicus	*chiriquinus*			x	N	F			x	
	Eptesicus	*furinalis*	x	x	x	N	F			x	
	Lasiurus	*atratus*	x	x	x	N	F			x	
	Lasiurus	*brossevillii*			x	N	F			x	
	Lasiurus	*ega*	x	x		N	F			x	
	Lasiurus	*egregius*	x	x	x	N	F			x	
	Myotis	*albescens*	x	x		N	F			x	
	Myotis	*nigricans*	x	x	x	N	F			x	
	Myotis	*oxiotus*	x			N	F			x	

continued

Appendix 4.2. *continued*

Order	Family	Genus	Species	G	S	FG	Active Period	Habit	Diet										
									Fr	S	L	Po	E	Fu	Pu	In	V	Bl	Fish
		Myotis	*riparius*	x		x	N	F								x			
		Rhogeessa	*hussoni*		x		N	F								x			
		Rhogeessa	*tumida*	x	x		N	F								x			
	Molossidae	*Cynomops*	*greenhalli*	x	x	x	N	F								x			
		Cynomops	*planirostris*	x	x	x	N	F								x			
		Eumops	*auripendulus*	x	x	x	N	F								x			
		Eumops	*bonariensis*	x			N	F								x			
		Eumops	*hansae*	x	x	x	N	F								x			
		Eumops	*maurus*	x	x		N	F								x			
		Eumops	*trumbulli*	x	x		N	F								x			
		Molossops	*abrasus*	x	x	x	N	F								x			
		Molossops	*neglectus*	x	x		N	F								x			
		Molossops	*paranus*	x	x		N	F								x			
		Molossops	*temminckii*	x			N	F								x			
		Molossus	*ater*	x	x	x	N	F								x			
		Molossus	*barnesi*			x	N	F								x			
		Molossus	*coibensis*	x			N	F								x			
		Molossus	*molossus*	x	x	x	N	F								x			
		Molossus	*pretiosus*	x			N	F								x			
		Molossus	*rufus*	x	x	x	N	F								x			
		Molossus	*sinaloae*	x	x	x	N	F								x			
		Neoplatymops	*mattogrossensis*	x			N	F								x			
		Nyctinomops	*laticaudatus*	x	x	x	N	F								x			
		Nyctinomops	*macrotis*	x	x	x	N	F								x			
		Promops	*centralis*	x	x	x	N	F								x			
		Promops	*nasutus*	x	x		N	F								x			
		Tadarida	*aurispinosa*	x	x	x	N	F								x			
		Tadarida	*laticaudata*	x	x	x	N	F								x			
		Tadarida	*macrotis*	x	x	x	N	F								x			
Primates	Callithricidae	*Saguinus*	*midas*	x	x	x	D	A	x				x			x			
	Cebidae	*Alouatta*	*seniculus*	x	x	x	D	A	x		x	x							
		Ateles	*paniscus*	x	x	x	D	A	x		x	x							

		Cebus	olivaceus	x	x	x	D	A	x	x				x	x		
		Chiropotes	satanus	x	x	x	D	A	x	x		x			x		
		Pithecia	pithecia	x	x	x	D	A	x	x	x						
		Saimiri	sciureus	x	x	x	D	A	x			x			x		
Carnivores	Canidae	Cerdocyon	thous	x	x		D	T	x						x	x	
		Speothos	venaticus	x	x	x	D	T								x	
	Procyonidae	Bassaricyon	gabbii	x			N	A	x			x			x		
		Nasua	nasua	x	x	x	D	B	x						x	x	
		Potos	flavus	x	x	x	N	A	x			x			x		
		Procyon	cancrivorous	x	x	x	N	T							x	x	x
	Mustelidae	Eira	barbara	x	x	x	D	B	x						x	x	
		Galictis	vittata	x	x	x	D	T	x							x	
		Lontra	longicaudis	x	x	x	B	Aq							x		x
		Pteronura	brasiliensis	x	x	x	D	A							x	x	x
	Felidae	Herpailurus	yaguarondi	x	x	x	B	T								x	
		Leopardus	pardalis	x	x	x	B	T								x	
		Leopardus	tigrinus	x	x	x	B	T								x	
		Leopardus	wiedii	x	x	x	B	T								x	
		Panthera	onca	x	x	x	B	T								x	x
		Puma	concolor	x	x	x	B	T								x	
Perissodactyla	Tapiridae	Tapirus	terrestris	x	x	x	B	T	x		x			x			
Artiodactyla	Tayassuidae	Tayassu	pecari	x	x	x	D	T	x	x				x	x	x	
		Pecari	tajacu	x	x	x	D	T	x	x				x	x	x	
	Cervidae	Odocoileus	virginianus	x	x	x	B	T	x		x	x		x			
		Mazama	americana	x	x	x	B	T	x		x	x	x	x			
		Mazama	gouazoubira	x	x	x	B	T	x		x	x		x			
Rodentia	Sciuridae	Sciurus	aestuans	x	x	x	D	A	x	x				x			
		Sciurus	gilvigularis	x			D	A	x	x							
		Sciurus	igniventris	x			D	A	x	x							
		Sciurellus	pusillus	x	x	x	D	A						x			
	Muridae	Neacomys	dubostii		x	x	N	T	x	x				x	x		
		Neacomys	guianae	x	x	x	N	T	x	x				x	x		
		Neacomys	melanius	x	x	x	N	T	x	x					x		
		Neacomys	paracou	x	x	x	N	T	x	x				x	x		
		Nectomys	squamipes	x	x	x	N	T	x	x				x	x		
		Neusticomys	oyapocki			x	N	Aq							x		
		Neusticomys	venezuelae	x			N	Aq							x		

continued

Appendix 4.2. *continued*

Order	Family	Genus	Species	G	S	FG	Active Period	Habit	Diet										
									Fr	S	L	Po	E	Fu	Pu	In	V	Bl	Fish
		Oecomys	*auyantepui*	x		x	N	T	x	x						x			
		Oecomys	*bicolor*	x	x	x	N	T	x	x						x			
		Oecomys	*concolor*	x			N	T	x	x						x			
		Oecomys	*rex*	x		x	N	T	x	x						x			
		Oecomys	*roberti*	x			N	T								x			
		Oecomys	*rutilus*	x	x	x	N	T								x			
		Oecomys	*trinitatis*	x			N	T								x			
		Oligoryzomys	*fulvescens*	x	x	x	N	T	x	x						x			
		Oligoryzomys	*fulvescens*	x	x	x	N	T	x	x						x			
		Oryzomys	*macconnelli*	x	x	x	N	T	x	x						x			
		Oryzomys	*megacephalus*	x	x	x	N	T	x	x						x			
		Oryzomys	*yunganus*	x		x	N	T	x	x						x			
		Rhipidomys	*leucodactylis*	x		x	N	T											
		Rhipidomys	*macconnelli*	x			N	A											
		Rhipidomys	*nitela*	x		x	N	A											
		Sigmodon	*alstoni*	x	x	x	N		x	x					x				
		Zygodontomys	*brevicauda*	x	x	x	N	T	x	x						x			
	Erethizontidae	*Coendou*	*melanurus*	x	x	x	N	A											
		Coendou	*prehensilis*	x	x	x	N	A	x	x	x				x				
	Caviidae	*Cavia*	*aperea*	x	x		N	T		x	x								
	Hydrochaeridae	*Hydrochaeris*	*hydrochaeris*	x	x	x	N	T			x				x				
	Cuniculidae	*Cuniculus*	*paca*	x	x	x	N	T	x		x				x				
	Dasyproctidae	*Dasyprocta*	*fuliginosa*	x			D	T	x	x									
		Dasyprocta	*leporina*	x	x	x	D	T	x	x									
		Myoprocta	*acouchy*	x	x	x	D	T	x	x									
	Echimyidae	*Echimys*	*chrysurus*	x	x	x	N	A	x		x								
		Isothrix	*sinnamariensis*	x	x	x	N	A	x	x	x								
		Makalata	*didelphoides*	x	x	x	N	A	x	x	x			x		x			
		Mesomys	*hispidus*	x		x	N	A	x	x	x			x		x			
		Proechimys	*cayennensis*	x	x	x	N	T	x	x	x			x		x			
		Proechimys	*cuvieri*	x	x	x	N	T	x	x	x			x		x			
		Proechimys	*hoplomyoides*	x			N	T	x	x	x			x		x			
		Proechim[illegible]	*warreni*	x			N	T	x	x	x			x		x			

5 Folivorous Insects in the Rainforests of the Guianas

Yves Basset,[1] Neil D. Springate[2] and Elroy Charles[3]
[1]*Smithsonian Tropical Research Institute, Balboa, Ancon, Panama;* [2]*Department of Entomology, The Natural History Museum, London, UK;* [3]*Faculty of Agriculture/Forestry, University of Guyana, Turkeyen, Georgetown, Guyana*

Introduction

Unsurprisingly, insect–plant interactions in the neotropical region have been best studied where long-term taxonomic and ecological programmes exist at well-endowed research stations. In countries near the Guianas, this is evident in Costa Rica (e.g. Guanacaste Conservation Area: Janzen, 1983, 1988; La Selva: Marquis and Braker, 1993), Panama (e.g. Barro Colorado Island: Leigh, 1996), Puerto Rico (e.g. Luquillo Station: Reagan and Waide, 1996), Venezuela (Rancho Grande: Beebe and Crane, 1947; Huber, 1986), Trinidad (Simla Research Station, Arima Valley: Beebe, 1952) and Brazil (National Institute for Amazonian Research, INPA, at Manaus: Adis and Schubart, 1984; Adis, 1997). In contrast, long-term ecological programmes in French Guiana (e.g. Les Nouragues: Charles-Dominique, 1995; Poncy *et al.*, 1999), Suriname and Guyana (Mabura Hill: ter Steege *et al.*, 1996) rarely had the opportunity to generate substantial insect collections from rainforest habitats and concomitant studies on the ecology of insect herbivores. A notable exception may be the insect collections made at the Kartabo field station in Guyana which were, unfortunately, discontinued as early as 1924 (Beebe, 1925). As a consequence, the literature relating to insect–plant interactions in the rainforests of the Guianas is limited and, in addition, scattered in various sources dealing with insect, vertebrate or plant ecology and biology. Thus, the present compilation is unavoidably selective and represents only a starting point for more elaborate literature searches on specific insect–plant interactions.

Traditionally, insect–plant interactions are classified in the categories of either primary consumption ('herbivory') or mutualism, which largely overlap (e.g. Whitham *et al.*, 1991). Others have argued that moderate insect damage to plants can be beneficial in promoting growth and nutrient recycling (e.g. Owen, 1980). These views are rather phytocentric and may be of secondary interest to entomologists interested in elucidating patterns of host use by Guianan insects. The present review takes an entomocentric approach in order to stimulate deeper analyses of patterns of host use, as more and better data become available. The review concentrates on leaf and sap resources provided by vascular plants and used by folivorous insects (leaf-chewing and sap-sucking insects) in the rainforests of the Guianas (French Guiana, Suriname and Guyana). Other plant resources used by insects, such as epiphylls, flowers (see Chapter 6), seeds, extrafloral nectaries, fruit bodies, stems (including ants nesting in stems, ant-gardens and phytotelmata), wood and roots, are not treated here. Papers focusing mainly on insect taxonomy or crop pests are not accounted for, unless reporting general patterns of host use or host records.

A brief outline of entomological activities in the rainforests of the three countries is essential to appreciate the degree to which our present knowledge of these insect–plant interactions may be biased towards particular insect groups and/or plant resources. Information available on the main groups of sap-sucking and leaf-chewing insects is then detailed. Particular sections are dedicated to leaf damage (herbivory) and to leaf-cutting ants, given the substantial information available on these popular subjects. We conclude by discussing several contentious issues particularly worthy of further investigation in the Guianas.

Rainforest Entomology in the Guianas: a Brief Outline

French Guiana

Entomological investigations in the rainforests of French Guiana have often been associated with the activity of isolated amateurs and professionals based in metropolitan France (e.g. Balachowsky, 1970). Many specimens from Maroni River and similar localities are also common in many major collections, including the Smithsonian and British museums. These mostly came via French dealers such as Le Moult (1955).

Recently, the laboratories of the Institut Français de Recherche Scientifique pour le Développement en Coopération (ORSTOM) at Cayenne and of the Institut National de Recherche Agronomique (INRA) at Kourou involved several entomological projects. Of relevance to this review, recent studies concentrated on the taxonomy of grasshoppers (Orthoptera) and on seedling attack by leaf-chewing insects. The 'Canopy Raft', a platform made of inflated beams and netting which allows access to the canopy, had two scientific missions in French Guiana, in 1989 and 1996 (see Hallé and Blanc, 1990; Hallé, 1998). Thus, there is some information on insect–plant interactions in the canopy (e.g. Delvare and Aberlenc, 1990; Lechat *et al.*, 1990; Sterck *et al.*, 1992; Dejean *et al.*, 1998; Lowman *et al.*, 1998). In addition, the grasshopper fauna of the forest canopy in French Guiana is well-known (see review in Amédégnato, 1997). In contrast, canopy collections and data are virtually non-existent for Suriname and Guyana, with the exception of samples obtained by the National Zoological Collection (see below) with yellow pan traps set up 10–25 m high in the canopy of Akintosoela, 80 km SSE of Paramaribo, as well as samples obtained from felled trees in a central rainforest of Guyana (Basset *et al.*, 1999).

The Department of Entomology of ORSTOM at Cayenne has one collection of Cerambycidae, including about 8000 specimens and 1500 species. The Laboratoire d'Environment Hydreco, created in 1990 with private funding from Electricité de France (EDF), owns a collection of about 1000 insect species and nearly 6000 specimens identified. However, many of the specimens collected in French Guiana are deposited in a few private collections and at the Museum National d'Histoire Naturelle, Paris, which is to establish an annexe near ORSTOM-Herbarium at Cayenne (see reviews in Amédégnato, 1997, 2003). Tavakilian (1993) reviews the state of knowledge of entomological research in French Guiana.

Suriname

The first scientific study of insect–plant interactions in the Americas may well be that of Maria Sybilla Merian (1705, 1719; Valiant, 1992), who was a remarkable naturalist and made beautiful illustrations of several Surinamese insects. She spent most of her time in coastal plantations and many of her rearings and illustrations concern crop insects, particularly moths and butterflies (Geijskes, 1951). Unfortunately, her nomenclature pre-dated that of Linnaeus, so that host records are difficult to extract from her studies.

More recently, entomological research in Suriname has been dominated by the studies of the resident Dutch entomologist Dirk Cornelis Geijskes, who was active from 1940 to 1986 (for a review of entomological activities in Suriname, see Geijskes, 1951, 1957). Although Geijskes was mainly interested in the taxonomy of dragonflies (Odonata), he made some general insect collections in rainforests (e.g. Geijskes, 1968) and studied some aspects of the biology of the leaf-cutting ants. Nowadays, a programme in bee ecology exists at the University of Suriname and studies of agricultural insect pests have been and are still prominent in the country.

The country has two sizeable insect collections. The National Zoological Collection of Suriname at the University includes a strong invertebrate section with about 1400 insect species, 900 of which are identified, and close to 10,000 individuals. Bee specimens are particularly well-represented. The collection is linked to a database which includes nearly 500 records on insect–plant interactions (Hiwat, personal communication). An additional 40,000 insect specimens collected from Atkintosula, 80 km SSE of Paramaribo (De Dijn, 2003), as well as a rainforest on laterite, have been sorted to various taxonomic levels and will be incorporated gradually in the general collections (De Dijn, 2003). In addition, information sources for entomologists at the library of the University are very good. The second collection is at the Agricultural Field Station of the Ministry of Agriculture and Fisheries, and holds approximately 4000 insect specimens. A few entomological monographs have been published in the journal *Studies on the Fauna of Suriname and other Guyanas.*

Guyana

The British Guiana Tropical Research Station at Kartabo was founded by the New York Zoological Society and was operated by William Beebe from 1916 to 1924 (Beebe, 1925). Extensive insect collections were made in the rainforest there (e.g. Beebe and Fleming, 1945; Fisher, 1944; Fleming, 1945, 1949, 1950). Miller (1994) considers Kartabo as one of the nine places in the neotropical region that had been sampled for moths fairly intensively during most months over a period of years. In addition, there have been several expeditions by entomologists based in the UK (e.g. O.W. Richards, Oxford University expedition, 1929), in the USA (e.g. University of Michigan expedition; M. Collins, T.M. Forbes, P. Spangler, W. Steiner, F.X. Williams) or in Canada (H.S. Parish), so that the entomological fauna of Guyana is relatively well-known, although collections are scattered overseas. For example, a relatively comprehensive catalogue of butterflies was compiled for Guyana as early as 1940 (Hall, 1940).

Today, pest crops are the target of most entomological studies in Guyana and there is substantial information on leaf-cutting ants. Two sizeable insect collections exist in Guyana. The National Insect Collection is housed at the National Agricultural Research Institute, Mon Repos, East Coast Demerara. It includes 160 insect drawers, mostly including insect pests and associated enemies, but rainforest specimens are poorly represented (Munroe, 1993). The Center for Biodiversity at the University of Guyana houses a small collection of butterflies and about 21,000 specimens of rainforest insect herbivores collected at Mabura Hills, central Guyana. Entomological information relevant to Guyana can often be found in the journals *Timehri* and *Zoologica* (New York).

In short, the state of entomological knowledge in the Guianas cannot compare with that of other countries in Central and South America, particularly Costa Rica, Panama or Brazil. Overall, the insect fauna of Guyana is probably the best known of the three countries, but the best insect collections relevant to the Guianas may be found in Suriname, French Guiana, at the Museum d' Histoire Naturelle (Paris), the Natural History Museum (London), the

Smithsonian Museum (Washington), the American Museum of Natural History (New York), the Zoological Museum (Leiden), the Museo del Instituto de Zoologia Agricola (Maracay) or the Museo E. Goeldi (Belem). Some taxonomic information exists for certain conspicuous, traditionally well-collected, insect taxa such as butterflies, moths and longicorn beetles. However, the paucity of taxonomic as well as ecological information on the main groups of rainforest folivores, Cicadellidae, Fulgoroidea, Chrysomelidae and some Curculionidae, is striking. Arguably, these groups, together with wood-boring insects (mainly Cerambycidae, Curculionidae and Scolytinae), represent an appreciable part of insect–plant interactions and biodiversity in the rainforests of the Guianas.

Sap-sucking Insects

Thysanoptera and Heteroptera

Rainforest host-plant records in the literature for sap-sucking insects in the Guianas are probably very incomplete and all originate from Guyana (Table 5.1). Although sap-sucking insects (Thysanoptera and Hemiptera) are the main sap consumers in rainforests, leaf-cutting ants (see later) often ingest directly the sap from cut leaves (Littledyke and Cherrett, 1976) and studies by Cherrett (1980) showed that this represents an appreciable part of energy requirements of a colony. Information about Thysanoptera is limited to a general account originating from Suriname, without host-plant records (Priesner, 1923). In the rainforests near Kartabo, Beebe (1925) considered the Pentatomidae, Coreidae and Lygaeidae dominant among herbivorous heteropterans. In Suriname, van Doesburg (1966) treated the families Largidae and Pyrrhocoridae, detailed some host records and, in particular, indicated that some species of largids feed on *Inga* spp. (Mimosaceae).

Auchenorrhyncha and Stenorrhyncha

In Kartabo, the most abundant homopterans include Cicadellidae, Membracidae, Coccoidea and Cercopidae (Beebe, 1925), while Cicadellidae, Membracidae, Achilidae and Ciixidae are particularly abundant in the understorey near Mabura Hills (Charles, 1998; Basset and Charles, 2000). Limited information is available on the mealybugs collected near Kartabo, particularly the species feeding on ant plants (Morrison, 1922). Some species appear to be wide generalists (Table 5.1). Another source of information for the mealybugs of the Guianas is Williams and Willink (1992), although this targets mostly pests of various crops and economic plants (but see one record in Table 5.1).

Metcalf (1945) reported about 39 species of Fulgoroidea collected in the rainforests near Kartabo. Many species are also found elsewhere in Central and South America, but no host records are available. Similarly, Metcalf (1949) reported on 23 species of xylem-feeding Cicadellidae (Tettigellidae and Gyponidae, now in subfamily Cicadellinae) collected at Kartabo, without mention of host records. Haviland (1925) collected 75 species of Membracidae collected at Kartabo during a 5-month period. The salient features of this fauna include the wide geographical distribution of the species, preference of most species for clearings over deep shade forest, many species being attended by ants but, when unattended, being solitary as adults. Haviland provided only two host records (Table 5.1), but suggested that many species were monophagous, or at least restricted to a few species of plants. However, published host records of sap-sucking insects in Mabura Hill, Guyana, indicated a larger range of plants used, particularly for Cicadellidae, Membracidae and Cixiidae (Basset and Charles, 2000; Table 5.1). The studies of Haviland (1925) and Funkhouser (1942) stress that many Membracidae found in Guyana are widely distributed in South America.

Table 5.1. Host-plant records indicated in the literature for sap-sucking insects in the Guianas. All records originate from Guyana. See additional information in the text.

Insect species	Family	Host-plant(s)[a]	Source
Herpis vittata F.	Derbidae	A, B, C, D, E	Basset and Charles, 2000
Plectoderes collaris F.	Achilidae	A, B, C, D, E	Basset and Charles, 2000
Sevia bicarinata F.	Achilidae	A, B, C, D, E	Basset and Charles, 2000
Sevia consimile Fennah	Achilidae	A, B, C, D, E	Basset and Charles, 2000
Taosa muliebris Walker	Dictyopharidae	B, E	Basset and Charles, 2000
Toropa ferrifera (Walker)	Dictyopharidae	A, B, C, D, E	Basset and Charles, 2000
Toropa picta Walker	Dictyopharidae	B, E	Basset and Charles, 2000
Nogodina reticulata F.	Nogodinidae	A, B, C, D, E	Basset and Charles, 2000
Boethoos globosa Haviland	Membracidae	*Vismia ferruginea* Kunth (Clusiaceae)	Haviland, 1925
Darnis lateralis F.	Membracidae	A, B, C, D, E	Basset and Charles, 2000
Endoastus productus Osborn	Membracidae	*Tachigali paniculata* Aubl. (Caesalpiniaceae)	Osborn, 1921
Potnia gladiator Walker	Membracidae	A, C, E	Basset and Charles, 2000
Stegaspis fronditia L.	Membracidae	B, E	Basset and Charles, 2000
Tragopa guianae Haviland	Membracidae	*Vismia ferruginea* Kunth	Haviland, 1925
Acrocampsa pallipes F.	Cicadellidae	A, B, C, D, E	Basset and Charles, 2000
Amblyscarta invenusta Young	Cicadellidae	A, B, C, D, E	Basset and Charles, 2000
Baluba parallela Nielson	Cicadellidae	A, B, C, D, E	Basset and Charles, 2000
Barbatana extera Freytag	Cicadellidae	A, B, D	Basset and Charles, 2000
Cardioscarta quadrifasciata L.	Cicadellidae	A, B, C, E	Basset and Charles, 2000
Dasmeusa basseti Cavichioli and Chiamolera	Cicadellidae	A, B, C, D, E	Basset and Charles, 2000
Dasmeusa pauperata Young	Cicadellidae	A, B, C, D, E	Basset and Charles, 2000
Docalidia o'reilly Nielson	Cicadellidae	D, E	Basset and Charles, 2000
Gypona bulbosa DeLong and Freytag	Cicadellidae	A, B, C, D, E	Basset and Charles, 2000
Gypona flavolimbata Metcalf	Cicadellidae	A, C, D	Basset and Charles, 2000
Gypona flavolimbata Metcalf	Cicadellidae	A, C, E	Basset and Charles, 2000
Gypona funda DeLong	Cicadellidae	B, C	Basset and Charles, 2000
Gypona glauca F.	Cicadellidae	B, C, E	Basset and Charles, 2000
Gypona offa DeLong and Freytag	Cicadellidae	C	Basset and Charles, 2000
Joruma coccinea McAtee	Cicadellidae	A, B, C, E	Basset and Charles, 2000
Ladoffa aguilari Lozada	Cicadellidae	A, B, C, D, E	Basset and Charles, 2000
Ladoffa comitis Young	Cicadellidae	A, B, C, D, E	Basset and Charles, 2000
Ladoffa ignota Walker	Cicadellidae	A, B, C, D, E	Basset and Charles, 2000

continued

Table 5.1. *continued*

Insect species	Family	Host-plant(s)[a]	Source
Macugonalia moesta (F.)	Cicadellidae	A, B, C, D, E	Basset and Charles, 2000
Mattogrossus colonoides (Linnavuori)	Cicadellidae	A	Basset and Charles, 2000
Planocephalus flavicosta (Stal.)	Cicadellidae	A, E	Basset and Charles, 2000
Poeciloscarta cardinalis F.	Cicadellidae	A, B, C, D, E	Basset and Charles, 2000
Soosiulus fabricii Metcalf	Cicadellidae	A, B, C, D, E	Basset and Charles, 2000
Soosiulus interpolis Young	Cicadellidae	A, B, C, D, E	Basset and Charles, 2000
Xedreota tuberculata (Osborn)	Cicadellidae	A, B, C, D, E	Basset and Charles, 2000
Xestocephalus desertorum (Berg)	Cicadellidae	A, E	Basset and Charles, 2000
Cryptostigma quinquepori (Newstead)	Coccidae	*Cecropia*, *Ficus*, *Pithecellobium*, *Microlobium*	Newstead, 1917a; Morrison, 1922
Akermes secretus Morrison	Coccidae	*Inga*, *Haematoxylum*, *Triplaris*	Morrison, 1922
Ceroplastes cirripediformis Comstock	Coccidae	*Argyreia nervosa*, *Hura crepitans* (generalist)	Newstead, 1917b
Eucalymnatus decemplex (Newstead)	Coccidae	*Lecythis* sp. (Lecythidaceae)	Newstead, 1920
Saissetia hurae (Newstead)	Coccidae	*Hura crepitans* L. (Euphorbiaceae)	Newstead, 1917a
Lecanodiaspis ingae Howell and Kosztarab	Lecanodiaspididae	*Inga* spp. (Mimosaceae)	Howell and Kosztarab, 1972
Cataenococcus rotundus (Morrison)	Pseudococcidae	*Cecropia angulata* I.W. Bailey (Cecropiaceae)	Williams and Willink, 1992
Dysmicoccus probrevipes (Morrison)	Pseudococcidae	*Cordia*, *Tachigali*	Morrison, 1929
Farinococcus multispinosus Morrison	Pseudococcidae	*Triplaris*, *Cecropia*	Morrison, 1922
Pseudococcus bromeliae Bouché	Pseudococcidae	*Tachigali*, *Cecropia*, *Cordia*, *Ananas*	Wheeler, 1921; Morrison, 1922
Trionymus petiolicola (Morrison)	Pseudococcidae	*Tachigali paniculata* Aubl.	Morrison, 1922
Arocera equinoxia (Westwood)	Pentatomidae	B, C, D, E	Basset and Charles, 2000
Mormidea ypsilon (L.)	Pentatomidae	E	Basset and Charles, 2000

[a]A=*Chlorocardium rodiei* (Scomb.), B=*Mora gonggrijpii* (Kleinh.) Sandw., C=*Eperua rubiginosa* Miq., D=*Pentaclethra macroloba* (Willd.) O. Kuntze, E=*Catostemma fragrans* Benth.

Recently, the sap-sucking insects feeding on the seedlings of five common tree species have been studied near Mabura Hill, Guyana, on a relatively large scale: *Catostemma fragrans* Benth. (Bombacaceae), *Chlorocardium rodiei* (Scomb.) (Lauraceae), *Eperua rubiginosa* Miq., *Mora gonggrijpii* (Kleinh.) Sandw. (both Caesalpiniaceae) and *Pentaclethra macroloba* (Willd.) O. Kuntze (Mimosaceae) (Basset 1999, 2000; Basset and Charles, 2000). Monthly surveys of almost 10,000 seedlings were performed over a 2-year period in a forest plot of 1 km^2. Collections included over 24,000 specimens and 425 species. The most speciose families were Cicadellidae (including many Cicadellinae, Coelidiinae and Idiocerinae), Derbidae, Membracidae (particularly Smiliinae), Achilidae and Cixiidae. The most abundant families were Psyllidae, Cicadellidae, Cixiidae, Derbidae and Pseudococcidae. Plataspididae were the only conspicuous family of Heteroptera, being abundant but not speciose. A small fraction of the material was identified to species level (Table 5.1). Although feeding records are difficult to ascertain in most cases, the magnitude of collections allows inference of some degree of host specificity. Some species are almost certain to be wide generalists. For example, among Cicadellidae, a striking pattern was the high proportion of Cicadellinae in the collections, which are all xylem-feeders and often highly polyphagous (Basset, 1999, 2000; Basset and Charles, 2000; Basset *et al.*, 1999, 2001). Many such species were collected from the five hosts studied and observed feeding *in situ*. On *E. rubiginosa*, the dominant sap-sucking insect was an unidentified species of *Isogonoceraia* (Psyllidae) (Hollis, personal communication). Similarly, Gombauld (1996) observed that in French Guiana, among sap-sucking insects feeding on the seedlings of *Eperua grandiflora* (Aubl.), Psylloidea were the dominant group.

Herbivory

In the understorey, Newbery and de Foresta (1985) observed that the percentage of leaf area lost to herbivores was greater in the primary forest than in pioneer vegetation at La Piste de St Elie in French Guiana. In total, it averaged 5.5% of leaf area lost. Mature leaves of the shaded forest understorey were more heavily grazed than those on pioneer trees and those on small trees which grew in large, well-illuminated gaps. In the canopy, Sterck *et al.* (1992) measured herbivory among trees, lianas and epiphytes at the stations of Petit Saut and Les Nouragues in French Guiana. Overall damage levels of both canopies were just over 5%, individual samples ranging from 0.8% to 12.8% damage, without clear differences between life forms. Similarly, Lechat *et al.* (1990) measured leaf damage in a transect from the ground to the canopy at Petit Saut. Individual samples ranged from 0% to 20%, but 90% of the samples had below 10% damage and 60% below the 5% damage level. Leaf damage was not correlated with the height of the samples. On average, preliminary results of Lowman *et al.* (1998) indicated 4.7% damage in the canopy near Paracou, with trees supporting vines averaging over twice as much herbivory as trees devoid of vines.

Gombauld and Rankin de Merona (1998) measured leaf damage on *Eperua falcata* (Aubl.), *E. grandiflora, Dicorynia guianensis* Amshoff (Caesalpiniaceae), *Goupia glabra* Aubl. (Celestraceae) and *Qualea rosea* Aubl. (Vochysiaceae) by tagging leaves at Paracou (and see Gombauld, 1996). For *E. falcata*, in non-limiting light conditions (as in tree fall gaps), low levels of insect damage on leaves are correlated with increased height and stem diameter growth, whereas high levels of damage are correlated with a decrease in these parameters. *E. grandiflora*, which depends on cotyledon reserves for growth, is not influenced by damage by leaf-eating insects (Gombauld, 1996). Among the five species studied, *D. guianensis* is unique in maintaining leaf production throughout the dry season. In this case, the relative impact of herbivory is reduced during the period when groundwater deficits create conditions unfavourable for plant growth. *E. falcata*, *E. grandiflora* and *Q. rosea* experience high levels of herbivory during the dry season and display significantly lower mean height growth during the rainy season than during the dry season. In contrast, *D. guianensis* and *G. glabra* have similar height growth regardless of the season. Despite the rainy season being the more favourable period for the activity of leaf-eating insects (Gombauld, 1996), ratios of damage (leaf area eaten/total leaf area) do not differ significantly between the rainy and the dry season, with the exception of *E. grandiflora*. This pattern confirms that the proportion of leaf area produced and eaten varies similarly during the year (Gombauld and Rankin de Merona, 1998).

In Guyana, Isaacs *et al.* (1996) measured apparent leaf damage on *Dicymbe altsonii* Sandw. (Caesalpiniaceae), which forms monodominant stands on bleached sand soils (albic arenosols). Across four transects, damage ranged from 10.7% to 12.9% and leaf-cutting ants accounted for about half of the leaf area lost. Ter Steege (1990) mentioned that leaf damage was <10% for most seedlings of another Caesalpiniaceae, morabukea (*Mora gonggrijpii*) at Mabura Hill, Guyana, though the level of apparent damage, due mainly to leaf-scraping Chrysomelidae and leaf-chewing Tettigoniidae, increases considerably in large treefall gaps (Hammond, unpublished data), possibly due to an increase in young leaf availability. On Maracá Island, in Roraima, Brazil (bordering Guyana), Nascimento and Proctor (1994) measured herbivory on *Peltogyne gracilipes* Ducke (Caesalpiniaceae). *P. gracilipes* forms monodominant stands on this river island and related species of *Peltogyne* are common in the Guianas (Hammond *et al.*, 1996). Apparent leaf damage on *P. gracilipes* amounted to 11.4%, but in 1992 severe defoliation occurred. Nearly 60% of trees showed heavy and extreme damage (from 50% and greater of the crown defoliated). The insect responsible was *Eulepidotis phrygionia* Hampson, a generalist moth (Noctuidae), which is widespread in Brazil.

Coley and Aide (1991) reviewed herbivory in temperate and tropical forests and found that annual rates of herbivory in the latter amounted to 10.9%. Therefore, available data for the Guianas suggest that levels of herbivory there may not be extremely different from those elsewhere in the tropics, with perhaps a tendency to be lower. Both Newbery and de Foresta (1985) and Sterck *et al.* (1992) remarked that, being measured in the dry season, level of leaf damage may have been underestimated. Indeed, leaf damage in French Guiana is often higher in the late wet and early dry seasons when new leaves are produced than during the dry season (Gombauld, 1996; on Panama, see also Aide, 1988). This also correlates with the seasonality of large leaf-chewing insects, such as grasshoppers, whose densities are highest during that period of the year (Amédégnato, personal communication). A second complication is that all the values reported for the Guianas, with the exception of Gombauld (1996), concern apparent leaf damage (percentage of area lost), which typically does not account for leaves eaten entirely and therefore underestimates leaf damage (e.g. Lowman, 1984). This could be a serious bias when densities of large-bodied leaf-chewing insects, such as grasshoppers, are locally high. In these instances, whole herbaceous plants can be consumed within 2 h (Amédégnato, personal communication). Monitoring grazing rates by tagging leaves is an alternative to this problem (Lowman, 1984).

Leaf-chewing insects

Leaf-chewing insects include mostly grasshoppers, stick insects, beetles, moths, butterflies, sawflies and leaf-cutting ants (see below). Host records for these rainforest insects in the Guianas are summarized in Table 5.2.

Table 5.2. Host-plant records indicated in the literature for leaf-chewing insects in the Guianas. See additional information in the text.

Insect species	Family	Host-plant(s)	Country	Source
Colpolopha spp.	Romaleidae	Dicotyledons (generalist)	French Guiana	Descamps, 1978, 1979[a]
Maculiparia guyanensis Amedegnato	Romaleidae	Monocotyledons (specialist)	French Guiana	Descamps, 1978, 1979[a]
Phaeoparia lineaalba L.	Romaleidae	Monocotyledons (specialist)	French Guiana	Descamps, 1978, 1979[a]
Abacris flavalineata (De Geer)	Acrididae	*Eperua grandiflora* (Aubl.) (Caesalpiniaceae)	French Guiana	Gombauld, 1996
Clematodina sastrei Amedegnato & Descamps	Acrididae	Dicotyledons (generalist)	French Guiana	Descamps, 1978, 1979[a]
Copiocera spp.	Acrididae	*Euterpe oleracea* Mart. (Arecaceae)	French Guiana	Amédégnato, 1996
Cryptocloeus fuscipennis Bruner	Acrididae	*Loreya mespiloides* Miq. (Melastomataceae)	French Guiana	Descamps, 1978, 1979[a]
Ommatolampis perspiscillata Johanssen	Acrididae	Dicotyledons (generalist)	French Guiana	Descamps, 1978, 1979[a]
Prionolopha serrata (L.)	Acrididae	*Eperua falcata* (Aubl.), *E. grandiflora*	French Guiana	Gombauld, 1996
Schistocerca pallens (Thunberg)	Acrididae	*Eperua falcata*	French Guiana	Gombauld, 1996
Arawakella unca Rehn and Rehn	Eumastacidae	*Vismia* sp. (Clusiaceae)	French Guiana	Descamps, 1978, 1979[a]
Eneoptera guyanensis Chopard	Gryllidae	*Eperua grandiflora*	French Guiana	Gombauld, 1996
Lutosa brasiliensis (Brunner)	Tettigoniidae	*Eperua falcata* (Aubl.) (Caesalpiniaceae)	French Guiana	Gombauld, 1996
Huradiplosis surinamensis Nijveldt	Cecidomyiidae	*Hura crepitans* L. (Euphorbiaceae)	Suriname	Nijveldt, 1968
Asphaera abbreviate F.	Chrysomelidae	*Dicorynia guianensis*, *Eperua falcata*	French Guiana	Gombauld, 1996
Asphaera nobilitata F.	Chrysomelidae	*Eperua falcata*, *E. grandiflora*	French Guiana	Gombauld, 1996
Chalcophyma collaris Lef.	Chrysomelidae	*Eperua falcata*, *E. grandiflora*	French Guiana	Gombauld, 1996
Chalcophyma fulgida Lef.	Chrysomelidae	*Eperua falcata*	French Guiana	Gombauld, 1996
Chalcophyma tarsalis Baly	Chrysomelidae	*Eperua falcata*	French Guiana	Gombauld, 1996
Coelomera cajennensis F.	Chrysomelidae	*Cecropia peltata* L. (Cecropiaceae)	Guyana	Bodkin, 1919
Cryptocephalus esuriens Suffrian	Chrysomelidae	*Mora*, *Eperua*, *Pentaclethra*, *Catostemma*	Guyana	Basset and Charles, 2000
Naupactus albulus Boheman	Curculionidae	*Eperua falcata*	French Guiana	Gombauld, 1996
Naupactus bellus Hustache	Curculionidae	*Dicorynia*, *Eperua*, *Goupia*, *Qualea*	French Guiana	Gombauld, 1996
Naupactus ornatus Schönherr	Curculionidae	*Eperua falcata*	French Guiana	Gombauld, 1996
Plectrophoroides unicolor Chevrolat	Curculionidae	*Eperua falcata*	French Guiana	Gombauld, 1996
Eurytides ariarathes Esper	Papilionidae	*Rollinia* sp. (Annonaceae)	French Guiana	Brévignon, 1990
Eurytides protesilaus (L.)	Papilionidae	*Duroia eriopila L.* (Rubiaceae)	Suriname	Merian, 1705
Papilio torquatus Cramer	Papilionidae	*Moniera trifolia* L. (Rutaceae)	French Guiana	Brévignon, 1990
Phoebis sennae (L.)	Pieridae	*Inga ingoides* (L.C. Richard) Will. (Mimosaceae)	Suriname	Merian, 1705
Lycorea ilione Cramer	Nymphalidae	*Ficus* spp. (Moraceae)	Guyana	Ackery, 1988
Napeocles jucunda Hübner	Nymphalidae	*Ruellia inflata* Rich. (Acanthaceae)	French Guiana	Brévignon, 1990
Chalybs janais (Cramer)	Lycaenidae	*Pentaclethra macroloba* (Mimosaceae)	Guyana	Basset and Charles, 2000
Thestius phloeus (Cramer)	Lycaenidae	*Vouacapoua americana*, *Catostemma fragrans*	F. Guiana, Guyana	Joly, 1996; Basset and Charles, 2000

continued

Table 5.2. *continued*

Insect species	Family	Host-plant(s)	Country	Source
Cremna thasus Stoll	Riodinidae	*Tilliandsia bulbosa* Hook (Bromeliaceae)	French Guiana	Brévignon, 1992
Euselasia thusnelda Möschler	Riodinidae	*Caraipa* (Clusiaceae)	French Guiana	Brévignon, 1995, 1997
Euselasia arcana Brévignon	Riodinidae	*Clusia* (Clusiaceae)	French Guiana	Brévignon, 1995, 1997
Euselasia euryone (Hewitson)	Riodinidae	*Mahurea* (Clusiaceae)	French Guiana	Brévignon, 1995, 1997
Euselasia midas F.	Riodinidae	*Tovomita* (Clusiaceae)	French Guiana	Brévignon, 1995, 1997
Napaea beltiana Bates	Riodinidae	*Catasetum barbatum* Lindl. (Orchidaceae)	French Guiana	Brévignon, 1992
Arsenura armida (Cramer)	Saturniidae	*Erythrina fusca* Loureio (Fabaceae)	Suriname	Merian, 1705
Syssphinx molina Cramer	Saturniidae	*Inga* sp. (Mimosaceae)	Suriname	Robinson and Kitching, 1997
Aleuron carinata Walker	Sphingidae	*Ambelania tenuiflora* Muell. Arg. (Apocynaceae)	French Guiana	Robinson and Kitching, 1997
Oryba kadeni Schaufuss	Sphingidae	*Palicourea* sp. (Rubiaceae)	French Guiana	Robinson and Kitching, 1997
Thysania agrippina (Cramer)	Noctuidae	*Bursea simaruba* (L.) (Burseraceae)	Suriname	Merian, 1705
Atomacera pubicornis F.	Argidae	*Ipomoea* sp. (Convolvulaceae)	Guyana, Suriname	McCallan, 1953; Smith, 1992
Hemidianeura leucopoda Cameron	Argidae	*Inga* sp. (Mimosaceae)	Guianas	Smith, 1992
Ptilia brasiliensis Lepeletier	Argidae	*Lecythis* sp. (Lecythidaceae)	French Guiana	McCallan, 1953; Smith, 1992
Ptilia peletieri Gray	Argidae	*Cnestidium rufescens* Planch. (Connaraceae)	Guyana, Suriname	Benson, 1930; Smith, 1992
Themos surinamensis Klug	Argidae	*Ceiba pentandra* (L.) Gaertn. (Bombacaceae)	Guianas	Smith, 1992

[a]And C. Amedegnato, Paris, 1998, personal communication.

Phasmids and Orthoptera

Host records for rainforest Phasmida in the Guianas proved difficult to find. In French Guiana, the communities of arboricolous grasshoppers (Orthoptera: Acridoidea) do not appear to be very different from elsewhere in Amazonia (Amédégnato and Descamps, 1980). Typically, these communities include a high number of very closely related sympatric species which often live on the same tree (Amédégnato, 1997). Descamps (1978) recorded at least 44 genera and 57 species in the families Romaleidae, Acrididae, Eumastacidae and Proscopiidae (in order of decreasing importance) from these communities. Grasshopper density in tree crowns appears relatively low: on average, 22 individuals and 10 species per tree (i.e. about 0.2–0.5 individual per m^2 of leaf area) in the Amazon, reaching 16.5 individuals per tree in French Guiana in particular. Grasshopper species richness is lowest on nutrient-poor soils in French Guiana (Amédégnato, personal communication). Densities also appear to be higher in the upper canopy than in the mid canopy or understorey (Amédégnato, 1997, 2003). Most grasshoppers either feed on tree foliage, or bark and small epiphytes, or on larger epiphytes.

However, host-plant records are rare (Table 5.2). Many species are rather polyphagous, such as most Eumastacidae. Of particular interest is the specialization on palms of a number of species within the Copiocerinae (Acrididae). In the Guianas, this concerns about 10 species in the genera *Copiocerina*, *Copiocera* and *Eumecacris* (Amédégnato, 1996). Usually, Lauraceae, Combretaceae and Myristicaceae support a rich and abundant grasshopper fauna, in contrast with Leguminosae and Lecythidaceae (Amédégnato, 1997).

Coleoptera and Diptera

Beebe (1925) commented on the richness of the phytophagous beetle families Chrysomelidae and Curculionidae in Kartabo, stating that the variety of the former was ‘unbelievable’. Chrysomelidae were the most species-rich and abundant leaf-chewing taxa feeding on seedlings in central Guiana (Basset and Charles, 2000). Unfortunately, the present state of taxonomic knowledge of the rainforest material in the Guianas, particularly Chrysomelidae, precludes any useful analysis beyond Beebe’s enthusiastic statement (Jolivet, personal communication). However, one particular genus of Galerucinae has been relatively well studied. At least 35 species of *Coelomera* (Chrysomelidae, Galerucinae) are known to feed on *Cecropia* spp. and other Cecropiaceae, and are widely distributed in the neotropical region (Jolivet, 1987). Unfortunately, the confusing state of the taxonomy of both *Coelomera* and *Cecropia* does not allow investigation as to whether there is a one-to-one correspondence between the beetle and the plant species (Jolivet, 1987). Jolivet and Salinas (1993) described oviposition by *Coelomera cajennensis* F. inside the internodes of *Cecropia peltata* L. in Venezuela, in a fashion similar to that of the ants inhabiting the hollow twigs. The outcome of this behaviour appears to be an improved protection of the egg masses. The larvae are free-living and often considerably damage the leaves of their host plant. Adults avoid ants by reflex bleeding, or thanatosis, and their larvae by enteric or buccal discharge. In particular, the larvae of some species exhibit a peculiar form of defence, cycloalexy (Vasconcellos-Neto and Jolivet, 1994). They form a circle, head to head, and their supra-anal shields and enteric secretions at the periphery of the circle protect them against ant or bug attack. Depending on local conditions, ants and beetles may cohabit or exclude themselves on the foliage, but this situation is not well understood (Jolivet, 1987, 1989).

Records of rainforest Curculionidae and Diptera feeding on leaves in the Guianas were even more difficult to extract from the literature (Table 5.2). In particular, literature on insect galling and leaf-mining in the Guianas is very limited, other than Nijveldt’s (1968) description of one gall midge from Suriname (Table 5.2).

Lepidoptera

Although there is considerable taxonomic information about the moth fauna of the Guianas, the larval biology of most species is unknown. For example, in French Guiana, Haxaire and Rasplus (1986, 1987) list 50 species of Sphingidae and de Toulgoet (1987) lists 200 species of Arctiidae, but no host-plant information is provided. De Jong (1983) lists 426 reported species of Hesperiidae in Suriname and stresses the wide distribution of most species in South America. Similarly, Lindsey (1928), Beebe and Fleming (1945) and Fleming (1945, 1949, 1950) provide species lists for several moth and butterfly families at Kartabo. Beebe (1925) further reported that Noctuidae, Geometridae and Pyralidae collected at lights at Kartabo represent about 55% of the species richness in macrolepidopteran moths, so it is probable that these families also contribute in large part to the caterpillar fauna in rainforests, although care must be taken extrapolating from such data. Further, Beebe (1925) considers Ithomiinae and Heliconinae (both in Nymphalidae) to be typical butterfly taxa belonging to the rainforest habitat, and Pieridae and Hesperiidae to be also common butterfly families in the rainforests near Kartabo.

The few host records available in the butterfly catalogue for Guyana (Hall, 1940) include records from crops and orchard trees. The information more particularly relevant to rainforest butterflies often needs to be tracked in the databases and rearing reports of keen professional and amateur lepidopterists (Table 5.2). Of interest, the few host records available involving epiphytic plants mainly concern riodinid butterflies (e.g. Brévignon, 1992; Table 5.2).

Hymenoptera

Knowledge of the phytophagous sawflies (Hymenoptera, 'Symphyta') of the Guianas is extremely limited. At least 60 species have been recorded from the Guianas in the families Argidae, Pergidae and Tenthredinidae, although the exact number is likely to be many times greater (Benson, 1930; McCallan, 1953; Smith, 1988, 1990, 1992). The faunas are best known from Guyana, but very little is known about their ecology, particularly the host plants of the larvae (Table 5.2). Of the recorded species, the majority are found within the genera *Scobina, Hemidianeura, Manaos, Ptilia* (all Argidae), *Perreyiella, Decameria, Aulacomerus* (all Pergidae), *Stromboceros, Adiaclema* and *Waldheimia* (all Tenthredinidae).

It is probable that phytophagous species of the parasitoid superfamilies Cynipoidea and Chalcidoidea are more common in the Guianas than the data in the collections and the literature suggest. For example, the genus *Eschatocerus*, of the gall-forming Cynipidae (Cynipoidea), is rather diverse and, usually, host-specific on certain Leguminosae (Fergusson, personal communication). Leaf-cutting ants of the tribe Attini (Formicidae) are discussed later.

Studies of Particular Host Plants

Leaf-chewing insects feeding on the seedlings of several tree species have been relatively well studied in French Guiana. Mature leaves of *Eperua* spp. are not greatly damaged by insects, and herbivory does not appear to cause major seedling mortality in comparison with the effect of vertebrate herbivores. In contrast with young leaves of *E. falcata*, those of *Eperua grandiflora* are damaged frequently by invertebrate herbivores, and are often defoliated totally, leading to high seedling mortality (Forget, 1992). Gombauld (1996) studied in more detail the insect herbivores feeding on the seedlings of *E. falcata* and *E. grandiflora* at Paracou, French Guiana. Patterns of attack and consequences for seedling growth and survival were contrasted between different treatments and canopy openness. A total of 16 leaf-chewing species were collected from the families Tettigoniidae (genus *Lutosa*), Acridoidea (*Abacris, Prionolopha, Schistocerca, Colpolopha*), Gryllidae (*Eneoptera*), Chrysomelidae (Eumolpinae:

Chalcophyma; Alticinae: *Asphaera*), Curculionidae (Entiminae: *Naupactus*, *Plectrophoroides*) and Saturniidae. Few species appeared to be host specific and Gombauld noted that *Asphaera* and *Naupactus*, for example, were feeding on other tree species (*Goupia*, *Dicorynia*, *Qualea*). Leaf damage appeared occasional, without permanent setting on the seedlings. Chrysomelids were the dominant group of leaf-chewing insects on *Eperua*, but it was not known whether herbivores were restricted to the understorey or fed also on *Eperua* leaves in the canopy. Rates of attack on *Eperua* seedlings were correlated positively with leaflet production and sometimes increased at higher seedling densities. However, damage remained low in the primary forest, presumably since secondary hosts for eumolpine beetles were less common there. Gombauld also stressed that *Eperua* seedlings represent a low resource for herbivorous insects and that the energy needed for insect dispersal and feeding may be considerable. This pattern was similar for insects feeding on seedlings in Mabura Hill (Basset, 1999, 2000). The case of *E. grandiflora* is interesting; since its seeds are dispersed by various rodents (Forget, 1992), seedlings may sometimes establish at some distance from the parent trees. In this situation, Gombauld (1996) predicted that populations of specialist insects would become more fragmented (and less successful) with increasing distance between individuals.

Forget (1994, 1996) studied the seedling dynamics of *Vouacapoua americana* (Caesalpiniaceae) at Piste de St Elie, and assumed that a majority of seedlings died due to development of fungi on stems observed during the establishment phase. Complementary observations were made by Joly (1996) in Les Nouragues, where young red leaves of sprouting seedlings of *V. americana* were heavily attacked (up to 100%) by *Thestius pholeus* (Cramer) (Lycaenidae) in June–July, i.e. in the late wet season. This species appears to be rather host-specific and it did not consume either mature leaves or other leaves sampled at random in the understorey (Joly, 1996; but consider an observation made in Guyana below). In addition, the intensity of attack by the caterpillar was not evenly distributed, and only *V. americana* seedlings growing in areas with high density of conspecific trees were damaged (Joly, personal communication). It is possible that the occurrence of the caterpillar is related to the density of *V. americana* and/or overall tree diversity at the community level, both of which may vary widely between forest areas, thus having different impact between sites. Given that *V. americana* may form dense patches, as well as other similarly large-seeded species in the Guianas (e.g. *Chlorocardium rodiei*) that are subjected to insect herbivory, it would be interesting to quantify the impact of such herbivores on juvenile recruitment in light of tree patch density.

Although many species of *Cecropia* are inhabited by ants, this genus has been deemed ‘the most hospitable tree of the tropics’ (Skutch, 1945). Many animal species feed on its leaves, nectar and inflorescences, food bodies and on the ants themselves. The insect fauna feeding on *Cecropia* spp. is reasonably well-known (e.g. Fiebrig, 1909; Wheeler, 1942; Jolivet, 1987) and includes many species of Aleyrodidae, Chrysomelidae, Curculionidae and Nymphalidae, but few records originate from the Guianas.

The leaf-chewing insects feeding on the seedlings of various tree species were studied extensively at Mabura Hill, Guyana (Basset, 1999; Basset and Charles, 2000). Collections included over 3100 specimens, 179 species and 16 insect families. The most abundant and speciose families included Chrysomelidae (particularly Galerucinae, Eumolpinae, Alticinae and Cryptocephalinae) and Curculionidae (particularly Entiminae). The majority of the remainder included 13 families of Lepidoptera, with Gelechiidae dominating. Many of these species, particularly among Eumolpinae and Entiminae, were able to feed on the seedlings of the five hosts studied. Interestingly, the lycaenid *Thestius pholeus*, which severely damages the seedlings of *V. americana* in French Guiana

(Joly, 1996), also attacks the young foliage of seedlings of *C. fragrans* at Mabura Hill (but more commonly saplings of the same species), and that of an unknown vine (Basset, personal observation). With the exception of Orthoptera (the collections targeted diurnal insects), the insects feeding on *E. rubiginosa* appear similar at the higher taxa level than those feeding on other *Eperua* spp. in French Guiana (Gombauld, 1996), although species may be different. In particular, at least 16 species of chrysomelids (mostly Eumolpinae), nine species of Curculionidae (Entiminae), plus various species of moths (Geometridae and Tortricoidea, notably), and a Cecidomyiidae, which induces bud galls, feed on *Eperua* seedlings at Mabura Hill.

Leaf-cutting ants

All leaf-cutting ants are members of Attini of the subfamily Myrmicinae (Hymenoptera: Formicidae). The tribe includes 190 species, confined to the nearctic and neotropical regions (Cherrett *et al.*, 1988). Weber (1949) gives detailed accounts of the biology of most species of leaf-cutting ants in Guyana, where eight genera are recorded: *Cyphomyrmex* (three species), *Myocepurus* (one species), *Myrmicocrypta* (six species), *Apterostigma* (six species), *Sericomyrmex* (five species), *Trachymyrmex* (nine species), *Acromyrmex* (four species) and *Atta* (three to six species). However, the main species of leaf-cutting ants which harvest fresh vegetable substrates are species of *Atta* and *Acromyrmex* (Cherrett *et al.*, 1988). In the Guianas, three species of *Atta* are present (*A. cephalotes* L., *A. sexdens* L. and *A. laevigata* [Smith]) and four of *Acromyrmex* (*A. octospinosus* (Reich.), *A. landolti* (For.), *A. hystrix* (Latreille) *and A. coronatus* [F.]).

In the neotropical region, leaf-cutting ants have been considered to be 'dominant invertebrates' (Wheeler, 1907) and the most serious general insect pests of agriculture (Cherrett, 1968). They cut sections of leaves and, to a lesser extent, flowers and fruits, transport them to the nest chambers underground, where they excrete on them and inoculate them with a mutualistic fungus species, *Attamyces bromatificus* Kreisel (Basidiomycetes). This species of fungus has never been found outside the ant nests (Cherrett *et al.*, 1988). The ants feed on the fungus and discard all detritus into refuse dumps. The fungi require careful gardening to be retained in monocultures and garden temperatures are regulated. Large underground nests of *Atta* may be over 100 m^2 in surface area and include millions of individuals (Cherrett, 1982). Leaf-cutting ant colonies may be long-lived, often persisting for ten years or more (Cherrett, 1986). In Suriname, Stahel and Geijskes (1939, 1940, 1941) described the organization of the nest of *Atta cephalotes* and *A. sexdens*, while Geijskes (1953) observed nuptial flights of *Atta*.

Grazing damage due to leaf-cutting ants may be considerable and the ants attack a variety of crops (e.g. Buckley, 1982). In particular, citrus, cocoa, pastures and coffee suffer most and the nomadism of some Amerindian tribes is said to have been a response to upsurges in *Atta* populations after forest clearing (Cherrett, 1982). Average figures suggest that in tropical rainforests leaf-cutting ants may be harvesting 17% of total leaf production (Cherrett *et al.*, 1988). Cherrett (1972b) studied the substrate being carried into a nest of *A. cephalotes* in a forest near Bartica, Guyana. He estimated that during a 24 h period, approximately 700 g of fresh vegetable substrate was carried into the nest, about 60% of it as leaves, the rest being flowers. Ants carrying leaf fragments were significantly larger than ants carrying flower fragments. A general bibliography of leaf-cutting ants has been compiled by Cherrett and Cherrett (1989).

The biology of leaf-cutting ants has been best studied in Guyana by Cherrett and his co-workers (e.g. Cherrett, 1968, 1972a,b, 1980, 1982, 1983, 1986; Cherrett and Peregrine, 1976). Although leaf-cutting ants such as *A. cephalotes* defoliate a wide range of plant species, including some ant plants such as *Tachigali paniculata* (Wheeler, 1921), they forage selectively and the impact on particular plant species may be much greater than on the plant community as a whole (Buckley, 1982). For example, *A. cephalotes* tends to damage broad-leaved plants only, whereas *A. laevigata* cuts both grasses and broad-leaved plants (Cherrett, 1972a). Thorough studies of the foraging patterns of *A. cephalotes* by Cherrett, based on relatively long-term observations of a nest near Bartica in Guyana (e.g. Cherrett, 1968, 1972b; and see Cherrett *et al.*, 1988), showed that the ants are mostly nocturnal. *A. cephalotes* cut leaf sections from 36 out of 72 available plant species in the study area, concentrating on a few of these. Most foraging activity took place up to 30–45 m away from the nest and, in the canopy, above 12 m. The ants prefer young leaf material, flowers and buds, and, in particular, plant tissues both less tough and dense and those with high moisture. Plants are particularly at risk when flowering or flushing. The less dense the plant material, the more successful the small workers will be in cutting and carrying it back to the nest.

Several other studies led Cherrett and his co-workers (see references above) to conclude in substance that:

1. Whilst the interaction between leaf-cutting ants and the fungi that they cultivate is highly specialized, it is relatively unspecialized with regard to the use of plant species;
2. Only 5% of the energy requirements of an ant colony is provided by the fungus, the rest being supplied directly by plant sap; however, the fungus provides essential nutrients;
3. The outcome of the mutualistic relationship between the ants and the fungus is a most unusual degree of ecological dominance in diverse tropical vegetation, brought about by wide polyphagy; in short, both ant and fungus can utilize a far wider range of host plants than either could alone and, in doing so, they attain large population sizes (Cherrett *et al.*, 1988);
4. *A. cephalotes* has developed a conservative grazing system which prevents it from over-exploiting, and hence destroying, the vegetation in the area around the nest; and
5. The introduction of agriculture disrupts the pattern in (4) above, and may contribute to the pest status of this species.

Discussion

Although the overall body of published information on insect–plant interactions in the rainforests of the Guianas may appear considerable, it concerns mainly pollinators, leaf-cutting ants, ant gardens and other ant–plant interactions (Table 5.3). Most other interactions have been neglected, particularly those involving folivorous insects, including leaf-mining and gall-making species. Despite these caveats, it is of interest to discuss whether leaf and sap resources and insect–plant interactions in the Guianas may be different than at other, better studied locations in tropical rainforests. We concentrate on monodominance of tree species and host specificity of folivorous insects in order to extend the debate to the diversity of insect–plant interactions in the rainforests of the Guianas.

Monodominance of tree species and host specificity of folivorous insects

As emphasized elsewhere in this book, tree diversity in the rainforests of the Guianas depends mostly on local conditions such as soil type, drainage class and topographical features. Floristically-rich mixed forests tend to occur on well-drained soils, whereas on poorly drained soils dominance of one or a few species of Leguminosae is commonplace. Typically, these monodominant stands exhibit a much lower vegetational diversity than mixed forests (e.g. Davis and Richards, 1933, 1934; ter Steege *et al.*, 1996). Monodominant species of Leguminosae are found in *Eperua*, *Mora*, *Dicymbe*, *Peltogyne*, *Dicorynia*, *Dimorphandra* and *Vouacapoua*. In addition, there is often a trend for certain species in genera such as *Chlorocardium*, *Carapa* (Meliaceae), *Eschweilera* spp. and *Lecythis* spp. (Lecythidaceae), *Catostemma* spp.,

Table 5.3. Assessment of current scientific knowledge about particular insect–plant interactions in the Guianas, with an emphasis on rainforest insects.

Interaction	French Guiana	Suriname	Guyana
Pollination	XX	XX	X
Flower consumers	0	0	X
Seed predation	X	0	X
Seed dispersal	X	0	X
Sap-sucking insects	0	0	X
Herbivory	XX	0	X
Leaf-chewing insects	X	0	X
Leaf-cutting ants	X	X	XX
Extrafloral nectaries and food bodies	XX	0	X
Stem-boring insects	0	0	X
Ants nesting in stems	XX	0	XX
Ant gardens	XX	0	X
Phytotelmata	0	0	X
Wood-eating insects	XX	X	XX

0=nil, anecdotal, restricted or difficult to obtain.
X=some information available, sometimes unpublished.
XX=relatively extensive information.

Pentaclethra, *Triplaris* (Polygonaceae), *Hura* (Euphorbiaceae) and *Alexa* (Papilionaceae) to achieve co-dominance locally in the Guianas (Whitton, 1962; ter Steege *et al.*, 1993; Forget, 1994).

What are the likely results of these mono- or co-dominance patterns for insect herbivores, particularly for free-living folivorous insects? Price (1992) considered the influence of the resource base on the community structure of tropical insect herbivores to be paramount. A review of the information available from different tropical locations showed that this is credible (Basset, 1996). The following hypotheses can be put forward, following Price's (1992) rationale:

1. Reduced vegetational diversity in monodominant stands may reduce local insect diversity (the number of available niches is likely to decrease, particularly in the canopy; see Connell and Lowman, 1989), but particular insect taxa may be rather specialized and diverse in monodominant stands, since both the resource base and the predictability of resources provided by monodominant stands may be high.
2. For various reasons, monodominant tree species may be unpalatable to insect herbivores; the low resource base provided by such trees may then locally promote low insect species richness and low numbers of specialist species.

Although both hypotheses overall predict reduced insect diversity, they differ with regard to insect–host specificity. What is the evidence? Only circumstantial data exist in order to approach this problem. One example worth mentioning pertains to aphids, which are less diverse than psyllids in tropical habitats. Dixon *et al.* (1987) convincingly explained differences in host use for tropical psyllids and aphids. Both groups are rather host-specific, but aphids are much shorter-lived than psyllids and, consequently, have more difficulties to disperse efficiently on their hosts in a mosaic of diverse tropical vegetation in comparison with psyllids. Aphids are well-represented on legumes (e.g. van Emden, 1972) and should be also relatively well-represented in monodominant legume stands in the Guianas, provided that their resource base is large enough (i.e. in this case, that their hosts are palatable enough). However, Beebe (1925) commented on the scarcity of aphids in the rainforests near Kartabo. Further, seedlings of the legumes *Eperua* spp., *Mora* spp. and *Pentaclethra macroloba* rarely support aphid species, but the two former genera often support high loads of psyllids (Gombauld, 1996; Basset and Charles, 2000). This observation suggests that monodominant stands may provide a low resource base for aphids, which is in support of hypothesis 2.

Forget (1992; see also Richards, 1996) suggested that chemical and structural defences in *Eperua* spp. leaves, in addition to understorey tolerance at all stages of development, are likely to participate in promoting tree dominance of this species group in the forests of the Guianas, consistent with hypothesis 2. As emphasized previously, monodominant stands occur mostly on nutrient-poor or poorly drained locations (e.g. Beard, 1946; ter Steege *et al.*, 1993). The leaves of slow-growing monodominant trees may not only be well-protected chemically and physically, they may also be relatively nutrient-poor (e.g. total nitrogen in sapling leaves of mono- or co-dominant species at Mabura Hill do not appear to exceed 1.8% of dry weight: Raaimakers *et al.*, 1994, Table 2.1). It is well known that herbivores of nutrient-impoverished plants tend to be more polyphagous than those feeding on nutrient-rich plants (Mattson and Scriber, 1987). An analogous situation to that in the Guianas could well be the dominant forests of dipterocarps in South-east Asia, which are low in nutrients and support very few lepidopteran defoliators, most of them being polyphagous (Holloway, 1989). Data from Barro Colorado Island, Panama, also suggest that Membracidae feeding on slow-growing, shade-tolerant hosts are less diverse than those feeding on fast-growing hosts, but, consistent with hypothesis 1, they appear to be rather specialized (Loye, 1992).

Further, many species in several insect groups (e.g. Acridoidea, Fulgoroidea, Membracidae, Cerambycidae, Scolytidae, Platypodidae, Hesperiidae, etc.) appear to exhibit a relatively wide geographical distribution, unconfined to the Guiana Shield. Generalist insect herbivores tend to be larger than specialists, and larger herbivores tend to have a wider geographical range (e.g. Gaston and Lawton, 1988). Thus, a wide geographical range of many species of insect herbivore would tend to support hypothesis 2. An alternative explanation is that insects are relatively specialized (monophagous or oligophagous), but their hosts are widely distributed. De Granville (1988) considered that about 35% of plant species (within certain plant taxa) are endemic to the Guianas. Further, many mono- or co-dominant species present in the Guianas are near-endemic and do not extend much beyond that region (e.g. ter Steege, 1990). For example, Nascimento and Proctor (1994) commented that monodominant stands are rare in Brazil (for example, *Peltogyne gracilipes* on Maracá Island). Thus, the relatively high level of endemism of several tree species in the Guianas would not suggest a pattern of high level of insect specialization on widely distributed host trees in Amazonia.

Overall, there appears to be more circumstantial evidence to accept hypothesis 2, although hypothesis 1 may be more appropriate for certain insect taxa. However, at present, we are reluctant to accept hypothesis 2 before two further points can be clarified. First, tropical insects which have been formally identified and whose geographical range can be inferred are more likely to be better known generalists, these often being insect pests. Although there has been speculation over the tremendous local endemism in Amazonian canopy insects (e.g. Erwin, 1983), the impression of high insect endemism in the lowland tropics may be merely a consequence of poor sampling (e.g. Gaston *et al.*, 1996). For example, before exchanging information with his colleagues in French Guiana, the present first author thought that the lycaenid *Thestius pholeus* was relatively specialized, feeding on *Catostemma* at Mabura Hill, Guyana. In fact, this species also feeds on *Vouacapoua* in French Guiana, with apparent host-specificity (Joly, 1996). This case of apparent local specialization, but with rather polyphagous habits over the entire geographical range of the species, has been documented in several cases (e.g. Fox and Morrow, 1981) and may indeed be relatively common in tropical insect herbivores (e.g. Janzen, 1981). Further, poor sampling increases the difficulty of recognizing morphological variation associated with many widespread species (Gaston *et al.*, 1996).

Secondly, most insect collecting in the Guianas has been from the understorey. The abundance and the diversity of insect herbivores are typically higher in the canopy of rainforests than in their understorey, because, as can be again argued, the resource base is higher in the former (e.g. Basset *et al.*, 1992, 1999, 2003). Similarly, understorey insects may be less specialized (and have a wider geographical range) than in the canopy. We have limited evidence that some free-living insects feeding on seedlings in the understorey appear to be rather generalist (Gombauld, 1996; Basset, 1999).

In summary, we believe that future studies upon the host-specificity of insect herbivores foraging in the canopy will help greatly to evaluate whether hypothesis 2 is generally correct and not an artefact of the limited information presently available. These studies should be supported by adequate sampling effort (sufficient numbers of insects collected from sufficient numbers of host plants) and adequate taxonomic effort.

Diversity of insect–plant interactions in the rainforests of the Guianas

It is well known that mammalian diversity in Amazonia is least in the Guiana subregion (Voss and Emmons, 1996). What of insect–plant interactions, and particularly insect herbivores, in this regard? First, entomological knowledge in arguably the best studied of the three Guianas, Guyana, is notoriously low (e.g. Munroe, 1993; Funk, 1997) and this greatly impedes analyses of host specificity and endemism. Second, generalizations are difficult since, for example, patterns of mammalian diversity do not follow those of butterflies worldwide (Robbins and Opler, 1997). Third, patterns may be dissimilar for different insect taxa belonging to different feeding guilds. For example, whereas folivorous insects could be less diverse in monodominant stands (see discussion above), bee diversity appears to be poorer in diverse forests (Roubik, 1990). Flowers represent a much less protected resource, both physically and chemically, than leaves and, as such, patterns of host use may well be very different between respective insect feeding guilds using these different resources.

There are no suitable data (i.e. sample size large enough and comparable in different locations) to compare the fauna of insect herbivores in the Guianas with those elsewhere, particularly in Amazonia. However, it may be argued that the interactions in ant gardens are clearly less diverse in the Guianas than in Peru (Davidson, 1988). Further, circumstantial evidence suggests that the diversity of insects, particularly of herbivores, may indeed be relatively low in the Guianas. For example, Tavakilian (1993) concluded, from extensive observation, that light trapping of Coleoptera in French Guiana yields poorer results than in Africa or Asia. Incidentally, the present first author was also surprised at the low occurrence, all year long, of insects attracted to light at Mabura Hill, Guyana, in comparison with other locations in the tropics.

Further, leaf-cutting ants are abundant in the neotropical region and common in the Guianas. Yet, despite this, leaf damage in the rainforests of the Guianas appears relatively low in comparison with elsewhere in the tropics (see Coley and Aide, 1991). This might be a consequence of low grazing rates on leaf resources low in nutrients, or an indication of the low abundance and diversity of insect herbivores other than leaf-cutting ants. It is well known that forests growing on nutrient-poor white sands, for example, are less prone to attack by free-living insect herbivores (e.g. Janzen, 1974). The data of Isaacs *et al.* (1996) in Guyana indeed suggest that apparent leaf damage unrelated to leaf-cutting ants amounted to about 5%, a rather low value.

However, care must be exercised in this type of extrapolation, since it is notoriously difficult to relate insect abundance and diversity to apparent leaf damage (e.g. Marquis, 1991; Basset and Höft, 1994).

These patterns of relatively low diversity, if actually correct, may be partly related to monodominance of certain host plants, as discussed previously for folivorous insects. Since monodominance patterns may increase with respect to soil types from the series French Guiana–Suriname–Guyana, to culminate in the latter country, it would be interesting to test whether the diversity of particular insect taxa also decreases along this series, as more and better data become available. One related question would be to assess local differences in insect species richness between well-drained and poorly drained forests. The extent of structural and vegetational differences in these two habitats is such (e.g. ter Steege *et al.*, 1993) that this is likely to affect many taxa of insect herbivores. For example, De Dijn (2004) reported such differences for bee assemblages in Suriname.

To conclude, it is clear that there are severe gaps in our knowledge of insect–plant interactions in the Guianas. We believe that it may be difficult, if not futile, to discuss possible patterns and speculate further beyond that which we have suggested above without additional data. We hope that this review will stimulate entomologists and ecologists to undertake much needed studies of insect–plant interactions in the rainforests of the Guianas. The Guianan Shield, with its unique large and often undisturbed forest formations, including some peculiar monodominant stands, represents a potential wealth of information for scientists, which may challenge some commonly held views in tropical ecology. Our immediate responsibilities are to study thoroughly these habitats before they disappear, and to conserve most of them for future generations.

Acknowledgements

Many colleagues helped collating the information presented in this review, provided us with unpublished information, often with great rapidity, or discussed various sections of the text: H.-P. Aberlenc, J. Adis, C. Amédégnato, H. Barrios, G. Beccaloni, V. Becker, C. Brévignon, G. Delvare, P.-M. Forget, D.S. Hammond, H. Hiwat, P. Jolivet, I. Kitching, R.L. Kitching, D. Matile, J. Miller, S.E. Miller, M.P.A. Quik-Stregels, G.S. Robinson, L.S. Springate, H. ter Steege. In particular, J. Adis, C. Amédégnato, G. Delvare, B. De Dijn, G.W. Fernandes, P. Gombauld, S.E. Miller and S.P. Ribeiro commented on the full version of earlier drafts. Studies on insect herbivores in Guyana by the first and third authors were funded by the Darwin Initiative for the Protection of the Species (UK) and the DfID (UK), respectively, and were logistically supported by the Tropenbos-Guyana Programme and Demerara Timbers Ltd.

References

Ackery, P.R. (1988) Hostplants and classification: a review of nymphalid butterflies. *Biological Journal of the Linnean Society* 33, 95–203.

Adis, J. (1997) Terrestrial invertebrates: survival strategies, group spectrum, dominance and activity patterns. In: Junk, W.J. (ed.) *The Central Amazon Floodplain. Ecology of a Pulsing System.* Springer, Berlin, pp. 318–330.

Adis, J. and Schubart, H.O.R. (1984) Ecological research on arthropods in Central Amazonian forest ecosystems with recommendation for study procedures. In: Cooley, J. and Golley, F.B. (eds) *Trends in Ecological Research for the 1980s.* Plenum Press, New York, pp. 111–144.

Aide, T.M. (1988) Herbivory as a selective agent on the timing of leaf production in a tropical understory community. *Nature* 336, 574–575.

Amédégnato, C. (1996) L'acridofaune palmicole neotropicale: diversité et origine. In: Couturier, G. (ed.) *La faune entomologique des palmiers dans les écosystèmes forestiers de l'Amazonie péruvienne.* Programme SOFT, Convention ORSTOM 224. Rapport Ministère de l'Environnement, Paris, pp. 1–28.

Amédégnato, C. (1997) Diversity in an Amazonian canopy grasshoppers community in relation to resource partition and phylogeny. In: Stork, N.E., Adis, J. and Didham, R.K. (eds) *Canopy Arthropods.* Chapman and Hall, London, pp. 281–319.

Amédégnato, C. (2003) Microhabitat distribution of forest grasshoppers in the Amazon. In: Basset, Y., Novotny, V., Miller, S.E. and Kitching, R.L. (eds) *Arthropods of Tropical Forests. Spatio-temporal Dynamics and Resource Use in the Canopy.* Cambridge University Press, Cambridge, UK, pp. 237–255.

Amédégnato, C. and Descamps, M. (1980) Etude comparative de quelques peuplements acridiens de la forêt neotropicale. *Acrida* 9, 171–216.

Balachowsky, A.S. (1970) La mission d'exploration entomologique du Muséum National d'Histoire Naturelle en Guyane française (octobre–décembre 1969). Introduction. *Annales de la Société Entomologique de France (N.S.)* 6, 563–570.

Basset, Y. (1996) Local communities of arboreal herbivores in Papua New Guinea: predictors of insect variables. *Ecology* 77, 1906–1919.

Basset, Y. (1999) Diversity and abundance of insect herbivores foraging on seedlings in a rain forest in Guyana. *Ecological Entomology* 24, 245–259.

Basset, Y. (2000) Insect herbivores foraging on seedlings in an unlogged rain forest in Guyana: spatial and temporal considerations. *Studies on Neotropical Fauna and Environment* 35, 115–129.

Basset, Y. and Charles, E. (2000) An annotated list of insect herbivores foraging on the seedlings of five forest trees in Guyana. *Anais da Sociedade Entomológica do Brasil* 29, 433–452.

Basset, Y. and Höft, R. (1994) Can apparent leaf damage in tropical trees be predicted by herbivore load or host-related variables? A case study in Papua New Guinea. *Selbyana* 15, 3–13.

Basset, Y., Aberlenc, H.-P. and Delvare, G. (1992) Abundance and stratification of foliage arthropods in a lowland rain forest of Cameroon. *Ecological Entomology* 17, 310–318.

Basset, Y., Charles, E.L. and Novotny, V. (1999) Insect herbivores on parent trees and conspecific seedlings in a Guyana rain forest. *Selbyana* 20, 146–158.

Basset, Y., Charles, E., Hammond, D.S. and Brown, V.K. (2001) Short-term effects of canopy openness on insect herbivores in a rain forest in Guyana. *Journal of Applied Ecology* 38, 1045–1058.

Basset, Y., Novotny, V., Miller, S.E. and Kitching, R.L. (2003) Conclusion: arthropods, forest types and interpretable patterns. In: Basset, Y., Novotny, V., Miller, S.E. and Kitching, R.L. (eds) *Arthropods of Tropical Forests. Spatio-temporal Dynamics and Resource Use in the Canopy.* Cambridge University Press, Cambridge, UK, pp. 394–405.

Beard, J.S. (1946) The Mora forests of Trinidad, British West Indies. *Journal of Ecology* 33, 173–192.

Beebe, W. (1925) Studies of a tropical jungle; one quarter of a square mile of jungle at Kartabo, British Guiana. *Zoologica* 6, 4–193.

Beebe, W. (1952) Introduction to the ecology of the Arima Valley, Trinidad, B.W.I. *Zoologica* 37, 157–183.

Beebe, W. and Crane, J. (1947) Ecology of Rancho Grande, a subtropical cloud forest in Northern Venezuela. *Zoologica* 32, 43–61.

Beebe, W. and Fleming, H. (1945) The Sphingidae (moths) of Kartabo, British Guiana, and Caripito, Venezuela. *Zoologica* 30, 1–6.

Benson, R.B. (1930) Sawflies collected by the Oxford University Expedition to British Guiana, 1929. *Annals and Magazine of Natural History* 6, 620–621.

Bodkin, G.E. (1919) Notes on the Coleoptera of British Guiana. *The Entomologist's Monthly Magazine* 55, 264–272.

Brévignon, C. (1990) Quelques élevages guyanais: *Napeocles jucunda* Hübner (Lep. Nymphalidae), *Papilio torquatus* Cramer, *Papilio anchisiades* Esper, *Eurytides ariarathes* Esper (Lep. Papilionidae). *Sciences Nat* 68, 19–21.

Brévignon, C. (1992) Elevage de deux Riodininae guyanais, *Napaea beltiana* Bates et *Cremna thasus* Stoll. I. A propos de la myrmécophilie des chenilles. *Alexanor* 17, 403–413.

Brévignon, C. (1995) Description de nouveaux Riodinidae de Guyane française (Lepidoptera). *Lambillionea* 95, 553–560.

Brévignon, C. (1997) Notes sur les Nemeobiinae de Guyane française. II – Le groupe de *Euselasia euryone* (Hewitson, 1856) (Lepidoptera Riodinidae). *Lambillionea* 97, 116–120.

Buckley, R.C. (1982) Ant–plant interactions: a world review. In: Buckley, R.C. (ed.) *Ant–Plant Interactions in Australia.* W. Junk, The Hague, The Netherlands, pp. 111–162.

Charles, E.L. (1998) The impact of natural gap size on the communities of insect herbivores within a rain for-

est of Guyana. MSc thesis, The University of Guyana, Georgetown.
Charles-Dominique, P. (1995) Les Nouragues, une station de recherche pour l'étude de la forêt tropicale. In: Legay, J.M. and Barbauld, R. (eds) *La Révolution Technologique en Ecologie.* Masson, Paris, pp. 183–199.
Cherrett, J.M. (1968) The foraging behavior of *Atta cephalotes* (Hymenoptera Formicidae). I. Foraging pattern and plant species attacked in tropical rain forest. *Journal of Animal Ecology* 37, 387–403.
Cherrett, J.M. (1972a) Chemical aspects of plant attack by leaf-cutting ants. In: Harborne, J.B. (ed.) *Phytochemical Ecology.* Academic Press, London, pp. 13–24.
Cherrett, J.M. (1972b) Some factors involved in the selection of vegetable substracts by *Atta cephalotes* (L.) (Hymenoptera, Formicidae) in tropical rain forest. *Journal of Animal Ecology* 41, 647–660.
Cherrett, J.M. (1980) Possible reasons for the mutualism between leaf-cutting ants. *Biologie et Ecologie mediterranéenne* 7, 113–122.
Cherrett, J.M. (1982) The economic importance of leaf-cutting ants. In: Breed, M.D., Michener, C.D. and Evans, H.E. (eds) *The Biology of Social Insects.* Westview Press, Boulder, Colorado, pp. 114–118.
Cherrett, J.M. (1983) Resource conservation by the leaf-cutting ants and *Atta cephalotes* in tropical rain forest. In: Sutton, S.L., Whitmore, T.C. and Chadwick, A.C. (eds) *Tropical Rain Forest Ecology and Management.* Blackwell, Oxford, UK, pp. 253–265.
Cherrett, J.M. (1986) The biology, pest status and control of leaf-cutting ants. *Agricultural Zoology Reviews* 1, 1–37.
Cherrett, J.M. and Cherrett, F.J. (1989) A bibliography of the leaf-cutting ants, *Atta* and *Acromyrmex* ssp. up to 1975. *ODNRI Bulletin* 14, 1–58.
Cherrett, J.M. and Peregrine, D.J. (1976) A review of the status of leaf-cutting ants and their control. *Annals of Applied Biology* 84, 124–128.
Cherrett, J.M., Powell, R.J. and Stradling, D.J. (1988) The mutualism between leaf-cutting ants and their fungus. In: Wilding, N., Collins, N.M., Hammond, P.M. and Webber, J.F. (eds) *Insect Fungus Interactions.* Academic Press, New York, pp. 93–120.
Coley, P.D. and Aide, T.M. (1991) Comparison of herbivory and plant defenses in temperate and tropical broad-leaved forests. In: Price, P.W., Levinsohn, T.M., Fernandes, G.W. and Benson, W.W. (eds) *Plant–Animal Interactions: Evolutionary Ecology in Tropical and Temperate Regions.* John Wiley & Sons, New York, pp. 25–49.
Connell, J.H. and Lowman, M.D. (1989) Low-diversity tropical rain forests: some possible mechanisms for their existence. *The American Naturalist* 134, 88–119.
Davidson, D.W. (1988) Ecological studies of neotropical ant gardens. *Ecology* 69, 1138–1152.
Davis, T.A.W. and Richards, P.W. (1933) The vegetation of Moraballi Creek, British Guiana. An ecological study of a limited area of tropical rain forest I. *Journal of Ecology* 21, 350–384.
Davis, T.A.W. and Richards, P.W. (1934) The vegetation of Moraballi Creek, British Guiana. An ecological study of a limited area of tropical rain forest II. *Journal of Ecology* 22, 106–155.
De Dijn, B. (2004) Stingless bee communities: are they poorer in more constrained ecosystems? IBRA-congress, Costa Rica, 1996.
De Dijn, B.P.E. (2003) Vertical stratification of flying insects in a Surinam lowland rainforest. In: Basset, Y., Novotny, V., Miller, S.E. and Kitching, R.L. (eds) *Spatio-temporal Dynamics and Resource Use in the Canopy.* Cambridge University Press, Cambridge, UK, pp. 110–122.
De Granville, J.-J. (1988) Phytogeographical characteristics of the Guianan forests. *Taxon* 37, 578–594.
de Jong, R. (1983) Annotated list of the Hesperiidae (Lepidoptera) of Surinam, with description of new taxa. *Tijdschrift voor Entomologie* 126, 233–268.
de Toulgoet, H. (1987) Description de nouvelles arctiides d'Amérique latine, suivie de la liste des espèces les plus marquantes recoltées à la Guyane française en janvier–février 1986 (23e note) (Lepidoptera, Arctiidae). *Nouvelle Revue d'Entomologie* 4, 233–245.
Dejean, A., Orivel, J., Corbara, B., Delabie, J. and Teillier, L. (1998) La mosaïque des fourmis arboricoles. In: Hallé, F. (ed.) *Biologie d'une canopée de forêt équatoriale – III. Rapport de la mission d'exploration scientifique de la canopée de Guyane,* octobre–décembre 1996. Pro-Natura International and Opération Canopée, Paris/Lyon, pp. 140–153.
Delvare, G. and Aberlenc, H.-P. (1990) Des entomologistes sur la canopée. In: Hallé, F. and Blanc, P. (eds) *Biologie d'une canopée des forêt équatoriale.* Rapport de Mission. Radeau des Cimes Octobre–Novembre 1989, Guyane Française. OPRDC, Paris, pp. 211–221.
Descamps, M. (1978) Etude des écosystèmes guyanais, III. Acridomorpha dendrophiles (Orthoptera Caelifera). *Annales de la Société Entomologique de France (N.S.)* 14, 301–349.

Descamps, M. (1979) Eumastacoidea néotropicaux. Diagnoses, signalisations, notes biologiques. *Annales de la Société Entomologique de France (N.S.)* 15, 117–155.

Dixon, A.F.G., Kindlmann, P., Leps, J. and Holman, J. (1987) Why there are so few species of aphids, especially in the tropics. *The American Naturalist* 129, 580–592.

Erwin, T.L. (1983) Beetles and other insects of tropical forest canopies at Manaus, Brazil, sampled by insectidal fogging. In: Sutton, S.L., Whitmore, T.C. and Chadwick, A.C. (eds) *Tropical Rain Forest: Ecology and Management.* Blackwell, Oxford, UK, pp. 59–75.

Fiebrig, K. (1909) *Cecropia peltata* und ihr Verhältnis zu *Azteca alfari*, zu *Atta sexdens* und anderen Insekten, mit einer Notiz über Ameisen-Dornen bei *Acacia cavenia. Biologische Centralblatt* 29, 1–16, 33–55, 65–77.

Fisher, W.S. (1944) New American Cerambycidae (Coleoptera) from British Guiana and Costa Rica. *Zoologica* 29, 1–2.

Fleming, H. (1945) The Saturnioidea (moths) of Kartabo, British Guiana, and Caripito, Venezuela. *Zoologica* 30, 73–80.

Fleming, H. (1949) The Pericopidae (moths) of Kartabo, British Guiana, and Caripito, Venezuela. *Zoologica* 34, 19–20.

Fleming, H. (1950) The Eurochromiidae (moths) of Kartabo, British Guiana, and Caripito, Venezuela. *Zoologica* 35, 209–216.

Forget, P.-M. (1992) Regeneration ecology of *Eperua grandiflora* (Caesalpiniaceae), a large-seeded tree in French Guiana. *Biotropica* 24, 146–156.

Forget, P.-M. (1994) Regeneration pattern of *Vouacapoua americana* (Caesalpiniaceae), a rodent-dispersed tree species in French Guiana. *Biotropica* 26, 420–426.

Forget, P.-M. (1996) Removal of seeds of *Carapa procera* (Meliaceae) by rodents and their fate in rainforest in French Guiana. *Journal of Tropical Ecology* 12, 751–761.

Fox, L.R. and Morrow, P.A. (1981) Specialization: species property or local phenomenon? *Science* 211, 887–893.

Funk, V.A. (1997) Using collection data and GIS to examine biodiversity information levels in Guyana. In: Hoagland, K.E. and Rossman, A.Y. (eds) *Global Genetics Resources: Access, Ownership, and Intellectual Property Rights.* Association of Systematics Collections, Washington, DC, pp. 117–128.

Funkhouser, W.D. (1942) Membracidae (Homoptera) from British Guiana. *Zoologica* 27, 125–129.

Gaston, K.J. and Lawton, J.H. (1988) Patterns in body size, population dynamics and regional distribution of bracken herbivores. *The American Naturalist* 132, 662–680.

Gaston, K.J., Gauld, I.D. and Hanson, P. (1996) The size and composition of the hymenopteran fauna of Costa Rica. *Journal of Biogeography* 23, 105–113.

Geijskes, D.C. (1951) General entomological research in Surinam up to 1950. In: Nederlandse Entomologische Vereniging (ed.) *Entomology in the Netherlands and their Overseas Territories.* The Netherlands Entomological Society, Amsterdam, pp. 46–54.

Geijskes, D.C. (1953) Nuptial flighttime of Atta-ants in Suriname. *Tijdschrift over Plantenziekten* 59, 181–184.

Geijskes, D.C. (1957) The zoological exploration of Suriname. *Studies on the Fauna of Suriname and other Guyanas* 1, 1–12.

Geijskes, D.C. (1968) Insect collecting in Suriname with the help of 'Malaise' traps. *Studies on the Fauna of Suriname and other Guyanas* 10, 101–109.

Gombauld, P. (1996) Variabilité de la phyllophagie par les insectes chez deux arbres de la forêt guyanaise, *Eperua falcata* et *E. grandiflora* (Caesalpiniaceae): impact des diminutions de surface foliaire et du microclimat sur la croissance et la survie des plantules. PhD thesis, Université Paris 6, Paris, France.

Gombauld, P. and Rankin de Merona, J.M. (1998) Influence de la saison sur la phénologie et la phyllophagie par les insectes de jeunes individus de 5 espèces d'arbre de forêt tropicale en Guyane (Fr.). *Annales des Sciences Forestières* 55, 715–725.

Hall, A. (1940) Catalogue of the Lepidoptera Rhopalocera (butterflies) of British Guiana. *British Guiana Department of Agriculture Entomological Bulletin* 3, 1–88. Reprinted from *Agricultural Journal of British Guiana* 10, 25–41, 96–104, 146–169, 215–252 (1939).

Hallé, F. (ed.) (1998) *Biologie d'une canopée de forêt équatoriale – III. Rapport de la mission d'exploration scientifique de la canopée de Guyane, octobre–décembre 1996.* Pro-Natura International and Opération Canopée, Paris/Lyon.

Hallé, F. and Blanc, P. (eds) (1990) *Rapport de Mission. Radeau des Cimes Octobre–Novembre 1989, Guyane Française.* OPRDC, Paris.

Hammond, D.S., Gourlet-Fleury S., Hout, P. van der, Brown, V.K. and ter Steege, H. (1996) A compilation of known Guianan timber trees and the significance of their dispersal mode, seed size and taxonomic affinity to tropical rain forest management. *Forest Ecology and Management* 83, 99–116.

Haviland, M.D. (1925) The Membracidae of Kartabo, Bartica District, British Guiana. *Zoologica* 6, 229–290 + errata.

Haxaire, J. and Rasplus, J.-Y. (1986) Contribution a la connaissance des Sphingidae de Guyane Francaise. 1re partie. (Lep.). *Bulletin de la Société Entomologique de France* 91, 275–285.

Haxaire, J. and Rasplus, J.-Y. (1987) Contribution à la connaissance des Sphingidae de Guyane Francaise, 2e Partie (Lep.). *Bulletin de la Société Entomologique de France* 92, 45–55.

Holloway, J.D. (1989) Moths. In: Lieth, H. and Werger, M.J.A. (eds) *Tropical Rain Forest Ecosystems. Biogeographical and Ecological Studies.* Elsevier, Amsterdam, pp. 437–453.

Howell, J.O. and Kosztarab, M. (1972) Morphology and systematics of the adult females of the genus *Lecanodiaspis* (Homoptera: Coccoidea: Lecanodiaspididae). *Research Division Bulletin Virginia Polytechnic Institute and State University, Blacksburg* 70, 1–248.

Huber, O. (1986) *La Selva Nublada de Rancho Grande, Parque Nacional 'Henri Pittier'.* Fondo Editorio Acta Cientifica Venezolana, Caracas.

Isaacs, R., Gillman, M.P., Johnston, M., Marsh, F. and Wood, B.C. (1996) Size structure of a dominant neotropical forest tree species, *Dicymbe altsonii*, in Guyana and some factors reducing seedling leaf area. *Journal of Tropical Ecology* 12, 599–606.

Janzen, D.H. (1974) Tropical blackwaters rivers, animals, and mast fruiting by the Dipterocarpaceae. *Biotropica* 6, 69–103.

Janzen, D.H. (1981) Patterns of herbivory in a tropical deciduous forest. *Biotropica* 13, 271–282.

Janzen, D.H. (1983) *Costa Rican Natural History.* The University of Chicago Press, Chicago, Illinois.

Janzen, D.H. (1988) Ecological characterization of a Costa Rican dry forest caterpillar fauna. *Biotropica* 20, 120–135.

Jolivet, P. (1987) Remarques sur la biocénose des *Cecropia* (Cecropicaeae). Biologie des *Coelomera* Chevrolat avec la description d'une nouvelle espèce du Brésil (Coleoptera Chrysomelidae Galerucinae). *Bulletin mensuel de la Société Linnéenne de Lyon* 56, 255–276.

Jolivet, P. (1989) The Chrysomelidae of *Cecropia* (Cecropiaceae): a strange cohabitation. *Entomography* 6, 391–395.

Jolivet, P. and Salinas, P.J. (1993) Ponte de *Coelomera cajennensis* (Fabricius, 1787) dans la tige des *Cecropia* (Col. Chrys.). *Bulletin de la Société Entomologique Française* 98, 472.

Joly, A. (1996) Dynamique de régéneration de deux espèces de Caesalpiniaceae, *Vouacapoua americana* et *Eperua falcata*, en Guyane. Diplome d'Etude Approfondies, Université Pierre et Marie Curie (Paris VI), Paris.

Lechat, F., Limier, F., Salager, J.L. and Roy, J. (1990) Profils de masse surfacique, teneur en azote et dégâts des herbivores dans une canopée tropicale. In: Hallé, F. and Blanc, P. (eds) *Biologie d'une canopée de forêt équatoriale. Rapport de Mission. Radeau des Cimes Octobre–Novembre 1989, Guyane Française.* OPRDC, Paris, pp. 178–189.

Leigh, E.G.J. (1996) Epilogue: Research on Barro Colorado Island, 1980–94. In: Leigh Jr, E.G., Rand, A.S. and Windsor, D.M. (eds) *The Ecology of a Tropical Forest,* 2nd edn. Smithsonian Tropical Research Institute, Washington, DC, pp. 469–503.

Le Moult, E. (1955) *Mes chasses aux papillons.* Editions Pierre Horey, Paris.

Lindsey, A.W. (1928) Hesperioidea from Kartabo district of British Guiana. *Journal of the Scientific Laboratories/Denison University (Denison, Iowa)* 23, 231–235.

Littledyke, M. and Cherrett, J.M. (1976) Direct ingestion of plant sap cut leaves by the leaf cutting ants *Atta cephalotes* (L.) and *Acromyrmex octospinosus* (Reich) (Form. Attini). *Bulletin of Entomological Research* 66, 205–217.

Lowman, M.D. (1984) An assessment of techniques for measuring herbivory: is rainforest defoliation more intense than we thought? *Biotropica* 16, 264–268.

Lowman, M.D., Foster, R., Wittman, P. and Rinker, H.B. (1998) Herbivory and insect load on epiphytes, vines, and host trees in the rain forest canopy of French Guiana. In: Hallé, F. (ed.) *Biologie d'une canopée de forêt équatoriale – III. Rapport de la mission d'exploration scientifique de la canopée de Guyane, octobre–décembre 1996.* Pro-Natura International and Opération Canopée, Paris/Lyon, pp. 116–128.

Loye, J.E. (1992) Ecological diversity and host plant relationships of treehoppers in a lowland tropical rainforest (Homoptera: Membracidae and Nicomiidae). In: Quintero, D. and Aiello, A. (eds) *Insects of Panama and Mesoamerica. Selected Studies.* Oxford University Press, Oxford, UK, pp. 280–289.

McCallan, E. (1953) Sawflies (Hym., Tenthredinidae and Argidae) from Trinidad, British Guiana and Venezuela. *The Entomologist's Monthly Magazine* 99, 126.

Marquis, R.J. (1991) Herbivore fauna of *Piper* (Piperaceae) in a Costa Rican wet forest: diversity, specificity and impact. In: Price, P.W., Levinsohn, T.M., Fernandes, G.W. and Benson, W.W. (eds) *Plant–Animal Interactions: Evolutionary Ecology in Tropical and Temperate Regions.* John Wiley & Sons, New York, pp. 179–208.

Marquis, R.J. and Braker, E.H. (1993) Plant–herbivore interactions: diversity, specificity, and impact. In: McDade, L.A., Hespenheide, H.A. and Hartshorn, G.S. (eds) *La Selva, Ecology and Natural History of a Neotropical Rain Forest.* University of Chicago Press, Chicago, Illinois, pp. 261–281.

Mattson, W.J. and Scriber, J.M. (1987) Nutritional ecology of insect folivores of woody plants: water, nitrogen, fiber and mineral considerations. In: Slansky, F. and Rodriguez, J.G. (eds) *Nutritional Ecology of Insects, Mites, Spiders and Related Invertebrates.* John Wiley & Sons, New York, pp. 105–146.

Merian, Maria Sybilla (1705) *Metamorphosis Insectorum Surinamensis.* Ofte Verandering der Surinaamsche Insecten. Priv. Ed., Amsterdam. [Reprinted by Pion Books, London, 1982].

Merian, Maria Sybilla (1719) *Dissertatio de Generatione et Metamorphosibus Insectorum Surinamensium.* Joannes Oosterwyk, Amsterdam.

Metcalf, Z.P. (1945) Fulgoroidea (Homoptera) of Kartabo, Bartica District, British Guiana. *Zoologica* 30, 125–143.

Metcalf, Z.P. (1949) Tettigellidae and Gyponidae (Homoptera) of Kartabo, Bartica District, British Guiana. *Zoologica* 34, 259–279.

Miller, S.E. (1994) Systematics of the neotropical moth family Dalceridae (Lepidoptera). *Bulletin of the Museum of Comparative Zoology* 153, 301–495.

Morrison, H. (1922) On some trophobiotic Coccidae from British Guiana. *Psyche* 29, 132–152.

Morrison, H. (1929) Some neotropical scale insects associated with ants (Hemiptera – Coccidae). *Annals of the Entomological Society of America* 22, 33–60.

Munroe, L.A. (1993) The status of biosystematics in Guyana. In: Jones, T. and Cook, M.A. (eds) *Proceedings of the Caribbean LOOP Planning Meeting,* Santo Domingo, 26–27 November 1993. CAB International, Wallingford, UK, pp. 22–26.

Nascimento, M.T. and Proctor, J. (1994) Insect defoliation of a monodominant Amazonian rainforest. *Journal of Tropical Ecology* 10, 633–636.

Newbery, D.M. and Foresta, H.D. (1985) Herbivory and defense in pioneer gap and understory trees of tropical rain forest in French Guiana. *Biotropica* 17, 238–244.

Newstead, R. (1917a) Observations on scale-insects (Coccidae) – III. *Bulletin of Entomological Research* 7, 343–380.

Newstead, R. (1917b) Observations on scale-insects (Coccidae) – V. *Bulletin of Entomological Research* 8, 125–134.

Newstead, R. (1920) Observations on scale-insects (Coccidae) – VI. *Bulletin of Entomological Research* 10, 175–207.

Nijveldt, W. (1968) Two gall midges from Surinam (Diptera, Cecidomyiidae). *Studies on the Fauna of Suriname and other Guyanas* 10, 61–66.

Osborn, H. (1921) Two *Tachigalia* membracids. *Zoologica* 3, 233–234.

Owen, D.F. (1980) How plants may benefit from the animals that eat them. *Oikos* 35, 230–235.

Poncy, P. Riéra, B., Larpin, L., Belbenoit, P., Jullien, M., Hoff, M. and Charles-Dominique, P. (1999) The permanent field station 'Les Nouragues' in the tropical rain forest of French Guiana: current projects and preliminary results on tree diversity, structure and dynamics. In: Dallmeier, F. and Comiskey, J.A. (eds) *Forest Biodiversity in North, Central and South America, and the Carribean: Research and Monitoring.* Man and the Biosphere Series 21. UNESCO and the Parthenon Publishing Group, Carnforth, UK, pp. 385–410.

Price, P.W. (1992) The resource-based organization of communities. *Biotropica* 24, 273–282.

Priesner, H. (1923) Ein Beitrag zur Kenntnis der Thysanopteren Surinams. *Tijdschrift voor Entomologie* 66, 88–111.

Raaimakers, D., den Ouden, F., van der Marel, M. and Boot, R. (1994) N and P as possible limiting factors for tree growth on acid sandy soils in tropical rain forest in Guyana. In: Raaimakers, D. (ed.) *Growth of Tropical Rain Forest Trees as Dependent on Phosphorus Supply. Tree Saplings Differing in Regeneration Strategy and their Adaptations to a low Phosphorus Environment in Guyana.* Tropenbos Series 11. The Tropenbos Foundation, Wageningen, The Netherlands, pp. 9–21.

Reagan, D.P. and Waide, R.B. (1996) *The Food Web of a Tropical Rain Forest.* University of Chicago Press, Chicago, Illinois.

Richards, P.W. (1996) *The Tropical Rain Forest, an Ecological Study*, 2nd edn. Cambridge University Press, Cambridge, UK.
Robbins, R.K. and Opler, P.A. (1997) Butterfly diversity and a preliminary comparison with bird and mammal diversity. In: Reaka-Kudla, M.L., Wilson, D.E. and Wilson, E.O. (eds) *Biodiversity II.* Joseph Henry Press, Washington, DC, pp. 69–82.
Robinson, G.S. and Kitching, I. (1997) Database 'Hosts'. The Natural History Museum, London. Records as from April 1997.
Roubik, D.W. (1990) Niche preemption in tropical bee communities: a comparison of neotropical and Malesian faunas. In: Sakagami, S.F. and Ohgushi, R.I. (eds) *Natural History of Social Wasps and Bees in Equatorial Sumatra.* Hokkaido University Press, Sapporo, pp. 245–257.
Skutch, A.F. (1945) The most hospitable tree (*Cecropia*). *The Scientific Monthly (New York)* 60, 1–17.
Smith, D.R. (1988) A synopsis of the sawflies (Hymenoptera: Symphyta) of America south of the United States: introduction, Xyelidae, Pamphilidae, Cimbicidae, Diprionidae, Xiphydriidae, Siricidae, Orussidae, Cephidae. *Systematic Entomology* 13, 205–261.
Smith, D.R. (1990) A synopsis of the sawflies (Hymenoptera: Symphyta) of America south of the United States: Pergidae. *Revista Brasileira de Entomologia* 34, 7–200.
Smith, D.R. (1992) A synopsis of the sawflies (Hymenoptera: Symphyta) of America south of the United States: Argidae. *Memoirs of the American Entomological Society* 39, 1–201.
Stahel, G. and Geijskes, D.C. (1939) Ueber den Bau der Nester von *Atta cephalotes* L. und *Atta sexdens* L. (Hym. Formicidae). *Revista de Entomologia* 10, 27–78.
Stahel, G. and Geijskes, D.C. (1940) Observations about temperature and moisture in *Atta*-nests. *Revista de Entomologia* 11, 766–775.
Stahel, G. and Geijskes, D.C. (1941) Weitere Untersuchungen über Nestbau und Gartenpilz von *Atta cephalotes* L. und *Atta sexdens* L. (Hym. Formicidae). *Revista de Entomologia* 12, 243–268.
Sterck, F., van der Meer, P. and Bongers, F. (1992) Herbivory in two rain forest canopies in French Guyana. *Biotropica* 24, 97–99.
Tavakilian, G. (1993) L'entomofaune de la forêt Guyanaise. In: SEPANGUY (ed.) *Gestion de l'écosystème forestier et aménagement de l'espace régional* (Congrès SEPANGUY, Cayenne 1990). SEPANGUY, Cayenne, pp. 125–130.
ter Steege, H. (1990) *A monograph of Wallaba, Mora and Greenheart.* Tropenbos Technical Series 5. The Tropenbos Foundation, Wageningen, The Netherlands.
ter Steege, H., Jetten, V., Polak, M. and Werger, M. (1993) Tropical rain forest types and soils of a watershed in Guyana, South America. *Journal of Vegetational Science* 4, 705–716.
ter Steege, H., Boot, R.G.A., Brouwer, L., Caesar, J.C., Ek, R.C., Hammond, D.S., Haripersaud, P., van der Hout, P., Jetten, V.G., van Kekem, A.J., Kellman, M.A., Khan, Z., Polak, A.M., Pons, T.L., Pulles, J., Raaimakers, D., Rose, S.A., van der Sanden, J.J. and Zagt, R.J. (1996) *Ecology and Logging in a Tropical Rain Forest in Guyana, with recommendations for Forest Management.* Tropenbos Series 14. The Tropenbos Foundation, Wageningen, The Netherlands.
Valiant, S.D. (1992) Questioning the caterpillar. *Natural History* 12/92, 46–59.
van Doesburg Jr, P.H. (1966) Heteroptera of Suriname. I. Largidae and Pyrrhocorridae. *Studies on the Fauna of Suriname and other Guyanas* 9, 1–60.
van Emden, H.F. (1972) Aphids as phytochemists. In: Harborne, J.B. (ed.) *Phytochemical Ecology.* Academic Press, London, pp. 25–42.
Vasconcellos-Neto, J. and Jolivet, P. (1994) Cycloalexy among chrysomelid larvae. In: Jolivet, P.H., Cox, M.L. and Petitpierre, E. (eds) *Novel Aspects of the Biology of Chrysomelidae.* Kluwer Academic, Amsterdam, pp. 303–309.
Voss, R.S. and Emmons, L.H. (1996) Mammalian diversity in neotropical lowland rainforests: a preliminary assessment. *Bulletin of the American Museum of Natural History* 230, 1–115.
Weber, N.A. (1949) The biology of the fungus-growing ants. Part IX. The British Guiana species. *Revista de Entomologia* 17, 114–172.
Wheeler, W.M. (1907) The fungus-growing ants of North America. *Bulletin of the American Museum of Natural History* 23, 669–807.
Wheeler, W.M. (1921) A study of some social beetles in British Guyana and of their relation to the ant-plant *Tachigalia. Zoologica* 3, 35–126.
Wheeler, W.M. (1942) Studies of neotropical ant-plants and their ants. *Bulletin of the Museum of Comparative Zoology, Harvard University* 90, 1–263.
Whitham, T.G., Maschinski, J., Larson, K.C. and Paige, K.N. (1991) Plant responses to herbivory: the contin-

uum from negative to positive and underlying physiological mechanisms. In: Price, P.W. Lewinsohn, T.M., Fernandes, G.W. and Benson, W.W. (eds) *Plant–Animal Interactions: Evolutionary Ecology in Tropical and Temperate Regions.* Wiley, New York, pp. 227–256.

Whitton, B.A. (1962) Forests and dominant legumes of the Amatuk Region, British Guiana. *Caribbean Forester* 23, 35–57.

Williams, D.J. and de Willink, M.C.G. (1992) *Mealybugs of Central and South America.* CAB International, Wallingford, UK.

6 Flower-visiting Insects in Guianan Forests: Pollinators, Thieves, Lovers and Their Foes

Bart P.E. De Dijn
National Zoological Collection, University of Suriname, Leysweg, Paramaribo, Suriname

Introduction

The importance of insects as pollen vectors or 'flying (plant) genitalia' has been known since the late eighteenth century (Barth, 1991), and the broad topic of pollination has been reviewed repeatedly and popularized (e.g. Knuth, 1906–9; Meeuse, 1961; Faegri and van der Pijl, 1979; Proctor *et al.*, 1996). Several recent reviews have provided general coverage of plant–pollinator interactions in tropical rainforests (Bawa, 1990) and in the context of certain neotropical forest regions (Amazonia by Prance, 1985; Central America by Schatz, 1990). From these reviews it is clear that most rainforest plants are pollinated by insects, and that neotropical pollination biology is more about insect–flower interactions than anything else. Literature from earlier in the last century in particular (e.g. Knuth, 1906–9) focused on the 'harmonious' co-adaptation between pollinators and flowers. This concept of harmony between flowering plants and archetypical pollinators lives on in an important part of the modern scientific community, as exemplified by records of flower visitation in which the visitors, certainly when they are bees, are automatically assumed to be pollinators. There is, however, more to flower visitation than pollination. Certainly in the neotropics not all the flower visitors are pollinators, and not all flowers visited are pollinated. To complicate matters further, certain flower visitors – not least the many stingless bees (Apidae: Meliponinae) – may be best characterized as thieves and robbers. These visitors will often destructively collect the floral reward 'intended' for pollinating visitors, failing to pollinate the host flower themselves and deterring actual pollinators in the process (Roubik, 1989). Still other visitors which are consistently associated with flowering plant parts have interests other than the flowers themselves; they are there primarily to find mates or animal prey. Despite the non-pollinating purposes that insects may have for visiting flowers, many neotropical insects do pollinate. In most cases, these pollinators are cross-pollinating as evident from the fact that most of the tree species in a neotropical forest are self-incompatible and require an animal pollen vector (Bawa *et al.*, 1985a,b). A pertinent first question to address in relation to the Guiana Shield rainforest is thus which flowers are visited by whom, why and whether visits by insects result in (cross-)pollination.

To begin to describe and quantify these basic components of flower–insect interactions requires a considerable length and breadth of field work. Currently, there are insufficient baseline data, let alone comprehensive studies, on flower visitation and pollination in Guiana Shield forests. Much of what will be presented below is based on extrapolation from studies done elsewhere (e.g. Central America and Central Amazonia). Fortunately, field observations suggest that most studies from other neotropical regions appear to be highly relevant to the rainforest of the Guiana Shield (De Dijn, personal observation). From the scientific point of view, however, the putative correspondence between Guianan and other neotropical pollinator systems must be regarded as a preliminary hypothesis in need of testing, viz. 'Are insect pollinators associated in the same way with the same or related host-plant species in the Guiana Shield, Central America and the Amazon Basin?' In essence, this is a question about regional diversity shifts in Amazonia and the neotropics. This chapter focuses on the flower-visiting insects in Guiana Shield rainforests. However, insects visiting flowering plant parts to feed on leaves (e.g. Lepidoptera larvae), young fruits (e.g. Hemiptera) or plant sap (e.g. Homoptera) will not be discussed here, as they are dealt with in Chapter 5 of this volume. Flower-visiting vertebrates are discussed in Chapter 4. The subject will be approached in the first instance by looking at the rewards available at flowers; most of these rewards are substances produced by the plant for the specific purpose of enticing insects to visit more flowers. The pollination syndrome and flower type concept of Faegri and van der Pijl (1979) will be used here to more comprehensively outline the various kinds of flower–insect interactions. The rewards at flowering plant parts are resources for insect visitors in the ecological sense, often being essential for their survival, which they access opportunistically and use selfishly to increase their fitness and reproductive success. This means that in addition to the archetypical pollinators, there are a whole host of other visitors which typically parasitize on the mutualism associated with a particular pollination syndrome. Once the scope of the mutualism we are dealing with is clear, I attempt to assess the diversity of flower-visiting insects in the Guiana Shield, as well as pollination systems, and based on these attributes, discuss several issues in relation to flowering plants and flower-visiting insects in Guiana Shield rainforests.

Rewards, Pollination Syndromes and Flower Types

Most insects visit flowers to obtain the rewards available, which in Amazonia can be pollen, nectar, oil, resin, fragrances or flower tissue (Prance, 1985). Some insects frequent flowers to find mates or prey (Gentry, 1978; De Dijn, personal observation; see also Roubik, 1989). In an attempt to impose some order on the variety of flower morphologies, visitor taxa, pollinating behaviours, etc., Faegri and van der Pijl (1979) introduced the concept of 'pollination syndromes', which – at least where pollination by animals is concerned – can be regarded as stereotypical mutual adaptations of flowering plants and pollinators. The syndromes are most often named after the legitimate pollinator taxa involved, such as moth, bird or bee, and can be subdivided further on the basis of flower type. Flower shape, colour and time of anthesis are part of a syndrome; each syndrome has its stereotypes, some of which are listed in Table 6.1 (for detailed descriptions and exceptions, see Faegri and van der Pijl, 1979, and Proctor *et al.*, 1996). The description of syndromes gravitates towards the more specific, 'specialist' ones. This is misleading in the sense that it creates the impression that most interactions between flowers and animals are essentially very specific and that other, non-pollinating animals have no business at specialist flowers, which is not the case (Janzen, 1983; De Dijn, personal observation). Many of the specialist syndromes are defined largely by the kind of animals that are being prevented – often very imperfectly (e.g. Roubik *et al.*, 1985) – from illegitimately (i.e. by damaging or destroying) accessing the floral rewards. Another reality is that a specialist flower will often also be visited by a wide range of legitimate (non-destructive) visitors, e.g. a hummingbird flower also being

Table 6.1. Floral rewards, pollination syndromes, and plant and insect taxa associated with them in the Guiana Shield rainforest.

Main reward	Syndrome	Typical characteristics of flower or inflorescence: Morphology	Colour	Scent	Timing	Plant families pollinated **(entire family)**	Insect visitors: **Primary** (P) or **secondary** (s) **pollinators**; non-pollinating visitors (n)
Pollen only	Wind	Exserted anthers/stamens; small flowers	Green, dull	None	Day	**Poaceae, Cyperaceae,** Piperaceae?, Moraceae?	**Apoidea**: P?/s/n; Coleoptera, Diptera & Dermaptera: n?
	Bee	Pollen hidden in tubular anthers with pores	Yellow, bright	Sweet subtle	Morning	Caesalpiniaceae, Solanaceae, Melastomataceae	**Apidae, Anthophoridae & Halictidae**: P/s/n; Diptera: n
	Bee	Many anthers forming exposed bundle/sphere	Bright	Sweet subtle	Morning	Mimosaceae, Myrtaceae?	**Apidae, Halictidae & Anthophoridae**: P,s
Nectar and pollen	Insect	Shallow flower with fully exposed pollen and nectar	White, yellow, bright	Sweet subtle	Day or (&) night	**Burseraceae, Sapindaceae, Malvaceae, Anacardiaceae, Dilleniaceae,** Euphorbiaceae, Palmae, Lauraceae?	**Hymenoptera & Diptera**: P/s/n; **Lepidoptera, Thysanoptera**, **Blattodea & Coleoptera:** s?/n
	Insect	Small flower with nectar in short-narrow corolla	White, yellow, bright	Sweet subtle	Day	**Meliaceae, Sapotaceae?**, Humiriaceae, Asteraceae, Rubiaceae, Boraginaceae	**Apoidea, Vespoidea & Sphecoidea**: P/s/n; **Lepidoptera & Diptera**: s?/n
	Bee (large)	Large flower with access to nectar (and pollen) mechanically obstructed	Yellow, white, bright	Sweet subtle	Day	**Convolvulaceae, Labiatae, Papilionoidea, Scophulariaceae**, Rubiaceae, Verbenaceae, Orchidaceae, Bignoniaceae, Maranthaceae, Lecythidaceae, Passifloraceae	**Apidae**: P/s/n; **Anthophoridae, Megachilidae**: P,s; Lepidoptera: n?
Nectar	Moth	Slender corolla with hidden nectar; many with exserted anthers	White	Sweet heavy	Night	**Amaryllidaceae?**, Mimosoidea, Apocynaceae, Bignoniaceae, Orchidaceae, Rubiaceae	**Sphingidae**: P/s/n; **other Lepidoptera**: s/n?
	Butterfly	Slender corolla with hidden nectar	Red, purple	Sweet subtle	Day	Rubiaceae, Cucurbitaceae, Verbenaceae	**Nymphalidae, Hesperiidae, Lycaenidae, Papilionidae & Pieridae**: P/s/n?
	Bird	Long corolla with hidden nectar; most with exserted anthers	Red	None	Day	**Bromeliaceae, Marcgraviaceae**, **Musaceae,** Passifloraceae, Zingiberaceae	Apidae: Meliponinae & Anthophoridae: Xylocopini: n
	Bat	Large, broad and thick-walled corolla	White	Fruity fetid	Night	Cactaceae, Bombacaeae, Lecythidaceae	Apidae: Meliponinae & Lepidoptera: n
Oil	Bee	Oil glands or oil hairs	Yellow, bright	Sweet subtle	Day	Malphigiaceae, Guttiferae? Orchidaceae?	**Anthophoridae**: P/s; **Apidae: Meliponinae**: s?/n
Resin	Bee	Resin glands	White, yellow	Sweet subtle	Day	Clusiaceae, Euphorbiaceae	**Apidae:Euglossinae:** P/s, **Apidae: Meliponinae**: P?/s/n
Scent	Bee	Scent glands (strongly fragrant)	(Dark)	Etherical	Day	Orchidaceae, Araceae, Solanaceae	**Apidae: Euglossinae**: P/s/n (male bees only)
Feeding tissue	Beetle	Flower with feeding tissue and holding chamber	Green, white	Fruity subtle	Night	**Nymphaceae**, Araceae, Annonaceae	**Scarabeidae: Dynastinae**: P; Nitidulidae: s?/n
Brood site	Wasp	Globular synconia	Green	None	?	Moraceae	**Agaonidae**: P
	Beetle: brood fl.	Small fleshy flowers in compact inflorescence	White, dull	Fruity subtle	Day and night	Palmae, Cyclanthaceae	**Curculionidae, Nitidulidae**: P,s; **Staphylinidae, Diptera**: s?/n
None (deceit)	**Fly**	Large trap-inflorescence or small ordinary florescence	Dark, green	Putrid subtle	?	**Aristolochiaceae**, Orchidaceae, Sterculiaceae?	**Diptera**: P/s/n

Timing=time of anthesis; ?=uncertain.

visited by butterflies and bees, which may also be pollinators (cf. case studies: Willmer and Corbet, 1981; Roubik, 1982a; Roubik *et al.*, 1985). Thirdly, many plants, including those in Guiana Shield forests (cf. below), would seem 'intentionally' unspecialized (i.e. generalist) as to the kind of visitors they attract as pollinators. This is particularly true in relation to insect pollination, but poorly investigated, at least in Amazonia, as remarked by Prance (1985). Based on the literature (cf. below), miscellaneous field observations in the Guianas (De Dijn, personal observation) and published descriptions and drawings of the regional flora (Lindeman and Mennega, 1963; Wessels Boer *et al.*, 1976; Gentry, 1993; Polak, 1992), an attempt is made in Table 6.1 to list important Guiana Shield plant families and flower visitors as a function of floral resources, pollination syndromes and flower types.

Pollen: from ancestral reward to specialist syndromes

It seems safe to assume that pollen was the main ancestral reward for the earliest pollinating flower visitors (beyond leaf-like flower parts). One can imagine a situation in Mesozoic times (cf. Proctor *et al.*, 1996) where phytophagous and scavenging beetles, cockroaches and other insects visit flowers and eat pollen at easily accessible generalist flowers, not unlike they do today in the Guiana Shield (see below). Although various flower-visiting insect taxa, such as beetles, wasps and moths, consume pollen to some extent, bees are strictly dependent on pollen for their survival. Bee larvae require pollen as a protein source, except those of a few divergent neotropical carrion-feeding species which also occur in the Guiana Shield (*Trigona hypogea* group; Roubik, 1982b). The females of most bee species will collect pollen at flowers and transfer it to their nest (Roubik, 1989). On the other hand, throughout the neotropics, *Lestrimelitta* (Apidae: Meliponinae) robs what has been collected by other social bees, while a whole range of bees, such as *Coelioxys* (Megachilidae), parasitize ready-made nests. In the Guiana Shield, flowers offering pollen as the sole reward are largely limited to those with tubular, poricidal anthers (so-called 'buzz' flowers; De Dijn, personal observation). These flowers have special adapted 'closed' anthers that are specifically adapted to buzz pollination by specialist bees of the common neotropical taxa *Eulaema*, *Euglossa*, *Melipona*, *Bombus* (all Apidae), *Augochloropsis* (Halictidae), *Centris*, *Xylocopa* and *Exomalopsis* (Anthophoridae) (Wille, 1963; Buchmann, 1985; De Dijn, personal observation). The pollen remains hidden inside the anther and will only dislodge via a narrow pore when vibrated properly, or buzzed, by specialist bees. The buzzing bees are dusted with pollen and pollinate by rubbing their pollen-dusted bodies against the floral stigmas. Neotropical plants also found in the Guianas that require buzzing include many *Solanum* s.l. (Solanaceae), *Cassia* s.l. (Caesalpiniaceae), and various Melastomataceae (Buchmann, 1983). These flowers are also visited by a host of non-buzzing bees, some of which are robbers that chew open the anthers, especially *Trigona* s.s. (Apidae: Meliponinae; Renner, 1983), while others are 'gleaners' that mop up pollen dusted on plant parts (e.g. some *Trigona* s.l.; Wille, 1963). Remarkably, some flowers with ordinary anthers (not poricidal) also offer only pollen and have been observed to be buzzed, especially those referred to as 'shaving brush' flowers, i.e. those having many stamens protruding to form a compact bundle (Buchmann, 1985). This flower type can be expanded to include the globular *Mimosa* (Mimosoidea) inflorescences buzzed by *Melipona* (Apidae: Meliponinae) in French Guiana and Suriname (Roubik, 1996a; De Dijn, personal observation). Other non-buzzing visitors that collect pollen at such flowers are equally non-destructive (e.g. *Trigona*, *Apis mellifera*). Grasses and sedges growing along creeks and in swamp forest and along forest-edge habitats in the Guiana Shield are also visited by insects such as bees (De Dijn, personal observation; see also Thomas, 1984).

These monocots produce only pollen and are generally assumed to be primarily wind-pollinated (Proctor *et al.*, 1996). Bees as well as flies and beetles may actually be pollinators of many neotropical grasses and sedges, such as *Pariana* (Poaceae) in Venezuelan cloud forest (Ramirez, 1989). Two very important Guiana Shield plant taxa that also only offer pollen from their 'catkin'-like flowers should be mentioned: *Piper* (Piperaceae) and *Cecropia* (Moraceae) (De Dijn, personal observation). *Cecropia* is assumed to be primarily wind pollinated (Ramalho *et al.*, 1990; Kress and Beach, 1994); and the same assumption can be made for *Piper* on the basis of its similar inflorescence shape (De Dijn, personal observation; Proctor *et al.*, 1996). *Cecropia* pollen is heavily collected by stingless bees and *Apis* in many neotropical forests (Ramalho *et al.*, 1990). In Suriname, both *Cecropia* and to a lesser extent *Piper* are important pollen sources for stingless bees (Biesmeijer *et al.*, 1992) and *Apis* (Kerkvliet and Beerlink, 1991). Whether these plants are also pollinated by these bees is unknown.

Nectar: from all-taxa to specialist reward

Floral nectar as a reward is the rule with the majority of angiosperms in Amazonia, although there are many striking and important exceptions (Prance, 1985). Nectar is an important currency used by plants to pay their pollinating flower visitors, which is easily understood when considering the relationship: nectar = dissolved sugars = energy to burn (Heinrich and Raven, 1972). Nectar composition (mainly dissolved sugars, but also amino acids), as well as sugar concentration and volume (or rate of secretion) will influence the kind of visitor (e.g. Proctor *et al.*, 1996), the rate of visitation (e.g. Heinrich and Raven, 1972) and ultimately plant reproductive output. Nectar is important to a variety of adult insects to fuel their daily activities, not least energy-consuming flight. In an evolutionary context, nectar can be regarded as a reward which has developed to attract the more energetic and therefore energy-demanding flying insects that tend to visit more flowers and through their longer flying distances prove to better cross-pollinators. Most female bees use the nectar they collect as the main carbohydrate source for their progeny (Roubik, 1989). Ants may also collect floral nectar to feed their larvae, but apparently to a lesser extent – possibly a consequence of toxins or repellents in the nectar (Haber *et al.*, 1981) or the fact that ants are the main visitors to extra-floral nectaries on plants where these occur. Below, it will be attempted to (re)define some syndromes and flower types that are primarily related to nectar, in a way that makes sense from a Guiana Shield perspective.

Generalist insect syndromes

Several common neotropical plant taxa, such as Sapindaceae and Anacardiaceae, have small, shallow 'open' flowers which offer uninhibited nectar and pollen access to a wide variety of insects, such as cockroaches, beetles, bees, wasps, flies, butterflies and moths (Janzen, 1983). The diversity of insect species visiting a single tree with a generalist flower type can be amazing, in contrast to the usually limited set of taxa visiting a patch of specialist flowers. Beetle diversity can be particularly striking, as there may be many species of Chrysomelidae, Curculionidae, Nitidulidae, Cantharidae and Mordellidae present, as well as some Dermestidae, Cerambicidae, Brentidae and Scarabaeidae (De Dijn, personal observation). All of these varied visitors may be legitimate pollinators, although this largely remains an assumption given the scarcity of direct observational studies on the pollination of plants with generalist flowers in the neotropics, or at least in Amazonia (Prance, 1985). There are detailed studies of Amazonian palms visited mainly by bees and a variety of beetles. These are reviewed by Henderson (1995); a more recent study on *Euterpe precatoria* suggests this palm is also heavily visited by bees (Kuchenmeister *et al.*, 1997). It may not be strictly possible to distinguish between what can be called 'shallow' (small, open) flower pollination and 'small-tube' flower pollination by insects. Both flower types conform to the generalist model in the sense that they are visited by a large number of insects belonging to a variety of taxa. Visiting taxa, however, are often restricted access to one of the two types. The small-tube flowers, such as those found in the family Meliaceae, hide their nectar and pollen inside a short, tubular (fused) corolla. This type of flower allows larger insects with a well-developed proboscis easy access to floral nectar, while for those large insects without a proboscis proper, access is nearly impossible (De Dijn, personal observation). While access to shallow flowers is unrestricted, it is restricted to aculeate Hymenoptera (wasps and bees), long-tongued flies (e.g. Syrphidae) and Lepidoptera (e.g. Hesperiidae and Lycaenidae) in small-tube flowers; very minute insects can also access the resources of small-tube flowers, such as small beetles, the ubiquitous Thysanoptera, and minute *Plebeia* and *Trigonisca* (Apidae: Meliponinae).

Large bee pollination syndrome

A more specialized syndrome, involving flowers with a large, tubular corolla, and large, mostly long-tongued bees as the main pollinators, appears to be distinctly neotropical (Frankie *et al.*, 1983; Janzen, 1983; De Dijn, personal observation). A typical large bee flower type is the 'large-tube' flower, a flower with a long (often fused) corolla which is usually narrow at the flower base, but may be very broad distally. Large bee flowers tend to be large, conspicuous and zygomorphic; they can only be legitimately visited by forceful bees which are able to overcome the physical and mechanical barriers which separate the floral nectar from the outside world (see Barth, 1991). Barriers include such simple devices as dense patches of hairs or an internal ring of stamens blocking the narrow tubular part of the corolla. The corolla base which holds the nectar is usually either thick-walled and tough, or is protected on the outside by a tough calyx. In the Guiana Shield, this type of design can be found within some *Psychotria* (Rubiaceae), *Stachytarphetum* (Verbenaceae) and Scrophulariaceae (De Dijn, personal observation). Flowers with larger, broader corollas tend to have different built-in barriers: a tough constriction at the corolla base requiring forceful insertion of the proboscis (e.g. with Convolvulaceae and many Bignoniaceae), or asymmetrical corolla 'deformations' or lips partially hiding and closing off the corolla mouth (e.g. with many Maranthaceae and Labiatae) or the nectar spur (e.g. with many Orchidaceae). A different flower design with the same effect, the flag-blossom, is found within most Papilionaceae. These flowers have tough, asymmetrical petals that hide nectar and pollen, and require cunning and force to push apart (Proctor *et al.*, 1996). A uniquely neotropical design is found with most Lecythidaceae: a fleshy asymmetrical hood serves to cover or hide that part of the flower which holds nectar and pollen and a strong and agile visitor is required to lift it (Mori and Boeke, 1987). Although legitimate access to the nectar of most of these flower types is obtained by a variety of Lepidoptera in the Guiana Shield, large bees, such as *Euglossa*, *Eulaema*, *Bombus*, *Melipona*, *Centris* and *Xylocopa*, as well as *Ptiloglossa* (Colletidae) and *Megachile* (Megachilidae) always seem to be the main pollinators (De Dijn, personal observation; Janzen, 1983; see also visitation records in Roubik (1979) and Heithaus (1979), and the discussion on Bignoniaceae in Gentry (1974), and on Lecythidaceae in Mori and Boeke (1987)). Shorter-tongued *Xylocopa*, and also the small *Trigona* are common robbers of these big-tube flowers; they perforate the corolla base from the outside and in so doing illegitimately access the floral nectar. The large bees appear to have the required behavioural repertoire and learning capacity, as well as the large body, to visit the flowers legitimately; they will often insert not just their proboscis, but their entire head or body in the corolla to gain access to the nectar. In the Guiana Shield, many smaller, shorter-tongued bees, such as *Augochloropsis* and *Trigona*, also legitimately visit a variety of large bee flowers, but for pollen which, as opposed to nectar, tends to be easily accessible (De Dijn, personal observation).

Butterfly and moth syndromes

Typical butterfly- or moth-pollinated flowers have a narrow tubular corolla or nectar spur (Haber and Frankie, 1989; Proctor *et al.*, 1996), and are thus similar in general shape to that of flowers pollinated by large, long-tongued bees. However, they generally lack the obstructive morphological features found inside large bee flowers (De Dijn, personal observation). Many flowers pollinated by Lepidoptera have relatively short corollas, shorter than one might expect on the basis of the often long proboscis of Lepidoptera, and have long filaments and styles that effectively position the anthers and styles well beyond the corolla proper. This is referred to as a 'brush' design, but a much more unusual one than the earlier mentioned shaving brush flower. The 'hairs' of the brush will often hinder the flower-visiting moth or butterfly, and force the latter to keep their bodies away from the corolla. These flower visitors will consequently use their long tongue 'from a distance' to probe for nectar, and in the process collect pollen dust on their body. Bird- and bat-pollinated flowers may have a similar brush design (see Proctor *et al.*, 1996, and Chapter 4 and Table 6.1, this volume). Flower-visiting Nymphalidae, Hesperiidae and Lycaenidae position their bodies just above or to the side of the corolla proper (De Dijn, personal observation), while hawkmoths (Sphingidae) will usually remain in hovering flight, directly in front or well above the corolla (Proctor *et al.*, 1996). The narrow corolla is an adaptation to the long, thin tongue of the majority of flower-visiting Lepidoptera. As in other neotropical areas (Haber and Frankie, 1989), the moth-flower design is taken to the extreme in some plant families in the Guiana Shield, such as the Rubiaceae, Amaryllidaceae and Orchidaceae, which each have at least one species with a corolla or nectar spur exceeding 10 cm in length (De Dijn, personal observation). While flower visitation in a neotropical habitat has been investigated at least once in a comprehensive manner in relation to hawkmoths (dry forest in Costa Rica: Haber and Frankie, 1989), there is no equally comprehensive data on the rest of the Lepidoptera at that location, nor on Lepidoptera pollination at other neotropical locations (but see Janzen, 1983).

Feeding tissue and brood-flowers

Several neotropical plants occurring in the Guiana Shield, e.g. *Nymphaea* (Nymphaeaceae), *Annona* (Annonaceae) and *Philodendron* (Araceae), have floral parts that attract large beetles during the female phase of flowering (Gottsberger, 1986). Beetles entering the large inflorescence at night gnaw at special feeding tissue (fleshy and nutritive flower parts) and become entrapped when the flower or inflorescence closes towards the morning. The next night – when the inflorescence is already in its male phase – the beetles get dusted with pollen and are allowed to escape (Proctor *et al.*, 1996). These inflorescences heat up in their female phase and produce a smell that attracts the beetles, most conspicuously Dynastinae (Scarabaeidae) of the genus *Cyclocephala*; the latter were observed visiting *Nymphaea* in Suriname (Cramer *et al.*, 1975). At least some Annonaceae may only attract smaller beetles (e.g. Nitidulidae) and flies, as observed in Guyana (van Tol and Meijendam, 1991) and elsewhere (e.g. Maas and Westra, 1992). Small Nitidulidae and Curculionidae are attracted by the hundreds or thousands to certain palm species, as recorded at several places in Amazonia (Henderson, 1995), and also observed with *Astrocaryum* in Suriname (De Dijn, personal observation). Liestabarth (1996) observed that *Phyllotrox* (Curculionidae) and *Epurea* (Nitidulidae) visit the inflorescences of certain Amazonian palm species and in the process pollinate; he also observed that they oviposit in the flowers. The minute beetle larvae develop rapidly in the flower tissue; a single palm inflorescence is a major breeding site for these beetles. Several *Phyllotrox* have been recorded to visit and oviposit at the inflorescences of five species of Cyclanthaceae in Venezuelan cloud forest (Seres and Ramirez, 1995). A characteristic of the beetle pollination syndrome would appear to be the sharing of several plant species by just a few pollinator species (Gottsberger, 1986; Liestabarth, 1996). Another type of brood-flower pollination syndrome occurring in the neotropics is distinctly one-on-one species-specific, and involves figs (*Ficus*: Moraceae) and fig wasps (Agaonidae). This fig wasp syndrome has been reviewed repeatedly (e.g. Janzen, 1979; Berg, 1990). It is quite complex and involves (chronologically, as the fig synconia develop):

1. Attraction of parent female wasps to female(-stage) fig synconia, resulting in pollination;
2. Oviposition and the development of wasp larvae in part of the fig florets;
3. Copulation of offspring wingless males and winged females in the synconia; and
4. Escape of pollen-laden female wasps from male(-phase) synconia.

Oils: food or an aid for bees

Most neotropical Malpighiaceae and many Melastomataceae have flowers with well-developed epithelial oil glands (Renner, 1989; Vogel, 1990). Other neotropical plants, such as the Commelinaceae, have flowers with fields of 'oil-hairs' (Faden, 1992). Epithelial flower oil is used in preference to nectar by some specialized neotropical bees, most notably female *Centris* (Buchmann, 1987). In the Guiana Shield, Malpighiaceae with oil flowers such as *Byrsonima* and *Stigmaphyllon* are quite common, as are oil-flower-visiting Centridini and Exomalopsini (Anthophoridae; e.g. *Centris* and *Paratetrapedia*; De Dijn, personal observation). These bees rub the oil from flowers they visit with their legs and have a peculiar toolkit, such as bristles to scratch open oil-bearing tissue, and brushes to mop up and transport the exudate oil (Neff and Simpson, 1981). Although it is quite clear that *Centris* is a good oil-flower pollinator, and that it uses oil as the main energy-rich food component for its larvae (Vinson *et al.*, 1989), the situation is less obvious for other oil-collecting anthophorids and oil-flower-visiting bees without special oil-collecting devices, such as Meliponinae. The latter are assumed to use the oil as a nest-building component or as pollenkit (a substance which makes pollen grains stick to their bodies; Roubik, 1989). There are, however, indications that at least two stingless bee species have a tight association with oil flowers in Suriname, and may thus act as important pollinators: (i) *Trigona kaieteurensis*, which has been found on numerous flowering *Stigmaphyllon* and never on other flowers; and (ii) *Cephalotrigona capitata*, which is commonly found on the oil-hair-covered flowers of *Vismia cayennensis* (together with *Exomalopsis*), but with difficulty on other flowers (De Dijn, personal observation).

Resin: building material for bees

Resin is offered by neotropical *Clusia* (Clusiaceae) and *Dalechampia* (Euphorbiaceae) flowers as a reward that is used by bees as a high-quality nest construction material (Armbruster, 1984). *Clusia* is diverse and common in the Guiana Shield, and is visited by a variety of Euglossinae and Meliponinae (Apidae), many of which collect floral resin, while some collect only pollen (Bittrich and Amaral, 1996, 1997; De Dijn, personal observation). There appears to be a great diversity of floral designs and breeding systems within *Clusia*, which have only recently begun to be investigated from a pollination biology perspective. The larger resin-flower-visiting bees, such as *Euglossa* and *Eufriesea* (Euglossinae), *Melipona* and larger *Trigona* (Meliponinae), seem to have the body size required for efficient pollination, while smaller Meliponinae may be infrequent pollinators or not pollinate at all. There are indications of a particularly tight association between at least one bee species and *Clusia* species in the Guiana Shield (De Dijn, personal observation; Roubik, personal communication) and presumably in Amazonia in general (cf. visitation recorded by Mesquita and Franciscon, 1995; Bittrich and Amaral, 1996, 1997). The large and enigmatic stingless bee *Duckeola ghilianii* has been found frequently on *Clusia* flowers, but not anywhere else. In addition, the otherwise quite elusive small bees of the *Trigona* subgenus *Frieseomelitta* can commonly be found on *Clusia* flowers, and would also appear to interact tightly, possibly as pollinators.

Scents: reward for male orchid bees

Orchid bees (Apidae: Euglossinae) only occur in the neotropics, and a limited set of neotropical plant taxa attract the males of these bees with very specific scents produced by 'fragrant', etherical chemicals, such as vanillin, cineole and eugenol (Dressler, 1982). Each orchid species appears to produce a distinct cocktail of fragrant chemicals, and different cocktails appear to attract different sets of male orchid bee species. The reward for the male bees are the fragrant chemicals themselves; these chemicals are mopped off the flower with the front legs and transferred to a glandular storage organ in the large hind tibia (Barth, 1991). Away from the flowers, male orchid bees release the evaporating chemicals present in their hind tibial glands throughout their territorial domain. This peculiar behaviour appears to have a function in mating, but the exact role remains unclear (Dressler, 1982; Roubik, 1989). In Suriname, a whole range of orchid species, such as *Catasetum* and *Gongora*, attract euglossine males, commonly *Eulaema*, *Euglossa* and *Eufriesea* (Werkhoven, 1986; De Dijn, personal observation). Orchids are, however, by no means the only Guiana Shield plants using chemicals to attract male euglossine bees. In French Guiana, at least two species of Lecythidaceae produce chemicals that attract male euglossines (Knudsen and Mori, 1996), as does one species of *Cyphomandra* (Solanaceae; Gracie, 1993). In Suriname, euglossines have in addition been observed on strongly fragrant *Spathiphyllum* (De Dijn, personal observation), in agreement with observations elsewhere (Williams and Dressler, 1976).

Fly pollination syndrome: deception

Flies that breed in rotting organic material and fungi are attracted to flowers producing smells which are unpleasant (to humans), such as those of *Aristolochia* (Aristolochiaceae; Proctor *et al.*, 1996). Flies are lured into the oil lamp-shaped inflorescence of *Aristolochia* when it is in its female phase. At that time most visitors fail to find the escape route out of the flower because they try to exit at the base, where the reproductive parts are, and where the surrounding plant tissue is translucent (suggestive of an exit-opening). In the subsequent male phase, the flies get dusted with pollen while the tissue near the inflorescence base darkens; this enables the trapped visitors to find the actual opening at the opposite end of the inflorescence. What deceives the flies in the first place is a smell resembling that of decaying plant material or fungi emanating from inside the flower. *Aristolochia* does, however, offer some nectar or other secretion upon which the trapped visitors appear to feed. Some Surinamese orchids (Werkhoven, 1986), and some Amazonian Sterculiaceae (Prance, 1985) also appear to selectively lure flies to their flowers by means of olfactory deception.

Rewards for insects at non-insect flowers

There are a variety of angiosperms in the Guiana Shield with flowers adapted for pollination by birds and bats (see Chapter 4, this volume, and Prance, 1985). While these typically large flowers (red ones for birds; white ones for bats; see Table 6.1) reward their vertebrate visitors with large amounts of dilute nectar, the flowers are also visited by insects. In Suriname, many of the insect visitors to these flowers are essentially scavengers, collecting pollen or surplus nectar, and do not appear to pollinate at all (De Dijn, personal observation). Several of these insects, most notably *Trigona*, will actually chew a hole in the flower base to get at the nectar which is difficult to access through the normal perianth opening, as has been observed in Guyana with bird-pollinated *Passiflora* (Dalessi, 1993), and in Suriname with *Norantea* (Marcgraviaceae) and *Symphonia* (Guttiferae) (De Dijn, personal observation). Another common illegitimate visitor at hummingbird flowers is *Xylocopa*, a big bee that makes slits in the base of corollas to access the hidden nectar (Roubik, 1989). Yet other visitors are secondary robbers, mainly small stingless bees, which rob nectar via the perforations made by the primary robbers.

Non-floral rewards: mates and animal prey

When the attractiveness of flowers to visitors is discussed, there rarely is explicit mention of mates and prey as rewards. A crucial attraction for some visitor taxa, certainly for many bee species, is a high concentration of suitable mates. This is most obvious when observing male bees at flower patches that only offer pollen and therefore do not offer any resource for male bees. Male bees in the genus *Centris* have been observed patrolling at *Cassia* flowers in Suriname (De Dijn, personal observation). Male *Ancyloscelis* (Anthophoridae) can be observed patrolling *Ipomaea* (Convolvulaceae) flowers and accosting, in flight, anything resembling a conspecific female or a competing male. Various Halictidae can be observed doing the same at, for example, inflorescences of Compositae. It is obvious that brood-flowers and trap-flowers act as important sites where both sexes can meet and mate (Liestabarth, 1996; Proctor *et al.*, 1996).

For many predators, flowers are ideal sites for finding prey. The main arthropod predator taxa at flowers in the Guiana Shield are assassin bugs (Reduviidae), robber flies (Asilidae), dragonflies (Odonata) and spiders (Araneae); preying mantises (Mantodea) and predatory carabid beetles have also been observed occasionally (De Dijn, personal observation). Reduviids are often found at flowers, trying to ambush small bees or wasps. Asilids can be seen most of the time staking out potential prey on a prominent flower or vegetative part in the immediate vicinity of a larger patch of flowers. Dragonflies occasionally do the same but appear to spend most of their time in flight right above the flowering vegetation. Some Reduviidae and Asilidae are near-perfect colour mimics of similar-sized common bees. This resemblance may serve either as a protection against predators or to mislead their prey (Roubik, 1989). Flower-visiting crab spiders (Thomisidae) in Suriname are particularly common on yellow or white flowers and often have a body with the same colour as the flower, which strongly suggests a camouflage function. These cryptic-coloured predators are rarely encountered anywhere else but on flowers (De Dijn, personal observation), indicating that a consistent association has developed over evolutionary time between these predators, the host-plant flowers and their visiting prey. Predatory sphecid wasps, as well as parasitic wasps and bees – most commonly *Coelioxys* (Megachilidae) – are often found on flowers, but all these appear to be there to forage for nectar, not to locate a host (De Dijn, personal observation). Parasitic insects actually using flowers to locate hosts for their offspring have rarely been observed in the Guiana Shield, although there is at least one record of a rhipiphorid beetle (Coleoptera) larvae (and adults) on *Borreria* in French Guiana (observation by Falin and De Dijn).

Diversity of Host Plants, Flower Visitors and Pollination Systems

One basic fact that needs to be established in relation to diversity in the Guiana Shield is how many flower-visiting insects are there compared to how many other flower visitors and flowering plants. It is obvious that '(insects) have to be somewhere' (E.O. Wilson, cited by C.D. Michener, personal communication), which includes flowering plant parts, and that consequently the diversity of insects at flowers will tend to be a function of general insect diversity. The type and extent of floral rewards that can be found in the natural world indicates that many insect taxa have compelling reasons to spend a larger proportion of their time visiting flowering plant parts than anywhere else. Arguably more meaningful diversity facts to establish are: how many flower visitors are actual pollinators and how many plants are being pollinated by any particular insect taxon. The latter, referred to as 'pollination systems', is discussed in much of the recent literature (e.g. Kress and Beach, 1994).

Host-plant diversity

On the basis of recent data on the Guiana Shield flora, there are at least 7493 species of flowering plants in the three Guianas (Guyana, Suriname and French Guiana; Boggan *et al.*, 1997) and 8622 in Venezuelan Guayana (Berry *et al.*, 1995). Each of the Guianas harbours more or less 5000 plant species (more in Guyana). Based on estimates of subregional levels of endemism (23% for Venezuelan Guayana, 5% for Guyana), taking into account that additional species (endemics and widespread species) should occur in the Brazilian part of the Guiana Shield, and assuming a bit less than 40% overall regional endemism (Berry *et al.*, 1995), there ought to be at least 10,000–11,000 native angiosperm species in the entire Guiana Shield rainforest. Another important figure to know is not regional but 'on the spot' (alpha) plant species richness in the Guiana Shield. Based on data from Ek (1997) obtained at 1-ha forest plots in Central Guyana, one plot will have some 70–100 tree species, 40–70 liana species and 15–25 herb species. This may add up to some 125–195 species of flowering plants, not counting epiphytes (which may be at least a few dozen species/ha).

Alpha diversity of flower visitors

Estimating the number of flower-visiting insect species (alpha species richness) at a Guiana Shield rainforest site is tricky, given that a comprehensive data set on a neotropical flower visitor community has to be puzzled together using data on a forest habitat block in Guanacaste province, Costa Rica. There, in a 'complex' lowland deciduous forest, Heithaus (1974, 1979) recorded a total of 230 flowering plant species and 586 flower visitor species, of which 573 were insects (170 were bees). Unfortunately, these figures are for diurnal visitors only, and data on predatory spiders, bugs and the like appear to be excluded; also, fig wasps do not seem to be included (fig trees may have been absent from the study site; if not, the three fig species in Guanacaste would contribute at least as many fig wasp species; based on Janzen, 1983). From other studies, there are some data on important nocturnal flower-visiting taxa: 65 hawkmoth species (Sphingidae) were recorded at the same general locality as used by Heithaus (based on Haber and Frankie, 1989), and Guanacaste has five common species of Glossophaginae and five other phyllostomatid bats (Chiroptera) which are known to visit flowers (based on Janzen, 1983). There are no comprehensive data whatsoever on nocturnal beetles and non-sphingid moths visiting flowers in Guanacaste. Based on this still incomplete data set, more than 95% of the flower visitor species are insects, and about 25% are bees. This high proportion of insects seems to be in line with the overwhelming richness of insect species in any tropical forest. The proportions of some insect taxa are not in line though; bees, for instance, surely make up much less than 25% (i.e. their proportional species richness based on the Guanacaste forest site flower visitation data) of the animal species at any neotropical forest site. Given that angiosperm species richness at the Guanacaste forest site would appear to be comparable to that at a Central Guyana forest plot (cf. above), flower visitor alpha species richness may be comparable too.

Regional diversity of flower visitors and pollinators

Given the paucity of data on the Guiana Shield insect fauna, assessing the flower visitor diversity at a (sub)regional scale is fraught with problems. Nevertheless, data on the diversity and natural history of some taxa do exist, if only for other forested subregions in the neotropics which are – hopefully – comparable to the Guiana Shield (Table 6.2). The diversity figures compiled here – although surely very incomplete – suggest that there may be thousands of flower-visiting insect species in the Guiana Shield, most of which are Coleoptera, Lepidoptera, Diptera and Hymenoptera. However, the majority of these species probably do not pollinate or at best are part of a larger pool of generalist pollinators, each of which may be quite inefficient at pollen transfer in most situations. A few insect families seem to contain efficient specialist pollinators, such as Agaonidae and Sphingidae, while with other families, just a small fraction of the species are such specialists, e.g. Scarabaeidae with specialist *Cyclocephala* and Anthophoridae with specialist *Ancyloscelis*. There seem to be many efficient pollinators amongst the generalist bees (e.g. Halictidae and Meliponinae; cf. discussion below).

Host-plant diversity vs. visitor and pollinator diversity

In each one of the three Guianas, there may thus be less than 1000 efficient pollinating insect species for some 5000 flowering plant species. For the Guiana Shield as a whole, the ratio of efficient pollinating insects to angiosperm species richness is probably even lower, given that the majority of these pollinator species are wide-

Table 6.2. Regional species richness, and quantity of species visiting flowers and pollinating, for insect taxa in the Guiana Shield and some other forested neotropical regions.

Taxon	Visitation type	Expected species richness	No. or % of species		Source
			Visiting	Pollinating	
Thysanoptera	sca/gen?	?	?	?	–
Blattodea	sca/gen?	SU: >48	>2?	none?	De Dijn, pers. obs.
Mantodea	pre	SU: >10	>2?	none?	De Dijn, pers. obs
Odonata	pre	SU: >260	?	none?	Geijskes, 1967
Dermaptera	sca/gen?	SU: >31	2	2?	Brindle, 1968
Coleoptera					
Carabidae	pre	SU: >31	>2	0?	De Dijn, pers. obs.
Staphylinidae	sca/pre/gen?	SU: > ?	few?	few?	Ashe?
		CR: >845	?	?	Ashe, 2000
Cantharidae	sca/gen?	SU: >4	all?	?	De Dijn, pers. obs
Mordellidae	sca/gen?	SU: >5	all?	?	De Dijn, pers. obs.
Scarabaeidae	sca/fts/gen?	SU: >>16	>5?	few?	De Dijn, pers. obs. Krikken?
Bruchidae	sca/gen?	SU: >>3	>2?	none?	De Dijn, pers. obs
Nitidulidae	sca/brd/gen?	SU: ?	>3?	few?	De Dijn, pers. obs. & Falin?
Elateridae	sca?	SU: ?	>2?	few?	De Dijn, pers. obs
Buprestidae	sca?	SU: >>5	>1?	none?	De Dijn, pers. obs
		CR: >245	?	?	Janzen, 1983
Chrysomelidae	sca/gen?	SU: ?	>10?	many?	De Dijn, pers. obs.
		VE: ?	?	?	Joly?
Cerambicidae	sca/gen?	FG: >1500	5 %?	<5 %?	Hequet, 1996
		VE: ?	?		Joly?
Curculionidae	sca/brd?	SU: ?	>11?	>3?	De Dijn, pers. obs.
& Brentidae	gen?	VE: ?			Joly?
Diptera					
Syrphidae	sca/gen	SU: >140	all?	many?	van Doesburg, 1966
Asilidae	pre	SU: >6	many?	none?	De Dijn, pers. obs.

Sepsidae & others	dct	SU: ?	?	?	?
		BR: ?	?	?	Brantjes, 1980
Calyptrata	sca/gen?	SU: ?	many?	few?	?
Lepidoptera					
Papilionidae	sca/nec/gen	FG: 37	all	few–many?	SEPANGUY, 1995–7
		CR: 40	all		Janzen, 1983
Pieridae	sca/nec/gen	SU: >12	all?	few–many?	De Dijn, pers. obs.
		CR: 70	all		Janzen 1983
Nymphalidae s.l.	sca/nec/gen	FG: >314	50%	few–many?	SEPANGUY, 1995–7
		CR: 433	60%		Janzen, 1983
Hesperiidae	sca/nec/gen	SU: > 45	all?	few–many?	De Dijn, pers. obs.
		CR: 250	all		Janzen, 1983
Lycaenidae s.l.	sca/nec/gen	SU: ?	all?	few–many?	De Dijn, pers. obs.
		CR: 250	all		Janzen, 1983
Sphingidae	nec	FG: ?	?	?	HYDRECO
		VE: 95	?	?	Chacin and Clavijo, 1995
		CR: 126	all?	many?	Janzen, 1983
Small moths	sca/nec?	SU: ?	?	few?	De Dijn, pers. obs.
Hymenoptera					
Vespoidea	sca/pre	SU: >150	many	many?	De Dijn, pers. obs.
Spheciformes	sca/pre	SU: >33	most	many?	De Dijn, pers. obs.
Apiformes (=Apoidea s.s.)	sca/nec/pol	FG: 245	all	most?	references for the
		BR: 250	all	most?	first three areas in
		CR: 200	all	most?	Roubik, 1992
		PA: 353	all	most?	Michener, 1954
Agaonidae	brd	SU: >10	all	all	Weiblen, pers. com.
		FG/SU: 38?	all	all?	Boggan *et al.*, 1997
		CR: 65?	all	all?	Janzen, 1983

SU=Suriname; FG=French Guiana; BR=Brazil, around Belem; PA=Panama; CR=Costa Rica; VE=Venezuela, Amazonas state; sca=scavenging (non-pollinating); pol=pollen only foraging (pollinating); gen=generalist (pollinating); nec=nectar foraging (pollinating); fts=feeding tissue (pollinating); brd=brood flower (pollinating); dct=deceit (pollinating); pre=predating (non-pollinating); ?=unknown/uncertain. De Dijn, pers. obs. is based on inspection of the NZCS collection, and unpublished observations in Suriname.

spread (De Dijn, personal observation). It is thus tempting to agree with the conclusions of Roubik (1992), expand a bit upon them, and conclude in analogy that in a tropical forest there are 'so few efficient pollinators (predominantly insects) and so many angiosperms'. There is no evidence pointing at non-insects, e.g. birds and bats, adding significantly to the number of efficient pollinators (but see Chapter 4). The above conclusion would, however, appear to be valid only at a broad scale, and not for any particular forest habitat at any particular study site (e.g. a Guanacaste dry forest site), nor during any particular short period of time. At any particular site and time, there would appear to be a better balance between the number of plant species actually flowering and the number of efficient pollinating species. This and related matters have been discussed in detail by Roubik (1992, 1996) for tropical bees. His conclusions are that, at smaller spatial and temporal scales, a generalist bee species will visit a limited set of flowering plant species, and that an individual generalist bee will tend to visit the flowers of one plant species for a prolonged period. This would elegantly explain how abundant, generalist bees might be responsible for the pollination of numerous plant species in the Guiana Shield rainforest and in other botanically diverse tropical forests.

Diversity of pollination systems

Several authors have described the 'pollination systems' of neotropical plant communities (Ramirez, 1989; Bawa, 1990; Kress and Beach, 1994; Seres and Ramirez, 1995). The pollination system concept would appear to be similar to that of pollination syndromes in that the systems are named after the main pollen vectors involved, most often animal taxa, and that the different classes are mutually exclusive. The figures listed are usually percentages (of plant species assigned to a pollination system class or type), thus indicating the relative importance of various pollen vector types in a plant community (Table 6.3). There is an impressive data set on the angiosperms of the rainforest at La Selva in Costa Rica (Bawa *et al.*, 1985b, Bawa, 1990; Kress and Beach, 1994), which is, however, still incomplete and biased against non-woody plants (most notably epiphytes). There are also data on the angiosperms of a high-altitude shrubland habitat ('El Jardin') in Venezuelan Guayana (Ramirez, 1989; the only data set from the Guiana Shield) and on the monocotyledons of the cloud forest at Rancho Grande in Venezuela (Seres and Ramirez, 1995). The La Selva data may be the most relevant in relation to the lowland forests of the Guiana Shield, the other data being representative of similar highland ecosystems in the Guiana Shield. One problem with the data is that the authors used different methods to determine the pollination system and used different classes. The biggest problem lies in the comparison of data on plants pollinated by 'small insects'; these plants are put in a separate class for La Selva, while for the Venezuelan localities they are always assigned to a variety of other classes. At an ecosystem level, pollination by insects, mostly bees, appears to be the general rule (see Table 6.3). With monocotyledons, beetle pollination would appear to surpass that by bees, mainly because of the many beetle-pollinated palms, Cyclanthaceae and Araceae (not just at Rancho Grande, but also at La Selva, based on inspection of the published tables). These plants belong mainly to the understorey, which may explain the relatively high occurrence of beetle pollination at this forest stratum. Bee pollination seems to be particularly important in the rainforest

Table 6.3. Pollination systems (Ps) in a lowland and highland neotropical plant community.

Location	La Selva, Costa Rica					Guayana, Venezuela		
Ecosystem	Lowland rainforest					High altitude shrubland		
Community	Trees (partial)		Angiosperms (partial)			Angiosperms		Monocotyledons
Stratum	Canopy	Sub-canopy	Canopy	Sub-canopy	Understorey	All strata	Various strata	Mainly understorey
Small insect Ps	23	8	27	14	5	11	?	?
Small bee Ps	8	17	4	18	16	14	56	16
Large bee Ps	44	22	37	20	22	24		
Wasp Ps	4	2	6	5	0	3		0
Butterfly Ps	2	5	4	4	5	4	11	0
Moth Ps	14	7	12	14	4	11		0
Beetle Ps	0	16	2	12	17	13	3	34
Fly Ps	0	0	0	0	3	2	10	11
Bird Ps	2	18	4	4	24	15	12	24
Bat Ps	4	4	4	8	1	4	0	5
Wind Ps	0	3	0	1	4	3	8	8
No. plant spp.	52	220	51	74	151	276	55	33
Source	Bawa, 1990		Kress and Beach, 1994			Ramirez, 1989		Seres and Ramirez, 1995.

canopy, where a much larger proportion of the plants (mainly trees) are pollinated by big bees than in lower strata.

Structure and diversity of flower visitor communities

The Guiana Shield has been, and still is, an important region for the study of bee community structure and diversity, and the impact of the exotic Africanized honeybee (*Apis mellifera*) on native bees and plants (see below). These studies are highly relevant for understanding the diversity and structure of entire flower visitor communities, which mainly consist of insects. It is logical that research has focused on strongly flower-dependent taxa, such as the bees, if only because narrower communities may be easier to comprehend. The diversity of flower visitor communities involved in one-on-one or few-on-few interactions with plants is logically a function of the number of plant species. Thus, the diversity of fig wasps is a reflection of the diversity of fig trees (see Janzen, 1983). For other more 'loosely' interacting, generalist visitor taxa, often referred to as the product of 'diffuse' coevolution, plant and visitor diversity should not necessarily be so tightly linked. The most striking example of this may be the relationship between plants and stingless bees in the neotropics. Similar-sized areas of humid lowland forest in the interior of the Guiana Shield, Central Amazonia, the Andean lowlands and Central America all appear to hold about 35–50 stingless bee species (Roubik, 1989), despite the fact that plant composition and diversity are quite different (Gentry, 1990; ter Steege *et al*., 2000). This fact is of great interest to ecologists who investigate the forces which shape community structure and diversity. A high diversity of exceedingly generalist flower visitors is difficult to explain using 'classical' ecological equilibrium concepts – where competition is invoked as the structuring force – given that all the species appear to fill the same food niche, and thus must be regarded as strong competitors which ought not to coexist. Roubik (1992) proposes that tropical forests are too heterogeneous and unpredictable over space and time for competitive processes to continue until an equilibrium is reached, and that this is the explanation for the high social bee diversity in the neotropics. De Dijn (2000) records significantly fewer stingless bee species in a swamp forest in Suriname, which he sees as the result of low woody plant diversity and tree polydominance. Johnson (1983) explains the coexistence of many species of stingless bees in part as a consequence of the different foraging styles of the species. The views of the latter two authors would appear to be more in line with the equilibrium theory, at least to some extent.

The overall community structure of flower visitors may best be understood in terms of guilds (as the bee guilds defined by Roubik, 1992) or functional groups (not unlike the ant groups of Andersen, 1997). This is something which remains to be elaborated further. The presence of certain groups or guilds, e.g. oil bees, and the diversity of groups and guilds, may be a function of general habitat characteristics and plant composition. Each guild or group has its own 'internal' diversity, which may be strongly influenced by plant diversity or spatio-temporal forest heterogeneity. One guild or group may have a strong impact on another one. For instance, the presence of social flower visitors, such as stingless bees, may lower the diversity of other visitor taxa, such as solitary bees. This impact of one flower visitors group on another has been a topic addressed repeatedly by Roubik (e.g. 1989, 1992); he concludes that highly social bees pre-empt feeding resources and thus limit opportunities for the establishment and speciation of more specialized solitary bee species. This offers an elegant explanation why bee species diversity is generally lower in tropical forests than, for instance, in (sub)tropical and Mediterranean-type habitats (Roubik, 1990, 1996). Support for the 'dominance' of stingless bees over other neotropical bee taxa, and arguably also over other flower visitor taxa, comes from data on their high ecosystem-wide abundance at flowers in neotropical ecosystems (Heithaus, 1974,

1979; Roubik, 1979) and data on the wide range of flowering plants they visit in the neotropics (Ramalho *et al.*, 1990; Biesmeijer *et al.*, 1992).

A recently introduced flower visitor species in the neotropics is the Africanized honeybee, *Apis mellifera*. Its impact on native stingless bees has been investigated in French Guiana by Roubik (1978). He failed to find convincing evidence of colony-level competition between *Apis* and *Melipona* (Roubik, 1982c), but he did observe some changes in the visitation and seed-set with *Mimosa pudica* (Mimosoidea), which he links to invasion by feral *Apis* (Roubik, 1996). The limited impact of an exotic flower visitor and the large 'redundant' diversity of generalist flower visitors, such as stingless bees, suggests that the pollination in forest ecosystems of the Guiana Shield may be quite insensitive to disturbances. What has surfaced to date, however, is only a small part of the puzzle; more case studies on other visitor and plant taxa need to be done, and more long-term data need to be collected. This topic is important in relation to forest conservation. Very tight pollination mutualisms are surely vulnerable to forest decline. Even some generalist pollination in the neotropics may decline seriously as a consequence of chronic forest degradation, as is probably happening in the Costa Rican dry forest (Frankie *et al.*, 1997). A breakdown of pollination mutualisms will have very serious consequences for plant reproduction and forest maintenance. One extreme consequence would be that many plant species would no longer reproduce and would slowly disappear from their natural habitats. Such a scenario is not improbable for valuable timber species, where a decrease in density due to over-harvesting will increase inter-individual distances, and could hamper cross-pollination.

Pollination Research in the Guiana Shield: Pitfalls and Priorities

Much of the information reviewed above (Tables 6.1–6.3) is based on extrapolation, and many of the data sets used are incomplete and probably biased; worse, some data sets may contain errors because of the many pitfalls related to recording and interpreting the roles of visitors, and assessing their impact on pollination. One needs to be cautious when using the data, and not take the conclusions based on them for granted. Pitfalls in relation to pollination studies in the neotropics could be summarized by the following:

1. Visitor vs. pollinator. Visitors are not necessarily pollinators, and this is also the case for archetypical pollinators such as bees (see discussion and examples above).
2. Pollen carrying vs. inter-plant pollen transfer. Even visitors with pollen 'at the right place' on their body, and observed to contact the style with parts of their body dusted with pollen, are not necessarily pollinators. Given that most neotropical plants may require cross-pollination (Bawa, 1985a), they will only be pollinators when in addition they move from one plant individual to the next. The latter will thus have to be demonstrated, at least for self-incompatible plant species (as done by Frankie *et al.*, 1976).
3. Occasional vs. consistently efficient pollinators. Visitors may be very abundant on a flowering plant (like stingless bees often are), but may only occasionally move from one conspecific plant to the next, and in so doing cross-pollinate. This is very much related to the amount of reward available on one plant, the size and flight range of the visitor and the distance between conspecific plants, as discussed in relation to nectar by Heinrich and Raven (1972).
4. Pollinators vs. non-pollinating visitors. Flower visitors that (partially) deplete the resources attracting legitimate pollinators may positively or negatively affect pollination. For clover in temperate ecosystems, an increase of seed set has been linked to the presence of non-pollinating visitors (reviewed in Heinrich and Raven, 1972), while in a neotropical setting, seed set with a treelet has been shown to decrease as a result of nectar robbing (Roubik *et al.*, 1985).

5. Pollinators vs. aggressive non-pollinating visitors. Many of the flower visitors that behave aggressively at flowers may not act as pollinators themselves and in addition chase away the actual pollinators; hence the term 'anti-pollinators', coined by Gentry (1978). Such anti-pollinators, however, may just as well promote cross-pollination by occasionally chasing or scaring away pollinators while these are in the process of visiting flowers (Roubik, 1989). Aggressors that may cause such events are territorial male bees, resource-monopolizing *Trigona* and various predators (De Dijn, personal observation).

One can safely conclude that there is a serious shortage of high-quality studies on flower visitation and pollination in the neotropics in general (let alone the Guiana Shield); few studies go 'all the way' from recording visitors (and accurately interpreting their roles) to assessing the contribution of individual visitor species or classes to pollination and seed production. Notable exceptions are:

1. A series of studies by Frankie and collaborators in Costa Rica (mainly on the canopy tree *Andira inermis*; reviewed in Frankie *et al.*, 1990, 1997);
2. A study of the pollination of the understorey shrub *Pavonia dasypetala* in Panama (Roubik, 1982a);
3. A study on flower visitors and reproduction of the treelet *Quassia amara* (Roubik *et al.*, 1985); and
4. A study on social bee visitation and seed set of the creeping herb *Mimosa pudica* in French Guiana (Roubik, 1996a).

It is obvious that an important research priority would be to undertake similar in-depth studies on the variety of key plant species in the Guiana Shield. Highest priority would need to be given to:

1. Representatives of diverse plant families and genera (see ter Steege, 2000);
2. Representatives of abundant plant families and genera (see data on trees in ter Steege *et al.*, 2000);
3. Plant species of great ecological importance, such as pioneer species (e.g. *Cecropia*) and locally dominant species (e.g. *Euterpe* and *Eperua*); and
4. Plant species of economic importance, such as timber tree species (e.g. *Cholorocardium rodiei* and *Vouacapoua americana*), and plants yielding important non-timber forest products (e.g. many larger palm species).

Another obvious Guiana Shield research priority is to undertake some ecosystem-wide studies on flower visitors and their roles (also plant breeding system), in which taxonomic and growth form bias is avoided as much as possible. Diverse 'mixed' lowland forest and the typical Guiana Shield species-poor forest types would deserve high priority here. Ecosystem-wide studies may need to precede the studies on individual plant species, in the spirit of the exemplary research strategy followed by Frankie and collaborators in Costa Rica (as explained in Frankie *et al.*, 1990). Along the lines of this strategy, the next step or research priority would be to study the ecology of some of the most important pollinator taxa, which will surely include most of the bees. An ambitious priority research programme for the Guiana Shield could have an even wider scope than the one in Costa Rica, and investigate plant reproduction in a holistic manner. This would mean: (i) initially combining ecosystem-wide studies on flower visitation and pollination with studies on seed set and seed dispersal; (ii) then focusing on representative and important plant species; and (iii) ultimately investigating representative and important animal taxa involved in flower visitation and pollination.

Questions to address in relation to flower visitation in the Guiana Shield

Flower visitation and the pollination mutualism in tropical forests have long been topics of considerable interest to ecologists (e.g. Heithaus, 1979; Janzen, 1983), morphologist-taxonomists and evolutionary biologists (e.g. Gentry, 1974; Neff and

Simpson, 1981). Although it is an ideal topic for crossover plant–animal studies, most of the hypotheses formulated and in part also investigated to date have been either 'botanical' ones, e.g. on plant breeding systems and effective pollination, or 'zoological' ones, e.g. on floral resource partitioning and flower visitation. Some of the main hypotheses are summarized here; the one on flower visitor community structure and diversity will be discussed further below, given that it has already been the subject of several influential studies conducted in the Guiana Shield.

A comprehensive review of some predominantly botanical questions in relation to pollination in tropical rainforests can be found in Bawa (1990); they can be summarized as follows:

1. How is adequate pollen flow realized in heterogeneous forests?
2. Do longevity and environmental pressures require rainforest tree populations to be heterogeneous?
3. What is the link between the various pollination syndromes and plant speciation?
4. What factors promote coevolution, e.g. of specialist plants and pollinators?
5. To what extent do plant–pollinator interactions shape (plant) community structure?
6. Do the mutualistic plant–pollinator interactions make forests stable and resilient?

Some pertinent 'zoological' questions have been formulated in relation to tropical bees by Roubik (1989, 1990, 1992, 1996) and can be restated more generally as follows:

1. How do flower visitors partition the resources at different spatial and temporal scales?
2. Are flower visitor communities mainly shaped by long-lived and highly social species?
3. Are flower visitor communities in competitive equilibrium or in constant flux?
4. Do many-on-many pollinating interactions make forests less sensitive to disturbance?

These research questions need to be addressed in the Guiana Shield to further wider comparisons of the results from different tropical forest regions in South America, the neotropics and the tropics in general. The Guiana Shield is an important component in any meta-analysis of pollination. Have the many unique or rare (pre)historical and ecological characteristics of the region similarly affected the way that plants and pollinating insects interact today? Some of the region's unique attributes might include:

1. The forests cover one of the Earth's most ancient and stable geological formations (see Chapters 1 and 2).
2. There are extensive tree-species-poor forests, such as xeromorphic forests on excessively drained sands, and hydrophytic forests in the poorly drained coastal plain (Chapters 2 and 7).
3. There are isolated table mountain, rock outcropping and savanna ecosystems within a sea of highly heterogeneous lowland rainforest (Berry *et al.*, 1995) (Chapter 2).
4. Tree species diversity in comparable forest ecosystems is lower than in Central Amazonia and the Andean foothills (ter Steege *et al.*, 2000) (Chapter 7).

The fact that the Guiana Shield is one of the largest, relatively unmodified forest areas remaining on Earth is another strong point of this region. Its intact food chains and communities allow for the investigation of ecosystems which have optimal functional integrity, i.e. where all the interacting rainforest organisms which define the system under study are still present. Such intact ecosystems allow for more meaningful research and ultimately a better understanding of nature; also, they are of great importance to elucidate the impact of human activities on nature, and provide clues on how to restore disturbed habitats (see Chapter 9).

Acknowledgements

Thanks to David Hammond and Yves Basset for timely feedback; George Weiblen,

Meindert Hielkema and Phillippe Cerdan for information on fig wasps; Paul Ouboter, Harrold Sijlbing and Usha Raghoenandan for facilitating the review work and Anil Gangadin and Helene Hiwat for assistance and discussions in the field (and comments on this manuscript).

References

Andersen, A.N. (1997) Functional groups and patterns of organization in North American ant communities: a comparison with Australia. *Journal Biogeography* 24, 433–460.

Armbruster, W.S. (1984) The role of resin in angiosperm pollination: ecological and chemical considerations. *American Journal Botany* 71, 1149–1160.

Ashe, J.S. (2000) Rove beetles (Staphylinidae) of the Monteverde area. In: Nadkarni, N. and Wheelwright, N. (eds) *Monteverde: Ecology and Conservation of a Tropical Cloud Forest.* Oxford University Press, New York, pp. 108–111.

Barth, F.G. (1985) *Insects and Flowers: the Biology of a Partnership* (transl. M.A. Biederman-Thorsen). Princeton University Press, Princeton, New Jersey.

Barth, F.G. (1991) *Insects and Flowers: the Biology of a Partnership.* Princeton University Press, Princeton, New Jersey.

Bawa, K.S. (1990) Plant–pollinator interactions in tropical rain forests. *Annual Review Ecology and Systematics* 21, 399–422.

Bawa, K.S., Perry, D.R. and Beah, J.H. (1985a) Reproductive biology of tropical lowland rain forest trees. I. Sexual systems and incompatibility mechanisms. *American Journal Botany* 72, 331–345.

Bawa, K.S., Bullock, S.H., Perry, D.R., Coville, R.E. and Grayson, M.H. (1985b) Reproductive biology of tropical lowland rain forest trees. II. Pollination systems. *American Journal Botany* 72, 346–351.

Berg, C.C. (1990) Reproduction and evolution in *Ficus* (Moraceae): traits connected with the adequate rearing of pollinators. *Memoirs New York Botanical Garden* 55, 169–185.

Berry, P.E., Huber, O. and Holst, B.K. (1995) Floristic analysis and phytogeography. In: Berry, P.E., Holst, B.K. and Yatskievych, K. (eds) *Flora of the Venezuelan Guayana,* Vol. 1: *Introduction.* Missouri Botanical Garden, St Louis, Missouri, pp. 161–191.

Biesmeijer, J.C., van Marwijk, B., van Deursen, K., Punt, W. and Sommeijer, M.J. (1992) Pollen sources for *Apis mellifera* (Hym, Apidae) in Surinam, based on pollen grain volume estimates. *Apidologie* 23, 245–256.

Bittrich, V. and Amaral, C.E. (1996) Flower morphology and pollination biology of *Clusia* species from the Gran Sabana (Venezuela). *Kew Bulletin* 51, 681–694.

Bittrich, V. and Amaral, C.E. (1997) Flower biology of some *Clusia* species from Central Amazonia. *Kew Bulletin* 52, 617–635.

Boggan, J., Funk, V., Kelloff, C., Hoff, M., Cremers, G. and Feuillet, C. (1997) *Checklist of the Plants of the Guianas (Guyana, Surinam, French Guiana),* 2nd edn. Centre Study Biological Diversity, Georgetown.

Brantjes, N.B.M. (1980) Flower morphology of *Aristolochia* species and the consequences for pollination. *Acta Botanica Neerlandica* 29, 121–213.

Brindle, A. (1968) The Dermaptera of Surinam and other Guyanas. *Studies Fauna Suriname and Other Guyanas* X, 1–60.

Buchmann, S.L. (1983) Buzz-pollination in angiosperms. In: Jones, C.E. and Little, R.J. (eds) *Handbook of Experimental Pollination Biology.* Van Nostrand Reinhold, New York, pp. 73–114.

Buchmann, S.L. (1985) Bees use vibration to aid pollen collection from non-poricidal flowers. *Journal Kansas Entomological Society* 58, 517–525.

Buchmann, S.L. (1987) The ecology of oil flowers and their bees. *Annual Review Ecology and Systematics* 18, 343–369.

Chacin, M.E. and Clavijo, J.A. (1995) Sphingidae (Insecta: Lepidoptera) del estado Amazonas, Venezuela. *Boletin Entomologia Venezolana* 10, 7–24.

Cramer, J.M., Meeuse, A.D.J. and Teunissen, P.A. (1975) A note on the pollination of nocturnally flowering species of *Nymphaea. Acta Botanica Neerlandica* 24, 489–490.

Dalessi, D.L.W.M. (1993) Flower visiting behaviour of Trigona (T.) williana on *Passiflora coccinea* and *Passiflora glandulosa* in Mabura, Guyana. MSc research report, Utrecht University, Utrecht, The Netherlands.

De Dijn, B.P.E. (2000) Stingless bee communities: are they poorer in more constrained ecosystems? In:

Proceedings Sixth International Conference on Apiculture in Tropical Climates. IBRA, Cardiff, UK, pp. 200–210.

Dressler, R.L. (1982) Biology of the orchid bees (Euglossini). *Annual Review Ecology Systematics* 13, 373–394.

Ek, R.C. (1997) Botanical diversity in the tropical rain forest in Guyana. PhD thesis, Utrecht University. Tropenbos-Guyana Series 4. Tropenbos-Guyana, Georgetown.

Engel, M.S. and Dingemans-Bakels, F. (1980) Nectar and pollen resources for stingless bees (Meliponinae, Hymenoptera) in Surinam (South America). *Apidologie* 11, 341–350.

Faden, R.B. (1992) Floral attraction and floral hairs in the Commelinaceae. *Annals Missouri Botanical Garden* 79, 46–52.

Faegri, K. and van der Pijl, L. (1979) *The Principles of Pollination Ecology,* 3rd edn. Pergamon Press, Oxford, UK.

Frankie, G.W., Opler, P.A. and Bawa, K.S. (1976) Foraging behavior of solitary bees: implications for outcrossing of a Neotropical forest tree species. *Journal of Ecology* 64, 1049–1057.

Frankie, G.W., Haber, W.A., Opler, P.A. and Bawa, K.S. (1983) Characteristics and organization of the large bee polllination system inthe Costa Rican dry forest. In: Jones, C.E. and Little, R.J. (eds) *Handbook of Experimental Pollination Biology.* Van Nostrand Reinhold, New York, pp. 411–448.

Frankie, G.W., Vinson, S.B., Newstrom, L.E., Barthell, J.F., Haber, W.A. and Frankie, J.K. (1990) Plant phenology, pollination ecology, pollinator behaviour, and conservation of pollinators in a neotropical dry forest. In: Bawa, K.S. and Hadley, M. (eds) *Reproductive Ecology of Tropical Forest Plants.* Man and the Biosphere Series, Vol. 7. UNESCO, Paris and Parthenon, Park Ridge, New Jersey, pp. 37–47.

Frankie, G.W., Vinson, S.B., Rizzardi, M.A., Griswold, T.L., O'Keefe, S. and Snelling, R. (1997) Diversity and abundance of bees visiting a mass flowering tree species in disturbed seasonal dry forest, Costa Rica. *Journal Kansas Entomological Society* 70, 281–296.

Geijskes, D.C. (1967) De insektenfauna van Suriname, ook vergeleken met die van de Antillen, speciaal wat betreft de Odonata. *Entomologische Berichten (Amsterdam)* 27, 69–72.

Gentry, A.H. (1974) Coevolutionary patterns in Central American Bignoniaceae. *Annals Missouri Botanical Garden* 61, 728–759.

Gentry, A.H. (1978) Anti-pollinators for mass-flowering plants? *Biotropica* 10, 68–69.

Gentry, A.H. (1990) Floristic similarities and differences between Southern Central America and Upper and Central Amazonia. In: Gentry, A.H. (ed.) *Four Neotropical Rainforests.* Yale University Press, New Haven, Connecticut, pp. 141–157.

Gentry, A.H. (1993) *A Field Guide to the Families and Genera of Woody Plants of Northwest South America (Colombia, Ecuador, Peru), with Supplementary Notes on Herbaceous Taxa.* University of Chicago Press, Chicago, Illinois.

Gottsberger, G. (1986) Some pollination strategies in neotropical savannas and forests. *Plant Systematics Evolution* 152, 29–45.

Gracie, C. (1993) Pollination of *Cyphomandra endopogon* var. *endopogon* (Solanaceae) by *Eufriesea* spp. (Euglossini) in French Guiana. *Brittonia* 45, 39–46.

Haber, W.A. and Frankie, G.W. (1989) A tropical hawkmoth community: Costa Rican dry forest Sphingidae. *Biotropica* 21, 155–172.

Haber, W.A., Frankie, G.W., Baker, H.G., Baker, I. and Koptur, S. (1981) Ants like flower nectar. *Biotropica* 13, 211–214.

Heinrich, B. and Raven, P.H. (1972) Energetics and pollination ecology. *Science* 176, 597–602.

Heithaus, E.R. (1974) The role of plant–pollinator interactions in determining community structure. *Annals Missouri Botanical Garden* 61, 675–691.

Heithaus, E.R. (1979) Flower visitiation records and resource overlap of bees and wasps in northwest Costa Rica. *Brenesia* 16, 9–52.

Henderson, A. (1995) *The Palms of the Amazon.* Oxford University Press, New York.

Hequet, V. (1996) *Longicornes de Guyane.* ORSTOM and Silvolab, Cayenne.

Janzen, D.H. (1979) How to be a fig. *Annual Review Ecology and Systematics* 10, 13–51.

Janzen, D.H. (ed.) (1983) *Costa Rican Natural History.* University of Chicago Press, Chicago, Illinois.

Janzen, D.H. (1984) Two ways to be a tropical big moth: Santa Rosa saturniids and sphingids. *Oxford Surveys in Evolutionary Biology* 1, 85–140.

Johnson, L.K. (1983) Foraging strategies and the structure of stingless bee communities in Costa Rica. In: Jaisson, P. (ed.) *Social Insects in the Tropics,* Vol. 2. Universite Paris-Nord, Paris, pp. 31–58.

Johnson, L.K. (1987) Communication of food source location by the stingless bee Trigona fulviventris. In:

Eder, J. and Rembold, H. (eds) *Chemistry and Biology of Social Insects.* Verlag J. Peperny, München, Germany, pp. 698–699.
Kerkvliet, J.D. and Beerlink, J.G. (1991) Pollen analysis of honeys from the coastal plain of Surinam. *Journal Apicultural Research* 30, 25–31.
Knudsen, J.T. and Mori, S.A. (1996) Floral scents and pollination in neotropical Lecythidaceae. *Biotropica* 28, 42–60.
Knuth, P. (1906–9) *Handbook of Flower Pollination* (transl. J.A. Davis). Clarendon Press, Oxford.
Kress, W.J. and Beach, J.H. (1994) Flowering plant reproductive systems. In: McDade, L.A., Bawa, K.S., Hespenheide, H.A. and Hartshorn, G.S. (eds) *La Selva. Ecology and Natural History of a Neotropical Rain Forest.* The University of Chicago Press, Chicago, Illinois, pp. 161–182.
Kuchenmeister, H., Silberbauer-Gottsberger, I. and Gottsberger, G. (1997) Flowering, pollination, nectar crop, and nectaries of *Euterpe precatoria* (Arecaceae), an Amazonian rain forest palm. *Plant Systematics Evolution* 206, 71–97.
Liestabarth, C. (1996) Pollination of *Bactris* by *Phyllotrox* and *Eperua*: implications of the palm breeding beetles on pollination at the community level. *Biotropica* 28, 69–81.
Lindeman, J.C. and Mennega, W.H.A. (1963) *Bomenboek voor Suriname.* Dienst LBB, Paramaribo.
Maas, P.J.M. and Westra, L.Y.T. (1992) *Rollinia.* Flora Neotropica, Monograph 57. New York Botanical Garden, New York.
Meeuse, B.J.D. (1961) *The Story of Pollination.* Ronald Press, New York.
Mesquita, R. de C.G. and Franciscon, C.H. (1995) Flower visitors of *Clusia nemorosa* G.F.W. Meyer (Clusiaceae) in an Amazonian white-sand campina. *Biotropica* 27, 254–257.
Michener, C.D. (1954) Bees of Panama. *Bulletin American Museum Natural History* 104, 1–175.
Mori, S.A. and Boeke, J.D. (1987) Pollination. In: Mori, S.A. and collaborators *The Lecythidaceae of a Lowland Neotropical Forest: La Fumee Mountain, French Guiana. Memoirs New York Botanical Garden* 44, 137–155.
Neff, J.L. and Simpson, B.B. (1981) Oil-collecting structures in the Anthophoridae (Hymenoptera): morphology, function, and use in systematics. *Journal Kansas Entomological Society* 54, 95–123.
Polak, A.M. (1992) *Major Timber Trees of Guyana. A Field Guide.* Tropenbos Foundation, Wageningen, The Netherlands.
Prance, G.T. (1985) The pollination of Amazonian plants. In: Prance, G.T. and Lovejoy, T.E. (eds) *Amazonia.* Pergamon Press, Oxford, UK, pp. 166–191.
Proctor, M., Yeo, P. and Lack, A. (1996) *The Natural History of Pollination.* Timber Press, Portland, Oregon.
Ramalho, M., Kleinert-Giovannini, A. and Imperatriz-Fonseca V.L. (1990) Important bee plants for stingless bees (Melipona and Trigona) and Africanized honeybees (*Apis mellifera*) in neotropical habitats. *Apidologie* 21, 469–488.
Ramírez, N. (1989) Biología de polinización en una comunidad arbustira tropical de la Alta Guayana Venezolana. *Biotropica* 21, 319–330.
Renner, S. (1983) The widespread occurrence of anther destruction by *Trigona. Biotropica* 15, 251–256.
Renner, S. (1989) A survey of reproductive biology in neotropical Melastomataceae and Memecylaceae. *Annals Missouri Botanical Gardens* 76, 496–518.
Roubik, D.W. (1978) Competitive interactions between neotropical pollinators and Africanized honey bees. *Science* 201, 1030–1032.
Roubik, D.W. (1979) Africanized honey bees, stingless bees, and the structure of tropical plant-pollinator communities. In: Caron, D. (ed.) *Proceedings IV International Symposium on Pollination.* Maryland Agricultural Experiment Station Miscellaneous Publications 1, pp. 403–417.
Roubik, D.W. (1982a) The ecological impact of nectar-robbing bees and pollinating hummingbirds on a tropical shrub. *Ecology* 63, 354–360.
Roubik, D.W. (1982b) Obligate necrophagy in an social bee. *Science* 217, 1059–1060.
Roubik, D.W. (1982c) Seasonality in colony food storage, brood production and adult survivorship: studies of *Melipona* in tropical forest (Hymenoptera: Apidae). *Journal Kansas Entomological Society* 55, 789–800.
Roubik, D.W. (1989) *Ecology and Natural History of Tropical Bees.* Cambridge University Press, Cambridge, UK.
Roubik, D.W. (1990) Niche preemption in tropical bee communities: a comparison of neotropical and Melanesian faunas. In: Sakagami, S.F., Ohgushi, R. and Roubik, D.W. (eds) *Natural History of Social Wasps and Bees in Equatorial Sumatra.* Hokkaido University Press, Sapporo, pp. 245–257.
Roubik, D.W. (1992) Loose niches in tropical communities: why are there so few bees and so many trees?

In: Hunter, M.D., Ohgushi, T. and Price, P.W. (eds) *Effects of Resource Distribution on Animal–Plant Interactions.* Academic Press, London, pp. 327–354.

Roubik, D.W. (1996) African honey bees as exotic pollinators in French Guiana. In: Matheson, A., Buchmann, S.L., O'Toole, C., Westrich, P. and Williams, I. (eds) *The Conservation of Bees.* IBRA, Cardiff, UK, pp. 173–182.

Roubik, D.W. (1996b) Measuring the meaning of honey bees. In: Matheson, A., Buchmann, S.L., O'Toole, C., Westrich, P. and Williams, I. (eds) *The Conservation of Bees.* IBRA, Cardiff, UK, pp. 163–172.

Roubik, D.W., Holbrook, N.M. and Parra, G.V. (1985) Roles of nectar robbers in reproduction of the tropical treelet *Quassia amara* (Simaroubaceae). *Oecologia (Berlin)* 66, 161–167.

Schatz, G.E. (1990) Some aspects of pollination biology in Central American forests. In: Bawa, K.S. and Hadley, M. (eds) *Reproductive Ecology of Tropical Forest Plants,* pp. 69-84. UNESCO, Paris.

SEPANGUY (1995–7) *Papillons Diurnes de Guyane.* SEPANGUY, Kourou, 9 posters.

Seres, A. and Ramírez, N. (1995) Biología floral y polinización de algunas monocotiledoneas de un bosque nublado venezolano. *Annals of the Missouri Botanical Garden* 82, 61–81

ter Steege, H. (2000) A perspective on Guyana and its plant richness. In: ter Steege, H. (ed.) *Plant Diversity in Guyana.* Tropenbos Series 18, Tropenbos Foundation, Wageningen, The Netherlands, pp. 11–17.

ter Steege, H., Sabatier, D., Castellanos, H., van Andel, T., Duivenvoorden, J., de Oliveira, A.A., Ek, R., Lilwah, R., Maas, P. and Mori, S. (2000) A regional perspective: analysis of Amazonian floristic composition and diversity that includes the Guiana Shield. In: ter Steege, H. (ed.) *Plant Diversity in Guyana.* Tropenbos Series 18. Tropenbos Foundation, Wageningen, The Netherlands, pp. 19–34.

Thomas, W.W. (1984) The systematics of *Rhynchospora* sect. Dichromena. *Memoirs New York Botanical Garden* 37, 1–116.

van Doesburg Jr, P.H. (1966) Syrphidae from Suriname. *Studies Fauna Surinam and Other Guyanas* IX, 61–107.

van Tol, I.A.V. and Meijendam, N.A.J. (1991) Field research on pollination and seed dispersal of Annonaceae (soursop family). Doctoral report 91191, Herbarium, Utrecht University, Utrecht, The Netherlands.

Vinson S.B., Williams H.J., Frankie, G.W. and Shrum, G. (1997) Floral lipid chemistry of *Byrsonima crassifolia* (Malpighiaceae) and a use of floral lipids by *Centris* bees (Hymenoptera: Apidae). *Biotropica* 29, 76–83.

Vogel, S. (1990) History of the Malpighiaceae in the light of pollination ecology. *Memoirs New York Botanical Garden* 55, 130–142.

Werkhoven, M.C.M. (1986) *Orchids of Suriname.* Vaco Press, Paramaribo.

Wessels Boer, J.G., Hekking, W.H.A. and Schulz, J.P. (1976) *Fa joe kan tak' mi no moi.* Stinasu, Paramaribo.

Wille, A. (1963) Behavioural adaptations of bees for pollen collecting from *Cassia* flowers. *Revista Biologia Tropical* 205–210.

Williams, N.H. and Dressler, R.L. (1976) Euglossine pollination of *Spathiphyllum* (Araceae). *Selbyana* 1, 349–356.

Willmer, P.G. and Corbet S.A. (1981) Temporal and microclimatic partitioning of the floral resources of *Justicia Aurea* amongst a concourse of pollen vectors and nectar robbers. *Oecologia (Berlin)* 51, 67–78.

7 Guianan Forest Dynamics: Geomorphographic Control and Tropical Forest Change Across Diverging Landscapes

David S. Hammond
Iwokrama International Centre for Rain Forest Conservation and Development, Georgetown, Guyana. Currently: NWFS Consulting, Beaverton, Oregon, USA

Geomorphic and Geographic Controls

The geological world consists of many regions with tremendous topographic relief and many regions of tremendous geological age, but few areas with both. Simply stated, this mutual exclusivity is a consequence of the regional balance between (epeirogenic) forces raising crustal mass and surface processes countering these forces. Over time, local diastrophic forces invariably quiesce and topographic relief declines as uplifting becomes less frequent and surface processes continue to eat away at the remaining exposed, and most resistant, features.

More specifically, diastrophic processes that regulate the rate and distribution of tectonic and volcanic activity slowly grow mountains and move continents (see Chapter 2). They also deliver pyroclastic flows, lahars, lavas and seismic events over smaller spatial scales that catastrophically deflect forest community processes within and near these areas. Expansion of ash clouds and aerosol trains over much larger areas influence radiative forcing effects on climate and deliver punctuated nutrient loads to affected forest areas. Volcanic eruptions and crustal uplifting create new weathering surfaces and reinvigorate depleted erosion bevels. These changes lead to injection of new materials through accelerated erosion, exposure of intrusive lithologies, outgassing and magmatic flow. More importantly, adjacent lowland regions act as the first-stop depositories of material conveyed downslope.

In contrast, where diastrophic activity is relatively quiescent, as in the case of Precambrian shield regions, these effects no longer dominate the regional landscape. Ancient phases of uplifting, volcanism and erosion no longer act as significant providers of new materials and life in these regions becomes largely dependent on internal recycling, atmospheric deposition and biological fixation routes to material acquisition.

Together, tectonic and volcanic activity have modulated the rate and spatial distribution of material flow into, within and out of the biosphere over geological time. They have also created structural barriers that alter the size and shape of contiguous climatic zones and force anisotropies in the delivery of forest-sustaining rainfall patterns (see Chapter 2). As a consequence, climates that sustain modern tropical forest communities have also varied tremendously.

Diastrophic effects, however, delimit

only part of the story. Long-term changes in the amount of incoming short-wave radiation (ISR), how this is distributed spatially and when it is delivered throughout the year form the other part. Sunspot and Milankovitch cycling, volcanic outgassing, fossil-fuel burning, meteoric impacts and geomagnetic reversals are leading external factors believed to force climate change by altering the planet's energy budget (see Chapter 2). Changes to the planetary budget translate into different regional climate responses as the net increase or decrease of global energy is variously transformed and transferred through the global atmospheric and oceanic circulation (see 'Climate', Chapter 2). In turn, the paths of oceanic and atmospheric currents change with a slowly shifting surface topography driven by the spatial distribution of diastrophic activity. Thus, the geographic position of tropical forests in a way reflects the interplay of long-term internal and external forcing factors and constitutes a major role in determining how these regions have been shaped by these changes.

In the case of the neotropics, the gradual widening of the Atlantic as part of the Gondwana break-up and final uplifting of the Panamanian land bridge fundamentally altered how climate was delivered to the Guiana Shield through their effects on regional oceanic circulation (Pitman *et al.*, 1993; Coates and Obando, 1996) (see Chapter 2). More importantly, these changes were largely concurrent with the known earliest phases of angiosperm and later mammalian radiations and extinctions across the neotropics (Fig. 7.1) (Raven and Axelrod, 1974; Friis *et al.*, 1987).

Geomorphographic Control?

Together, diastrophic, or geomorphic, variation and geographic position exert a pronounced control on the main abiotic features shaping tropical forest environments over a wide range of timescales. This geomorphographic control defines a set of constraints exerted by the spatial distribution of age-related geological features and geographic location of these features on habitat change trajectories.

Large-scale, structural constraints on the trajectory of biological change and measured community diversity have been proposed before, particularly as a means of reconciling discordance in palaeontological and palaeoclimatic records (Ricklefs, 1987; Jackson, 1994), but also in pursuit of an explanation for floristic differences across the neotropics (Gentry, 1982).

Most efforts to understand the factors underlying changes in tropical forest diversity, structure and distribution through structural controls have arisen as a consequence of concepts delimiting *Pleistocene refugia* (e.g. Haffer, 1969; Vuilleumier, 1971; Prance, 1973, 1981). The idea of the changing Pleistocene climate altering tropical forests was an incredibly important step forward (van der Hammen, 1972, 1974) but the basis for spatially delimiting refugia across the neotropics based on zoological and botanical specimen collections has proved unconvincing, particularly in light of the built-in bias associated with these approaches in resolving spatial patterns (Nelson *et al.*, 1990).

Almost as a counterweight to this structural explanation, recent efforts have attempted to explain patterns through more general dynamics linking biological behaviour at population, community and metacommunity levels with biogeographic changes in tropical forest composition and distribution over contemporary timescales (e.g. Hubbell, 2001). While these efforts have and will continue to merit considerable attention, the scale of questions addressed and mechanisms proposed need to be clearly delimited. Exclusively seeking contemporary climate and/or edaphic correlations with standing tree diversity may be discounting important differences in historical timelines and structural controls. These may be constraining local community dynamics across the tropics in different ways. These structural differences can create significant non-stationarity when analysing spatial patterns of standing diversity, confounding attempts to realistically interpolate point patterns. This may arise as

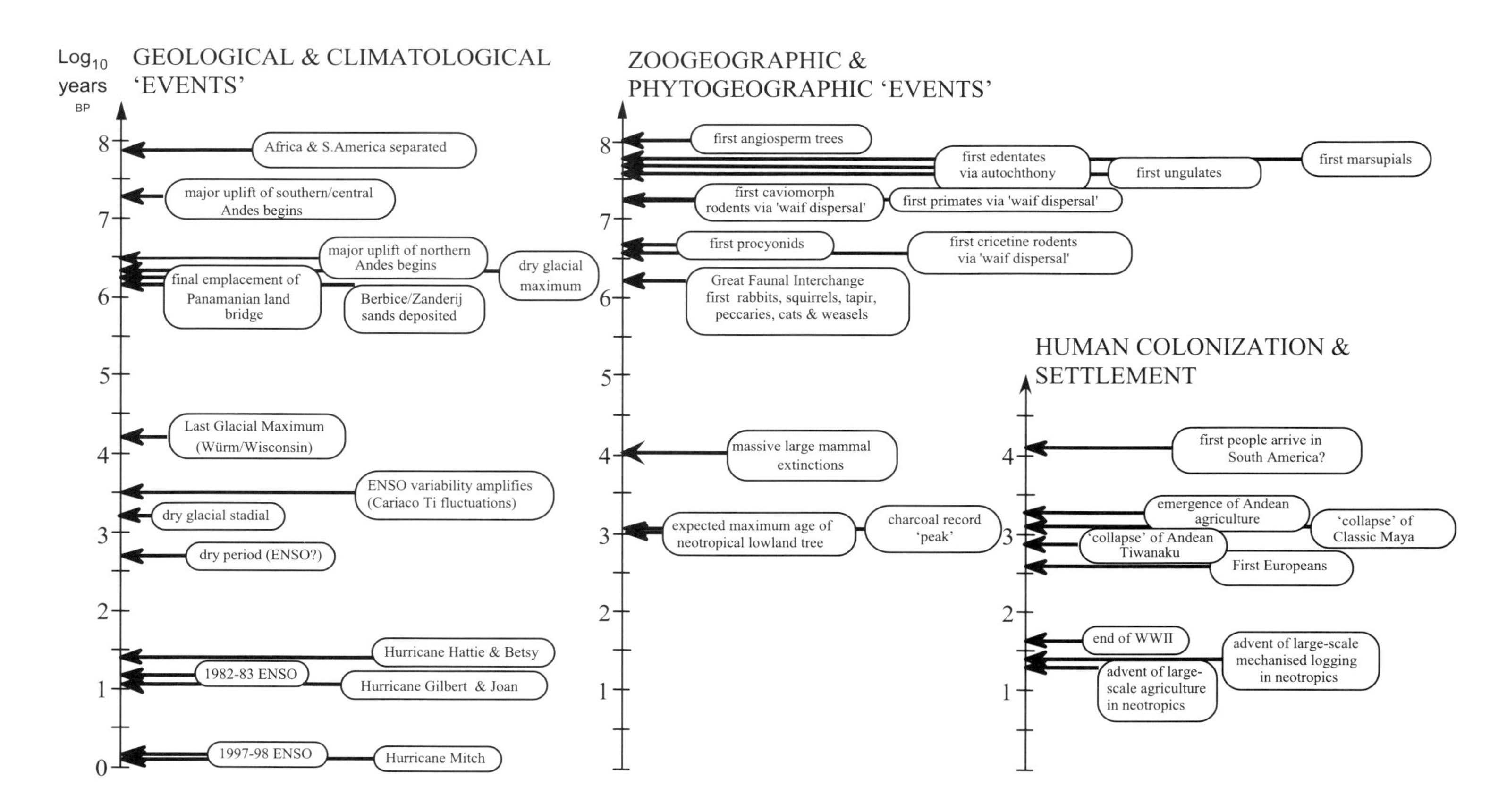

Fig. 7.1. A putative chronology of significant events shaping modern neotropical rainforests. Note the timing of the Würm dry glacial maximum, large mammal extinction and the earliest undisputed record of humans in South America at the Pleistocene–Holocene boundary (≈ 10,000–15,000 years BP).

a consequence of anisotropic variation in patch scales that implicitly underlie concepts of standing beta-diversity.

In effect, geomorphography controls the trajectory on which smaller-scale mechanisms shape forest variation. If this is true, then scaling-up continuously proximate state measures of these mechanisms over larger spatial areas will lead to diverging outcomes. These divergences should define the transition across different geomorphographic regions. The hypothetical consequences of geomorphographic variation on neotropical forest variation, and in particular its role in the Guiana Shield, are discussed later.

Spatial Segregation of Effects in the Neotropics

Visiting the Andean piedmont, the flooded Amazon varzea, the volcanic landscapes of Central America and the Guiana Shield interior conjures up distinctly different notions of tropical forests. While forests in these contrasting landscapes share many general attributes that traditionally have characterized tropical forests since the earliest descriptions of the great naturalists, the physical foundations underpinning them are rarely the same. Rock, soil and the surface waters that shape each of these show important variation across the neotropics. Granted this variation may seem insignificant when contrasting neotropical with nearctic realms. But within respective realms, the geological underpinnings continue to play an important role in defining the plant growth environment and, more specifically, how and at what rate environments are changing. The role of geology in forest formation extends well beyond the most commonly addressed question of parent material influence over soil fertility that has sequestered the view that the intervening depth of lowland soils has separated rock from playing any role in tropical forest growth. At wider scales, it is the indirect effects of juxtaposed geologies, and their inter-related ages, origins and geographic positions that constitute a discernible effect on forest patterns.

The contrast between the geomorphology of Precambrian shield areas and Phanerozoic, often Cenozoic, mountain belts and sedimentary platforms across the tropics could not be more striking and reflects efforts to distinguish these broad differences as hydrogeochemical and geological provinces (Fittkau, 1974; Putzer, 1984; Räsänen, 1993). Shield regions are senescent geological features. Remaining rock formations are those most resistant to erosion, and weathering rates are slow and driven by chemical acidity. Conversely, the Cenozoic belts defining the Andean, Central American and Caribbean volcanic arcs are dominated by young geological features. Rock formations are continuously growing through uplift or volcanism, but also easily eroded through mechanical action, releasing elevated concentrations of non-siliceous material that form hydrological and soil conditions that are only weakly acidic and of higher nutrient status (see 'Soils', Chapter 2). The interlink between rock age, geomorphology and acidity defines an important axis of variation for neotropical forests.

Landscapes with contrasting geomorphologies are also influenced by their geographic distribution. Within the tropical belt, geographic location links a forest landscape with the oceanic and atmospheric drivers that regulate rainfall and its variation within and across years (see 'Climate', Chapter 2). Over much longer periods, geographic position defines the periodicity, extent and duration of rainfall decline that can precipitate large-scale contraction of tropical forest cover. Latitudinal distance from the long-term average position of the inter-tropical convergence zone (ITCZ), or meteorological equator, is relevant, but a position along the eastern or western rim of an oceanic basin is also important. Western rim regions currently receive considerable rainfall throughout the year, but suffer anomalous rainfall decline, often precipitously, when cross-basin oceanic and atmospheric circulatory features weaken or shift latitudes. In contrast, eastern rim regions are characterized by a regime of low rainfall, punctuated by periods of tremendous precipitation. The precipitation response to this oscillatory feature, known as the El Niño Southern Oscillation, or ENSO, is strongly attached to geographic position (see 'Climate', Chapter 2).

Interestingly, the effects of geographic position on ENSO response will also influence ISR received at the forest canopy. Surface ISR is strongly correlated with cloud cover and two cloud formation mechanisms are believed to control large-scale cloud cover over neotropical forest regions in particular. The first of these, convective and wave train formations, are associated with the movement of the ITCZ. Changes in ITCZ behaviour are the most prominent feature delivering ENSO rainfall anomalies, largely as a forced response to changes in the annual distribution of sea surface temperatures and atmospheric pressure gradients that govern their movement. The second is linked to cloud deck formation along mountain slopes through orographic uplift. ISR substantively declines along slopes below cloud deck formation. The link between geographic location, rainfall variation and ISR defines another important axis of variation in the neotropics.

In a modern context, the relative impact of *geomorphic control* in the neotropics is spatially segregated along passive and active plate margins (Fig. 7.2A), while *geographic control* is principally linked to juxtaposition of forest and oceanic basins (Fig. 7.2B). The intersection of the two controls delimits broad zones varying in long-term constraints on primary productivity due to relative deficiencies in light, moisture and nutrient availability. Across the eastern Guiana Shield, permanent nutrient deficiencies are exacerbated by fluctuation in ENSO-modulated rainfall, high levels of both ISR and OLR and hydrological isolation from the Andes. In comparison, the western Amazon of southern Colombia, Ecuador and northern Peru experience localized, short-term nutrient deficiencies, consistently high moisture availability and relatively low ISR and OLR levels (see 'Soils' and 'Climate', Chapter 2). Southwestern Amazon, Central America and some parts of the Caribbean are generally characterized by long-term renewal of nutrient availability from relatively young lithologies actively exposed to mechanical weathering, long-term instability of rainfall either driven by ENSO (Panama, some parts of Caribbean), easterly wave dynamics (that influence large hurricane formations) and/or more permanent adjustments to ITCZ ranging (southwestern Amazon) and intermediate levels of ISR and OLR (see Chapter 2).

Ramping and Dampening Ecosystems – Hypothetical Considerations

Geomorphographic control exerts a profound influence on regional differences in forest change trajectories for the simple reason that it embraces the principal mechanisms through which external energy flows into the system. Flat-line constancy of internal planetary and solar processes and the astronomical relationships between them would eliminate all discernible variation, but does not represent a plausible state condition capable of supporting life (see 'Longer-term climate forcing factors', Chapter 2). Variation in the conformation of these influences, however, is a plausible and potent driver of change rates that are also spatially heterogeneous. This broad classification clearly requires further explanation.

Geological age in the tropics, holding climate constant, is strongly correlated with a host of physical attributes, including topographic diversity, rock mineralogy, weathering rates, soil nutrient and metal toxicity status, soil depth and physicochemical characteristics of draining waterways (see Chapter 2).

Climate variation in the tropics is strongly correlated with geographic position through both a latitudinal and longitudinal component. Latitude effects are linked to the ranging of the ITCZ as part of the global circulation and how this is affected by changes in external forcing effects. Regions along the broadest limits of ITCZ latitudinal ranging are likely to experience more frequent fluctuations in rainfall

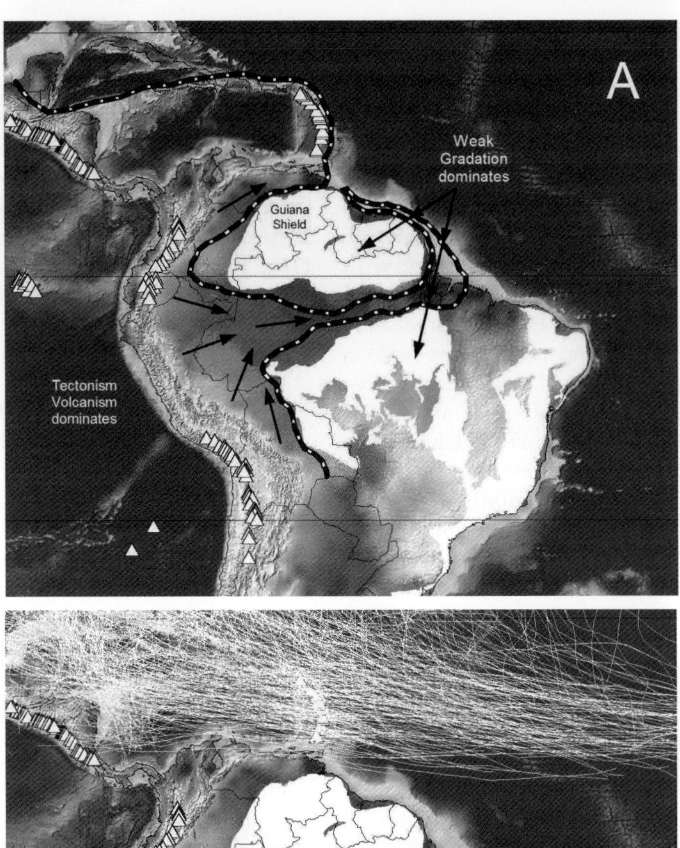

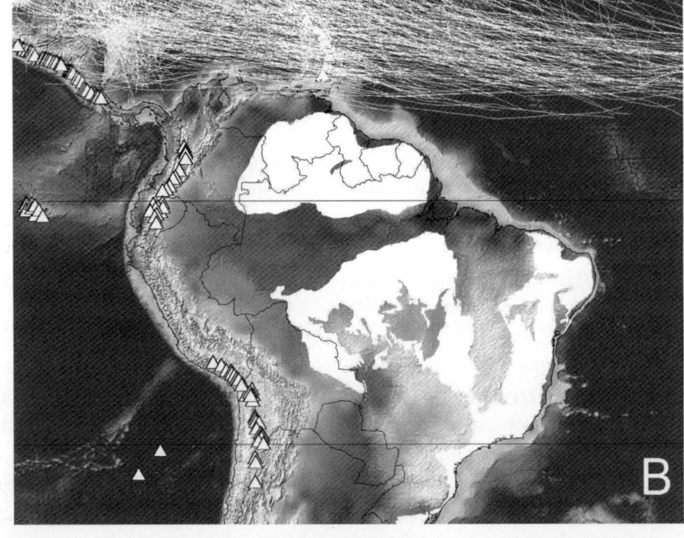

Fig. 7.2. (A) Large-scale diastrophic and gradation-dominated zones across the neotropics based on distribution of volcanic (triangles), tectonic (solid lines on ocean floor), faults (short lines on land) and Precambrian shield features. Ramping and dampening systems left and right of dividing line, respectively. (B) Overlay of hurricane storm tracks (1900–1998) in relation to position of exposed Precambrian of Guiana Shield region.

as ITCZ narrows and extends along these margins in response to more modest shifts in energy distribution. Only local maxima and minima along the most substantive phase variations in ITCZ ranging (e.g. during glacial maxima) would reach regions closer to the meteorological equator (see 'Prehistoric climates of the Guiana Shield', Chapter 2).

Longitudinal effects, however, can alter

rainfall delivery at these low latitudes while extending rainfall along the ITCZ range limits. Wave harmonics across both Pacific and Atlantic equatorial regions trigger changes in trans-basin thermoclines and shift migratory behaviour of warm sea surface temperatures. Consequently, these precipitate anomalous shifts in the oceanic ITCZ position and reduce rainfall along western basin perimeters as the ascending branch of the zonal Walker circulation retracts eastward (see Climate, Chapter 2).

Simultaneously varying the distribution of geological formations and anomalous behaviour of Walker- and Hadley-modulated rainfall over the neotropics produces a series of distinct systems, which can be discriminated by a series of biogeochemical attributes, and are coherently linked through a series of testable cause-and-effect relationships. These can be simplified by collapsing ranges into two over-arching types: ramping and dampening systems, based on age of the underlying geology and long-term precipitation changes.

Ramping systems

The eastern slopes, piedmont and geosynclinal depression of the western Amazon Basin form a high-energy, ramping system. Volcanic islands along the active tectonic margin of the Caribbean Plate and in Central America are also classified as ramping systems. Tropical forests of these regions are strongly shaped by very active tectonic and volcanic land-building processes and tremendous gradation. Diastrophic activity, however, exceeds gradation effects and topographic relief is increasing. This creates a continuous supply of new, young weathering surfaces. Rapid mechanical weathering of these surfaces, particularly in the Andean highlands, translates into elevated suspended solids and dissolved mineral concentrations in the major river systems (see Chapter 2). Soils are generally being eroded faster than they are formed along the slopes. In the subtending depression, they are being reworked and mixed vigorously through river meandering (Salo *et al.*, 1986). Steeper slopes combined with regular, heavy rainfall and high specific discharge rates (see Chapter 2) increase the likelihood of mass wasting events through landslip (Garwood *et al.*, 1979) or chronic hydrological disturbances (Gullison *et al.*, 1996). In the case of Central America and the Caribbean, tectonically active regions are also affected by catastrophic hurricane events (Fig. 7.2B). All of these broad mechanisms, inherently linked to geomorphography, play a significant role in the import and re-distribution of nutrients across the landscape.

Dampening systems

In contrast, the Guiana and Brazilian Shields are low-energy, dampening systems. Forests in these regions are shaped by a 70+ million year absence of significant tectonic or volcanic activity combined with modest gradation. Diastrophic activity is of such little effect that even the extremely low gradation rates dominate surface processes of the region and the landscape relief is slowly declining towards a peneplain. Soils are thickening faster than they are being eroded and major sedimentary depressions are dominated by quartzic sands slowly re-worked by low-energy waterways, often with courses constrained by structural features of the underlying geology (see 'Geology', Chapter 2). Mechanical weathering through hydrological disturbances and landslides is relatively rare, as reflected in the extremely low suspended solid and nutrient concentrations of the region's endemic waterways (see 'River, Lake and Tidal Systems', Chapter 2). Hurricanes over the last century at least have consistently tracked northwards of the Guiana Shield (Fig. 7.2B) in line with geographic control over the major rainfall-delivering mechanisms (see 'Climate', Chapter 2). In fact, forests of the shield regions are rarely affected by any geological or climatic phenomena that can regularly damage forests at the stand level. Only

ENSO-driven fire and modern human activity appear as plausible stand-level catastrophes affecting forests of the Guiana Shield (see below).

Fire

Fire may be the only significant forcing factor, apart from modern human industry, capable of delivering sweeping change to forests of the Guiana Shield region. Periodic fires across many forest regions of the Guiana Shield have been well-documented since the mid-18th century (McTurk, 1882; Hohenkerk, 1922; Oliphant, 1938; Hughes, 1946; Fanshawe, 1954; Schulz, 1960; Vink, 1970; Bubberman, 1973; Saldarriaga, 1994; Hammond and ter Steege, 1998). Foresters recognized early on the threat these posed to the commercial, mainly timber and balata, value of forests in the region (Wood, 1926; British Guiana Forest Department, 1935) and fire-abatement strategies formed a serious part of early efforts to manage forests for timber production.

Early foresters also noticed that fire events in closed-canopy forests of the interior shield region were largely restricted to 'drought' years. We now know that these 'drought' years are in fact severe El Niño phases of the Southern Oscillation and that tropical forests on western rims of the equatorial Atlantic and Pacific are susceptible to anomalous collapse of precipitation, particularly during the normal seasonal dry periods (see 'Climate', Chapter 2). Virtually every year documented to coincide with forest fire events in Guyana and Suriname was a strong to severe warm phase of ENSO (Fig. 7.3). More recent satellite monitoring of these events suggests that other regions of Venezuela and Roraima state in Brazil are also highly susceptible to fire during these periods (Table 7.1).

The charcoal evidence also indicates a long history of prehistoric fires affecting almost every forest area at some stage over the Holocene (see 'Prehistoric climates of

Table 7.1. The estimated number of pixels associated with fire events from June 1997 to June 2003 across the Guiana Shield and selected parts of the wider neotropics. Pixel counts were obtained via NOAA's AVHRR-3 sensor. Data source: http://www.cptec.inpe.br, except Guianas estimated by author based on AVHRR saturated pixel counts.

Area of coverage		Fire pixels (AVHRR-3)		
		All areas	Closed forest	% in forest
Venezuela		49,086	147	0.3
	Bolivar	7,606	111	1.5
	Amazonas	998	36	3.6
	Delta Amacuro	917	–	0.0
Guyana		450	36	8.0
Suriname		230	35	15.2
French Guiana		12	1	8.3
Brazil		752,891	131,199	17.4
	Amapa	3,097	1,104	35.6
	Roraima	9,735	3,236	33.2
	Para	74,714	74,563	99.8
	Para in GS	536	536	100.0
	Amazonas	6,190	4,154	67.1
	Amazonas in GS	1,460	950	65.1
	Maranhao	32,864	32,864	100.0
Peru		5,031	6	0.1
Bolivia		80,560	260	0.3
Total Guiana Shield		25,041	6,045	24.1

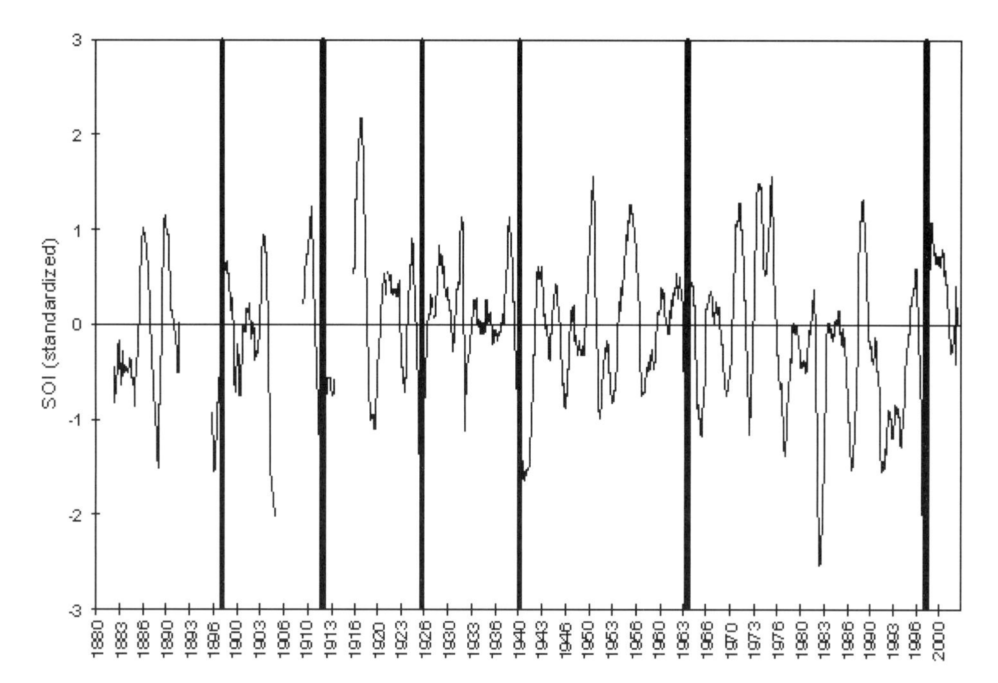

Fig. 7.3. Major historical forest fire years (solid vertical lines) in Guyana and Suriname and their association with 12-month moving averages of standardized SOI scores (see Fig. 2.19). See text for references used to identify fire years.

the Guiana Shield', Chapter 2). The extent, magnitude and ignition source of these events, however, remain poorly resolved. On the one hand, ubiquitous soil charcoal could result from repeated ground fires of limited spatial extent. On the other, they could be associated with stand replacement events that swept through vast areas of forest cover, but only at multi-millennial scales. In all likelihood, a regime of high-frequency, spatially limited events compounded by low-frequency massive change appears the most plausible course, given the interaction between seasonal variation in rainfall and inter-annual deepening of these into severe drought phases during warm ENSO phases. Extension of ENSO drought by weeks or months (Meggers, 1994), combined with spatially delimited human activity, would make some forest regions more susceptible to large-scale stand replacement events than others.

While the occurrence of fire in tropical forests of the Guiana Shield is irrefutable, the role of fire in preventing or facilitating stand mono-dominance is not as clear cut. Some large-seeded dominants, such as *Attalea speciosa*[1] and its frequent co-dominant, *Lecythis lurida*, in the Babassu forests of southeastern Amazon, clearly profit from occasional fire (Anderson *et al.*, 1991). Many canopy palms of the *terra firme* forest also respond vigorously to fire events while their dicotyledonous sympatrics suffer. Like these, many of the large-seeded dominants in the Guiana Shield exhibit classic fire-response attributes (after Kauffman and Uhl, 1990), such as epicormic sprouting, thick bark and coppicing (Table 7.2). However, others, such as *Chlorocardium rodiei*, show few of these traits (e.g. coppicing) and are incapable of tolerating even modest fire events either as an adult or seed (D. Hammond, personal observation). At the same time, other large-seeded associates, such as *Mora gonggrijpii*, have been observed to respond well to fires where they ultimately assume dominance after neighbouring greenheart (*Chlorocardium rodiei*) populations are killed (Wood, 1926).

Table 7.2. Examples of dominant species of the Guiana Shield with fire-type adaptations.

Adaptation to fire	Species
Thick bark	*Manilkara bidentata*
	Lecythis zabacujo
	Cedrelinga caeteniformis
	Tabebuia insignis
	Dimorphandra conjugata
Epicormic sprouting	*Dimorphandra conjugata*
Coppicing	*Dicymbe* spp.
	Mora gonggrijpii
	Pentaclethra macroloba
	Attalea regia
	Astrocaryum spp.
Root suckering	*Dimorphandra* spp.
	Guadua spp.

Remote sensing of modern-day fires also shows considerable spatial variation in both intensity and frequency of occurrence. In Brazil, South Pará alone accounted for more than 50% of all AVHRR-saturated pixels attributed to forest burning between 1997 and 2002 (CPTEC, 2003). Forest burning in the Guiana Shield contributed only 2–3% of the total fire pixels registered for Brazil and Venezuela. Fire across the three Guianas, even during peak ENSO events in early 1998 and 2002, remained an order of magnitude lower than those spread across the southeastern Amazon basin. Yet, relative to closed-canopy forests of the western Amazon in Peru and Bolivia, modern burning across the Guiana Shield is elevated (Table 7.1).

The growth of South Pará as a continental centre of forest fire is not surprising given the geographic location of the region in relation to Walker and Hadley circulatory behaviour and the very different socio-economic and demographic changes taking place in the region in comparison to the Guiana Shield. Nonetheless, the role of fire in forest systems of the shield region clearly represents one of very few forcing factors that catalyse change in an otherwise dampening system. Ironically, while active disturbance may invigorate the transfer of weathered mineral nutrients across many ramping systems, fire events only appear to further the pace of dampening by acting as a major export route for preciously scarce nutrients from a relatively closed and isolated regional nutrient cycle (see below).

Nutrient Balance and Migration

Fire plays such an important role not simply because it can destroy or degrade forest. Rather, it is the disruption of the critical, internal nutrient recycling process that, from all available evidence, is disproportionately important to forests in the Guiana Shield. The region has very few autochthonous sources of nutrients compared to ramping system regions across the neotropics, and this makes it particularly vulnerable. Long-standing acidity, relatively modest topographic relief over much of the region and hydrogeochemical isolation from the younger, steeper ranges that typify much of western South America, Central America and the Caribbean, have left most nutrients critical for plant productivity constrained to the biomass itself or locked up as inaccessible stores (e.g. soil phosphorus under low pH conditions) (see Chapter 2). Most available nutrients have been shown to be rapidly recycled at the soil surface (Jordan and Herrera, 1981). In fact, the classic perception of tropical forests as 'closed' nutrient systems was largely influenced by work done within some of the poorest and wettest forests in the Guiana Shield (Richards, 1952; Herrera *et al.*, 1978; Jordan, 1987). While tropical forests are increasingly being viewed as open, or at least as leaking closed systems, it is clear that recycling dominates the mass balance of nutrients in most forest systems of the Guiana Shield. Relative to other regions, however, this appears to represent the extreme low end of a much broader spectrum defining the relative role of internal nutrient recycling to forest system maintenance.

A number of potential import pathways can contribute to the nutrient balance of tropical forests, including atmospheric deposition, weathering of rocks either at the site (autochthonous) or elsewhere (allochthonous) and biological fixation (Fig.

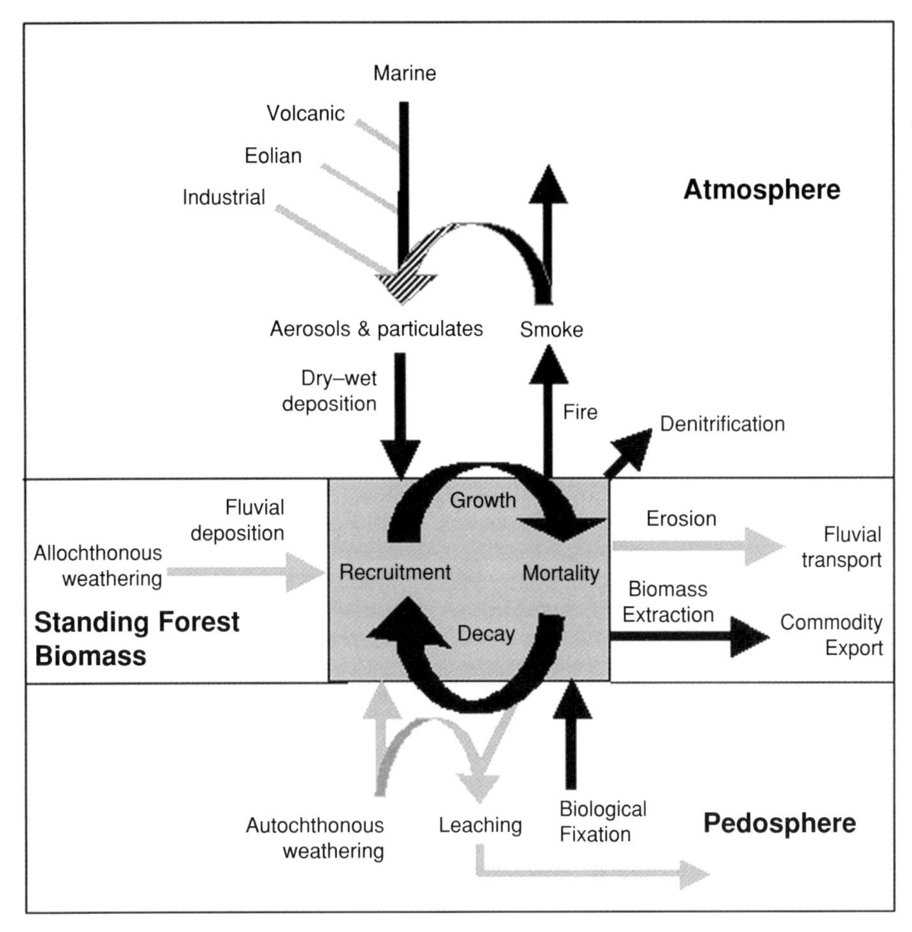

Fig. 7.4. Schematic of the major pathways (arrows) influencing the regional nutrient balance of Guiana Shield forests (grey box), emphasizing those making minimal (grey) and major (black) contributions. Striped arrow indicates an event-based contribution made through migrating fire emission particulates.

7.4). These do not act as equal pathways for all of the mineral and organic nutrients needed to sustain tropical plant growth. Rhizobial bacteria and mycorrhizal fungi solely fix di-nitrogen from the atmosphere into nitrate, while rock weathering provides little nitrogen, but acts as a major source of minerals, such as calcium, magnesium and potassium.

Atmospheric deposition

The abundance of aerosol particulates available for deposition is strongly influenced by the geographic location of sources. Several major sources include volcanoes, deserts (aeolian sediments), industrial centres and fires (Fig. 7.4).

The geographic location of the Guiana Shield places it largely southeastward of the major volcanic chains of the Caribbean, Andes and Central America (see Fig. 7.2A). The only source of aeolian particulates is North Africa, but effective transport of substantive quantities across the Atlantic is strongly linked to easterly wave phenomena that tend to concentrate in the subequatorial regions north of the Guiana Shield (Jones *et al.*, 2003).

Few industrial regions are within the Guiana Shield region and emissions from these areas are nominal (Holzinger *et al.*, 2001). The northern rim is, however, occupied by heavy industry, particularly around the Ciudad Guayana–Lago Guri area, and more industrialized areas can at least create hydrocarbon plumes that extend over remote areas of Venezuelan Guayana (Holzinger *et al.*, 2001). Generally, however, industrial impacts are relatively weak. This is well-illustrated by the very low atmospheric sulphur aerosol concentrations measured in the region (see 'Longer-term climate forcing factors', Chapter 2).

Monitoring of mineral import via atmospheric deposition suggests that this route has a nominal effect on the nutrient mass balance of forests in the Guiana Shield, except to elevate concentrations of sodium and chlorine as a consequence of ocean-sourced moisture (Haripersad-Makhanlal and Ouboter, 1993; Brouwer, 1996). At Mabura Hill, internal recycling of key plant nutrients, such as P, Ca, K and Mg, was measured at 20–85 times greater than contributions via atmospheric deposition (Brouwer, 1996). Some events, however, such as large fires, could lead to downwind migration of important nutrients within the region (Fearnside, 1990, p. 114) and act as important routes through which fire-susceptible edaphic habitats, such as *campinarana*, *caatinga*, dakama bush and muri scrub, are formed and diminish through repeated export of fossil nutrients (Anderson, 1981).

Autochthonous weathering

Nutrients made available through weathering of underlying rock surfaces can be substantial where forests are juxtaposed downstream from extensive highland areas subject to the action of snow and glacial ice on exposed rock faces. Across the Guiana Shield, few areas are located in such a way. Where upstream highlands are present, they consist largely of conglomerate sandstones or silica-rich metamorphics (see 'Geology', Chapter 2) rather than younger formations with greater calcium and magnesium concentrations. Thicker soil profiles are typically associated with these older erosion bevels, creating additional barriers to autochthonous nutrient contributions to forest growth and maintenance. In some instances, significant rooting depths combined with localized mafic lithologies (e.g. dolerite intrusive) might elevate contributions from *in situ* rock weathering of some minerals, such as magnesium (Brouwer, 1996). Generally speaking, however, this forms a nominal pathway to import most mineral nutrients in the Guiana Shield (Fig. 7.4).

Fluvial deposition (allochthonous weathering)

Modest autochthonous weathering probably characterizes most sedimentary depressions forming the bulk of the lowland tropics, but the depth and extent of overlying sediments and soils would invariably modulate the ability of forest plants to tap into mineralization zones through deep roots prior to leaching. Far more important, however, is the arrival of these nutrients via fluvial deposition from distant weathering surfaces. Based on physico-chemical attributes of draining waterways, this pathway represents an insignificant route for both import and redistribution of most nutrients across the Guiana Shield (see 'River, Lake and Tidal Systems', Chapter 2) (Fig. 7.4), particularly in comparison to other ramping system regions of the neotropics connected to highland regions undergoing significant weathering of young lithologies and tremendous site turnover through relatively high river migration rates (Kalliola *et al.*, 1999).

Biological fixation

Import of nitrogen into tropical forests occurs mainly through fixation of di-nitrogen by bacterial or mycorrhizal associates. This is believed to represent a mere 2–5% of the nitrogen available for uptake via internal recycling of organic matter, a level comparable to that deposited from atmospheric sources (Perreijn, 2002). Nonetheless, at the individual or forest stand level, it may be an important contributor to the nitrogen economy. Many tree species found in relatively high densities across parts of the Guiana Shield have been confirmed as having rhizobial root nodules, including *Chaemacrista* spp., *Dimorphandra* spp., *Inga* spp., *Pentaclethra macroloba*, *Clathrotropis* spp., *Diplotropis purpurea*, *Ormosia* spp. and *Swartzia* spp. (Norris, 1969; Raaimakers, 1995; Perreijn, 2002). Other dominants, including *Mora* spp., *Dicymbe* spp., *Eperua* spp. and *Vouacapoua* spp., *Chlorocardium rodiei*, *Catostemma fragrans* and *Eschweilera sagotiana*, have been associated with mycorrhizal, mainly arbuscular but some ectotrophic, fungi (Moyersoen, 1993; Béreau and Garbaye, 1994; Perreijn, 2002; T. Henkel, personal communication). Many of these taxa form both rhizobial and mycorrhizal associations.

Perreijn (2002) showed that both import of nitrogen via biological fixation and export via the main route of nitrogen loss from the system (denitrification) can vary considerably between forest types in the Mabura Hill area. The type of mycorrhizal associates formed, namely AM (arbuscular mycorrhizae) or ECM (ectomycorrhizae), and their association with rhizobia may also be linked to soil type (Moyersoen, 1993).

Long-term mineral emigration

The ratio of nutrient import to standing biomass in the Guiana Shield is likely to be one of the lowest in the world. With only small quantities of nutrients to be gained through atmospheric deposition, rock weathering, fluvial deposition and biological fixation, conservation through adept internal cycling appears the only plausible mechanism sustaining standing forest biomass across most parts of the Guiana Shield. The importance of this internal cycling is likely to be greater in these forests than those on sites with greater exposure to autochthonous and/or allochthonous nutrient sources (Cuevas, 2001) as a consequence of geomorphography.

The fragility of the nutrient cycle depends on its ability to recover after major losses. In the case of systems acutely dependent on internal recycling, catastrophic biomass loss will have a substantively greater impact on forest recovery times. Responses to these events across forested regions of the Guiana Shield are thus invariably more acute than in adjacent ramping system regions, where recycling efficiency is often offset by relatively active nutrient importation. Even putative small and brief transitions associated with shifting cultivation are unlikely to recover to pre-clearance biomass levels within a century in Venezuelan Amazonas (e.g. Saldarriaga, 1987). Repeated loss of all or most of the standing biomass over a relatively brief period of time will eventually lead to degraded woody plant communities and grass-dominated savannas as a natural consequence of a system moving along a dampening phase of the geomorphic cycle.

A long-term process of mineral emigration from the Guiana Shield to adjacent regions is taking place. Large or frequent fires extensively disrupt local nutrient recycling in the region and significant nutrient migration occurs eastward through river leachates and westward through wind-driven smoke as a consequence. Since the 18th century, extraction of wood and non-wood products and subsequent export from forests of the region have also contributed to significant nutrient emigration, particularly of mineral nutrients with few import sources, such as magnesium and calcium (see Chapter 9). As a consequence, periodic large-scale losses of sparse nutrients may have catalysed the dampening of ecosystem productivity across the Guiana Shield and selected for species capable of achieving large growth gains using very small quantities of nutrients (see Chapter 3, this volume; Raaimakers, 1995).

Functional Consequences of Geomorphographic Control

The way in which geomorphography describes a basket of inter-related system attributes suggests a number of functional consequences linking productivity and diversity of forest systems to underlying geology and geographic position.

Productivity

The consequences of energy state to forest change is non-trivial. Long-term site stability and individual tree longevity are expected to be, on average, relatively low and stand turnover relatively high in ramping systems. Yet, net ecosystem productivity (NEP) is also expected to be relatively higher than in dampening systems. This may in part reflect the high import–export fluxes that characterize ramping systems over the long term. In effect, they are open systems that continuously receive and lose material, primarily through hydrological, but also atmospheric, transport. In contrast, dampening systems conform to the conventional view of tropical forests as closed systems that rely principally on material recycling and nutrient conservation in order to maintain productivity. Spatial anisotropies in nutrient flux effects driven by geomorphographic controls across the neotropics complicate a picture of tropical forests generally characterized at smaller spatial scales by a high rainfall–low nutrient–high productivity relationship (Huston, 1994). Again, import flux and recycling effects co-occur, but fluxes should play a more important role in the long-term maintenance of higher productivity levels that have been shown to characterize ramping system regions. In contrast, dampening systems such as the Guiana Shield are in a state of chronic nutrient decline as export through leaching, fire and, more recently, human appropriation, far exceed import from weathering and atmospheric deposition (see below). In fact, major lateral redistribution of nutrients in the Guiana Shield may arise principally through low-frequency fire and at smaller continuous scales, through plant–animal relationships.

Scale of edaphic effects

Meaningful fractionation of soils in the western Amazon Basin based on their chemical and physical attributes has often proven difficult over large areas of *terra firme* forest (e.g. Duivenvoorden and Lips, 1995). Geomorphographic controls would suggest differences in edaphic patch size distribution among regions. In ramping regions, the effects of high-energy hydrological systems reworking thick sediments over a relatively recent geological timescale should work to decrease edaphic patch size over the long term. Conversely, the autochthonous processes dominating edaphic change in dampening systems exposed to a much longer history of geological change would lead to maintenance of patch sizes governed by geology rather than hydrology. At smaller scales, the most drastic edaphic transitions in both regions are, however, governed primarily by water table dynamics and position along the catena rather than nutrient availability (ter Steege, 1993; ter Steege and Hammond, 2001). This smaller-scale hydrological, rather than larger-scale nutrient, gradient best discriminated differential mortality responses to ENSO drought at a 50 ha scale on Barro Colorado Island (Condit *et al.*, 1996).

Disturbance and turnover

High NEP requires spare productive capacity (SPC). The route to high SPC depends on the long-term stability of the system. Where abiotic conditions are not limiting, varying frequency and amplitude of disturbance (after Huston, 1994) will trigger concomitant wave-like shifts in SPC. During interstitial phases characterized by modest (low amplitude) disturbance, SPC will approach zero and variation in NEP will become strongly attached to the rate of turnover in standing biomass and how this varies in proportion to microbial activity. Hence, regions with high NEP could be: (i) in a non-equilibrium state, moving along a larger-scale response trajectory initiated by a disturbance phase or event of proportional amplitude that inflated SPC; or (ii) a dynamic equilibrium state, responding to a high frequency–low amplitude regime of disturbance that creates continuous, low-level SPC, primarily through small-scale stand turnover. Neotropical forests are likely a composite of both, but forest areas following a non-equilibrium trajectory at any given time are hypothesized to account for a relatively larger proportion of ramping regions in comparison to those classified as dampening systems.

Attribute selection

If stand turnover rates are relatively high, this would place a premium on trees that function as *r*, rather than *K*, strategists. Short-lived species with higher growth and dispersal rates and early reproductive maturity are known to benefit under a high turnover environment. In tropical forests, where the average tree longevities are considerably greater, the principal connection between life-history and population attributes has to be viewed at longer timescales. One proximate measure of longer-term compositional responses to differences in forest change regimes is the relative frequency, and thus success, conferred upon taxa by their life-history attributes.

Small seed size in tropical forests is generally associated with greater dispersal distances after accounting for plant size effects. This is achieved largely as a trade-off against early low-light survivorship nearer to parent trees. Other selective forces that enhance survivorship (e.g. ingestion, burial, chemical defence) can also exert pressure on seed size. But the main effect rests with the probability of achieving the critical level of cohort survivorship needed to stabilize or expand a species population and whether this is greater: (i) when encountering elevated light levels or escaping distance-dependent predation through enhanced dispersal; or (ii) through longer post-germination residence times in low-light environments nearer to parent trees. In this general case, elevated turnover should confer a comparative advantage to small-seeded individuals over the long term, as highly clumped, large-seeded populations are more susceptible to extinction in the face of externally driven disturbance. This outcome would be consistent with a system driven at larger scales by density-independent effects (after Huston, 1994), such as catastrophic geological or climatic events. These clearly figure more prominently in the history of ramping systems, such as western Amazonia and Central America in comparison to the Guiana Shield.

Phylogenetic inertia

If higher forest turnover rates result in asymmetric survivorship of smaller seeds, then contribution of novel alleles among small-seeded cohorts will over time exceed those of larger seeded taxa that are more susceptible to local extinction and declining genetic diversity as a consequence of limited dispersal. Consequently, it is plausible that smaller-seeded taxa will diversify at rates that outstrip extinction and account for an increasing fraction of the standing diversity in ramping as opposed to dampening systems. Taxa that benefit from or depend upon chronic, widespread disturbance should generally occur more frequently in ramping system environments because opportunities for population persistence are greater.

Phylogeographic responses

Ramping systems are more recent landscapes borne along active tectonic plate margins. Prior to uplifting, many of these were dominated by shallow marine or epicontinental environments and offered relatively little, if any, area for terrestrial species to accumulate and evolve. Angiosperm evolution clearly pre-dates commencement of the Andean orogeny (about 20 Ma BP) and uplift of the

Panamanian land-bridge (14–18 Ma BP) by as much as 120 million years and considerable diversification is believed to have already led to the presence of most common neotropical plant families by the Eocene, approximately 20 Ma prior to the diastrophic upheaval that would reshape the region (Friis *et al.*, 1987; Romero, 1993). The recent shaping of ramping system regions intuitively suggests that these would support fewer archaic lineages and that their forests should be dominated by derived, rather than ancestral, clades, unless alternating periods of regional forest extinction have led to significant phylogeographic reversals.

Biological Consequences of Life in Ramping and Dampening Systems

Seed size

Few comparisons of community seed size distributions between neotropical sites have been made, the early work of Foster and Janson (1985) being a leading exception. General comparisons made by Hammond and Brown (1995) using seed size sets from Foster and Janson's western Amazon site (Manu, Peru), Foster's Panamanian site (Barro Colorado Island – BCI) (Foster, 1982) and their own site in Guyana (Mabura Hill) illustrated the high relative occurrence of large-seeded species in central Guyana relative to the other two sites. Based on forest compositional and seed size data, large-seededness appears to be a typical trait of forests across Venezuelan Guayana, Guyana, Suriname and French Guiana. Ter Steege and Hammond (2001) later showed that seed size weighted by relative stem density in clustered 0.1 ha plots enumerated across Guyana declined southward from a peak near the central part of the country. This cline was broadly associated with an increase in tree alpha-diversity and certain dispersal syndromes. Seed size, while commonly implicated as one of many factors regulating plant–animal relationships, had rarely been implicated directly as a factor influencing patterns of standing plant diversity in the tropics. This is despite the fact that the consequences of seed size to individual survivorship and fitness have been broadly demonstrated for quite some time (Harper *et al.*, 1970; Baker, 1972; Salisbury, 1974), albeit only more recently in relation to long-lived tropical woody plants (Foster, 1986). This paucity of attention combined with the noticeable absence of mechanistic explanations for described geographic variation in woody plant diversity within the tropical belt underscores the potential significance of seed size as one important determinant of long-term success of long-lived woody plants under varying macro-environmental conditions.

Dispersal, survivorship and growth plasticity

One logical role of seed size rests with two processes that it clearly regulates, dispersal distance and early survivorship. Larger seeds generally confer an early survivorship advantage in limiting environments. This advantage has been shown to be considerable for species with super-sized seeds (see Chapter 3) (e.g. Foster, 1986; Boot, 1996; Dalling *et al.*, 1997; Zagt, 1997; Hammond and Brown, 1998; Hammond *et al.*, 1999; Rose, 2000). These species typically fall within the upper 90th percentile of a community's seed-size distribution (Hammond and Brown, 1995). Tree species with super-sized seeds also appear to suffer disproportionately low mortality during early recruitment stages (i.e. smaller size classes). *Prioria copaifera* (Fig. 26.4 in Hubbell and Foster, 1990), *Chlorocardium rodiei* and *Dicymbe altsonii* (Zagt, 1997, pp. 188, 197) are good examples.[2]

Very high per capita survivorship of large-seeded tropical species also appears to exact a trade-off with growth-rate plasticity during early recruitment phases (see Chapter 3) (Boot, 1996; Rose, 2000), which is consistent with much broader trends between large-seededness and relative growth rate (via specific leaf area) (Lambers and Poorter, 1992; Westoby *et al.*, 1997). Smaller-seeded tropical tree species, on average, express more plastic growth responses to increases in resource availability, but fail to maintain positive carbon balances under scarce resource conditions that are typically tolerated by many larger-seeded tree species. These distinctions, however, appear largely restricted to canopy trees. Many understorey specialists appear capable of tolerating low light levels but also have much smaller seed sizes. Adult plant size in this instance consistently explains a significant part of the difference across a wide range of taxa and habitats (Hammond and Brown, 1995; Leishman *et al.*, 1995), underlining the importance of segregating growth strategies against growth objectives.

Generally speaking, unparalleled survivorship of large-seeded juveniles comes at the cost of wider dispersability and lower achievable growth rates. Large-seeded species, in the absence of targeted movement by humans or water, are lousy travellers and the expected outcome of this sluggish state is a hyper-clumped distribution of reproductive adults borne from very low per capita mortality rates. In fact, most tropical trees show a clumped distribution (Hubbell, 1979), but most *very large*-seeded tree populations are distributed this way principally through dispersal limitation, rather than the combination of dispersal and suitable site availability, as is the case with smaller-seeded pioneers or narrow edaphic specialists. The scale and ubiquity of large-seeded hyper-clumping across the Guiana Shield is reflected in the very high relative abundances of these species recorded through small-scale plot sampling (e.g. Davis and Richards, 1933; Fanshawe, 1952; Schulz, 1960; Ogden, 1966; Maas, 1971; Bariteau, 1992; ter Steege *et al.*, 1993; Comiskey *et al.*, 1994; Coomes and Grubb, 1996; Dezzeo and Briceño, 1997; Ek, 1997; Salas *et al.*, 1997; ter Steege, 2000; van Andel, 2000) and the transition from one oligarchy to another seen through forest inventories carried out at larger landscape scales across Venezuela, Guyana, Suriname and French Guiana (Hughes, 1946; Rollet, 1969; de Milde and de Groot, 1970; Welch and Bell, 1971; de Milde and Inglis, 1974; ter Steege, 2000).

Seed size vs. stand evenness

Terra firme trees with large seeds may dominate most stands across the Guiana Shield, but this phenomenon dissipates somewhat in the central Amazon, declining further at sites located along the eastern Andean piedmont and foredeep and in many parts of wet tropical Central America, such as BCI, Panama. In these regions, evenness is typically greater and, on average, characterized by species with smaller seeds. Not all sites within the Guiana Shield are necessarily composed of large numbers of large-seeded species (Maas, 1971; Thompson *et al.*, 1992), nor have all sampled areas outside the shield region proven substantially more even in the abundance of canopy species (e.g. *Pentaclethra* dominance at La Selva – Hartshorn and Hammel, 1994). Where these occur, average seed size continues to shift with changing evenness. In fact, it is difficult to find a measured neotropical forest that is dominated by one or several small-seeded species to the same extent as that found associated with super-sized species.

Together, these plots show a consistent relationship between the number of species needed to account for half of the total plot stem count and the average of these species' seed size weighted by their respective relative abundances in the plot. A power function links the two (Fig. 7.5).[3] This relationship, if proven stable over a wider range of plot data, would suggest a power law describing the effects of seed size on species relative success, as measured by its relative abundance.

Fitted data suggest that a seed size-dominance effect is not perfectly segregated by geomorphographic region. But even so, why would some regions, such as the Guiana Shield, tend to have a larger portion of their forest cover, which tends to be characterized by lower alpha diversity (ter Steege *et al.*, 2000), dominated by relatively large-seeded species? And why would there necessarily be a link between seed size and the degree of local success certain tree taxa achieved relative to other sympatrics? To get at the heart of this role, we have to go

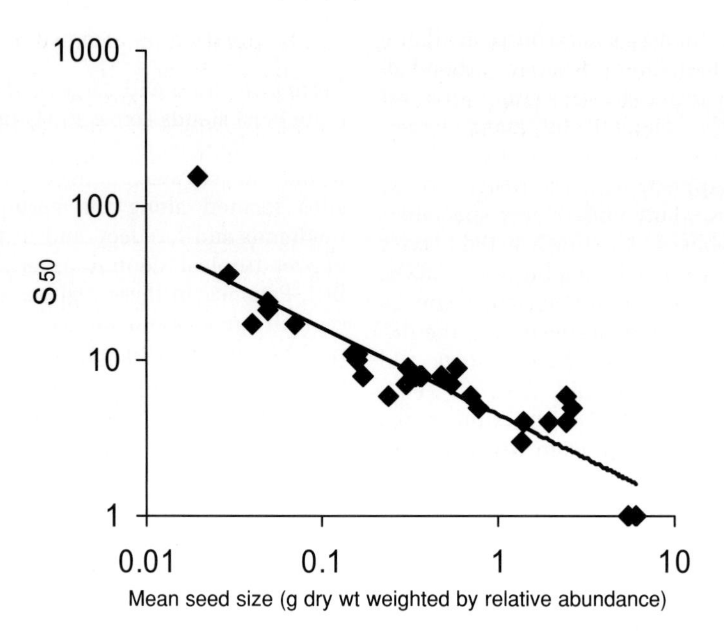

Fig. 7.5. Empirical relationship between the number of species accounting for half of all stems >10 cm dbh recorded in plots (S_{50}) and the weighted average seed mass of those species. Weighting based on relative abundance of stems. Relative abundance data: Davis and Richards (1933); Maas (1971); Comiskey *et al.* (1994); Hartshorn and Hammel (1994); Valencia *et al.* (1994); Dezzeo and Briceño (1997); Ek (1997); Ferreira and Rankin-de-Merona (1998); van Andel (2000). Seed mass data: Foster (1982); Foster and Janson (1985); Hammond and Brown (1995); D. Hammond (unpublished data). Seed mass was assigned based on volume-mass or intra-generic relationships where direct seed mass values were not available. Mass was assigned as minimum of quarter logarithmic intervals to reduce error effects.

back to basic concepts of population growth and the consequences of changes in reproductive output, early survivorship rates and their relationship with landscape-scale environmental stability.

When is it good to be big?

Generally, large-seeded offspring travel relatively short distances, survive more frequently and longer, but also grow more slowly. They generally tolerate biological attack better (Janzen, 1969; Dalling *et al.*, 1997; Hammond and Brown, 1998), but are also more frequently attacked, sometimes as a consequence of their poor dispersal and the effects of this as part of distance and density-dependent relationships with their main predators (Janzen, 1969; Hammond and Brown, 1998).

Environmental stability at small, ecological scales in wet tropical forestlands is generally characterized by relatively low rates of canopy turnover and long-term edaphic continuity. Forest gap dynamics are well explored as the leading mechanism responsible for driving local population recruitment and species diversity (e.g. Grubb, 1977; Orians, 1982) and have been variously implicated in a wide range of ecological phenomena believed to impart selective (short-term) advantages to species with contrasting recruitment and life-history attributes (e.g. Denslow, 1980; Schupp *et al.*, 1989).

But early perceptions of disturbance in ecology were in many ways driven by the relatively small range and scale of tropical field study. This was despite the well-documented record of cyclonic effects in the Caribbean and Australasian domains (Webb, 1958; Wadsworth, 1959; Whitmore, 1974) and the increasing assortment of evidence supporting a more prominent role for larger-scale disturbance mechanisms, including prehistoric human occupation (see Chapter 8), in characterizing the extent of environmental stability across some lowland forests (Garwood *et al.*, 1979; Saldarriaga and West, 1986; Salo *et al.*, 1986; Nelson *et al.*, 1994).

Large-seeded species disperse poorly, and in environments susceptible to these larger-scale disturbances they are more susceptible to extinction. Other adaptations (e.g. greater adult resistance to drought, fire, wind and pestilence) may offset some of this risk. But poor dispersal and the consequent increase in adult aggregation mean that, over time, compacted meta-populations of large-seeded taxa become more prone to changes in environmental stability across their range with little capacity to recolonize widely or quickly. High offspring survivorship around remnant adult clusters and/or vegetative regrowth can maintain the presence of these species, but the slow progression to dominance is in effect reset. Relatively high-frequency and/or large-scale disturbances can permanently reduce the representation of these taxa as fewer and fewer clusters remain and more agile taxa occupy canopy space.

Under high-frequency, but relatively small-scale disturbance regimes, large-seeded taxa will consistently be out-competed by smaller-seeded, more mobile species in the high light conditions generally attached to higher disturbance rates. Alternatively, low-frequency, but large-scale events precipitate the likelihood of catastrophic population loss due to hyper-aggregation of adults and a limited capacity to recolonize large disturbance patches. More reproductive adults, and thus reproductive output, are lost per unit area as aggregation increases under these circumstances. Ultimately, in this scenario large-seeded populations will dwindle in relative abundance until disturbance effects dissipate. During these relatively stable periods, population aggregates of large-seeded isolates can reconstitute and eventually begin to expand as a consequence of their high per capita survivorship and the advantage this brings during relatively stable environmental phases. In-breeding depression, low heterozygosity and high inter-patch variability should be detected in these isolates as sibling outcrossing rates bottleneck gene flow relative to more broadly dispersed, but locally rare, species.

Hence, if relatively stable environments persist in some regions much longer than others, the relative abundance of large-seeded upland taxa should increase in small sample plots as the aggregates expand across the full, tolerated edaphic continuum. In some instances, adaptations to juxtaposed edaphic conditions may create further opportunities for expansion of large-seeded dominance (e.g. *Mora* spp.; ter Steege, 1993) when the distinction between these facies remains relatively stable as a consequence of slow weathering rates and infrequent allochthonous influence on local substrate development. In regions experiencing more extensive and frequent substrate turnover due to mass wasting through landslips, erosion–deposition dynamics of river movement and flash-flooding events, opportunities for large-seeded, dicotyledonous trees to maintain and expand their ecological hegemony over forest communities would be curtailed and more broadly distributed, smaller-seeded species are predicted to increase in relative abundance as the comparative advantage of faster growth and greater mobility supersede those of high juvenile survival rates.

Large-seeded hegemonies of the Guiana Shield

Thus, large-seeded taxa can only proliferate and achieve relatively high local abundances over larger areas under relatively stable forest environments. While they exhibit elevated per capita survivorship under resource-limited environments relative to small-seeded species with comparable life-history objectives (canopy position, sexual reproduction), this advantage is lost when resources become less limiting and large-scale site stability decreases over longer timescales.

The predominant scale, or scalar range, of events affecting environmental stability contrasts considerably based on varying geomorphographic control effects. If these large-scale, regional differences play a significant role in determining the success of large-seeded species, then relative abundance and species turnover should be high in ramping relative to dampening systems. Across the Guiana Shield, turnover (or beta diversity) can be quite high across a landscape largely driven by long-standing autochthonous soil formation processes atop Proterozoic rock remnants. Fanshawe's (1952) system of forest associations–fasciations for Guyana intuitively mirrors this transition between hyperclumped distributions of edaphically segregated, and dispersal-limited, large-seeded species. Across the western Amazon lowlands, relatively recent allochthonous soil formation atop Quaternary sedimentary cover dominates. Here, the environment is changing more rapidly and relative residence time for substrate in any given area is, on average, lower (see 'River, Lake and Tidal Systems', Chapter 2). The patchiness that evolves in these regions is less pronounced than that seen on much older landscapes. The relatively low species turnover inferred through comparisons of plots across the western lowland landscape would be consistent with this longer-term influence (Condit *et al.*, 2002). The association between dominance and large-seededness and the relative frequency of these locations in the Guiana Shield would support a view of regional contrasts in site stability. If the incorporation of further data upholds the relationship and the most common species from plots in ramping and dampening regions cluster along the log and lag phases, respectively, this would suggest that species geographic ranges increase with higher regional site turnover rates.

Wood density

Another functional attribute that holds prospect as a surrogate for long-term selection under varying levels of environmental stability is wood density. Wood density describes a weight–volume relationship of lignin–cellulose under varying moisture content. Standardizing density measurements at a constant moisture content (12% is commonly employed) improves the comparison between individuals, species and across regions with varying moisture regimes. Growth in trees largely occurs through accrual of lignin–cellulose, so a loose relationship between measured average annual growth rates and wood densities (Brown and Lugo, 1990; ter Steege and Hammond, 2001) is not entirely unexpected. Arguably, the more pivotal question in regard to tropical forest spread and persistence rests in identifying the main pressures affecting wood density variation and, like seed size, its rise or decline at community scales.

Tropical hardwoods exhibit a broader range of wood densities than any other forest biome. Both the lightest and densest woods measured are known from tropical forest species (Williamson, 1984). Within this range, tree species in the Guiana Shield achieve, on average, the highest wood densities in the world (Fig. 7.6). The region may ultimately prove to be one of several areas with high spatial averages for wood density, but timber trees from the eastern Brazilian Amazon have been noted as having elevated wood densities relative to other regions (Whitmore and da Silva, 1990). This is not to say that species of low density do not occur, but rather heavy-hardwood taxa, and the relative abundance of these taxa, are elevated across the region relative to other areas. Light-wooded taxa (<0.5 g/cm^3) are widely distributed across the Guiana Shield (e.g. *Apeiba*, *Ceiba*, *Cedrela*, *Jacaranda*, *Parkia*, *Schefflera*, *Trattinickia*) and interestingly, are typified as pioneer or gap specialists with very small seed sizes. But why, like seed size, would tree species across the Guiana Shield tend to have higher wood densities, on average, than other tropical regions?

High wood density is believed to develop as a response to environmental limitations placed on growth through low light and/or low moisture availability (Howe, 1974; Zobel and van Buijtenen, 1989). Several studies within and between

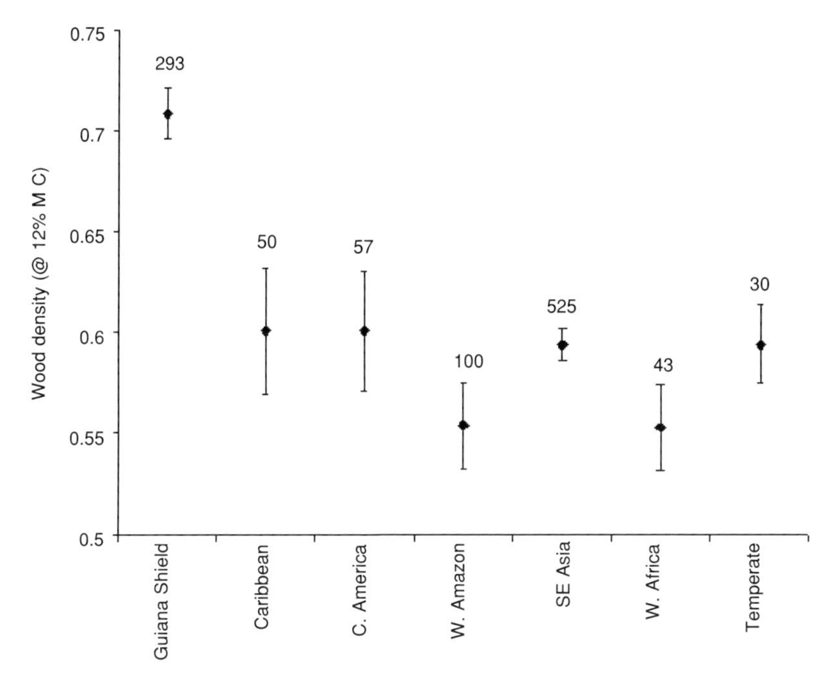

Fig. 7.6. Mean (± SE) wood density at 12% moisture content for species from the Guiana Shield and other neotropical regions. Values for West African, SE Asian and temperate hardwood species are also presented as a broader comparison. Values above SE bar are the number of species used to calculate mean. Species values based on one to six published sources. Data sources: Williams (1939); Record and Hess (1943); Suriname Forest Service (1955); van der Slooten and Martinez (1959); Longwood (1962); Hoheisel and Karstedt (1967); Hoheisel (1968); Mullins and McKnight (1981); Lavers (1983); CIRAD/CTFT (1989); INPA/CPPF (1991); Soerianegara and Lemmens (1993); Gerard *et al.* (1996); Brown (1997).

species suggest that tropical wood density increases in drier tropical forest conditions (Howe, 1974; Chudnoff, 1976; Wiemann and Williamson, 1989b; Gonzalez and Fisher, 1998). In a study of balsa (*Ochroma pyramidale*), arguably the least dense neotropical timber tree, Whitmore (1973) concluded that wood density decreased in Costa Rican life zones with shorter dry seasons.

However, sensitivity to environmental conditions is also modulated by inherent, or phylogenetic, constraints on growth plasticity. Different species under the same or very similar growth conditions have been shown to often have considerably different growth responses, limiting the range of measured growth rates (see Chapter 3) and, *inter alia*, wood densities. The average wood densities for SE Asian and Guiana Shield forests, for example, may reflect the high relative abundance of dipterocarps and caesalpinoid legumes, respectively. The question is whether wood densities, and other life-history attributes, differ among these tropical forest regions through the selective advantages they have provided in the face of different regional environmental histories or simply radiated as the result of periods of local isolation created through these histories (Whitmore and Prance, 1987).

At landscape scales, low light and low moisture effects are inversely related, since closed canopies that create low-light understorey conditions are less likely to be maintained in low moisture environments. The effects on wood density, however, could prove similar in both open-canopy forests exposed to chronic moisture deficits and closed-canopy forests exposed (in the subcanopy) to chronic light competition. In closed canopy forests that are also periodically exposed to severe moisture deficits, taxa with much slower growth rates and metabolic demands would achieve a premium in relatively dark, but periodically dry, conditions. In forests such as those of the Guiana Shield, where environmental stability is relatively high, but significant seasonal and inter-annual drought phases occur (see Chapter 2), slow growth would appear advantageous. The advantage to taxa exhibiting lower growth rates in the face of low light and moisture must come through survivorship, since their poor carbon accrual will ultimately make them losers where faster-growing species achieve similar recruitment rates.

Radial gradients in specific gravity (a correlate of wood density) provide some indication of the contrast in growth histories in relation to varying climates. Very little, if any, data on radial wood density gradients from species in the Guiana Shield were known to the author. However, examining radial gradients measured for long-lived hardwood species in various forest life zones in Costa Rica (Wiemann and Williamson, 1988, 1989a, 1989b) can help to circumstantiate growth-history patterns of Guiana Shield species that have been explored from a physiological perspective (see Chapter 3). Wieman and Williamson's assessment of the pith to bark changes in specific gravity for tree species growing in three tropical life zones illustrates how specific gravity of less dense species, particularly in lowland wet forests, shows greater increases in density as they get older. As initial specific-gravity increases, the change in density later in life approaches zero (Fig. 7.7). Dry forest trees appear to approach zero at lower densities than wet or montane species, logically as a consequence of greater moisture stress. Measured specific gravities of most dominant tree taxa in the Guiana Shield (Chrysobalanaceae: mean =

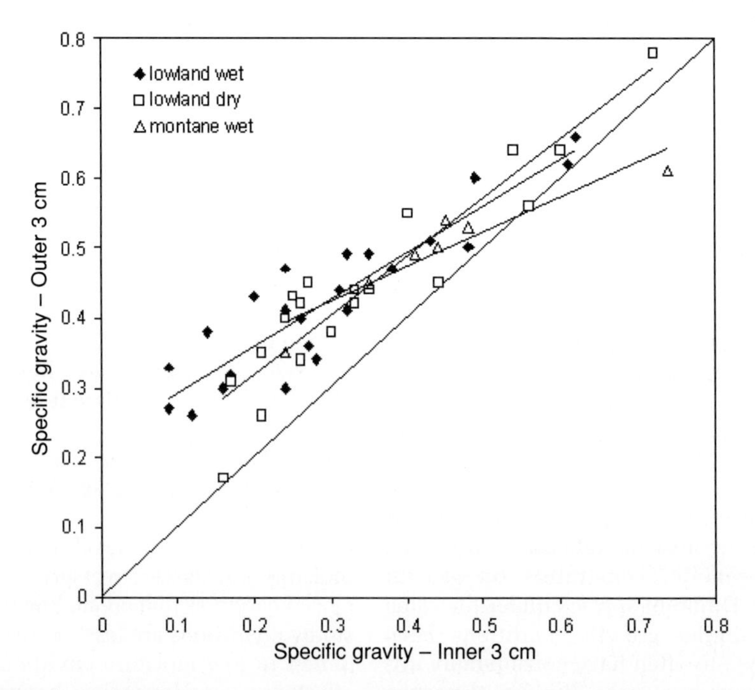

Fig. 7.7. Inner and outer specific gravity values for a range of Costa Rican lowland wet, tropical dry and montane forest tree species adapted from Wiemann and Williamson (1989a,b). Linear least-squares regression lines fitted to each life zone set. Diagonal line represents no radial change in specific gravity for different density values. Points above this line show an increase in density with age, those below a decrease.

0.82 (n=14); Lecythidaceae: mean = 0.7 (18); Caesalpiniaceae: mean = 0.74 (n=23)) are invariably at the upper end of the Costa Rican scale, tentatively suggesting very little increase or a decrease in wood density over their growth histories, but this needs to be tested.

When is it good to be dense?

The differences in wood densities between the Guiana Shield and other neotropical regions thus may be attributed to a number of divergent attributes of these forests. First, differences may simply be due to poor sampling of pith to bark gradients in wood density. If species in other forests are growing more slowly later in life and many common Guiana Shield species show relatively constant variation in wood density, due in part to more modest changes in growth rates (e.g. Zagt, 1997), then the way in which samples are taken could account for much of the difference between regional groupings. Secondly, patterns in radial wood density from Costa Rica and average wood densities from the various neotropical regions suggest that moisture may be more limiting than light in achieving higher wood densities. If so, then forests across many parts of the Guiana Shield would be most impacted by this effect. Work by Schulz (1960, Fig. 68) supports the view that anomalous amplification of dry-season conditions can lead to substantial decreases in growth rates, at least for some common trees in Suriname. The susceptibility of the Guiana Shield region to ENSO-induced precipitation anomalies and the relatively modest drought response associated with this signal in Costa Rica and western Amazon (see 'Climate', Chapter 2) could select for higher wood density either as a consequence of selection for lower metabolic cost requirements (low growth rates) or some other structural or resistance advantage. Thirdly, the greater mechanical investment that higher wood densities represent would only be advantageous to reproductive adult individuals with extended longevities. In forest regions more frequently disturbed, the advantage of investment would yield less return in terms of reproductive effort, particularly as seed crop-size decreases with increasing seed size. Saldarriaga (1987, Fig. 3.3) showed that average tree wood density (specific gravity) certainly remains significantly lower in areas up to 80 years after shifting cultivation relative to 'mature' forest in Venezuelan Amazonas, although overall densities remained comparatively high (consistent with Fig. 7.6). Ramping systems with higher site turnover rates should have more canopy trees with lower wood densities, holding moisture and light availability constant. Together, closed canopy, periodic drought and low site turnover mutually reinforce selection for higher wood densities across many forested regions of the Guiana Shield.

Phylogenetic position

During the earliest period of angiosperm evolution, most of the modern-day neotropical forestlands were believed to be epicontinental. That is, they were covered by a shallow marine environment. Considerable geological evidence supports this view (Harrington, 1962). In fact, many of the fossil hydrocarbon deposits currently being tapped across the sedimentary lowlands of Venezuela, Colombia, Brazil and Ecuador are formed in part from organic materials of shield origin deposited during this period (e.g. Ramón *et al.*, 2001).

This presents a rather simplistic view of a more complex palaeogeographic landscape, but also highlights the rational linkage between the incidence of tree species occupying basal positions among tropical angiosperm phylogenies and the occurrence of Precambrian geologies. Precambrian regions have a longer terrestrial legacy and thus have been capable of supporting evolution of arboreal plants over a much longer period.

This should not conjure up a view of uniform antiquity across the Guiana Shield specifically and Precambrian regions in general. Significant biogeographic changes have invariably led to considerable source–sink exchange of taxa through time between these and much younger adjoining regions (Raven and Axelrod, 1974). It should suggest, however, that upland forest tree taxa would have seen fewer opportunities for establishment in western and central tropical America independent of the Precambrian shield environment prior to the Tertiary period, 70 million years ago.

Many modern tree taxa are indeed not equally represented across geomorphographic regions within the neotropics, either as a function of taxonomic representation or abundance. Among these, the distribution of taxa within the Fabaceae (in broad terms), Lauraceae, Moraceae, Arecaceae, Lecythidaceae, Chrysobalanaceae and Bombacaceae are among the most strongly contrasting, both in the position they currently hold within their familial phylogeny, but also in their phenotypic attributes and the implications of these to their performance within ramping and dampening systems.

Caesalpinoids and Swartzieae

Within the legume family, the caesalpinoids are generally considered to be ancestral. Most of the super-dominant species across the Guiana Shield are caesalpinoids and, of these, a large fraction belong to the Amherstiae and Caesalpinieae. Their unusually high abundances within the forests of the Guiana Shield (Davis and Richards, 1933; Myers, 1936; Fanshawe, 1952; Lindeman and Molenaar, 1959; Whitton, 1962; Maas, 1971; Comiskey *et al.*, 1994; Huber, 1995; Johnston and Gillman, 1995; Coomes and Grubb, 1996; Ek, 1997; Toriola-Lafuente, 1997; Poncy *et al.*, 1998; ter Steege and Zondervan, 2000; Hollowell *et al.*, 2001), combined with a similar dominance, often of the same or closely related genera (e.g. *Cynometra*, *Microberlinia*, *Tetraberlinia*, *Gibertiodendron*, *Julbernardia*) across much of western tropical Africa (Hart *et al.*, 1989) suggests a Gondwanan origin. The tribe Swartzieae is considered intermediate between ancestral caesalpinoid and derived papilinoid groups. It is particularly well represented across many parts of the shield region through its main tree genus, *Swartzia* (but also *Aldina*), including a large number of species (currently) endemic to the area.

At the same time, caesalpinoid tree taxa (incl. *Swartzia*) are noticeable by their relatively poor representation among canopy stem counts in most sampled forests in western Amazon, Central America and some parts of the Caribbean (Fig. 7.8). In most western Amazon sites that have been inventoried, palms and moracs, rather than caesalpinoid legumes, achieve the highest relative densities (Fig. 7.8).

Chrysobalanaceae and Lecythidaceae

These distinct and speciose families of eastern Amazon trees are exceptionally well-represented in most upland forest types across the Guiana Shield and the Amazon Downwarp where they can achieve co-dominance, typically with each other or legumes, such as *Eperua* spp. or *Alexa* spp. (Fig. 7.8) (Fanshawe, 1954; Lindeman and Molenaar, 1959; Maas, 1971; Mori and Boom, 1987; Huber *et al.*, 1995; Toriola-Lafuente, 1997). These groups, however,

Fig. 7.8. Regional variation in relative representation of different neotropical families. Relative density of stems ≥10 cm dbh taken from selected plots located in Central America (CA), the western Amazon (WA), the Amazon Downwarp (AD) and the Guiana Shield (GS). Note difference in scales for more dominant Caesalpiniaceae and Arecaceae. Source data: La Selva: Hartshorn and Hammel (1994); BCI: Condit *et al.* (1996); Condit (1998) and CTFS-Panama (http://www.ctfs.si.edu/datasets); Cuyabeno: Balslev *et al.* (1987); Beni: Comiskey *et al.* (1998); Smith and Killeen (1998); Choco: Galeano *et al.* (1998); Jau: Ferreira and Prance (1998); Manaus: Ferreira and Rankin-de-Merona (1998); La Fumee: Mori and Boom (1987); Kwakwani: Comiskey *et al.* (1994); Winana, Snake and P. Jacob Creeks, Kamisa and Blanch Marie Falls: Maas (1971); Moraballi: Davis and Richards (1933); Pibiri: Ek (1997); Barama and Moruca: van Andel (2000); Mayaro: Beard (1946); Rio Chanaro: Dezzeo and Briceño (1997).

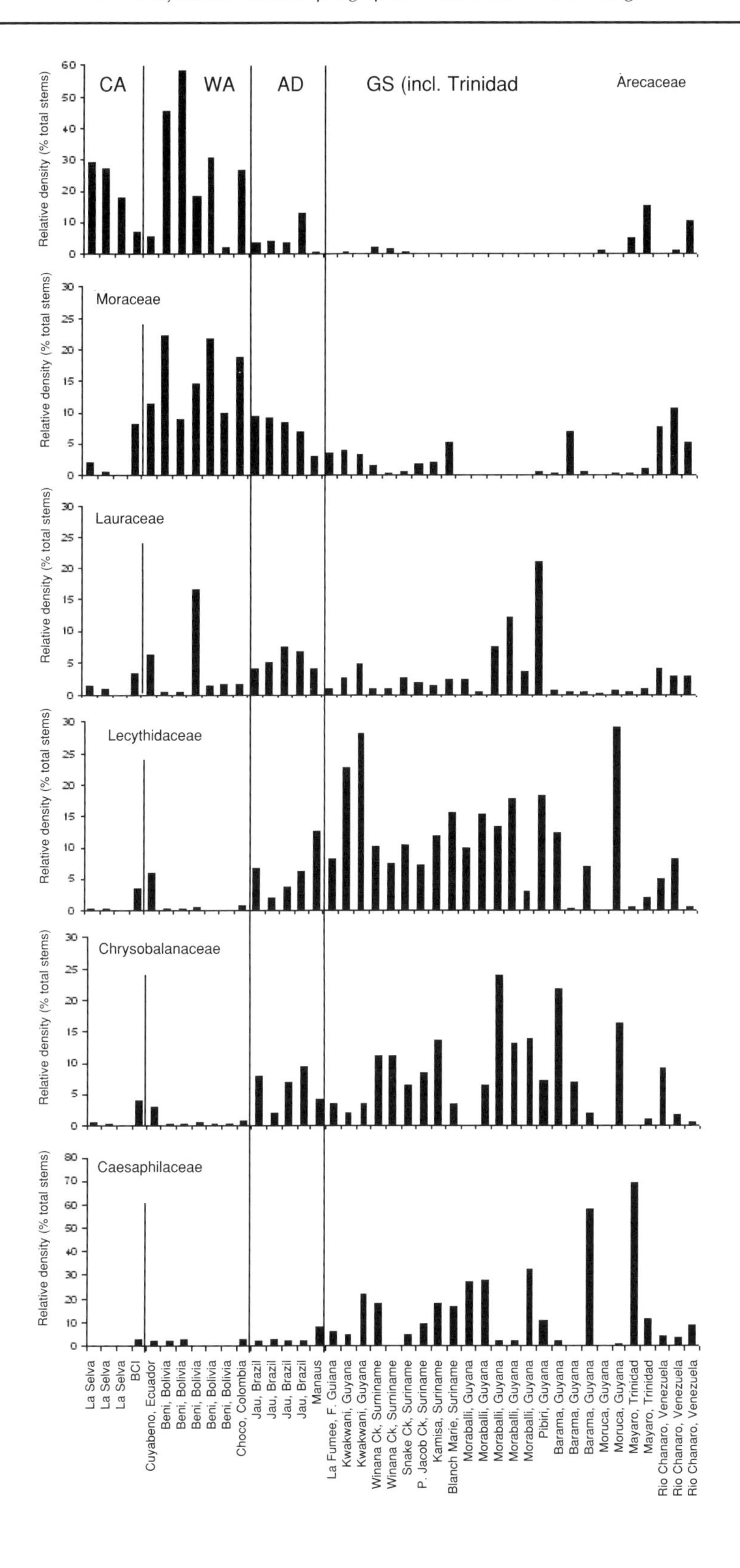
CA
WA
AD
GS (incl. Trinidad
Arecaceae
Moraceae
Lauraceae
Lecythidaceae
Chrysobalanaceae
Caesaphilaceae
Relative density (% total stems)
La Selva
La Selva
La Selva
BCI
Cuyabeno, Ecuador
Beni, Bolivia
Beni, Bolivia
Beni, Bolivia
Beni, Bolivia
Beni, Bolivia
Choco, Colombia
Jau, Brazil
Jau, Brazil
Jau, Brazil
Jau, Brazil
Manaus
La Fumee, F. Guiana
Kwakwani, Guyana
Kwakwani, Guyana
Winana Ck, Suriname
Winana Ck, Suriname
Snake Ck, Suriname
P. Jacob Ck, Suriname
Kamisa, Suriname
Blanch Marie, Suriname
Moraballi, Guyana
Moraballi, Guyana
Moraballi, Guyana
Moraballi, Guyana
Moraballi, Guyana
Pibiri, Guyana
Barama, Guyana
Barama, Guyana
Barama, Guyana
Moruca, Guyana
Moruca, Guyana
Mayaro, Trinidad
Mayaro, Trinidad
Rio Chanaro, Venezuela
Rio Chanaro, Venezuela
Rio Chanaro, Venezuela

are strikingly rare across most of the western Amazon and Central America (Fig. 7.8).

Both Chrysobalanaceae and Lecythidaceae (in the strict sense) are not known from fossil records prior to the Oligocene, 38 to 25 Ma BP, suggesting that these groups radiated after, and in part as a consequence of, significant separation between South America and Africa (Romero, 1993) over the Palaeogene. The widespread distribution of closely related taxa across the tropics, however, indicates that ancestral lines were in place before dispersal among regions was extinguished (Raven and Axelrod, 1974). The Gondwanan association of these species, combined with widespread fruit morphologies among canopy tree taxa, suggesting relatively sluggish dispersal, may have limited colonization of the western Amazon and Central America during early periods of forest formation across these ramping systems.

Lauraceae

The Lauraceae is a widespread tropical family with considerable taxonomic representation across the neotropics (Fig. 7.8). It is also one of the most phylogenetically basal tropical tree families, along with others included in the subclass Magnoliidae (Takhtajan, 1997). Greenheart (*C. rodiei*) and determa (*Sextonia rubra*), two very common tree species endemic to the Guianas, however, are believed to fall outside these three recognized tribes of the Lauraceae, based on morphological and molecular evidence (Chanderballi *et al.*, 2001). The substantial phylogenetic analyses of the tropical Lauraceae undertaken by Chanderballi *et al.* (2001) clearly show a complex formed by *Chlorocardium*, *Sextonia*, *Mezilaurus* and the monotypic *Anaueria* occupying the most basal position with the constructed lauraceous phylogeny. The authors view this complex as a Gondwanan relict. The ancestral position of the Guiana Shield dominants within the phylogeny of the Lauraceae is only paralleled by the far less dominating presence of *Chlorocardium venenosum* across parts of Colombia, Ecuador and Peru, *Mezilaurus* spp. across the Guiana Shield, Colombia, Peru and Bolivia, and the monotypic *Anaueria* genus in Brazil and Peru (Rohwer *et al.*, 1991; van der Werff, 1991).

Moraceae

Moraceae (*s.l.*) includes a group of pantropical taxa that are well-represented in the fossil record as early as the Late Cretaceous, prior to the break-up of Gondwana (Romero, 1993). Striking among the biogeographic patterns attached to this group across the neotropics is the widespread occurrence of *Poulsenia armata* and *Pseudomeldia* spp. throughout western Amazon (Balslev *et al.*, 1987; Foster, 1990a; Gullison *et al.*, 1996; Smith and Killeen, 1998) and Central America (Hubbell and Foster, 1990), where they achieve significant relative densities, often as consociates of *Iriartea deltoidea*, *Jessenia* spp. or other arborescent palm species (Fig. 7.8). Both genera occur throughout the Guiana Shield (Boggan *et al.*, 1998), but rarely achieve the densities and frequencies found across other geomorphographic regions. Commonly recognized pioneers, such as *Ficus* and *Cecropia*, contribute significantly to floodplain forests along upper Amazon tributaries (e.g. Foster, 1990b), but are noticeably absent from many seasonally or superannually inundated river edges in the Guiana Shield (Davis and Richards, 1933; Johnston and Gillman, 1995; Aymard *et al.*, 1998; van Andel, 2000). Two other genera, *Pourouma* and *Coussapoa* do not appear to contribute significantly to these differences and are typically rare throughout most longer-standing forests. Nonetheless, Moraceae are one of several 'oligarchic' groups exhibiting widespread local abundance across forests of the sub-Andean trough (Pitman *et al.*, 2001).

Arecaceae

The stilt-rooted *Iriartea deltoidea* is present as one of the most abundant arboreal species in plots across the western Amazon with surprising frequency. From the forests of lowland Bolivia (Gullison *et al.*, 1996; Smith and

Killeen, 1998), Peru (Gentry and Terborgh, 1990) and Ecuador (Balslev *et al.*, 1987) through to La Selva in Costa Rica (Hartshorn and Hammel, 1994) and Nicaragua, this canopy palm consistently is found in relatively high abundances across much of its range throughout western Amazon and Central America (Henderson *et al.*, 1995) (Fig. 7.9). It is noticeably absent from the Guiana and Brazilian shield regions and Smith and Killeen (1998) commented on the abrupt disappearance of the species when crossing from Quaternary to Precambrian geologies in the southeastern Amazon.

The dominant *Orbignya* (now *Attalea speciosa*) achieves hyper-dominance in vast aggregates across the Brazilian Shield from the Palaeozoic Tocantins basin to the Quaternary sediments of lowland Bolivia (Anderson *et al.*, 1991; Henderson *et al.*, 1995) (Fig. 7.9). This dominant species is virtually absent north of the Amazon main stem and found only in the southern region of Guyana and Suriname, where it does not form hyper-dominant stands, possibly as a consequence of human introduction (Kahn and Moussa, 1997). In this case, range and local abundance appear tied to geographic position, rather than underlying geology.

It is clear that the arborescent Arecaceae achieve higher taxonomic richness and greater relative densities in *terra firme* forests across the Amazon Downwarp, sub-Andean Foredeep and Central America land-bridge relative to the Guiana Shield (Kahn *et al.*, 1988) (Fig. 7.8). In part, this reflects the greater evenness among dicotyledonous tree families in western Amazon and Central America relative to many parts of the Guiana Shield. Palms are, however, well-represented in certain habitats within the Guiana Shield. Coastal swamp forests, savanna gallery forests and localized perched catchments are often, but not always, dominated by a single canopy palm species (e.g. *Euterpe*, *Mauritia*) (Fanshawe,

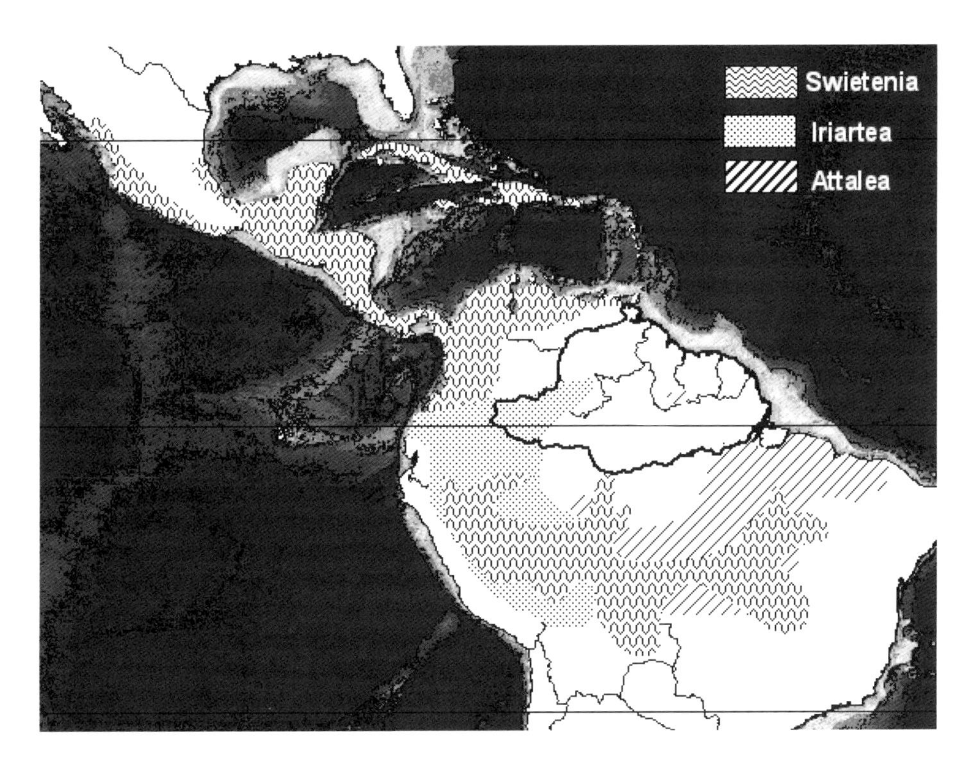

Fig. 7.9. The hemispheric distribution of several well-known disturbance-driven taxa. Distributional coverages derived from: *Swietenia* spp.: Pennington (1981) and Lamb (1966); *Iriartea deltoidea*: Henderson *et al.* (1995) and *Attalea speciosa*: Anderson *et al.* (1991).

1954; Lindeman and Molenaar, 1959; Huber, 1995; van Andel, 2000). Most upland forests are also typically characterized by a much higher abundance and richness of understorey palms (e.g. Ek, 1997), a trait they share with most other neotropical regions (Gentry, 1990; Kahn and de Granville, 1992). However, the greater presence of arborescent palms in these areas relative to the shield region is difficult to overlook, even when considering some of the most diverse forests identified in the Guiana Shield (e.g. French Guiana: Mori and Boom, 1987; Poncy *et al.*, 1998).

Disturbance, Mobility and Ranging Across the Neotropics

The likelihood of ancient clades that originated along the western rim of Gondwana (Raven and Axelrod, 1974) surviving *in situ* to modern times would be much smaller than along the eastern rim of modern South America, where Precambrian landscapes dominate. Consequently, taxa with life-history attributes supporting greater mobility and shorter inter-generational times should have radiated faster and/or more profusely (Marzluff and Dial, 1991). Where environmental stability was low, a premium would have resided with highly mobile taxa spread across larger geographic ranges as a buffer to extinction. Over time, these lineages would account for an increasing share of the standing diversity as a consequence of lower extinction rates in these areas.

In this case, taxa that are adapted to low stability environments in ramping system regions should also be associated with speciation centres as opportunities for mobile, well-distributed taxa propagate and those for species adapted to quiescent landscapes, such as the Guiana Shield, dwindle or go locally extinct. Two pertinent examples can be given to support this view.

One is the case of *Inga*, a widespread neotropical genus of trees with small to medium-sized seeds and sugar-rich aril attractive to a wide range of vertebrates. The colonization capacity of *Inga* spp. is well-documented from forest inventories in regions of known historic disturbance throughout the Guiana Shield, where relative density of the genus increases (van der Hout, 1999; van Andel, 2000; ter Steege *et al.*, 2002). *Inga* also contributes disproportionately to species richness of neotropical forests and high levels of alpha diversity in many forests of the western Amazon, but also across most regions (Gentry, 1990; Valencia *et al.*, 1994).

Richardson *et al.* (2001) identified the eastern slopes of the Andes as the centre of diversification for the genus, arguing that the geological history of the region led to explosive speciation, primarily due to life-history features of the taxon, especially its relatively short generation time. If true, then one very speciose neotropical tree genus can be associated with large-scale environmental instability typical of a ramping system.

Similarly, *Cecropia* represents another case of a disturbance-loving taxon believed to have speciated rapidly in the western Amazon as a consequence of Andean uplift and the environmental instability it created (Franco-Roselli and Berg, 2003). As a consequence, the eastern Andean slopes of Ecuador and western Colombia contain the highest species richness within this genus.

Circumstantial evidence highlighting the contrast between ramping and dampening systems and the role of geomorphographic control can also be gleaned from the contemporary distribution of tree species known to associate with large-scale or catastrophic disturbances. *Swietenia* spp., *Ochroma pyramidale, Attalea speciosa* and *Iriartea deltoidea* are common features of many forest habitats across much of Central America, the Caribbean and western Amazon. These taxa also have a strong association with areas affected by widespread, calamitous or heightened levels of disturbance, such as fire, hurricanes and hydrological events (Lamb, 1966; Foster, 1990b; Anderson *et al.*, 1991; Gullison *et al.*, 1996), yet are widely absent from the Guiana Shield region (Fig. 7.9). This is not to falsely imply that the shield region is devoid of disturbance-dependent taxa. Rather, it emphasizes the significance of near exclusion of these well-known indicators of event-based change and the congruence of their distributions with the physical features that distinguish ramping from dampening system regions. As a consequence, catastrophic disturbance impacts are not absent, but less pronounced across the Precambrian landscape of the Guiana Shield relative to adjoining regions. Ultimately, the relationship between landscape-scale variation in site fertility, predictability of moisture availability either through precipitation and/or (sub)surface flow, historical chronologies of human impact and habitat modification and the underlying life history attributes that shape opportunities for effectively competing and colonizing need to be considered concomitantly in assessing why forest communities vary at different spatial scales.

Geomorphographic Control – an Organizing Principle?

The concept of large-scale geomorphographic variation broadly distinguishing high- and low-energy ecosystems within the neotropics adheres to a basic dynamical flow concept that establishes the direction and magnitude of flux from a central pool, population or steady state. In population models these take the form of birth and death or immigration and emigration. In biogeography theory, speciation and species extinction identify processes provisioning gains and losses to a standing community. From a geological perspective, ramping and dampening systems are riding on different phases of the geomorphic cycle and forces that affect variation in climatic conditions. A series of logical biophysical distinctions can be attached to these different phases as testable hypotheses concerning the role of geomorphographic constraint on local and regional forest change (Fig. 7.10).

Geomorphographic control is not meant as a 'silver bullet' hypothesis or a unifying theory capable of fully explaining why there are so many species in tropical forests or how they function. Rather, it is presented as an organizing principle that

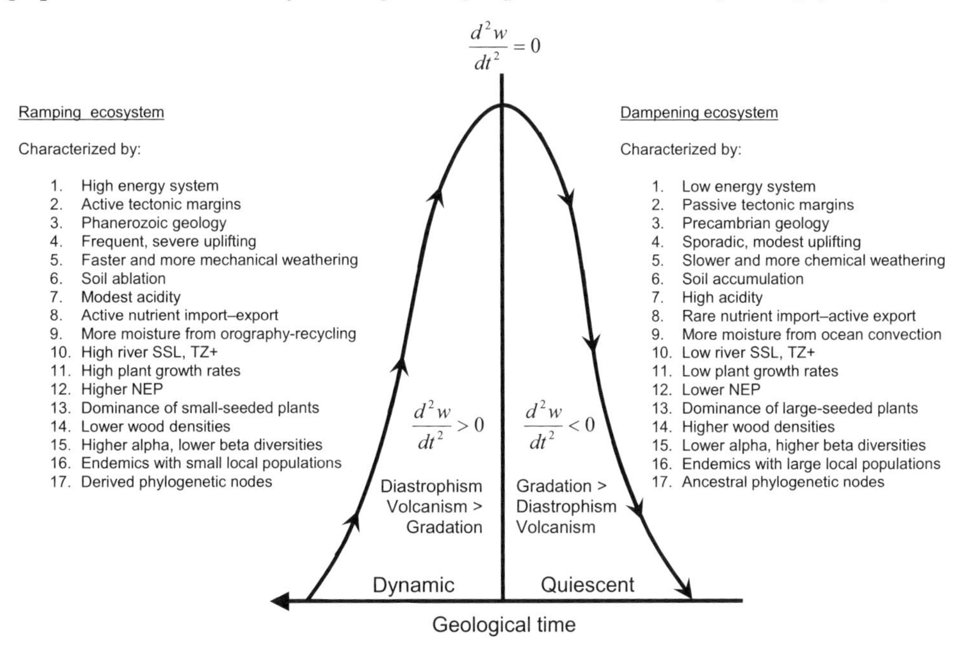

Fig. 7.10. Depiction of ramping and dampening systems along the geomorphic cycle and hypothesized attribute sets. System phases are defined by the change in the rate of weathering (w). The Guiana Shield is considered a dampening system, while Central America, western Amazon and the Caribbean (less Trinidad and Tobago) are considered ramping system regions. See text for further explanations.

may bear consideration when attempting to take on such a mighty task. Perhaps more importantly, it brings into question whether scaling-up interesting site-specific results from dampening (ramping) systems bears much relevance to ramping (dampening) systems and whether concepts, such as 'Amazonia' and 'tropical rainforest', constructively clarify a (statistical) region or merely perpetuate a perspective of tropical environmental uniformity that does not exist, and from all available evidence has never existed since the earliest known start of terrestrial angiosperm evolution. These wide-scale differences in standing forest composition and the large-scale factors forcing change in these forests suggest that applying a uniform conservation and management approach across these regions may lead to contrasting and unforeseen outcomes.

Notes

[1] Formerly *Orbignya phalerata.*

[2] The latter species have reached much greater relative abundances than the former at their respective study sites in Panama and Guyana, but it is interesting to note *Prioria's* slow, but steady, increase in stem number (>10 cm diameter breast height (dbh)) between 1980 and 2000 within the 50 ha forest plot monitored on BCI (CTFS-Panama).

[3] This power function evolved as a consequence of exploring the relationship between forest evenness (or dominance) and seed size by empirically fitting a function on to plot data. The plot data consisted of the number of stems ≥10 cm dbh per species as a proportion of total stems of this diameter or larger in a handful of 1 ha sites, mainly across the Guiana Shield, but also, where possible, from other neotropical regions. Time did not allow for the fitting of further data that exist. The general relationship follows as:

$$S_{50} \cong \lambda \left[\frac{\sum_{1}^{s} \frac{i_s}{i_t} w}{n_s} \right]^{-0.5}$$

where S_{50} is the number of most common species, S, needed to account for half of the total number of stems, w is the mean seed weight (g) for each species S, i is the total number of stems for each species S in plot t, n is the number of mean seed weight values for S species and λ is a seed size scaling coefficient.

The function fits the distribution of empirical data well (r^2=0.82) despite incorporating a wide range of plot data varying considerably in geographic location and soil types. It is important to note, however, that stratification across geomorphographic regions is unbalanced and the large number of sites from the Guiana Shield may reflect a spatially bound scale of applicability, rather than a universal relationship. Also, seed sizes can vary at considerable magnitudes within and between conspecific adults, particularly as minimum seed size increases. Thus again, this is a general trend excepted by local variation, but a decline in average seed size with decreasing dominance (or increasing evenness) appears sufficiently noteworthy to merit further exploration.

References

Anderson, A.B. (1981) White-sand vegetation of the Brazilian Amazonia. *Biotropica* 13, 199–210.

Anderson, A.B., May, P.H. and Balick, M.J. (1991) *The Subsidy from Nature – Palm Forests, Peasantry, and Development on an Amazon Frontier.* Columbia University Press, New York.

Aymard, G., Cuello, N. and Schargel, R. (1998) Floristic composition, structure, and diversity in the moist forest communities along the Casiquiare channel, Amazonas state, Venezuela. In: Dallmeier, F. and Comiskey, J.A. (eds) *Forest Biodiversity in North, Central and South America, and the Caribbean.* Parthenon Publishing, Paris, pp. 495–506.

Baker, H.G. (1972) Seed weight in relation to environmental conditions in California. *Ecology* 53, 997–1010.

Balslev, H., Luteyn, J., Ollgaard, B. and Holm-Nielsen, L.B. (1987) Composition and structure of adjacent unflooded and floodplain forest in amazonian Ecuador. *Opera Botanica* 92, 37–57.

Bariteau, M. (1992) Régénératoin naturelle de la forêt tropical humide de Guyane: étude de la répartition spatiale de *Qualea rosea* Aublet, *Eperua falcata* Aublet et *Symphonia globulifera* L. *Annales des Sciences Forestières* 49, 382–395.

Beard, J.S. (1946) The mora forests of Trinidad, British West Indies. *Journal of Ecology* 33, 173–192.

Béreau, M. and Garbaye, J. (1994) First observations on the root morphology and symbioses of 21 major tree species in the primary tropical rain forest of French Guyana. *Annales des Sciences Forestiere* 51, 407–416.

Boggan, J., Funk, V., Kelloff, C., Hoff, M., Cremers, G. and Feuillet, C. (1998) *Checklist of the Plants of the Guianas (Guyana, Suriname, French Guiana)*, 2nd edn. BDGP – NMNH Smithsonian Institution, Washington, DC.

Boot, R.G.A. (1996) The significance of seedling size and growth rate of tropical forest seedlings for regeneration in canopy openings. In: Swaine, M.D. (ed.) *The Ecology of Tropical Forest Tree Seedlings.* Parthenon, Paris, pp. 267–284.

British Guiana Forest Department (1935) *Forestry in British Guiana.* Supplementary statement prepared by the Forest Department for presentation to the Fourth British Empire Forestry Conference (South Africa), 1935. BGFD, Georgetown.

Brouwer, L. (1996) *Nutrient Cycling in Pristine and Logged Tropical Rain Forest – A Study in Guyana.* Tropenbos-Guyana Series 1. Tropenbos-Guyana, Utrecht, The Netherlands.

Brown, S. (1997) *Estimating Biomass and Biomass Change of Tropical Forests.* FAO Forestry Paper 134. FAO, Rome.

Brown, S. and Lugo, A.E. (1990) Tropical secondary forests. *Journal of Tropical Ecology* 6, 1–32.

Bubberman, F.C. (1973) De bosbranden van 1964 in Suriname. *Nieuwe West-Indische Gids* 49, 163–173.

Chanderballi, A.S., van der Werff, H. and Renner, S.S. (2001) Phylogeny and historical biogeography of Lauraceae: evidence from chloroplast and nuclear genomes. *Annals of the Missouri Botanical Garden* 88, 104–134.

Chudnoff, M. (1976) Density of tropical timbers as influenced by climatic life zones. *Commonwealth Forestry Review* 55, 203–217.

CIRAD/CTFT (1989) *Bois des DOM-TOM. Tome 1 – Guyane.* CIRAD-CTFT, Nogent-sur-Marne, France.

Coates, A.G. and Obando, J.A. (1996) The geologica evolution of the Central American isthmus. In: Coates, A.G. (ed.) *Evolution and Environment in Tropical America.* Chicago University Press, London, pp. 21–56.

Comiskey, J.A., Aymard, G. and Dallmeier, F. (1994) Structure and composition of lowland mixed forest in the Kwakwani region of Guyana. *Biollonia* 10, 13–28.

Comiskey, J.A., Dallmaier, F. and Foster, R.B. (1998) Forest structure and diversity in managed and unmanaged rainforest of Beni, Bolivia. In: Dallmaier, F. and Comiskey, J.A. (eds) *Forest Biodiversity in North, Central and South America, and the Caribbean.* Parthenon Publishing/UNESCO, Paris, pp. 663–680.

Condit, R. (1998) *Tropical Forest Census Plots.* Springer, Berlin.

Condit, R., Hubbell, S.P. and Foster, R.B. (1996) Changes in tree species abundance in a neotropical forest: impact of climate change. *Journal of Tropical Ecology* 12, 231–256.

Condit, R., Pitman, N.C.A., Leigh, E.G.J., Chave, J., Terborgh, J., Foster, R.B., Nuñez-Vargas, P., Aguilar, S., Valencia, R., Villa, G., Muller-Landau, H.C., Losos, E. and Hubbell, S.P. (2002) Beta-diversity in tropical trees. *Science* 295, 666–669.

Coomes, D.A. and Grubb, P.J. (1996) Amazonian caatinga and related communities at La Esmeralda, Venezuela: forest structure, physiognamy and floristics, and control by soil factors. *Vegetatio* 122, 167–191.

CPTEC (2003) Histogramas dos focos de calor em Ombrofila Densa – NOAA-12 for Amapa, Amazonas, Roraima and Para states, Brazil, and Bolivar and Amazonias states, Venezuelan Guayana, 1998–2003. http://www.cptec.inpe.br/products/queimadas

Cuevas, E. (2001) Soil versus biological controls on nutrient cycling in terra firme forests. In: McClain, M.E., Victoria, R.L. and Richey, J.E. (eds) *The Biogeochemistry of the Amazon Basin.* Oxford University Press, London, pp. 53–67.

Dalling, J.W., Harms, K.E. and Aizprúa, R. (1997) Seed damage tolerance and seedling resprouting ability of *Prioria copaifera* in Panama. *Journal of Tropical Ecology* 13, 481–490.

Davis, T.A.W. and Richards, P.W. (1933) The vegetation of Moraballi creek, British Guiana: an ecological study of a limited area of tropical rain forest. I. *Journal of Ecology* 21, 350–384.

de Milde, R. and de Groot, D. (1970) *Forest Industries Development Survey – Guyana. Reconnaissance Survey of the More Accessible Forest Areas.* UNDP/FAO, Georgetown.

de Milde, R. and Inglis, C. (1974) *Inventory of the Fallawatra, Nassau and Kabalebo Areas.* FAO, Paramaribo.

Denslow, J.S. (1980) Gap partitioning among tropical rainforest trees. *Biotropica* 12 (Suppl.), 47–55.

Dezzeo, N. and Briceño, E. (1997) La vegetacion en la cuenca del Rio Chanaro: medio Rio Caura. In: Huber, O. and Rosales, J. (eds) *Ecologia de la Cuenca del Río Caura, Venezuela II. Estudios Especiales.* Refolit CA, Caracas, pp. 365–385.

Duivenvoorden, J.F. and Lips, J.M. (1995) *A Land-ecological Study of Soils, Vegetation and Plant Diversity in Colombian Amazonia.* Tropenbos Foundation, Wageningen, The Netherlands.

Ek, R.C. (1997) *Botanical Diversity in the Tropical Rain Forest in Guyana.* Tropenbos-Guyana Series 5. Tropenbos-Guyana Programme, Wageningen, The Netherlands.

Fanshawe, D.B. (1952) *The Vegetation of British Guiana: a Preliminary Review.* Imperial Forestry Institute Paper 29. Imperial Forestry Institute, Oxford, UK.

Fanshawe, D.B. (1954) Forest types of British Guiana. *Caribbean Forester* 15, 73–111.

Fearnside, P.M. (1990) Fire in the tropical rain forest of the Amazon basin. In: Goldammer, J.G. (ed.) *Fire in the Tropical Biota.* Springer, Heidelberg, Germany, pp. 106–116.

Ferreira, L.V. and Prance, G.T. (1998) Species richness and floristic composition in four hectares in the Jaú National Park in upland forests in Central Amazonia. *Biodiversity and Conservation* 7, 1349–1364.

Ferreira, L.V. and Rankin-de-Merona, J.M. (1998) Floristic composition and structure of a one-hectare plot in terra firme forest in central Amazonia. In: Dallmeier, F. and Comiskey, J.A. (ed.) *Forest Biodiversity in North, Central and South America, and the Caribbean.* UNESCO MAB, Paris, pp. 649–662.

Fittkau, E.J. (1974) On the ecological classification of Amazonia 1. The geological development of Amazonia. *Amazoniana* 5, 77–134.

Foster, R.B. (1982) The seasonal rhythm of fruitfall on Barro Colorado Island. In: Leigh, E.G.J., Rand, A.S. and Windsor, D.M. (eds) *The Ecology of a Tropical Forest – Seasonal Rhythms and Long Term Changes.* Smithsonian Institution Press, Washington, DC, pp. 151–172.

Foster, R.B. (1990a) Floristic composition of the Manu floodplain forest. In: Gentry, A.H. (ed.) *Four Neotropical Forests.* Yale University Press, New Haven, Connecticut, pp. 99–111.

Foster, R.B. (1990b) Long-term change in the successional forest community of the Rio Manu floodplain. In: Gentry, A.H. (ed.) *Four Neotropical Forests.* Yale University Press, New Haven, Connecticut, pp. 565–572.

Foster, R.B. and Janson, C.H. (1985) The relationship between seed mass and establishment conditions in tropical woody plants. *Ecology* 66, 773–780.

Foster, S.A. (1986) On the adaptive value of large seeds for tropical moist forest trees: a review and synthesis. *Botanical Review* 52, 260–299.

Franco-Roselli, P. and Berg, C.C. (2003) Distributional patterns of *Cecropia* (Cecropiaceae): a panbiogeographic analysis. *Caldasia* 19, 285–296.

Friis, E.M., Chaloner, W.G. and Crane, P.R. (eds) (1987) *The Origins of Angiosperms and their Biological Consequences.* Cambridge University Press, Cambridge, UK.

Galeano, G., Cediel, J. and Pardo, M. (1998) Structure and floristic composition of a one-hectare plot of wet forest at the Pacific coast of Chocó, Colombia. In: Dallmaier, F. and Comiskey, J.A. (eds) *Forest Biodiversity in North, Central and South America, and the Caribbean.* Parthenon Publishing/UNESCO, Paris, pp. 551–568.

Garwood, N.C., Janos, D.P. and Brokaw, N. (1979) Earthquake-caused landslides: a major disturbance to tropical forests. *Science* 205, 997–999.

Gentry, A.H. (1982) Neotropical floristic diversity: phytogeographical connections between Central and South America, Pleistocene climatic fluctuations, or an accident of the Andean orogeny? *Annals of the Missouri Botanical Garden* 69, 557–593.

Gentry, A.H. (1990) Floristic similarities and differences between southern Central America and upper and central Amazonia. In: Gentry, A.H. (ed.) *Four Neotropical Forests.* Yale University Press, New Haven, Connecticut, pp. 141–157.

Gentry, A.H. and Terborgh, J. (1990) Composition and dynamics of the Cocha Cashu 'mature' floodplain forest. In: Gentry, A.H. (ed.) *Four Neotropical Forests.* Yale University Press, New Haven, Connecticut, pp. 542–564.

Gerard, J., Miller, R.B. and ter Welle, B.J.H. (1996) *Major Timber Trees of Guyana – Timber Characteristics and Utilization.* Tropenbos Series 15. Tropenbos Foundation, Wageningen, The Netherlands.
Gonzalez, J. and Fisher, R.F. (1998) Variation in selected wood properties of *Vochysia guatemalensis* from four site in Costa Rica. *Forest Science* 44, 185–191.
Grubb, P.J. (1977) The maintenance of species-richness in plant communities: the importance of the regeneration niche. *Biological Reviews* 52, 107–145.
Gullison, R.E., Panfil, S.N., Strouse, J.J. and Hubbell, S.P. (1996) Ecology and management of mahogany (*Swietenia macrophylla* King) in the Chimanes Forest, Beni, Bolivia. *Botanical Journal of the Linnean Society* 122, 9–34.
Haffer, J. (1969) Speciation in Amazonian forest birds. *Science* 165, 131–137.
Hammond, D.S. and Brown, V.K. (1995) Seed size of woody plants in relation to disturbance, dispersal, soil type in wet neotropical forests. *Ecology* 76, 2544–2561.
Hammond, D.S. and Brown, V.K. (1998) Disturbance, phenology and life-history characteristics: factors influencing distance/density-dependent attack on tropical seeds and seedlings. In: Newbery, D.M., Prins, H.T.T. and Brown, N.D. (eds) *Dynamics of Tropical Communities.* Blackwell Scientific, Cambridge, UK, pp. 51–78.
Hammond, D.S. and ter Steege, H. (1998) Propensity for fire in the Guianan rainforests. *Conservation Biology* 12, 944–947.
Hammond, D.S., Brown, V.K. and Zagt, R.J. (1999) Spatial and temporal patterns of seed attack and germination in a large-seeded neotropical tree. *Oecologia* 119, 208–218.
Haripersad-Makhanlal, A. and Ouboter, P.E. (1993) Limnology: physico-chemical parameters and phytoplankton composition. In: Ouboter, P.E. (ed.) *The Freshwater Ecosystems of Suriname.* Kluwer Academic, Dordrecht, The Netherlands, pp. 53–75.
Harper, J.L., Lovell, P.H. and Moore, K.G. (1970) The shape and size of seeds. *Annual Review of Ecology and Systematics* 1, 327–356.
Harrington, H.J. (1962) Paleogeographic development of South America. *Bulletin of the American Association of Petroleum Geologists* 46, 1773–1814.
Hart, T.B., Hart, J.A. and Murphy, P.G. (1989) Monodominant and species-rich forests of the humid tropics: causes for their co-occurrence. *American Naturalist* 133, 613–633.
Hartshorn, G.S. and Hammel, B.E. (1994) Vegetation types and floristic patterns. In: McDade, L.A., Bawa, K.S., Hespenheide, H.A. and Hartshorn, G.S. (eds) *La Selva: Ecology and Natural History of a Neotropical Rain Forest.* University of Chicago Press, London, pp. 73–119.
Henderson, A., Galeano, G. and Bernal, R. (1995) *Field Guide to the Palms of the Americas.* Princeton University Press, Princeton, New Jersey.
Herrera, R., Jordan, C.J., Klinge, H. and Medina, E. (1978) Amazon ecosystems – their structure and functioning with particular emphasis on nutrients. *Interciencia* 3, 223–231.
Hoheisel, H. (1968) *Identification of Some Colombian Wood Species and Their Possible Use on the Basis of Physical and Mechanical Properties.* IFLAIC, Merida.
Hoheisel, H. and Karstedt, P. (1967) *Identification of Ecuadorian Wood Species for Possibilities of Utilization on Basis of Technological Results.* Instituto Forestal Latino Americano, Merida.
Hohenkerk, L.S. (1922) A review of the timber industry of British Guiana. *Journal of the Board of Agriculture of British Guiana* 15, 2–22.
Hollowell, T., Berry, P.E., Funk, V. and Kelloff, C. (2001) *Preliminary Checklist of the Plants of the Guiana Shield.* Vol.1: *Acanthaceae-Lythraceae.* Biological Diversity of the Guianas Program Series. BDGP – Smithsonian Institution, Washington, DC.
Holzinger, R., Kleiss, B., Donoso, L. and Sanhueza, E. (2001) Aromatic hydrocarbons at urban, sub-urban, rural and remote sites iin Venezuela. *Atmospheric Environment* 35, 4917–4927.
Howe, J.P. (1974) Relationship of climate to the specific gravity of four Costa Rican hardwoods. *Wood and Fiber* 5, 347–352.
Hubbell, S.P. (1979) Tree dispersion, abundance and diversity in a tropical dry forest. *Science* 203, 1299–1309.
Hubbell, S.P. (2001) *The Unified Theory of Biodiversity and Biogeography.* Monographs in Population Biology 32. Princeton University Press, Oxford, UK.
Hubbell, S.P. and Foster, R.B. (1990) Structure, dynamics, and equilibrium status of old-growth forest on Barro Colorado Island. In: Gentry, A.H. (ed.) *Four Neotropical Forests.* Yale University Press, New Haven, Connecticut, pp. 522–541.
Huber, O. (1995) Vegetation. In: Berry, P.E., Holst, B.K. and Yatskievych, K. (eds) *Flora of the Venezuelan Guayana.* Timber Press, Portland, Oregon, pp. 97–160.

Huber, O., Gharbarran, G. and Funk, V. (1995) *Vegetation Map of Guyana* (preliminary version). Centre for the Study of Biological Diversity, University of Guyana, Georgetown.

Hughes, J.H. (1946) Forest resources. In: Roth, V. (ed.) *Handbook of Natural Resources of British Guiana.* The Daily Chronicle Ltd, Georgetown, pp. 47–191.

Huston, M.A. (1994) *Biological Diversity: the Coexistence of Species on Changing Landscapes.* Cambridge University Press, Cambridge, UK.

INPA/CPPF (1991) *Catálogo de Madeiras da Amazônia.* INPE/CPPF, Manaus.

Jackson, J.B.C. (1994) Constancy and change of life in the sea. *Philosophical Transactions of the Royal Society, London, Series B* 343, 55–60.

Janzen, D.H. (1969) Seed-eaters versus seed size, number, toxicity and dispersal. *Evolution* 23, 1–27.

Johnston, M. and Gillman, M. (1995) Tree population studies in low-diversity forests, Guyana. I. floristic composition and stand structure. *Biodiversity and Conservation* 4, 339–362.

Jones, C., Mahowald, N. and Luo, C. (2003) The role of easterly waves on African desert dust transport. *Journal of Climate* 16, 3617–3628.

Jordan, C.J. (ed.) (1987) *Amazonian Rain Forests – Ecosystem Disturbance and Recovery.* Ecological Studies. Springer, London.

Jordan, C.J. and Herrera, R. (1981) Tropical rainforests: are nutrients really critical? *American Naturalist* 117, 167–180.

Kahn, F. and de Granville, J.-J. (1992) *Palms in the Forest Ecosystems of Amazonia.* Springer, London.

Kahn, F. and Moussa, F. (1997) El papel de los grupos humanos en la distribución geográfica de algunas palmas en la Amazonía y su periferia. In: Rios, M. and Pedersen, H.B. (eds) *Uso y manejo de recursos vegetales.* Ediciones Abya-Yala, Quito, pp. 83–99.

Kahn, F., Mejia, K. and de Castro, A. (1988) Species richness and density of palms in terra firme forests of Amazonia. *Biotropica* 20, 266–269.

Kalliola, R.J., Jokinen, P. and Tuukki, E. (1999) Fluvial dynamics and sustainable development in upper Rio Amazonas, Peru. In: Henderson, A. (ed.) *Várzea – Diversity, Development and Conservation of Amazonia's Whitewater Floodplains.* NYBG Press, New York, pp. 271–282.

Kauffman, J.B. and Uhl, C. (1990) Interactions of anthropogenic activities, fire, and rain forests in the Amazon basin. In: Goldammer, J.G. (ed.) *Fire in the Tropical Biota.* Springer, Heidelberg, pp. 117–134.

Lamb, F.B. (1966) *Mahogany of Tropical America: Its Ecology and Management.* University of Michigan Press, Ann Arbor, Michigan.

Lambers, H. and Poorter, H. (1992) Inherent variation in growth rates between higher plants: a search for physiological causes and higher consequences. *Advances in Ecological Research* 23, 188–261.

Lavers, G.M. (1983) *The Strength Properties of Timber.* HMSO, London.

Leishman, M.R., Westoby, M. and Jurado, E. (1995) Correlates of seed size variation: a comparison among five temperate floras. *Journal of Ecology* 83, 517–530.

Lindeman, J.C. and Molenaar, S.P. (1959) *The Vegetation of Suriname.* Vol. 1. Part 2. Van Eedenfonds, Amsterdam.

Longwood, F.R. (1962) *Present and Potential Commercial Timbers of the Caribbean – with Special Reference to the West Indies, the Guianas, and British Honduras.* Agriculture Handbook No. 207. USDA-Forest Service, Washington, DC.

Maas, P.J.M. (1971) Floristic observations on forest types in western Suriname I. *Proceedings of the Konikl. Nederl. Akademie van Wetenschappen – Amsterdam, Series C* 74, 269–302.

Marzluff, J.M. and Dial, K.P. (1991) Life history correlates of taxonomic diversity. *Ecology* 72, 428–439.

McTurk, M. (1882) Notes on the forests of British Guiana. *Timehri* 1, 173–215.

Meggers, B.J. (1994) Archeological evidence for the impact of mega-Niño events on Amazonia during the past two millennia. *Climatic Change* 28, 321–338.

Mori, S.A. and Boom, B.M. (1987) Chapter II. The forest. *Memoirs of the New York Botanical Garden* 44, 9–29.

Moyersoen, B. (1993) *Ectomicorrizas y micorrizas vesiculo-arbusculares en caatinga amazónica del sur de Venezuela.* Scientia Guaianae, 3. Reflofit CA, Caracas.

Mullins, E.J. and McKnight, T.S. (1981) *Canadian Woods – Their Properties and Uses.* University of Toronto Press, Toronto, Canada.

Myers, J.G. (1936) Savanna and forest vegetation of the interior Guiana plateau. *Journal of Ecology* 24, 162–184.

Nelson, B.W., Ferreira, L.V., da Silva, M.F. and Kawaski, M.L. (1990) Endemism centres, refugia and botanical collection density in Brazilian Amazonia. *Nature* 345, 714–716.

Nelson, B.W., Kapos, V., Adams, J.B., Oliveira, W.J., Braun, O.P.G. and do Amaral I.L. (1994) Forest disturbance by large blowdowns in the Brazilian Amazon. *Ecology* 75, 853–858.

Norris, D.O. (1969) Observations on the nodulation status of rainforest leguminous species in Amazonia and Guyana. *Tropical Agriculture* 46, 145–151.

Ogden, J. (1966) Ordination studies on a small area of tropical rain forest. MSc Forestry, University of Wales, Bangor, UK.

Oliphant, F.M. (1938) *The Commercial Possibilities and Development of the Forests of British Guiana.* The Argosy Co. Ltd, Georgetown, Guyana.

Orians, G.H. (1982) The influence of tree-falls in tropical forests on tree species richness. *Tropical Ecology* 23, 255–279.

Pennington, T.D. (1981) *Meliaceae.* Flora Neotropica Monographs 28. NYBG, New York.

Perreijn, K. (2002) *Symbiotic nitrogen fixation by leguminous trees in tropical rain forest in Guyana.* Tropenbos-Guyana Series 11. Tropenbos-Guyana, Georgetown.

Pitman, W.C.I., Cande, S., LaBrecque, J.L. and Pindell, J.L. (1993) Fragmentation of Gondwana: the separation of Africa from South America. In: Goldblatt, P. (ed.) *Biological Relationships between Africa and South America.* Yale University Press, London, pp. 15–34.

Pitman, N.C.A., Terborgh, J., Silman, M.R., Nuñez-Vargas, P., Neill, D.A., Cerón, C.E., Palacios, W.A. and Aulestia, M. (2001) Dominance and distribution of tree species in upper Amazonian terra firme forests. *Ecology* 82, 2101–2117.

Poncy, O., Riéra, B., Larpin, D., Belbenoit, P., Jullien, M., Hoff, M. and Charles-Dominique, P. (1998) The permanent field research station 'Les Nouragues' in the tropical rainforest of French Guiana: current projects and preliminary results on tree diversity, structure and dynamics. In: Dallmaier, F. and Comiskey, J.A. (eds) *Forest Biodiversity in North, Central and South America, and the Caribbean.* Parthenon Publishing/UNESCO, Paris, pp. 385–410.

Prance, G.T. (1973) Phytogeographic support for the theory of Pleistocene forest refuges in the Amazon Basin, based on evidence from distribution patterns in Caryocaraceae, Chrysobalanaceae, Dichapetalaceae, and Lecythidaceae. *Acta Amazonica* 3, 5–28.

Prance, G.T. (1981) The changing forests. In: Prance, G.T. and Lovejoy, T.E. (eds) *Key Environments: Amazon.* Pergamon Press, London, pp. 146–165.

Putzer, H. (1984) The geological evolution of the Amazon basin and its mineral resources. In: Sioli, H. (ed.) *The Amazon: Limnology and Landscape Ecology of a Mighty Tropical River and its Basin.* Dr. W. Junk, Dordrecht, The Netherlands, pp. 15–46.

Raaimakers, D. (1995) *Growth of Tropical Rainforest Trees as Dependent on Phosphorus Supply.* Tropenbos Series 11. Tropenbos Foundation, Ede, The Netherlands.

Ramón, J.C., Dzou, L.I., Hughes, W.B. and Holba, A.G. (2001) Evolution of the Cretaceous organic facies in Colombia: implications for oil composition. *Journal of South American Earth Sciences* 14, 31–50.

Räsänen, M. (1993) La geohistoria y geologia de la Amazonia Peruana. In: Danjoy, W. (ed.) *Amazonia Peruana. Vegetatcion Humeda Tropical en el Llano Subandino.* PAUT/ONERN, Jyväskylä, Finland, pp. 43–65.

Raven, P.H. and Axelrod, D.I. (1974) Angiosperm biogeography and past continental movements. *Annals of the Missouri Botanical Garden* 61, 539–673.

Record, S.J. and Hess, R.W. (1943) *Timbers of the New World.* Yale University Press, New Haven, Connecticut.

Richards, P.W. (1952) *The Tropical Rain Forest.* Cambridge University Press, Oxford, UK.

Richardson, J.E., Pennington, R.T., Pennington, T.D. and Hollingsworth, P.M. (2001) Rapid diversification of a species-rich genus of neotropical rain forest trees. *Science* 293, 2242–2245.

Ricklefs, R.E. (1987) Community diversity: relative roles of local and regional processes. *Science* 235, 167–171.

Rohwer, J., Richter, H. and van der Werff, H. (1991) Two new genera of neotropical Lauraceae and critical remarks on their generic delimitation. *Annals of the Missouri Botanical Garden* 78, 388–400.

Rollet, B. (1969) *L'architecture des Forêts Denses Humides Sempervirentes de Plaine.* CTFT, Nogent Sur Marne, France.

Romero, E.J. (1993) South American paleofloras. In: Goldblatt, P. (ed.) *Biological Relationships between Africa and South America.* Yale University Press, New Haven, Connecticut, pp. 62–85.

Rose, S.A. (2000) *Seeds, Seedlings and Gaps – Size Matters.* Tropenbos-Guyana Series 9. Tropenbos Foundation, Wageningen, The Netherlands.

Salas, L., Berry, P.E. and Goldstein, I. (1997) Composición y estructura de una comunidad de árboles grandes en el valle del Río Tabaro, Venezuela: una muestra de 18.75 ha. In: Huber, O. and Rosales, J. (eds) *Ecologia de la Cuenca del Río Caura, Venezuela II. Estudios Especiales*. Refolit CA, Caracas, pp. 291–308.

Saldarriaga, J.G. (1987) Recovery following shifting cultivation. In: Jordan, C.F. (ed.) *Amazonian Rain Forests – Ecosystem Disturbance and Recovery*. Springer, London, pp. 24–33.

Saldarriaga, J.G. (1994) *Recuperación de la Selva de 'Tierra Firme' en el Alto Río Negro Amazonia Colombiana–Venezolana*. Estudios en la Amazonia Colombiana 5. Tropenbos Colombia Programme, Bogotá.

Saldarriaga, J.G. and West, D.C. (1986) Holocene fires in the northern Amazon basin. *Quaternary Research* 26, 358–366.

Salisbury, E.J. (1974) Seed size and mass in relation to environment. *Proceedings of the Royal Society* 186, 83–88.

Salo, J., Kalliola, R., Hakkinen, I., Makinen, Y., Niemela, P., Puhakka, M. and Coley, P.D. (1986) River dynamics and the diversity of Amazon lowland forest. *Science* 322, 254–258.

Schulz, J.P. (1960) Ecological studies on rain forest in Northern Suriname. *Verhandelingen der Koninklijke Nederlandse Akademie van Wetenschappen, Afd. Natuurkunde* 163, 1–250.

Schupp, E.W., Howe, H.F., Augspurger, C.K. and Levey, D.J. (1989) Arrival and survival in tropical treefall gaps. *Ecology* 70, 562–564.

Smith, D.N. and Killeen, T.J. (1998) A comparison of the structure and composition of montane and lowland tropical forest in the Serrania Pilón Lajas, Beni, Bolivia. In: Dallmaier, F. and Comiskey, J.A. (eds) *Forest Biodiversity in North, Central and South America, and the Caribbean*. Parthenon Publishing/UNESCO, Paris, pp. 681–700.

Soerianegara, I. and Lemmens, R.H.M.J. (eds) (1993) *Plant Resources of South-East Asia*. Pudoc, Wageningen, The Netherlands.

Suriname Forest Service (1955) *Surinam Timber*. Eldorado, Paramaribo.

Takhtajan, A. (1997) *Diversity and Classification of Flowering Plants*. Columbia University Press, New York.

ter Steege, H. (1993) *Patterns in Tropical Rain Forest in Guyana*. Tropenbos Series 3. Tropenbos Foundation, Ede, The Netherlands.

ter Steege, H. (2000) *Plant Diversity in Guyana*. Tropenbos Series 18. Tropenbos Foundation, Wageningen, The Netherlands.

ter Steege, H. and Hammond, D.S. (2001) Character convergence, diversity, and disturbances in tropical rain forest in Guyana. *Ecology* 82, 3197–3212.

ter Steege, H. and Zondervan, G. (2000) A preliminary analysis of large-scale forest inventory data of the Guiana Shield. In: ter Steege, H. (ed.) *Plant Diversity in Guyana*. Tropenbos Foundation, Wageningen, The Netherlands, pp. 35–54.

ter Steege, H., Jetten, V.G., Polak, A.M. and Werger, M.J.A. (1993) Tropical rain forest types and soil factors in a watershed area in Guyana. *Journal of Vegetation Science* 4, 705–716.

ter Steege, H., Sabatier, D., Castellanos, H., van Andel, T., Duivenvoorden, J.F., de Oliveira, A.A., Ek, R.C., Lilwah, R., Maas, P.J.M. and Mori, S.A. (2000) An analysis of Amazonian floristic composition, including those of the Guiana Shield. *Journal of Tropical Ecology* 16, 801–828.

ter Steege, H., Welch, I.A. and Zagt, R. (2002) Long-term effect of timber harvesting in the Bartica Triangle, Central Guyana. *Forest Ecology and Management* 170, 127–144.

Thompson, J., Proctor, J., Viana, V., Milliken, W., Ratter, J.A. and Scott, D.A. (1992) Ecological studies on a lowland evergreen rain forest on Maracá Island, Roraima, Brazil. I. Physical environment, forest structure and leaf chemistry. *Journal of Ecology* 80, 689–703.

Toriola-Lafuente, D. (1997) Regeneration naturelle en Guyane française: ARBOCEL, une juene forêt secondaire de 19 ans. Thèse de Doctorat. L'Université Paris 6. Paris.

Valencia, R., Balslev, H. and Paz y Miño, C.G. (1994) High tree alpha-diversity in Amazonian Ecuador. *Biodiversity and Conservation* 3, 21–28.

van Andel, T. (2000) *Non-timber Forest Products of the North West District of Guyana*. Part I. Tropenbos-Guyana Series 8A. Tropenbos-Guyana Programme, Georgetown.

van der Hammen, T. (1972) Changes in vegetation and climate in the Amazon basin and surrounding areas during the Pleistocene. *Geologie en Mijnbouw* 51, 641–643.

van der Hammen, T. (1974) The Pleistocene changes of vegetation and climate in tropical South America. *Journal of Biogeography* 1, 3–26.

van der Hout, P. (1999) *Reduced Impact Logging in the Tropical Rain Forest of Guyana. Ecological, Economic and Silvicultural Consequences*. Tropenbos-Guyana Series 6. Tropenbos-Guyana Programme, Georgetown.

van der Slooten, H.J. and Martinez, P. (1959) *Descripcion y Propiedades de Algunas Maderas Venezolanas.* Instituto Forestal Latino Americano, Merida.

van der Werff, H. (1991) A key to the genera of Lauraceae in the New World. *Annals of the Missouri Botanical Garden* 78, 377–387.

Vink, A.T. (1970) *Forestry in Surinam – a Review of Policy, Planning, Projects, Progress.* Suriname Forest Service (LBB), Paramaribo.

Vuilleumier, B.S. (1971) Pleistocene changes in the fauna and flora of South America. *Science* 1973, 771–779.

Wadsworth, F.H.G.H.E. (1959) Effects of the 1956 hurricane on forests in Puerto Rico. *Caribbean Forester* 20, 38–51.

Webb, L.J. (1958) Cyclones as an ecological factor in the tropical lowland rain forest, North Queensland. *Australian Journal of Botany* 6, 220–230.

Welch, I.A. and Bell, G.S. (1971) *Great Falls Inventory.* Guyana Forest Department, Georgetown.

Westoby, M., Leishman, M.R. and Lord, J. (1997) Comparative ecology of seed size and dispersal. In: Silvertown, J., Franco, M. and Harper, J.L. (eds) *Plant Life Histories: Ecology, Phylogeny and Evolution.* The Royal Society, London, pp. 143–162.

Whitmore, J.L. (1973) Wood density variation in Costa Rican balsa. *Wood Science* 5, 223–229.

Whitmore, T.C. (1974) *Change with Time and the Role of Cyclones in Tropical Rain Forest on Kolombangara, Solomon Islands.* Paper 46. Commonwealth Forestry Institute.

Whitmore, T.C. and da Silva, J.N.M. (1990) Brazil rain forest timbers are mostly very dense. *Commonwealth Forestry Review* 69, 87–90.

Whitmore, T.C. and Prance, G.T. (1987) *Biogeography and Quaternary History in Tropical America.* Oxford Science, Oxford, UK.

Whitton, B.A. (1962) Forests and dominant legumes of the Amatuk region, British Guiana. *Caribbean Forester* 23, 35–57.

Wiemann, M.C. and Williamson, G.B. (1988) Extreme radial changes in wood specific gravity in some tropical pioneers. *Wood and Fiber Science* 20, 344–349.

Wiemann, M.C. and Williamson, G.B. (1989a) Radial gradients in the specific gravity of wood in some tropical and temperate trees. *Forest Science* 35, 197–210.

Wiemann, M.C. and Williamson, G.B. (1989b) Wood specific gravity gradient in tropical dry and montane rain forest trees. *American Journal of Botany* 76, 924–928.

Williams, L. (1939) *Maderas Economicas de Venezuela.* Boletin Tecnica No. 2. Ministerio de Agricultura y Cria, Caracas.

Williamson, G.B. (1984) Gradients in wood specific gravity of trees. *Bulletin of the Torrey Botanical Club* 111, 51–55.

Wood, B.R. (1926) *Report by the Conservator of Forests on the Valuation of the Forests of the Bartica-Kaburi Area.* BGFD, Georgetown.

Zagt, R.J. (1997) *Tree Demography in the Tropical Rain Forest of Guyana.* Tropenbos-Guyana Series 3. Tropenbos-Guyana Programme, Georgetown.

Zobel, B.J. and van Buijtenen, J.P. (1989) *Wood Variation, its Causes and Control.* Springer, London.

8 Socio-economic Aspects of Guiana Shield Forest Use

David S. Hammond
Iwokrama International Centre for Rain Forest Conservation and Development, Georgetown, Guyana. Currently: NWFS Consulting, Beaverton, Oregon, USA

Lowland Tropical Forests, Limiting or Limited Environment?

Tropical forests are typically described and categorized by their plant community composition and structure and the animals that use them for shelter and food. The fact that people, communities and societies, with their own unique cultural and social dimensions, have inhabited much, if not most, neotropical forests at one point or another over the last 12,000–15,000 years is increasingly difficult to dispute (Bryan, 1973; Barse, 1990; Roosevelt, 1991a). Yet biophysical and anthropological perspectives on the forces driving change in both lowland forests and indigenous societies have largely remained separate. From the anthropological perspective, this pigeonholing has thrown discussion into a polemical, all-or-nothing, debate – either the prevailing environment has controlled socio-cultural change (Meggers, 1954, 1973, 1977, 1984, 1996; Steward, 1955; Lowenstein, 1973) or social structure has strongly influenced forest change (e.g. see Balée, 1992, 1998; Denevan, 1992; Stahl, 1996; Heckenberger *et al.*, 1999, 2003).

On one side is the concept of prevailing environmental conditions limiting how societies in lowland forest regions developed – what rituals they performed, how they went about meeting basic needs and the measure and magnitude of social organization they achieved. Given an extremely infertile tropical landscape, prehistoric cultures remained at the most basic levels of organization, practised few sophisticated rituals (such as elaborate burial) and rarely remained in one place through a generation. Their capacity to alter a 'pristine' forest was constrained as much as the forest constrained their capacity to achieve higher social complexity. However, by focusing largely on the role of low-amplitude, continuous environmental change in shaping prehistoric cultural development, environmental determinism fails to account for the role of social cooperation in overcoming environmental challenges (e.g. see arguments put forth by Erickson, 1999; Dillehay and Kolata, 2004). Cooperation, and its many benefits, is arguably one of the most important factors sustaining social gregariousness.

Conversely, assuming that environmental factors rarely limit sociocultural growth and change fails to appropriately weight the influence of punctuated, high amplitude events or phases that (continue to) test the resiliency of societies around the world. During these periods, the same gregariousness that assisted cooperation for mutual benefit also catalysed catastrophic loss by accelerating the impact of disease, famine and war under growing environmental limitations (and heightened social inequities as a consequence). Tremendous evidence has accrued to the fact that catastrophic loss and efforts to avoid these events were deeply inculcated in the spiritual and cultural beliefs that drove social thinking even in the most successful civilizations. Incipient environmental challenges can lead to subsequent cultural decline through dissipation of dominant social structures, as in the case of the Maya (Curtis *et al.*, 1996) and Tiwanaku (Binford *et al.*, 1997), particularly when sophisticated spiritual and cultural systems typical of strongly hierarchical societies fail to overcome these challenges. Environmental challenges may also plausibly retard growth in social cohesion and population within much smaller communities, particularly when their position within the sociocultural landscape renders them less capable of weathering catastrophic environmental change relative to stronger, adjacent communities and cooperation between these neighbours is weak or non-existent.[1]

Limiting and limited environment

The universal application of environmental limits as a homogeneous force acting upon all pre-Hispanic Amerindian communities inhabiting the lowland neotropics appears counterintuitive when considering the reality of biophysical variation across the region (see Chapter 2) (Roosevelt, 1998, p. 192; 1999, p. 372). At the same time, a belief that human inhabitation fundamentally altered most of 'Amazonia' and that this always led to an improvement in land capability ostensibly lacks the material evidence needed to support this sweeping counter-viewpoint (e.g. see Lavallée, 2000). Despite hyperbolic claims of forest manipulation, local examples of more advanced pre-Hispanic Amerindian settlements in the neotropical lowlands have been unearthed that clearly show land management techniques, such as raised fields, causeways, drainage canals and soil enrichment (Denevan, 1970; Parsons and Denevan, 1974; Spencer *et al.*, 1994, 1998; Erickson, 2000). These are akin to but arguably not as developed (or well preserved) as those employed by advanced highland civilizations in Mexico, Central America and the Andes (e.g. the *chinampas* of Lake Xochimilco and Chalco, Vallé de Mexico) (see later, 'Site types: earth engineering').

Acceptance of a uniformly limiting lowland environment would have to account first for the interaction between the historical geography of prehistoric colonization and social development and the geographic variation in biophysical dynamics across the region (see Chapter 2) in a way that maintains communities in a highly isolated state with little capacity or motivation to innovate, build, trade or identify more economic settlement locations. The colonization of all lowland regions of South America could not have occurred simultaneously. Available evidence suggests that this process occurred along lines of least risk and greatest familiarity, demarcated by the major waterways and coastlines (Rouse and Cruxent, 1963; Dillehay, 2000; Lavallée, 2000) as has been argued in the case of North America prehistory (Dixon, 1999). Common environmental changes, such as severe drought, flooding, fire, disease, mudslides, earthquakes, volcanic eruptions, hurricanes and other environmental tests of social resiliency have not affected all parts of the lowlands in the same way or to the same degree over the Quaternary (see Chapters 2 and 7), so the environmental limitation theory must assume these as redundant to high-frequency challenges presented by more protracted (or low amplitude) attributes of a changing landscape, such as soil, rock and topography. But these too are not uniform across the lowlands of South America, even by the strictest demarcation of the lowland forest ecosystem (see Chapter 2). More importantly, they could be engineered to suit human needs – high amplitude changes to the landscape conveyed by catastrophic weather and diastrophic events were (and remain) less easily managed. The interaction of these varying environmental conditions and human geographies supports a multiplex view of human–forest relationships – in some regions the relationship was more likely driven principally by sociocultural development and expansion, in others it was driven predominantly by biophysical limitations.

Assumption of pristine beginnings

Ecologists, by seeking exclusive biophysical explanations for spatial patterns of standing biological diversity, assume implicitly that past effects of human activity were spatially uniform (see Roosevelt, 1999, p. 385). By default the adopted view is one of an environment uniformly limiting prehistoric sociocultural growth across the lowland neotropics (Meggers, 1954) combined with an assumption that modern patterns of inhabitation and land use roughly follow prehistoric ones. By selecting remote locations, samples are stratified in relation to modern patterns and thus address the effects of broad-ranging variation in prehistoric human impact. But then an assumption of geological, pedological and climatic uniformity is made. Again, neither assumption is valid (see Balée, 1992, 1998; Chapter 2, this volume).

In the case of highly organized pre-Columbian societies, such as the Maya or Tiwanaku, both perspectives appear equally true – innovation and social structure drove novel means of extremely productive, well-managed land use in some areas (Siemens and Puleston, 1972; Matheny, 1976) (e.g. around lakes, floodplains, wetlands), but perhaps less so in others (e.g. rocky upland slopes, sandy plateaux). The role of intensified drought as a social dismantling factor is increasingly difficult to deny in the case of large Amerindian civilizations, such as the Maya (Curtis *et al.*, 1996; Gill, 2000; Hodell *et al.*, 2001; Haug *et al.*, 2003).

But how would drought resonate through such societies, leading to their dismantling and thus change in forest use practices? Plausible avenues of decline may simply mirror more recent agricultural practices. For indigenous communities working within the *ejido* system across southern Mexico, community growth leads to the formation of landholding and landless groups within the farming collective, a *minifundio*, despite intentions to the contrary (Simpson, 1937). In the case of the Zoque of Chiapas, landless *ejidatarios* are forced to cultivate marginal lands on the boundary of the community holding. The management of these rocky slopes for agricultural production is difficult, particularly during periods of extreme weather. In contrast, the more fertile valleys below are managed more easily and extensively by the few large, land-owning members (Hammond, 1991). Management tools in both past and present Maya agroforestry and religious practices have also modified forest composition through the selective retention of useful trees and groves (e.g. *Brosimum*, *Spondias*, *Manilkara*) (Puleston, 1982; Gómez-Pompa, 1987; Gómez-Pompa *et al.*, 1990; Hammond, 1991). Many of the trees retained by the Maya, Uto-Azteca, Zoque, Zapotecs and their descendants are also commonly found throughout much of the mature forest area across modern-day southern Mexico, Guatemela and Belize (e.g. Breedlove, 1981).

Thus, growing social inequalities, forest management decisions and accelerating phases of environmental change may have combined to drive the rate, intensity and type of forest use by prehistoric lowland Mayan communities, as they do today. Social and biophysical forces interacted dynamically to define the intensity and geographic extent of forest modification by pre-Hispanic Mayan societies and, in turn, the effects of environment on social change. In the lowland tropical forests of South America, large permanent structures like those that have assisted in defining the spatial distribution of Maya populations remain largely undiscovered or non-existent. Where structures are identified in lowland forest areas, they are typically linked to exposed rock formations or large river margins and yield less insight into use of the surrounding forests growing on deep soils of undulating *terra firme*. This unresolved footprint of prehistoric human impact confounds efforts to demonstrate causal linkages between patterns of standing forest diversity and biophysical attributes, particularly where these attributes and those characterizing past human forest use ultimately prove to be spatially concordant.

Do patterns of standing diversity purely reflect geographic shifts in climate, topography and soil attributes or the influence of people selectively manipulating forests along these physical gradients? What if a fraction of remote sample plots used to explore patterns of plant diversity were under field-fallow management or favoured as hunting grounds several thousand years ago? Would the inflection of this forest use still resonate in the standing diversity and structure registered today in remote forests? If so, then the biophysical trajectory defining forest change at a site may also be constrained by its position along an intensity gradient of prehistoric land use. Community, population and even (long-lived) individual responses measured in modern time may partially reflect historical modification of composition along a selection axis defined at one point by human resource decision-making, rather than by plant life history–population dynamics alone.

Evidence accruing through archaeological, ethnobotanical, phytolithic and charcoal studies increasingly supports this view. Many lowland neotropical forest sites previously considered as pristine are now in fact considered modified as a consequence of human use at various times throughout the Holocene (Eden *et al.*, 1984; Piperno, 1990, 1994; Kennedy and Horn, 1997; Wood and McCann, 1999; Piperno and Jones, 2003). Balée refers to these forests strongly shaped by past human use as 'artefactual landscapes' and suggests that they are the remnants of former agricultural regression–expansion phases driven by sociopolitical, rather than environmental, forces. War, disease and poor decision-making forced large, organized chiefdoms to dissolve into nomadic bands. The relative importance of foraging and horticultural practices in sustaining livelihoods, and thus impacts on forest cover and composition, oscillated with these changes in social structure. Again, however, explanations ascribing total control over lowland social structure to sociopolitical events fail to integrate the clear relationship between resource scarcity and societal focus. Spatial and temporal variation in resource scarcity would more likely have interacted with the strength of sociocultural cohesion and innovation to shape forest resource use patterns, as they do today (see Fig. 1.1, Chapter 1).

Modern land-use and forest change

The relationship between contemporary land-use patterns and forest change is, for obvious reasons of data availability, much clearer. The web of feedback linkages is far from fully resolved, particularly in the case of the lowland neotropics, but the relationship between socio-economic and biophysical drivers of forest change is well-documented (Moran, 1982; Hecht, 1985; Buschbacher, 1986). Interestingly, many causes of modern forest use in the Amazon appear to have very little to do with rational land use and everything to do with sociopolitics, a fact not to be overlooked while seeking to sculpt a robust interpretation of relationships between scattered archaeological evidence and the prehistoric record of environmental change.

Dialogue on modern lowland forest land-use patterns unfortunately also suffers from an inherent tendency to scale-up site-specific findings. Much like the debate over the prehistoric timing and extent of human lowland occupation, the geographic variation, scale-dependency and relative impact of modern lowland land-use practices often get lost in the drive to scale-up applicability. As a consequence, concepts of environmental and historical lowland uniformity are perpetuated (the 'Amazon'). Continuous impact gradients are transformed into binomial ones, despite considerable, high-quality multidisciplinary field evidence arguing to the contrary. Deforestation and defaunation (both forms of liquidation) define the extreme end of a resource-use gradient that equally embraces lower intensities of selective logging, hunting, field agriculture and plant extraction, through to a minimum allowable use regime defining traditional sacred and modern protected area concepts. The primacy of one end of the spectrum over the other varies geographically and with spatial scale as socio-economic and biophysical features intersect to define land use and its consequences.

About This Chapter

The Guiana Shield defines one end of this spectrum and the socio-economic and cultural influences on its past and future forest landscape are not necessarily consonant with other regions of the neotropics. Thus, this chapter aims to illustrate how the region fits within the broader socio-economic and (pre)historic human landscape. It briefly describes and explores the interaction between prehistoric, historic and modern socio-economic features and their relationship with forest resource use throughout the Guiana Shield as a conduit to further exploring the changing geographic extent and focus of human use.

Prehistoric archaeological evidence and the historical record of colonization can assist in delimiting spatial variation of anthropogenic impacts and the likely measure of forest modification that occurred as a consequence. It can also help establish how resource use objectives and methods have changed, and the implications these changes have had for forest conservation and sustainable use and the way in which people used and managed their local forest resources. The merging of immigrants, whether forced through slavery, offered through indenture or undertaken through the prospect of opportunity, and indigenous societies created a new forest-use dynamic that has in many areas changed the forest trajectory for the foreseeable future.

The chapter also broadly characterizes the modern historical trends in population growth and distribution, socio-economic conditions and forest resource use as key processes intertwining with biophysical features to shape tropical forest change across the Guiana Shield. It does not cover the wide-ranging political dimension of modern forest use in the region, including the implications of land ownership, multinational incorporation, public financial incentives (or subsidies) to forest-use industries, or international relations, among others, although these are clearly important and have been widely discussed elsewhere. History, health, education, law and other facets of our social systems are, and historically have been, fundamental drivers of forest conservation and use patterns (see Fig. 1.1, Chapter 1). There is good reason, and sufficient evidence, to believe that these also played an equally important role in pre-Columbian societies. Exploring biophysical explanations alone unintentionally dismisses the role of humans in shaping modern-day forests of the region prior to and following the arrival of the first Europeans. Given the vast and relatively accessible literature on modern history, an account of modern (post-1900) political change in the Guiana Shield countries has not been included here, although it clearly has driven much of the forest use and protection in the region.

As with all synthetic works of this kind, it is far from definitive and the reader is referred to the many references included in this and other chapters of the volume for root perspectives and results. The chapter does not address the taxonomic relationships between the various archaeological traditions, complexes and series that are typically employed to distinguish cultural 'phases' based on stylistic interpretation of artefacts. Based on the review conducted here, it appears that more often than not, new styles lead to new taxa, creating a proliferation of site-based classifications. These are less often explained as part of a larger, emerging picture of cultural development in South America (Lavallée, 2000, p. 107). Establishing a coherent chronology among the diverse artefacts at one site is typically beset by methodological issues, often steeped in controversy, let alone among the hundreds of excavated sites and solitary artefact finds located along most lowland rivers. While potentially resolving the long-term link between spatial and temporal patterns of forest use, an intensive look at prehistoric migration across the lowland forests of the Guiana Shield is beyond the scope of this chapter and references to the various traditions, phases and series have been avoided intentionally. A basic system comprised of Palaeo-Indian (earlier than 6.5 ka BP), Meso-Indian (6.5–3.0 ka BP), Neo-Indian (3.0–0.5) and Indo-Hispanic epochs (0.5–0.1) (after Rouse and Cruxent, 1963) has been adopted where larger time intervals require distinction, or evidence is assigned an age based on similarities with other styles rather than material dating.

Human Prehistory of the Guiana Shield

Entry, dispersal and population growth in the neotropics

Three facets of human prehistory in the lowland neotropics continue to remain unsatisfactorily resolved: (i) the timing and location(s) of initial arrival; (ii) the ensuing pattern of colonization across the region; and (iii) the maximum density the lowland populations achieved prior to the arrival of the first Europeans. All three have substantial bearing on the extent and intensity of prehistoric forest modification because together they define how and in what way forest resources were utilized, and managed, over a period that would have seen the standing forest biomass naturally turnover between 20 and 2400 times based on estimates for modern forest gap formation and stem mortality (see Chapter 7).

The peopling of the lowland neotropics is believed by many anthropologists, archaeologists and linguists to have commenced around 12,000 years ago as a southward movement of the initial Clovis migration across the Bering Strait land bridge and ice sheets. Many others, however, argue that evidence increasingly points to a pre-Clovis peopling of South America, either through a more rapid seaborne southward movement along the Pacific coastline or as an entirely separate event not linked to the Bering Strait route (Dixon, 1999; Dillehay, 2000; Lavallée, 2000). Radiocarbon-dated materials associated with several archaeological sites have been posited as support for an earlier inhabitation of lowland areas, although these thus far are clearly associated with coastal, rather than deep interior, regions and only adjust the earliest known entry by a millennium at most (Roosevelt *et al.*, 1996; Sandweiss *et al.*, 1998; Dillehay, 1997).[2] The exception is a series of cave sites near the São Francisco River valley in the eastern Brazilian Shield that have associated charcoal radiocarbon dated between 32 and 56+ ka BP (Guidon and Delibrias, 1986; Bahn, 1993; Santos *et al.*, 2003), based on new oxidation pretreatment techniques (Bird *et al.*, 1999) (Fig. 8.1).[3] At another site in the same region, submitted U-Th decay dates for putative quartzic artefacts range from 204 to 295 ka BP (de Lumley *et al.*, 1988). Ancillary approaches based on linguistic (Nichols, 1990) and molecular (Torroni *et al.*, 1994) analyses have also argued for a significantly earlier entry date, some 20–30 ka BP, assuming entry through a single group. Studies employing alternative linguistic, molecular and anthropological techniques, however, have also yielded results that fully support the original view of a later, Clovis entry (Nettle, 1999; Brace *et al.*, 2001; Zegura *et al.*, 2004). The growing number of near Clovis dates for South American locations combined with more extensive documentation and site validation (Dillehay, 1997) generally support the view of an entry into South America occurring much faster and earlier than previously thought.

The movement of prehistoric peoples across the neotropics has always remained clouded by shifting theoretical perspectives and the inability to resolve a clear spatial chronology from material finds that can be complicated by patterns of trade (Lathrap, 1973), nomadism (Rostain, 1994), commu-

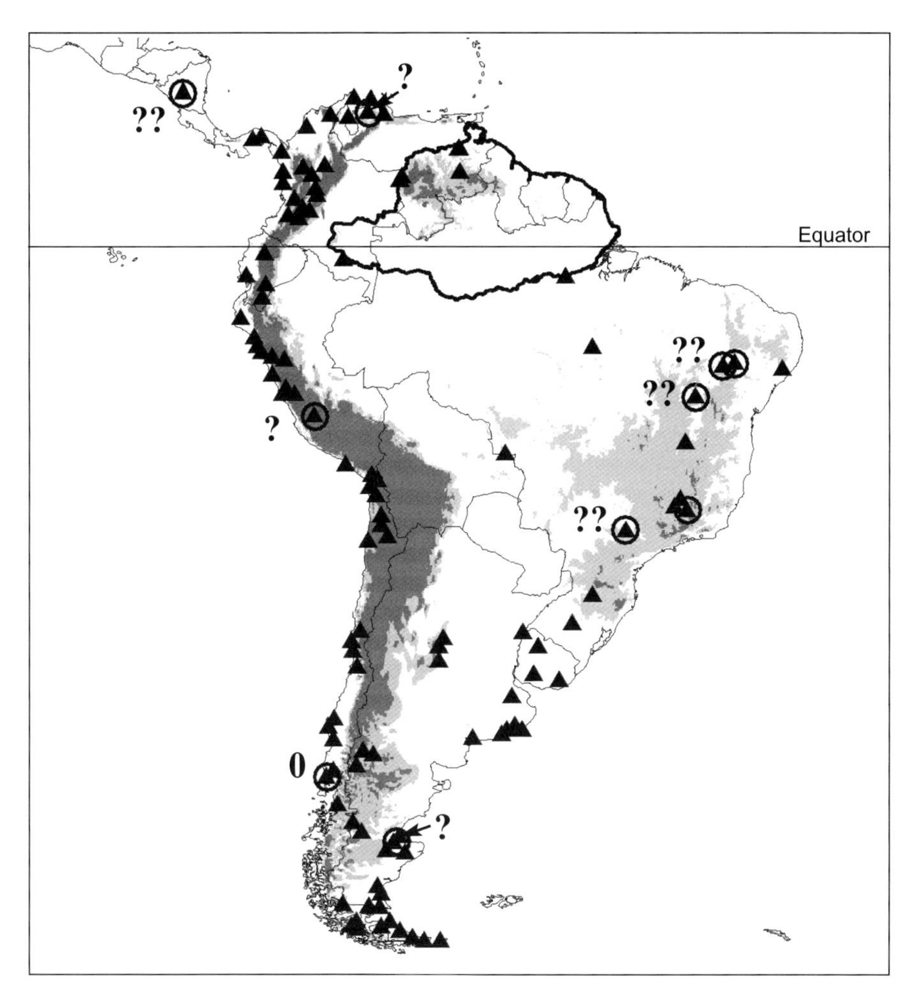

Fig. 8.1. Archaeological sites (solid triangles) proposed to reflect occupation by humans more than 8000 years BP (late Pleistocene–early Holocene). Circled locations are those proposed as pre-Clovis (c. 12,500 BP) inhabitation sites. Question mark number indicates relative lack of confidence/controversy in evidence supporting antiquity of the occupation from considerable (??) to very little (0). Site locations and age uncertainties adapted from assessments made by Lavallée (2000), Dillehay (2000) and Gassón (2002). Note the absence of any putative pre-Clovis sites from the Guiana Shield and larger lowland Amazonia.

nity fission–fusion (Meggers, 1995), migratory overlap, material and site recycling (DeBoer *et al.*, 1996; Meggers, 2003) and questions regarding assigned provenance of charcoal and other artefacts typically used to date occupation (Meltzer *et al.*, 1994). This latter source of error is particularly important in establishing the early Holocene occupations, since supporting evidence at virtually all of these sites is anchored by charcoal radiocarbon dates. The error arises from the fact that charcoal dated to the early Holocene is not always associated with archaeological remains (see Chapter 2) (Hammond *et al.*, unpublished results), is susceptible to size-dependent vertical and horizontal movement (e.g. in lake sediments: Whitlock and Millspaugh, 1996) and lacks the calibration that would allow anthropogenic charcoal to be distinguished as a matter of probability from that accumulated as a consequence of vegetation fires across the neotropics (e.g. eastern England: Bennett *et al.*, 1990).

Assuming earliest charcoal dates faithfully indicate timing of site occupation, a picture of wide-ranging colonization between 14 ka and 8 ka BP is apparent.[4] Both coastal and overland routes have been proposed at one time or another, but continue to shift as new sites are integrated and others removed from the accepted chronology (Meltzer, 1995; Dixon, 1999) (Fig. 8.1). Using the most recently proposed Pleistocene–early Holocene sites, the most favoured early route would be along the Pacific coastline, with subsequent movement eastward and down the Atlantic coast towards southeastern Brazil with colonization occurring along the Caribbean coast of Venezuela, the lower Amazon, and up the São Francisco watershed along the eastern rim of the Brazilian Shield. Sea level along both coasts would have been 20–100 m lower than present (see Fig. 2.36), depending on timing and extent of local isostatic, steric and epeirogenic adjustments (see 'Sea-level change – Neogene', Chapter 2). If early settlers were largely engaged in maritime livelihoods (Sandweiss *et al.*, 1998), then many of the earliest settlements may currently be submerged along key coastal stretches (Dillehay, 2000). This would seem all the more likely along coastal lengths where the exposed Pleistocene continental shelf was wide (e.g. along the Guiana coast between Isla de Margarita, Venezuela and Parnaiba River in Maranhão state, eastern Brazil) and less likely where there was only a narrow shelf exposure (south of the Parnaiba) (Fig. 2.37). Interestingly, the two oldest sites at Monte Verde, Chile and Pedra Pintura, Brazil (Fig. 8.1) are located adjacent to coastal regions with narrow shelves. No modern coastal sites of similar antiquity have been identified along the much wider tropical Atlantic shelf regions (see below).

Entry and early occupation in the Guiana Shield (12–7 ka BP)

If early movement and colonization along the eastern rim of the Guiana Shield occurred during the Pleistocene termination, site evidence of these is probably buried seaward of the current coastline. None of the existing archaeological, linguistic, molecular or anthropological data support an occupation of the large, central interior (>300 km inland) of Amazonia earlier than 9200 years BP. The earliest physical dates come from interior sites along the upper Orinoco (Barse, 1990), along the northern shield margin, at Monte Alegre (near the mouth of the Maicaru River, N. Pará), along the southern shield margin (Roosevelt *et al.*, 1996), in the Serra do Carajás region of central Brazil and along the western shield margin at Peña Roja (Gnecco and Mora, 1997). The location of all of these sites could also be seen as peripheral to the central Amazonian lowlands, given their proximity to extensive shrub or grasslands (upper Orinoco, Serra do Carajás) or coastal river mouths (Monte Alegre), except Peña Roja, where uncertainty remains over the provenance of the dated charcoal deposits. No material from sites within the central Guiana Shield thus far have been dated to this early period. The absence of materials in the region dating to the earliest millennia is, however, difficult to interpret in the context of entry into the Guiana Shield for several reasons.

First, considerable archaeological materials have been found throughout much of the Guiana Shield but anchored by few radiometric dates. Instead, seriation combined with stratigraphic techniques were principally used in an attempt to reconstruct cultural timelines, migration pathways and trade relationships across the region (e.g. Howard, 1947; Evans and Meggers, 1960). These are exclusively based on the stylistic and stratigraphical association of ceramic materials (see 'Material types: ceramics', below), an approach that governed archaeological focus during the peak period of investigation in the region from 1940 to 1980. Ancillary material amenable to direct dating, such as charcoal, pollen, seeds, phytoliths and bone material were less commonly the focus of investigations, although charcoal has long been the material of choice in anchoring timelines for ceramic and other midden remains despite changes in field archaeological methods (see above). Another potential bias reducing the likelihood of older finds in the Guiana Shield is linked to the heightened acidity of the region's soils and rivers (see Chapter 2). Human skeletal remains, the *sine qua non* of human presence, are composed largely of calcium carbonate and would dissolve at a much higher rate under these very acidic conditions relative to drier and/or colder conditions presented in the eastern Brazilian Shield and Andean regions, respectively. Where skeletons have been found in the Guiana Shield, they are often associated with raised, calcium-enriched shell middens (e.g. Williams, 1996a, 1998) or rock outcrops along the savanna periphery (e.g. Evans and Meggers, 1960). Both site types act to buffer the build-up of acidity that otherwise occurs in deep tropical forest soils or coastal areas with peat, cat clay or other stagnic properties (see 'Soils and Soil Fertility', Chapter 2). As a consequence, while occupation by prehistoric groups over most of the riverine and savanna regions along the periphery of the shield is probable, the earliest timing of their entry, duration of occupation and extent of forest use at the heart of the Guiana Shield remain unclear.

The finds along the upper Orinoco and at Monte Alegre and Carajás, along with those at Pedra, preliminarily suggest that some of the earliest occupations of the lowland interior occurred along shield peripheries. The advantages of long-term site stability and upland escape from pests afforded by rock formations along shield perimeters combined with continued access to the heightened productivity along major river mainstems (Orinoco, Amazon and Tocantins) would make these locations, in theory, ideally suited to a livelihood based predominantly on foraging, hunting and fishing (also proposed by Goulding *et al.*, 1996, p. 21). For the same reasons, however, the selective discovery of these ancient materials may reflect an inherent spatial bias in material dating. Akin to peat bogs and lakes that selectively offer superior pollen yields (see Chapter 2), habitation sites established along shield perimeter regions are inherently more likely to remain in a well-preserved condition relative to those established in depressions continuously affected by severe hydrological or diastrophic disturbances. The Amazon Depression, Sub-Andean Trough and Coastal Shelf regions have experienced significantly higher substrate turnover rates throughout the Quaternary over much of their area as a result of river wandering and sea-level change. As a consequence, site mixing or ablation in these areas is far more likely (Lathrap, 1968; Meggers, 1984, p. 642).

The earliest entry into the Guiana Shield would appear consistent with the Palaeo-Indian epoch, dating before 6.5 ka BP, but no earlier than 9 ka BP. This is based on the radiocarbon dates of pre-ceramic materials along the region's perimeter (mentioned above) and selective finds of lithic remains (particularly projectile points) in Venezuelan Guayana (Dupouy, 1956, 1960; Rouse and Cruxent, 1963), western and southern Guyana (Roth, 1924, 1929; Evans and Meggers, 1960) and along the Rio Negro (Roosevelt, 1999, p. 378). Published data, however, thus far suggest that the Guiana Shield was not a first point of settlement for populations that more plausibly moved southward along the Caribbean coast (Evans and Meggers, 1960), across the Venezuelan and Colombian llanos, along the Orinoco (Rouse and Cruxent, 1963; Gassón, 2002), and along the Amazon River between the Atlantic and eastern Andes (Evans and Meggers, 1968) (Fig. 8.1), even though evidence supporting these routes comes largely from assessments of more recent (<6 ka BP) prehistoric cultural connections. Whether these people in fact arrived as a southward extension of the Clovis migration along the spine of the Central and South American cordilleras, via a Pacific coastal route, by transoceanic voyage (e.g. waif arrival), or a combination thereof, current data would indicate that the Guiana Shield was not the first region to be colonized. The upper Orinoco–Rio Negro axis may have been used as a thoroughfare connecting the relatively productive northern llanos and southern Amazon Downwarp, but archaeological evidence from the watersheds of both the Amazon and Orinoco does not to date support an early, long-term occupation of the region. This early occupation is more plausibly assigned to other physiographic regions, such as the Pacific coastal lowlands, eastern Brazilian Shield, Central America or the Andean highlands. While long-term, sedentary occupation over most of the Guiana Shield interior by early foraging proto-horticulturalists appears improbable, this thesis still requires much more intensive study to further argue the case either way. Given the incredibly small search and sampling effort employed across the most remote parts of the shield region (Guayana Highlands, Tumucumaque Uplands), the initial date and extent of occupation by Palaeo-Indian societies remains highly speculative.

Artefact evidence supporting prehistoric occupation of the Guiana Shield

A wide array of artefacts, however, has been collected from various locations within the Guiana Shield (Fig. 8.2). Most of this material currently points to an Amerindian inhabitation dating back to at least 6 ka BP, although not all evidence has been easily integrated into a cultural timeline (e.g. ceramic styles), or typologically connected with other similar materials elsewhere (e.g. petroglyphs, stone implements). The main forms of evidence known from the Guiana Shield can be grouped into material artefact and site type categories. Material artefacts have been grouped here into: (i) lithic; (ii) ceramic; (iii) metalwork; and (iv) organic remains. Site type groups include: (i) caves, grottos and rockshelters; and (ii) earthworks.

Material types: lithic remains

Lithic remains include a diverse group of art and artisanal tools and products. Rock was often used as a canvas (petroglyphs, rock paintings), a medium (tools, stone arrangements, monoliths, jewellery) or a workbench (polissoirs).

Petroglyphs (roche gravées, arte rupestre, rotstekeningen)

DISTRIBUTION Carved rock art has been described from most parts of the world (Chippendale and Taçon, 1998) and in particular across the Americas (Mallery 1893; Rouse, 1949; Dubelaar, 1986). Petroglyphs represent some of the earliest documented evidence of prehistoric occupation, figuring prominently in the journals of early explorers, due largely to their relative visibility and permanence compared to other indicators of prehistoric human presence. They have been described from locations throughout the Guiana Shield over the last 130 years by a large number of archaeologists, anthropologists, geologists, foresters, surveyors and explorers (Brown, 1876; Im Thurn, 1883; Rodway, 1919; Cruxent, 1950; McKenna, 1959; Evans and Meggers, 1960; Guppy, 1961; Hurault *et al.*, 1963; Rouse and Cruxent, 1963; Bubberman, 1973; von Hildebrand, 1975; Dubelaar, 1976, 1981, 1986; Poonai, 1978; Williams, 1978b, 1979b; Dubelaar and Berrangé, 1979; Blair, 1980; Hilbert and Hilbert, 1980; Rostain, 1987, 1994; Rivas, 1993; Greer, 1995; Gassón, 2002) (Fig. 8.2, filled circles). Despite the geographical scale of their distribution across northeastern South America, petroglyph location in the shield and adjacent regions typically adheres to at least two of the following basic criteria, i.e: (i) the punctuated or isolated outcropping of intrusive rock formations; (ii) proximity to water; and (iii) an association with waterways or coastlines that link different parts of the broader region.

Petroglyphs have been found least frequently in savannas encircling the shield region, despite these areas being more amenable to surveying (Fig. 8.2). It is important to note that several significant petroglyph groups have, however, been recorded in the Sipaliwini (Bubberman, 1973), Parú (Frikel, 1969) and Rupununi

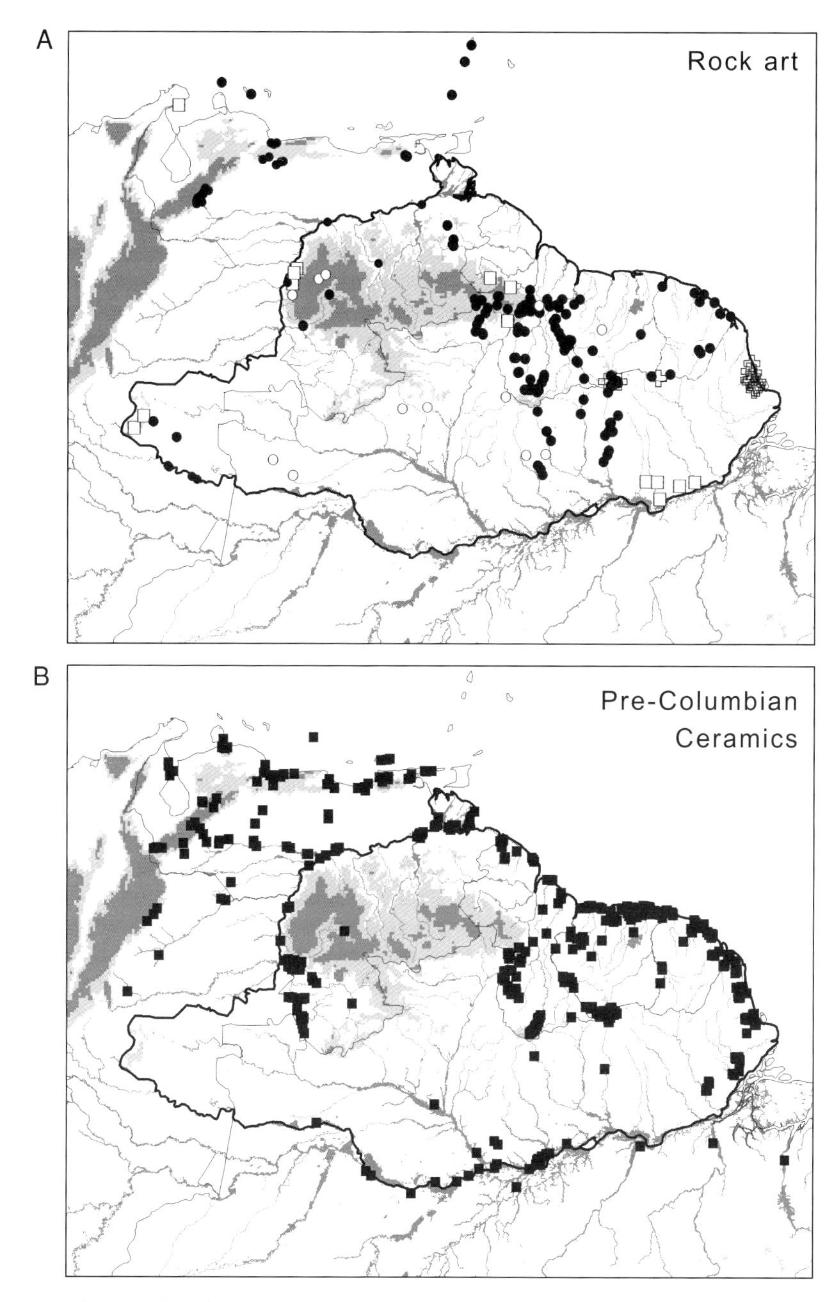

Fig. 8.2. Distribution of artefactual evidence of prehistoric occupation across the Guiana Shield. (A) Rock art: petroglyphs (solid circles), rock paintings (open squares), stone arrangements (crosses), locations of sandstone formations suitable for cave formation and rock painting (open circles). (B) Pre-Columbian ceramics (solid squares) identify sites ranging from Palaeo-Indian to Indo-Hispanic epochs. (C) Earthworks: shell mounds (solid triangles), earth mounds (solid circles), raised fields (solid squares), causeways (open squares with solid centre), highland terraced fields (open circles with solid centre). (D) *Terra preta*: site locations (solid squares) ranging in size from one to several hundred hectares. See Notes for source references.

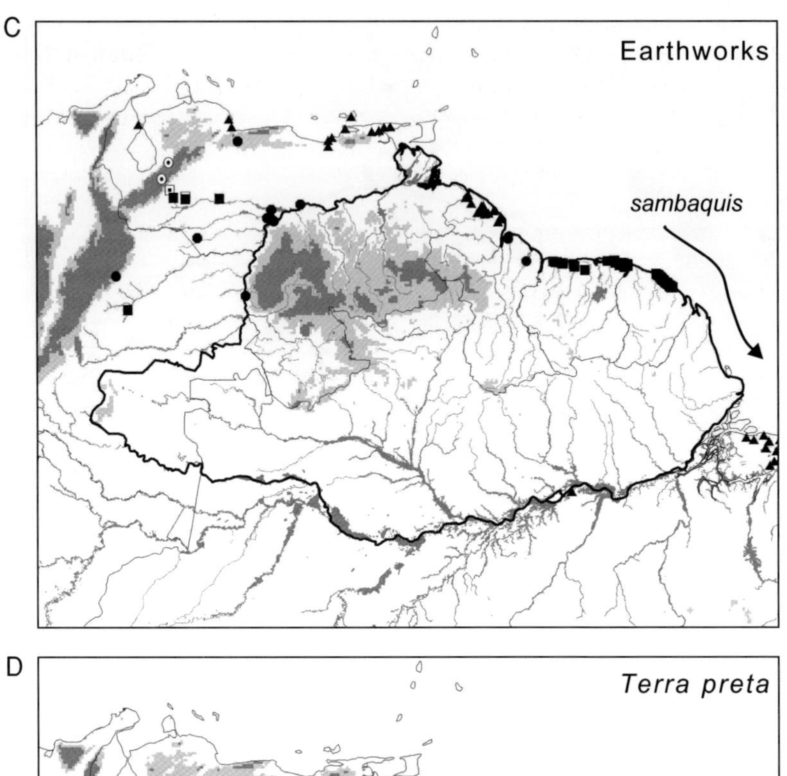

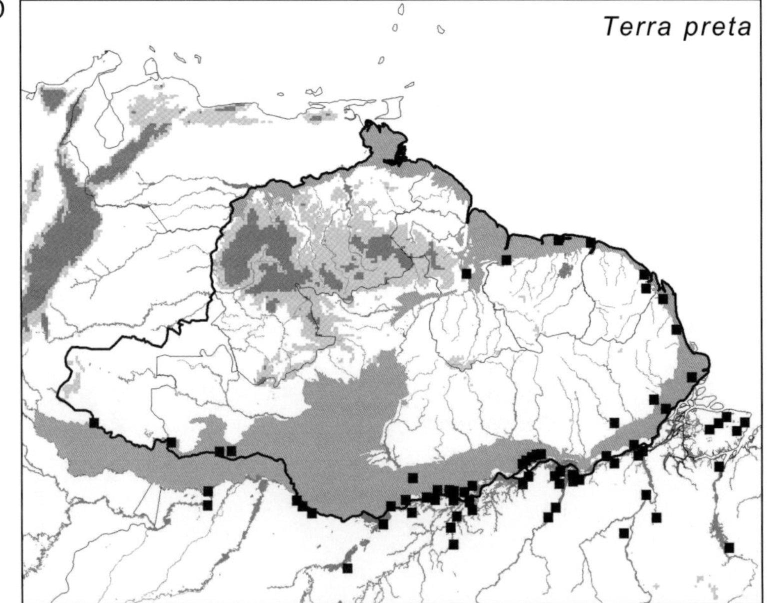

Fig. 8.2. *continued*

(Osgood, 1946; Hanif, 1967; Berrangé, 1977; Williams, 1979a) savannas in Suriname and Guyana, respectively. These locations also act as junctures between tributary headwaters connecting the central Amazon with the northwestern equatorial Atlantic and Caribbean (Evans and Meggers, 1960, p. 300; Dubelaar, 1981; Versteeg and Bubberman, 1992) (Fig. 2.30) and are subject to significant seasonal inundation that physically links or reduces the distance between divergent basins (Fig. 2.35). This would make local savanna rock outcrops functional homologues to the exposed dykes and sills that yield rock drawings more often registered along rivers (Dubelaar, 1981).

Glyphs associated with mountainous terrain are even more poorly known. Several glyphs have been tentatively identified along the rocky slopes of the Iwokrama Mountains, central Guyana (Plew, 2002), although these require further documentation and validation. They are believed to be virtually absent from the outcrops scattered across central Suriname and French Guiana (Dubelaar, 1986), but have been less comprehensively surveyed in the Pakaraima highlands of Guyana and Venezuela. Petroglyphs are much better known from various watersheds across the southern face of the Venezuelan Andes to the north (e.g. Novoa and Costas, 1998 in Gasson, 2002), and along the middle reaches of the Caura, Caroni, Ventauri, Cuyuni and Orinoco Rivers in Venezuelan Guayana (Cruxent, 1947; Evans *et al.*, 1959; Blair, 1980; Dubelaar, 1986; Silva, 1996). But here too their occurrence appears to reflect some geographic selection process. In most instances they are found near waterways, supporting the association of rock with water, but more intensive surveying at various locations along the eastern Andean slopes (Spencer, 1991) and southern Guyana (Berrangé and Johnson, 1972; Dubelaar and Berrangé, 1979; Williams, 1979a) confirms that not all waterways that offer suitable sites for petroglyph manufacture, or are connected with lowland areas, are necessarily worked.

They also appear sparse or absent over most of the coastal flatlands from the Orinoco to western French Guiana (Fig. 8.2). Their absence from these areas may be more easily understood, being consonant with the distribution of deep Quaternary sediments that dominate most of the coastal landscape. Where these deposits thin along the coast north of the Orinoco (Guarapiché/Paria Peninsula) and in eastern French Guiana (near Cayenne), Precambrian or Palaeozoic rock is exposed and petroglyphs are present (Rouse and Cruxent, 1963; Dubelaar, 1986; Rostain, 1987; Versteeg, 1995). Similarly, sediments dominating the geosynclinal trough southward of the Venezuelan and Colombian Andes offered very few rock outcrops to prehistoric petroglyph artists. As a consequence, examples of rock art are also largely absent from the extensive wet grasslands of the Venezuelan and Colombian llanos (Gassón, 2002) (Fig. 8.2).

Williams (1979a) also notes, however, that petroglyphs have yet to be recorded in the area of Guyana north of the Mazaruni River (Fig. 8.2), despite the presence of numerous exposed rock lines along the major rivers in this region. In the case of the upper river reaches and piedmonts of the adjacent Guayana Highlands, the little evidence of petroglyph writing available may also reflect the relatively low survey effort expended to date in this rugged region. Local residents typically have a far more comprehensive knowledge of site type and location that is rarely reflected in the literature.

PURPOSE The incidence of petroglyphs across the region is consistent with rock–water landscape controls on manufacturing opportunities, but their absence is not. Where petroglyphs occur, they are typically found clustered along river sections or in savannas, but numerous sites amenable to their manufacture have not been utilized. It is reasonable to propose that prehistoric petroglyph artists selected sites on the watershed scale, based on an approximate distribution of known carvings (Fig. 8.2), the distribution of suitable rock formations and considering studies where the delineated study area has been thoroughly surveyed (e.g. Berrangé and Johnson, 1972; Dubelaar and Berrangé, 1979).

The notion of watershed-based site selection would tentatively support the use of basin connectivity as a third criterion for selection and the view that petroglyphs may represent trade or travel route markers in certain instances (e.g. Redmond and

Spencer, 1990; Spencer, 1991). Efforts to attach them otherwise to existing ceramic chronologies have been attempted, although evidence is largely interpretive (Greer, 1995). Fishing guidelines (Williams, 1979b), spiritual icons and shamanistic practices (Rivas, 1993; Williams, 1993) have also been suggested as the purpose underlying petroglyph manufacture at various locations, but in many ways these theories lack the clear support required for critical consideration. As an iconographic system that still lacks a robust translator, petroglyphs remain open to wide-ranging interpretation. This fact is emphasized by the apparent similarity in symbol use (e.g. concentric circles) between shield and other sites on multiple continents (e.g. Europe, Africa) that have no apparent chronological, cultural or geographic affiliation within the timeframe of shield colonization and occupation over the Holocene.

TIMING OF MANUFACTURE Petroglyphs are not readily or reliably dated using radiometric techniques applied to other artefacts, in part because their manufacture involves the removal, rather than addition, of material. The absence of clear stylistic associations across the region and the possibility that many large, complicated works may in fact represent palimpsests further confounds efforts to figure these important remnants of prehistoric culture into a functional timeline and establish the ultimate purpose behind their creation. Several authors have suggested links with cultural phases ranging from late Prehispanic (500 years BP) through to early Mesoindian (7000 BP) periods. The wide range in time period offered illustrates the uncertainty over their origins in the absence of material dates.

Williams (1978b, 1979b) noted that a number of petroglyph clusters in Guyana along the Kassikaityu, Mazaruni and Berbice were only exposed during anomalously dry years (i.e. very strong El Niño years) when river stages were extremely low. This is also the case at sites along the middle Essequibo River (D. Hammond, personal observation). Williams widely interpreted this to suggest that river stages were permanently lowered in the past as a consequence of sustained rainfall decline ('mega-Niño events' in the parlance of Meggers (1994)). Surprisingly, trial recreations of petroglyph processing have not been undertaken to establish the amount of time and effort required to create existing images and whether these could be created under river stage extremes similar to those of today. It remains equally probable that images were simply manufactured during droughts of modern duration and accumulated through a succession of such periods out of opportunity, as part of a ritualized recording of these periods or through placement of instructions or markers as historical reference points denoting low river stage extremes (i.e. El Niño severity). Alternatively, rock faces may have simply been dismantled by river currents (or human action) over the intervening period, lowering petroglyph levels. Further study is needed to establish the most plausible route and functional significance, if any, of their emplacement level.

Stone arrangements (géoglyphes, assemblages de pierres, stone rows and circles)

DISTRIBUTION Stone arrangements have been documented from the Calcoene River area along the Amapá coast (Meggers and Evans, 1957), the Sipaliwini and Parú savannas (Frikel, 1961, 1969; Bubberman, 1973; Boomert, 1980b, 1981), the south (Williams, 1979c) and north (Schomburgk, 1836; Brown, 1876) Rupununi savannas, the Rio Branco savannas (Henderson, 1952), the south Pakaraimas/Ireng region near Kurukubaru (Brown, 1876; Roth, 1929), the lower Ventauri River region of Venezuelan Amazonas (Evans *et al.*, 1959) and the Tumucumaque Uplands along the southern border of French Guiana (Hurault *et al.*, 1963; Rostain, 1987, 1994) (Fig. 8.2, crosses). They have typically been found resting atop the smooth, convex surfaces of exposed outcrops of crystalline basement rock (granitoids, greenstones), often with a good view of the surrounding area (Meggers and Evans, 1957; Hurault *et al.*, 1963; Frikel, 1969; Williams, 1979c).

PURPOSE Stone alignments are widely interpreted as having a ceremonial, ritualistic and/or religious function in the belief system of prehistoric societies (Farabee, 1918; Meggers and Evans, 1957; Frikel, 1969), supposedly since their function cannot be interpreted as serving any other practical role. The variation in alignment geometry across shield-based examples supports the view that they are not monothetic cultural insignia, but probably developed by different societies in commemoration of unrelated events or beliefs (Boomert, 1981, p. 144).

TIMING OF MANUFACTURE Boomert (1981) suggests that many of these alignments are of post-Columbian age, but little material evidence is available that would improve confidence in any assigned age. Historical accounts of stone pile-making by Amerindians combined with transcription of early oral traditions concerning stone alignments provide some support for a more recent (post-Columbian) manufacture of these, at least in the Rupununi–Rio Branco region (Schomburgk, 1848; Brown, 1876; Roth, 1929). Disturbance at many sites (e.g. Amapá: Meggers and Evans, 1957) would also suggest the possibility of active re-working of arrangements that would erase any typological information present in the stone patterning.

Polissoirs (grindstones)

DISTRIBUTION Sites believed to be used for the manufacture of stone implements have been located along virtually all of the major rivers in the interior of the shield (Evans and Meggers, 1960; Rouse and Cruxent, 1963; Williams, 1979b, 1996b; Boomert, 1980b; Rostain, 1994; Gassón, 2002). Unlike petroglyphs, however, *polissoirs* are known to occur along much smaller waterways and those that do not link different basins (e.g. Fig. 66 in Rostain, 1994). Their association with water, while not exclusive (e.g. S. Rupununi: Williams, 1979c), is compelling based on studies documenting their occurrence across the shield region (Abonnenc, 1952; Evans and Meggers, 1960).

PURPOSE Compared with other lithic remains, the functional purpose of grindstone sites is relatively clear. Their proximity to water would provide easy access to sorted abrasives, such as riverine quartz sands, and small stone materials, that are not as easily obtained from the surrounding deep, forest soils while offering a more open, hospitable workshop location to the artisan. The use of these workbench sites for the processing of other wood, fibre and food materials is also likely.

TIMING OF MANUFACTURE Polissoirs remain poorly integrated into the cultural timeline of prehistoric settlement and resource use. Some archaeologists have assigned them to later Meso- or Neo-Indian (ceramic) epochs (Cruxent, 1947; Rouse and Cruxent, 1963; Williams, 1979a), albeit without radiometric dating of the lithic or other site-associated material. Like petroglyphs, their presence is established by the removal of material and this makes material dating equally difficult. Radiometric dating of other materials found at grindstone sites can be associated with these, but the possibility of site recycling invariably fogs establishment of earliest use without independent material dates.

Rock paintings (piedras pinturas, pictographs)

DISTRIBUTION Rock paintings, known from scattered locations across the region, are typically found within caves, grottos, rockshelters and along protected cliff walls. They appear to occur most frequently along the highland perimeters across the region adjacent to some of the larger rivers, including the upper and middle Orinoco (Cruxent, 1947, 1950; Perera and Moreno, 1984; Tarble, 1991, 1999; Greer, 1995), the eastern slopes of the Guayana Highlands near the Mazaruni and Ireng Rivers (Brown, 1876; Storer-Peberdy, 1948; Henderson, 1952; Attenborough, 1956; Poonai, 1974), in the Chiribiquete complex in south-central lowland Colombia along the Vaupés River and along the southern rim at Monte Alegre, adjacent to the lower Amazon River (Roosevelt *et al.*, 1996) and the lower Rio Negro (Wallace, 1853a) (Fig. 8.2, open squares). A vast number of painted cave sites along the middle and upper São Francisco River in the eastern Brazilian Shield region have also been documented (Prous *et al.*, 1984; Guidon and Delibrias, 1986; Lavallée, 2000).

Rock paintings are noticeable by their absence from French Guiana, Suriname, eastern Guyana and Amapá (Dubelaar, 1986). They are not expected from areas north of Monte Alegre and its neighbouring hills (Morro Grande, Serra Azul, S. Parauaguara) or the upper Rio Negro. This is based on an interesting lithological foundation that tentatively appears to link sites otherwise located in contrasting landscapes. Most if not all rock painting sites documented across eastern South America are associated with sedimentary rock types. The most detailed accounts, such as Monte Alegre and Pedra Furada, describe rock paintings sandwiched between a sedimentary base formed from Palaeozoic clastics and a siliceous or evaporitic precipitate cover (Roosevelt *et al.*, 1996). Rock paintings along the lower Rio Negro reported by Wallace (1853a) were also associated with sandstone escarpments that represent the westward extension of the Palaeozoic belt of epicontinental sedimentaries that form the Monte Alegre structures (Fig. 2.4). Sites in Colombia, although less intensively studied, are also associated with Palaeozoic sandstones forming the Chiribiquete complex. In the Pakaraima region of Guyana, sandstones providing the canvas for rock paintings belong to the Precambrian Roraima Formation. The absence of suitable sedimentary rock formations across Suriname (except Tafelberg, Emma Range), French Guiana and much of the southern shield area would support the absence of records from these areas. Notable exceptions in this region include isolated sites with sedimentary lithologies, such as the isolated Rio Novo mesa, Serra Araca (Jauri), Urupi Mountain, Ja and Uneuixi Mountains and Makari Mountain that form the Quasi-Roraima sedimentaries (see Fig. 8.2, open circles). This apparent lithological prerequisite would also suggest that the major massifs forming the Venezuelan Pantepui should eventually yield further sites, if the primacy of large river access seen to operate elsewhere did not exclude these headwater locations from the site selection process used by prehistoric, rock painting societies of the Guayana Highlands.

Potential taphanomic (bias) effects of lithology on the distribution of prehistoric rock paintings are substantial (Bednarik, 1994). Porous, sedimentary rocks may enhance pigment adsorption relative to more dense types. Sedimentary rocks form caves and rockshelters more readily as a consequence of their higher rates of chemical weathering (see 'River, Lake and Tidal Systems', Chapter 2) and paintings protected from exposure may remain visible longer. Chemical precipitates are also more likely to form on the inside surface of caves than on exposed rock surfaces, creating a protective cover over the dyed sandstone surfaces. Alternatively, prehistoric painters may have recognized the longevity offered by a sedimentary canvas and selected known sites for this reason. Alone, it remains unclear whether rock painting sites mark important prehistoric population centres, such as the Santarém or Puerto Ayacucho areas, or are the remnants of a much more widespread practice.

PURPOSE Painted caves and rockshelters are believed to have been used as both habitation (Roosevelt *et al.*, 1996) and sacred (Tarble, 1991, 1999) sites based on artefactual evidence recovered from floor deposits. Variation in pictographic styles and overlap of motifs suggest that these functions may have been employed by different groups during alternating periods of time (Greer, 1995), but too few sites have been studied in any detail to synthesize the role of cave art in regional prehistoric societies.

TIMING OF MANUFACTURE Unlike petroglyphs, polissoirs or stone arrangements, organic paint pigments offer an opportunity to employ radiometric techniques to materially date the age of rock paintings.

More intensively studied rock shelters and caves with paintings have been dated to the Pleistocene termination, around 10 ka BP (Guidon and Delibrias, 1986; Roosevelt *et al.*, 1996). On the other side of the shield, however, several sites along the upper and middle Orinoco region have been tentatively assigned to periods extending only to 2450 years BP (Greer, 1995; Gassón, 2002). Most known sites within the Guiana Shield remain poorly studied (e.g. Tramen Cliffs, Guyana) despite their intriguing topographic and artefactual attributes and despite local knowledge and historical reference to their existence. Herrera (1987, cited in Gasson, 2002) believes that pictographs from the Andean piedmont, as well as the Puerto Ayacucho region, are palimpsests of work carried out during different periods. If true, material dating of pigment samples from all images would be required to develop an accurate timeline of manufacture.

Stone implements

DISTRIBUTION Tools shaped from stone are some of the most common archaeological finds in the Guiana Shield and throughout the wider neotropics. They have been found at most archaeological sites in the shield region, including numerous sites along the Caribbean coast, Orinoco River and its major tributaries in Venezuela (Osgood and Howard, 1943; Rouse and Cruxent, 1963; Sanoja and Vargas, 1983; Wagner and Arvelo, 1986), along most of the coastal and interior rivers of Guyana (Brett, 1852; Im Thurn, 1883; Verill, 1918a; Roth, 1924; Carter, 1943; Osgood, 1946; Evans and Meggers, 1960; Poonai, 1978; Williams, 1978a, 1996a), Suriname (Ten Kate, 1889; Penard and Penard, 1917; Geijskes, 1960; Boomert, 1979, 1980a), French Guiana (Reichlen and Reichlen, 1944; Abonnenc, 1952; Rostain, 1994) and Amapá (Meggers and Evans, 1957), Pará, Roraima and Amazonas (Becker and de Mello Filho, 1963; Boomert, 1979; Hilbert and Hilbert, 1980) (Fig. 8.2). While stone implements have not been found at all excavated sites, a clustered pattern akin to that resolved for petroglyph locations is not apparent based on available survey results. In fact, stone implements are so widely dispersed across the region that the many isolated collections made over the past several centuries have contributed more than site excavations to typological comparisons (e.g. stone axes, Boomert, 1979, Table 1).

PURPOSE The functional purpose assigned to stone implements is wide-ranging. Size, form and edging characteristics are typically used to infer function of stone implements, along with wear patterns and association with other artefactual materials. Manufactured stone implements have been characterized principally as: (i) projectile points; (ii) cutting tools, such as axes, chisels and adzes; (iii) grinding and pounding tools; (iv) polishing tools; and (v) hammerstones.

Of all the implement types recovered, projectile points and axe heads are perhaps most indicative of direct forest resource use by prehistoric inhabitants. Projectile points also hold a particularly important position in the matrix of material evidence underpinning the prehistoric timing of Palaeo-Indian immigration, assignment of subsistence lifestyles and geographic connections (Dillehay, 2000; Lavallée, 2000). Williams (1998) assigns a spear-fishing function to points excavated from shell mounds along the Guyana coast, although the basic form of these and most other bifacial projectiles leaves these open to interpretation. Primary function is typically assigned based on site location or the composition of accompanying artefacts (e.g. animal remains). Similarly formed points were also found in the Ireng, Cuyuni and Mazaruni Rivers (Evans and Meggers, 1960; Williams, 1978a), near Salto de Hacha, Caroni river (Rouse and Cruxent, 1963), at Tapaquen (Cruxent, 1972) and along the Paragua (Dupouy, 1960), upper Orinoco rivers (Barse, 1990) and Gran Sabana (Dupouy, 1956) in Venezuelan Guayana.

Stone axe heads from the shield and adjacent regions have been typologically classified by ten Kate (1889) and Boomert (1979). More than 90% of the specimens classified are believed to have been primarily used as wood-working tools in the manufacture of house framing materials, agricultural plot establishment and corial (canoe) production. Smaller axe heads have been assigned a ceremonial function or viewed as tokens of social rank (Boomert, 1979).

TIMING OF MANUFACTURE The earliest assigned projectile point ages are associated with two sites on the shield periphery, one near Puerto Ayacucho on the upper Orinoco (Barse, 1990), and the other at Monte Alegre, near the lower Amazon (Roosevelt *et al.*, 1996). To date, no points excavated in the Guiana Shield have been given a materially dated age older than the 11 ka BP assigned to the stemmed, winged points found by Roosevelt *et al.* Cruxent (1972) considered points unearthed at Tapequén, Bolivar State, Venezuela to be at least 13,000 years old, but this antiquity has not been confirmed through radiometric dating. Barse assigned a later age of 7 ka BP to points associated with hearth charcoal at his study sites. Williams (1998) asserts that several isolated finds of projectile points from rivers indicate colonization of coastal Guyana earlier than 7.2 ka BP.[5] Axe heads and other stone implement finds from the Guianas have been almost entirely assigned to more recent, Neo-Indian (<3 ka BP) periods based on accompanying ceramic materials (Evans and Meggers, 1960; Versteeg and Bubberman, 1992), but a few locations associated with earlier periods suggest that their use pre-dates the manufacture of ceramics (Gassón, 2002). Many of the axe heads recovered along the Atlantic coast sites are shaped from mafic rock types that are found inland of the Quaternary sediments. This further complicates age assignments from coastal sites since it opens up the prospect of inter-generational reuse of stone implements and/or trade with interior communities prior to deposition at the excavated location.

Material types: ceramics

Distribution

Complete or fragmented remains of ceramic wares are associated with most excavated sites in the Guiana Shield (Fig. 8.2). Consequently, the distribution of ceramic artefacts across the region principally reflects spatial variation in sampling effort. Effort spent can be stratified at two spatial levels: (i) watershed; and (ii) river proximity. At the watershed level, the Cuyuni, Mazaruni, Caroni, Caura, Ventauri, Branco, middle Rio Negro, Uaupes/Vaupés, Uatuma, Jatapu, Mapuera and Jari have been sparsely explored, although existing evidence, both lithic and ceramic, indicates that many of these were important thoroughfares connecting people living along the Caribbean and Atlantic coast and the large floodplain corridor along the Amazon River prior to European arrival (Lathrap, 1973; Boomert, 1987; Rostain, 1994).

The distribution of ceramic finds also reflects on the substantial under-sampling of *terra firme* forest sites outside floodplains (Fig. 8.2). The depiction of the earliest lowland inhabitants as *varzea* dwellers (Roosevelt, 1998), while logical, has not been adequately tested by documenting the absence of similar inhabitation sites at upland forest locations throughout the shield region. Information available regarding early use of upland areas is almost entirely limited to savanna and rockshelter locations (e.g. see Rostain, 1994; Gassón, 2002). Given the record of artefact recovery from floodplain excavations to date, it is to be expected that most future floodplain excavations along major rivers will yield ceramics.

Purpose

Ceramic wares are generally believed to have coincided with a sedentary, if not agrarian, lifestyle. This relationship, if strictly true, is crucial in assessing both the timing and magnitude of changes in size, social organization and lifestyle of prehistoric populations in the lowland neotropics. Most excavated ceramic materials are the broken remnants (sherds) of vessels, probably used for carrying, processing, cooking and storing the basic staples of everyday life. Ceramic griddles, believed to have been used to make cassava and/or maize bread, have been found at various locations (see references in Gassón, 2002). Still other more elaborate, but less functional, finds are believed to have been used for ceremonial purposes or as trade objects. In many instances, vessels were used to house buried remains (e.g. Evans and Meggers, 1960).

Timing of manufacture

Ceramic artefacts excavated at Taperinha, along the south bank of the lower Amazon near Santarém, Brazil are believed to be the oldest thus far unearthed in the lowland neotropics (Roosevelt, 1991a). These materials, found in a shell midden, were radiometrically dated to more than 7.0 ka BP. Williams (1981a) also found ceramic sherds taken from a coastal Guyanese shell mound (Piraka) that he associated with charcoal radiocarbon-dated to 7.2 ka BP. Several other sites along the lower Amazon River (Marajó, Monte Alegre) and the Atlantic coast of the shield region (Venezuela, Guyana, Amapá) have also been assigned to periods earlier than 4.0 ka BP (summarized in Roosevelt, 1995). Among the best known of these is the sequence of pottery types excavated at Parmana and adjacent sites along the Orinoco immediately downstream from its confluence with the Apure (Gassón, 2002). Roosevelt (1980) assigned the earliest pottery types found at Parmana to a period dating back 4.45 ka BP, while Lathrap and Oliver found pottery at nearby Agüerito that they believe was manufactured more than 5.2 ka BP. Both early assignments, however, have been questioned on several accounts (Zucchi *et al.*, 1984; Barse, 2000).

The majority of ceramic artefacts either dated directly or by stratigraphic association are of late Meso-Indian, Neo-Indian or Indo-Hispanic age, less than 4.0 ka BP. Late Neo-Indian and Indo-Hispanic ceramic styles are particularly common. This includes virtually all of the sites excavated to date along the upper (Zucchi, 1991), middle (Simões, 1987) and lower (Simões, 1974; Heckenberger *et al.*, 1998) Rio Negro, southern Guyana and Suriname (Evans and Meggers, 1960; Boomert, 1981; Williams, 1981b), the lower Orinoco (see references in Gassón, 2002, pp. 288–291) and coasts of Suriname, French Guiana and Amapá (Meggers and Evans, 1957; Rostain, 1994).

Material types: metalworkings

Very few examples of worked metalcraft have been recorded for the Guiana Shield region. Whitehead (1990) mentions the discovery of a gold pendant on the upper Mazaruni River, but suggests that this is a trade object obtained through routes linking tribes of the region with highland metalworkers of Colombia. Compared to highland Andean sites, however, metalworked artefacts are virtually absent from the regional record until the appearance and proliferation of European trade goods, such as metal axeheads, in the 17th century.

Material types: organic remains

Organic remains are the most readily dated artefacts recovered from archaeological sites and often represent the only material dates available for anchoring sequences of excavated strata. Among the most commonly recovered organic remains are charcoal, woody endocarps, worked and unworked animal bones, pollen and/or phytoliths. Less commonly, human remains have been recovered at lowland neotropical sites.

Human remains, particularly when these are presented in a manner that indicates ceremonial burial, are unequivocal proof of human occupation. Radiometric dating of these finds directly places this occupation along an absolute timeline of human inhabitation and land use, rather than one defined purely by stratigraphy and interpretation. Remains, however, have been excavated at only a handful of the hundreds of sites studied across the Guiana Shield, Venezuelan llanos and Amazon Downwarp.

Along the coastal rim of Amapá, Meggers and Evans (1957) unearthed hundreds of funeral urns at 21 sites. Most contained cremated remnants, but several were found with uncremated remains, partially cremated remains or a mixture of both. In eastern French Guiana, remains of an adult and infant have been found interned in one of several funeral caves located in the Montagne Bruyere, between the Ouanary and Oyapock Rivers, although accompanying objects suggest they are of Indo-hispanic age (Cornette, 1985, cited in Rostain, 1994), as is the case for many of the other sites (Petitjean Roget, 1983). More than 300 archaeological sites have been studied across Suriname (Versteeg and Bubberman, 1992), but only a few coastal sites associated with the Hertenrits (including Buckleburg, Wageningen sites), Tingiholo and Barbaekoeba earth mound-drained field complexes have yielded human remains (Boomert, 1980a, 1993; Versteeg, 1985; Khudabux *et al.*, 1991). These finds are believed to coincide with an earlier, prehistoric period, probably dating to the Neo-Indian epoch. A similar situation exists along the Guyana coastlands where human remains have been recovered, but from shell mounds near the Waini and Pomeroon Rivers of the northwest district (Williams, 1981a, 1996a). Unlike Amapá, French Guiana or Suriname, however, few caches of badly decomposed or partially cremated human bones have been documented from burial jar sites examined in the interior. These are located in rockshelters and caves found along the piedmont of the Pakaraima Mountains of southwest Guyana (Evans and Meggers, 1960), where local people know of many other sites containing burial urns (J. James, personal communication). Accompanying materials of European and North American manufacture indicate, however, that most of these burials occurred during the 18th and 19th centuries. The Atlantic coast of Venezuela has also yielded several sites with human remains. Like those along the northwestern littoral of Guyana, these have largely been excavated with or without burial urns from the many shell mounds populating the coastal stretch between the Orinoco Delta and Maracaibo (Rouse and Cruxent, 1963) (Fig. 8.2). Sites in the interior region of Venezuelan Guayana are generally few, although a series of funeral sites are known from explorations of the Puerto Ayacucho cave complex (Perera, 1971) and remains recovered from sites further downstream at Parmana (Roosevelt, 1980; van der Merwe *et al.*, 1981). Sites excavated along the upper Rio Negro and headwater tributaries of the Orinoco have not included verifiable human remains (Evans *et al.*, 1959; Zucchi, 1991). Major sites along the southern rim of the shield region have also not included human remains among the recovered artefacts (Roosevelt, 1991a; Roosevelt *et al.*, 1996).

Animal remains are noticeably more common among artefacts excavated across the Guiana Shield, although these too are found more frequently and in greater amounts (or reported more often) at sites along the Atlantic coastline and lower Amazon (Verill, 1918a; Carter, 1943; Meggers and Evans, 1957; Evans and Meggers, 1960; Boomert, 1980a, 1993; Williams, 1981a, 1998; Versteeg, 1983; Versteeg and Bubberman, 1992; Rostain, 1994; Roosevelt *et al.*, 1996) than the interior (Cruxent, 1950; Evans *et al.*, 1959; Evans and Meggers, 1960; Frikel, 1969; Boomert, 1980b, 1981; Hilbert and Hilbert, 1980; Williams, 1981b; Zucchi, 1991). Remains from coastal and riverine sites emphasize many of the same aquatic resources currently harvested from the adjacent freshwater, estuarine and marine habitats, although changes in relative sea-level and discharge dynamics over the Holocene have also altered resource opportunities in many of these areas (e.g. molluscs, bivalve distributions) (e.g. Altena and van Regteren, 1975).

Site types: earth engineering

Earth engineering exemplifies, perhaps more than any other material evidence, the ability of prehistoric societies to modify their local environment as a means of improving their living conditions. Although known worldwide, studies documenting engineered earth sites in the lowland tropics are particularly significant as testimony to the innovative environmental solutions employed by prehistoric societies to expand otherwise limited human carrying capacities (Denevan, 1970; Parsons and Denevan, 1974; Williams, 1979a; Versteeg, 1983; Roosevelt, 1991b; Wood and McCann, 1999; Erickson, 2000; Mann, 2000a,b; Wood and Mann, 2000).

Raised fields (champs drainés, montones), causeways (calzados) and earth mounds

DISTRIBUTION Arguably the best known example of lowland earth engineering is the extensive network of raised fields and connecting causeways that are found across the seasonally flooded expanse of the Mojos plains in Bolivia (Denevan, 1970; Parsons and Denevan, 1974). A similar complex of drained fields and causeways exists within the seasonally flooded savanna along the Apure River in Venezuela linked to the golden age of prehistoric chiefdoms at El Gaván and El Cedral (Zucchi, 1972, 1973, 1978, 1984; Spencer, 1991; Spencer *et al.*, 1994, 1998) (Fig. 8.2) and extensive raised field complexes at Caño Ventosidad (Zucchi, 1985). A less extensive system was also discovered along the piedmont of the upper Meta River in Colombia (Reichel-Dolmatoff, 1974). Earth engineering of similar or greater magnitude is known from highland lake shores such as Titicaca in Bolivia and Xochimilco in Mexico, but no other systems of comparable range and complexity have been documented to occur in the lowland neotropics.

Other, more modest, systems that have been well-documented from the lowlands occur almost exclusively along the rim of the Guiana Shield (Fig. 8.2). These consist of raised field and earthen mound complexes along the Suriname and French Guiana coastlands (Boomert, 1980a, 1993; Versteeg, 1983, 1985; Rostain, 1994), scattered earthen mounds between the Abary and Canje Rivers in northeast Guyana (Im Thurn, 1884; Verill, 1918b; Osgood, 1946; Poonai, 1962; Boomert, 1978; Thompson, 1979) and mound clusters along the middle Orinoco, particularly at the regional centre of Parmana (Roosevelt, 1980; Gassón, 2002) (Fig. 8.2).

PURPOSE Raised field-mound complexes are believed to have created a productive capacity that exceeded basic subsistence requirements in the western Llanos de Venezuela. Maize is believed to have been the primary crop, unlike historical subsistence systems in the Guiana Shield that focused on manioc production. Connected by causeways, raised fields supported food surpluses that allowed for ritualistic feasting and other activities that maintained social alliances in an advanced political landscape of competing chiefdoms (Spencer, 1991; Spencer *et al.*, 1998; Redmond *et al.*, 1999).

In contrast, materials recovered from earthen mounds and raised field-mound complexes studied along the Orinoco and Guianan coastlands do not support assignment of a similar level of social complexity. Ceramic evidence from mounds in Suriname and French Guiana, such as griddles and smoothers, suggest manioc was the cultigen of choice (Boomert, 1993; Rostain, 1994),[6] although the presence of grinding stones (metates) at some sites suggest maize as an alternative staple (Boomert, 1980a). Along the Orinoco, it is believed that initial manioc cultivation was replaced by maize as this crop was introduced eastward through the Colombian/Venezuelan llanos (Roosevelt, 1980; Gassón, 2002). The size of field complexes and cultural artefacts found from within earthen mounds point to relatively large Amerindian populations (Marajó Island: Roosevelt, 1991b; 20,000+ in French Guiana: see Rostain, 1994; lower Rio Negro: Heckenberger *et al.*, 1999) living in communities bound by increasingly complex social interactions, trade and religious beliefs prior to final abandonment in the centuries preceding European arrival (Boomert, 1980a; Versteeg and Bubberman, 1992).

TIMING OF MANUFACTURE Raised field and earthen mound complexes spread between the northern llanos and French Guianan coastline have been assigned noticeably concurrent Neo-Indian dates, based on radiometric analysis of *in situ* charcoal, wood and peat materials. The Hertenrits mound of western Suriname is believed to have been constructed through successive layering over a period of 150 years, from 1.2 to 1.1 ka BP (Boomert, 1980a). Development of the Buckleburg mounds is believed to ante-date Hertenrits by several hundred years, being built from 1650 to 1350 years BP. The field-mound complexes of eastern Suriname and French Guiana that in part contain Barbakoeba-type pottery were used between 1800 and 750 years BP (Versteeg, 1983; Boomert, 1993; Rostain, 1994). Marajoara mound-builders at the mouth of the Amazon are believed to have been active from 1540 to 650 years BP (Meggers and Evans, 1957; Roosevelt, 1991b). More extensive complexes at Gaván and El Cedral in the Apure Basin of Venezuela are thought to have been constructed and intensively used from 1400 to 950 years BP (Spencer *et al.*, 1994, 1998), while those further downriver at Caño Ventosidad were active thereafter between 750 years BP and the arrival of Europeans, 450 years BP (Zucchi, 1985). No earth engineering of this type has been justifiably assigned to a period earlier than 2000 years BP and overlap with intensive use of adjacent shell mounds along the coastal region appears minimal.

Shell mounds (sambaquis)

DISTRIBUTION Many early inhabitants of the eastern neotropical rim are believed to have relied substantively on the estuarine and marine resources offered by the shallow, but rising, coastal waters of the Holocene (Fairbridge, 1976). Along certain coastal stretches of the eastern shield perimeter, the earthen mounds and raised fields disappear and mounds formed from the heaping of discarded crab, mollusc and snail shells speckle the landscape. The shell mounds spread in clusters along a coastal belt ranging from Panama and Colombia, across the Guajira Peninsula in north Venezuela and along the coasts of eastern Venezuela, northwestern Guyana, Marajó Island, Pará, Maranhao, Bahia and southward to Rio Grande do Sul. More than 1000 known shell mounds form this coastal belt, with over 90% of these occurring along the Brazilian coast south of the Rio Saõ Francisco. Far fewer mounds are documented along the lower Amazon (Marajó, Santarém) (Roosevelt, 1991a,b), northwest Guyana (Verill, 1918a; Osgood, 1946; Evans and Meggers, 1960; Williams, 1981a, 1996a), around the Guiria Peninsula (Sanoja and Vargas, 1983), along the Golfo Triste, and through to the Guajira Peninsula and the Caribbean coasts of Colombia and Panama (Rouse and Cruxent, 1963; Williams, 1996a). The documented site from northwest Guyana is the only mound cluster known to exist within the Guiana Shield.

Shell mounds are excellent point sources of alkaline agents in an otherwise acidic tropical environment. Early Europeans quickly recognized their utility for a wide range of applications that had similarly motivated prehistoric coastlanders to initially construct the mounds. As a result, documented sites probably represent only a subset of a much larger, original population diminished through selective mining as a source of agricultural fertilizer, liming agents, road-building materials and stable building sites, among other uses (Rouse and Cruxent, 1963, p. 75; Gaspar, 1998).

PURPOSE The distribution of shell mounds emphasizes the widespread technological adaptation of coastal societies along the western tropical Atlantic to the prevailing environmental conditions. The harvesting and processing of crustaceans, molluscs, snails and bonefish are clearly seen as principal activities from the preserved animal remains and various bone and wood fishing tools recovered (Evans and Meggers, 1960; Williams, 1981a). Yet, human remains in many mounds combined with a wide range of lithic and ceramic artefacts not related to resource use of the littoral environment indicate that the structures were also used at various stages as domiciles and burial grounds (Simões, 1961; Williams, 1993, 1996a; Gaspar, 1998).

TIMING OF MANUFACTURE Basal contents of numerous mounds have been dated radiometrically to give an approximate age of their earliest construction and use. The earliest dates assigned to shell mounds within the coastal belt range from 7500 to 1480 years BP. Among the earliest of these are those assigned to the Taperinha mound (>7.5 ka) near Santarém (Roosevelt, 1991a), the Piraka mound (>7.2 ka) in northwest Guyana (Williams, 1981a), Banwari Trace (>7.1 ka) in Trinidad (Table 10 in Williams, 1996a), Cerro Iguanas and Guayana (>5.5 ka) in Venezuela (Rouse and Cruxent, 1963; Williams, 1996a) and the Forte and Geriba II sites from southeastern Brazil (Gaspar, 1998). Numerous other mounds interspersed with these older sites have been assigned initial use dates as late as 700 years BP. The stratification of cultural artefacts and ceramic styles combined with anchor material dates of charcoal and pottery sherds indicate that many large mounds grew through successive, though possibly discontinuous, occupations over several millennia.

Terra preta (do índio) (Indian black earth, terre noire)

Terra preta are recognized by their dark, humus-rich horizons, intercalated charcoal, abundance of ceramic sherds and unusually high agricultural productivity (Sombroek, 1966; Glaser *et al.*, 2001). They also have been identified by a geochemical signature of high plant nutrient and anomalously low arsenic levels that is distinct from background ferrasolic soils (Lima da Costa and Kern, 1999).

Unlike the short-term spike in nutrients available immediately following slash-and-burn, areas of *terra preta* are consonant with sustained elevation of nutrient levels. Experiments suggest that the incorporation of massive charcoal layers assist in the selective retention of phosphorus, potassium, calcium and other micronutrients that are poorly retained in areas dominated by upland forest soils (Ferrasols, Oxisols, Podzols) otherwise susceptible to rapid leaching (see sections on 'Soils and Soil Fertility' and 'River, Lake and Tidal Systems', Chapter 2). Nutrient elevation is initiated by an enrichment process. Mass deposition of organic garbage, such as fish bones, freshwater snail and turtle shells (Roosevelt, 1999), provides a slow-release source of scarce micronutrients and pH buffers that reduce the loss of otherwise growth-limiting nutrients, such as phosphorus, and suppress dissolution of aluminium and iron that can be toxic to plants. Agricultural productivity is substantially increased as a consequence of these enrichment and buffering effects (Lehmann *et al.*, 2003).

Terra preta soils are widely distributed throughout the Amazon Downwarp, along major rivers draining the Brazilian Shield, such as the Tocantins and Xingu (Heckenberger *et al.*, 1999), and along the Quaternary sediment belt encircling the Guiana Shield. These latter sites include several known from the Atlantic coast of Amapá, French Guiana (Rostain, 1994) and Suriname (Versteeg and Bubberman, 1992; Boomert, 1993) and along the Caquetá River in Colombia (Eden *et al.*, 1984). Sites across the interior of the Guiana Shield are, by comparison, rare. Only a few or no sites have been located in the middle and upper Rio Negro basin (Sombroek *et al.*, 2002), interiors of French Guiana, Suriname, Roraima, Guyana and Venezuela Guayana. Williams (1994) noted a single site that is currently cultivated at Kurupukari, central Guyana along the southern limit of the Berbice Formation, but this site needs further documentation (Fig. 8.2).

Given the widespread role of tropical acidity in promoting agriculturally adverse nutrient leaching, heavy metal mobility and formation of chelated, metallo-organic complexes, the development of *terra preta* represents a significant feat of environmental engineering that arguably outstrips the success achieved by modern efforts to improve the agricultural productivity of tropical forestlands. This interpretation of *terra preta*, however, is not universal. Instead, some view the development of these soils as merely the consequence of continuous refuse disposal (Balée, 1992).

While the heightened nutrient retention properties of *terra preta* are well-documented, the limits to agricultural intensification on these putative anthrosolic soils remain unclear in the absence of controlled manipulative experiments. Sombroek *et al.* (2002) indicate that where modern agricultural use exceeds several consecutive years, crops begin to show signs of malnutrition. The extent to which these soils are linked to hypothesized prehistoric population densities depends entirely upon their ability to sustain agricultural production. The combination of total area and rate of productivity attached to *terra preta* provides a calculable basis for estimating prehistoric human population sizes in regions where these are known to exist.

Most *terra preta* deposits are associated with site artefacts dated between 900 and 400 years BP (Roosevelt, 1991b; Rostain, 1994; McCann *et al.*, 2001). The Kaurikreek site in Suriname, however, has been assigned a much earlier date, between 3.6 and 2.5 ka BP (Versteeg and Bubberman, 1992). Williams (1994) assigns a radiocarbon date of 3 ka BP to the Kurupukari site in central Guyana, making this arguably the oldest known *terra preta* site in the forested neotropics. The age of charcoal samples taken from other sterile arenosolic sites north of the area, a history of ENSO drought, the great longevity of many tree species from the region (see 'Prehistoric climates of the Guiana Shield', Chapter 2) and absence of any artefacts pointing to charcoal as an event-specific product of human activity, seriously questions the connection between charcoal age and human occupation in this instance.

Comparison with prehistory of other neotropical regions

Mounting archaeological evidence supports the view of a prehistoric neotropical lowland that was colonized at or near the same time that highland and temperate regions of the Americas began to house early arrivals from Asia. While certain assumptions regarding the interpretation of evidence still leave the possibility of a more recent, post-Clovis arrival of humans in South America, the growing geographic range of sites, the broad assortment of researchers and institutions exploring these sites and the use of a broadening spectrum of technologies for processing artefactual evidence, argues for a pre-Clovis colonization (for comprehensive assessment of arguments see Dillehay, 2000). This earliest colonization, in a manner probably repeated numerous times thereafter, remained close to the coast. The coastal fringe of the continent is the environment that has arguably experienced the longest, most continuous and most intense history of human impact.

The prehistoric timeline and magnitude of lowland tropical forest occupation in the interior is another matter. Archaeological evidence suggests quite different settlement and livelihood patterns for the main physiographic regions of the neotropics. Evidence of early Palaeo-Indian occupation is largely restricted to worked lithic remains excavated from prehistoric hearths. These have been found at several locations in the northern Guiana Shield, such as the Sipaliwini savanna (Boomert, 1980b), Guayanan Highlands and Gran Sabana (Dupouy, 1956, 1960; Rouse and Cruxent, 1963; Cruxent, 1972), and near Puerto Ayacucho (Barse, 1990). More substantial, but equally controversial, finds of early Palaeo-Indian occupation have been found along the Saõ Francisco River among the Palaeozoic sandstone canyons fringing the eastern rim of the Brazilian Shield (see overview by Lavallée, 2000). The Amazon Downwarp east of Manaus is also increasingly being viewed as a landscape of early occupation through site excavations at Taperinha and Monte Alegre (Roosevelt, 1991a; Roosevelt *et al.*, 1996).

The discovery of early lithic (and some ceramic) remains from these sites points to widespread migration within the lowland regions, but provides little room for additional, and plausible, interpretations. Palaeo-Indian sites, while in some cases stratigraphically superposed by complex cultural artefacts, collectively indicate simple, non-sedentary lifestyles based largely on foraging and hunting. In the absence of additional evidence, the most parsimonious interpretation is one where occupation of the Guiana Shield is transitory and small groups are moving frequently. This lifestyle, if proven to be correct, would suggest human impacts on the forest were primarily linked to forest fauna through hunting and fishing and, potentially, the use of fire as a tool in savanna-based hunting. The absence of any evidence pointing to an agrarian lifestyle before 8 ka BP would eliminate the role of slash-and-burn in modifying forests. More subtle management practices, such as distribution and enrichment planting of useful forest species (e.g. Brazil nut, palms), may have been employed.

The Meso- and Neo-Indian periods ushered in a gradual advancement of the way in which people of the lowland neotropics interfaced with their environment. From around 8 to 3 ka BP, the most striking evidence points to prehistoric communities assembling along the major rivers and coastlines of the region. Unlike their predecessors, however, they left a more substantive legacy of their occupation, and lifestyle – the shell mound, or *sambaqui* belt. The remnants of this belt attest to an attachment with the sea and the predictable resources the shallow littoral environment had to offer. Artefacts excavated from these sites suggest people made collecting trips to the interior, but again dates assigned to ceramics found along the upper reaches of ocean-draining basins are not generally concordant with dates assigned to early stages of mound construction.

Ceramic-making societies inhabiting the smaller tributary waters of the Guiana Shield and elsewhere were infrequent until around 2.5 ka BP, based on available evidence. At the same time, raised fields, earth mounds, *terra preta* and diverse and abundant ceramic traditions provide testimony to a golden age of lowland agrarian civilization that employed sophisticated techniques to overcome inherent environmental incompatibilities with (semi-)permanent agricultural production. Consistent evidence supports the view that prehistoric peoples of the lowland neotropics had undergone a growth in social complexity comparable, but different, to neighbouring highland societies over the several thousand years prior to European arrival. Archaeological evidence increasingly points to the sweep of European occupation and disease as the primary force dismantling prehistoric social complexity of these lowland Amerindian societies.

The fact that virtually all evidence points to a social complexity fundamentally linked to geomorphic features of the region further argues the case for a rapid attrition among the most sophisticated societies. Their ties to the relatively fertile *varzea* and coastal littoral landscapes placed the bulk of prehistoric population at the frontline of the epidemiological and cultural tsunami that swept the region with the influx of Europeans. The harsh interior of the shield regions inhibited sedentary lifestyles and population growth compared to that achieved along the Andes-linked waterways of the Amazon Downwarp and Sub-Andean Foredeep. The interior of the Guiana Shield, situated along the meteorological equator, would have presented the most severe limits to lowland agriculture, fishing and transportation. Rivers coursing (i) predominantly through Quaternary sediments, (ii) receiving water from a wider geographic area and (iii) with headwaters sourced in the Andes would have been (i) less susceptible to El Niño-driven drought effects, (ii) more capable of delivering seasonal sediment recharges that would support greater sedentary population growth and (iii) more amenable to transportation and travel (see 'River, Lake and Tidal Systems', Chapter 2). These large rivers, along with the coastlines, would have provided far more opportunities for regular trade and interaction among the many peoples inhabiting the region than oligotrophic waterways terminating in the forested Guayana Highlands and Tumucumaque Uplands. With the first Europeans arriving in the Americas, the golden age of Amerindian lowland societies along the rim of the Guiana Shield reached its terminus and a new age of human influence on the forests of the region began.

Colonial History: AD 1500–1900

Gold and the search for El Dorado

The Guiana Shield was from the beginning an important entry point in the exploration of the New World and consequent exploitation of its inhabitants and resources by the monarchies of Europe. On Columbus' third voyage, he and his crew travelled to the delta of the Orinoco, crudely charting its contours and channels in the process. Several years later, Alonso de Ojeda travelled further up the Orinoco and in 1500 Vicente Yáñez Pinzón charted all of the channels of the Orinoco delta. Less than 25 years after the arrival of Columbus in the Caribbean islands and his subsequent exploration of the Orinoco mouth, the Spanish had established settlements along the shoreline of present-day Venezuela and the nearby island of Margarita. By the early 1500s, Trinidad was being established as an important staging ground for Spain and its representatives to exert their claim over the lands of the northwestern Guiana Shield, a vast area that was virtually unknown to them. During the 1530s, the Spanish led a series of reconnaissance forays up the lower reaches of the Orinoco, under the command of the *doradista* Diego de Ordaz (Hemming, 1978b). From 1528 to 1566 the House of Welser, a German banking firm, received the 'right' to exploit large areas of western Venezuela from Emperor Charles V in payment of a debt owed to them by the Spanish Crown. Two German explorers, Phillip von Hutton and Nikolaus Federmann, were sent by the Welser group to establish the extent of their control and to search for gold in eastern Venezuela. While the Welser group were apparently driven in their exploitation of the people and the land during their 38-year lease, von Hutton and Federmann failed to find gold. The exploits of Ordaz, von Hutton and Federmann were the first in a subsequent chronology of encounters between the gold-seeking Europeans and the traditional inhabitants of these lands spurred by the competing regional interests of expanding European empires. Stories of a magnificent city accrued from subsequent expeditions up the Orinoco River by Diego Fernandez de Serpa, Pedro Maraver de Silva and, most famously, Antonio de Berrio, increased the anticipation of great wealth to be found in the hinterland of the shield (Ojer, 1960; Hemming, 1978a).

By the time the English explorer Sir Walter Raleigh arrived off the coast of Guiana in 1594, small gold artefacts and very large stories from the region had made their way into the hands and thoughts of Spanish, Dutch and German explorers. The stories, low on fact and rich in speculation, confirmed in Raleigh's mind the existence of a great kingdom, *Manoa*, located within the heart of the Guiana Shield. The object of the early *doradistas'* quest in the region, *El Dorado*,[7] or the Golden Man, was not without precedence and represented just another chapter in a long line of expeditions mounted in obsessive pursuit of ill-gotten wealth. The substantive collections of worked gold and silver that had been discovered and extricated from the temples and burial grounds of highly developed city-states in Colombia, Peru and Mexico catalysed a widely held belief that all indigenous societies of South America held in equal esteem and skill the working of gold and silver (Whitehead, 1997). Raleigh's record of his first trip to the Guiana coast in search of the mythic *El Dorado* exemplifies the magnitude of expectation borne from decades of nearly unfathomable effort made by Spanish *doradistas* in quest of the great treasure houses, real and fabled, of the pre-Columbian inhabitants of South and Central America:

> The Empyre of Guiana is directly east from Peru towards the sea, and lieth under the Equinoctial line, and it hath more abundance of Golde then any part of Peru, and as many or more great Cities then ever Peru had when it flourished most: it is governed by the same lawes, and the Emperour and people observe the same religion, and the same forme and policies in government as was used in Peru, not differing in any part: as I have beene assured by such of the Spanyardes as have seene *Manoa* the emperiall Citie of Guiana, with the Spanyardes cal *el Dorado*, that for the greatnes, for the riches, and for the excellent seate, it farre exceedeth any of the world, at least of so much of the world as is knowen to the Spanish nation: it is founded upon a lake of salt water of 200 leagues long like unto *mare caspium*.
> (Raleigh, 1596)

If there was an *El Dorado* in South America, he and his empire had long disappeared before European arrival in the Guiana Shield. The image of *Manoa* on the shores of Lake Parima, however, remained central to the European perceptions of the Guiana Shield region long after Raleigh's demise. Other explorers continued to chase Raleigh's dream (e.g. Nicolas Horstman in 1740 (Bancroft, 1769)) and as late as the mid-19th century, some European mapmakers were still showing the fabled lake as a prominent landscape feature in the region (Fig. 8.3). It wasn't until Richard and Robert Schomburgk, under commission by the Prussian Emperor, carried out an extensive survey of the boundaries of the growing British colony in the 1830s, that the existence of the lake and its great golden city were finally put to rest in most European minds (Schomburgk, 1848).[8] In its stead were the deflated reality of an overgrown, seasonally enlarged lake, called *Amacu*, located in the Rupununi savannas and the existence of a small Macushi village, *Pirara*, on the unflooded rise adjacent to its shores (Schomburgk, 1840). The early expectations of quick wealth, largely bred by the early Spanish and English explorers, gave way to the realization that if the Guiana Shield had something to offer to its constantly fighting European colonizers, then it would have to be pursued through more orthodox forms of production and trade.

Early trade and colonization

Gold wasn't the only potential that the early Europeans were exploring in the 17th-century Guiana Shield, albeit the most coveted. As early as the late 1500s, Dutch, French and English traders were seeking to develop commerce with Amerindians along the Guianan coast and up the Amazon and Negro Rivers, starting with the exchange of their simple manufactured goods for native plant products, foremost being dyewood (Brazilwood), annatto and tobacco (Edmundson, 1901, 1904a,b; Lorimer, 1979).

The historical record suggests that the Dutch had already established trading outposts and commenced with rudimentary cultivation of cash crops in the Corentyne, Essequibo and Pomeroon regions of Guyana by 1620 (Whitehead, 1988). In 1621, the Dutch monarchy began to formalize its commercial presence in the Guiana region by establishing an outpost for a new trading company, the Dutch West Indies Company, created to oversee its crown interests in the New World. This outpost, Fort *Kyk-over-al*, was situated on an island in the D'Essekebe, or Essequibo, River[9] downstream from the present-day town of Bartica and would coordinate Dutch trade in the region from this location and other satellite outposts up until the mid-1700s. Trade outposts were established along the Barima and Pomeroon Rivers and trade missions sent up the Mazaruni and Cuyuni Rivers with mixed success (Whitehead, 1988).

At the same time, several Englishmen, most notably Captain Charles Leigh and Robert Harcourt, attempted unsuccessfully to establish British colonies along the lower Oyapock River (Harcourt, 1613; Purchas, 1906). The French, also seeking to establish a foothold in the Guianas, suffered even greater difficulties in founding colonies along the coastal zone of French Guiana. Attempts by Gaspar de Sostelle to settle at the mouth of the Montsinery and Cayenne Rivers in the late 1500s ended in conflict

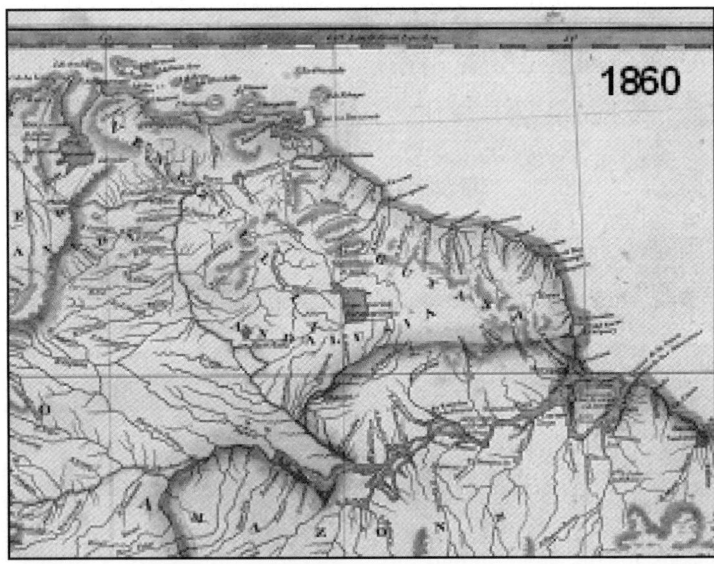

Fig. 8.3. Early cartographic representations of the Guiana Shield emphasizing the importance of Lake Parima in the geographical mind-set of early Europeans. Drafted in remote European offices based purely on the speculations of recent visitors, they illustrate how self-interest of early explorers fuelled the transformation of thinly supported assertions into generally accepted fact.

and failure. It wasn't until 1604 that Daniel de la Revardiere established a French outpost at the present-day location of Cayenne, only to be further built up by the French and subsequently destroyed by Dutch, English and Amerindian raiding parties (Harlow, 1925). Although a small village with a fort (Ceperou) would manage to remain situated at Cayenne throughout the late 17th and 18th centuries, attempts to establish Cayenne as a permanent and thriving settlement would continue to meet largely with failure up until the mid-19th century (Rodway, 1912).

Spanish settlements during the 1600s for the most part remained restricted to a few neglected and frequently overrun and rebuilt outposts, first at Santo Tomé and later at Moitaco, located along the lower Orinoco River (Fig. 8.4) (Ojer, 1966; Whitehead, 1988). The Kingdom of New Granada was considered by the Spanish to extend much further southeastward from this position, but the wealth flowing from the Spanish Main in Mexico, Colombia, Peru and Bolivia preoccupied resources and interest. The Spanish hold on the region throughout this period was largely affected by the frequent raiding and trading activities of Dutch and Caribs coming from the Essequibo (Whitehead, 1988). Deterred by a lack of crown resolve and frequent conflict with Carib inhabitants and European rivals, the Spanish colonial development and occupation of the modern-day Venezuelan Guayana was virtually absent up until the early 1700s (Ojer, 1966).[10]

The Portuguese, preoccupied with cash crop production along the Atlantic coast between present-day Belém and São Paulo and bound to the east by the Treaty of Tordasillas signed with Spain in 1494, had hardly ventured inland along the Amazon River and its northern tributaries during the 17th century. This left much of the trade in these parts to the Dutch West Indies Company, which established outposts along the lower Amazon and Negro Rivers (Edmundson, 1904a,b). The Portuguese did, however, establish a fort in 1616 along the lower Amazon at the present-day site of Belém in defiance of the treaty with Spain and had extended their reach along the southern rim of the Guiana Shield by establishing a fort at the confluence of the Amazon and Negro Rivers – called Manaos – by 1669 (Fig. 8.4). Up until the 18th century, however, the southern and western part of the Guiana Shield remained for all intents and purposes unaffected by the colonization sweeping the coastal shores and highlands of South America and the Caribbean. Trade in dyes, timber and tobacco were undertaken with the residents of these lands, but few permanent settlements were established. The interior of the southern shield area was virtually unknown to the Europeans, but the diseases they brought with them were rapidly establishing their presence among the indigenous peoples inhabiting the area.

Sugar and souls

The Dutch West Indies Company (DWIC), working from its base on the Essequibo, continued to expand and develop its commercial interests in dyes, tobacco, cotton, coffee and, increasingly, sugar. The company had begun to cultivate cane along the banks of the Essequibo and Pomeroon Rivers in the 1530s and by 1638 sugar became the major source of revenue for the company in its Essequibo and Demerara holdings. Cotton and tobacco were increasingly unable to compete with production from the more fertile lands of the British North American colonies and their cultivation was ultimately discontinued, leaving cane and coffee as the major agricultural products. The inland sugar estates of the Dutch continued to expand as far inland as the lower Mazaruni River, but were increasingly seen to be uneconomic as diminishing soil fertility reduced yields. Realizing the dwindling land capability of the interior, the DWIC relocated its commercial headquarters in 1738 to Fort Island (Zeelandia) at the mouth of the Essequibo River and concentrated sugar cane production in new plantations along the lower Essequibo islands and the lower Demerara River. The appointment of Laurens Storm van's Gravesvande as secretary, then commander and finally director general of the Essequibo and Demerara holdings from 1738 to 1772 led to a rapid expansion of the sugar estate between the Pomeroon and Berbice Rivers and saw a massive influx of (mainly) English planters leaving the fragile and densely packed croplands of the British Caribbean (Storm van's Gravesande, 1911). The rapid and successful development of the lower Demerara River became the engine behind Storm van's Gravesande's success and later the transition from Dutch to British colonial control.

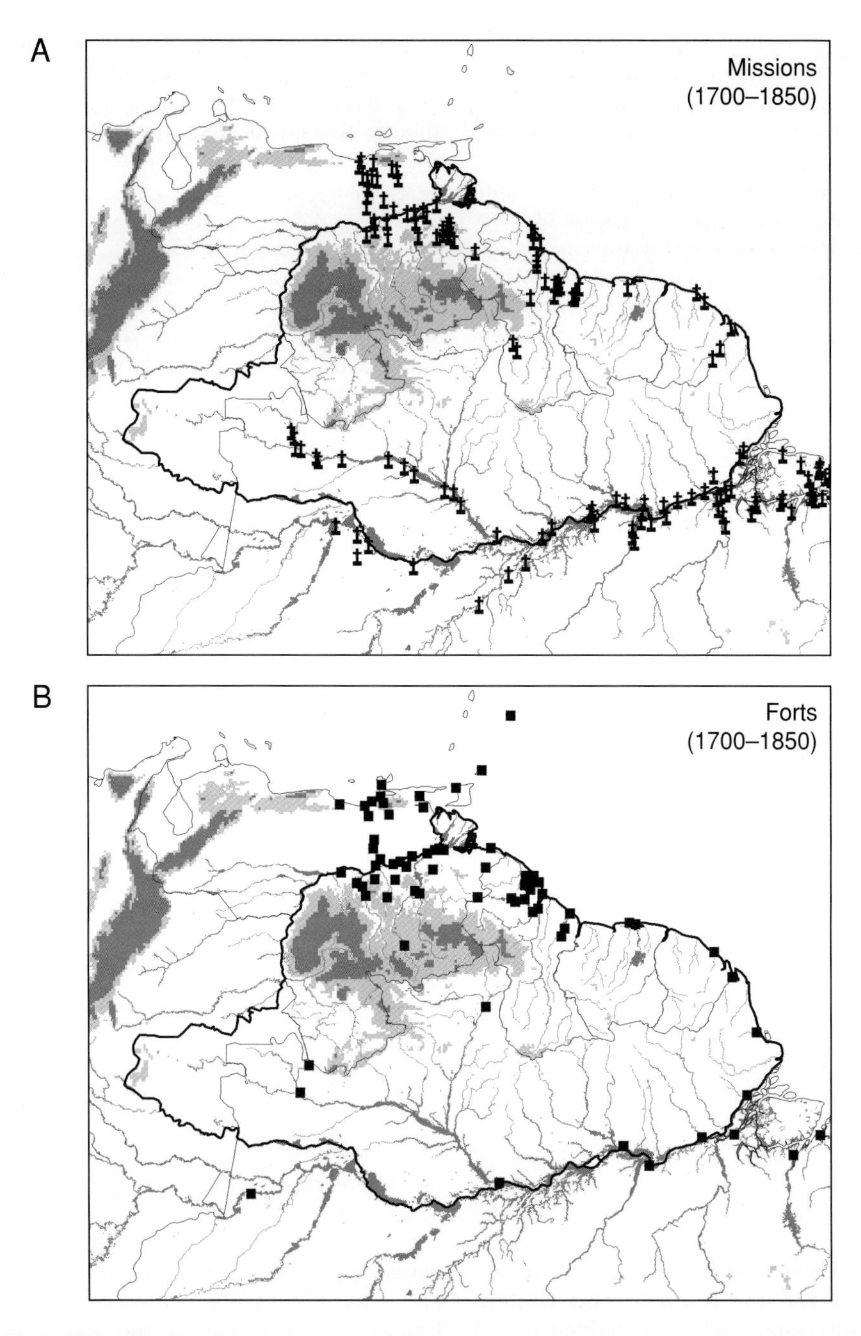

Fig. 8.4. Distribution of early European forts and missions across the Guiana Shield and its periphery. (A) Missions: sites include those established by Capuchin and Jesuits in Venezuela, Jesuits along the Rio Negro, lower Amazon and coasts of Amapá and French Guiana, Moravians in Guyana (Berbice) and Suriname, and various Protestant denominations in (southern and northwestern) Guyana. (B) Forts: sites include those established by the Spanish in Venezuela, Dutch and British in Guyana and Suriname, French in French Guiana and Portuguese in Amapá, along the Amazon River, Rio Negro and Rio Branco. Forts, however, changed hands among these four nations frequently during the period 1700–1850 (see Fig. 8.6). See Notes for source references.

During the same period, an English aristocrat, Francis Willoughby, successfully established a permanent settlement near to present-day Paramaribo in Suriname after disastrous attempts sponsored by the Duke of Corland in 1639 and Earl of Warwick in 1642 to settle the Pomeroon River (Edmundson, 1901). Willoughby's settlement eventually led to the creation of one of the first sugar plantocracies in the shield region and, importantly, the full-scale development of the deplorable practice of importing slave labour from Africa to work the cane fields that was first initiated by the Dutch in their Essequibo, Demerara and Berbice *patroons* (holdings). Within 20 years of founding the first permanent European settlement in Suriname, more than 500 plantations had been established. At the Treaty of Brede in 1667, the Suriname plantations and lands between the Courentyne and Maroni Rivers were permanently handed over by the English to the Dutch in exchange for the North American colony of New Amsterdam (Manhattan Island, New York) (Dalton, 1855). Combined with their *patroons* in Guyana, the Dutch effectively controlled the production of agricultural commodities over a 5000 km^2 area by the end of the 17th century (Fig. 8.5). Most of this area had previously been coastal swamp forest and estuarine plains.

Throughout the 1700s, the colonial development of the Guiana Shield would rest principally with the shifting political, religious, economic and military objectives shaping regional relationships between Spain and Holland. The majority of colonial activity would be concentrated along the coastal margins between the Orinoco and Maroni Rivers and sugar production would reign supreme among these considerations. Trade in other forest materials, such as letterwood (*Brosimum*) and annatto (*Bixa orellana*) also continued to influence

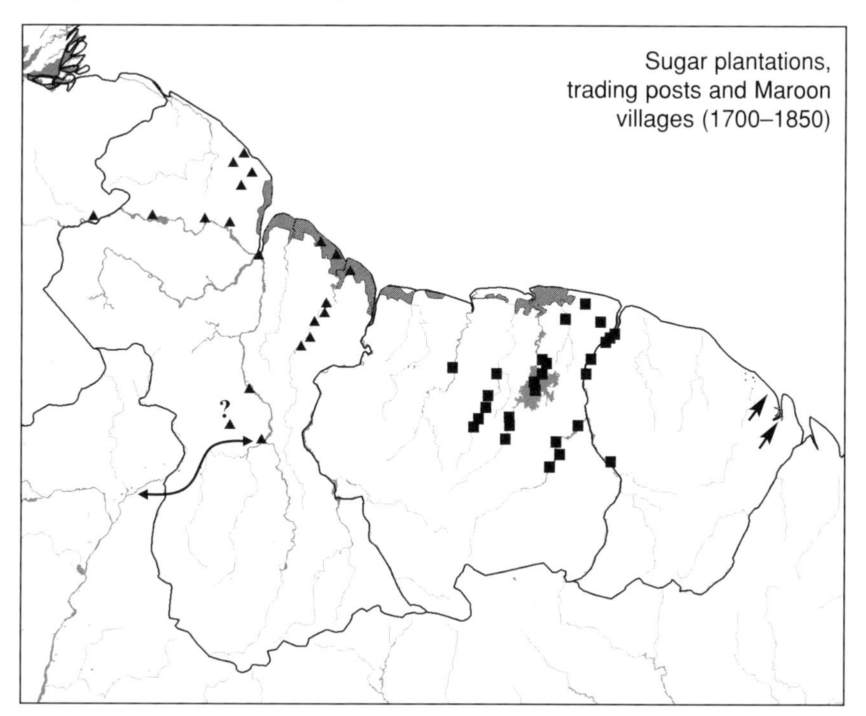

Fig. 8.5. Distribution of early Dutch trading posts in Guyana (solid triangles), distribution of sugar plantations at their peak across Guyana, Suriname (shaded areas) and French Guiana (arrows) and the region settled by Maroons in eastern Suriname–western French Guiana after fleeing coastal plantations (solid squares). Long arrow indicates possible 18th-century trade connection between Dutch and Rio Negro Amerindians through the upper Rio Branco-Rupununi. See Notes for source references.

the economic success and cultural impact of European colonization. Dutch trading outposts and forts were established throughout Guyana during the 1700s, most notably at Mahaicony, Abary, Wironie and Canje in the Berbice region, along the Cuyuni, Pomeroon and Maruca Rivers and much further inland at Arinda near the confluence of the Siparuni and Essequibo Rivers (Fig. 8.5) (Whitehead, 1988). A small, 60 km^2 French colony at Cayenne had several small plantations, but the exposure of the Precambrian basement rock up to the coast effectively limited opportunities for sugar production to the small stretches of riverine alluvium along the lower Approuague, Mahury, Cayenne and Montsinery waterways (Rodway, 1912). Instead, annatto and coffee were cultivated as principal cash crops.

While the Dutch, British and French continued to expand their imperial interests in the region through trade with Amerindians and the cultivation of sugar and coffee, Spain's effort to extend its colonial interests in Guayana were strongly linked to the activities of the Capuchin, Franciscan and Jesuit missions. Spain had undergone a tumultuous collapse during the late 17th century and much of its initial imperial grandeur had been taken away by raiding Dutch and English. When Philip of Anjou succeeded to the Spanish throne, he committed the resources of the Spanish crown to consolidating their control over their remaining American colonies in an attempt to prevent further economic erosion caused by the expanding Dutch and English settlements. Building upon a knowledgebase accrued through earlier unsuccessful expeditions deep into the Guayana region along the Orinoco, Caroni and Caura Rivers (Gumilla, 1741), the early missionaries rapidly expanded their network of forts and missions as Spain fully committed military support to their efforts to subdue the Carib population and eliminate the threat of further Dutch expansion from their established hold in the Essequibo. Despite considerable resistance from the Orinoco, Imataca and Barima Carib populations, the persistent efforts of the Capuchin and Franciscan missions with support from the Spanish crown had thoroughly eliminated the 'Dutch-Carib threat' by the end of the 18th century (Whitehead, 1988). By 1780, the Spanish missions had established over 35 new settlements, or reducciones,[11] along all of the main southern tributaries of the Orinoco, facilitated the foundation of a provincial capital of Guayana at Angostura (Ciudad Bolivar) and extended their *entradas* against the Caribs as far south as the Uraricoera River in present-day Roraima state and as far east as the Pomeroon and lower Cuyuni Rivers (Fig. 8.3). The economic spin-off of the missionary activity was the establishment of a sugar mill at the Caroni mission and a cattle herd numbering more than 145,000 head along the Yuruari River in the Imataca region (Whitehead, 1988). The Dutch, by now increasingly focused on returns from the production of sugar and coffee from their estates in Guyana and Suriname, and less so on the trade in dyes and other forest products, allowed their 150-year alliance with the Caribs against the Spanish to dissolve in an effort to avert Spanish reprisals against their burgeoning, and increasingly sedentary, economic interests in the region (Storm van's Gravesande, 1911).

The French and Portuguese were also busy consolidating and partitioning their control over parts of the southern Guiana Shield during the 18th century. Forcible pursuit of their respective claims came at a price – paid by the native inhabitants of Amapá. During the early years of the 18th century, the Portuguese furthered their hold over the region by eliminating tribes that did not support their claim. As a consequence, several thousand people were either forcibly removed to Marajó Island by the Portuguese or fled to the French-held part of the Oyapock watershed or the coastal region of Cayenne (Hurault, 1972).

Finally, at the Treaty of Utrecht in 1713 they agreed to define the Oyapock River as the boundary between French and Portuguese territorial interests in the southeastern corner of the shield. This expanded the Portuguese system of *donatários*, or captaincies (akin to the Dutch *patroon* system) created first at São Vicente in 1535, to an area north of the Amazon River's mouth. The French, however, would not gain complete control over French Guiana until 1816, when agreement, through the Treaty of Paris, was reached with Britain and Portugal over France's claim to the present-day extent of French Guiana. As a consequence, the British would hand back the colony after an 8-year period of occupation (Fig. 8.5).

Portuguese control over the great lowland forest area stretching across the western rim of the Guiana Shield was restricted to their early settlements at Belém and Manaus during much of the early 17th century, but this changed dramatically by 1750. Portugal agreed a new delimitation of their colonial interests in South America with Spain through the Treaty of Madrid. This redefined the boundary originally set in 1494 through the Treaty of Tordasillas. The Treaty of Madrid gave extensive claim rights to Portugal over the entire Amazon Basin, but not the southern part of the Guiana Shield extending north from the Amazon River (Fig. 8.3). Combined with the agreement reached with the French in 1713 through the Treaty of Utrecht, the Portuguese 'paper' control over the southern rim of the Guiana Shield area was now complete. Despite the vast expansion of their interests in South America under the Treaty of Madrid, the Portuguese monarchy would never see material control over the southern province of the Guiana Shield that was later to become the Brazilian states of Roraima, North Pará and Amapá. The Portuguese had, however, commenced with the colonization of the Amazon proper through two different routes, namely regular trading missions along the Amazon tributaries and the missionary activities of the Jesuits. Both were inextricably bound to one another. Trade with Amerindians continued along a similar line as that established by the Dutch in the 16th and 17th centuries. Cacao, or cocoa, was the major trade item, though other forest products were also sought, including oils from the copaiba (*Copaifera* spp.) and andiroba (*Carapa guianensis*) trees, vanilla and other aromatic plants. Cacao would account for the lion's share of trade from the Amazon during much of the mid-1700s.

While trade in forest plant products was increasing, the Jesuit mission was rapidly expanding its presence on the Amazon and Negro Rivers. By the mid-1700s, more than 60 major settlements had been established along the Amazon River and more than 50,000 Amerindians lived and worked within the confines of these Jesuit-run compounds, or *aldeias* (Fig. 8.4) (Hemming, 1978a). In many ways, the Jesuit *aldeias* were similar to the Capuchin *reducciones* that were being established concurrently in the northwestern part of the Guiana Shield along the Orinoco and its tributaries. Most (albeit exclusively European) historical accounts suggest that the approach taken by the Jesuits, however, relied less on forced resettlement through the use of the military than was typically used by the Capuchins, although their cooperation with 'ransom missions' aimed at procuring Amerindian slaves was far from benign (Hemming, 1978a). The Jesuits' missions would become the mainstay of European agricultural expansion along the Rio Negro region up until the late 1700s. The toll on the native inhabitants of the Rio Negro region was staggering. By 1750, Franciscan diaries estimate that nearly three million people had been exported from the region as slaves to work the coastal sugar estates (Hemming, 1978a). By 1785, more than 400,000 coffee and cacao trees had been planted in cleared forest areas along the lower and middle Rio Negro as part of the Jesuit network of Amerindian-fuelled agro-industry and the region had been widely depopulated (Reis, 1943). Cattle-ranching was promoted as a secondary use of cleared land no longer able to support intensive crop production, but this too quickly exhausted the meagre nutrition afforded livestock in a deforested landscape near the mission centres (Chernela, 1998).

Colonial consolidation and decline

Less than 50 years after the Spanish missions and Dutch West Indies Company had firmly established their control over the main ports and waterways in the region and were reaping profit from their activities, the tides of change were already beginning to rise. The Dutch colonial administration of their Demerera, Essequibo, Berbice and Suriname holdings were increasingly occupied with the control of ill-treated slaves on their sugar and cotton plantations (Collis, 1965). The pillaging of the Dutch holdings in Suriname by the French and the subsequent flight of many slaves to the forested interior during the chaos bolstered the African population living free of the plantations in Suriname. An African slave revolt at Berg-en-Daal in 1730 and a major rebellion that shook the Berbice estates in 1763 added to this growing population (Rodway, 1912). The Suriname plantations were thereafter continuously harried by rebellion and runaway slaves that escaped into the forest interior during the late 18th and early 19th century. These escaped slaves, called *Maroons* or Bushland Negroes, numbered more than 10,000 in Suriname, Berbice, Demerara and Essequibo by the 1770s (Fig. 8.5) (de Groot, 1977; Rose, 1989) and the plantation owners in Suriname were required to expend more and more effort in order to maintain their tenuous hold on a slave population that outnumbered them by more than 10 to 1 by the late 18th century (Whitehead, 1988). Eventually, peace treaties were signed between the Dutch and the Aucans and Saramaccans in 1761 (Rodway, 1912), but fighting continued with the Bonni and Baron groups (Collis, 1965). Contracts were established with Amerindians to recapture those slaves that managed to escape into the hinterland and defences to protect the plantation margins from marauding Maroons had to be constructed. The sum of these actions began to take their toll on the DWIC's ability to profitably operate their sugar estate holdings in the Guianas. The growing resistance to slavery among plantation workers preoccupied the Dutch to such an extent that they were also unable to respond in kind to the increasingly provocative Spanish movements eastward through Guayana during the 1700s. Trading outposts on the Barama, Moruca and upper Cuyuni Rivers were burned and briefly occupied in the 1760s and the extent of influence the Dutch enjoyed in the Guiana Shield during their 'golden age of trade' in the 17th and 18th centuries was eventually restricted to the coastal sugar estates by the early 1800s.

The age of Capuchin and Franciscan control over Venezuelan Guayana and its Carib inhabitants came to an abrupt end in 1817 as the Spanish empire reached the pinnacle of its expanse and then recoiled under the weight of Napoleonic France and revolution in its colonies. The war of independence, fought between 1810 and 1818 in Venezuela, led to the formation of Grancolombia, a vast nation centred on the spine of the Andes and extending between Guayana and Ecuador (Fig. 8.6). The Catholic missions that had virtually ruled parts of Guayana since the early 1700s provided assistance to the royalist forces during the struggle for independence. By 1818 they had been completely removed from power in the Guiana Shield as a result of these loyalist sympathies. No further policies were developed to subjugate the Carib inhabitants, though the effect of the mission on Carib regional domination was clear from the catastrophic decline in the population through disease and death in the *reducciones* and immigration to the Essequibo/Cuyuni area (Whitehead, 1988). Twelve years later, Grancolombia was dissolved and the republic of Venezuela was born. Through the break up of Grancolombia, Venezuela (and Colombia) became the first of the modern-day countries in the Guiana Shield to achieve independence (Fig. 8.6). Brazil, while severing links with the Portuguese royal family around the same time as Simon Bolivar and other revolutionaries were fighting for independence from Spain, would continue to be ruled through its own monarchy up until the conversion to a rudimentary republic in 1889 (Fig. 8.6).

The Dutch hold on the coastland of the Guianas collapsed altogether as they entered the 19th century. Napoleon's takeover of the Netherlands and subsequent war with Great Britain severely weakened

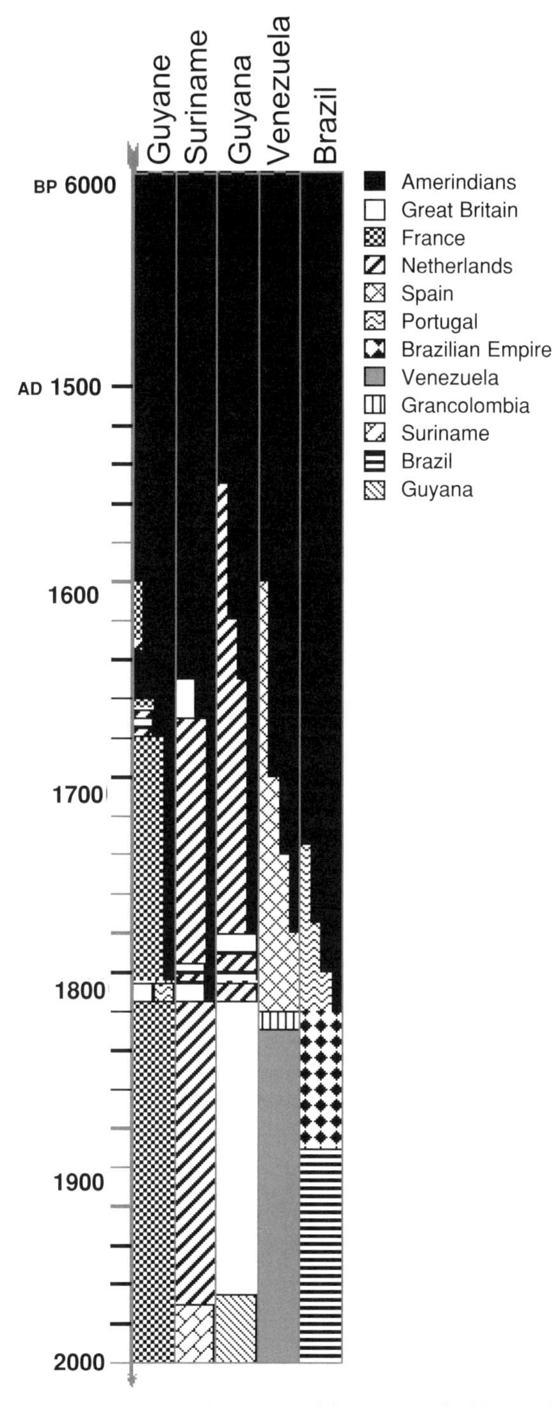

Fig. 8.6. Timeline of occupation and European colonization of the Guiana Shield according to country. Each country timeline (separated by vertical lines) consists of four levels of control represented by four vertical bars. One bar signifies trading missions, expeditions and the presence of a few scattered trading posts, two bars signifies the establishment of several forts and/or missions, three bars signifies the presence of towns and cites and four bars indicates complete political authority over the area.

the ability of the Dutch to administer their colonies. The Dutch West Indies Company was increasingly being seen to lack the skills necessary to improve the profitability of the sugar estates and maintain control over a growing resentment among smallholder planters and resistance from slaves. In 1796, the Dutch allowed the DWIC's contract to expire and assumed direct responsibility for the Demerara and Essequibo holdings. The change in administration proved to have little effect on the decline in Dutch control over the area. The French (and French-controlled Batavian Republic of The Netherlands) briefly controlled the Essequibo and Demerara region in 1782, establishing a new capital, Longchamps, at the mouth of the Demerara River. The Dutch regained full control in 1784, renaming the French settled town Stabroek and designating it as the new capital of their Guiana holdings (Fig. 8.6). British forces, already occupying Suriname since 1775, were sent in 1796 to occupy the Essequibo and Demerara holdings yet again. Six years later, the area was returned to Dutch control in accordance with the Treaty of Amiens (Fig. 8.6). Less than a year later, British troops returned to Stabroek and the British would retain lasting control over all of the sugar estates, though Dutch interests and properties were not forfeited. In 1814, the Dutch formally agreed to hand over all three of its former patroons, Berbice, Demerara and Essequibo, to the British through the Treaty of London. Two years later, the British returned the former colonial lands in Suriname to the Dutch and agreed through the Treaty of Paris to return Guyane to French control after 8 years of occupation in collaboration with the Portuguese (Fig. 8.6). By 1831, Stabroek was renamed Georgetown and the three colonies were united under a single colonial administration until Guyana's independence in 1966. Suriname, or Dutch Guiana, would remain a Dutch colony for the next 155 years until independence in 1975. Cayenne and the lands of present-day Guyane would remain as a French colony until 1945, when it would be given full department (state or county) status (Fig. 8.6).

By the end of the 18th century, the Jesuit missions of the Rio Negro region also had all but disappeared. As agricultural productivity was exhausted within the *aldeias* landholdings and the population of Amerindian workers dwindled in the face of disease, forced emigration and hardship, the focus of Jesuit and Portuguese exploitation shifted to the Solimões and its western tributaries (Reis, 1943). Over the next 50–80 years, the region would be virtually abandoned by the Portuguese and Spanish colonists, until global economic expansion would reignite interest in the region as the source of a traditional product being widely sought for its new industrial applications – rubber (Chernela, 1998).

Emancipation and new immigration

The period 1810–1840 proved to be one of the most important junctures in the history of the Guiana Shield. While the production of coffee and cotton in the plantations of the South American east coast from the Essequibo to Bahia largely declined over the first part of the nineteenth century, sugar exports rose precipitously. Plantations in the Demerara, Essequibo and Berbice colonies had only recently come under final control of the British and the sugar industry was expanding in these possessions while well-established plantocracies in the West Indies had reached their material limits to expansion. The prospects for a burgeoning agricultural estate in British Guiana that would bring considerable profits to the plantation owners appeared good and the drive for further slave labour to feed estate production and expansion was in full motion. By 1810, forced emigration of Africans as slaves had reached its highest rates and the standing population of slaves measured more than 1.4 million in the British colonies, 460,000 in the Dutch colonies and 1.8 million in Brazil (Curtin, 1969). Buoyant sugar prices, steady market demand and the free-flow of slave labour augured well for the continued dominance of the plantocracy in the life of colonial America.

Several factors, however, began to alter this prospect by 1840. Incessant war among the major regional empires since the 1790s began to take its toll on sugar returns through shipping and infrastructural losses. At the same time, a voice of social justice calling for the end of the slave trade, and eventually slavery, began to be heard in the halls of the British political system. Anti-slave abolitionists, led by missionary groups, began to exert enormous pressure on the political justification for this abhorrent practice. By 1807, the use of British ships to transport slaves was made a crown offence (Waddell, 1967).

In part driven by the constant fear of their enormous slave populations and the memory of past revolts, particularly the great Haitian rebellion of 1793 led by Toussaint L'Ouverture, the 1763 Berbice revolts and a growing Maroon population, plantation owners became even more repressive in their treatment of the sugar workers. Combined with increasing efficiency demands, declining supply of new slave labour and a dropping sugar price, owners placed an increasing burden on their existing slave populations (Higman, 1984). As a consequence, a series of revolts erupted in 1823 (Demerara, British Guiana) (da Costa, 1994), 1831 (Jamaica), 1835 (Salvador, Bahia) and 1838 (Rio de Janeiro, Maranhão) (Skidmore, 1999). Spurred on by the great Demerara slave rebellion in British Guiana and the vociferous foment of the Anti-slavery Society, all slaves within the British Atlantic colonies were emancipated with the passage of the Slave Emancipation Act in 1833 (Fig. 8.7).

First deprived of their inflow of new slave labour, and then plantation labour altogether, British sugar interests found themselves competitively disadvantaged. The gradual remission of the Sugar Act by 1850 increased the competing flow of sugar from outside the existing colonial plantations to the UK. British interests quickly redressed this imbalance by proactively promoting the widespread end of slavery throughout the sugar-producing regions of the Americas. Fifteen years later, slavery was similarly abolished in French colonies, in the Danish islands by 1848 after a series of revolts, in Venezuela by 1854, and across the Dutch West Indian holdings (including Guiana) in 1863 (Fig. 8.7). Brazil, the only slave-owning country in the Americas by 1865, would continue this deplorable practice until legally outlawed in 1888 through imperial decree, the *Lei Áurea*, after a series of earlier stop-gap actions that circumvented release of the labour force by providing first for the freedom of newborn children and then seniors exceeding the age of 65 rightfully failed to quell discontent (Toplin, 1972). The ending of the slave trade decimated a sugar industry that had seemingly passed its prime and was increasingly being challenged by other industrial sectors, including sugar beet production in Europe and North America. The advent of the industrial revolution in Europe also ushered sugar to the rank of staple commodity, which supported the development of more lucrative, manufacturing sectors rather than acted as a point source of profit on its own. By 1845, cotton was no longer exported from the Guianas and sugar receipts declined precipitously (Waddell, 1967) (Fig. 8.7).

The rise and fall of plantation slavery in the Guianas also had an impact on the forests of the near interior. During the expansionary years of the plantation system in the late 18th-century Guianas and Brazil, considerable amounts of timber and land were put to use in the construction of housing, millworks, sugar punts, bridges, canal revetments and empoldering that formed the infrastructural foundation of the plantocracy. These were continuously maintained through the extraction of timber from upland forests adjacent to the coastal swamplands. Constant military action between European powers in the region during the 18th and early 19th centuries required timbers for ship-building and fort construction. By the 1840s, nearly 2% of the population along the coast of British Guiana alone was believed to be employed in the extraction of wood products, either as independent suppliers or as part of nine registered businesses that would have provided materials to over 300 sugar, coffee

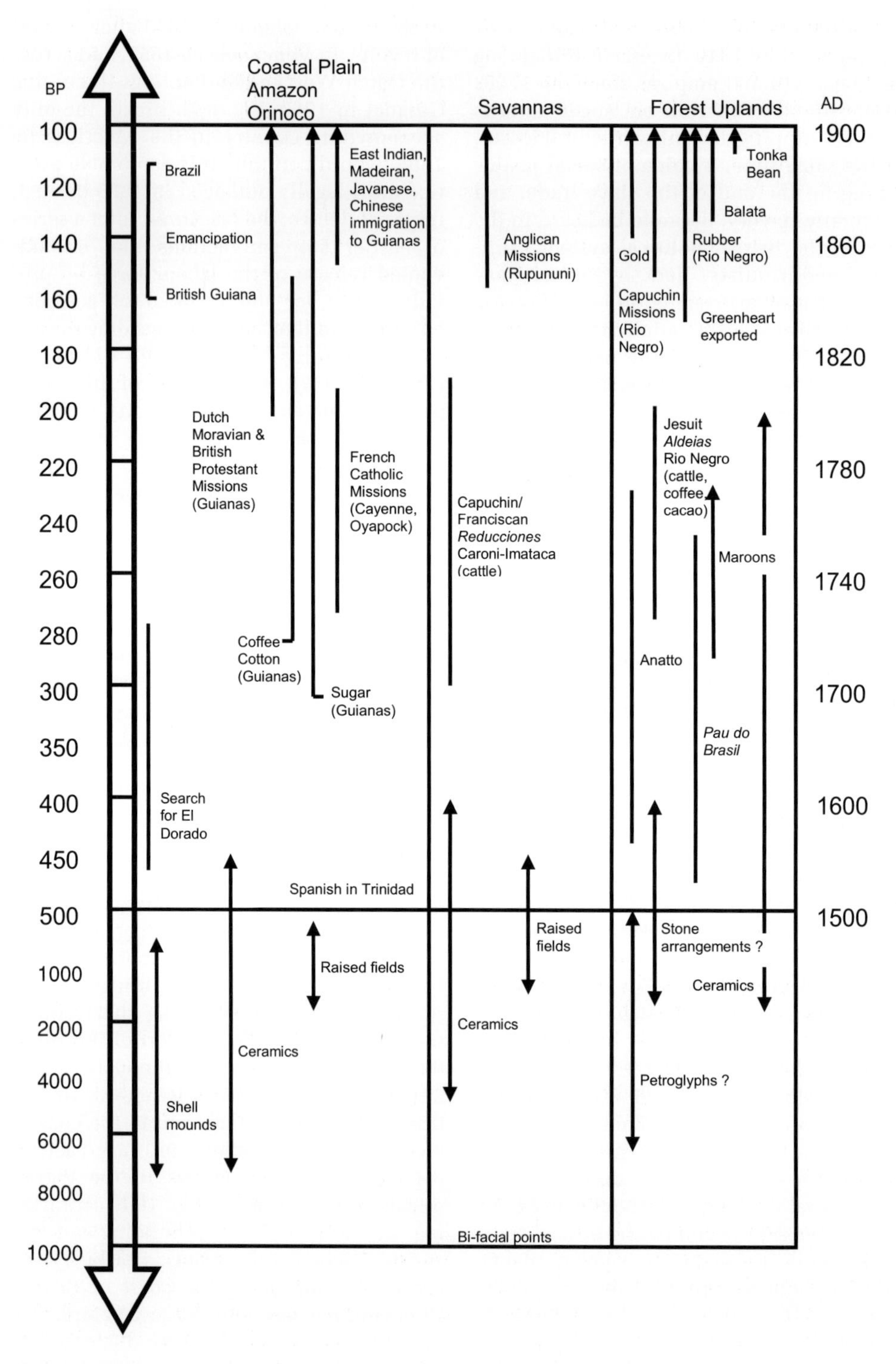

Fig. 8.7. Timeline of archaeological artefact dates and colonial phases across the three main ecosystems of the Guiana Shield up to the end of the 19th century. Colonial phases are designated by principal mission, trade, agriculture, mining and timber extraction periods.

and cotton plantations (Schomburgk, 1840). In addition to this trade, timber trees in the Guianas were increasingly recognized for their density, durability and strength in maritime and industrial applications overseas and a growing export trade had developed by the time the sugar industry was in post-emancipation decline. By 1850, nearly 70% of the sugar plantations operating at the turn of the century had closed and the export of timber (mainly greenheart) from British Guiana alone amounted to over 5000 m^3 (Fig. 8.6) (Schomburgk, 1840).

Africans fleeing slavery in the coastal plantations would also have begun to modify the forests of eastern Suriname's near interior as they started to utilize forest resources to sustain their growing populations (Fig. 8.5). Across the middle Maroni, Commewijne, Suriname, Nickerie, Courentyne and Berbice watersheds, Maroons were establishing their own renewed social order independent of the plantocracy. This independence was at its greatest in eastern Suriname and in all likelihood had the greatest impact on the forests of that region.

In contrast, the revolutionary fight against Spain across Venezuela, the expulsion of the Capuchin mission from Guayana and the collapse of the Jesuit system along the Rio Negro, combined with the devastation of disease and forced emigration upon the native populations, meant that much of the western Guiana Shield remained largely as it was prior to colonization or had effectively been emptied of its human inhabitants. In many areas, this is the forest that would ultimately be encountered by travelling natural historians documenting, and preparing, the misperception of a uniformly verdant and pristine system devoid of human influence.

By the 1850s, the sugar industry of the Atlantic coastline had diminished substantively due to a shortage of labour after emancipation. Dutch and British sugar interests began a mass programme of indebturement that would (in part) see the arrival of new sugar workers, mainly from India and Java (Indonesia). In the case of Guyana, this migration would see nearly 250,000 East Indians arrive between 1850 and 1920 (Bisnauth, 2000). The consequence of this and other immigration, mainly from Madeira (Portugal) and China, was to increase the coastal populations, reinvigorate the sugar industry, establish an expanding rice industry and invariably place greater demands on the forest resources of the near interior. As the century came to a close, coastal agriculture was in full swing, but it was the growth of forest-based industries that would rapidly extend the range of industrial forest use to the deepest reaches of the Guiana Shield.

The new El Dorado: rubber, timber and placer gold

In 1818 Charles Macintosh created a fashion revolution in England with his rain-repelling jackets made from a cloth impregnated with natural rubber. He would patent the process in 1837 and the demand for this new kind of outerwear would start a drive for natural rubber that lasted more than a century. The British fashion for practical outwear, however, would pale in comparison to the tremendous industrial demand that emerged as the Industrial Revolution moved into high gear. The properties of rubber, balata and other natural latex materials were soon recognized for their wide range of uses, particularly after a series of inventions that assisted in the development of new product lines.

In 1842, Charles Goodyear patented the vulcanization process and a vibrant export market for the raw material developed over the next several decades. Vulcanization extended rubber's natural elasticity over a wide range of temperatures, an important consideration when manufacturing goods for use in the harsh winters of Europe and North America. As the first cars and their pneumatic rubber tyres became a common sight in many parts of North America and Europe, demand boomed and the Amazon rubber industry was born. By the 1880s, *seringa* (or Indian) rubber (*Hevea brasiliensis*) production was operating across much of the Rio Negro basin and effort was underway to establish plantations in the Guianas and Venezuela. Brazil remained the sole supplier of natural rubber, a native plant of the Amazon lowlands, until 1876 when a British expatriate exported (many say illegally) seeds of *Hevea* to Trinidad. Soon thereafter, British and Dutch colonial companies shipped these to Malaya and Indonesia, where plantation production of rubber flourished for the next 100 years and effectively eliminated the Brazilian monopoly on supply during the peak decades of demand. Nonetheless, Brazil remained a major supplier of rubber throughout this period, particularly to the USA and Germany (Barham and Coomes, 1996).

At the same time, another natural rubber, balata, was increasingly recognized for its useful properties, particularly in the manufacture of machine belting, golf balls and later, telegraph cable insulation. A concurrent balata boom across the Guiana Shield created a drive into the forest interior that had up until the mid-1800s only experienced sporadic establishment of outposts trading for annatto, dyewood, tobacco and selected timbers and a few missions. By the close of the 19th century, thousands of balata bleeders were traversing the forested interior of the shield seeking new bulletwood trees (*purguo, nisperillo, boletri, balata franc, maçaranduba*) (*Manilkara bidentata, M. huberi*) to tap.[12]

Logging also was beginning to develop as an export industry in many parts of the shield during the latter half of the 19th century. In (the then) British Guiana, production of timber (mainly logs and squares) totalled more than 406,000 m^3 between 1850 and 1900 and led the region in hardwood exports. The bulk of timber extraction occurred along the northern strip of forests running parallel to the Coastal Plain, along the savannas and lower and middle reaches of the major Amazon and Orinoco tributaries below the first major impediments to transport, the falls. Nearly all timber was extracted via the rivers below these falls (e.g. McTurk, 1882).

The increase in the use of forests for latex and timber during the 19th century was only one part of the transformation experienced by the forested interior and the people living there. The intended purpose of the earliest shield explorers was to find hidden gold of a forgotten Indian empire. Nearly 400 years later, gold was extracted commercially, only as placer dust rather than the ornate workings of Manoan metalsmiths envisioned by Raleigh and others. Commercial working of gold commenced in a series of deposits throughout the Guianas and Venezuelan Guayana between 1860 and 1880, but as early as the 1840s in the case of the Imataca region of Venezuelan Guayana (e.g. Holmes and Campbell, 1858) (Fig. 8.6). By 1900, several tens of thousands of miners had moved into the forested shield interior of the Guianas and Venezuela. In French Guiana, gold miners alone accounted for nearly 25% of the total population (49,000, including 6500 convicts) by this time and mining was the principal economic activity undertaken in the territory (Rodway, 1912).

The Guiana Shield as a place of refuge, 1500–1900

The archaeological, ethnological and historical evidence indicates that the European colonization that followed Columbus' arrival must represent one of the most rapid and complete ethnic turnovers ever experienced in human history. Patterns of colonization along the shield rim reconstructed from archaeological and historical evidence point to parallels between the first Amerindians and Europeans. Both arrived by boat along the continental coastline and lower river reaches through a wave-like series of small settlements that had mixed, long-term success. As populations established a modicum of permanence, population centres grew and spread, innovated and explored new avenues of agricultural production and trade. War, hardship and political discordance forced populations to migrate and invariably clash with or be absorbed by resident societies upon their arrival. Consequently, the human history of the shield region is one defined by frequent cycles of migration, displacement and establishment.

The striking difference between cycles reconstructed for pre-Hispanic Amerindian societies from archaeological evidence and those historically documented for Europeans hinges on an important difference – the maintenance of an external, continuous, flow of immigrants that forever sought to benefit through familiar socio-economic and ideological links with Europe, often even after political independence was achieved. Evidence suggests that Amerindian societies, particularly those along major rivers and the coast, also maintained contacts with other groups, but these were largely limited to the same lowland environment (but see a narrative on trade links between lowland Manau (Manao) of the Rio Negro and highland Chibcha in Colombia in Hemming (1978a, p. 440)). It is not unreasonable to assume that they had in turn an independent and more perceptive understanding of the wider risks and opportunities associated with the lowland neotropical landscape and a better ability to identify the most suitable living locations from both a biophysical and socio-economic perspective. Tragically, what they shared with the first Europeans was an ignorance of the risks and opportunities that each other's cultures brought with contact. The immediate consequences conveyed through epidemic disease, displacement and enslavement established a trend of counterproductive land-use that has lasted for centuries.

By the time the majority of native inhabitants realized the dangers of contact, the topology of turnover had transformed from points, to lines and finally areas of impact, especially as initial points sparked secondary spread of disease and slave-export among Amerindian communities themselves (Colchester, 1997a). Flight became the *sine qua non* of the Amerindian response. As a result, the shield interior would become a regional refuge for Amerindian groups fleeing the plantations, *encomiendas*, missions, *aldeias*, *reducciones* and slave-raiding parties forming the colonial Guiana Shield (Evans and Meggers, 1960; Riviére, 1969; Hemming, 1978a; Boomert, 1981). For much of the 18th century and early 19th century, this proved for some to be an effective strategy for cultural survival. However, with the renewed push into the interior forestlands and savannas by late 19th-century natural resource companies, missions and government outposts, only the most remote forests of the Tumucumaque Uplands and Guayana Highlands remained at a distance from the management of commercial and political interests still largely aimed at fostering wealth through overseas export and cultural change through import.

The flight of Amerindian groups into the heart of the Guiana Shield appears across virtually every part of the region, although the movement out of the Rio Negro, North Pará and Amapá regions of Brazil was without doubt the greater (Reis, 1943; Hemming, 1978a; Skidmore, 1999). The intense and sustained enslavement of Amerindians from these regions for 300 years between AD 1600 and 1900 compelled many of the remaining groups to move into the relative protection of the remote interior regions of the Guianas and southeastern Venezuela. Hurault (1972) details the diaspora of the Palikur fleeing Portuguese slave-raiders along the Amapá coast and the arrival of the Wayana and Wayapi moving upstream along the northern tributaries of the Amazon in Pará only to arrive at their new home along the upper Oyapock. Maroons fleeing sugar plantations in eastern Dutch Guiana also pushed Amerindian groups, such as the Emerillon, further up the Maroni, Suriname, Saramacca and Tapanahoni Rivers. Archaeological and ethnohistorical evidence also supports the migration of the Taruma from the middle Rio Negro across the Acarai Mountains into southern Guyana, a feat subsequently repeated by the Atorai, Wai-Wai (Evans and Meggers, 1960), Wapishana and Macushi (Macuxi). A similar geography of flight developed from the eastern part of Venezuelan Guayana into western Guyana and the Caura basin as Carib (Kari'ña) communities fled the 18th century Spanish military and internment in the Capuchin *reducciones* (Whitehead, 1988). Later, during the late 19th century, the Yekuana and Sanema (Yanomami) migrated northward from their ancestral lands along the upper Orinoco and Parima Rivers towards the upper Ventauri, Erebato and Caura watersheds in order to avoid capture by other groups slave-trading with the Portuguese or to obtain forest products for exchange with Dutch and British traders (Colchester, 1997a, pp. 114–119). The headwaters of the major rivers draining the shield continue to act as a place of refuge for Amerindian groups seeking to escape the pressures exerted by a spreading *mestizo* (*caboclo*) lifestyle, where they have traditionally been forced to the bottom of the developing social class system (e.g. upper Caura River: Silva, 1996).

Commercial Use of Guiana Shield Forests, 1900–2000

The earliest decades of the 20th century witnessed the beginnings of modern commercial forest transformation across much of the Guiana Shield. Many of the Amerindian societies once firmly established along the rim of the shield had migrated into the deep interior, fleeing the sweeping changes that had already altered the social and economic dynamic of the Atlantic coastal plains and floodplains of the Amazon, Negro and Orinoco Rivers. The Industrial Revolution created a global surge of new demand for natural products as the steam engine and line assembly catalysed the manufacturing processes once performed by hand.

The growth of commercial forest use in the Guiana Shield widely preceded a similar rise in our understanding of what it contained, how it worked and its role in modulating local, regional and global change. These invisible services were understood but rarely articulated fully by people who acquired many years, if not a lifetime, of observations in the forested regions. This included several early European tropical foresters, natural historians and, in a different way, Amerindian inhabitants. To most others, the vastness of the forest frontier appeared immeasurable when considered in proportion to the state of technological capacity. Most early commentators believed the main limitation to development was the lack of technological capacity to exploit this vastness, rather than the sustainability of this exploitation. Focus was dominated by a product–by–product dissection of the forest system and the relative, saleable value of each, typically without thought to the mounting opportunity costs transferred on to other forest users.

In a manner widely duplicated throughout the remaining natural forests of the world, the business of locating, quantifying, removing and refining the natural resources of forestlands proceeded at a pace that was principally driven by human labour at first. By 1920, more than 60,0000 people were engaged in the extraction of timber, gold, bauxite, diamonds, balata, seringa rubber and other minor forest products across the interior of the Guiana Shield. At the same time that employment in the forest sector was booming, the commercial extraction of forest products was conceded to a few, large companies operating under licence from the colonial governments (e.g. timber, rubber: British Guiana Lands and Mines Dept, 1920) or was controlled by an autochthonous class of middlemen that ruthlessly managed trade between producers and buyers. In the case of rubber-tapping, this further promoted the formation of an underclass from the large, mainly Amerindian, population of extractivists (e.g. the rubber 'patrões' of the Rio Negro) (Barham and Coomes, 1996; Chernela, 1998).

As the enormous surge towards the interior grew, a number of botanists and foresters in (the then) British Guiana began to question the wisdom of unmitigated extraction and the absence of any effort to manage resources as a means of securing their future contribution to economic growth (McTurk, 1882; Jenman, 1885; Davis, 1933). At the same time, other natural historians and anthropologists recognized the proactive collusion of regional administrators and local power figures in forcibly pressing Amerindian inhabitants along the Rio Negro into extractivist work (Wallace, 1853a; Koch-Grünberg, 1909; Nimuendajú, 1950) and inculcating both local and migrant labourers in a complex exchange, or *aviamento*, system through which rubber was traded to isolated tappers for goods sold at prices sufficiently inflated in order to maintain indebtedness (Bunker, 1985). The growth and development of forest-based commodities in the region during the late 19th century has increasingly dominated the regional economies, as they once did in the early 16th century. Overlapping boom-and-bust cycles of commodity production dominate the history of forest use and forest livelihoods across many parts of the region over the last 100 years.

Commercial mining

The Guiana Shield has figured prominently in the global production of several metal and gemstone commodities over the past century. More than 60 metal and non-metal mineral products have been extracted from the region in various quantities since 1900, but most are restricted to only a few sites (e.g. manganese) and/or are of low-grade content. These are, for purposes here, considered of marginal significance both in terms of global trade and regional socio-economic and environmental impacts. Among the minerals that have fundamentally shaped the 20th-century economies of the region, gold, bauxite, diamonds and iron rank far above the others.

Gold

This precious metal has been known to occur in commercial quantities in the Guianas since as early as the 1820s (Rodway, 1912), but was first brought to the attention of Europe's political and business sectors through discoveries at Tapaquen (El Callao) in Venezuelan Imataca in 1849, on the Mana River in French Guiana in 1864, and along the Suriname and Saramacca Rivers of Suriname in 1876 (Rodway, 1888). The earliest commercial gold-mining activity of any significance began in 1849 at El Callao, the oldest active gold mine in the shield region (Gibbs and Barron, 1993). In 1882, a large expansion of production commenced in (the then) British Guiana when placer (alluvial) deposits of commercial grade were discovered on the Cuyuni River. Placer, or alluvial, gold deposits were soon being worked along most of the major rivers draining the greenstone belt running behind the northern coastal rim of the shield region (see Chapter 2) (Gibbs and Barron, 1993). Gold production from the Guiana Shield, however, has consistently amounted to less than 1% of total global output (Table 8.1). Spurred on by the output of several large mines, the contribution of the region to global gold production exceeded 1% for the first time in 1995 and has continued to expand.

Greenstones and gold

The distribution of greenstone formations is the key to understanding the spatial pattern of historic, present and future gold mining and its associated impacts across the region. Representing one of the most widely distributed and most well-established sources of auriferous ore across the global Precambrian landscape (Goldfarb *et al.*, 2001), more than 90% of the major gold deposits worked in the region over the past century are closely associated with the spatial distribution of the greenstone belt (Fig. 8.8) and are cited as first-order exploration targets for large tonnage–low grade gold deposits (Voicu *et al.*, 2001). Greenstones are unusually widespread across the Guiana Shield and represent the largest areal exposure of this Precambrian formation in South America (see 'Greenstone belts', Chapter 2). They also occur across the Brazilian (Guapore) Shield (e.g. Bahia, Goias, Minas Gerais), where they have been mined extensively since the early 18th century and continue to account for the major share of Brazil's declared gold production (US Bureau of Mines, 1923–2002).

Table 8.1. Fraction of total estimated global production derived from the Guiana Shield expressed as 10-year averages from 1930 to 2000. Prior to 1979, bauxite was only produced in Guyana and Suriname. Values in parentheses after 1980 are for Guyana and Suriname only. More than 90% of gold and 95% of diamonds produced in Brazil originate from deposits outside the Guiana Shield. Sources: US Bureau of Mines (1932–2002), DNPM (1995–2001), British Guiana Lands and Mines Department (1910–1945), British Guiana Geological Survey Department (1957–1963), Guyana Geological Survey Department (1964–1970), Guyana Geological Surveys and Mines Department (1971–1979), Guyana Geology and Mines Commission (1980–2002).

	% Global production		
Interval (years)	Gold*	Bauxite**	Diamonds***
1930–1940	0.6	18.5	0.6
1940–1950	0.4	38.3	0.5
1950–1960	0.2	33.6	0.5
1960–1970	0.1	17.8	0.9
1970–1980	0.1	11.6 (11)	1.9
1980–1990	0.2	12.4 (5.9)	0.4
1990–2000	0.9	16.2 (4.7)	

*Venezuela, Guyana, Suriname, French Guiana only.
**Venezuela, Guyana, Suriname, North Pará only.
***Venezuela and Guyana only.

Virtually all of the historic production across the Guiana Shield is associated with these greenstone formations and their (palaeo) placer deposits. More than half of the total gold produced in Suriname since 1885 originates from the Lawa (upper Maroni/Marowijne) river greenstones and associated (palaeo) placers as well as a major share of that produced in French Guiana from the St Elie, Paul Isnard, Adieu Vat, Sophie and Repentia greenstones of the Marowijne Supergoup (US Bureau of Mines, 1923–2002; Gibbs and Barron, 1993, Fig. 16.3). Greenstones and their placers along the Mazaruni and Potaro districts have accounted for more than 95% of gold production in Guyana since 1890 (British Guiana Lands and Mines Dept, 1910–1945; Macdonald, 1968). Virtually all of the major gold-producing areas, including those along the Yuruari (El Callao, Lo Increible), Botanamo and other (Las Cristinas-Kilometro 88) tributaries of the upper Cuyuni River in eastern Bolivar state are attached to greenstones of the Carichapo and Botanamo Groups (Gibbs and Barron, 1993; USGS and CVGTM, 1993).

Historical influences on production

Placer gold extraction was, and remains, a relatively inexpensive and simple procedure when compared to more sophisticated operations required to process greenstone and associated rock containing gold-quartz veins. Early peaks in gold production history of Guyana, Suriname and French Guiana (black bars 1 and 2, Fig. 8.9) in part reflect the availability of untapped riverine placer deposits, particularly those located in the relatively accessible near interior. Many of these deposits were worked several times by different outfits prior to abandonment and the combined extraction from these areas declined considerably thereafter (British Guiana Lands and Mines Department, 1910–1945) (Fig. 8.9).

Other major world events, such as the onset of the Second World War, as well as local labour strikes, economic recession, unemployment spikes, environmental disasters (e.g. at Omai in August 1995) also influenced annual fluctuations in gold production within the region (Fig. 8.9). The influence of these events, migration of

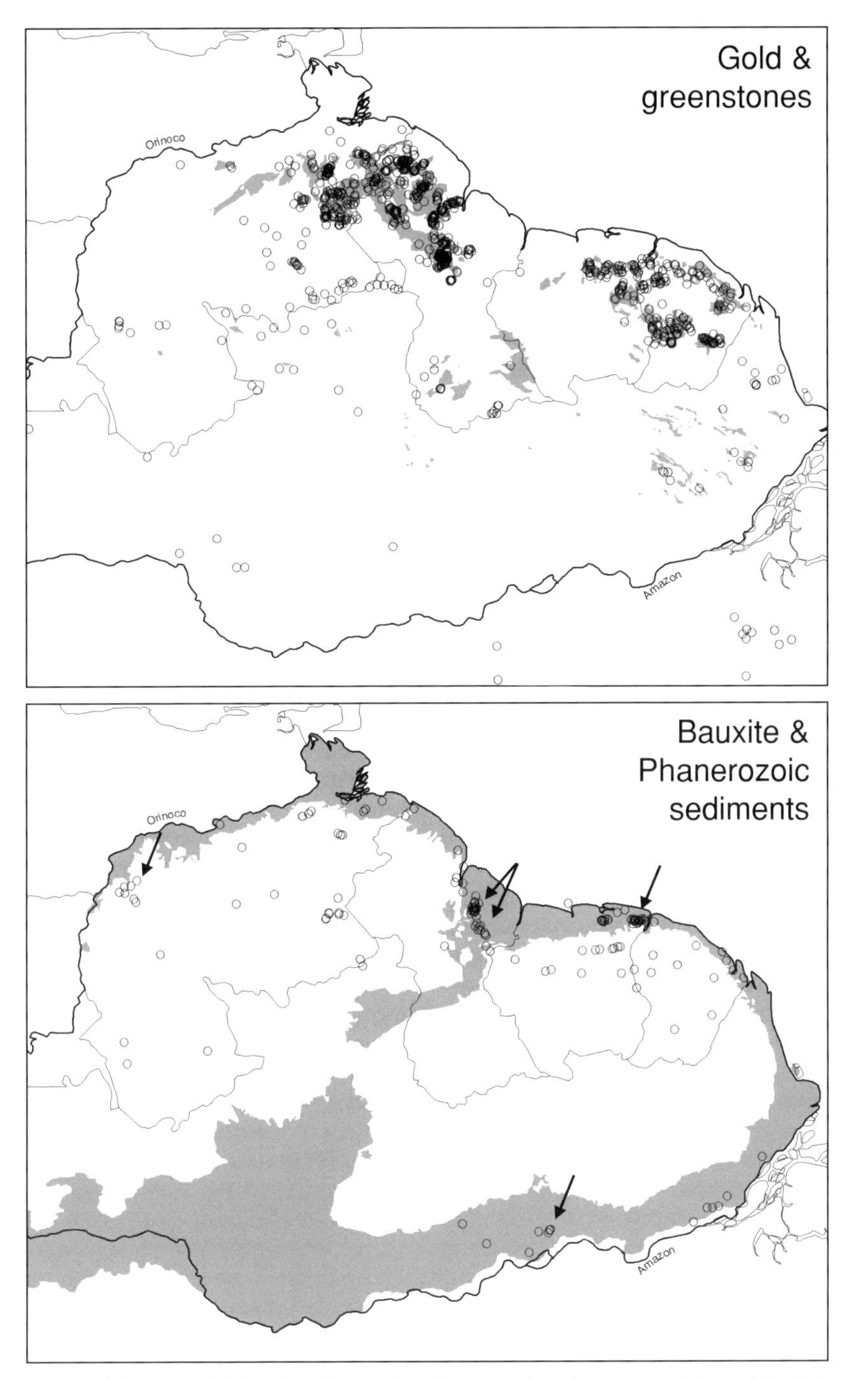

Fig. 8.8. Commercial-grade gold, bauxite, diamond and iron ore deposits (empty circles) of the Guiana Shield and their affiliated geological formations (grey areas). Deposit locations taken from USGS and CVGTM (1993). Geological coverages adapted from Gibbs and Barron (1993). Arrows indicate location of working bauxite and iron mines that have supplied the major share of production depicted in time series of Fig. 8.9. General region of greenstone (and placer) gold deposits being heavily worked (illegally) by Brazilian garimperos in Guyana, Suriname and French Guiana are indicated by open stars.

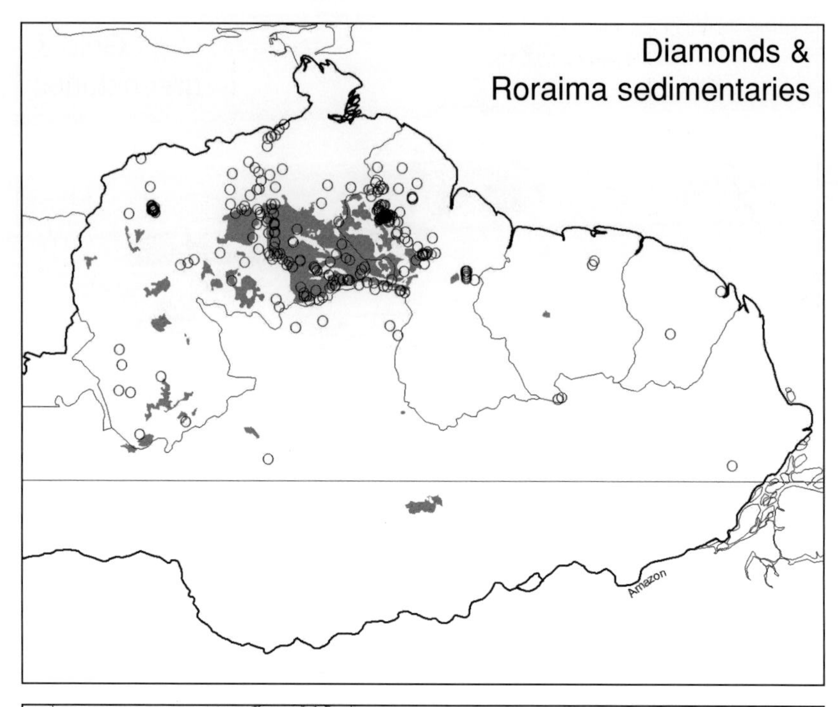

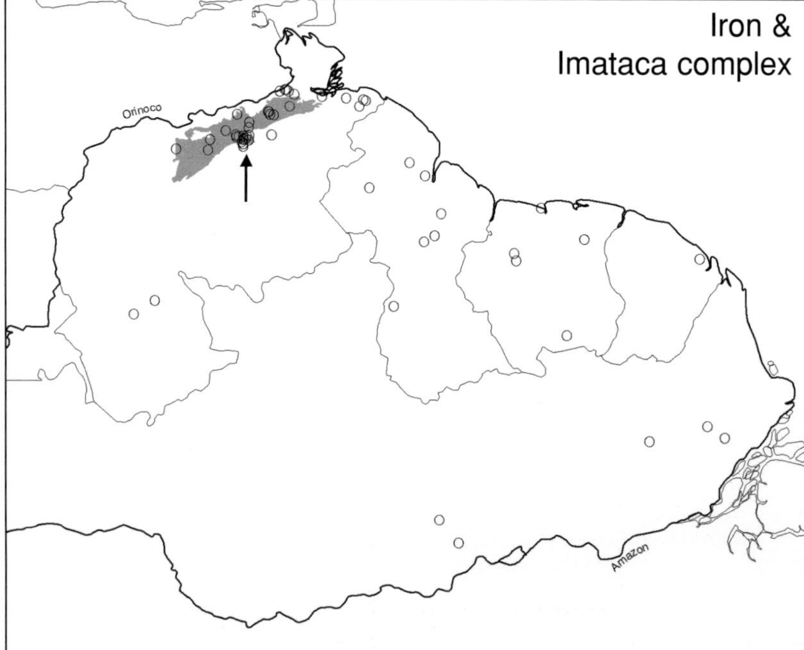

Fig. 8.8. *continued*

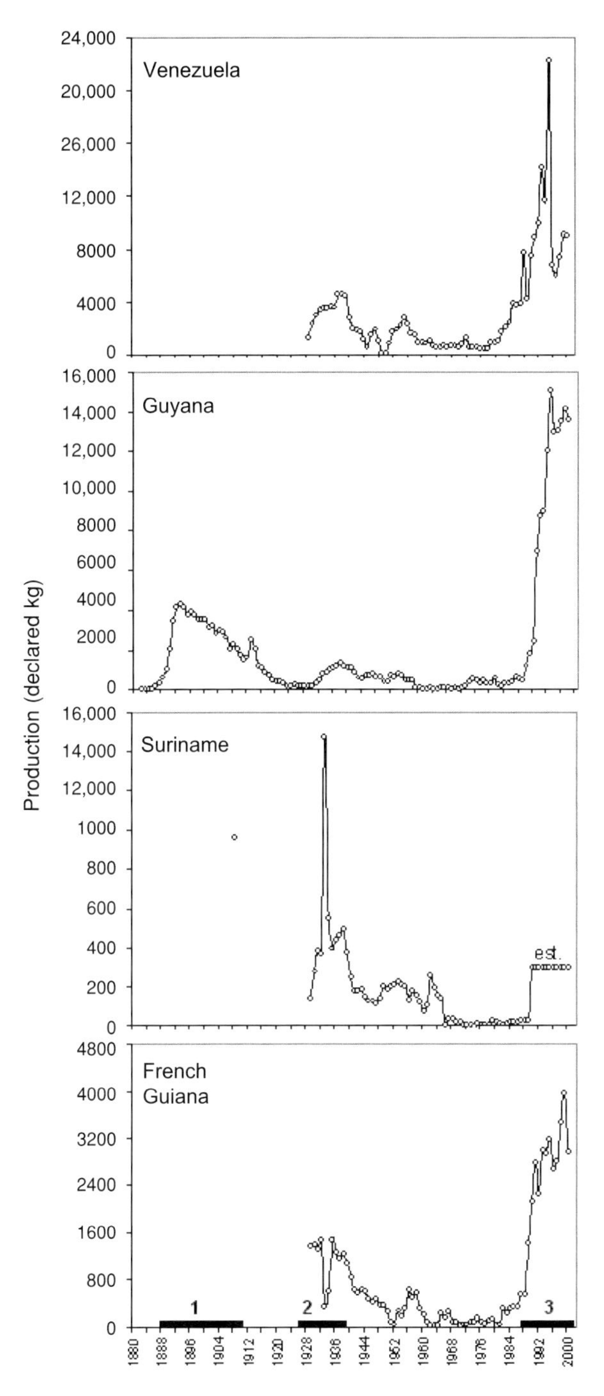

Fig. 8.9. Patterns of 20th-century gold production in the Guiana Shield according to country. Based on official, declared production only. Undeclared production can equal or exceed declared amounts. Note difference in production scale for each country. Source: Gibbs and Barron (1993), USGS and CVGTM (1993), British Guiana Lands and Mines Dept (1910–1945), Departmento Nacional de Produção Mineral (1999, 2001), Guyana Geology and Mines Commission (1980–2002), US Bureau of Mines (1923–2002).

labour between countries and the strong political and economic ties binding most of the region to Europe and North America are reflected in the synchronicity of production spikes across the region (Figs 8.9 and 8.10). Similarly, when the Bretton Woods Agreement was signed in 1944, fixing the world gold price at US$35/oz, production declined and remained flat in all of the gold-producing countries forming the Guiana Shield (Fig. 8.10). When the Bretton Woods fixed gold price was abandoned in 1972, a new gold industry emerged throughout the region as previously uneconomic reserves could be profitably extracted at the rapidly inflating unit prices. In 1980, gold reached its historical maximum at US$850/oz, as gold production rates remained flat but demand by speculators and for new electronic applications surged upward shortly after the end of the Bretton Woods price control (Fig. 8.10). By the early 1980s, production had begun to increase worldwide, the increase in supply and speculator sales began to meet more of the early demand and the unit price declined to under US$400/oz. The beginning of the 1990s witnessed exponential growth in declared gold production in Venezuela, Guyana and French Guiana, albeit at varying magnitudes (Fig. 8.10). Spurred on by political change, attractive incentives to foreign investment, improved exploration and extraction technologies and the prospects of a post-Bretton Wood unit price of US$300+, several large mines were commissioned or planned across the greenstone belts of the north Guiana Shield by the end of the century (Fig. 8.11).

Large-scale operations

By 2000, several large mining ventures had become fully operational in Guyana (Omai Phase II) and Venezuela Guayana (La Camorra, Las Cristinas-Kilometro 88) (Fig. 8.11) and several well-established sites in Venezuela (El Callao) had increased processing rates, sending declared production to record levels (Fig. 8.9). For example, in its first year of operation in 1993, Omai alone accounted for 72% of total declared gold produced nationally (Guyana Geology and Mines Commission, 1994) and represented one of the largest open-cast mines in the Americas. These large operations process vast quantities of low-grade, auriferous ore using heavy machinery extracting 10,000 (330) to 50,000 (1660) g (troy oz.) per day. Given the immense capital investment needed to process greenstone ore, the profitability of large-scale operations is highly sensitive to the relationship between ore grade, recovery rates and the world gold price. In 1999, when gold unit prices dipped below US$270, large-scale mining of lower-grade deposits operated along very narrow margins, particularly where recovery rates declined or ore grades failed to meet expectations. Operations were temporarily suspended in some instances (e.g. construction of Placer Dome's Las Cristinas mine (Torres, 1999)). Sound investment strategies in the gold futures market[13] are purported to have offset thin margins at one of these large operations during this period, but this strategy only further emphasizes the external socio-economic controls on commodity production, and forest use, in the region. Without flotation of the world gold price after Bretton-Woods, most of the existing (Omai, Konawaruk, El Callao Phase II, Las Cristinas) or planned (Gross Rosebel, Suriname; Paul Isnard Phase II, French Guiana; Las Brisas, Venezuela) lower-grade ore mining ventures would not be economical (Fig. 8.11).

Medium-scale operations

While a few large operations account for the lion's share of declared production, medium-scale operations that employ an intermediate level of capital account for the largest operational area of forestland. Capital normally takes the form of diver-operated, bucket and/or 'missile' dredges that are designed to process smaller quantities of unlithified sediment and soil from river channels and banks to produce 1000 (33) to 10,000 (330) g (troy oz.) per day. During the early 20th century, dredges were the most capital-intensive form of mining and accounted for the largest fraction of annual production in Guyana (British

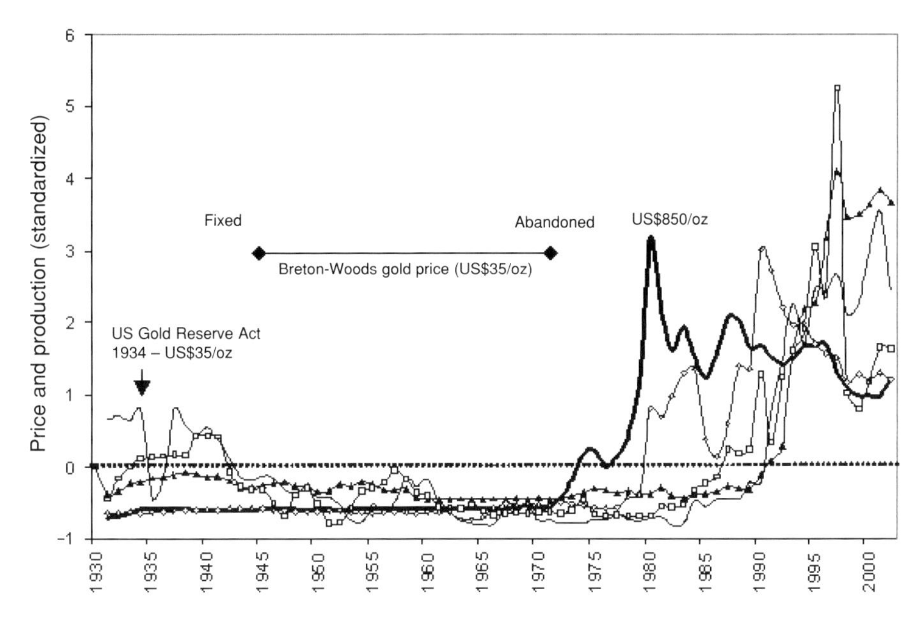

Fig. 8.10. Declared gold production in Venezuela (empty squares), Guyana (solid triangles), Suriname (empty circles) and French Guiana (thin solid line) vs. world gold price (thick solid line) with Gold Reserve Act, Bretton-Woods fixed gold (and exchange rate) price period and historic peak gold price indicated. Gold prices from World Gold Council and www.kitco.com based on London PM fix.

Guiana Lands and Mines Department, 1910–1945). As gold prices increased, the reworking of placer deposits at lower grades also became profitable. The use of 'missile' dredges to work palaeo-placer (old river course) deposits has increased substantially since their initial introduction in riverbed operations. In Guyana, the number of licences issued to dredging operations increased from 1057 in 1991 to 1682 in 1994 (Guyana Geology and Mines Commission, 1994) and the total number of claims increased steadily from 2195 claim licences in force in 1972 (Guyana Geological Surveys and Mines Dept, 1974) to nearly 15,000 in 1994 (Guyana Geology and Mines Commission, 1994).

Small-scale operations

Historically, small-scale operations have accounted for the largest number of persons employed in gold mining. Licensed and unlicensed artisanal miners, *garimpeiros* (Brazil) or porkknockers (Guyana) literally define a gold rush as they quickly move *en masse* to excavate ore-bearing reserves with a minimal capital investment. Traditionally using pans, sifters, bateaux and quicksilver (mercury) to separate gold dust from sediment, but increasingly employing single, small-engine missile dredges, their combined effort can produce a significant quantity of gold. This, however, rarely forms part of declared annual production. Where declarations have increased (e.g. Guyana, Brazil), estimates can be substantial. In Brazil, *garimpo* (*garimpeiro* collectives) production constituted nearly 62% of total gold production in 1991, declining steadily to 16% in 2002 as prices declined, higher-grade, shallow deposits were depleted and the Brazilian government began to exercise greater environmental standards over small-scale operators (Departmento Nacional de Produção Mineral, 1999). In Suriname, undeclared production by small-scale operators is believed to have exceeded 15,000 kg in 1997 (Heemskerk, 2001) and 30,000 kg (100× the declared amount) in 2001 (Szczesniak, 2001). Increasing quantities of gold are also known to have been

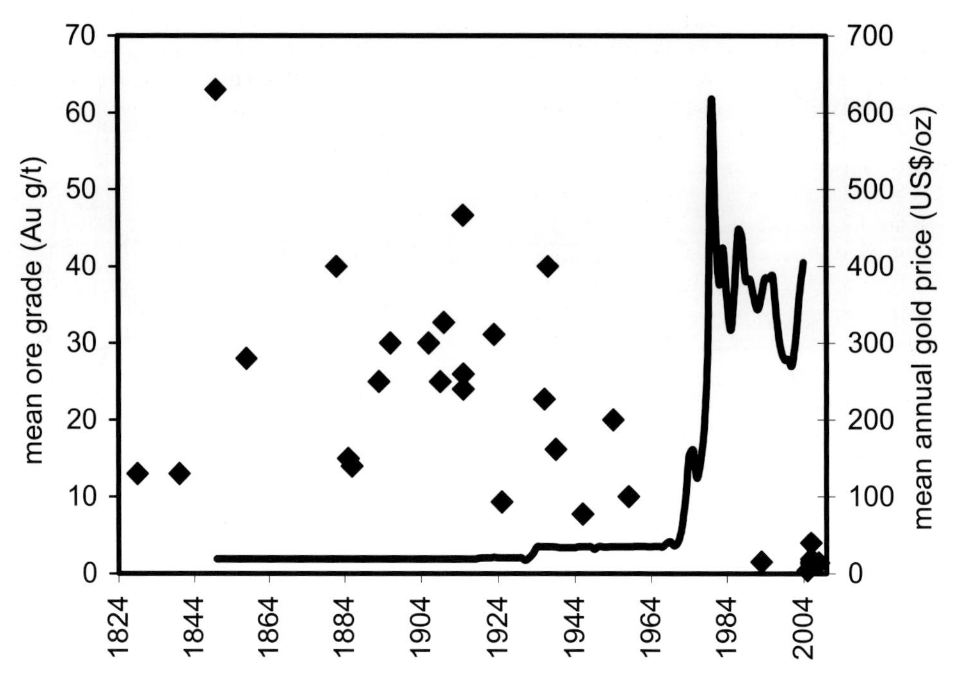

Fig. 8.11. Relationship between ore grade of gold deposits (solid diamonds, left axis) in the Guiana Shield and world gold price (solid line, right axis) since 1825. Gold prices: World Gold Council and www.kitco.com based on London PM fix. Mean ore grades derived from Gibbs and Barron (1993), Guyana Geology and Mines Commission (1994), USGS and CVGTM (1993), Cambior (2002).

smuggled by unlicensed Brazilian miners from southern greenstone regions of French Guiana, Guyana and Venezuela since the early 1990s (Fig. 8.6, empty stars), but quantified estimates are not available.

Social impacts

The socio-economic impacts of mining in the Guiana Shield vary with operational scale. Large-scale operations are almost exclusively operated through multinational business partnerships between companies registered in Canada, Venezuela, the UK, the USA, South Africa, Australia, Brazil and France, with subsidiary partners drawn from both the domestic private-sector and government. Despite a complex and very active process of continuous trading, purchasing and selling of various interests in different exploration and operational rights among the major global mining companies, most are listed on major stock-markets and thus remain relatively visible compared with smaller-scale operators. Operations are under greater scrutiny, inaugurated with great fanfare and heralded as important contributors to the national economies (which they often are). Working conditions are strongly regulated, well-equipped and generally superior to those experienced at medium- or small-scale operations. Wage rates are typically higher and a workforce with more substantial skill sets is selected.

Large gold-mining operations also generate considerable revenue to government through royalties, area fees, fuel taxes and employee income taxes. From the period 1991 to 1999, the Government of Guyana received over US$63 million in various direct cash payments from Omai Gold Mines Ltd, amounting to nearly 10% of estimated gross operational revenue and accounting for 3.7% of total government revenue over the period, a substantial contribution from a single industrial entity.[14]

The greatest socio-economic drawback to most of the large-scale mining operations is their relatively short operational life. Operational life expectancy of most large-scale operations has been and continues to be projected at less than 15 years (e.g. Omai, 1993–2005), although there are exceptions (e.g. El Callao). Large operations processing low-grade deposits are also most susceptible to the vagaries of the international gold market. Low gold prices typically curtail further exploration and mining at this scale. Combined with dwindling ore grades at operational sites, the Guiana Shield region is exposed to economic rippling as successive large-scale operations commence and then close, but with insufficient overlap to maintain employment levels and benefits. Government revenue flow, personal disposable income and local economic liquidity all suffer through these boom-and-bust cycles.

Considerable manpower is also mobilized and concentrated around large-scale enterprises during the period of their operation. For example, the Omai operation was employing more than 1000 skilled staff by its 10th year of operation. Closure creates a tremendous economic vacuum in nearby communities that have rapidly geared to provide goods and services during the waxing years of mineral production. With little demand for these post-closure, much of the infrastructure becomes redundant, unsustainable and what wealth that has accrued departs in search of other opportunities.

Medium- and small-scale operators tend to be owned by an individual or family from within the country, increasingly with Brazilian involvement, and typically maintain a lower operational profile. Small artisanal miners are normally single individuals or groups that are seeking to achieve some measure of cash income that otherwise is unattainable due to slow or no demand for their acquired skills, their limited formal education and training and inability to emigrate. During periods of economic recession, Heemskerk (2001) suggests that artisanal mining may be modulating the impacts of low wage-earning opportunities by offering an outlet to the unemployed during periods of high economic volatility in Suriname. This too is clearly the case in Guyana (D. Hammond, personal observation). Increasingly, this outlet in the Guianas is being used by Brazilians *garimpeiros* that have seen their opportunities diminish in Roraima, Pará and Amapá as the national government increases its enforcement of federal laws protecting the environment and Amerindian lands.

But this open-ended opportunity to earn higher wages through small-scale mineral extraction in the interior is not without substantive social costs (Heemskerk, 2001). Among the greatest concerns attached to small-scale operations are health issues related to sexually transmitted diseases, mercury poisoning, occupational accidents and insect vector-borne infections, such as malaria, dengue and yellow fever. HIV prevalence among small-scale miners in Guyana can be very high (Palmer *et al.*, 2002), as is their susceptibility to insect-borne diseases once a camp is infected. Accident rates are also comparably high among medium- and small-scale operators where safety guidelines, if existing, are rarely enacted and access to health care is inadequate. Prostitution services are frequently located near mining camps and HIV infection among these women is invariably high. The return of miners to their home villages leads to further spread of these infections to their families and other community members.

Cultural attrition in gold-mining frontiers across the Guiana Shield is also commonplace and has led extensively to violence, abuse and exploitation of traditional Amerindian communities across many parts of the region when ore-bearing deposits overlap with existing Amerindian lands and traditional resource space (Forte, 1996, p. 96; Tierney, 2000).

Environmental impacts

FOREST LOSS Mining by definition involves the rapid removal of the soil, subsoil and/or upper parent rock strata. This alone constitutes a significant and unrecoverable change to the local environment that is not consonant with changes occurring to the surrounding landscape. Large rock-crushing operations effectively create a quarry site that cannot be restored to its previous forest type, although there are prospects for the establishment and growth of some native plants (D. Hammond, personal observation). In most instances of dry land mining, large areas are not effectively levelled and filled, creating ponding, vertical cliffs and tailing piles that present difficulties for most native forest plant species. Instead, ruderal plants invade and achieve population densities uncharacteristic of forest ecosystems. In Guyana, many former mine sites have remained deforested (but vegetated) nearly a half century after abandonment (D. Hammond, personal observation). Ponding creates swamp-like environments ideally suited for mosquito breeding.

MERCURY CONTAMINATION The poorly regulated use of mercury as a cheap gold amalgamating agent by tens of thousands of small artisanal miners is widespread in the neotropics, making it one of the major global sources of biospheric mercury (Nriagu and Pacyna, 1988; Nriagu *et al.*, 1992). The major share of anthropogenic mercury enters the environment through atmospheric emissions emanating from amalgam burning and bullion smelting (Pfeiffer *et al.*, 1993; Drude de Lacerda, 2003), consequent deposition, uptake by forest plants and translocation via litterfall to the soil matrix (Mélières *et al.*, 2003).

Mercury also enters aquatic systems directly through mining effluents and the extent and breadth of its inculcation into the forest riverine habitats of the Guiana Shield is considerable. Elevated mercury concentrations have been documented in alluvial sediments of the Essequibo and Mazaruni Rivers in Guyana (Miller *et al.*, 2003), some parts of the Sinnamary watershed in French Guiana (Richard *et al.*, 2000), discharged mine water and sediment in Suriname (Gray *et al.*, 2002), concentrated as methylmercury in floating meadows along the Rio Negro (Guimarães *et al.*, 2000) and at high concentrations (6–32.6 μg/g) in the hair and fish consumed by about one-quarter of women sampled along the upper Rio Negro. Concentrations in carnivorous fish were higher than in non-carnivorous fish in French Guiana (Richard *et al.*, 2000), presumably reflecting trophic bioaccumulation, but only a small fraction were considered a health hazard.

The biophysical evidence linking mercury transport from mine waste to effluent to river sediments to fish and, ultimately, to humans and other birds and mammals in the Guiana Shield is apparent. But few studies have comprehensively dealt with the entire source-to-sink chain (e.g. using tracers), and the anthropogenic connection often remains unclear in the face of other biochemical pathways that naturally elevate Hg concentrations across some lowland regions. Volcanic emissions, weathering of Hg-bearing lithologies and air–ocean exchange processes are natural sources of Hg accumulation in forests. Massive forest fires release naturally occurring mercury from biomass and soils into the atmosphere and these events have been proposed as a major transit source of environmental Hg in the Amazon basin (Veiga *et al.*, 1994). An upper Rio Negro sediment core extending back 41 ka BP lends some credence to this view (Santos *et al.*, 2001). The core shows elevated mercury accumulation rates since 18 ka BP and Santos *et al.* suggest this change may reflect an increase in forest fire activity. Forest land-use that leads to serious soil ablation or erosion (e.g. roads, land dredging) can also mobilize naturally occurring mercury (Roulet *et al.*, 1999). Global and regional transport of atmospheric mercury is also believed to contribute substantially to levels measured in Amazonian lakes and other aquatic systems (Melack and Forsberg, 2001). High concentrations of mercury derived from natural sources have also been measured in lowland Amazonian soils. Together, uncertainty concerning the relative influence of natural sources, atmospheric depositional and rock weathering pathways and variation in standing levels of mercury complicate standard perceptions that contamination is solely related to gold mining.

However, the geographic position and geomorphic attributes of the Guiana Shield suggest that diastrophic sources, such as volcanoes, are not major contemporary contributors to standing mercury levels in the region. Active chains are located north or west of the shield region and sub-stratospheric aerosol emissions move northwestward, reducing the likelihood of direct deposition from these sources. Physical weathering of substrate is probably lower over (most parts of) the Guiana Shield than any other neotropical region so would not act as a significant contemporary contributor to elevating mercury levels, although chemical weathering pathways would be substantial (as for aluminium and iron). The absence of any commercial mercury deposits in the region suggests concentrated point-source lithologies no longer exist or never existed (Macdonald, 1968; Gibbs and Barron, 1993; USGS and CVGTM, 1993). The length and extent of past weathering, however, could have released mercury from previous depositional episodes that would have been stored in the deep soil substrates and overlying vegetation and periodically released through fires. Over time, repeated fires during dry periods and accelerated chemical weathering during wet periods would have eventually exported most mercury westward through prevailing trade winds or eastward through the major waterways, respectively.

However, the location and geology of the shield also makes it a potent bio-accumulating region once mercury is imported. Mercury complexes readily with both dissolved and particulate organic matter (Roulet *et al.*, 1999) in lowland forest environments. Aquatic conditions of low pH and high DOC favour methylmercury production and the consequent bioaccumulation through the food chain. The Guiana Shield represents the largest assemblage of low pH/high DOC waterways in South America (and arguably the tropical world) (see 'River, Lake and Tidal Systems', Chapter 2). High-level human mercury contamination has been positively correlated with low river pH and high DOC fractions, independent of mining activities in the Amazon (Silva-Forsberg *et al.*, 1999). The prevalence of both predatory and benthic-feeding fish species in the aquatic systems of the Guiana Shield would concentrate effects as carbon-complexed mercury is accumulated through the food chain. The impacts of repeated gold-mining activities with little pre-effluent settling and liberal use of liquid mercury in blackwater-dominated watersheds are compounding background mercury concentrations in a system with potentially severe downstream public and environmental health consequences.

SEDIMENT LOADING Rivers draining the Guiana Shield have naturally low levels of suspended sediment (see 'River, Lake and Tidal Systems', Chapter 2) and the magnitude of sediment discharge from modern land-dredging operations is unprecedented. While sediment levels do vary among rivers draining the region, most are considerably lower than those documented for waterways draining Andean areas dominated by active physical weathering (see Fig. 2.34). The very recent influx of massive sediments associated with gold mining is invariably altering long-standing attenuation of the aquatic systems to relatively slow rates of biophysical change. Aquatic and terrestrial habitats of the shield are highly acidic environments dominated by chemical, rather than physical, weathering processes. The massive rise in suspended sediment resulting from mining in the region is comparable to natural mass wasting processes that characterize intense physical weathering of the Andean piedmont and is unprecedented in the last several millennia of the shield interior (see 'River, Lake and Tidal Systems' and 'Prehistoric Climates of the Guiana Shield', Chapter 2). Forest life that is adapted to relatively static patterns of changing acidity, sediment loading and deposition will be competitively disadvantaged under the alien conditions created by hyper-rising sediment influx created through poorly managed gold mining. Ruderal and alien species better suited to sediment-driven river systems are likely to benefit if mining activities sustain elevated sediment loads in these regions.

Bauxite

In contrast to a gold-mining environment embracing a wide range of operational scales and controls, bauxite is mined exclusively at very large operational scales. Its formation through chemical reduction, particularly along ancient laterite surfaces buried by deep Quaternary sediments, concentrates aluminum in a high clay matrix as iron dissolves under the extreme acidity. These coastal plain bauxites (after Gibbs and Barron, 1993) produced some of the purest (claimed) bauxite and are strongly correlated with the distribution of Phanerozoic sediments along the rim of the Guiana Shield (Fig. 8.8). The largest bauxite mines in Guyana (Linden, Intuit, Kwakwani: Fig. 8.8, A) and Suriname (Moengo, Onverdacht, Paranam and Lelydorp: Fig. 8.8, B) form part of this group.

Other, lower-grade deposits occur at higher elevations. The very large mines at Trombetas in Brazil (Fig. 8.8, D) are exploiting one of these, associated with a series of erosional surfaces extending to the Cretaceous (Gibbs and Barron, 1993). Los Pijiguaos, sole producer of bauxite in Venezuela, is located at an even higher elevation, and these deposits belong to a group of much older, high-elevation or plateau, bauxites (after Gibbs and Barron, 1993). They are formed on dissected planation surfaces of Mesozoic–Cenozoic age atop the rapakivi granites forming the Parguaza batholith in north Amazonas state (Soler and Lasaga, 2000) (Fig. 8.8, C). Other upland commercial deposits are found in the Paru-Jari River area and in lower grades (higher iron content) at Nuria and Los Guaicos, Venezuela; Kopinang, Guyana; Bakhuys and Brownsberg Mountains in Suriname; and Monts Kaw, Tortue, Française and Lucifer in French Guiana.

Historical influences on production

Bauxite was first produced in 1917 in (British) Guiana and shortly thereafter in 1922 production commenced in (Dutch) Suriname (Fig. 8.12) and was maintained in both at comparably low levels until the Second World War. The extremely high demand for aluminium in the manufacturing of armaments and military supplies brought a rapid and unprecedented upsurge in mining activity to the relatively secure Guianas from 1939 to 1945 (Fig. 8.12). By the end of the war, British and Dutch Guiana were the largest and longest producers of bauxite in a world with a growing demand for aluminium products. They accounted for nearly 40% of total global bauxite production during this period (Table 8.1). Production in both countries peaked immediately following independence (1966–1975) and has experienced a decline since then (Fig. 8.12), in part due to rapid growth in competition from Jamaica, Australia, Brazil and Venezuela after 1950 (US Bureau of Mines, 1923–2002) and in part due to increases in unit production costs due to the age of the deposit, management inefficiencies, social unrest and worker strike actions (Table 8.1). Despite this, they remain significant sources of the more valuable refractory-grade and calcined bauxite. Mining of lower-grade deposits over a 228 km^2 concession embracing the Kaw, Rouri and Mahury mountains of French Guiana (near Cayenne) was planned from 1968 to 1972, but failed to materialize when Alcoa (CMAG) (USA) and Péchiney (France) allowed their mining rights to lapse in 1973 (US Bureau of Mines, 1968–1973) as bauxite prices dropped and remained low after the post-war surge in demand was supplied from other existing suppliers, making the prospect uneconomic. By 1983, production in Brazil exceeded both Guyana and Suriname, principally through activation of its Trombetas operation in 1979, but also from later activation of mines along the Brazilian Shield periphery in the states of Minas Gerais and Santa Catarina (Fig. 8.12). Bauxite production in Venezuela commenced in 1986 and has risen steadily through the massive-scale production at the Los Pijiguaos mine in the Parguaza district of Venezuelan Guayana (Fig. 8.12). By the 1990s, the Guianas contributed a mere 5% to global production, while deposits in

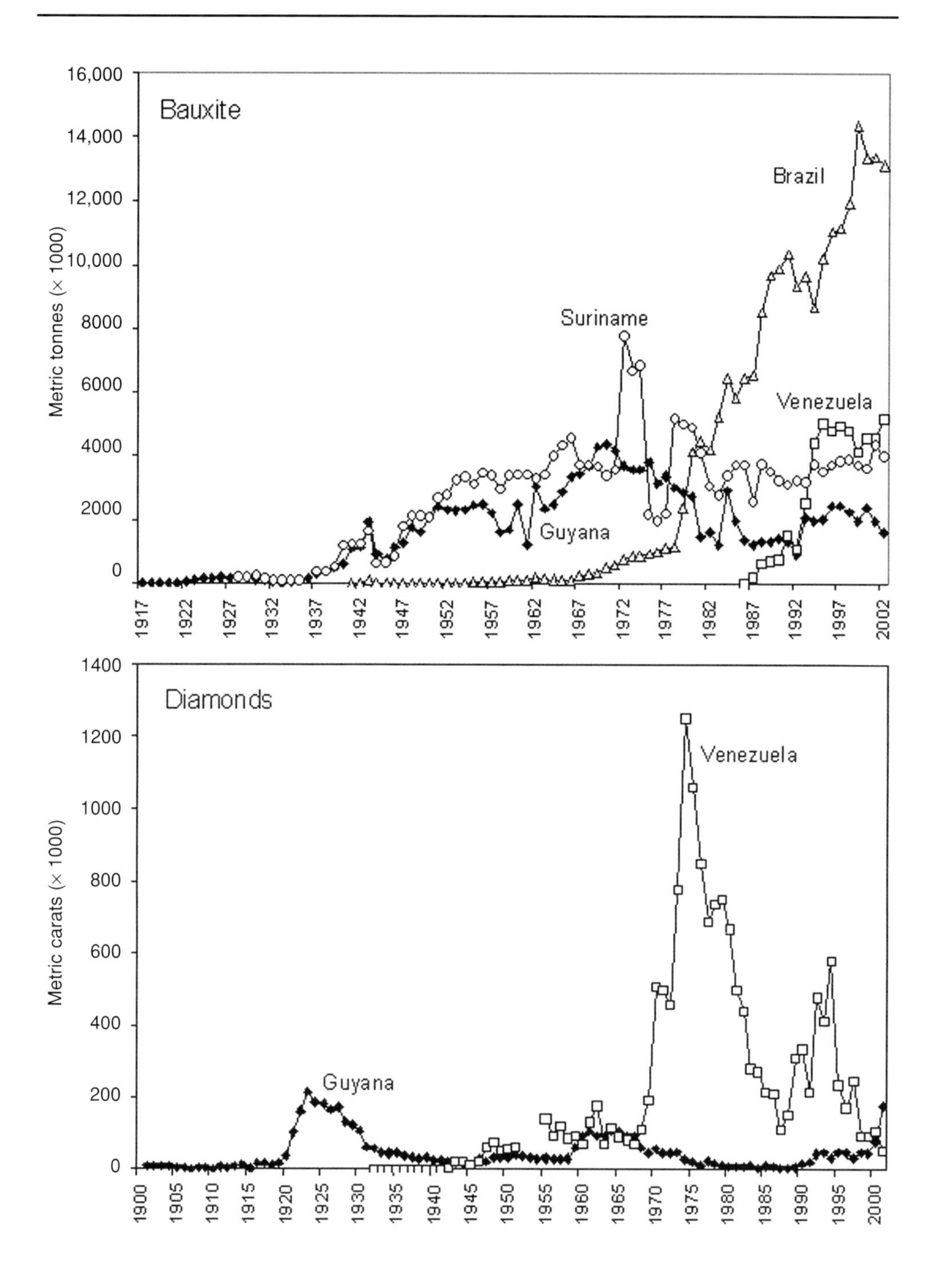

Fig. 8.12. Twentieth-century bauxite and diamond production trends in the Guiana Shield according to country. Only Venezuela and Guyana support significant diamond mining in the Guiana Shield. Source: British Guiana Lands and Mines Dept (1910–1945), Departmento Nacional de Produção Mineral (1999, 2001), Guyana Geology and Mines Commission (1980–2002), US Bureau of Mines (1923–2002).

Brazilian and Venezuelan parts of the shield combined to account for nearly 12%, although little or none of this was of calcined or refractory grade.

Social impacts

Bauxite has dominated the 20th-century economies of Guyana and Suriname. Exports have accounted for unrivalled fractions of national GDP, foreign exchange earnings and government revenues since the 1930s. In 1980, bauxite earnings accounted for more than 30% and 20% of Suriname's and Guyana's GDPs (at factor cost), respectively (Bank of Guyana, 1985). These contributions declined to 15% and 4% by 2000, but still accounted for a significant portion (70%) of foreign-exchange earnings in Suriname, while declining to a little above 15% in Guyana (Bank of Guyana, 2000), mainly due to growth in gold mining and the agriculture sector. The historic pre-eminence of bauxite production in the economies of Guyana and Suriname has clearly acted as an important source of foreign exchange and government revenue. Yet one could argue that the disproportionate weight of any single commodity has deterred economic diversification and innovation as the wider public and private sectors maintain a focus on the needs of one, or a few, dominant extractive industries. These needs are heavily affected by fluctuating world market prices that vary independent of mine performance. These prices are in part a reflection of downstream demand, but also changes in global competition from regions where bauxite production plays a relatively minor role in shaping the health of its parent economy, while being produced at a much lower production cost (e.g. Australia, China). Unit production costs of bauxite from Guyana and Suriname are some of the highest in the industry and unit export prices have exceeded world unit prices in only eight of the last 20 years (US Bureau of Mines, 1923–2002; Perez *et al.*, 1997). As a consequence, the industry has contracted significantly, leaving communities established to service mine complexes with few employment opportunities and stagnant economic development. Often these communities turn to artisanal (chainsaw) logging, wildlife collecting, gold mining and smuggling to achieve a measure of household income when the structured prospects for employment disappear in the mining industry.

Environmental impacts

Like large gold-mining operations, bauxite production in the Guiana Shield is opencast, stripping soil and sediment from an area of several square kilometres and creating a new, more dramatic landscape of hills and canyons formed from excavated and redistributed sediment. The physiographic and edaphic consequences of mining have historically involved little mitigation or restoration effort, although the policy and regulatory trends across the region are moving towards requirements for post-mining landscape renewal (US Bureau of Mines, 1923–2002). Forest restoration practices have been documented for the large Porto Trombetas (North Pará) operation along the south rim of the shield region. Results suggest that vegetation re-establishment is possible, but the likelihood of early cohort recovery of many important long-lived woody taxa (e.g. Annonaceae, Chyrsobalanceae, Lauraceae) is low due to poor dispersal and establishment success and greater management intervention may be necessary over a period extending decades (Parrotta *et al.*, 1997; Parrotta and Knowles, 1999).

Diamonds

Production from Minas Gerais, Brazil and then the Kimberley fields of South Africa has dominated diamond production since the early 1700s (Svisero, 1995). Alluvial deposits of diamonds were encountered in western Guyana as early as 1887, but the first declared commercial production in the Guiana Shield was not recorded until 1901 (Fig. 8.12). Ten years later, production from Tepequen Mesa in northernmost Roraima state and the Caroni rivers commenced. Further finds in the Paragua, Cuchivero and Guaniamo Rivers draining the Guayana Highlands and throughout the North Pakaraima mountains in Guyana led to a series of production peaks over the last century (Fig. 8.12), but the region has generally failed to supply more than 1% of total global production (Table 8.1), although brief peaks in the 1920s (in Guyana) and the mid-1970s (in Venezuela) saw production reach 4% and 2% of global output, respectively. More than 95% of production in the region is restricted to Venezuela and Guyana. Production in Suriname, French Guiana and Brazilian parts of the southern shield area is virtually non-existent, principally due to the absence of Proterozoic sedimentary lithologies associated with the Roraima Supergroup that are strongly correlated with alluvial deposits (Fig. 8.8) (see 'Prominent geological regions of the Guiana Shield', Chapter 2). Between 25% and 60% of recovered diamonds are typically of gemstone quality, with the remainder graded for industrial or bort use (mainly for industrial abrasive and cutting applications) (US Bureau of Mines, 1923–2002).

Roraima sedimentaries and diamonds

Major African diamond fields are associated with Precambrian shield areas and kimberlite pipes (or diatremes) in particular, where ore grades can reach commercially viable levels. Kimberlites of commercial quality are not believed to occur widely across the Guiana Shield (Janse and Sheahan, 1995), probably due to excessive erosion of their parent rock that released the more friable diamond-containing kimberlite. The 1989 find at Quebrada Grande in Venezuelan Guayana (CVG-TECMIN, 1991; Gibbs and Barron, 1993; USGS and CVGTM, 1993) suggests, however, that pipe remnants may be rare, rather than entirely absent from the region. Virtually all of the historic production from the region has drawn from placer deposits along the major streams draining the Roraima Supergroup of eastern Venezuela/western Guyana. Diamond deposits have not been successfully located from within the remnants of the Roraima sedimentaries, but most alluvial deposits are believed to have formed through repeated redistribution of material from palaeo-placers associated with these deposits into more recent alluvial traps (e.g. behind falls) (Briceño, 1984).

Historical influences on production

The discovery of placer diamond deposits along rivers in the region and subsequent working of these deposits to exhaustion is probably the primary pace-setter of historic production from the region. Relatively rare commercial deposits are sufficiently scattered to effectively prevent any sustained, long-term trends in output. Most deposits, once discovered, appear to have a life expectancy of between 5 and 15 years (Fig. 8.8). For example, the 1968 discovery of the Guaniamo diamond placers in Venezuela led to the largest and longest production spike in the history of the shield region and accounted for 85% of Venezuela's declared historical output (USGS and CVGTM, 1993, p. 87).

After the 1870 discovery of the vast Kimberley fields in South Africa, diamonds became relatively common and the carat price dropped to less than US$50 for a clear, cut one-carat stone. Realizing their investment was at risk, the DeBeers syndicate or cartel was formed in 1889, 2 years before major commercial production in (the then) British Guiana to restrict the global diamond supply and sustain the appearance, if not the reality, of diamonds as a rare and valuable mineral commodity. Despite continued expansion of diamond production, unit prices reached nearly US$800 per carat (as above) in 1922 under the management of global trade by DeBeers (Ball, 1934, Fig. 106). The rapid rise in diamond prices catalysed commercial production in (British) Guiana from 1922 to 1927 (Fig. 8.12). As prices fell to US$500 per carat shortly after the stock market crash and onset of the Great Depression, production in (British) Guiana dropped to pre-1922 levels (Ball, 1934).

Like gold, diamonds held value during periods of economic uncertainty when paper money and stock investments rapidly devalued. During periods of prosperity, gold and diamonds were bought as investments. When economic depression restricted income, gold and diamonds were sold, driving up supply in a market with diminishing demand. The valuation of diamonds, however, is more complex than the global supply–demand relationship governing gold prices and the quantity of production alone is not a good indicator of its value. Clarity, colour, size (caratage) and whether stones are cut or uncut (rough) play significantly on their value. The ratio of gem to industrial grade diamond production has traditionally been higher in Guyana than Venezuela or Brazil, although total output has remained considerably lower (US Bureau of Mines, 1923–2002).

By 1950, DeBeers had inculcated the largest growing economy, the USA, with the perception that diamonds were an integral part of courtship and marriage through a clever and sustained advertising campaign that shaped the post-war generation's perception of diamonds as heirlooms not to be re-sold. A soaring demand and tight control on the release of supply sent prices upwards, sparking further investment in diamond exploration and extraction in Guyana and Venezuela (Fig. 8.12). Combined with the location of placer deposits, DeBeers' cunning control over the diamond supply and a spreading perception of diamonds as investments, tokens of commitment and family keepsakes, has sustained production in the Guiana Shield in the face of significant supplies emanating from south and west-central Africa, southern Brazil and most notably, Australia. As the main Guaniamo fields in Venezuela reached their productive lifespan, Guyana became the largest producer of diamonds in the Guiana Shield in 2001, a position it had not held since 1967 (Fig. 8.12).

Social impacts

Diamonds are rarely mined through large-scale operations in the Guiana Shield and the socio-economic costs and benefits attached to small- and medium-scale gold-mining and diamond operations are similar. In many instances, operations target both commodities where placers contain both. By far the largest placer deposits are found in Guyana and Venezuelan Guayana and these have attracted Brazilian *garimpeiros* as the few deposits known from Brazilian Roraima (e.g. Tépequem, upper Cotinga and Ireng (Mau, Mutum) Rivers) near commercial depletion or have been fully claimed.

Environmental impacts

Extraction of diamonds from alluvial sediments is done through a physical, rather than chemical, sorting process. Without any mercury to manage, vast quantities of sediment are processed and flushed directly (back) into waterways without further consideration, although diamonds can be associated with conglomeratic deposits with higher gravel content than those typically worked for gold and the change in suspended sediment loads and other physicochemical attributes of waterways are probably, on average, less than those associated with gold mining. In many instances, however, both commodities are mined from the same (palaeo)placers.

Commercial NTFP extraction

From a commercial perspective, the golden age of the non-timber forest product trade commenced in the late 1800s and lasted until the 1950s. During this time, a wide range of natural forest products were in high demand as new manufacturing industries required vast supplies of natural materials, but synthetics had not yet reached their wide-ranging and encompassing dominance as the materials of choice in the manufacture of working parts. Advances in transportation made the connection between remote interior supply and overseas manufacturing centres economical. As a consequence, a commercial drive of unduplicated magnitude swept the Guiana

Shield when businesses quickly organized to supply raw forest materials to burgeoning European and North American industrial sectors. Among these, the extraction of tropical tree latexes exceeded all others in terms of both socio-economic and environmental consequences to forests and forest-based livelihoods of the 20th-century Guiana Shield.

Balata and Pará rubber (borracho, seringa) (Fig. 8.14)

Unlike most production of mineral commodities, rubber is largely now only a chapter in the history of forest use in the Guiana Shield. Rubber production continues in western Amazon in part due to structured government subsidies and international fair-trading networks and environmental organizations attempting to generate some livelihood support for rubber-tappers (*extractivistas*) in the wake of their well-known struggle to retain their livelihood in the face of dominating interests, a period that was highly publicized after the murder of Chico Mendes. By 1995, production of both balata (*Manilkara bidentata*) and Pará rubber (*Hevea brasiliensis*) had slowed to a trickle of its former documented levels as new synthetics eclipsed the last (e.g. golf balls) of their formerly wide-ranging applications and government subsidy programmes terminated (Fig. 8.13).

The history of Pará rubber collection is dominated by large-scale extractive activities in Brazil, Peru and Colombia. Very little commercial-scale production was recorded from the Guianas or Venezuelan Guayana, in part due to the relatively low density of *Hevea* in the north Guiana Basin and the failure of early efforts to establish plantations in the Guianas. Production in Brazil was first recorded shortly after the European recognition of the elastic properties of *gutta percha* and its potential applications. Large populations of *Hevea brasiliensis*, the Pará rubber tree, across the lowland Amazon made the industrial-scale extraction from natural forests possible, although the rapid rate of patented inventions that characterized the late 19th century continued to create a demand that outstripped the natural supply. Unit prices soared. Imports to the USA alone reached 258,000 tonnes, ten times recorded exports from Brazil, during the peak period of production between 1910 and 1920. Pará rubber production in Brazil peaked at 43,000 tonnes in 1914, spurred on by inventions of the pneumatic tyre, telegraphic and other mass-manufacturing applications (Fig. 8.13). But extracting rubber from such an extensive area added costs and though the rubber tree was amenable to near-continuous tappings, supply became constrained by the species' natural abundance. Plantation trials in Brazil largely failed as trees succumbed to the rubber-tree fungus, *Dothidella ulei* (Barham and Coomes, 1996; Goulding *et al.*, 1996). The rubber tree's escape from this fungus in Asia allowed the British and Dutch to establish widespread plantations by 1925. Tapping of these soon thereafter ended the Brazilian monopoly, flooded a market already dwindling in the face of further inventions, such as the wireless telegraph, which reduced mass-scale demand for cable-insulating materials (Fig. 8.13). With the contraction of industry during the Great Depression, the Brazilian rubber extractivist economy was effectively dead. Production in Brazil recovered temporarily during the Second World War, when most Japanese-occupied Asian plantations were unable to supply a growing war-time demand in the USA and Great Britain, but this abruptly declined at the end of the war in 1945 (Fig. 8.13). By the end of the Great Amazon Rubber Boom in 1945, nearly 1.3 million tonnes of rubber had been produced from Brazil alone. Perhaps one-fifth of this was produced from the forests of the southern Guiana Shield along the Jari River and the Rio Negro.

The story of balata production is largely derived from the Guianas, although relatively small commercial quantities were harvested up until the 1930s from north Pará, Amapá and the Rio Negro basin (Pinton and Emperaire, 1992) and continued to be collected sporadically from these areas for domestic consumption and handi-

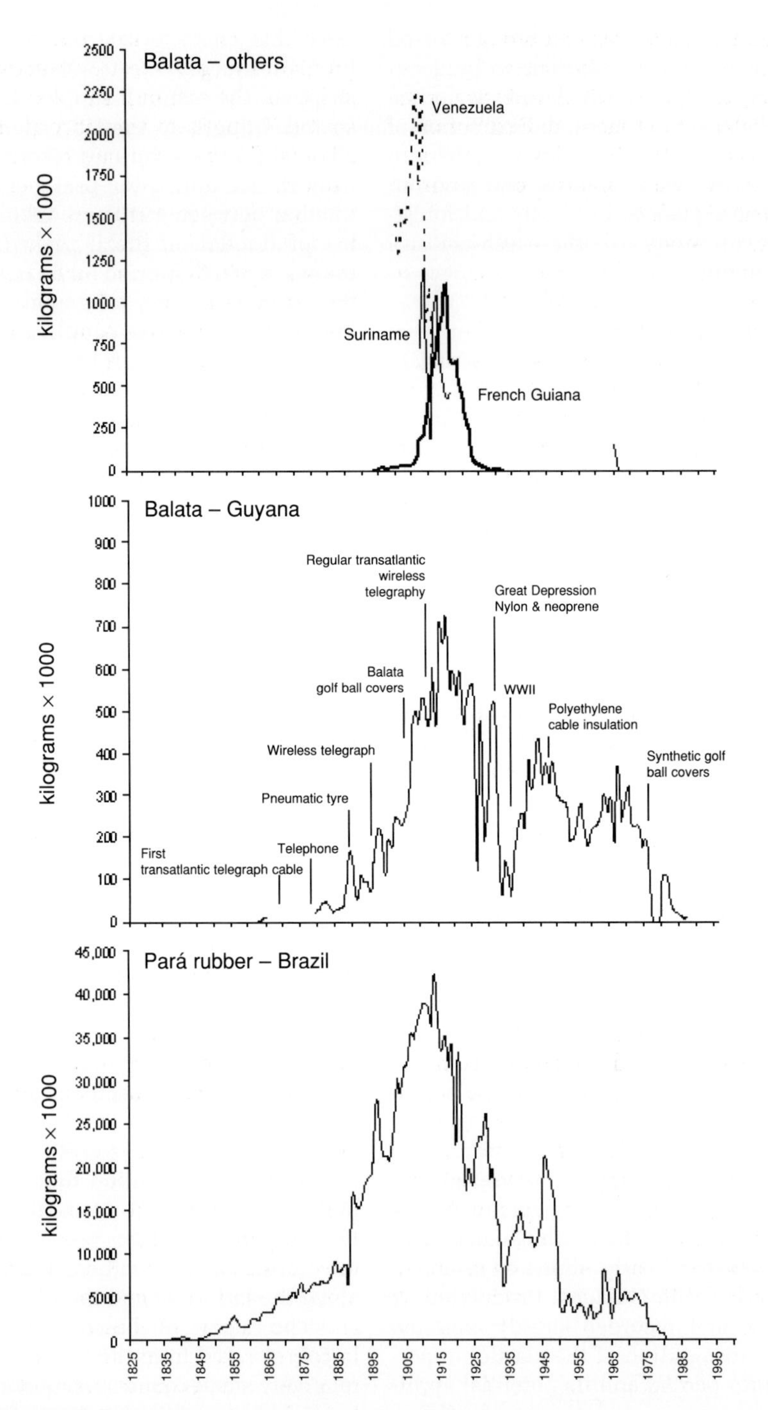

Fig. 8.13. Historical balata and Pará rubber (borracha, seringa) production in the Guiana Shield and Brazilian Amazon (including the Guiana Shield) with important events shaping the boom-and-bust character of both industries highlighted. Note difference in balata and rubber production scales. Source: IBGE (2003a,b), British Guiana Lands and Mines Dept (1910–1945), Châtelain (1935), Bruleaux (1989), Jenman (1885).

craft exports. Balata was first brought to the attention of French industry in 1855 (Bruleaux, 1989) and British industry at the Crystal Palace International Exhibition of 1862 (Davis, 1933). The following year saw the first commercial-quantity exports from the Guiana Shield.

(British) Guyana was the earliest and longest commercial producer of balata in the region. From 1863 to 1988, nearly 28,000 t of balata were produced from this country, compared to 9300 t from French Guiana over the period 1898–1936 (Fig. 8.13). Commercial production began in Venezuelan Guayana and (Dutch) Suriname in 1894 (Gonggryp, 1923) and similarly proved to have a shorter lifespan, although Suriname exports continued into the 1970s (Vink, 1970). Balata proved to have superior insulating properties and it was used widely in the production of submarine telegraph cable covering until the advent of commercial wireless telegraphy, and eventually phone services, rapidly curbed demand (British Guiana Forest Department, 1935a). Remaining demand was further eroded after replacement by polyethylene in the early 1950s (Fig. 8.13). The intermediate elasticity of balata (more rigid than Pará rubber, but more flexible than *gutta percha*) also made it invaluable to the early manufacture of machinery drive belts and, later, in the production of dimpled golf ball covers and shoes (British Guiana Balata Committee, 1912; Coomes, 1995). The gradual accumulation of cheaper and more reliable synthetic alternatives eroded balata's usefulness in a manufacturing process that increasingly sought cost reductions, supply certainty and technical improvement. Balata no longer had any viable market niche and commercial production ceased altogether in Suriname by 1975 and in Guyana by 1982 (Fig. 8.13).

Social impacts

The economic value of the rubber and balata industries over the peak period of demand from 1890 to 1920 was tremendous when considering the relative modern-day contribution of non-timber extractive industries to government revenue, employment, export earnings and per capita GDP. By 1910, the Brazilian rubber industry was estimated to have employed nearly 150,000 (Bunker, 1985; Barham and Coomes, 1996). The balata industries in British, Dutch and French Guianas employed more than 8300, 4000 and 500 the same year, while the number of persons actively engaged in the industry in Venezuela was estimated at 10,000 (British Guiana Balata Committee, 1912; Brett, 1916; Gonggryp, 1923). By 1930, nearly half of the working Amerindian male population of the Rupununi district in (British) Guiana was thought to have been employed as balata bleeders (Davis, 1933).

Government, business and personal income also rose substantively in rubber- and balata-bleeding regions during the peak production years. Gross regional income from 1910 rubber sales in the Brazilian Amazon were comparable, in real terms, to income levels in 1960 and real per capita income was nearly twice that estimated for 1970 (Bunker, 1985). In Guyana, balata revenues collected in the form of royalties, export duties, rents (area fees) and survey fees accounted for 43% (1925) to 80% (1905) of total annual government forest product revenue between 1900 and 1925 (British Guiana Lands and Mines Dept, 1910–1945; British Guiana Balata Committee, 1912). By the 1950s, this contribution was reduced to a mere 2–3%.

Yet, the expected injection of great wealth delivered by the booming rubber and balata years was quickly consolidated under *patrões* (patrons) and export businesses that absorbed much of the benefit resulting from the labour of rubber-tappers, most of whom were Amerindians or former subjects and descendants of the plantocratic economy. Middlemen and licence-holders established themselves as sole proprietors of goods to their field labourers typically working in remote frontier regions in British Guiana (British Guiana Balata Committee, 1912) and Venezuelan Amazonas (Sizer, 1991, cited in Richards, 1993). Equally, in Brazilian Amazonas, hyper-inflated provision prices would quickly eat up earnings, and many rubber-tappers remained in debt servitude as supply prices were increased in line with earnings (Bunker, 1985; Barham and Coomes, 1996).

The overseas demand for natural latex at the beginning of the 20th century was of such magnitude that it quickly outstripped labour capacity of the meagre native population remaining in the Amazon region after 300 years of enslavement, disease and forced emigration. Vast numbers of former sugar workers from Bahia in Brazil and Berbice in (British) Guyana moved to rubber and balata production regions as a consequence. Where Amerindian populations remained, such as the upper Rio Negro and Uaupes/Vaupés River, they were often forced into service or cruelly treated by rubber traders (Nimuendajú, 1950; Chernela, 1993; Saldarriaga, 1994) and then forced to reintegrate into (Salesian) missionary compounds as the only means of escaping continuous harassment (Chernela, 1998). Working crews in British and (particularly) Dutch Guiana appear to have been employed on more reasonable terms, but systems of extending credit against provisions appear to have been a commonplace method of ensuring tappers received the least benefit from the balata production process (British Guiana Balata Committee, 1912).

With the collapse of the rubber trade in Brazil, the vast majority of non-resident *mestizo* (Brazil, *caboclo*; Guyana, coastlander) tappers remained in the Amazon where many fell into a continuation of the *aviamento* system, but increasingly based on other extractive commodities, such as fish, skins, bushmeat, Brazil nuts, spices and plant fibres and oils (Santos, 1968). Several major regional market centres emerged during the rubber boom period in the Brazilian part of the Guiana Shield, most notably at Manaus and Barcelos (Lescure and de Castro, 1990). This post-rubber period established the foundation for the modern-day, mixed-market rural livelihoods commonly documented throughout the Rio Negro, Amazon, Orinoco and Solimões basins.

Environmental impacts

Rubber and balata bleeding satisfied somewhat different industrial demands, were pursued around separate geographic centres of peak production and employed different tapping techniques (Fig. 8.14 for balata). *Hevea* was amenable to a nearly continuous regime of bleeding, as vessels distributing latex around the tree's sub-bark were connected. In contrast, once *Manilkara* was initially bled, further bleeding of the same area did not yield any appreciable quantity of latex. In *Manilkara*, sub-bark latex storage occurs in separate, vertical vessels that are not refilled, and latex is only produced when new ones are formed (Gonggryp, 1923). Davis (1933) and others anticipated a 15+ year recharge time for a tapped tree to again yield commercial quantities of latex.

This distinction affected the way in which latex was collected between the two sources and, ultimately, the impact latex-collecting would have had on the metapopulations of these species across the Guiana Shield over the last 100–130 years. While Pará rubber collecting worked continuously around the same *Hevea* stands, balata collectors had to initially roam over vast areas of forest within each concession in order to locate additional trees that had yet to be tapped. During the early years of industrial balata collecting, latex was collected by felling the entire tree. In effect, the resource was liquidated in an effort to achieve the maximum volume of latex at the current market price. The timber value of the tree (at the time 20× that of balata) was left to rot in the forest.[15] As relatively accessible stands dissolved or disappeared, tapping practices were placed under increasing scrutiny (Jenman, 1885), and felling of bulletwood trees was banned in British and Dutch Guiana by 1895. Recommendations by Jenman (1885) that only half of the bark be bled over a single cycle were eventually incorporated into the British Guiana Crown Lands Regulations of 1919. This reduced the amount of latex collected per tree during the initial tapping (Hohenkerk, 1919), but was intended to

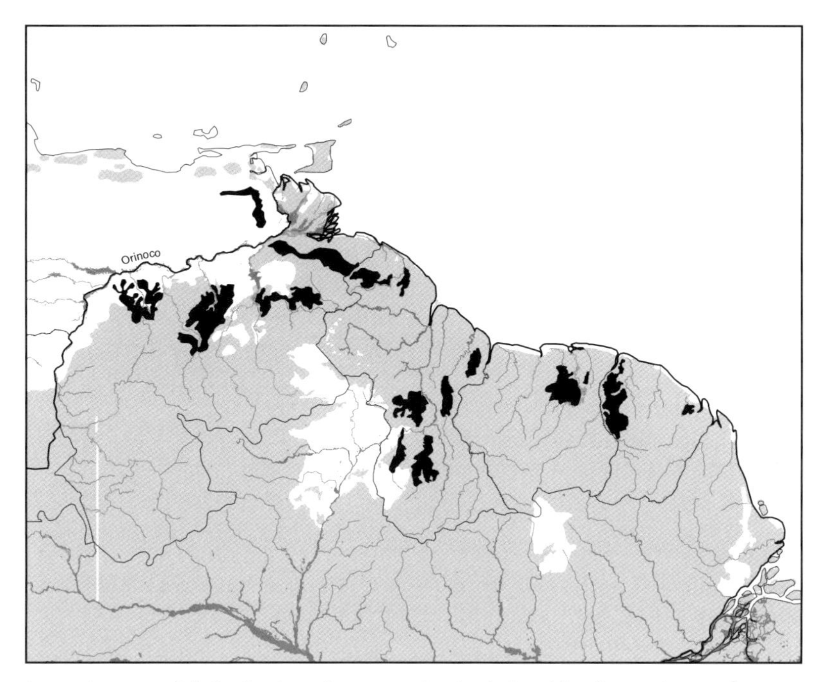

Fig. 8.14. Approximate spatial distribution of most productive balata-bleeding regions and concessions exploited throughout the late 19th and early 20th century. Based on descriptions in Brett (1916), Gonggryp (1923), Jenman (1885), Châtelain (1935) and Guyana Forestry Commission (1985). Maps of Suriname balata concessions in Struycken de Roysoncour and Gonggrijp (1912) and produced for Ordinance of 1914 were not consulted. Areas in Suriname are thus less accurately located.

increase the proportion of trees surviving until the next extraction cycle. Tapping the entire circumference of *Manilkara* trees (20–90 cm dbh) typically resulted in 8–50% mortality, while bleeding only half of the bark area killed only 4–14% of exploited individuals (Davis, 1933). Full circumference tapping typified extraction from both Dutch and French Guiana throughout the period, leading to larger volumes being harvested from fewer trees over a shorter period relative to British Guiana. Invariably, mortality was higher too. In Venezuela, virtually the entire supply of balata originating from forests of the Guayana region came from felled trees (Brett, 1916; Gonggryp, 1923). Stocking rates of bulletwood (*purguo*) were considerably higher across the main bleeding regions between Maturin and the upper Cuyuni, between Ciudad Bolivar and Tumeremo, around La Peragua on the Caroni and Caicara on the Orinoco River. As a consequence, the production peak was early, intense and brief as the extractable reserves diminished with each felled tree (Fig. 8.13). Felling restrictions were not put in force until 1918, since the balata industry believed felling was the most efficient method of extracting latex from a tree that would die regardless of the technique employed (Brett, 1916). Some early experimental trials indicated that latex yields were lower on felled trees compared to other methods that left trees standing (Anderson, 1914; Bancroft and Bayley, 1914), but experimental design was poorly constructed and results were not entirely in line with the conclusions (Gonggryp, 1923). Balata and maçaranduba (*Manilkara huberi*) trees were also bled in small commercial quantities across parts of North Pará and Roraima states in Brazil as late as the 1980s (Lescure and de Castro, 1990). As in Venezuela, *Manilkara* latex was commonly collected through tree-felling across many parts of the Brazilian Amazon, a technique that remained in practice as late as the 1990s in South Pará (D. Hammond, personal observation).

The estimable impacts of different latex production strategies on bulletwood populations across the Guiana Shield are staggering, even when adopting conservative mortality figures and liberal latex production rates.[16] During the peak period of balata production from 1910 to 1920 (1905–1915 in Venezuela), between 4.2 and 13.2 million *Manilkara* trees are estimated to have been killed either through felling or tapping across the North Guiana Basin region of the shield, based on mortality and yield rates documented at the time (Jenman, 1885; Hohenkerk, 1919; Gonggryp, 1923; Davis, 1933; Châtelain, 1935). Commercial bleeding in Guyana alone is estimated to have killed between 900,000 and 4.8 million *Manilkara* trees across the major producing regions of the country during the entire 125-year history of balata production.

Other latexes, oils, resins, fibres and foods

Pará rubber and balata extraction dominated the early 20th-century forest economies in the Guiana Shield, but a huge variety of natural forest products commonly forming an integral part of Amerindian traditional culture also began to balloon into fully fledged regional, national and international trade networks. Of these, a dozen or so reached significant levels of production during the 20th century. These can be divided into two groups, namely: (i) latexes, oils and resins; and (ii) fibres and foods.

Generally speaking, the latex–oil–resin group characterized a pattern of boom-and-bust production similar to rubber and balata. Their production peaked with the first half of the 20th century and diminished substantially as demand outstripped supplies of naturally sourced materials and synthetic alternatives were successfully developed to overcome these supply constraints on demand by the 1950s. As with rubber and balata, small quantities continued to be produced for local and national consumption, but these represent an ever-diminishing source of income for rural inhabitants across the Guiana Shield and other regions of the neotropics (e.g. Anderson and Ioris, 1992). It is not unreasonable to say that their status as commercial NTFPs was founded simply on the lag between accelerating rates of innovation in manufacturing (1860–1920) and chemical (1920–1970) engineering. Rosewood oil (*Aniba* spp.), cumarin (*Dipteryx* spp.) and tanning bark (*Rhizophora mangle*) evolved international markets only to decline as alternative sources were developed and commercially accessible quantities declined through destructive harvesting (*Rhizophora* bark) or competing uses (e.g. *Dipteryx* lumber) (Homma, 1992).

Commercial rosewood oil production in the Guiana Shield was driven mainly by extraction from forests in French Guiana and Brazil from 1880 to 1930 to feed a growing perfume industry. Commercial extinction of rosewood trees due to destructive harvesting of the oil and competition from synthetics (linalool) and other exporting countries (e.g. Mexico) led to a price decline and eventual closure of distillation operations, leading to the collapse of the rosewood oil industry by the 1930s (Homma, 1992). Production continued sporadically up until the 1960s in French Guiana (Bruleaux, 1989) and small production quantities are still reported from Brazil (IBGE, 2000), but the process of mining slow-growing *Aniba* trees for their oil in the states of Pará and Amazonas has led to deep contraction of the industry since its peak in the 1960s.

The seeds of the *sarrapia*, or tonka bean, tree were also in great commercial demand during the early 1900s for use in cigarettes and perfume (Anon., 1936; Fanshawe, 1950). More than 1600 tonnes of the bean were produced in Venezuelan Guayana from 1926 to 1936, accounting for the bulk of the export from the shield region. Nearly all of this volume was collected from the middle Caura and Cuchivero watersheds, where relatively high stocking rates of *Dipteryx* were encountered. Annual production fluctuated by two orders of magnitude, due to interannual variation in fruit crop size. Synthesis of coumarin in the 1930s effectively eliminated supply variations due to the vagaries of natural phenology and the demand for natural sources declined precipitously. Cumarin is still produced in Brazil, with nearly 90% of the 10–50 annual tonnes reported to originate from the North Pará region of the Guiana Shield between 1990 and 2002 (IBGE, 2003b).

The story of food and fibre products followed a different course over the 20th century and many have retained or grown into viable national and international markets. Most prominent within this group is the production of the heart and fruits of the *açai* (pinot, manicole, cabbage, manaka) palm, Brazil nuts and *piaçaba* (*chiquichiqui*) fibre.

Heart-of-palm

Commercial palm-heart production was recorded from Guyana as early as 1920 (British Guiana Lands and Mines Dept, 1910–1945). In the 1960s, palm-heart production moved from southern Brazil to Pará and Amapá as stocks of *Euterpe edulis* declined below commercial quantities as a consequence of overharvesting (Richards, 1993). The Amazon estuary has been a particularly important centre of palm-heart production since this time, generating nearly US$300 million per annum and employing 30,000 (Pollak *et al.*, 1995).

All of the modern production of the Guiana Shield is drawn from operations in Amapá (IBGE, 2003b), French Guiana (Ricci, 1989), northwestern Guyana (van Andel, 2000) and the Delta Amacuro region based on both *E. oleracea* and *E. precatoria* harvesting. Although levels have been declining due to unsustainable harvesting rates and lower market prices, palm-heart production still ranks highly in contribution to total revenue generated from forest extractive industries in both Guyana (van Andel, 1998, 2000) and Brazil (IBGE, 2003b). Over the last decade, however, production in Amapá has been declining, while increasing in Guyana (Fig. 8.15), making this country the largest producer in the Guiana Shield by the end of the millennium.

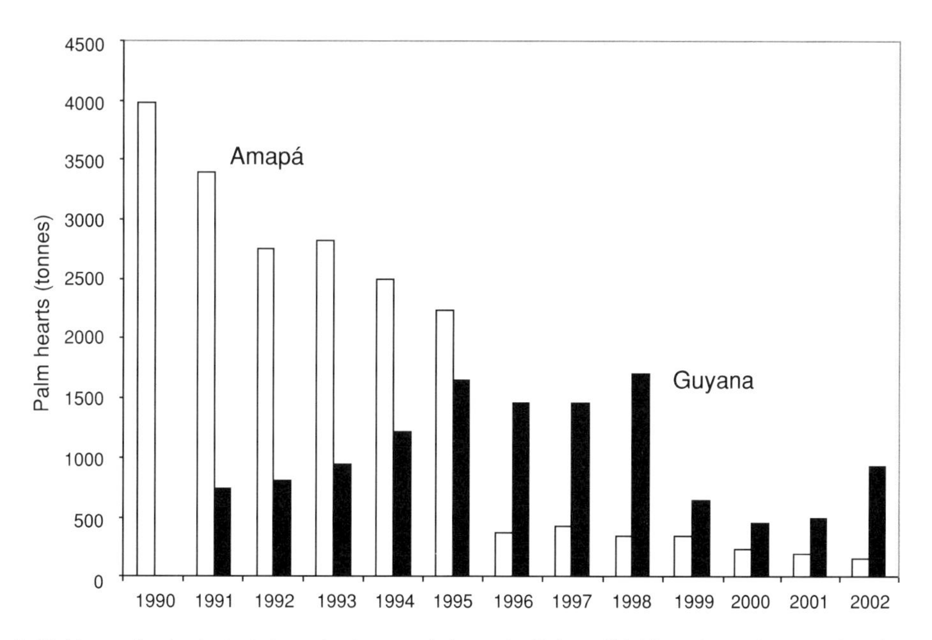

Fig. 8.15. Heart-of-palm (palmito) production trends for main Guiana Shield producers in Amapá and Guyana at the end of the 20th century. Production reflects commercial harvesting of *Euterpe* spp. only. Sources: van Andel (2000), Guyana Forestry Commission (1980–2002), IBGE (2003b).

Açai fruit

Commercial harvest of fruit from the *Euterpe* palm is a relatively modern development based on production almost exclusively carried out in the eastern Brazilian Amazon with only a small percentage (1–2.5%) derived from the Guiana Shield (Amapá). Virtually all of the commercial produce in Brazil is destined for consumption in Macapá and Belem with little or no export (Strudwick and Sobel, 1988) and provides the largest contribution of any NTFP to rural household incomes in the lower Amazon area (Anderson and Jardim, 1989) and second largest revenue-earner at the national level (IBGE, 2003b). Production in Amapá has been declining in contrast to the expanding production from municipios along the Amazon River and its major southern tributaries (Table 8.2).

Brazil nuts

When the rubber boom busted in the 1920s, many of the former rubber tappers sought other work in extracting Brazil nuts for a growing international market. The centre of Brazilian production originated in South Pará, moving westward to Amazonas and finally Acré, where the majority of nuts were produced up until the early 1990s (along with neighbouring Bolivian and Peruvian lowland regions). By the end of the 20th century, IBGE records indicate a steady decline in declared production from Acré and (southern) Amazonas state is now the top contributor to national production. Nationally, production has nearly halved over the period 1990 (51.2 kt) to 2002 (27.4 kt) and a mere quarter of peak production achieved in 1976 (104.5 kt), mainly due to a decline in production from Acré (IBGE, 1985, 2003b). Despite this decline, Brazil nut production remains the second most significant contributor to the extractive forest economy in the post-rubber Amazon basin. The Brazil nut industry has (thus far) avoided the more severe boom-and-bust life history typifying many other international commodities (Homma, 1992), despite having a mixed history of recurrent inequities very similar to the oligopolistic *aviamento* system that structured the rubber economy throughout the early 20th-century lowland neotropics.

Declared production from *municipios* in the Guiana Shield account for a significant part of modern Brazil nut production, ranging between 16% (1998) and 55% (1995) during the 1990s (IBGE, 2003b).[17] Small quantities of Brazil nuts were recorded as having been shipped for sale in Guyana and Venezuela during the first half of the 1900s (Fanshawe, 1950), but the extremely low stocking rates or absence of *Bertholletia* over most of the Guianas and Venezuelan Guayana has precluded any serious commercial extraction from the North Guiana Basin region of the shield. This absence and the fact that most production emanates from *terra firme* forests in *municipios* along or adjacent to the Amazon Downwarp or other large rivers lends some support, when compared with the archaeological record (see 'Human prehistory', above), that this species owes much of its current distribution and local abundance (in groves, or *castanhais*) to past human manipulation (Balée, 1989).

Piaçaba (chiqui-chiqui) fibre

Piaçaba fibre is collected from the leaves of the palm *Leopoldinia piassaba* and manufactured into a wide range of basic household consumables (see Table 8.2) for sale in regional markets in Venezuelan and Brazilian Amazonas states and Colombia (Putz, 1979; Saldarriaga, 1994; Narváez and Stauffer, 1999). During the late 19th–early 20th century, *piaçaba* was exported for broom manufacturing in Europe (Wallace, 1853b), but is now largely restricted to regional markets (Puerto Ayacucho, Barcelos) that are within close proximity to raw materials sources. *Leopoldinia piassaba* (and *L. major*) is narrowly restricted to the Rio Negro white sand–blackwater habitat (Goulding *et al.*, 1988; Henderson *et al.*, 1995) that dominates the faulted depression crossing the central shield region (see 'Major faults,

Table 8.2. Selected past and present non-timber forest products extracted for commerce from the Guiana Shield region, the peak decade of production and their commercial applications. (+) denotes small quantities with no or unknown peak period.

		Country or state[a]										
Scientific name	Common names from region	DA	BOL	VE-AM	GUY	SUR	FG	RR	AP	PA	BR-AM	General commercial use
Euterpe spp.	acai uassai, guassai, baboen pine, pina			+					1980s	2000s	2000s	Juice, ice cream flavourings
Euterpe spp.	palmito, cabbage	+			1990s		+		1080s			Palm heart
Carapa spp.	andiroba, crabwood, krabba				+					+		Soap, candles
Copaifera spp.	copaiba		+	+	+					+	1970s	Scented soaps and oil
Heteropsis spp.	nibbi, cipo titica			+	1990s	+						Furniture, handicraft
Clusia spp.	kufa				1990s							Furniture
Mauritia flexuosa	Ite, miriti, buriti, moriche	2000s	+	2000s	+				+	+	+	Handicraft, flour drink
Bertholletia excelsa	Brazil nut, castanha, noix du Bresil			+	+			+	1960s	1970s	1970s	Nuts
Rhizophora mangle	tanning bark				1925/1945							Leather tanning
Couma utilis	sorva, leche caspi				+	+				+	1925	Chewing gum, paints, varnish
Leopoldinia piassabe	piacava, chiqui-chiqui			2000s							2000s	Handicrafts, rope, brooms, mats, brushes
Dipteryx spp.	cumaru, sarrapia, tonka bean		1940s		1930s					1940/1980s	1940s	Tobacco flavouring, vanilla substitute
Orchidaceae	tropical orchids				1920s							Horticulture trade
Aniba spp.	Pau rosa, bois de rose, rosewood						1920s			1960s	1960s	Eau de toilette, perfume
Hymenaea spp.	locust resin, copal jatoba				1925					1960s		Faux amber, varnish

Source: IBGE, GFC, Fanshawe (1950), Richards (1963), van Andel (2000), Bruleaux (1989), Homma (1992), Lescure & de Castro (1990).

[a]DA, Delta Amacuro; BOL, Bolivar; VE-AM, Venezuelan Amazonas; GUY, Guyana; SUR, Suriname; FG, French Guiana; RR, Roraima; AP, Amapá; PA, Pará; BR-AM, Brazilian Amazonas.

downwarps, rift valleys and geosynclines' and 'Upland and sedimentary plain soils', Chapter 2). Production increased significantly in the Rio Negro basin at the close of the 20th century (from 1200 to 8000 tonnes per annum), but still only accounted for a relatively small fraction of total national production (<10%) (IBGE, 2003b), due to the widespread harvesting of a similar, rament-like fibre (also called *piaçava*) from *Attalea funifera* along the Atlantic coast of northeastern Brazil, especially in Bahia. Despite potential inter-regional competition, *piassaba* production provides an important contribution to rural incomes along the upper Rio Negro region. The non-destructive harvesting of the rament from each adult plant dampens the likelihood that commercial extinction of the base resource will curtail long-term, regional demand as in the case of other NTFPs. It is easily transported and does not require major capital investments in order to expand marketing reach that otherwise limits more perishable products, such as *açai* fruit (Richards, 1993).

Commercial timber extraction

The commercial extraction of tropical timber in the countries of the Guiana Shield dates back to the late 1500s, but only began to compete in economic significance with the long-standing agricultural sector as the 19th century concluded. In many areas, the demise of the early NTFP industries sparked a revaluation of the forest resources. In the Guianas, this shifted principally along a view that timber had failed to reach its full potential and that the Guianas held unsurpassed quantities of superior timbers (Rodway, 1912; Hohenkerk, 1922; Furse *et al.*, 1924; Valeix and Mauperin, 1989). In Brazil and Venezuela, a focus on alternative forms of NTFP extraction and traditional agricultural and livestock-rearing dominated. Timber production, apart from high-graded *Swietenia macrophylla* extraction, remained of relatively minor regional importance. Through a series of technological, policy and economic transitions over the course of the 20th century, timber extraction has grown from modest beginnings into a recognized economic sector throughout most of the region. Paralleling this growth in production is a growing recognition of the considerable socio-economic and biological consequences of this form of forest land-use, although timber production in the Guiana Shield has taken many forms that, like the scale of mining operations, vary in their socio-economic and environmental consequences.

Production trends

Up until the end of the 19th century, timber production had been restricted to only a handful of species (e.g. *Brosimum*, *Cedrela*, *Chlorocardium*, *Carapa*, *Manilkara*, *Mora*, *Ocotea*) extracted in small quantities measuring no more than several thousand cubic metres per year (Schomburgk, 1840; Malfoy, 1989; Valeix and Mauperin, 1989). Efforts to establish commercial plantations of these and other native and exotic species during the early 20th century largely failed due to biological invasion of the monocultures and higher-than-expected costs of plantation maintenance (Dawkins and Philip, 1998). In the Guianas, concepts of economic development remained squarely fixed on the timber-driven largesse bestowed upon North American economies, that were nonetheless rapidly exhausting their hardwood supplies by the early 1900s (Williams, 1989). The increase in North American demand, combined with continued favourable performance of heavy hardwoods in maritime applications (shipbuilding, coastal defence works, canal locks, wharves) in Europe and the West Indies, sparked increased demand for timber products from the region. But the early 20th century timber industries of the Guianas were poorly geared to fulfil the market demand for industrial heavy hardwoods in The Netherlands, Great Britain, France, Martinique, Belgium, the USA, Canada and Bermuda, among others (e.g. Furse *et al.*, 1924). By the time investment mobilized to increase production, the Great Depression hit most national economies, weakening commercial investment in and market demand for timber products from the region (Oliphant, 1938), but increasing domestic and regional demand in the Caribbean (British Guiana Forest Department, 1935b).

The greatest increase in global tropical timber production occurred from 1940 to 1980 in the aftermath (and in part as a consequence) of the Second World War and output from the Guiana Shield paralleled this worldwide trend (Fig. 8.16). Timber production from the Guianas increased rapidly with the initiation of reconstruction in Europe and economic overdrive of a post-war North America, reaching new historical highs in annual production in Guyana, Dutch Guiana and French Guiana by 1965 (Fig. 8.16). The establishment of the ill-fated Jari paper pulp plantation project over 16k km^2 of lower Amapá boosted total wood production, but low growth rates and poor performance of the ill-advised selection of a fast-growing exotic, *Gmelina arborea*, thwarted efforts to recapture the massive financial investment (Rollet, 1980; Palmer, 1986). With independence in Guyana and Suriname, production increased further only to decline across the Guianas in the early 1980s as worldwide recession hit wood product demand. Production during the 1980s remained depressed and erratic. But by 1990, estimated timber production in the Guianas since the end of the Second World War totalled nearly 17.9 million cubic metres. Over the course of the early 1990s, considerable foreign investment in Guyana's timber industry propelled annual production above 500,000 m^3/year (Fig. 8.16), a level never previously achieved in the Guiana Shield. Production from Amapá during the early 1990s also peaked during a period of timber extraction from national forest to offset shortfall in the Jari project, now under new management and objectives

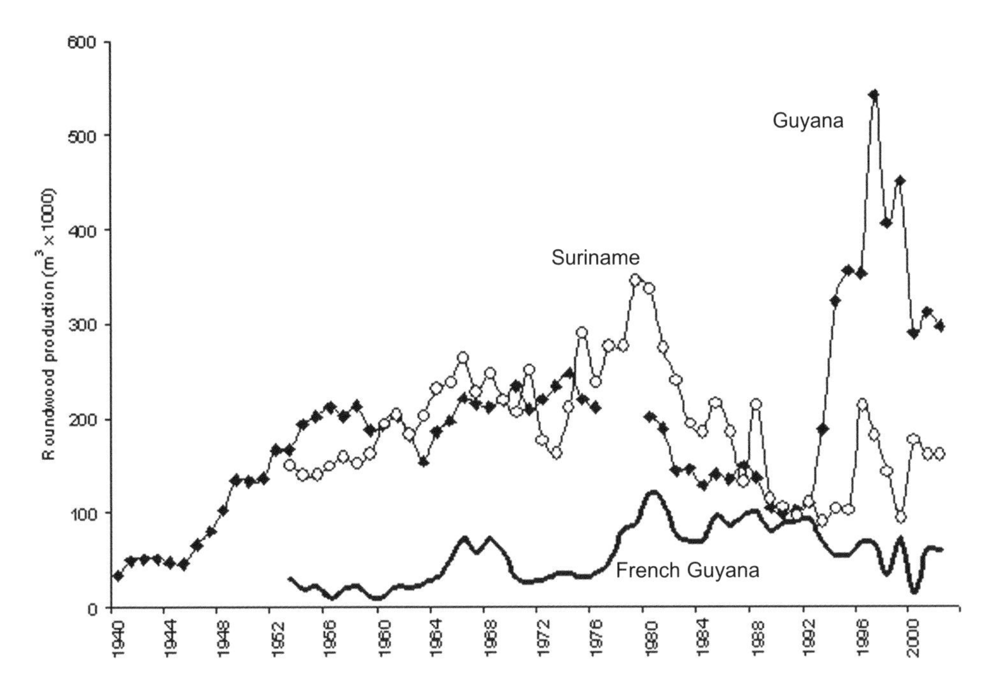

Fig. 8.16. Tropical timber production trends in the Guianas 1940–2002 for wood extracted from natural forest sources only. Sources: British Guiana Forest Dept (1940–1965), Guyana Forest Department (1966–1976), Vink (1970), Guyana Forestry Commission (1980–2002), Office National des Forêts (1992, 1997a), IBGE (2003a,b), ITTO (2003), FAOSTAT (2003).

(Table 8.3). At the end of the 20th century, Guyana had become the region's largest commercial producer of wood products, with Suriname, Bolivar state in Venezuela and North Pará (less Jari) trailing (Table 8.3). From 1990 to 2002, an estimated 14.1 million cubic metres of wood was extracted from natural forests in the Guiana Shield, nearly 80% of the volume produced over the preceding 40 years. Placing this volume in a larger geographic perspective, however, it represents a mere 4% of the total tropical roundwood production reported for Brazil from 1990 to 2002 (Table 8.3). This nominal figure reflects both the inaccessibility created by the geomorphology of the region and the traditional avoidance of forest conversion approaches to timber production that have been more commonly employed across the southern Amazon region of Brazil (see Chapter 9).

Forest area exploited or managed for timber production

Virtually all of the timber extracted from forests of the region prior to 1950 was floated or loaded on punts (*ballahoos* in Guyana) or barges for downstream transport to sawmills and loading docks. As a consequence, the rapids interrupting virtually every major waterway in the Guiana Shield delimited the region of major production prior to road-building. Commercial extinction of forests below the falls over centuries of repeated creaming was inevitable (McTurk, 1882; Hohenkerk, 1922). Im Thurn (1883) described the late 19th century perception of this limitation in (the then) British Guiana:

> Next to [the sugar tract] is the timber tract, from which alone timber has as yet been remuneratively brought to market. This extends toward the interior as far as the lowest cataracts on the various rivers. It is presently impossible to cut timber profitably beyond these cataracts, owing to the difficulty which there would be in carrying any cut beyond that to market; so that an imaginary line, roughly parallel to the sea-coast, and cutting each of the great rivers at their lowest cataracts, marks the further limit from the coast of this tract.

Im Thurn's imaginary line extends around the rim of the Guiana Shield (e.g. Goulding *et al.*, 1988, p. 25), effectively creating a barrier to transport of heavy consignments, particularly the dense woods that typified most of the commercial timber volumes sought for industrial applications. In Suriname, this imaginary line formed a 620,000 ha exploitable forest belt, and similar problems of rapid overharvesting and commercial depletion of merchantable species below the cataract line (Vink, 1970).

As the limitation of navigable waterways in the region was overcome through a series of road-building campaigns (e.g. British Guiana Forest Department, 1935a, 1940–1965; Valeix and Mauperin, 1989), the exploitable forest estate expanded and new forest reserves dedicated for timber or multiple use purposes were established (Fig. 8.17). By 2000, an estimated 12.4% of the Guiana Shield had been allocated to timber production (Table 8.4), although the status of forests within these areas varied from heavily exploited and commercially depleted through to largely unexploited (Guyana Forestry Commission, 1980–2002; Office National des Forêts, 1997b; Miranda *et al.*, 1998).

Technological and transport transitions

The expansion of commercial timber production forests in many ways accompanied technological advances that conditionally rendered road-building and longer transport distances economically feasible.

Up until the 1950s, felling was carried out exclusively by hand-axe and many operations still transported logs manually over rolling sticks ('grey stick method' in Guyana) and using draft animals (British Guiana Forest Department, 1940–1965; Vink, 1970). Logs of low-density wood (e.g. *Cedrela*, *Carapa*) were 'worked' over cataracts by hand and floated to the mill, while high-density timbers were assisted using a punt or barge (Viera, 1980). In other

Table 8.3. Estimated annual timber production from natural forest sources in the Guiana Shield, 1990–2002. Sources: Guyana Forestry Commission (1980–2002), IBGE (2003b), ITTO (2003), FAOSTAT (2003).

A.		Non-coniferous, tropical roundwood in $m^3 \times 1000$												
Country		1990	1991	1992	1993	1994	1995	1996	1997	1998	1999	2000	2001	2002
Venezuela		878	750	1016	960	769	591	599	779	684	754	664	695	647
	Delta Amacuro*	18	15	20	19	15	12	12	16	14	15	13	14	13
	Bolivar**	176	150	203	192	154	118	120	156	137	151	133	139	129
	Amazonas***	9	8	10	10	8	6	6	8	7	8	7	7	6
Guyana		98	104	112	189	326	358	354	544	408	453	289	312	297
Suriname		106	97	112	90	104	103	213	182	143	94	176	162	162
French Guiana		89	91	93	70	55	55	68	66	35	72	15	60	60
Brazil		25,087	26,859	26,938	27,014	27,090	27,164	27,238	27,312	27,384	28,452	29,444	29,444	29,444
	Amapa	340	353	317	333	330	352	76	57	73	83	84	71	78
	Para (no Jari)	44	41	50	48	86	65	65	67	84	130	139	153	119
	Para (Jari)	910	1.1013	884	924	1,004	891	809	796	799	772	790	758	743
	Roraima	34	36	38	–	17	–	17	17	20	27	27	25	75
	Amazonas	452	14	19	92	169	185	14	69	92	93	76	90	85
Total Guiana Shield		1,366	909	974	1,043	1,263	1,255	944	1,181	1,012	1,124	960	1,033	1,026
B.		As % of total Guiana Shield												
Venezuela														
	Delta Amacuro	1.3	1.7	2.1	1.8	1.2	0.9	1.3	1.3	1.4	1.3	1.4	1.3	1.3
	Bolivar	12.9	16.5	20.9	18.4	12.2	9.4	12.7	13.2	13.5	13.4	13.8	13.5	12.6
	Amazonas	0.6	0.8	1.0	0.9	0.6	0.5	0.6	0.7	0.7	0.7	0.7	0.7	0.6
Guyana		7.2	11.4	11.5	18.1	25.8	28.5	37.5	46.1	40.3	40.3	30.1	30.2	28.9
Suriname		7.8	10.7	11.5	8.6	8.2	8.2	22.6	15.4	14.1	8.4	18.3	15.7	15.8
French Guiana		6.5	10.0	9.5	6.7	4.3	4.4	7.2	5.6	3.5	6.4	1.6	5.8	5.8
Brazil		1,837	2.956	2,765	2,589	2,145	2,164	2,884	2,312	2,705	2,532	3,068	2,849	2,870
	Amapá	24.9	38.9	32.5	31.9	26.1	28.1	8.0	4.9	7.2	7.4	8.8	6.9	7.7
	Pará (no Jari)	3.2	4.5	5.1	4.6	6.8	5.2	6.9	5.7	8.3	11.5	14.5	14.8	11.6
	Pará (Jari)	66.6	111.5	90.8	88.6	79.5	71.0	85.6	67.4	78.9	68.7	82.4	73.4	72.4
	Roraima	2.5	4.0	3.9	–	1.3	–	1.8	1.4	1.9	2.4	2.8	2.4	7.3
	Amazonas	33.1	1.6	2.0	8.9	13.4	14.8	1.5	5.8	9.1	8.3	8.0	8.7	8.3

*Calculated at 2% of total production. **20% of total production. ***1% of total production.

Table 8.4. Estimated area allocated for timber production in the Guiana Shield including private plantations (Jari), national forest area allocated for sustainable use, long-term concessions and short-term cutting leases. Amerindian and agricultural lands are not included. Sources: Guyana Forestry Commission (1980–2002), Vink (1970), G. Zondervan (personal communication), Office National des Forêts (1997b), Rollet (1980), IBAMA, MARNR.

Country	Area (km^2)	% of natural area in GS	% of GS
Venezuela	103,348	22.8	4.5
Guyana	55,565	25.8	2.4
Suriname	13,700	8.8	0.6
French Guiana	3,930	4.5	0.2
Brazil	107,036	10.3	5.7
Colombia	–	–	–
Total	283,579		12.4

instances, timber production above rapids remains largely a local subsistence activity. Few roads had been developed to provide access to upland timber stocks and virtually all timber extracted during this period was taken from an area bound by the course of the major waterways. In Suriname, ditch-blasting was approached as a means of accessing valuable peeler log species (mainly *Virola*) growing in difficult swamp locations (Bubberman and Vink, 1966), but with unexpected consequences.

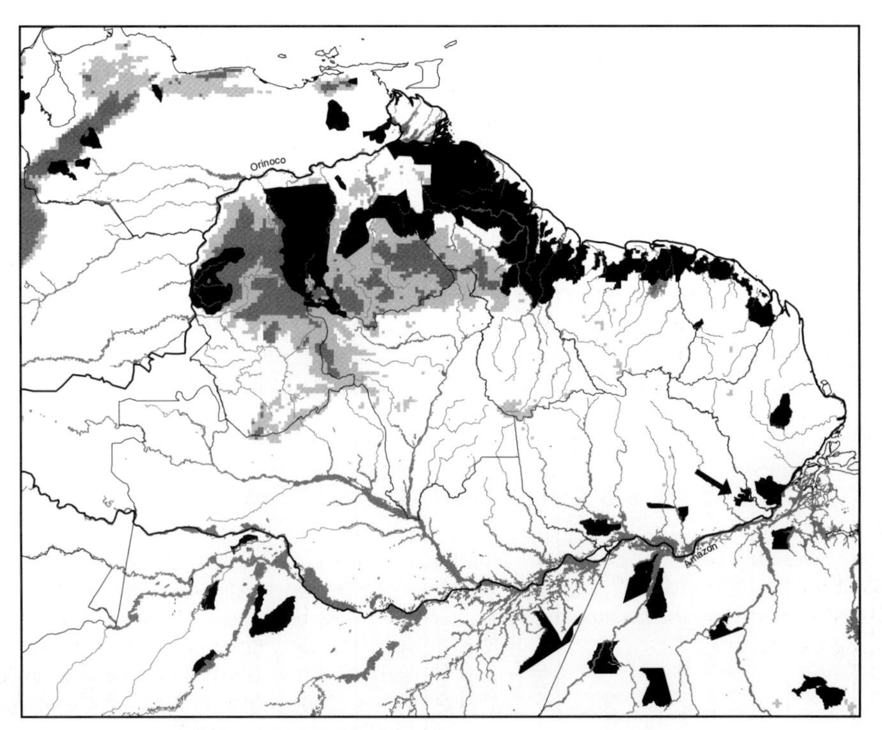

Fig. 8.17. Cumulative distribution of official timber producing regions (solid black) across the Guiana Shield by end of the 20th century. Only one large plantation (Jari, arrow) has been established in the region. Levels of commercial stocking and use vary within allocated areas. GIS coverage sources: GFC, MARNR, WRI-GFW, IBAMA, LBB, Office National des Forêts (1997b), Iwokrama International Centre.

Table 8.5. Installed electricity generation capacity (kW) in the Guiana Shield by 2003. Thermal electric is based on hydrocarbon (mainly diesel) use only. Power from geothermal, solar, wind or nuclear sources was negligible. Hydroelectric power in French Guiana and Amapá includes non-public generational capacity, mainly for Kourou and Jari facilities, respectively. Per capita generation in watts (W) based on population presented in Table 8.6. Sources: ANEEL (2003), US Energy Information Administration (2003).

Country/State	Hydroelectric	Thermal electric	Total	% hydro	W per capita
Venezuela Guayana	15,461.0	–	15,461.0	100.0	10.008
Guyana	<0.01	0.8	0.8	<0.01	0.001
Suriname	0.2	0.2	0.4	48.6	0.001
French Guiana	116.0	0.5	116.5	99.6	0.741
Amapá	68.0	138.7	206.7	32.9	
North Pará	–	22.7	22.7	–	
North Amazonas	250.0	66.0	316.0	79.1	
Roraima	5.0	189.7	194.7	2.6	0.187*
	15,900.2	418.5	16,318.7	97.4	4.157

*Based on total Brazilian population in Guiana portions of states.

small plants have been developed in the past to provide power to mining operations and small towns (e.g. Moco-Moco complex providing electricity to Lethem, Rupununi, south Guyana).

The Guri (Raul Leoni) facility in Venezuelan Guayana dominates both hydroelectric generation and installed capacity from the Guiana Shield region (Table 8.5). French Guiana, and Amapá and Amazonas states in Brazil also derive a significant fraction of their generating capacity from hydroelectric facilities. Guyana, North Pará and Roraima currently have little or no significant generating capacity. Further facilities are under construction or planned in Brazil along the Jari River (Santo Antonio), Oiapoque (Salto Cafesoca) (ANEEL, 2003) and in Venezuela downstream of existing facilities along the lower Caroni River (US Energy Information Administration, 2003) and along the Kuribrong River in the Pakaraima Mountains of western Guyana.

Renewable sources

Hydroelectric generation from tidal energy is another approach that could prospectively be employed at small scales along the coast of Guyana, Suriname, French Guiana and Amapá. In fact, tidal energy was harnessed by small-scale facilities in Suriname, French Guiana and the Amazon estuary, possibly as early as the 1700s, and has the potential to contribute to local generation (Anderson *et al.*, 1999), particularly where tidal gates are already employed to protect reclaimed coastal lowlands during peak daily and seasonal stages. Wind generation is also a traditional technology that could be employed at commercial scales along many parts of the shield coastline (Persaud *et al.*, 1999). Solar cell units are currently employed throughout many parts of the interior that are not connected to more traditional thermal or hydroelectric sources as part of the national transmission grid.

Seasonal and interannual variation in source energies

A steady, uninterrupted supply of fuels and electricity brings tremendous socio-economic benefit, but the heavy reliance of many parts of the shield region on a single energy source continues to expose regional economies to growth-dampening fluctuations and foreign debt accrual.

The list of factors potentially influencing world oil market prices is wide-ranging and expanding. Guyana and French Guiana are the largest per capita importers of petroleum products in the shield region and this reliance is increasing (Fig. 8.19). In contrast, Suriname's new petroleum production industry, combined with the hydroelectric facility at Afobakka, has steadily dampened its per capita requirement for overseas oil products, making the country a net exporter for the first time briefly in 1999. Suriname's transition mimics that achieved, albeit at a much greater scale, by Venezuela in its installation of hydroelectric facilities along the Caroni River and development of extensive oil fields that has made it the largest exporter of energy in South America. Currently it exports excess hydroelectric capacity to Colombia and Roraima, Brazil through a growing national transmission grid and has consistently exported between 30 and 50 barrels of oil per person each year since 1980 (US Energy Information Administration, 2003).

Installation of oil production facilities in Suriname and Guyana could bode well for their economies in the future as the global peak in hydrocarbon production is passed (between 2010 and 2020) and the world petroleum economy begins its descent towards commercial extinction. Bringing their reserves on-line during this late post-peak phase should see considerably higher world market oil prices and greater revenues from the relatively small proven offshore reserves of the Guyana–Suriname Basin and revise views on how best to achieve benefit from their relatively large per capita forest cover (see Chapter 1).

Brazil generates a massive 80%+ of its national electricity requirement through hydroelectric facilities, mainly located south of the Amazon River, while maintaining a relatively stable per capita reliance on overseas oil imports (Table 8.4). The large hydroelectric capacity assists in buffering the Brazilian economy from the vagaries of the world oil market, but exposes it to rainfall variations that commonly affect the region as a consequence of ENSO. Facilities planned or located along rivers draining the Guiana Shield and northeastern Brazil are particularly susceptible to large-scale seasonal failure in rainfall (see 'Climate', Chapter 2). Countering this interannual failure in generating capacity is the fact that

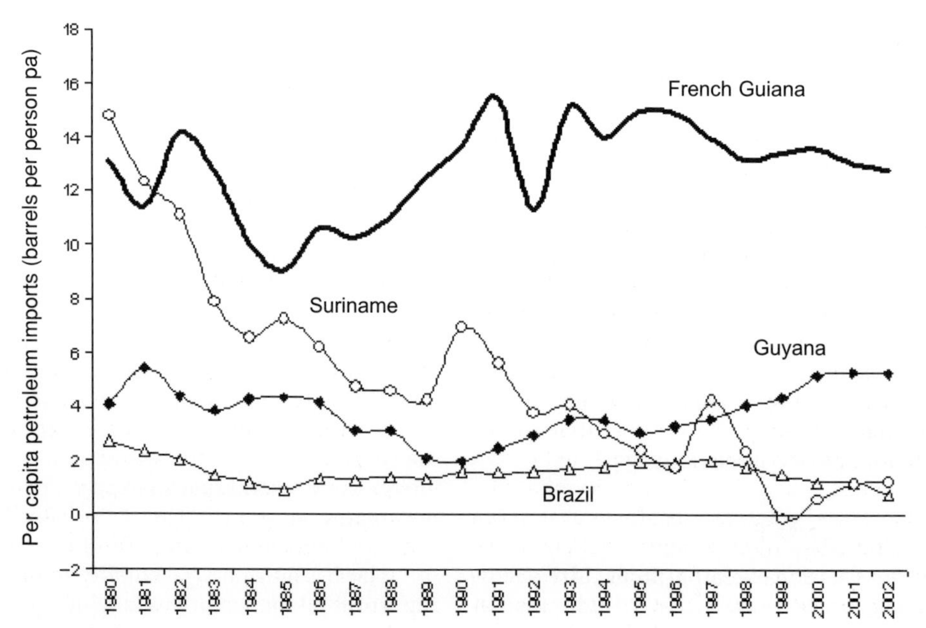

Fig. 8.19. Variation in the amount of oil imported in each of the net importing countries of the Guiana Shield region. Calculated as difference between annual hydrocarbon consumption and net production divided by population. Brazilian imports are for entire country. Data source: US Energy Information Authority (2003).

many of the rivers draining the Guiana Shield are relatively sediment-free (without upstream mining) and the life expectancy of the reservoirs (embalses), such as Guri or Brokopondo should be greater than locations elsewhere in the neotropics.

The relatively high number of sunshine days experienced along the Savanna Trough of the central shield region (see 'Climate', Chapter 2) would auger well for larger-scale solar generation as part of a hybrid facility that could assist in modulating seasonal and interannual variation in hydroelectricity generating capacity. Rainfall and total sunshine hours are inversely related and hydroelectric and solar hybrid systems would reach peak generation during opposing seasonal and interannual phases. The relatively low population growth in the region, however, makes future energy requirements prospectively less than in other tropical lowland countries.

Population Trends in the Guiana Shield

The very small population of the Guiana Shield continues to play an important role in buffering large-scale degradation of tropical forests in the region. The region's population, however, has a long (pre-)history dominated by successive periods of colonization, abandonment and immigration borne from the recurrent realization that the biophysical environment of the region has few quick rewards on offer. The region's natural population growth, borne through increasing reductions in the infant mortality rate over the past century, has continuously been sapped by the mass emigration of many working age people to southern Brazil, French Guiana, Venezuela, the Caribbean, North America and Europe (Fig. 8.20). The effect of this exodus is even more striking when one considers that countries such as Guyana and Suriname have actually recorded negative population growth rates over several brief periods in the last 50 years. Countries such as French Guiana and Venezuela have shown consistent population growth in the shield region, particularly linked to large-scale industrial and technology centres at Kourou and Ciudad Bolivar. During the 19th century, (British) Guiana held the largest recorded population in the region and maintained this lead up until the 1980s (Fig. 8.20). The growth of Ciudad Guayana into one of the largest cities in Venezuela after installation of several large-scale mining and industrial processing industries in the region sparked an unprecedented level of population immigration to the region (McDonald, 1969). By 1984, Bolivar state alone had the largest recorded population in the Guiana Shield. At the same time, population growth in Guyana and Suriname stalled and then actually declined as birth rates failed to keep up with the 1–3% of the population emigrating each year (Fig. 8.21).

Population growth in the region has historically remained one of the lowest in the world despite several hundred years of structured immigration to the region to serve the needs of the main agricultural and forest-based industries, establish geopolitical sovereignty over remote frontiers, push industrial development and to offset ethnic dominance (Friedman, 1969; Dew, 1978; Bisnauth, 2000). As a result, the immense pressure exerted by rapid population growth on natural resources and social infrastructure in many other parts of the world has been largely indiscernible in the region over the last 500 years. By the year 2000, the population of the shield had only managed to grow to a size equivalent to one-quarter that of The Netherlands, in an area that is 55 times larger (see Table 8.6). Standing population densities remain some of the lowest in the world (World Bank, 2001), averaging less than 3 persons/km^2 across the area (Table 8.6) and falling to less than 1 person/km^2 in the vast, inaccessible forests of southern French Guiana, Guyana, Suriname and certain parts of the Brazilian states of Amapá, Pará, Roraima and Amazonas and the Venezuelan state of Amazonas.

In fact, a far more compelling issue constraining sustainable development in many parts of the Guiana Shield is not population growth, but the emigration of young

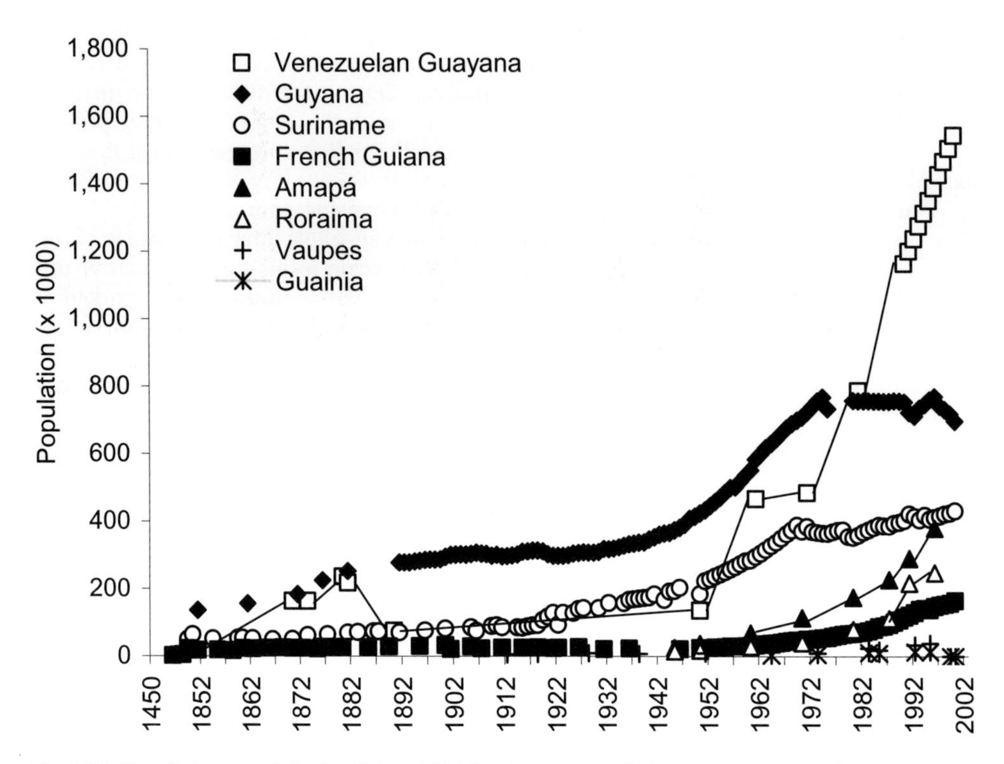

Fig. 8.20. Population growth in the Guiana Shield region since 1700 by country or state. Data sources: INSEE-Guyane, OCEI-Venezuela, IGBE-Brazil, GBS-Guyana, ABS-Suriname, Abonnenc (1951), Hurault (1972), Rowland (1892), Schomburgk (1840).

adults to the cities and from these cities to other countries. In many instances, a chronic lack of jobs that meet wage expectations and the opportunity to follow in the footsteps of thousands of relatives and friends has prolonged a mass exodus of skilled and educated labour that commenced with periods of political tension and uncertainty in many countries of the shield area. Movement to the large cities and then often onward to Europe, North America and, increasingly, to the Caribbean has become a standard engine behind the unusually low population growth recorded in many parts of the Guiana Shield. Guyana and Suriname, in particular, have been affected by this pattern (see Fig. 8.21). French Guiana in turn has received many of these emigrants, particularly from Suriname, Haiti and Brazil (Frouté, 2000), adding to the arrivals entering on work attached to the Kourou Space Centre.

The significance of modest population growth in the Guianas is reflected in both the structure of their domestic economies and the rate and method of forest resource use. A historic reliance on foreign capital-driven commodity extraction, political instabilities, difficulty in achieving wage competitiveness with more populous LDCs, uncertain job tenure across consecutive commodity extraction cycles and chronic limits to domestic consumer demand for most refined goods and services are both a cause and consequence of slow net population growth, creating a chronic cycle of economic under-development affected by high rates of emigration. Repatriation of funds from emigrants in the case of the Guianas is believed to play a significant role in modulating the impact of global economic vicissitudes on the very small, commodity-driven economies of the Guiana Shield region, but periods of economic downturn

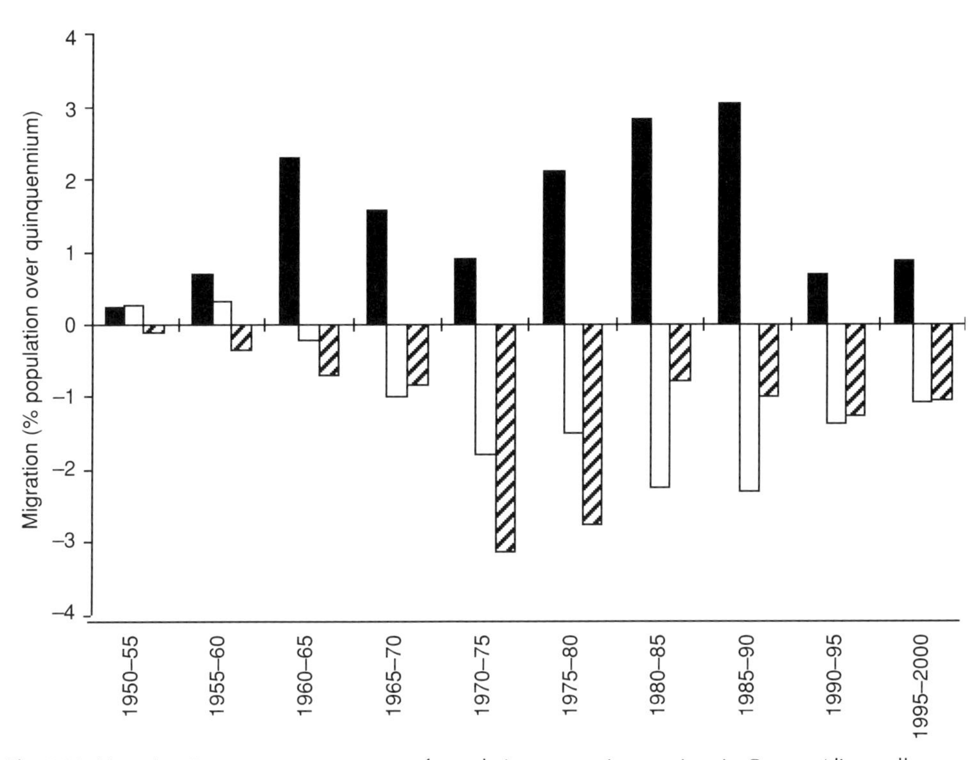

Fig. 8.21. Net migration rate as a percentage of population over quinquennium in Guyana (diagonally hatched columns), Suriname (empty), and French Guiana (solid). Negative and positive values reflect net emigration and immigration, respectively. Data source: UN-CEPAL (2003).

still drive many of the unemployed to engage in forest extraction activities (e.g. Heemskerk, 2001). Only Venezuelan Guayana has been able to overcome these limitations, principally through liquidation of forestland resources across eastern Bolivar state and downstream refinement into energy, metals and other durable goods for both domestic and overseas use. The modern development of the Guiana Shield forestlands, however, has not been pursued without significant social and environmen-

Table 8.6. Population, land area and population density of the Guiana Shield. Values are based on census numbers for administrative districts falling within the shield area. Data sources: CIA World Fact Book (2000), OCEI (2000), IBGE (2003a), Guyana Bureau of Statistics (2000), Algemeen Bureau voor de Statistiek (2000), Frouté (2000).

			Population		
Country	Population (no. persons)*	Land area (km^2)	Density (no. km^2)	As % contribution to total Guiana Shield	As % of country's
Brazil	1,041,595	1,204,279	0.9	26.5	0.6
Colombia	53,650	170,500	0.3	1.4	0.1
French Guiana	157,213	88,150	1.8	4.0	100.0
Guyana	697,286	214,980	3.2	17.8	100.0
Suriname	431,303	156,000	2.8	11.0	100.0
Venezuela	1,544,915	453,950	3.4	39.4	6.4
Guiana Shield	3,925,962	2,287,859	1.7	100.0	

*Year 2000.

tal costs (Sizer and Rice, 1995; Miranda *et al.*, 1998).

Amerindian populations and titled lands

The fact that a disproportionate bulk of the social costs associated with forestland use in the Guiana Shield has been borne by the original inhabitants is irrefutable. From the earliest waves of disease, enslavement, debt servitude and forced emigration, pre-Columbian populations along the shield rim declined from reasonable estimates in the millions to a sparse and scattered population measured in thousands. Estimates of Amerindian populations in French Guiana during the 17th century consistently exceeded 30,000, but by 1750 had declined to less than 3000 due to waves of epidemic disease exacerbated by concentration of populations in missions (Hurault, 1972; Zonzon and Prost, 1996). A nearly identical chronology of decline along the rim and subsequent flight to the remote interior of the shield region has been conveyed for Venezuela Guayana (Whitehead, 1988) and the Rio Negro basin (Hemming, 1978a; Chernela, 1998). In most instances, the number of distinct sociolinguistic cultures disappearing over the first 200 years of contact with Europeans is simply impossible to estimate, although the demise of many groups (e.g. the Norak along the Approuague in French Guiana) is clearly documented (Hurault, 1972).

Nineteenth-century indigenous people continued to struggle under the sweeping cultural, economic and epidemiological changes advanced by continued European colonization. Little population growth was achieved over most of the 19th and early 20th centuries and many groups (e.g. Taruma, Atorais), and their unique cultural contributions, disappeared altogether during this period. Schomburgk (1840) estimates that a mere 7000 Amerindians associated with 10 groups lived in the 19th century boundaries of (the then) British Guiana. This figure changed little up until the end of the 1800s, although arguably based on relatively little effort in accounting for people living in the far interior, where many had fled from the Rio Negro–Branco region and the coastlands of the Guianas. The plight of indigenous peoples remained subject to disparate colonial perspectives on their role in the rapidly changing society, represented by comments made by Schomburgk (1840):

> History informs us that the discoverers of South America found the continent densely peopled by Indians. What then has become of the millions of aborigines who once inhabited these regions? Driven from their lands, now in possession of the Europeans and their descendants, they have wandered from their ancient homes, strangers in their own country; and diseases and vices introduced by the settlers, and feuds among themselves, have all but annihilated the rightful owners of the soil. It is a melancholy fact, but too well founded, that wherever Europeans have settled, the extermination of the native tribes has succeeded their arrival. (Schomburgk, 1840, pp. 48–49)

and almost unbelievably later by Rowland (1892):

> The Aborigines number on the schedules 7,463 or 3,917 male and 3,546 females. To the total population this is 4 per cent. The Registrar-General gives 10,000 more of this race as estimated to be wandering about the interior of the colony. The number on the schedule shows a decrease on the figures of 1881, when 7,762 were returned. This race is of little or no social value and their early extinction must be looked upon as inevitable in spite of the sentimental regret of Missionaries. At the same time it is unnecessary to hasten the process in any way, for in this matter, nature, as ever, is much more gentle than man. (Rowland, 1892, pp. 55–56)

While commentaries at the time clearly varied between sincere empathetic and outright racist perspectives on the Amerindian condition, both seemed agreed on the inevitable loss of Amerindian society from British Guiana. By the 1930s, however, many communities had begun to expand and Amerindian societies across much of the shield region entered a period of sustained growth that continues into the 21st century. By 1960, estimates of the Amerindian population in British Guiana had increased to nearly 23,000 (Anon., 1960), although in part this also probably reflects the use of more accurate assessment techniques. At the end of the 20th century, nearly 53,000 Amerindians contributed to modern Guyanese society.

Nearly a quarter of a million Amerindians are estimated to have lived within various regions of the Guiana Shield by the end of the millennium,[18] largely in the Guayana Highlands, Gran Sabana, Rupununi and Sipaliwini savannas and the rivers draining the forested Tumucumaque Uplands. Belonging to at least six major linguistic groups, Amerindian peoples inhabit virtually all parts of the shield region, but the upper Corentyne and lower Amapá, Rio Negro and Amazon regions are notable by their relatively sparse coverage (Fig. 8.22). In part, the modern distribution is a reflection of long-standing historical migration towards the central highlands of the shield, but also shaped by the distribution of (nearly) exclusive legal title to variously sized fragments of their former homelands (Fig. 8.23). Only Brazil and Guyana had provided some form of official land title to Amerindian communities by the close of the 20th century (Fig. 8.23), although infringements upon these lands, overlapping allocation of commercial land-use driven by disparate ministerial priorities and legal exclusions to land use rights remain commonplace in many areas. Location of Amerindian communities in areas allocated for commercial timber and/or mining have been particular points of conflict in Guyana (e.g. Forte, 1996; Colchester, 1997b) and Brazil (e.g. Tierney,

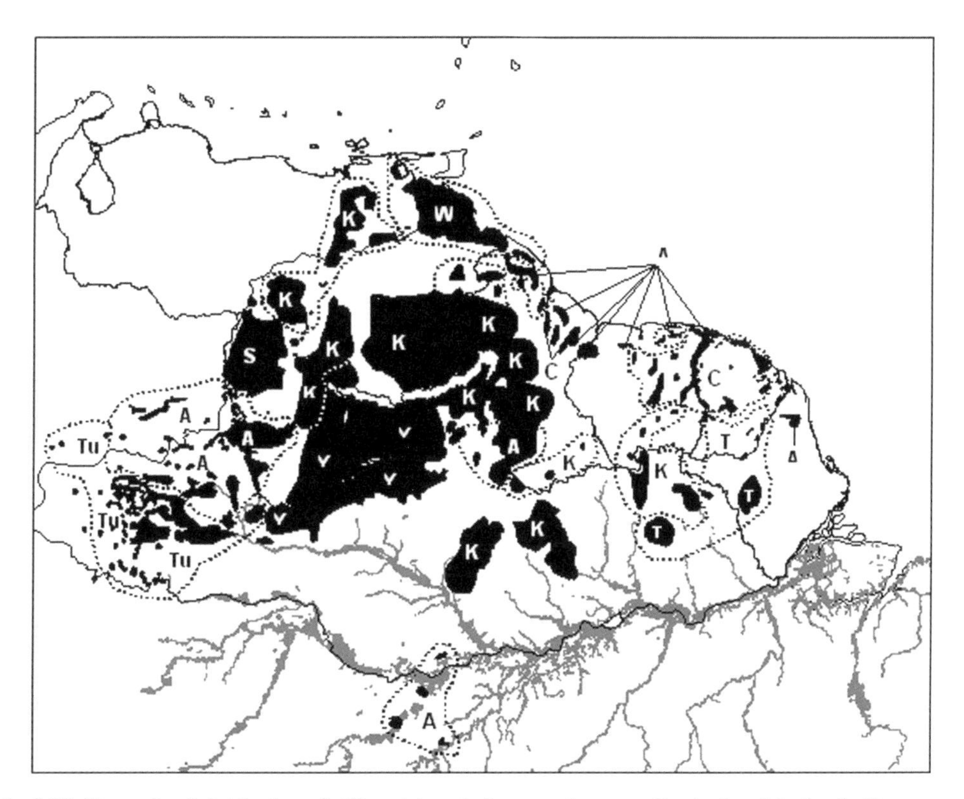

Fig. 8.22. Generalized distribution of different Amerindian peoples according to linguistic family (A: Arawak (Lokono), C: Creole (Maroon), K: Karib (Kariña), T: Tupi, Tu: Tukano, Y: Yanomami, S: Saliva, W: Warrau). Adapted from Queixalos and Renault-Lescure (2000).

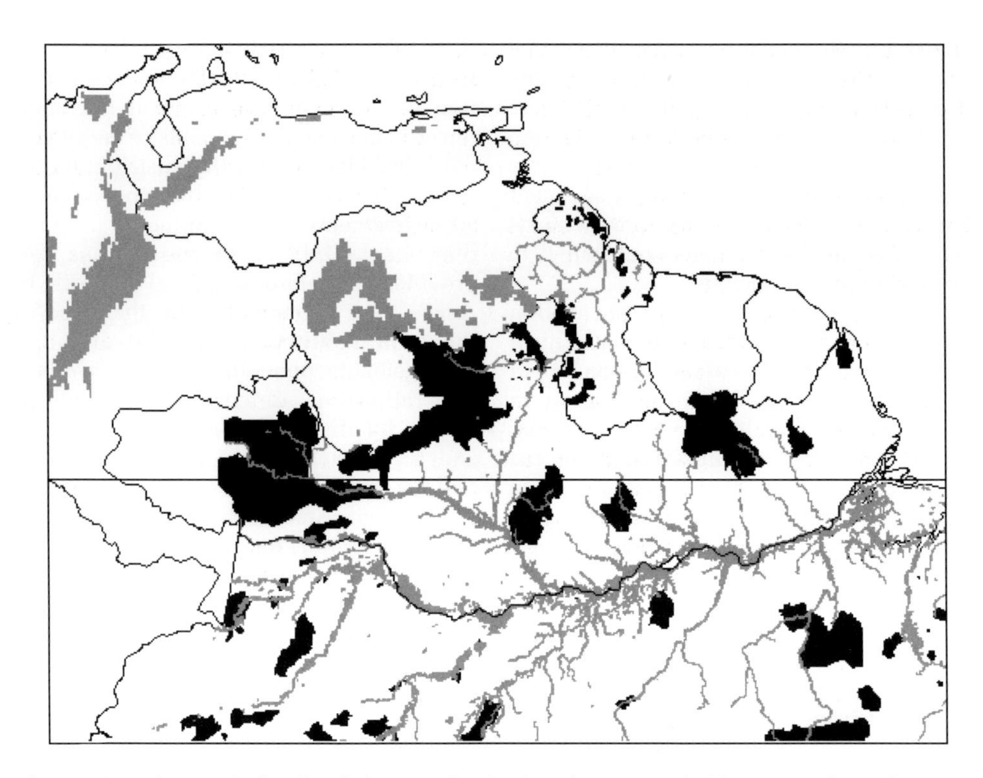

Fig. 8.23. Distribution of officially titled Amerindian lands in the Guiana Shield. Note overlap with commercial mining and timber areas in Figs 8.8 and 8.17. Solid grey areas represent elevations >1100 m asl.

2000) during the last century of the millennium. In Guyana, recent government efforts have worked to improve the legal status and distribution of Amerindian land titles and rational expansion of these exclusive zones, as has the Brazilian government through further tightening of environmental controls over widespread unlicensed mining operations in the north Amazon.

Notes

[1] Erickson (1999), in his rebuff of environmental determinism, indirectly argues for cooperation as one route to weathering periods of environmental hardship. The example of upland Andean Uru society 'moving in and out' with their Aymara and Quechua neighbours in response to environmental vicissitudes around Lake Titicaca emphasizes the need for support during these periods, though without knowledge of individual movement patterns, this too seems an oversimplified explanation. Small, isolated communities in the forested lowlands are less likely to have such support at hand. Particularly in *terra firme* areas not receiving the seasonal delivery of sediment from the Andes (see Chapter 2) nor the expanded livelihood opportunities afforded by settlement along savanna and coastal ecotones, the motivation to move back to a prior location would be diluted by new-found opportunities (for hunting, fishing, farming and escaping parasites and pests) and tempered by relations with neighbouring tribes.

[2] Considerable criticism has been levied against the methods used to assess many of the several dozen putative non-Clovis sites. Lithic material interpretation and identification, stratigraphic

assignment and radiocarbon dating methods and analyses rank high among the criticisms (Meltzer, 1995; Haynes, 1997; Reanier, 1997; Barse, 1997; Dillehay, 2000; Lavallée, 2000). Nonetheless, physical dating of several of the earliest sites appears to have measured up to these criticisms (e.g. Monte Verde in Chile: Dillehay (1997); Quabrada Jaguay, south coast of Peru: Sandweiss *et al.* (1998)). None of these are found in the lowland forests of Amazonia and point to a more specific early peopling of the Pacific coast and Western Cordillera of South America that is consistent with a coastal route to early southward migration.

[3] All before present (BP) dates mentioned in the text are uncalibrated (i.e. ^{14}C dates) unless indicated otherwise (calibrated).

[4] The very early date (>30 ka BP) of charcoal associated with the Pedra Pintura site in eastern Brazil remains highly isolated, particularly when considering the Clovis/pre-ceramic radiocarbon dates determined for associated site materials that are without doubt indicative of a human presence, such as rock painting pigment (12 ka to 6 ka BP), teeth (12.21±0.04 ^{14}C ka BP: Peyre *et al.* (2000)) and skeletal remains (9.67±0.14 ^{14}C ka BP: Peyre (1993)). This alone does not refute the possibility that people inhabited South America before (the end of) the LGM, but additional sites with similar dates would be needed to support the notion. Even presupposing the link between charcoal and human site use is correct, other concurrent site ages would be needed to indicate a colonization process that was more than an isolated waif dispersal event, perhaps along ENSO-enhanced equatorial currents, that failed to lead to widespread population growth.

[5] It is important to note that all ages are inferred by radiocarbon dating of material stratigraphically associated with the points. This approach to dating lithic artefacts, while commonplace, cannot be considered unequivocal proof of age and rarely receive widespread concurrence (Meltzer *et al.*, 1994; Barse, 1997; Haynes, 1997; Dillehay, 2000) due to uncertainty concerning stratigraphic interpretation and the role of human vs. natural depositional processes.

[6] Perry (2002) analysed starch grains adhered to ceramic griddle fragments found at Pozo Azul Norte I, an archaeological site near Puerto Ayacucho on the Orinoco where some of the earliest signs of human occupation have been assigned (Barse, 1990). Her results indicate ceramic griddles were used to process a wide variety of starchy flours, including those made from palms, arrowroot, maize and manioc.

[7] El Dorado is used today to generally describe a land of great untapped wealth, prospect or potential. The real El Dorado, however, had arguably already been discovered by the time that Cortez, Pizarro and de Quesada established Spain's colonial grip on the New World by dispossessing the Aztec (1521), Inca (1530) and Chibcha (1533) empires of the gold and silver they had accrued. It is not surprising that the flame of anticipation surrounding the location of another El Dorado in the Guiana Shield was fuelled by stories of a supreme ruler, a city and a lake. Previous successes of the Spanish conquistadores in Mexico, Peru and Colombia included a ruler (Moctezuma, Atahualpa and Zipa), a city (Tenochtitlan, Cuzco, Bogota), and a nearby lake (Texcoco, Titicaca, Guatavita). The existence of the city Manoa, called Meta by the Spanish, on the shores of a Lake Parima (or Parime), and ruled by an Amerindian emperor that dusted his body in gold each day, had considerable inspiration.

[8] It has been commonly suggested that to the native American perpetuating tales of wealth to be found just over the next ridge, river or mountain was a means of ensuring that the unwelcome attentions and demands of the doradistas did not stay long in their villages (Whitehead, 1997).

[9] The root of the word Essequibo has been suggested by many 19th-century authors as the name of one of Cristobal Columbus' lieutenants, Juan de Esquivel (=Esquibel=Essequibel=Essequibo) (Schomburgk, 1840; Dalton, 1855). Kirke (1898), Rodway (1911) and more recently, Benjamin (1982) suggest, however, that the name in fact is derived from an Arawak (Lokono) word. The etymological roots of many names of prominent physical and biological features of the Guiana Shield are equally difficult to identify and may never be resolved (e.g. greenheart).

[10] An interesting, although undoubtedly dramatized, account of the Spanish efforts to control the Orinoco and Trinidad from 1592 to 1813 is given by Naipaul (1969).

[11] A reduccion, or compound, was a Capuchin mission where Caribs were forcibly resettled

through the use of the military. In this way, the missions could maintain central control over the widely scattered Amerindians.

[12] Bulletwood trees do not continuously produce balata latex. The latex is stored in elongated cells wedged between the corky and vascular cambia that run parallel to the main trunk axis. When cut, the latex flows downwards from that portion of the cell chamber above the cut. These sections empty through the cut before coagulated latex closes the opening but do not immediately refill with latex. Studies conducted during the peak balata production period in the early 20th century indicate that these areas of bark would require 10–15 years before they were again able to produced latex in commercial quantities (Hohenkerk, 1919; Gonggryp, 1923).

[13] Basically, predicting the future gold unit price at time *t* and being paid (or paying) the difference between predicted and actual price at *t*. In this instance, the producer is selling the gold at the futures price, which was predicted higher than actual.

[14] Data source: Omai Gold Mines Limited (2000) and Bank of Guyana (2000). Gross revenues at Omai estimated based on annual average world gold price multiplied by total annual production (futures dealings may alter this relationship as would held reserves). Government share of this gross revenue based on value of 5% royalties, income taxes, duty and consumption taxes, and payments made to government corporations and departments from 1991 to 1999.

[15] In Trinidad and Tobago, where timber resources were scarce, bleeding of bulletwood trees for their balata latex was prohibited entirely as a consequence of its effects on tree survivorship (Gonggryp, 1923).

[16] Estimates of tree mortality associated with latex bleeding are based on the following:

	Guyana	Suriname	Venezuela	French Guiana
Minimum impact scenario				
Mortality, % (M)	4	8	90	8
Latex yield, kg (Y)	1.4	2.3	6.4	2.3
Maximum impact scenario				
Mortality, %	20	50	100	50
Latex yield, kg	0.9	1.6	2.3	1.6

Source: mortality – Davis 1933; yield – Gonggryp 1923, Brett 1917, Hohenkerk 1919.

where $\sum_{1}^{c}\left(\left[\frac{P_{yc}}{Y_{sc}}\right]\right)M_{sc}$

P being the declared production, *Y* being the yield, *M* being the mortality rate, *S* being the impact scenario, C being the country.

[17] Production from municipios (Pará: Prainha; Amazonas: Coari, Codajás, Itacoatiara, Manacapuru) spread across both the Amazon Downwarp and Guiana Shield are included here but accounted for no more than 9% of the amount attributed to shield forests over the period 1990–2002 (IBGE, 2003).

[18] Taken from censii of various national statistical offices, including IBGE (Brazil), DANE (Colombia), OCEI (Venezuela), and INSEE (French Guiana).

References

Abbonenc, E. (1951) *Aspects Démographiques de la Guyane Française. 1. Historique, 2. Démographie actuelle, 3. Avenir de la population.* Institut Pasteur de la Guyane et du Territoire de l'Inini, Cahors (FRA).

Abonnenc, E. (1952) Inventaire et distribution des sites archéologiques de Guyane française. *Journal de la Société des Américanistes* XLI, 43–63.

Algemeen Bureau voor de Statistiek (2000) *Statistich jaarboek van Suriname.* ABS, Paramaribo, Suriname.

Altena, C. and van Regteren, O. (1975) The marine Mollusca of Suriname (Dutch Guiana), Holocene and Recent. Part 3, Gastropoda and Cephalopoda. *Zoologie Verhandelening* 139, 1–104.

Anderson, A.B. and Ioris, E.M. (1992) The logic of extraction: resource management and income generation by extractive producers in the Amazon. In: Padoch, C. (ed.) *Conservation of Neotropical Forests – Working from Traditional Resource Use.* Columbia University Press, New York, pp. 175–199.

Anderson, A.B. and Jardim, M.A.G. (1989) Costs and benefits of floodplain forest management by rural inhabitants in the Amazon estuary: a case study of Açai palm production. In: Browder, J.O. (ed.) *Fragile Lands of Latin America: Strategies for Sustainable Development.* Westview Press, London, pp. 114–129.

Anderson, C.W. (1914) Report on experimental tappings of balata trees. *Journal of the British Guiana Board of Agriculture* 7, 34–39.

Anderson, S.D., Marques, F.L.T., Fernandes, M. and Nogueira, M. (1999) The usefulness of traditional technology for rural development: the case of tide energy near the mouth of the Amazon. In: Henderson, A. (ed.) *Várzea: Diversity, Development, and Conservation of Amazonia's Whitewater Floodplains.* NYBG, New York, pp. 329–344.

ANEEL (Agência Nacional de Energia Eléctrica) (2003) Capacidade de geração do Brasil (http://www.aneel.gov.br). ANEEL.

Anon. (1936) Notes – the tonka bean. *Agricultural Journal of British Guiana* 6, 115–116.

Anon. (1960) *Report on British Guiana for the Year 1960.* British Guiana Government, Georgetown.

Attenborough, D. (1956) *Zoo Quest to Guiana.* Lutterworth Press, London.

Bahn, P.G. (1993) 50,000-year-old Americans of Pedra Furada. *Nature* 362, 114.

Balée, W. (1989) The culture of Amazonian forests. *Advances in Economic Botany* 7, 1–21.

Balée, W. (1992) People of the fallow: a historical ecology of foraging in lowland South America. In: Redford, K.H. and Padoch, C. (eds) *Conservation of Neotropical Forests: Working from Traditional Resource Use.* Columbia University Press, New York, pp. 35–57.

Balée, W. (1998) *Advances in Historical Ecology.* Columbia University Press, New York.

Ball, S.H. (1934) Precious and semiprecious stones (gem minerals). In: Kiessilng, O.E. (ed.) *Mineral Yearbook 1934.* US Bureau of Mines, Washington, DC, pp. 1079–1096.

Bancroft, C.K. and Bayley, S.H. (1914) Report on the experimental bleeding of balata trees at Onderneeming. *Journal of the Board of Agriculture of British Guiana* 7, 20–33.

Bancroft, E. (1769) *An Essay on the Natural History of Guiana in South America.* T. Becket and P.A. De Hondt, London.

Bank of Guyana (1985) *Annual Report and Statement of Accounts 1990.* Bank of Guyana, Georgetown.

Bank of Guyana (2000) *Half Year Report and Statistical Bulletin 2000.* Bank of Guyana, Georgetown.

Barham, B.L. and Coomes, O.T. (1996) *Prosperity's Promise: the Amazon Rubber Boom and Distorted Economic Development.* Dellplain Latin American Studies. Westview Press, Boulder, Colorado.

Barse, W.P. (1990) Preceramic occupation in the Orinoco river valley. *Science* 250, 1388–1390.

Barse, W.P. (1997) Dating a paleoindian site in the Amazon in comparison with Clovis culture. *Science* 275, 1949–1950.

Barse, W.P. (2000) Ronquin, AMS dates, and the middle Orinoco sequence. *Interciencia* 25, 337–341.

Becker, M.D.C. and de Mello Filho, D.P. (1963) Ensaio de tipologia lítica brasileira. *Revista Museu Paulista* 14, 439–453.

Bednarik, R. (1994) A taphonomy of paleoart. *Antiquity* 68, 68–74.

Benjamin, J. (1982) The naming of the Essequibo river. *Archaeology and Anthropology (W. Roth Museum – Guyana)* 5, 29–66.

Bennett, K.D., Simonson, W.D. and Peglar, S.M. (1990) Fire and man in the post-glacial woodlands of eastern England. *Journal of Archaeological Science* 27, 635–642.

Benoist, R. (1931) Bois de la Guyane française. *Archiv Botanique 5, Memoir 1,* 1–229.

Berrangé, J.P. (1977) *The Geology of Southern Guyana, South America.* Overseas Memoir 4. HMSO, London.

Berrangé, J.P. and Johnson, R.L. (1972) A guide to the upper Essequibo river, Guyana. *The Geographical Journal* 138, 41–52.

Binford, M.W., Kolata, A.L., Brenner, M., Janusek, J.W., Seddon, M.T. and Curtis, J.H. (1997) Climate variation and the rise and fall of an Andean civilization. *Quaternary Research* 47, 235–248.

Bird, M.I., Ayliffe, L.K., Fifield, L.K., Turney, C.S.M., Cresswell, R.G., Barrows, T.T. and David, B. (1999) Radiocarbon dating of 'old' charcoal using a wet oxidation-stepped combustion procedure. *Radiocarbon* 41, 127–140.

Bisnauth, D. (2000) *The Settlement of Indians in Guyana 1890–1930.* Peepal Tree, Leeds, UK.

Blair, D. (1980) Notes of an expedition from Georgetown to the gold diggings on the borders of Venezuela (from original notes of 1857). *Archaeology and Anthropology (W. Roth Museum – Guyana)* 3, 1–63.

Boomert, A. (1978) Prehistoric habitation mounds in the Canje river area? *Archaeology and Anthropology (W. Roth Museum – Guyana)* 1, 44–51.

Boomert, A. (1979) The prehistoric stone axes from the Guianas: a typological classification. *Archaeology and Anthropology (W. Roth Museum – Guyana)* 2, 99–124.

Boomert, A. (1980a) Hertenrits: an arauquinoid complex in north west Suriname. Part 1. *Archaeology and Anthropology (W. Roth Museum – Guyana)* 3, 68–103.

Boomert, A. (1980b) The Sipaliwini complex of Surinam: a summary. *Nieuwe West-Indische Gids* 54, 94–107.

Boomert, A. (1981) The Taruma phase of southern Suriname. *Archaeology and Anthropology (W. Roth Museum – Guyana)* 4, 104–157.

Boomert, A. (1987) Gifts of the Amazons: 'green stone' pendants and beads as items of ceremonial exchange in Amazonia and the Caribbean. *Antropologica* 67, 33–54.

Boomert, A. (1993) The Barbakoeba archaeological complex of northeast Suriname. *Tijdschrift voor Surinaamse Taalkunde, Letterkunde, Cultuur en Geschiedenis* 12, 198–222.

Brace, C.L., Nelson, A.R., Seguchi, N., Oe, H., Sering, L., Qifeng, P., Yongyi, L. and Tumen, D. (2001) Old world sources of the first new world human inhabitants: a comparative craniofacial view. *Proceedings of the National Academy of Sciences USA* 98, 10017–10022.

Breedlove, D.E. (1981) *Flora of Chiapas. Part 1: Introduction to the Flora of Chiapas.* California Academy of Sciences, San Francisco, California.

Brett, H. (1916) *Venezuelan Production of Balata.* British Consulate, La Guaira, Venezuela.

Brett, W.H. (1852) *The Indian Tribes of Guiana: Their Condition and Habits, with Researches into their Past History, Superstitions, Legends, Antiquities, Languages, etc.* Bell and Daldy, London.

Briceño, H.O. (1984) Genesis de yacimientos minerales venezolanos II – placeres diamantíferos de San Salvador de Paúl. *Acta Cientifica Venezolana* 36, 154–158.

British Guiana Balata Committee (1912) *Report, Minutes of the Proceedings and Report of Evidence.* BG Governors Office, Georgetown, Guyana.

British Guiana Forest Department (1935a) *Forestry in British Guiana. Supplementary Statement Prepared by the Forest Department for Presentation to the Fourth British Empire Forestry Conference (South Africa), 1935.* BGFD, Georgetown, Guyana.

British Guiana Forest Department (1935b) *Timber Supply, Consumption, and Marketing in British Guiana. Special Statement Prepared for Presentationa to the 4th British Empire Forestry Conference (South Africa), 1935.* BGFD, Georgetown, Guyana.

British Guiana Forest Department (1940–1965) *Annual Report of the Forest Department.* Department of Lands and Mines, Georgetown, Guyana.

British Guiana Geological Survey Department (1957–1963) *Annual Report on the Geological Survey Department.* Geological Survey Department, Georgetown, Guyana.

British Guiana Lands and Mines Department (1910–1945) *Annual Report on the Lands and Mines Department.* Lands and Mines Department, Georgetown, Guyana.

British Guiana Lands and Mines Department (1920) *Report of the Land and Mines Department for the Year 1920.* Lands and Mines Department, Georgetown, Guyana.

Brown, C.B. (1876) Canoe and camp life in British Guiana. *Journal of the Anthropological Institute of Great Britain and Ireland* 2, 254–257.

Bruleaux, A.-M. (1989) Deux productions passées de la forêt Guyanaise. *Bois et Forêts des Tropiques* 219, 99–113.

Bryan, A. (1973) Paleoenvironments and cultural diversity in the late Pleistocene South America. *Quaternary Research* 3, 237–256.

Bubberman, F.C. (1973) Rotstekeningen in de Sipaliwinisavanne. *Nieuwe West-Indische Gids* 49, 129–142.

Bubberman, F.C. and Vink, A.T. (1966) Ditch-blasting as a method of opening up the tropical swamp forest in Suriname. *Netherlands Journal of Agricultural Science* 14, 10–15.
Bunker, S.G. (1985) *Underdeveloping the Amazon – Extraction, Unequal Exchange, and the Failure of the Modern State.* University of Illinois Press, Chicago, Illinois.
Buschbacher, R.J. (1986) Tropical deforestation and pasture development. *Bioscience* 36, 22–28.
CAMBIOR (2002) *Rosebel Project Technical Report.* Cambior, Paramaribo.
Carter, J.E.L. (1943) An account of some recent excavations at Seba, British Guiana. *American Antiquity* 9, 89–99.
Central Intelligence Agency (2000) *The World Factbook.* CIA, Washington, DC.
Châtelain, G. (1935) La balata et la gomme de balata en Guyane française. *l'Agronomie Coloniale* 209, 148–151.
Chernela, J.M. (1993) *The Wanano Indian of the Brazilian Amazon: A Sense of Space.* University of Texas Press, Austin, Texas.
Chernela, J.M. (1998) Missionary activity and indian labor in the upper Rio Negro of Brazil, 1680–1980: a historical–ecological approach. In: Balée, W. (ed.) *Advances in Historical Ecology.* Columbia University Press, New York, pp. 313–333.
Chippendale, C. and Taçon, P.S.C. (1998) *The Archaeology of Rock Art.* Cambridge University Press, London.
Colchester, M. (1997a) Ecología social de los Sanema. In: Rosales, J. (ed.) *Ecología de la Cuenca del Río Caura, Venezuela II. Estudios Especiales.* Refolit, C.A., Caracas, pp. 111–140.
Colchester, M. (1997b) *Guyana: Fragile Frontier.* Earthscan Publications, London.
Collis, L. (1965) *Soldier in Paradise: the Life of Captain John Stedman, 1744–1797.* Michael Joseph, London.
Coomes, O.T. (1995) A century of rain forest use in western Amazonia: lessons for extraction-based conservation of tropical forest resources. *Forest and Conservation History* 39, 108–120.
Corothie, H. (1948) *Maderas de Venezuela.* Impreta Nacional, Caracas.
Cruxent, J. (1947) Pinturas rupestres de El Carmen, en el Rio Parguaza, Estado Bolivar, Venezuela. *Acta Venezolana* 11, 1–60.
Cruxent, J. (1950) Archaeology of Cotúa Island, Amazonas Territory, Venezuela. *American Antiquity* 16, 10–16.
Cruxent, J. (1972) Tupuquén: un yacimiento con lítica de tipo Paleo-Indio. *Acta Cientifica Venezolana* 23, 1–17.
Curtin, P.B. (1969) *The Atlantic Slave Trade: A Census.* University of Wisconsin, Madison, Wisconsin.
Curtis, J.H., Hodell, D.A. and Brenner, M. (1996) Climate variability on the Yucatan peninsula (Mexico) during the past 3500 years, and implications for Maya cultural evolution. *Quaternary Research* 46, 37–47.
CVG-TECMIN (1991) *Proyecto Inventario de los Recursos Naturales de la Región Guayana.* CVG-TECMIN, Ciudad Bolivar, Venezuela, pp. 1–15.
da Costa, E.V. (1994) *Crowns of Glory, Tears of Blood – the Demerara Slave Rebellion of 1823.* Oxford University Press, Oxford, UK.
Dalton, H.G. (1855) *The History of British Guiana.* Longman, Brown and Green, London.
Davis, T.A.W. (1933) *Report on the Balata Industry with Special Reference to the Rupununi District.* British Guiana Legislative Council No. 4. Georgetown, Demerara, British Guiana.
Dawkins, H.C. and Philip, M.S. (1998) *Tropical Moist Forest Silviculture and Management – a History of Success and Failure.* CAB International, Wallingford, UK.
de Lumley, H., de Lumley, M.A., Beltrão, M.C., Yokoyama, Y., Labeyrie, J., Danon, J., Delibrias, G., Falguers, C. and Bischoff, J. (1988) Decouverte d'outils taillès associes a des faunes du Pleistocene moyen dans la Toca da Esperança, Etat de Bahía, Bresil. *Comptes Rendus de l'Académie des Sciences de Paris* 306, 307–317.
de Groot, S.W. (1977) *From Isolation Towards Integration – the Surinam Maroons and their Colonial Rulers.* Martinus Nijhoff, The Hague, The Netherlands.
DeBoer, W.R., Kintigh, K. and Rostoker, A.G. (1996) Ceramic seriation and site reoccupation in lowland South America. *Latin American Antiquity* 7, 263–278.
Denevan, W.D. (1970) Aboriginal drained field cultivation in the Americas. *Science* 169, 647–654.
Denevan, W.M. (1992) The pristine myth: the landscape of the Americas in 1492. *Annals of the Association of American Geographers* 82, 369–385.
Departmento Nacional de Produção Mineral (1999) *Indicadores da Produção Mineral 1998.* Divisão de Economia Mineral, Brasilia.
Departmento Nacional de Produção Mineral (2001) *Anuário Mineral Brasiliero 2001.* DNPM, Brasilia.

Detienne, P. and Chanson, B. (1996) L'éventail de la densité du bois des feuillus. *Bois et Forêt des Tropiques* 250, 19–30.

Dew, E. (1978) *The Difficult Flowering of Suriname – Ethnicity and Politics in a Plural Society.* Martinus Nijhoff, The Hague, The Netherlands.

Dillehay, T.D. (1997) *Monte Verde: A Late Pleistocene Settlement in Chile,* Vol. 2: *the Archaeological Context and Interpretation.* Smithsonian Institution Press, Washington, DC.

Dillehay, T.D. (2000) *The Settlement of the Americas – a New Prehistory.* Basic Books, New York.

Dillehay, T.D. and Kolata, A.L. (2004) Long-term human response to uncertain environmental conditions in the Andes. *Proceedings of the National Academy of Sciences* 101, 4325–4330.

Dixon, E.J. (1999) *Bones, Boats and Bison: Archaeology and the First Colonization of Eestern North America.* University of New Mexico Press, Albuquerque, New Mexico.

DNPM (Departmento Nacional de Produção Mineral) (1995–2001) *Anuário Mineral Brasileiro.* DNPM, Brasilia.

Drude de Lacerda, L. (2003) Updating global Hg emissions from small-scale gold mining and assessing its environmental impacts. *Environmental Geology* 42, 308–314.

Dubelaar, C.N. (1976) Jeannine Sujo Volsky, el estudio del arte rupestre en Venezuela. *Journal de la Société des Américanistes* 52, 322–327.

Dubelaar, C.N. (1981) Petroglyphs in Suriname: a survey. *Archaeology and Anthropology (W. Roth Museum – Guyana)* 4, 64–80.

Dubelaar, C.N. (1986) *The Petroglyphs in the Guianas and Adjacent Areas of Brazil and Venezuela: An Inventory.* Monumenta Archaeologica 12. Institute of Archaeology, University of California, Los Angeles, California.

Dubelaar, C.N. and Berrangé, J.P. (1979) Some recent petroglyph finds in southern Guyana. *Archaeology and Anthropology (W. Roth Museum – Guyana)* 2, 61–78.

Ducke, A. (1943) The most important woods of the Amazon valley. *Tropical Woods* 74, 1–15.

Dupouy, W. (1956) Dos piezas de tipo paleolítico de la Gran Sabana, Venezuela. *Boletín del Museo de Ciencias Naturales, Caracas* 2/3, 95–102.

Dupouy, W. (1960) Tres puntas líticas de tipo paleo-indio de la Paragua, Estado Bolivar, Venezuela. *Boletín del Museo de Ciencias Naturales, Caracas* 5/6, 7–14.

Eden, M.J., Bray, W., Herrera, L. and McEwan, C. (1984) Terra preta soils and their archaeological context in the Caquetá basin of southeast Colombia. *American Antiquity* 49, 125–140.

Edmondson, J.R. (1949) Reaction of woods from South America and Caribbean areas to marine borers in Hawaiian waters. *Caribbean Forester* 10, 37–41.

Edmundson, G. (1901) The Dutch in western Guiana. *English Historical Review* 16, 640–675.

Edmundson, G. (1904a) The Dutch on the Amazon and Negro in the 17th century. Part 1. Dutch trade on the Amazon. *English Historical Review* 18, 642–663.

Edmundson, G. (1904b) The Dutch on the Amazon and Negro in the 17th century. Part 2. Dutch trade in the Rio Negro basin. *English Historical Review* 19, 1–25.

Erickson, C.L. (1999) Neo-environmental determinism and agrarian 'collapse' in Andean prehistory. *American Antiquity* 73, 634–642.

Erickson, C.L. (2000) An artificial landscape-scale fishery in the Bolivian Amazon. *Nature* 408, 190–193.

Evans, C. and Meggers, B.J. (1960) *Archeological Investigations in British Guiana.* Bulletin of the Bureau of American Ethnology, 177. Smithsonian Institution Press, Washington, DC.

Evans, C. and Meggers, B.J. (1968) *Archeological Investigations on the Rio Napo, Eastern Ecuador.* Smithsonian Contributions to Anthropology, 6. Smithsonian Institution Press, Washington, DC.

Evans, C., Meggers, B.J. and Cruxent, J. (1959) Preliminary results of archeological investigations along the Orinoco and Ventuari rivers, Venezuela. *Actas del XXXIII Congreso Internacional de Americanistas,* San José, Costa Rica.

Fairbridge, R.W. (1976) Shellfish-eating Indians in coastal Brazil. *Science* 191, 353–359.

Fanshawe, D.B. (1950) *Forest Products of British Guiana II. Minor Forest Products.* Forest Department, British Guiana, Georgetown, Guyana.

FAOSTAT (2003) *FAOSTAT – Forestry Data* (http://www.apps.fao.org/faostat). FAO, Rome.

Farabee, W.C. (1918) *The Central Arawaks.* Anthropological Publications, 10. The University Museum, Philadelphia.

Fimbel, R.A., Grajal, A. and Robinson, J.G. (eds) (2001) *The Cutting Edge – Conserving Wildlife in Logged Tropical Forests.* Biology and Resource Management Series. Columbia University Press, New York.

Findley, W.P.K. (1938) The natural resistance to decay of some empire timbers. *Empire Forestry Journal* 17, 249–259.

Forte, J. (1996) *Thinking about Amerindians.* Janette Forte, Georgetown, Guyana.

Friedman, J. (1969) The changing pattern of urbanization in Venezuela. In: Rodwin, L. (ed.) *Planning Urban Growth and Regional Development: the Experience of the Guayana Program of Venezuela.* MIT Press, Cambridge, Massachusetts, USA, pp. 40–59.

Frikel, P. (1961) Fases culturais e aculturaçao intertribal no Tumucumaque. *Boletin de Museu Paraense Emilio Goeldi. Nova série-Antropologia* 16.

Frikel, P. (1969) Tradition und archäologie im Tumuk-Jumak/Nordbrasilien. *Zeitschrift für Ethnologie* 94, 103–130.

Frouté, O. (2000) Les Dom: une population encore jeune, mais de moins en moins. *INSEE Premiere* 747, 1–4.

Furse, R.D., Troup, R.S. and Ainslie, J.R. (1924) *Report of a Committee of the Empire Forestry Conference in Canada Appointed to Enquire into the Present Forestry Situation in British Guiana, and to Make Proposals for Improvements.* British Guiana Combined Court – First Special Session, Georgetown.

Gaspar, M.D. (1998) Considerations of the sambaquis of the Brazilian coast. *Antiquity* 277, 592–615.

Gassón, R.A. (2002) Orinoquia: the archaeology of the Orinoco river basin. *Journal of World Prehistory* 16, 237–311.

Geijskes, D.C. (1960) History of archeological investigations in Surinam. *Berichten van de Rijksdienst voor het Oudheidkundig Bodemonderzoek* 10, 70–77.

Gibbs, A.K. and Barron, C.N. (1993) *The Geology of the Guiana Shield.* Oxford University Press, Oxford, UK.

Gill, R.B. (2000) *The Great Maya Droughts: Water, Life and Death.* University of New Mexico Press, Albuquerque, New Mexico.

Glaser, B., Guggenberger, G., Haumaier, L. and Zech, W. (2001) Persistence of soil organic matter in archaeological soils (terra preta) of the Brazilian Amazon region. In: Rees, R.M., Ball, B.C., Campbell, C.D. and Watson, C.A. (eds) *Sustainable Management of Soil Organic Matter.* CAB International, Wallingford, UK, pp. 190–194.

Gnecco, C. and Mora, S. (1997) Late Pleistocene/Early Holocene tropical forest occupations at San Isidro and Peña Roja, Colombia. *Antiquity* 71, 683–690.

Goldfarb, R.J., Groves, D.I. and Gardoll, S. (2001) Orogenic gold and geologic time: a global synthesis. *Ore Geology Reviews* 18, 1–75.

Gómez-Pompa, A. (1987) On Maya silviculture. *Estudios Mexicanos* 3, 1–19.

Gómez-Pompa, A., Flores, J.S. and Fernandez, M.A. (1990) The sacred groves of the Maya. *Latin American Antiquity* 1, 247–257.

Gonggryp, J.W. (1923) On bleeding balata. *The Official Gazette* 2014, 1–12.

Goulding, M., Carvalho, M.L. and Ferreira, E.G. (1988) *Rio Negro: Rich Life in Poor Water.* SPB Academic, The Hague, The Netherlands.

Goulding, M., Smith, N.J.H. and Mahar, D.J. (1996) *Floods of Fortune – Ecology and Economy Along the Amazon.* Columbia University Press, New York.

Gray, J.E., Labson, V.F., Weaver, J.N. and Krabbenhoft, D.P. (2002) Mercury and methylmercury contamination related to artisanal gold mining, Suriname. *Geophysical Research Letters* 29, 2105–2112.

Greer, J. (1995) El arte rupestre del sur de Venezuela: una síntesis. *Boletín de la Sociedad de Investigación del Arte Rupestre de Bolivia SIARB* 11, 38–52.

Guidon, N. and Delibrias, G. (1986) Carbon-14 dates point to man in the Americas 32,000 years ago. *Nature* 321, 769–771.

Guimarães, J.R.D., Meili, M., Hylander, L.D., de Castro e Silva, E., Roulet, M., Mauro, J.B.N. and Alves de Lemos, R. (2000) Mercury net methylation in five tropical flood plain regions of Brazil: high in the root zone of floating macrophyte mats but low in surface sediments and flooded soils. *Science of the Total Environment* 261, 99–107.

Gumilla, P.J. (1741) *El Orinoco Ilustrado y Defendido, Historia Natural, Civil y Geographica de Este Gran Rio y de sus Caudalosos Vertientes.* Manuel Fernandez, Madrid.

Guppy, N. (1961) *Wai-Wai – Through the Forests North of the Amazon.* John Murray, London.

Guyana Bureau of Statistics (1990–2000) *Guyana Statistical Bulletin.* Bureau of Statistics, Georgetown, Guyana.

Guyana Forest Department (1966–1976) *Annual Report of the Guyana Forest Department.* Guyana Forest Department, Georgetown, Guyana.

Guyana Forestry Commission (1980–2002) *Annual Report of the Guyana Forestry Commission.* GFC, Georgetown, Guyana.
Guyana Forestry Commission (1985) *An Introduction to the Balata Industry in Guyana.* GFC, Georgetown, Guyana.
Guyana Geological Survey Department (1964–1970) *Annual Report of the Geological Survey Department.* GSD, Ministry of Agriculture and Natural Resources, Georgetown, Guyana.
Guyana Geological Surveys and Mines Department (1971–1979) *Annual Report of the Geological Surveys and Mines Department.* GSMD, Ministry of Energy and Natural Resources, Georgetown, Guyana.
Guyana Geological Surveys and Mines Department (1974) *Annual Report 1974.* GSMD, Ministry of Energy and Natural Resources, Georgetown, Guyana.
Guyana Geology and Mines Commission (1980–2002) *Annual Report of the GGMC.* GGMC, Georgetown, Guyana.
Guyana Geology and Mines Commission (1994) *Annual Report of the GGMC.* GGMC, Georgetown, Guyana.
Hammond, D.S. (1991) Restoration of tropical dry forest after milpa agriculture, Chiapas, Mexico. PhD thesis. School of Environmental Sciences, University of East Anglia, Norwich, UK.
Hanif, M. (1967) Petroglyphs in the Rupununi. *Timehri* 43, 19–27.
Harcourt, R. (1613) *A Relation of a Voyage to Guiana.* John Beale, London.
Harlow, V.T. (1925) *Colonising Expeditions to the West Indies and Guiana, 1623–1667.* Hakluyt Society, London.
Haug, G.H., Günther, D., Peterson, L.C., Sigman, D.M. and Hughen, K.A. (2003) Climate and collapse of Maya civilization. *Science* 299, 1731–1735.
Haynes, C.V.J. (1997) Dating a paleoindian site in the Amazon in comparison with Clovis culture. *Science* 275, 1948.
Hecht, S.B. (1985) Environment, development and politics: capital accumulation, and the livestock sector in Amazonia. *World Development* 13, 663–684.
Heckenberger, M.J., Neves, E.G. and Peterson, J.B. (1998) De onde surgem os modelos? As origens e expansões Tupi na Amazônia central. *Revista de Antropologia* 41, 70–96.
Heckenberger, M.J., Peterson, L.C. and Neves, E.G. (1999) Village size and permanence in Amazonia: two archaeological examples from Brazil. *Latin American Antiquity* 10, 353–376.
Heckenberger, M.J., Kuikuro, A., Kuikuro, U.T., Russell, J.C., Schmidt, M., Fausto, C. and Franchetto, B. (2003) Amazonia 1492: pristine forest or cultural parkland? *Science* 301, 1710–1714.
Heemskerk, M. (2001) Do international commodity prices drive natural resource booms? An empirical analysis of small-scale gold mining in Suriname. *Ecological Economics* 39, 295–308.
Hemming, J. (1978a) *Red Gold – the Conquest of the Brazilian Indians 1500–1760.* Harvard University Press, Cambridge, Massachusetts.
Hemming, J. (1978b) *The Search for El Dorado.* Macmillan, London.
Henderson, A., Galeano, G. and Bernal, R. (1995) *Field Guide to the Palms of the Americas.* Princeton University Press, Princeton, New Jersey.
Henderson, G. (1952) Stone circles and tiger's lairs. *Timehri* 31, 62–66.
Herrera, L. (1987) Apuntes sobre el estado de la investigación arqueológica en la Amazonia Colombiana. *Boletín de Antropología (Universidad de Antioquia)* 6, 21–61.
Higman, B. (1984) *Slave Populations of the British Caribbean, 1807–1834.* John Hopkins University Press, Baltimore, Maryland.
Hilbert, P.P. and Hilbert, K. (1980) Resultados preliminares da pesquisa arqueologica nos rios Nhamunda e Trombetas, baixo Amazonas. *Boletin de Museu Paraense Emilio Goeldi. Nova série-Antropologia* 75, 1–11.
Hodell, D.A., Brenner, M., Curtis, J.H. and Guilderson, T. (2001) Solar forcing of drought frequency in the Maya lowlands. *Science* 292, 1367–1370.
Hohenkerk, L.S. (1919) Experiments to determine the best method of bleeding balata trees. *The Official Gazette* 11343, 1–8.
Hohenkerk, L.S. (1922) A review of the timber industry of British Guiana. *Journal of the Board of Agriculture of British Guiana* 15, 2–22.
Holmes, W. and Campbell, W.H. (1858) Report of an expedition undertaken to explore a route by the rivers Waini, Barama and Cuyuni to the gold fields of Caratal, and thence by Upata to the river Orinoco. *Proceedings of the Royal Geographical Society* 2, 154–157.
Homma, A.K.O. (1992) The dynamics of extraction in Amazonia: a historical perspective. *Advances in Economic Botany* 9, 23–31.

Horn, E.F. (1948) Teredo resistant timbers of the Amazon valley. *Tropical Woods* 93, 35–40.

Howard, G.D. (1947) *Prehistoric Ceramic Styles of Lowland South America, their Distribution and History.* Yale University Publications in Anthropology. Yale University Press, New Haven, Connecticut.

Hughes, J.H. (1946) Forest resources. In: Roth, V. (ed.) *Handbook of Natural Resources of British Guiana.* Daily Chronicle Ltd, Georgetown, Guyana, pp. 50–111.

Hurault, J.M. (1972) *Français et Indiens en Guyane 1604–1972.* 1018, Union Générale D'Éditions, Paris.

Hurault, J.M., Frenay, P. and Raoux, Y. (1963) Petroglyphs et assemblages de pierres dans le Sud-Est de la Guyane française. *Journal de la Société des Américanistes* 52, 133–156.

IBGE (1985) *Produção Extrativa Vegetal – 1984.* Vol. 11. IBGE, Rio de Janeiro.

IBGE (2003a) *Estatísticas do Século XX.* IBGE, Rio de Janeiro.

IBGE (2003b) *SIDRA – Produção Extrativa Vegetal.* IBGE, Rio de Janeiro.

Im Thurn, E.F. (1883) *Among the Indians of Guiana.* Kegan Paul, Trench and Co., London.

Im Thurn, E.F. (1884) Artificial mound behind plantation Leonora. *Timehri* 3, 373.

ITTO (2003) *Annual Review and Assessment of the World Timber Situation.* ITTO, Yokohama, Japan.

Janse, A.J.A. and Sheahan, P.A. (1995) Catalogue of world wide diamond and kimberlite occurrences: a selective and annotative approach. *Journal of Geochemical Exploration* 53, 73–111.

Jenman, G.S. (1885) Balata and the balata industry. *Timehri* IV, 153–233.

Kennedy, L.M. and Horn, S.P. (1997) Prehistoric maize cultivation at the La Selva biological station, Costa Rica. *Biotropica* 29, 368–370.

Khudabux, M.R., Maat, G.J.R. and Versteeg, A.H. (1991) The remains of prehistoric amerindians of the 'Tingi Holo Ridge' in Suriname: a physical anthropological investigation of the 'Versteeg collection'. *Comptes Rendus du XIIéme CIAC,* Cayenne, F. Guiana, AIAC.

Kirke, H. (1898) *Twenty-five Years in British Guiana.* Sampson, Low, Marston and Co., London.

Koch-Grünberg, T. (1909) *Zwei Jahre unter den Indianern: Reisen in Nordwest-Brasilien 1903/1905.* E. Wasmuth, Berlin.

Lathrap, D.W. (1968) Aboriginal occupation and changes in river channel in the central Ucayali, Peru. *American Antiquity* 33, 62–79.

Lathrap, D.W. (1973) The antiquity and importance of long-distance trade relationships in the moist tropics of pre-Columbian South America. *World Archaeology* 5, 170–186.

Lavallée, D. (2000) *The First South Americans – the Peopling of a Continent from the Earliest Evidence to High Culture.* University of Utah Press, Salt Lake City, Utah.

Lawrence, S. and Coster, P. (1965) Petroleum potential of offshore Guyana. *Oil and Gas Journal* 67, 1–25.

Lehmann, J., Pereira da Silva Jr, J., Steiner, C., Nehls, T., Zech, W. and Glaser, B. (2003) Nutrient availability and leaching in an archaeological Anthrosol and a Ferrasol of the central Amazon basin: fertilizer, manure and charcoal amendments. *Plant and Soil* 249, 343–357.

Lescure, J.P. and de Castro, A. (1990) *L'extractivisme en Amazonie Centrale. Aperçu des Aspects Economiques et Botaniques.* ORSTOM, Cayenne.

Lima da Costa, M. and Kern, D.C. (1999) Geochemical signatures of tropical soils with archaeological black earth in the Amazon, Brazil. *Journal of Geochemical Exploration* 66, 369–385.

Lorimer, J. (1979) The English tobacco trade in Trinidad and Guiana, 1590–1617. In: Hair, P.E.H. (ed.) *The Westward Enterprise: English Activities in Ireland, the Atlantic, and America, 1480–1650.* Wayne State University Press, Detroit, Michigan, pp. 124–150.

Lowenstein, F.W. (1973) Some considerations of biological adaptation by aboriginal man to the tropical rain forest. In: Meggars, B.J., Ayensu, E.S. and Duckworth, W.D. (eds) *Tropical Forest Ecosystems in Africa and South America: a Comparative Review.* Smithsonian Institution Press, Washington, DC, pp. 293–310.

Macdonald, J.R. (1968) *A Guide to Mineral Exploration in Guyana.* Geological Survey of Guyana, Ministry of Agriculture and Natural Resources, Georgetown, Guyana.

Malfoy, L.J. (1989) De la marine royale aux meubles dits 'de port' ... ou la 'petite' histoire des bois de Guyane au XVIIIe siécle. *Bois et Forêts des Tropiques* 220, 99–104.

Mallery, G. (1893) Picture writing of the American Indians. *Tenth Annual Report of the Bureau of American Ethnology.* Smithsonian Institution, Washington, DC.

Mann, C.C. (2000a) Earthmovers in the Amazon. *Science* 287, 786–789.

Mann, C.C. (2000b) The good earth: did people improve the Amazon basin. *Science* 287, 788.

Matheny, R.T. (1976) Maya lowland hydraulic systems. *Science* 193, 639–646.

McCann, J.M., Wood, W.I. and Meyer, D.W. (2001) Organic matter and anthrosols in Amazonia: interpreting the Amerindian legacy. In: Rees, R.M., Ball, B.C., Campbell, C.D. and Watson, C.A. (eds)

Sustainable Management of Soil Organic Matter. CAB International, Wallingford, UK, pp. 180–189.

McDonald, J.S. (1969) Migration and the population of Ciudad Guayana. In: Rodwin, L. (ed.) *Planning Urban Growth and Regional Development: the Experience of the Guayana Program of Venezuela.* MIT Press, Cambridge, Massachusetts, pp. 109–125.

McKenna, S.J. (1959) Discovery of rock carvings. *Timehri* 22, 17.

McTurk, M. (1882) Notes on the forests of British Guiana. *Timehri* 1, 173–215.

Meggers, B.J. (1954) Environmental limitation on the development of culture. *American Anthropologist* 56, 801–824.

Meggers, B.J. (1973) Some problems of cultural adaptation in Amazonia, with emphasis on the pre-European period. In: Meggars, B.J., Ayensu, E.S. and Duckworth, W.D. (eds) *Tropical Forest Ecosystems in Africa and South America: a Comparative Review.* Smithsonian Institution Press, Washington, DC, pp. 311–320.

Meggers, B.J. (1977) Vegetational fluctuation and prehistoric cultural adaptation in Amazonia: some tentative correlations. *World Archaeology* 8, 287–303.

Meggers, B.J. (1984) The indigenous peoples of Amazonia, their cultures, land use patterns and effects on the landscape and biota. In: Sioli, H. (ed.) *The Amazon – Limnology and Landscape Ecology of a Mighty River and its Basin.* Dr. W. Junk Publishers, Dordrecht, The Netherlands, pp. 627–647.

Meggers, B.J. (1994) Archeological evidence for the impact of mega-Niño events on Amazonia during the past two millenia. *Climatic Change* 28, 321–338.

Meggers, B.J. (1995) Archaeological perspectives on the potential of Amazonia for intensive exploitation. In: Nishizawa, T. and Uitto, J.I. (eds) *The Fragile Tropics of Latin America.* United Nations University Press, New York, pp. 68–93.

Meggers, B.J. (1996) *Amazonia – Man and Culture in a Counterfeit Paradise.* Revised edition. Smithsonian Institution Press, London.

Meggers, B.J. (2003) Revisiting Amazonia circa 1492. *Science* 302, 2067–2068.

Meggers, B.J. and Evans, C. (1957) *Archeological Investigations at the Mouth of the Amazon.* Bulletin of the Bureau of American Ethnology, 167. Smithsonian Institution, Washington, DC.

Melack, J.M. and Forsberg, B.R. (2001) Biogeochemistry of the Amazon floodplain lakes and associated wetlands. In: McClain, M.E., Victoria, R.L. and Richey, J.E. (eds) *The Biogeochemistry of the Amazon Basin.* Oxford University Press, Oxford, UK, pp. 235–274.

Mélières, M.-A., Pourchet, M., Charles-Dominique, P. and Gaucher, P. (2003) Mercury in canopy leaves of French Guiana in remote areas. *Science of the Total Environment* 311, 261–267.

Meltzer, D. (1995) Perspectives on the peopling of the New World. *Annual Review of Anthropology* 24, 21–45.

Meltzer, D., Adovasio, J. and Dillehay, T.D. (1994) On a Pleistocene human occupation at Pedra Furada, Brazil. *Antiquity* 68, 695–714.

Miller, J.R., Lechler, P.J. and Bridge, G. (2003) Mercury contamination of alluvial sediments within the Essequibo and Mazaruni river basins, Guyana. *Water, Air and Soil Pollution* 148, 139–166.

Miranda, M., Blanco-Uribe, A., Hernández, L. and Ochoa, J.G. (1998) *All that Glitter is Not Gold: Balancing Conservation and Development in Venezuela's Frontier Forests.* WRI, Washington, DC.

Moran, E. (1982) *Developing the Amazon.* University of Indiana Press, Bloomington, Indiana.

Naipaul, V.S. (1969) *The Loss of El Dorado.* André Deutsch, London.

Narváez, A. and Stauffer, F. (1999) Products derived from palms at the Puerto Ayacucho markets in Amazonas state, Venezuela. *Palms* 43, 122–129.

Nettle, D. (1999) Linguistic diversity of the Americas can be reconciled with a recent colonization. *Proceedings of the National Academy of Sciences USA* 96, 3325–3329.

Nichols, J. (1990) Linguistic diversity and the first settlement of the New World. *Language* 66, 475–521.

Nimuendajú, C. (1950) Reconhecimento dos rios Içana, Ayari, e Uaupés: relat'orio apresentado ão Serviço de Proteção äos Índios do Amazonas de Acre, 1927. *Journal de la Société des Américanistes* 39, 125–182.

Novoa, P. and Costas, F. (1998) *Arte Rupestre del Estado Barinas.* Editorial Litho-Centro, Mérida.

Nriagu, J.O. and Pacyna, J.M. (1988) Quantitative assessment of worldwide contamination of air, water and soils by trace metals. *Nature* 333, 134–139.

Nriagu, J.O., Pfeiffer, W.C., Malm, O., Souza, C.M.M. and Mierle, G. (1992) Mercury pollution in Brazil. *Nature* 356, 389.

OCEI (now INE: Instituto Nacional de Estadistica) (2000) *Población – Magnitud y Estructura (1990–2015).* OCEI (http://www.ocei.gov.ve/ine/poblacion).

Office National des Forêts (1992) *Rapport du Gestion 1992 – Dirrection Régionale de l'ONF – Guyane*. ONF-Guyane, Cayenne.

Office National des Forêts (1997a) L'exportation des bois guyanais. *Bois et Forêts de Guyane* 5, 6–7.

Office National des Forêts (1997b) Schéma d'aménagement forestier en Guyane. *Bois et Forêts de Guyane* 5, 2.

Ojer, P. (1960) *Don Antonio de Berrio – Gobernador del Dorado*. Universidad Católica Andrés Bello, Caracas.

Ojer, P. (1966) *La formación del oriente Venezolano. I. Creación de las gobernaciones*. Universidad Católica Andrés Bello, Caracas.

Oliphant, F.M. (1938) *The Commercial Possibilities and Development of the Forests of British Guiana*. The Argosy Co. Ltd, Georgetown, Guyana.

Omai Gold Mines Limited (Cambior) (2000) *Statement of Total Investment, Royalties and Taxes Paid, Purchase of Local Goods and Services and Infrastructural Development over the period 1991–2000*. Public Communications Department, OGML, Georgetown, Guyana.

Osgood, C. (1946) *British Guiana Archaeology to 1945*. Yale University Press, New Haven, Connecticut.

Osgood, C. and Howard, G.D. (1943) *An Archaeological Survey of Venezuela*. Yale University Press, New Haven, Connecticut.

Palmer, C.J., Validum, L., Loeffke, B., Laubach, H.E., Mitchell, C., Cummings, R. and Cuadrado, R.R. (2002) HIV prevalence in a gold mining camp in the Amazon region, Guyana. *Emerging Infectious Diseases* 8, 330–331.

Palmer, J.R. (1986) Lessons for land managers in the tropics. *Revue Bois et Forêts des Tropiques* 212, 16–27.

Parrotta, J.A. and Knowles, O.H. (1999) Restoration of tropical moist forests on bauxite-mined lands in the Brazilian Amazon. *Restoration Ecology* 7, 103–116.

Parrotta, J.A., Knowles, O.H. and Wunderle, J.M.J. (1997) Development of floristic diversity in 10-year-old restoration forests on a bauxite mined site in Amazonia. *Forest Ecology and Management* 99, 21–42.

Parsons, J.J. and Denevan, W.D. (1974) Pre-Columbian ridged fields. In: Zubrow, E.B.W., Fritz, M.C. and Fritz, J.M. (eds) *New World Archaeology: Theoretical and Cultural Transformations*. WH Freeman and Co., San Francisco, California, pp. 240–248.

Penard, A.P. and Penard, T.E. (1917) Popular notions pertaining to primitive stone artifacts in Surinam. *The Journal of American Folk-Lore* 30, 251–261.

Perera, M.A. (1971) Contribución al conocimiento de la espeleología histórica en Venezuela. Parte II. La arqueología hipogea del Orinoco medio, Territorio Federal Amazonas. *Boletín de la Sociedad Venezolana de Espeleología* 3, 151–163.

Perera, M.A. and Moreno, H. (1984) Pictografías y cerámica de dos localidades hipogeas en la penillanura del norte, Territorio Federal Amazonas y Distrito Cedeño del Estado Bolivar. *Boletín de la Sociedad Venezolana de Espeleología* 21, 21–32.

Perez, L., van Beek, F., Justice, G., Shah, S. and Zandamela, R. (1997) *Recent Economic Developments and Selected Issues*. IMF, Washington, DC.

Perry, L. (2002) Starch analyses reveal multiple functions of quartz 'manioc' grater flakes from the Orinoco basin, Venezuela. *Interciencia* 27, 635–639.

Persaud, S., Flynn, D. and Fox, B. (1999) Potential for wind generation on the Guyana coastlands. *Renewable Energy* 18, 175–189.

Petitjean Roget, H. (1983) Evolution et décadence de l'art funéraire des sites pré et post-colombiens de la baie de l'Oyapock. *Compte Rendu du IX^{éme} CIECPPA*, Santo Domingo, D.R. CRC, Université de Montréal.

Peyre, E. (1993) Nouvelle découverte d'un homme préhistorique américain: une femme de 9,700 ans au Brésil. *Comptes Rendus de l'Académie des Sciences de Paris* 316 (série II), 839–842.

Peyre, E., Guérin, C., Guidon, N. and Coppens, Y. (2000) Resultados da datação de dentes humanos da Toca do Garrincho, Piaui, Brésil. *Anais da X Reunião Cientifica da SAB*, Recife, Brazil, Universidade Federal de Pernambuco.

Pfeiffer, W.C., Drude de Lacerda, L., Salomons, W. and Malm, O. (1993) Environmental fate of mercury from gold mining in the Brazilian Amazon. *Environmental Review* 1, 26–37.

Pinton, F. and Emperaire, L. (1992) L'extractivisme en Amazonie brésilienne: un système en crise d'identité. *Cahiers de Sciences Humaines* 28, 685–703.

Piperno, D. (1990) Fitolitos, arqueología y cambios prehistóricos de la vegetación en un lote de cincuenta hectáreas de la isla de Barro Colorado. In: Leigh, E.G.J., Rand, A.S. and Windsor, D.M. (eds) *Ecología de un Bosque Tropical*. STRI, Balboa, Panama, pp. 153–156.

Piperno, D.R. (1994) Phytolith and charcoal evidence for prehistoric slash-and-burn agriculture in the Darien rain forest of Panama. *The Holocene* 4, 321–325.

Piperno, D.R. and Jones, J.G. (2003) Paleoecological and archaeological implications of a Late Pleistocene/Early Holocene record of vegetation and climate from the Pacific coastal plain of Panama. *Quaternary Research* 59, 79–87.

Plew, M.G. (2002) *A Report on an Archaeological Survey in the Iwokrama Mountains with Recommendations Regarding the Development of Protocols Relating to Archaeologcial Sites.* Iwokrama International Centre, Georgetown, Guyana.

Pollak, H., Mattos, M. and Uhl, C. (1995) A profile of palm heart extraction in the Amazon estuary. *Human Ecology* 23, 357–386.

Poonai, N.O. (1962) Archaeological sites on the Corentyne coast. *Timehri* 33, 52–53.

Poonai, N.O. (1974) Recollections of a naturalist. *Timehri* 44, 31–44.

Poonai, N.O. (1978) Stone age Guyana. *Archaeology and Anthropology (W. Roth Museum – Guyana)* 1, 5–23.

Prous, A., Junqueira, P.A. and Malta, I.M. (1984) Arqueología do alto médio São Francisco – região de Janúaria e Montalvânia. *Revista de Arquelogia (Museu Paraense Emilio Goeldi)* 2, 59–72.

Puleston, D.E. (1982) The role of Ramón in Maya subsistence. In: Flannery, K.V. (ed.) *Maya Subsistence.* Academic Press, New York, pp. 353–365.

Purchas, S. (1906) *Hakluytus Posthumus or Purchas his Pilgrimes, 6.* James Maclehose and Sons, Glasgow, UK.

Putz, F.E. (1979) Biology and human use of *Leopoldinia piassaba. Principes* 23, 149–156.

Queixalos, F. and Renault-Lescure, O. (2000) *As Linguas Amazonicas Hoje.* Instituto Socoiambiental, São Pãulo.

Raleigh, W. (1596) *The Discoverie of the large, rich, and bewtiful empyre of Guiana, with a relation of the great and golden citie of Manoa (which the Spanyards call El Dorado) and of the provinces of Emeria, Arromaia, Amapaia, and othe countries, with their rivers, adjoyning.* Robert Robinson, London.

Reanier, R.E. (1997) Dating a paleoindian site in the Amazon in comparison with Clovis culture. *Science* 275, 1948–1949.

Redmond, E.M. and Spencer, C.S. (1990) Investigaciones arqueologicas en el piedemonte y los llanos altos de Barinas, Venezuela. *Boletin de la Asociacion Venezolana de Arqueologia* 5, 4–24.

Redmond, E.M., Gassón, R.A. and Spencer, C.S. (1999) A macroregional view of cycling chiefdoms in the western Venezuelan llanos. In: Bacus, E. and Lucero, L. (eds) *Complex Polities in the Ancient Tropical World.* AAA, Arlington, Virginia, pp. 109–129.

Reichel-Dolmatoff, A. (1974) Un sistema de agricultura prehistórica de los Llanos Orientales. *Revista Colombiana de Antropologia* 17, 253–262.

Reichlen, H. and Reichlen, P. (1944) Contribution à l'archéologie de la Guyane française. *Journal de la Société des Américanistes* 35, 1–24.

Reis, A.C.F. (1943) *O Processo Histórico da Economia Amazonense.* Editora Paralelo, Rio de Janeiro.

Ricci, J.-P. (1989) Les pinotières. *Revue Bois et Forêts des Tropiques* 220 (spécial Guyane), 55–63.

Richard, S., Arnoux, A., Cerdan, P., Reynouard, C. and Horeau, V. (2000) Mercury levels of soils, sediments and fish in French Guiana, South America. *Water, Air and Soil Pollution* 124, 221–244.

Richards, E.M. (1993) *Commercialization of Non-timber Forest Products in Amazonia.* NRI, Chatham, UK.

Rivas, P. (1993) Estudio preliminar de los petroglifos de Punta Cedeño, Caicara del Orinoco, estado Bolivar. In: Fernández, F. and Gassón, R.A. (eds) *Contribuciones a la Arqueología Regional de Venezuela.* Fondo Editorial Acta Científica, Caracas, pp. 165–196.

Riviére, P.G. (1969) *Marriage Among the Trio, A Principle of Social Organisation.* Oxford University Press, Oxford, UK.

Rodway, J. (1888) The gold industry in Guiana. *Timehri* 1, 75–96.

Rodway, J. (1911) Our river names. *Timehri* 1, 53–56.

Rodway, J. (1912) *Guiana: British, Dutch and French.* South American Series. T. Fisher Unwin, London.

Rodway, J. (1919) Timehri or pictured rocks. *Timehri* 6, 1–11.

Rollet, B. (1980) Jari: succès ou échec? Un exemple de développement agro-sylvo-pastoral et industriel en Amazonie brésilienne. *Bois et Forêts des Tropiques* 192, 3–34.

Roosevelt, A.C. (1980) *Parmana. Prehistoric Maize and Manioc Subsistence Along the Amazon and Orinoco.* Academic Press, New York.

Roosevelt, A.C. (1991a) Eighth millennium pottery from a prehistoric shell midden in the Brazilian Amazon. *Science* 254, 1621–1624.

Roosevelt, A.C. (1991b) *Moundbuilders of the Amazon: Geophysical Archaeology on Marajo Island, Brazil.* Academic Press, San Diego, California.
Roosevelt, A.C. (1995) Early pottery in Amazonia: twenty years of obscurity. In: Barnett, W. and Hoopes, J. (eds) *The Emergence of Pottery.* Smithsonian Institution Press, Washington, DC, pp. 115–132.
Roosevelt, A.C. (1998) Ancient and modern hunter-gatherers of lowland South America: an evolutionary problem. In: Balée, W. (ed.) *Advances in Historical Ecology.* Columbia University Press, New York, pp. 190–212.
Roosevelt, A.C. (1999) Twelve thousand years of human–environment interaction in the Amazon floodplain. In: Padoch, C., Ayres, J.M., Pinedo-Vasquez, M. and Henderson, A. (eds) *Várzea: Diversity, Development, and Conservation of Amazonia's Whitewater Floodplains.* NYBG, New York, pp. 371–392.
Roosevelt, A.C., Lima da Costa, M., Machado, C.L., Michab, M., Mercier, N., Vallada, H., Feathers, J., Barnett, W., da Silveira, M.I., Henderson, A., Sliva, J., Chernoff, B., Reese, D.S., Holman, J.A., Toth, N. and Schick, K. (1996) Paleoindian cave dwellers in the Amazon: the peopling of the Americas. *Science* 272, 373–384.
Rose, J.G. (1989) Runaways and Maroons in Guyana history. *History Gazette (University of Guyana)* 4, 1–14.
Rostain, S. (1987) Roche gravées et assemblages de pierres en Guyane française. *Equinox* 24, 35–69.
Rostain, S. (1994) L'occupation amerindienne ancienne du littoral de Guyane. Thèse de Doctorat. Universite de Paris I-Pantheon/Sorbonne, Paris.
Roth, W.E. (1924) *An Introductory Study of the Arts, Crafts and Customs of the Guiana Indians.* Annual Report of the Bureau of American Ethnology, 38. Smithsonian Institution, Washington DC.
Roth, W.E. (1929) *Additions Studies of the Arts, Crafts and Customs of the Guiana Indians.* Bulletin of the Bureau of American Ethnology, 91. Smithsonian Institution, Washington DC.
Roulet, M., Lucotte, M., Farella, N., Serique, G., Coelho, H., Sousa Passos, C.J., Jesus da Silva, E., Scavone de Andrade, P., Mergler, D., Guimaraes, J.R.D. and Amorim, M. (1999) Effect of recent human colonization on the presence of mercury in Amazonian ecosystems. *Water, Air and Soil Pollution* 112, 297–313.
Rouse, I. (1949) Petroglyphs. In: Steward, J.H. (ed.) *Handbook of South American Indians,* Vol. 5. US GPO, Washington, DC, pp. 493–502.
Rouse, I. and Cruxent, J. (1963) *Venezuelan Archaeology.* Caribbean Series, 6. Yale University Press, London.
Rowland, E.D. (1892) The census of British Guiana, 1891. *Timehri* 6, 40–68.
Saldarriaga, J.G. (1994) *Recuperación de la Selva de 'Tierra Firme' en el Alto Río Negro Amazonia Colombiana–Venezolana.* Estudios en la Amazonia Colombiana 5. Tropenbos Colombia Programme, Bogotá.
Sandweiss, D.H., McInnis, H., Burger, R.L., Cano, A., Ojeda, B., Paredes, R., del Carmen Sandweiss, M. and Glascock, M.D. (1998) Quebrada Jaguay: early South American maritime adaptations. *Science* 281, 1830–1832.
Sanoja, M. and Vargas, I. (1983) New light on the prehistory of eastern Venezuela. *Advances in World Archaeology* 2, 205–244.
Santos, G.M., Bird, M.I., Parenti, F., Fifield, L.K., Guidon, N. and Hausladen, P.A. (2003) A revised chronology of the lowest occupation layer of the Pedra Furada rock shelte, Piauí, Brazil: the Pleistocene peopling of the Americas. *Quaternary Science Reviews* 22, 2303–2310.
Santos, G.M., Cordeiro, R.C., Silva Filho, E.V., Turcq, B., Lacerda, L.D., Fifield, L.K., Gomes, P.R.S., Hausladen, P.A., Sifeddine, A. and Albuquerque, A.L.S. (2001) Chronology of the atmospheric mercury in Lagoa da Pata basin, upper Rio Negro region of Brazilian Amazon. *Radiocarbon* 43, 801–808.
Santos, R. (1968) O equilíbrio da firma aviadora e a significação econômica institutional do aviamento. *Pará Desenvolvimento* 3, 7–30.
Schenk, C.J., Higley, D.K. and Magoon, L.B. (2000) Region 6 – Central and South America. In: USGS World Energy Assessment Team (ed.), *World Petroleum Assessment 2000.* USGS, Washington, DC.
Schenk, C.J., Viger, R.J. and Anderson, C.P. (1999) *Maps Showing Geology, Oil and Gas Fields and Geologic Provinces of South America.* 97-470D. USGS, Denver, Colorado.
Schomburgk, R. (1836) Report of an expedition into the interior of British Guayana in 1835–1836. *The Journal of the Royal Geographic Society of London* 6, 224–284.
Schomburgk, R.H. (1840) *A Description of British Guiana.* Simpkin, Marshall and Co., London.
Schomburgk, R.H. (1848) *Reisen in British Guiana den Jahren 1840–1844.* J.J. Weber, Leipzig.
Siemens, A.H. and Puleston, D.E. (1972) Ridge fields and associated features in southern Campeche: new perspectives on lowland Maya. *American Antiquity* 37, 228–239.

Silva, N. (1996) Etnografia de la cuenca del Caura. In: Rosales, J. and Huber, O. (eds) *Ecología de la Cuenca del Río Caura, Venezuela. I. Caracterización General.* Refolit C.A., Caracas, pp. 98–105.

Silva-Forsberg, M.C., Forsberg, B.R. and Zeidemann, V.K. (1999) Mercury contamination in humans linked to river chemistry in the Amazon Basin. *Ambio* 28, 519–521.

Simões, M.F. (1961) Coletores-pescadores ceramistas do litoral do Salgado (Para). *Boletín Museu Paraense Emilio Goeldi, Antropologia,* 1–26.

Simões, M.F. (1974) Contribuição à arqueologia dos arredores do baixo Río Negro, Amazonas. In: Simões, M.F. (ed.) *Programma Nacional de Pequisas Arqueológicas.* Museu Paraense Emílio Goeldi, Belem, pp. 165–188.

Simões, M.F. (1987) Pesquisas arqueológicas no Médio Rio Negro (Amazonas). *Revista de Arquelogia (Museu Paraense Emilio Goeldi)* 4, 83–116.

Simpson, E.N. (1937) *The Ejido – Mexico's Way Out.* University of North Carolina Press, Chapel Hill, North Carolina.

Sizer, N. and Rice, R. (1995) *Backs to the Wall in Suriname: Forest Policy in a Country in Crisis.* World Resources Institute, Washington, DC.

Skidmore, T.E. (1999) *Brazil – Five Centuries of Change.* Oxford University Press, New York.

Soler, J.M. and Lasaga, A.C. (2000) The Los Pijiguaos bauxite deposit (Venezuela): a compilation of field data and implications for the bauxitization process. *Journal of South American Earth Sciences* 13, 47–65.

Sombroek, W. (1966) *Amazon Soils: A Reconnaissance of the Soils of the Brazilian Amazon Valley.* Pudoc, Wageningen, The Netherlands.

Sombroek, W., Dirse, K., Rodrigues, T., d. S. Cravo, M., Jarbas, T.C., Wood, W.I. and Glaser, B. (2002) *Terra preta and Terra mulata: Pre-Columbian Amazon Kitchen Middens and Agricultural Fields, Their Sustainability and their Replication.* 17th WCSS, Bangkok, Thailand.

Spencer, C.S. (1991) The coevolution and the development of Venezuelan chiefdoms. In: Rambo, A.T. and Gillogly, K. (eds) *Profiles in Cultural Evolution: Papers from a Conference in Honor of Elman Service.* University of Michigan Press, Ann Arbor, Michigan, pp. 137–165.

Spencer, C.S., Redmond, E.M. and Rinaldi, M. (1994) Drained fields at La Tigra, Venezuelan llanos: a regional perspective. *Latin American Antiquity* 5, 119–143.

Spencer, C.S., Redmond, E.M. and Rinaldi, M. (1998) Prehispanic causeways and regional politics in the llanos of Barinas, Venezuela. *Latin American Antiquity* 9, 95–110.

Stahl, P.W. (1996) Holocene biodiversity: an archaeological perspective from the Americas. *Annual Review of Anthropology* 25, 105–126.

Steward, J.H. (1955) *Theory of Culture Change: the Methodology of Multilinear Evolution.* University of Illinois Press, Urbana, Illinois.

Storer-Peberdy, P. (1948) Discovery of Amerindian rock-paintings. *Timehri* 28, 54–58.

Storm van's Gravesande, L. (ed.) (1911) *The Rise of British Guiana.* Hakluyt Society, London.

Strudwick, J. and Sobel, G.L. (1988) Uses of *Euterpe oleracea* in the Amazon estuary, Brazil. *Advances in Economic Botany* 6, 225–253.

Struycken de Roysancour, C.A.J. and Gonggryp, J.W. (1912) *Het balata vraagstuk in Suriname.* Department van den Landbouw, Paramaribo, Suriname.

Svisero, D.P. (1995) Distribution and origin of diamonds in Brazil: an overview. *Journal of Geodynamics* 20, 493–514.

Szczesniak, P.A. (2001) The mineral industry of Suriname. *USGS Minerals Yearbook – 2001.* USGS, Washington, DC, pp. 16.1–16.3.

Tarble, K. (1991) Piedras y potencia, pintura y poder: estilos sagrados en el Orinoco medio. *Antropologica* 75, 141–164.

Tarble, K. (1999) Style, function and context in rock art of the middle Orinoco area. *Boletín de la Sociedad Venezolana de Espeleología* 33, 17–33.

Ten Kate, H.F.C. (1889) On West Indian stone implements and other indian relics. *Bidragen tot de Taal, Land en Volkenkunde van Nederlansch Indië* 38, 153–160.

Theofilos, N.G. (1975) Other areas of South America. *Mineral Yearbook Area Reports: International 1975.* US Bureau of Mines, Washington, DC, pp. 1273–1281.

Thompson, A. (1979) Discovery of a new mound with remains of aboriginal inhabitants of the Abary area. *Archaeology and Anthropology (W. Roth Museum – Guyana)* 2, 149–150.

Tierney, P. (2000) *Darkness in El Dorado – How Scientists and Journalists Devastated the Amazon.* W.W. Norton and Company, New York.

Toplin, R.B. (1972) *The Destruction of Brazilian Slavery, 1850–1888.* University of California Press, Berkeley, California.

Torres, I.E. (1999) The mineral industry of Venezuela. In: *Mineral Yearbook Vol. III – Area Reports: International and Commodity Summaries.* US Bureau of Mines, Washington, DC, pp. 27.1–27.5.
Torroni, A., Neel, J.V., Barrantes, R., Schurr, T.G. and Wallace, D.C. (1994) Mitochondrial DNA 'clock' for the Amerinds and its implications for timing their entry into North America. *Proceedings of the National Academy of Sciences USA* 91, 1158–1162.
UN-CEPAL (United Nations Comisión Económica para América Latina y el Caribe) (2003) *Anuario Estadístico de América Latina y el Caribe 2003.* CEPAL, Santiago.
US Bureau of Mines (1923–2002) *Mineral Yearbook* Vol. III – *Area Reports: International and Commodity Summaries.* USGS, Washington, DC.
US Energy Information Administration (2003) *World Electricity Generation, Consumption and Installed Capacity by Country, 1980–2002.* EIA-DOE, Washington, DC.
USGS and CVGTM (1993) *Geology and Mineral Resource Assessment of the Venezuelan Guayana Shield.* USGS Bulletin 2062. US GPO, Washington, DC.
USGS World Energy Assessment Team (2000) *World Petroleum Assessment 2000.* USGS, Washington, DC.
Valeix, J. and Mauperin, M. (1989) Cinq siécles de l'histoire d'une parcelle de forêt domaniale de la terre ferme d'Amerique du Sud. *Bois et Forêts des Tropiques* 219, 13–29.
van Andel, T. (1998) Commercial exploitation of non-timber forest products in the north-west district of Guyana. *Caribbean Journal of Agriculture and Natural Resources* 2, 15–28.
van Andel, T. (2000) *Non-timber Forest Products of the North West District of Guyana.* Part I. Tropenbos-Guyana Series 8A. Tropenbos-Guyana Programme, Georgetown.
van der Merwe, N.J., Roosevelt, A.C. and Vogel, J.C. (1981) Isotopic evidence for prehistoric subsistence change at Parmana, Venezuela. *Nature* 292, 536–538.
Veiga, M.M., Meech, J.A. and Onate, N. (1994) Mercury inputs from forest fire in the Amazon. *Nature* 368, 816–817.
Velasco, P. and Ensminger, R.H. (1984) Other areas of South America. In: *Mineral Yearbook Area Reports: International.* US Bureau of Mines, Washington, DC, pp. 1057–1067.
Verill, A.H. (1918a) Prehistoric mounds and relics of the northwest district of British Guiana. *Timehri* 5, 11–20.
Verill, A.H. (1918b) A remarkable mound discovered in British Guiana. *Timehri* 5, 22–25.
Versteeg, A.H. (1983) Raised field complexes and associated settlements in the coastal plain of western Suriname. Drained field agriculture in Central and South America – *44th International Congress of Americanists,* Manchester, UK.
Versteeg, A.H. (1985) The prehistory of the young coastal plain of west Suriname. *Berichten van de Rijksdienst voor het Oudheidkundig Bodemonderzoek* 35, 653–750.
Versteeg, A.H. (1995) The occurrence of petroglyphs in the Lesser Antilles and the Virgin Islands. In: Dubelaar, C.N. (ed.) *The Petroglyphs of the Lesser Antilles, the Virgin Islands and Trinidad.* Foundation for Scientific Research in the Caribbean Region, Amsterdam, pp. 22–25.
Versteeg, A.H. and Bubberman, F.C. (1992) Suriname before Columbus. *Mededelingen Surinaams Museum* 49, 3–65.
Viera, V.S. (1980) *Logging in Guyana and Considerations for Improvements.* Guyana Forestry Commission, Georgetown.
Vink, A.T. (1970) *Forestry in Surinam – A Review of Policy, Planning, Projects, Progress.* Suriname Forest Service (LBB), Paramaribo.
Voicu, G., Bardoux, M. and Stevenson, R. (2001) Lithostratigraphy, geochronology and gold metallogeny in the northern Guiana Shield, South America: a review. *Ore Geology Reviews* 18, 211–236.
von Hildebrand, E. (1975) Levantamiento de los petroglifos del rio Caquetá entre La Padrera y Araracuara. *Revista Colombiana de Antropologia* 19, 303–370.
Waddell, D.A.G. (1967) *The West Indies and the Guianas. Modern Nations in Historical Perspective.* Prentice-Hall, Englewood Cliffs, New Jersey.
Wagner, E. and Arvelo, L. (1986) Monou-Teri: un nuevo complejo arqueológico en el alton Orinoco, Venezuela. *Acta Cientifica Venezolana* 37, 689–696.
Walker, W. (1878) *British Guiana at the Paris Exhibition: catalogue of exhibits, to which are prefixed some illustrative notices of the colony.* Paris Exhibition, London.
Wallace, A.R. (1853a) *A Narrative of Travels on the Amazon and Rio Negro.* Lockhart Publishers, London.
Wallace, A.R. (1853b) *Palm Trees of the Amazon and their Uses.* Van Hoorst, London.
Whitehead, N. (1990) The Mazaruni pectoral: a golden artefact discovered in Guyana and the historical sources concerning native metallurgy in the Caribbean, Orinoco and Northern Amazonia. *Archaeology and Anthropology (W. Roth Museum – Guyana)* 7, 19–38.

Whitehead, N.L. (1988) *Lords of the Tiger Spirit – a History of the Caribs in Colonial Venezuela and Guyana 1498–1820.* Foris Publications, Dordrecht, The Netherlands.

Whitehead, N.L. (1997) *The Discoverie of the Large, Rich, and Bewtiful Empyre of Guiana.* Manchester University Press, Manchester, UK.

Whitlock, C. and Millspaugh, S.H. (1996) Testing the assumptions of fire-history studies: an examination of modern charcoal accumulation in Yellowstone National Park, USA. *The Holocene* 6, 7–15.

Williams, D. (1978a) A Mazaruni-type handaxe? *Archaeology and Anthropology (W. Roth Museum – Guyana)* 1, 32–33.

Williams, D. (1978b) Petroglyphs at Marlissa, Berbice river. *Archaeology and Anthropology (W. Roth Museum – Guyana)* 1, 24–31.

Williams, D. (1979a) Controlled resource exploitation in contrasting neotropical environments evidenced by meso-indian petroglyphs in southern Guyana. *Archaeology and Anthropology (W. Roth Museum – Guyana)* 2, 141–148.

Williams, D. (1979b) Preceramic fishtraps on the upper Essequibo: a survey of unusual petroglyphs on the upper Essequibo and Kassikaityu rivers, 12–28 March 1979. *Archaeology and Anthropology (W. Roth Museum – Guyana)* 2, 141–148.

Williams, D. (1979c) A report on preceramic lithic artifacts in the south Rupununi savannas. *Archaeology and Anthropology (W. Roth Museum – Guyana)* 2, 10–53.

Williams, D. (1981a) Excavation of the Barabina shell mound, North West District: an interim report. *Archaeology and Anthropology (W. Roth Museum – Guyana)* 2, 125–140.

Williams, D. (1981b) Three sites of the Taruma phase in southeast and east Guyana. *Archaeology and Anthropology (W. Roth Museum – Guyana)* 4, 81–103.

Williams, D. (1993) The forms of the shamanic sign in the prehistoric Guianas. *Archaeology and Anthropology (W. Roth Museum – Guyana)* 9, 3–21.

Williams, D. (1994) Annex C. Archaeology and Anthropology. *Environmental Impact Assessment – the Linden Lethem Road.* ERM, Georgetown.

Williams, D. (1996a) The origin, characterization and chronology of the Mabaruma subseries of the barrancoid tradition. *Archaeology and Anthropology (W. Roth Museum – Guyana)* 11, 3–53.

Williams, D. (1996b) *Prehistoric Cultures of the Iwokrama Rain Forest.* GNRA, Georgetown, Guyana.

Williams, D. (1998) The archaic colonization of the Western Guiana Littoral and its aftermath. *Archaeology and Anthropology (W. Roth Museum – Guyana)* 12, 22–41.

Williams, L. (1939) *Maderas Economicas de Venezuela.* Ministerio de Agricultura y Cria, Caracas.

Williams, M. (1989) *Americans and their Forests.* Cambridge University Press, Cambridge.

Wood, W.I. and Mann, C.C. (2000) Earthmovers of the Amazon. *Science* 287, 786–789.

Wood, W.I. and McCann, J.M. (1999) The anthropogenic origin and persistence of Amazonian dark earths. *Yearbook of the Conference of Latin American Geographers* 25, 7–14.

World Bank (2001) *World Bank Development Indicators database.* World Bank, Washington, DC.

Zegura, S.L., Karafet, T.M., Zhivotovsky, L.A. and Hammer, M.F. (2004) High resolution SNPs and microsatellite haplotypes point to a single, recent entry of Native American Y chromosomes into the Americas. *Molecular Biology and Evolution* 21, 164–175.

Zonzon, J. and Prost, G. (1996) *Histoire de la Guyane.* Maison Neuve and Larose, Paris.

Zucchi, A. (1972) La prehistoria de los llanos occidentales: investigaciones recientes. *Acta Cientifica Venezolana* 23, 185–187.

Zucchi, A. (1973) Prehistoric human occupations of the western Venezuelan llanos. *American Antiquity* 38, 182–190.

Zucchi, A. (1978) La variabilidad ecologica y la intensificacion de la agricultura en los llanos venezolanos. In: Wagner, E. and Zucchi, A. (eds) *Unidad y Variedad Ensayos en Homenaje al Dr. J.M. Cruxent.* IVIC-CEA, Caracas, pp. 349–374.

Zucchi, A. (1984) Alternative interpretations of pre-Columbian water management in the western llanos of Venezuela. *Indiana* 9, 309–327.

Zucchi, A. (1985) Recent evidence for pre-Columbian water management systems in the western llanos of Venezuela. In: Farrington, I.S. (ed.) *Prehistoric Intensive Agriculture in the Tropics.* Oxford University Press, Oxford, UK, pp. 167–180.

Zucchi, A. (1991) El Negro–Casiquiare–Alto Orinoco como ruta conectiva entre el Amazonas y el Norte de Suramerica. *Comptes rendus du XII*, AIAC, Cayenne.

Zucchi, A., Tarble, K. and Vaz, J. (1984) The ceramic sequence and new TL and C14 dates for the Agüerito site of the middle Orinoco. *Journal of Field Archaeology* 11, 155–180.

9 Forest Conservation and Management in the Guiana Shield

David S. Hammond
Iwokrama International Centre for Rain Forest Conservation and Development, Georgetown, Guyana. Currently: NWFS Consulting, Beaverton, Oregon, USA

Exposed Precambrian Landscapes – a Pretext for Caution

More than 550 million years of weathering, with few or no major diastrophic events, can take its toll on the roots of landscape carrying capacities. Most available information for the Guiana Shield converges along a central theme – ancient surface geology creates difficult conditions for human societies. These difficult conditions are not created as a result of catastrophic impacts commonly associated with risk and calamitous loss, such as volcanic eruptions, earthquakes, hurricanes and landslides. Rather, it is their absence that created conditions challenging early human society and its advance towards more sedentary and permanent lifestyles.

The nexus of debate concerning the prehistory of human settlement across the Amazon in many ways reflects the transition across geomorphographic regions. Accrued artefactual evidence supporting complex social units and long-term inhabitation in the lowland neotropics is drawn almost exclusively from ramping system regions (see 'Ramping and dampening systems', Chapter 7), while depictions of pre-Columbian lowland forest inhabitants as small bands of para-nomadic wanderers appears to have been strongly shaped by evidence derived from dampening systems typified by Precambrian shield interiors (see Chapters 7 and 8). Hence, it is not unreasonable to state that existing archaeological evidence points to a 'ring of early lowland civilization' encircling the Guiana Shield along the major sedimentary basins separating it from the Andean highlands and other shield regions south of the equator (see 'Human Prehistory of the Guiana Shield', Chapter 8). Ceramic evidence is widely distributed across the region, but most submitted dates for those located in the interior are of more recent, even Indo-Hispanic ages (see 'Material types: ceramics', Chapter 8). Together, this reflects on the existing debate over levels and limits on pre-Columbian, lowland cultural development and sophistication in one important way. It suggests that the upland *terra firme* forest regions may have exerted a relatively pronounced limiting effect on human social structures, but that this was likely consequent, not parallel, to the development of sophisticated, sedentary social centres in the main sedimentary depressions across the neotropics. Evidence also indicates that longer-term occupation of the shield interior, and its more difficult environmental conditions, occurred as groups fled intertribal or inter-clan conflict and later, the sweeping spread of colonizing Europe. Thus, historical anthropological characterizations post-AD 1500 probably reflected upon only a very small remnant of the cultural depth and complexity of lowland societies that existed along the shield perimeter more than 600 years ago.

Globally, most known centres of early sedentary human civilization are also noticeably absent from the Precambrian shield regions of the world, having established instead in the more suitable Phanerozoic sedimentary basins or mountainous landscapes. In the adjacent Brazilian Shield region, putative pre-Clovis artefacts have been discovered principally along the scattered Palaeozoic sedimentary cliffs delimiting the São Francisco watershed, not in the heart of the adjoining shield region (see Fig. 8.1). During the later Neo-Indian period, signs of advanced Amerindian civilization are best known from the raised field complexes of sediment-filled wetlands, the Llanos de Mojos of Bolivia (Denevan, 1970), Llanos de Barinas of Venezuela (Spencer *et al.*, 1998) and the sedimentary coasts along the Atlantic (see 'Site types – earth engineering', Chapter 8). But was such a vast area immune from advanced human colonization because of a simple contrast between soil productivity conditions attached to *terra firme* and *varzea* forests? Probably not. Tropical forests in Precambrian landscapes may have been less susceptible to early human colonization for a number of reasons, of which soil fertility counts importantly. However, not all factors limiting human colonization across shield regions are immediately attached to a general, but hopefully dwindling (Moran, 1995), misconception of uniform upland soil infertility across the lowland tropics (Meggers, 1996).

Soil fertility

Soil fertility throughout the wet tropics is, however, relatively low in comparison to drier or colder regions of the planet due to much higher acidity and chemical weathering rates, but relevant differences in soil nutrient content exist across geomorphographic regions (see 'Soils and soil fertility', Chapter 2). In part, this is due to a much higher fraction of area affected by high-energy fluvial systems, hurricanes, volcanic ejecta and landslips across ramping system regions, such as Central America and western Amazon (Räsänen *et al.*, 1987; Kalliola *et al.*, 1999) compared to shield regions. However, highly infertile and relatively fertile soil facies can be found across all regions. The difference is met when considering the relative distribution of facies across the fertility spectrum. In this case, soils of the Guiana Shield are highly right-skewed towards uniformly low nutrient status, relative to other regions (see Fig. 2.15). Even floodplain soils offer relatively fewer incremental gains in soil fertility, in contrast to the well-established differences between *varzea* soils and adjacent areas of *terra firme* in the downwarp and sub-Andean sedimentary basins, as well as across parts of Central America (see Fig. 2.12).

Acidity

From a hydrochemical perspective, the Amazon basin is a very large, spatially anisotropic, acid–base titration. Acidity is regulated westward by the relatively cold, alkaline base-rich contribution of the Andean slopes, but on average becomes increasingly acidic eastward across the lowlands, particularly in the Precambrian shield regions (see Chapter 2). A large part of this change-over comes with the duration of exposure to much higher temperatures in the lowlands. The interesting prospect of lower ambient temperatures across the Last Glacial Maximum (Liu and Colinvaux, 1985; Colinvaux *et al.*, 1996) (also see 'Prehistoric climates of the Guiana Shield', Chapter 2) would suggest that these acidity-buffering contributions would be sustained in much higher concentrations further downstream during colder phases, creating with it expanded zones of more modest acidity and higher soil fertility across western Amazon and the downwarp. Anticipating this type of extension across most of the eastern shield regions would appear much less plausible.

The southward location of the Brazilian Shield creates a drier seasonal environment that modulates to some extent the impacts of high rainfall on acidity under closed-forest conditions. Development of the Amazon Downwarp surfaces from Andean-derived materials also assists in buffering these areas against high acidity that is both spatially widespread and persistently low across the Guiana Shield. Similarly, across many regions of Central America and the Caribbean, more seasonal climate combined with contemporary weathering of wide-ranging clastic, chemical (e.g. oolitic limestones) and organic-based (e.g. fossiliferous limestones) sedimentary features work to counter widespread tropical acidity.

Prospects for long-term, prehistoric occupation at the centre of neotropical terrestrial acidity, the interior Guiana Shield, would seem counterintuitive when more amenable environments exist along the periphery. Acidity is not only linked to terrestrial productivity through regulation of available phosphorus and the concentration of heavy-metal concentrations that can create further physiological challenges to plant survival and growth (see Chapter 3). It also regulates aquatic productivity and structures the trophic assemblage of primary producers, consumers and predators.

Aquatic productivity is arguably the most important determinant of long-term human occupation in tropical environments. As the main source of protein, fish play a pivotal role in sustaining subsistence and buffering more tenuous productivities attached to upland tropical soils. Across the *varzea* regions of the Amazon and its major tributaries, high aquatic productivity is the one attribute that distinguishes these areas from others. Perhaps only the presence of engineered earth in the form of *terra preta* strikes equivalence with relative fertility of the *varzea* across the eastern Amazon. Yet, these too appear largely restricted to the Phanerozoic sediment cover of the Amazon Downwarp and Atlantic rim of the Guiana Shield see (Fig. 8.2).

Across the Guiana Shield, *varzea*-like features are also largely restricted to the periphery. In the interior, they are virtually absent, being replaced by some of the most acidic fluvial systems known from the tropics. Combined with high humic acid contents in many rivers, acidity and low oxygenation causes aquatic productivity to plummet and with it the carrying capacities it may have offered to sedentary, prehistoric societies (see 'River, lake and tidal systems', Chapter 2). Only near the turbulence created by rock features cross-cutting river channels is productivity believed to increase significantly. These 'falls' may have provided the greatest opportunities for long-term occupation in an environment otherwise disfavouring sedentarism.

Access

Perhaps of equal or greater importance is the likelihood that forest-covered Precambrian areas posed more substantial challenges to early settlers due to difficulties in access. Across the lowland neotropics, people have traditionally traded, migrated and communicated along rivers. Yet, rivers draining shield regions become some of the most difficult to navigate as sedimentary cover thins and exposed Proterozoic structures create tremendous barriers to movement. The role of the 'falls' has been one of historically curtailing more rapid use of resources in the Guiana Shield interior and concentrating development along the margins (e.g. McTurk, 1882; Im Thurn, 1883; Goulding *et al.*, 1988).

Control

Yet, today those same barriers are increasingly proving surmountable. Road construction and an expanding network of aerodromes is increasingly connecting the shield interior and perimeter and with this, both the prospects and problems of modern forestland use. Improved and expanded access to the interior forested regions of the Guiana Shield has brought forest fire ignition sources (e.g. Hammond and ter Steege, 1998), an upsurge in unregulated hunters and wildlife collectors, illicit smugglers, socially transmitted diseases (e.g. Palmer *et al.*, 2002), pollution (see Chapter 8), cultural transformation and social upheaval to regions previously insulated from the widespread effects of frontier natural resource use by the constraints on river navigability.

Isolation

From many perspectives, the Precambrian geology of the Guiana Shield has been isolating. Whether viewed as isolation from the sediments of the Andes, from the fertility of the *varzea*, through the geological control on river navigability or from the sociocultural complexity that appears to have and continues to typify the shield perimeter, the Guiana Shield interior appears as an isolating environment. It is precisely this long-standing isolation that characterizes the relatively higher risk attached to uncontrolled and poorly managed forestland use in the Guiana Shield compared to other regions arguably more strongly shaped by prehistoric human occupation and catastrophic events.

Modern forest cover and deforestation rates

Perhaps as a consequence of these factors, even today deforestation rates and forest area coverages remain some of the lowest and highest in the world, forming a significant and growing part of the standing tropical forest area in South America and the world (Table 9.1). As deforestation trends continue elsewhere, the Guiana Shield will increasingly represent a greater share of the remaining closed forest cover, unless land-use practices and patterns change dramatically.

The purpose of this chapter is to review and synthesize available information on the conservation and management of the shield region. It characterizes the history, effort and modern approaches to habitat protection in the region and offers a view on landscape-scale conservation gaps based on this synthesis. The chapter also provides a brief overview of approaches assessed and employed in the quest for sustainable forest management, including reduced impact logging techniques, extractive reserves and concession and community-based management. It explores some of the key challenges in achieving adequate conservation and habitat protection across the region and those facing efforts to sustainably utilize various forest resources towards further economic development. The chapter concludes with a perspective on forest management that highlights briefly how lack of individual opportunity and a mismatch of objectives with biophysical processes can limit success in achieving sustainable management.

Conservation Patterns and Approaches

Areas designated for forest protection

The Guiana Shield contains a world-class system of protected areas that has grown from one of the first areas officially designated for protection in South America (Kaieteur National Park, Guyana in 1929) through to the largest protected area ever established in the neotropics in 2002 (Tumucumaque National Park, Brazil) (Table 9.2). The region boasts three major World Heritage Sites, including the Central Suriname Reserve (designated 2000), the Central Amazon Conservation Complex[1] in Brazil (2003) and the oldest, Canaima National Park in eastern Venezuela (1994). Three coastal areas have also been designated as Wetlands of International Importance (Ramsar), two at Basse-Mana (established 1993) and Les Marais de Kaw (1993) along the French Guiana coast and another covering the tropical mangrove system at Coppenamemonding in Suriname since 1985.

The rate of growth in area allocated to protection in the Guiana Shield has matched or exceeded the accrual rate globally (Fig. 9.1). The establishment of new protected areas around the world has

Table 9.1. General status of tropical forest cover in the Guiana Shield in relation to continental and global standing area. Source: Land and total forest area – FAO (2000); land area in Guiana Shield – OCEI/MARNR (2000), IBGE (2000), FAO (2000); tropical forest area in Guiana Shield – FAO (2000), except Venezuela based on Huber (1995) and Brazil based on IBGE municipio estimates.

	Land Area (km^2)		Intact Forest Area (km^2)			Deforestation (%)
	Total	in GS	Total	Tropical	Tropical in GS	1990–2000
French Guiana	88,150	88,150	79,260	29,260	79,260	–
Suriname	156,000	156,000	141,130	141,130	141,130	–
Guyana	214,980	214,980	168,790	163,726	163,726	2.8
Venezuela	882,060	453,950	495,060	346,542	287,630	4.2
Brazil	8,456,510	1,204,279	5,439,050	4,133,678	945,573	4.1
Colombia	1,038,710	170,500	490,601	412,105	170,500	3.7
Guiana Shield	10,836,410	2,287,859	6,813,891		1,342,246	
South America	17,547,510		8,856,180	5,490,832	26.9%	
World	148,000,000		38,694,550	10,060,583	13.2%	

Table 9.2. The 35 largest protected areas legislatively established in the five countries of the Guiana Shield. Only publicly owned areas restricted to low-impact use are included. IUCN category: Ia – Strict Nature Reserve managed for science, Ib – Wilderness Area managed mainly for wilderness protection, II – National Park managed for ecosystem protection and recreation, III - Natural Monument managed for conservation of specific natural features, IV – Habitat/Species Management Area managed for conservation through management intervention.

Rank	Protected area	Designation	Country	IUCN	Year Est	Area (km²)
1	Montanhas de Tumucumaque	National Park	Brazil	II	2002	38,670
2	Formaciones de Tepuyes	Natural Monument	Venezuela	III	1990	34,200
3	Parima-Tapirapecó	National Park	Venezuela	II	1991	33,100
4	Canaima	National Park	Venezuela	II	1962	30,000
5	Pico da Neblina	National Park	Brazil	II	1979	22,000
6	Jau	National Park	Brazil	II	1980	22,720
7	Central Suriname Reserve	Nature Reserve	Suriname	II	1998	16,000
8	Serrania La Neblina	National Park	Venezuela	II	1978	13,600
9	Serrania de Chiribiquete	National Park	Colombia	II	1989	12,800
10	Puianawai	Nature Reserve	Colombia	III	1989	10,920
11	Nukak	Nature Reserve	Colombia	III	1989	8,550
12	Cavo Orange	National Park	Brazil	II	1980	6,190
13	Delta de Orinoco	National Park	Venezuela	II	1991	5,698
14	Uatuma	Biological Reserve	Brazil	Ia	1990	5,600
15	Viruá	National Park	Brazil	II	1998	2,270
16	Rio Treombetas	Biological Reserve	Brazil	Ia	1979	3,850
17	Lago Piratuba	Biological Reserve	Brazil	Ib	1980	3,570
18	Anavilhanas	Ecological Station	Brazil	Ia	1981	3,500
19	Jau Sarisariñama	National Park	Venezuela	II	1978	3,300
20	Yapacana	National Park	Venezuela	II	1978	3,200
21	Niquia	Ecological Station	Brazil	Ia	1985	2,866
22	Jari	Ecological Station	Brazil	Ia	1982	2,271
23	Serra da Mocidade	National Park	Brazil	II	1998	3,610
24	Duida-Marahuaca	National Park	Venezuela	II	1978	2,100
25	Iwokrama	Wilderness Preserve	Guyana	Ib	1997	1,800
26	Monte Roraima	National Park	Brazil	II	1989	1,160
27	Maracá	Ecological Station	Brazil	Ia	1981	1,013
28	Sipaliwini	Nature Reserve	Suriname	IV	1972	1,000
29	Nouragues	Nature Reserve	French Guiana	Ia	1995	1,000
30	Marais de Kaw-Roura	Nature Reserve	French Guiana	IV	1998	947
31	Caracarai	Ecological Station	Brazil	Ia	1982	806
32	La Trinité	Nature Reserve	French Guiana	Ia	1996	760
33	Maracá-Jipioca	Ecological Station	Brazil	Ia	1981	720
34	Kaieteur (extended)	National Park	Guyana	II	1929 (98)	630
35	Forêt de Saül	Prevectorially Decreed Biotope	French Guiana	IV	1995	600

tapered off since 1994, after experiencing a 22-year span of consistent annual increase in area since 1972. The establishment of Canaima National Park in 1962 represented one of the earliest efforts to protect large areas of tropical forest in the world and the most significant contiguous area designated in the Guiana Shield until the 2002 establishment of the Montanhas de Tumucumaque (Table 9.2). The annualized rate of protection since national system establishment (in the region) varies considerably by country, but several distinct patterns are discernible. First, the rates of countries with only a part of their national territory in the region, viz. Colombia, Venezuela and Brazil, is considerably higher than those with their entire (departmental) area in the region, i.e. French Guiana, Suriname and Guyana (Fig. 9.2). The difference is logical, given the need for the smaller, shield-bound countries to consider all of their potential land-use options, while their larger neighbours have other territory to consider in meeting their national economic goals. Nonetheless, the rate of protected area accrual in Guyana remains well below that achieved in Suriname or French Guiana since establishment of the inaugural protected area in each country. Guyana's rate of accrual is clearly affected by the very early establishment of Kaieteur National Park, but even if adjusted to the year of independence in both Guyana (1966) and Suriname (1975), Guyana's rate remains much lower than the 400 km^2 per annum that would place it in line with other countries in the region. The development of a national protected area system, after having legislated for the 1600 km^2 Iwokrama Forest Wilderness Preserve in 1997, was imminent by 2003 and has been forming around a series of designated sites since the early 1990s (Ramdass and Haniff, 1990; Agriconsulting, 1993). Hopefully this will alter the country's current standing as the smallest contributor to the regional network of protected areas.

By 2003, this network amounted to 1.6% of the global area protected, based on IUCN-WCMC calculations of 18.763 million km^2 of area officially designated worldwide as protected. This equates to 4.6% of IUCN Category I–III areas registered globally by 2003 (Chape *et al.*, 2003). Nearly 13% of Guiana Shield forests are estimated to be under formal protection through six national systems, although the proportion of each country's territory under protection in the region varies from just over 1% in Guyana to nearly 28% of Venezuelan Guayana (Table 9.3). On average, there are

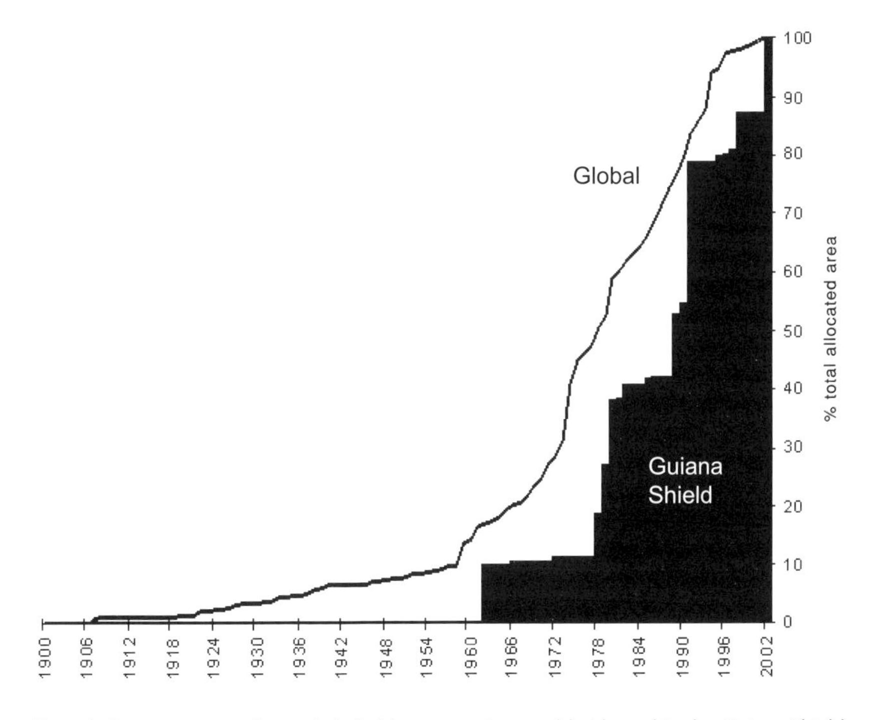

Fig. 9.1. Twentieth-century growth trends in habitat protection worldwide and in the Guiana Shield depicted as percentage of all officially recognized protected area in 2002. Sources: global (http://sea.unep-wcmc.org/wdbpa, but see Chape *et al.*, 2003), Guiana Shield (http://sea.unep-wcmc.org/wdbpa cross-referenced with size and establishment data from national park and protected area agencies and administrators – INPARQUES, IBAMA, EPA-Guyana, INDERENA). Figures do not include areas proposed but without legislative mandate, designated Biosphere Reserves (except where these include national system units), or Amerindian reserves, resguardos, titled or ancestral land areas (see Chapter 8).

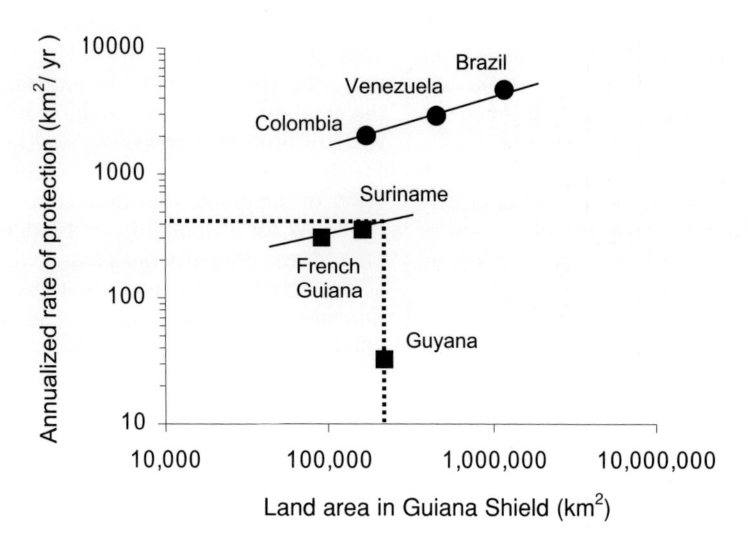

Fig. 9.2. Relationship between rate of area protection ($\log_{10}$) and total area ($\log_{10}$) within Guiana Shield for each of the six countries forming the region since inception of their respective protected area systems.

nearly 7.5 ha of forest under protection for every person living in the shield region, based on 2000 population estimates (Table 9.3).

The nearly 290,000 km^2 (almost the size of Guyana and French Guiana combined) of protected area already established across the region covers an estimated 12.3% of tropical humid forest area remaining globally (Chape *et al.*, 2003). This is a substantial figure and likely to rise when processes in Guyana and French Guiana lead to the formal expansion of their national (departmental) systems of protection that currently lag other regional commitments (Table 9.3). At the same time tropical wet forest area elsewhere continues to decline at a faster pace, pushing upwards even further the fractional contribution of Guiana Shield areas to global protection of this general forest biome.

The vast area under protection within the shield is distributed over 51 distinct units exceeding 10 km^2 (35 largest in Table 9.1) (Fig. 9.3). Numerous units with an area less than 10 km^2 also contribute to forest protection and research under systems of ecological reserves established directly by national forestry services or as part of required management planning in timber concessions (e.g. biological reserves in timber concessions, Guyana). Of course, simple measures of proportional forest area receiving legal protection do not adequately address whether the spatial distribution of protected area: (i) captures the main features of the forest landscape; (ii) is likely to retain long-term conservation value; or (iii) is functionally performing according to the conservation objectives established through legislative mandate. It does, however, identify a continuing commitment to forest protection in the region that exceeds that typically encountered in most other regions of the world.

Landscape conservation assessments

Several large global conservation and environmental organizations have increasingly taken on the self-designated task of assessing and classifying the terrestrial and marine regions of the planet according to their conservation value, immediacy of the threats confronting their persistence and priorities for investment in their protection. The Guiana Shield has formed part of the area assessed through these approaches. The results and relevance of these global

Table 9.3. General status of habitat protection in the Guiana Shield.

	Venezuela	Colombia	Guyana	Brazil	Suriname	French Guiana	Guiana Shield
Total area in GS (km^2)	453,950	170,500	214,980	1,204,279	156,000	88,150	2,287,859
Population (yr 2000)	1,544,915	53,650	697,286	1,041,595	431,303	157,213	3,925,962
Total PA (km^2)	125,243	32,270	2594	120,816	16,359	4970	302,252
% of GS area in PAs	27.6	18.9	1.2	10.0	10.5	5.6	13.2
Year initial PA established	1962	1989	1929	1979	1961	1989	
Protected area per person (ha)	8.1	60.1	0.4	10.4	3.8	3.2	7.4

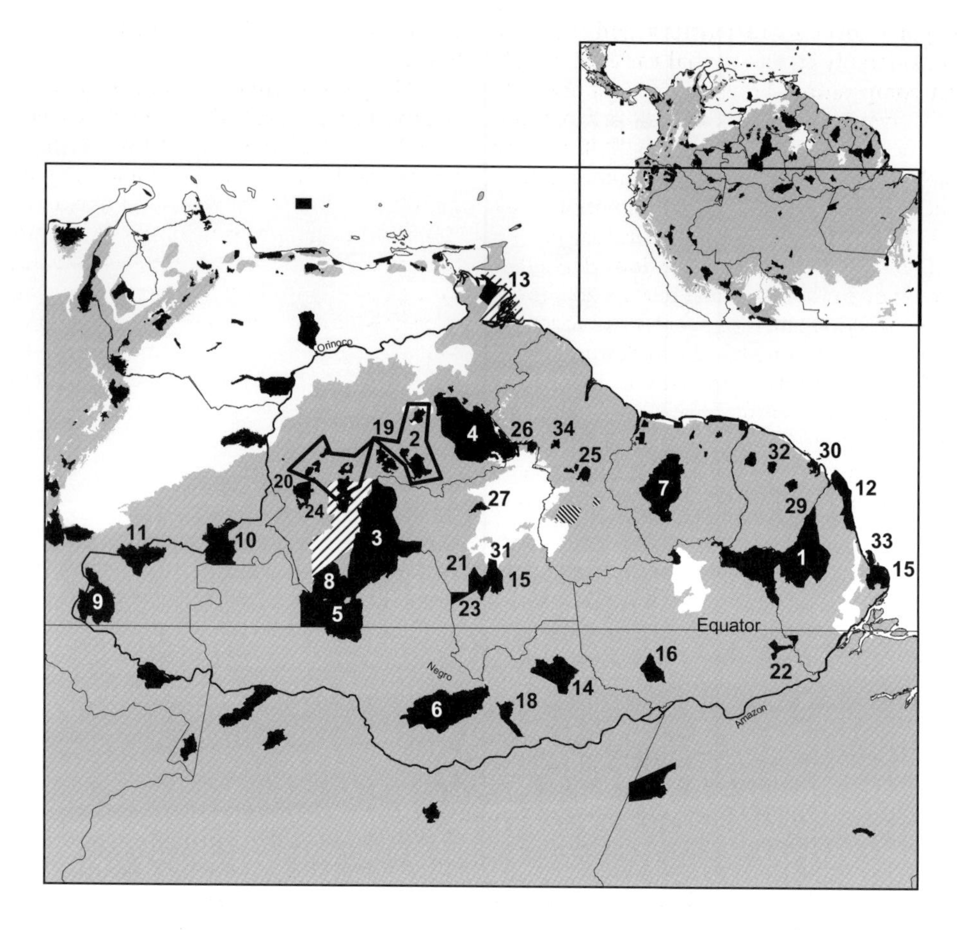

Fig. 9.3. Spatial distribution of official protected area (solid black) across the Guiana Shield and wider neotropics (inset). Numbered units are consistent with rank in Table 9.2. Designated Biosphere Reserves extending beyond boundaries of protected areas (right diagonal hatching) and proposed conservation units in Guyana (left diagonal hatching) are also presented. Forest cover is solid grey. Spatial data sources: Brazil: IBAMA; Venezuela: INPARQUES and WCPA/WCMC; Guyana: Iwokrama, CI-Guyana, EPA; Suriname: CELOS; French Guiana and Colombia: WCPA/WCMC.

initiatives to shield features are summarized here.

Ecoregions

Olson *et al.* (2001) fractionated the planetary terrestrial surface area into 867 ecoregions based on a combination of landforms, vegetation types and variation in ecological processes.[2]

As part of this global classification system, the Guiana Shield is dissected into 14 ecoregions, the seven largest being endemic to the shield area (as defined in this volume) (Fig. 9.4). The four largest ecoregions effectively dissect the shield area into quadrants that embrace an archipelago formed from the three smaller, edaphic-based units. Other smaller units largely reflect variation in coastal vegetation between the Orinoco delta and Amazon mouth and along the upper Orinoco, Solimões and Rio Negro.

While savanna, *campinarana* and *tepui* formations define distinct, often abrupt, edaphic transitions, the four larger ecoregions cover more territory and embrace considerably greater spatial variation in forest composition. Forest-type classifications in Guyana and the Guianas comprehensively analysed by ter Steege (1998) and ter Steege and Zondervan (2000) more coherently delimit compositional transitions that are not consistent with margins of the Guiana Moist Forest ecoregion. This discrepancy can for the most part be attributed to geology. Transitions between areas of exposed crystalline basement dominated by TATE granitoids, greenstone belts and Phanerozoic sands (of the Berbice Formation) (see 'Greenstone belts', Chapter 2) provide a spatial delimiter of forest type associations that dissect Olson *et al.*'s region into at least three distinct ecoregions (dashed lines in GMF ecoregion of Fig. 9.4). The Guayana Highland ecoregional area is consistent with geological transitions associated with distributions of Roraima sedimentaries, Uatuma volcanics and Parguaza granites that dominate Venezuelan Guayana, although east to west precipitation gradients associated with the Savanna Trough are not fully delimited. The change from this unit to the adjacent Negro-Branco Moist Forest unit overlying the Casiquiare Rift reflects an important landscape forest transition. Perhaps more importantly, the geology also spatially defines the largest and growing threat to forest integrity and function: unregulated mining.

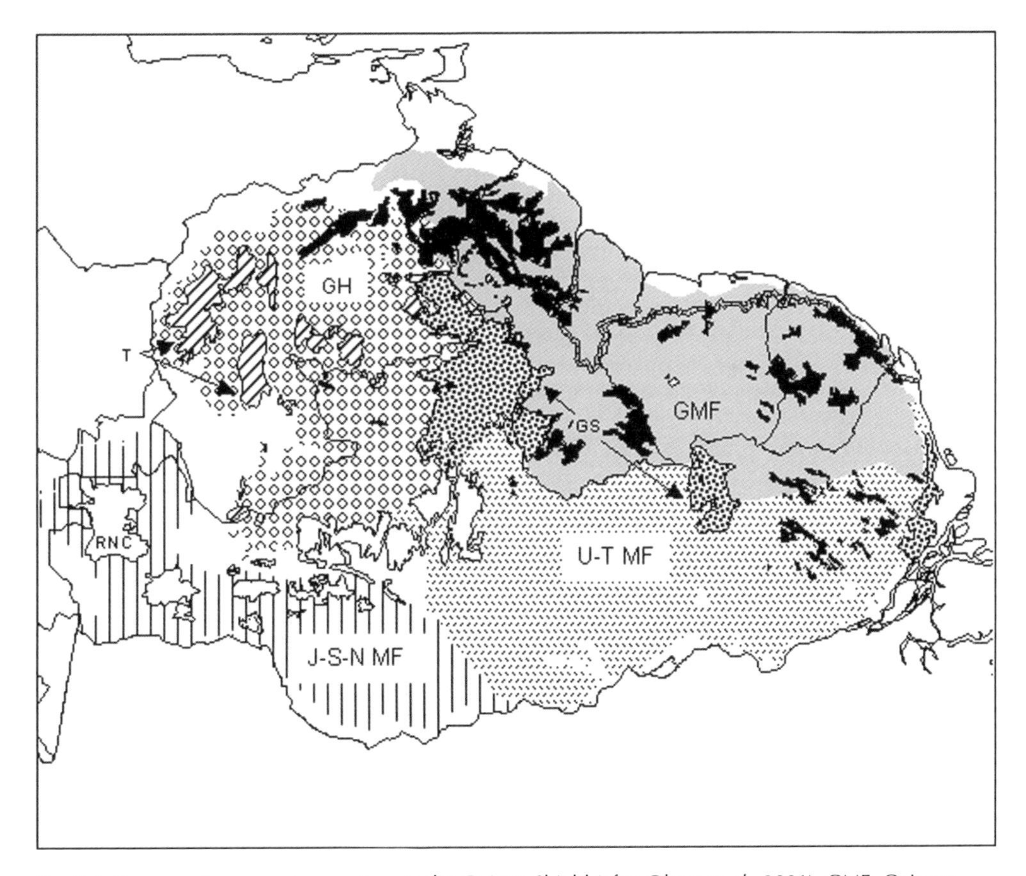

Fig. 9.4. Seven major ecoregions unique to the Guiana Shield (after Olson *et al.*, 2001). GMF: Guianan moist forests; U-T MF: Uatuma-Trombetas moist forests; GS: Guyanan savannas; GH: Guayanan highlands moist forest; T: Tepuis; RNC: Rio Negro campinarana; J-S-N MF: Japurá-Solimões-Negro moist forest. Ecoregion spatial coverage: WWF-USA. Thick line demarcates southern boundary of Berbice Formation, a unique ecoregion candidate currently subsumed within the GMF.

Forest frontiers

A four-point classification of frontier forest threat status by Bryant *et al.* (1997) identifies forests under medium or high threat of degradation along most of the perimeter area of the Guiana Shield (Fig. 9.5A). Three core areas of low threat were assigned to the main upland 'islands' forming the Guiana Shield, i.e. the Tumucumaque Uplands, Guayana Highlands and Chiribiquete Plateau (and eastern Colombian lowlands) (see 'Shield macro-features', Chapter 2). The distribution of threat magnitude assigned by Bryant *et al.* appears largely linked to areas allocated to selective logging (Fig. 8.17) and, to a lesser degree, agriculture. Low threat status is widely assigned to major mining regions containing some of the largest greenstone belts in the world, including areas in Suriname, French Guiana and Amapá (Fig. 9.5B). Mining activity throughout the valleys and along the periphery of the Roraima sedimentaries in Venezuelan Guayana also suggests that assignment of low risk status is not appropriately weighting the history and future of unregulated mining activity. Both geological groups are strongly associated with gold and diamond-bearing substrates that have attracted vast numbers of small-scale miners seeking subsistence incomes (see 'Commercial mining', Chapter 8).

The pressures to keep valuable greenstone formations open to mining are high and areas that have already been subjected to mining are not normally considered for conservation. It is not surprising, therefore, that countries with large greenstone belts and a large number of registered mines adjoining these areas have relatively little of this area allocated for habitat protection (Fig. 9.6) and widely overlap with areas allocated for timber production (Fig. 9.7), although these two resource-use practices and the regulations designed to moderate their impacts are often highly incompatible. For example, the Guyana Forestry Commission's Code-of-Practice establishes a minimum creek buffer zone applicable in all commercial forestry operations. This sensibly mitigates the impact of logging on water quality and sedimentation and is widely implemented as part of sound forestry practices globally (Dykstra and Heinrich, 1996). Yet, placer gold deposits are located in these very same creeks and rivers making them subject to massive sediment influxes as a consequence of current extraction techniques (see 'Commercial mining', Chapter 8), considerably reducing the functional value of buffer zoning.

Hotspots

Areas of high conservation value that are disproportionately threatened with loss or degradation have been variously classified as biodiversity hotspots (Davis *et al.*, 1997; Myers *et al.*, 2000). None of these is located in the Guiana Shield region. Areas that have extensive forest cover and face few imminent threats are collectively included as a Wilderness Area. Virtually the entire Amazon Downwarp, Sub-Andean Foredeep and Guiana Shield comprise the largest tropical Wilderness Area identified.

Hotspots are delimited by the estimated fraction of 'original' vegetation cover lost combined with the estimated fraction of endemics composing this original cover. While relatively little forest cover has been lost across the Guiana Shield, several important features suggest that an assessment of this kind based on recent population growth and deforestation trends alone may not sufficiently embrace future vulnerabilities. The low-energy attributes that characterize modern forest processes in the Guiana Shield (see Chapters 2, 3 and 7) suggest that this region's susceptibility to catastrophic change is disproportionately higher than other regions of the neotropics. The large number of endemic plants in relatively high local abundances but restricted geographic ranges within the shield region (ter Steege, 2000) also argues for a rethink on how methods for assessing conservation value cope with geomorphography and its influence on the way climate, substrate, phylogeny and people combine to influence forests across these regions. Across the neotropics, geomorphographic control on forest processes renders the Guiana Shield

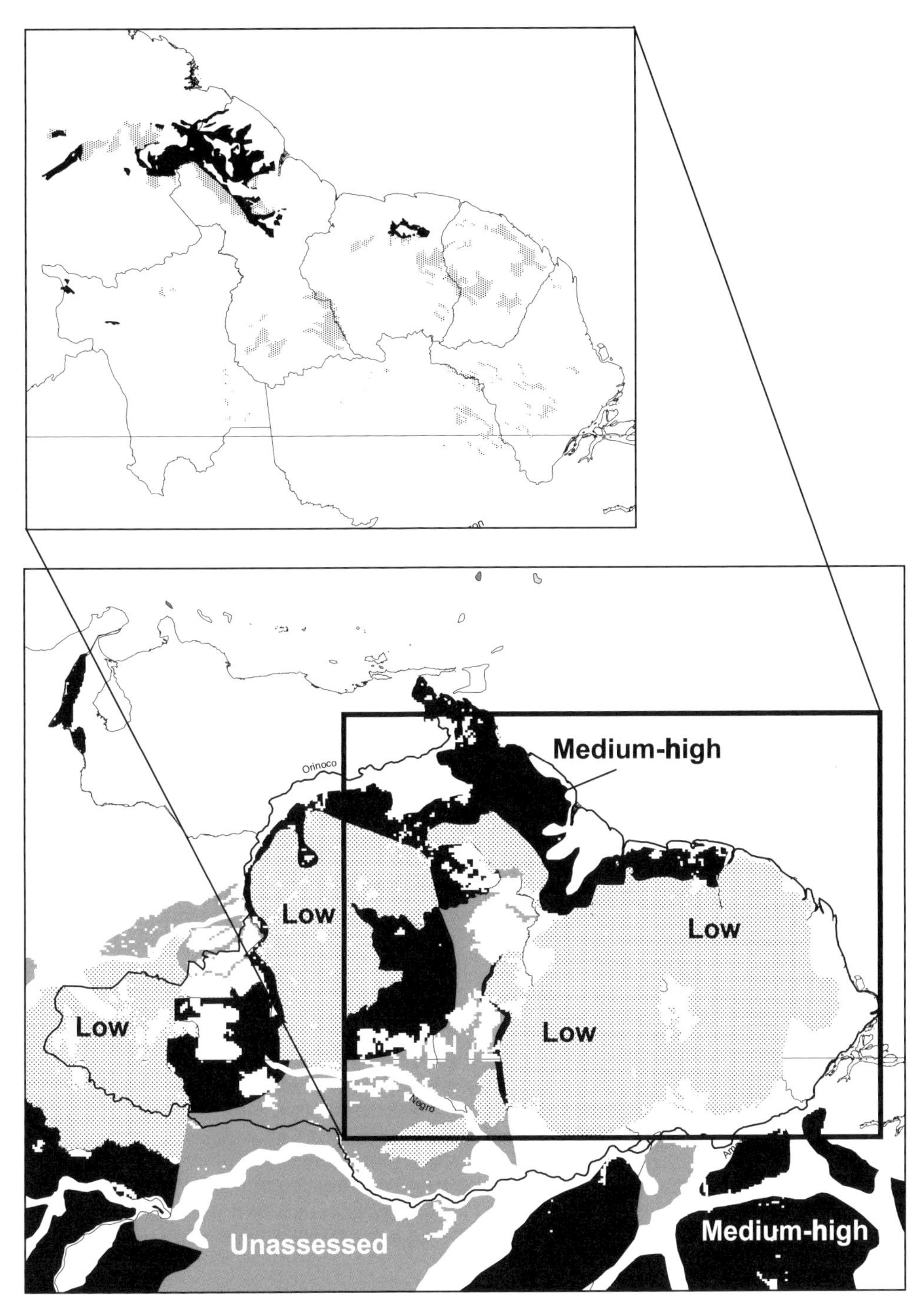

Fig. 9.5. Two-point spatial threat assessment of regions within the Guiana Shield (after Bryant *et al.*, 1997). Threat assessment spatial coverage: WRI.

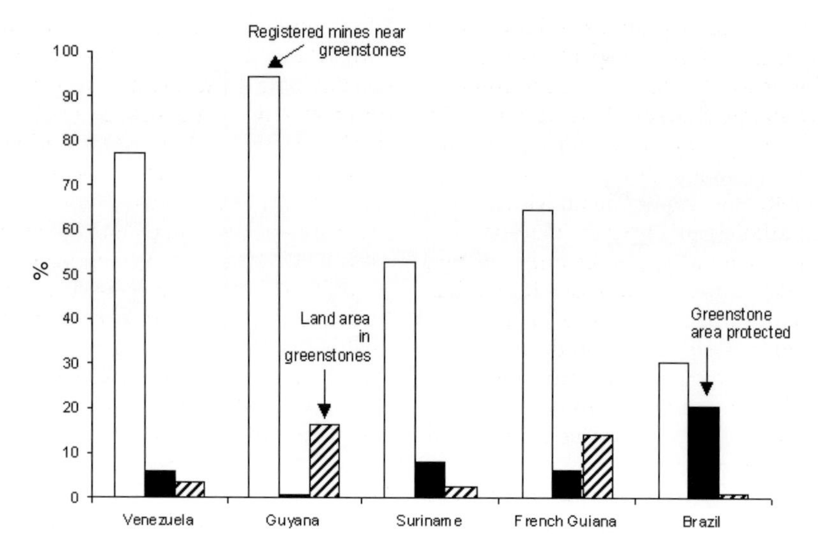

Fig. 9.6. Greenstone belts, gold mines and protected areas. Percentage of land area covered by greenstones, the percentage of registered gold mines located within 10 km of the nearest greenstone formation and percentage of these greenstones located within existing protected areas (IUCN I–IV categories) (see Fig. 9.2). Greenstone distribution based on Gibbs and Barron (1993). Mine locations based on USGS and CVGTM (1993).

particularly vulnerable to widespread change at comparably modest levels of human intervention, particularly when this is poorly managed.

Habitat and forest types as conservation units

Recognizing aquatic habitat groups based on salinity and dissolved organic carbon content helps to separate marine and estuarine habitats along the Atlantic coast from freshwater systems characterized by widely varying DOC, TSS and TZ^+ (Furch, 1984) (see 'Hydrology', Chapter 2). Discrimination of terrestrial habitats according to vegetation type strongly shadows variation in edaphic attributes (Davis and Richards, 1933; Fanshawe, 1952; Richards, 1952; Schulz, 1960; Heyligers, 1963; Cooper, 1979; Lescure and Boulet, 1985; ter Steege *et al.*, 1993; Duivenvoorden and Lips, 1995; Coomes and Grubb, 1996; ter Steege, 2000; ter Steege and Hammond, 2001), particularly in relation to surface and soil moisture status, but also parent material provenance. Rainfall and temperature variation, in relation to topography and geographic location, exert a larger control over soil–vegetation relationships (see 'Soils and soil fertility' and 'Climate and weather sections', Chapter 2), as may historical patterns of human influence (see Chapter 8).

Dissecting the Guiana Shield by the spatial distribution of these aquatic and terrestrial habitats arguably defines an optimum scale for discriminating distributions of all but the largest (e.g. jaguar, puma, tapir) or most mobile (e.g. migrants) of species (see Chapter 4). There are several reasons for this optimum.

First, geomorphographic control delimits, at the widest scale, the range of possible edaphic and aquatic attributes. For example, the absence of significant sources of mineral calcium in the shield, and thus certain edaphic and aquatic conditions, reflects both the geographic position (rainfall and temperature patterns) and weathering age of the region.

Secondly, the underlying Precambrian geomorphology exerts a pronounced influence over drainage dynamics and the extent and magnitude of hydrological disturbance over the lowland edaphic environment. The low-energy system characterizing the surface drainages of the shield region promote autochthonous over allochthonous pathways to soil development. Thus, longstanding differences in parent material and climate resonate more significantly through the soil development process. In regions where widespread import of externally sourced materials is delivered through relatively high-energy fluvial systems, more frequent soil turnover would reduce edaphic patchiness and increase variation at much smaller mixing scales.

Thirdly, habitats in the Guiana Shield represent one of the oldest existing terrestrial regions in the neotropics. During the late Cretaceous and prior to the uplift of both the Andean mountains and the Panamanian land bridge, terrestrial and freshwater life were already evolving on a shield landscape that was dominated by slow gradation and weak diastrophism. In comparison, the western South American environment was believed to be largely epicontinental (shallow marine) (see Chapters 2 and 7). This would suggest a much longer legacy of continuous *in situ* terrestrial and freshwater evolution across the major shields relative to other regions of the neotropics. While proxy measures suggest climate fluctuated significantly over at least the last 80 million years, no evidence currently points to a wholesale extinction across the Guiana Shield. This cannot be discounted, but neither can it be conclu-

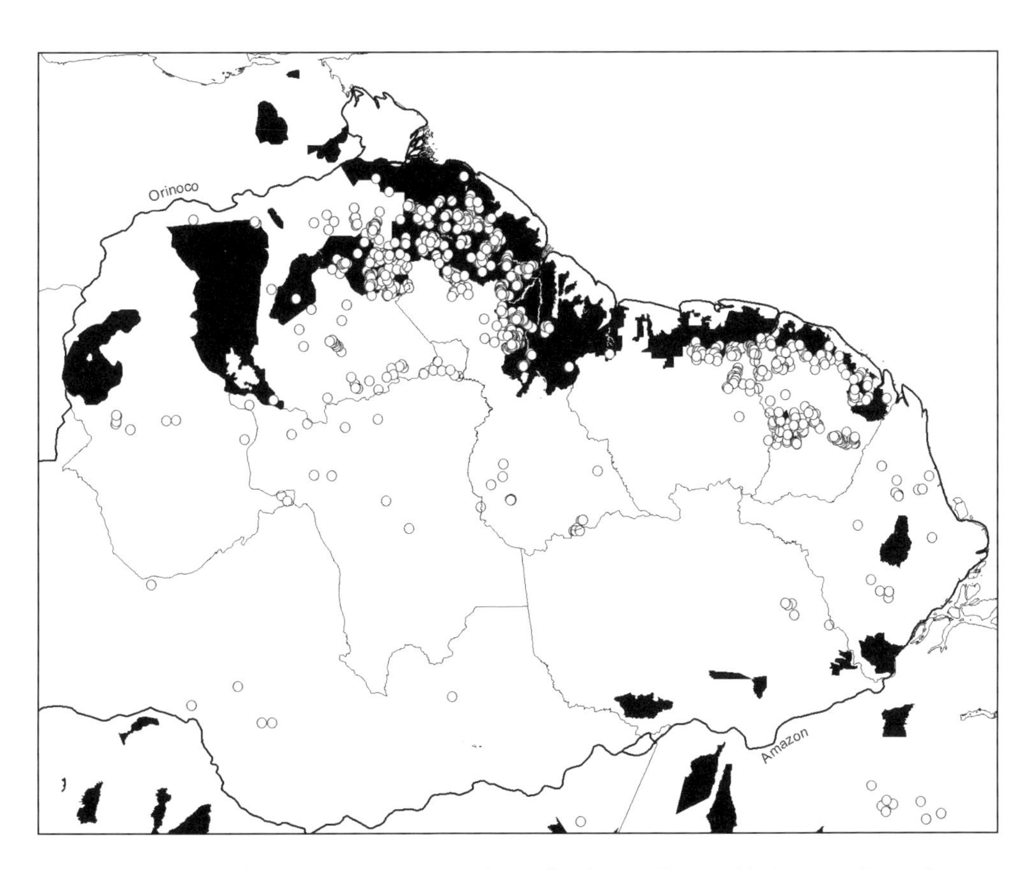

Fig. 9.7. Zones of overlapping mining (empty circles) and timber production (black) across the north Guiana Shield. See Figs 8.8 and 8.17 for data sources.

sively supported from palaeontological evidence derived from other parts of the neotropics and then scaled up across very large regions that have contributed few, if any, fossilized remains.

While some lineages undoubtedly disappeared from the region during unfavourable shifts in climate, it is difficult to counter a much earlier start to terrestrial evolution of life in the shield. Given the relative stability of the region's geographic position over most of the Cenozoic, it would have supported the accumulation of biotic attributes that bring a comparative advantage in low-energy environments. The primacy of meso-scale edaphic specialization in defining opportunity in this competitive environment translates into a large number of modern forest types dominated by one or few endemic species that are hyper-abundant within their highly restricted geographic ranges and express attributes that place a premium on high per capita survival, not dispersal.

At larger classificatory scales, much of this important habitat variation is lost. Clear distinctions identified by foresters, ecologists, botanists and natural historians over the last century or more are merged into single associations. The drawback at these large scales is that the aggregation of wide-ranging patchiness simply reinforces notions of tropical environmental uniformity. Ironically, viewing systems at this large scale supports the primacy of smaller-scale processes, that strongly influence alpha-diversity levels (e.g. Hubbell *et al.*, 1999), as the nexus of conservation decision-making. The important role of the physical environment in structuring variation at meso-scales is removed. Meso-scale, habitat or beta-diversity, is lost in the analysis.

Across much of the Guiana Shield where alpha diversity is relatively low (ter Steege *et al.*, 2000), and large populations of narrow-ranging endemics are common (Fanshawe, 1952; Richards, 1996), the collapse of habitat variation into larger units fails to emphasize the value of this uncommon feature. Endemic species in the shield region are typified by a very high abundance to range ratio (e.g. *Alexa* spp., *Dicymbe* spp., *Eperua* spp., *Chlorocardium rodiei, Dicorynia guianensis, Catostemma* spp., *Micrandra* spp., *Mora gonggrijpii*). Common species in Central America and western Amazon appear to have much larger geographic ranges with much lower relative abundances (e.g. *Iriartea deltoidea, Poulsenia armata, Pseudomeldia* spp.). Uniqueness, an important component of conservation value, thus equates differently in the Guiana Shield than it does in other neotropical regions affected by different geomorphographic controls. These regions are also likely to express components of conservation value differently.

In the Guiana Shield, delineation of conservation units based on forest type characterization creates greater opportunities for successfully meeting the challenges that confront representative conservation. Unique habitats or forest types across the northern shield region work well as working units because these are strongly linked to transitions between relatively large-seeded endemics with rapidly diminishing representation at larger scales. Refined soil-type classification cross-referenced with geographic ranging of endemics, in this instance, would yield an optimum conservation topology if the objective is representative forest system protection.

Management Approaches and their Applicability

Mainly protection

As a management approach, protected areas (after IUCN categories I–III) aim to carry out a series of functionally important roles, at least in theory. According to the IUCN, protected forest areas should: (i) maintain forest ecological and genetic processes; (ii) minimize artificial disturbances; (iii) provide opportunities for low-impact research, education and recreation; and (iv) protect outstanding natural features and scenic areas of national or international significance. In practice, several of these broad-based, generic objectives can prove difficult to achieve without substantial tailoring at the conservation unit level for a number of reasons.

Ecological and genetic processes

One difficulty is defining measurable indicators of forest ecological and genetic health that are appropriate to the system being protected. For example, protection of forests in the western United States over the last half century has implicitly involved thwarting significant disturbance from fire, largely as a consequence of the catastrophic forest and human loss caused by the Great Fires of 1910 (Pyne, 2001). Massive, stand-replacement fires at the turn of the 21st century have now prompted a re-think about how best to define 'natural' processes and how best to 'protect' these. Fires are seen now as an inimical part of most forest ecological processes and subduing their influence over decadal scales has created conditions literally fuelling catastrophic stand replacement events at landscape scales.

Across the neotropics, regional forest systems are also subject to different large-scale influences on local ecological processes. While these invariably overlap, the Guiana Shield area is clearly exposed to far fewer of these (Fig. 7.2; see Chapters 2 and 7). Charcoal and historical evidence point to fire as a widespread natural agent of catastrophic change across the shield region, other parts of the eastern Amazon and, possibly, parts of Panama and Central America. In contrast, the western Amazon has yet to be characterized as a fire-affected system. Rather, it appears more strongly controlled by high-energy hydrological disturbances (Räsänen *et al.*, 1987) (see Chapters 2 and 7). Equally, the relative impact of pre-Columbian indigenous societies appears to vary spatially (see Chapter 8). The relative impacts of these different trajectories on smaller-scale ecological processes are not readily thrown under the same definition or monitored using the same criteria because they may:

1. Alter the standing forest stock of life-history attributes (Hammond and Brown, 1995; ter Steege and Hammond, 2001);
2. Affect how these bring comparative advantage to different plant and animal taxa under prevailing conditions;
3. Shape the relative influence of deterministic (e.g. historical human forest use) and stochastic (e.g. density-dependent mortality) processes on forest change; and
4. Regulate how fast the balance of standing taxa and biomass is changing as a consequence.

'Artificial' disturbances

Consequently, separating disturbances that form part of the 'natural' ecological process from those that are 'artificial' can also prove difficult, if forests being protected prove to have been subject to widespread, long-lasting and/or intensive prehistoric use. Many small protected areas in Central America previously believed to have been 'pristine' have subsequently proven affected in part by pre-Columbian agricultural use (e.g. Barro Colorado Island, Panama: Piperno, 1990; La Selva, Costa Rica: Kennedy and Horn, 1997) as have many protected forest areas in the Guiana Shield (e.g. Kanuku Mountains, Guyana (proposed): Evans and Meggers, 1960; e.g. Iwokrama, Guyana: Williams, 1994; Nouragues, French Guiana: Ledru *et al.*, 1997). Again, the problem arises in discriminating the relative effects of these chronological series of events from other stochastic processes at work. If the accumulated archaeological evidence reflects the true distribution of prehistoric human impacts, then chronologies of human impact across the interior of the Guiana Shield should be compressed relative to Central America, the Amazon Downwarp and Sub-Andean Foredeep regions.

Research, education and recreation

Unlike the more diffuse realities of environmental protection, these real-time activities are more amenable to practical management. The main difficulty in generalizing arises with the thresholding definition of low-impact. Take the example of fire. Across many protected areas in the neotropics the creation of managed fire treatments as part of research would not be considered among the low-impact, allowable uses. Yet, ecotonal forests in some protected areas along the perimeter of tropical savannas are frequently affected by fire (Thompson *et al.*, 1992). If the objective of restricting use is to minimize human influence over 'natural' system processes, then the allowable range of activities associated with these general forms of low-impact use are again subject to local considerations of suitability. In the Guiana Shield many locations would require significantly lower impact allowances than parallel communities within ramping system regions.

Protected area management in the Guiana Shield

Strategies and methods for prioritizing and selecting protected areas across the Guiana Shield have dominated processes associated with protected area management (Hoosein, 1996; Rodríquez and Rojas-Suárez, 1996; Huber, 1997; Stattersfield *et al.*, 1998; ter Steege, 1998; Funk *et al.*, 1999; ter Steege, 2000; Mittermeier *et al.*, 2001; Funk and Richardson, 2002). Exploring appropriate objectives and methods of managing established areas has received less region-wide attention, although this is arguably the major constraint facing functional conservation in most areas. Bruner *et al.* (2001) concluded that the number of park guards was the best determinant of protection effectiveness in tropical parks based on questionnaire surveys of park managers and park staff (56%), NGOs and researchers (30%) and protected area agencies (14%).[3] Defending protected areas through enforcement and patrolling invariably plays an important role in deterring illegal activity. But taken as a top priority, enforcement and patrolling is a blunt instrument that is symptomatic of failure, rather than success, in effectively demonstrating protected area benefits to local communities to the same or greater degree than has been done at international levels.

Significant areas across the shield region have been established for habitat protection and this coverage already rivals all other tropical forest regions for the title of 'best protected'. Only Guyana (and to a lesser extent French Guiana) has yet to legislate for areas that have been under consideration since the 1990s, but this is likely to change. The region boasts nearly 30% of remaining closed tropical forest cover in South America and 13% of that estimated to occur globally (Tables 9.1 and 9.2). Approximately 20% of lowland tropical forest in the region is protected and this is likely to rise to 25% by 2010.

With one-quarter of the area protected and all countries in the region already or expected to have more than 10% of their land area committed to strict habitat protection, tailoring objectives and identifying practices that will lead to relevant conservation in these areas would appear crucial. Moreover, articulating these through well-considered management plans and coordinated actions has remained noticeably absent from many designated areas (e.g. Huber, 1995, 2001). In part this is due to a shortfall in resources needed to effectively manage for conservation in a region being increasingly used for its mineral, timber and wildlife resources and weighed down by excess external debt burdens, poor access to education and health care, bouts of social unrest, high rates of emigration and mounting costs of infrastructural maintenance (see Chapter 8).

Mainly timber

Management for timber production has been a long-standing focus across many regions of the Guiana Shield since the early 1900s. Regional notions of sustainability, however, have broadened beyond maintenance of harvestable volume to include many non-commodity benefits (e.g. formulated in the Tarapoto Agreement). Nonetheless, timber remains the largest forest sector contribution to national GDP, apart from mining, and efforts to improve timber production and its management have explored several avenues (e.g. King, 1963). These have focused principally on plantation and natural forest management approaches to timber production and the potential for supplying material to paper pulp, (peeled and sliced) veneer, plywood, sawn split and hewn wood markets (e.g. Vink, 1970; Welch, 1975a).

Plantation forestry

The prospect of meeting domestic demand and supplying export markets with timber grown in a plantation setting has largely proven unworkable over most parts of the Guiana Shield where trials have been conducted (Vink, 1970; de Graaf, 1986). A wide range of native commercial timber species have been examined, including *Simaruba*, *Carapa*, *Cedrela*, *Peltogyne*, *Dipteryx*, *Centrolobium*, *Caryocar*, *Anacardium*, *Virola*, and *Dicorynia* with little success (British Guiana Forest Department, 1940–1965; Welch, 1975a; de Graaf, 1986). The potential of faster-growing exotics such as *Pinus caribaea*, *Eucalyptus* spp. and *Gmelina arborea*, as well as valuable hardwoods, such as *Swietenia*, *Khaya* and *Tectona* was also explored with mixed success due to the high costs of pest and weed management combined with disappointing growth increments (Vink, 1970). For example, one trial of *P. caribaea* grown on ferrasolic soils in Guyana achieved an annual average increment of 1.8 m^3/ha over a 20-year period, a figure only marginally improving on estimated volume accrual in many natural forests of the region. Perhaps the most outstanding example of plantation failure is the case of the pulpwood operation established along the lower Jari River in south Amapá/east Pará states (Rollet, 1980; Palmer, 1986). Monocultured *Gmelina* and *Pinus* were battered by fungal and insect damage and rapid declines in productivity, principally through disregard for the role of internal nutrient cycling (Russell, 1987) and density-dependent attack in maintaining productivity. Over the last 20 years, however, some productivity gains have been made, primarily by addressing these shortfalls and refining and redirecting pre-site management effort (McNabb and Wadouski, 1999), but nutrient emigration through biomass removal, particularly scarce calcium, continues to proceed at very high rates in stands of *Eucalyptus urograndis* (Spangenberg *et al.*, 1996).

By the close of the 20th century, very few plantations were operating within the Guiana Shield region. Although Brazil produced between 32 and 46 million m^3 of plantation wood fibre alone, a mere 4–5% of this originated in the Guiana Shield and virtually all of this was associated with the Jari pulpwood operation (IBGE, 2002). As of 2000, no other commercial-scale plantations are producing significant quantities in the region.

Plantations hold tremendous prospect in providing a reliable supply of light hardwood and wood fibre to supply local downstream industries and market demand (F. Wadsworth, personal communication). Little success has been met across the Guiana Shield in establishing commercially viable plantations of both natural and exotic species. The principal difficulty arising is the cost-effectiveness of producing wood through this approach. Relatively slow growth increments combined with elevated pest and weed management costs and stagnant prices for low-end tropical timber and fibre work to reduce or eliminate profit margins. Improving growth, cost-effectiveness of management or timber prices would potentially make plantation forestry a more attractive investment. In many parts of the Guiana Shield where forests have been degraded through high-frequency selective logging, plantations under these circumstances might offer a means of improving landscape fertility (Lugo, 1995), increasing wood supply and offering greater flexibility in allocating natural forest areas to lower-impact direct and indirect uses. It is unlikely, however, that the accelerated growth required from plantation-grown timber would foster the same strength and durability properties (see 'Wood density', Chapter 7) that have typified the main line of exported timber products from the Guiana Shield for centuries (e.g. Mackay, 1926).

Natural forest management

Managing standing mixed forests of the interior for long-term production of these heavy hardwood products holds considerably greater promise if natural regeneration is employed as the main avenue to restocking. Systematic selective logging has been the main route to timber production in the Guianas since the early 20th century (McTurk, 1882; Hohenkerk, 1922), although pro-active effort to manage forests for timber was not taken up until the 1950s.

A number of silvicultural approaches have been explored in an attempt to direct natural regeneration towards greater timber-tree stocking. Among the most widely known are thinning (e.g. through poisoning and girdling) (King, 1965; Jonkers and Schmidt, 1984; de Graaf *et al.*, 1999), enrichment planting, climber and liana elimination (e.g. Putz, 1991), seed tree retention (e.g. Plumptre, 1995) and stand damage control (e.g. van der Hout, 2000). Combined, these techniques form a process that aims to modify over time the existing mix of species and size classes in a way that improves timber value and manageability of the stand, often referred to as 'domestication' of natural forest (Dickinson *et al.*, 1996; de Graaf, 2000). Employed *à la carte*, they can act as effective tools in ensuring that forest functioning remains largely unimpeded, although (at least transitory) shifts in composition are inevitable at the harvesting intensities required to support capital-intensive, industrial approaches.

Reduced-impact logging

Stand damage control is arguably the *raison d'être* for employing reduced-impact logging (RIL) techniques (Hendrison, 1989). However, reducing the unit costs of extraction is commonly submitted as the financial pay-off associated with controlling damage (Holmes *et al.*, 1999; van der Hout, 1999, 2000). This pay-off is intimately associated with the up-front investment in preharvest planning of felling and extraction (e.g. Hammond *et al.*, 2000). As a consequence, operational efficiency is improved and wood waste reduced in comparison with unplanned approaches (Boltz *et al.*, 2003). Yet, many operators have been slow to implement these practices, in part because the management of more complex approaches to harvesting bear with them additional costs beyond forest operational considerations (Hammond *et al.*, 2000). The financial benefits of employing RIL techniques are often slim, neutral or slightly negative (Barreto *et al.*, 1988; Winkler, 1997; van der Hout, 1999, 2000; Armstrong and Inglis, 2000), suggesting even short-term lapses in management rigour could create further financial losses. These 'other' costs become even more significant when it is apparent that they may not lead to any immediate financial benefits from a market that is slow to pay higher tropical timber prices (Barbier *et al.*, 1994), despite efforts to structure incentives for better forest stewardship through certification approaches. Operators are largely unwilling to bear greater up-front costs and greater financial risk in the face of longer-term political instability and uncertain land tenure (Pearce *et al.*, 2003). In effect, growing and harvesting trees through natural forest management approaches in the tropics requires adoption of an investment horizon that extends well beyond the limits of normal risk tolerance. Across the Guiana Shield, this horizon may extend beyond the lifetime of the investor for a number of reasons linked to underlying geomorphographic controls and the need to maintain competitiveness in a market characterized by a wide range of interchangeable material options.

Sustainability of harvest rates and intensities

A competitive real rate of financial return on an investment is rarely characterized by a minimum 50+ year period of maturity. Yet, this may be what is required in order to sustain timber yields in the Guiana Shield through up-front investments in refined management practices under operational and market conditions at the turn of the 21st century. Several factors come into play in defining why the term of investment should be of such length.

First, stand-level increments for many heavy hardwood timber trees are generally some of the lowest in the industry (Prince, 1971; Veillon, 1985; Silva *et al.*, 1995; Zagt, 1997). Even after considerable logging intensities, that should have strong liberating effects, models indicate that forest stocking is unlikely to reach preharvest levels within 40 years at sites in western tropical Venezuela (Kammesheidt *et al.*, 2001), Guyana (Zagt, 1997; ter Steege *et al.*, 2002) and Suriname (de Graaf, 2000). On this basis, a cutting cycle of at least 60 years has been recommended in order to maintain forest biomass (Zagt, 1997; Kammesheidt *et al.*, 2001), although the CELOS system in Suriname originally recommended a 20–25-year cycle (de Graaf, 1986).

Secondly, hidden defects such as hollowing or decay are very common in commercial size-classes, at least in most timber trees of Suriname and Guyana (de Milde, 1970; de Milde and Inglis, 1974; D. Hammond, personal observation). While not normally considered in assessments of forest stand responses to silvicultural intervention, defect rates among many slower-growing, heavy hardwood species (e.g. *Chlorocardium*, *Mora*, *Swartzia*) are generally greater than in faster-growing, long-lived colonizers that dominate timber industries in many other parts of the neotropics (e.g. *Swietenia*, *Cedrela*, *Cedrelinga*). Defective logs, once mistakenly felled, are normally left in the forest or at the consolidation market or mill yard once detected. Most modern gang-saw operations do not easily cope with these types of defects. Thus, the 'processable' volume can be considerably lower than the standing commercial volume often used as a measuring stick to gauge initial extraction opportunity and later postharvest stand performance.

Thirdly, investment in good forest management practices, such as RIL techniques, can only yield a return below a threshold of logging intensity. Above this threshold, the relative value of employing RIL techniques diminishes rapidly as the density of felling and extraction overrides canopy and potential crop-tree conservation effects typically improved by RIL at lower intensities (Sist *et al.*, 1998; van der Hout, 2000). Consequently, a considerable volume of harvestable timber is foregone over the short-term as an investment in longer-term timber quality and sustained volumes takes precedence. This creates an almost irreconcilable predicament for large-scale operators that work to achieve the economies of scale necessary in recouping hefty capital investments in equipment and downstream processing facilities.

Relatively high work rates must be achieved in order to financially warrant the use of heavy machinery and large sawmills. In Guyana, initial efforts to mechanize the timber industry focused precisely on the higher work rates that could be achieved (Grayum, 1971). Expressed on a unit basis, costs were expected to be lower than more labour-intensive methods (Fig. 9.8A). However, expressed as a function of time, increased use of machinery demanded higher production rates, because daily costs were considerably higher (Fig. 9.8B).

In forests across the Guiana Shield, slow growth rates, high defect rates, sizeable areas of non-commercial forest (e.g. *bana*, *muri* scrub, palm swamp) (Fanshawe, 1954; Vink, 1970; Welch, 1975b), poor timber price growth and increasing transport distances combine to challenge the financial sustainability of capital-intensive operations. Yet, views persist that much shorter cutting cycles and higher intensities should be employed across some remote areas of the region (J. Leigh, personal communication).

In reality, commercial logging intensities have traditionally been self-limiting across many forest areas because of these factors. Extraction rates of 6–10 m^3/ha are not uncommon in many modern, large-scale operations (van der Hout, 1999; Armstrong and Inglis, 2000). Yet, even modest increases to 20 m^3/ha recommended as part of the CELOS Management

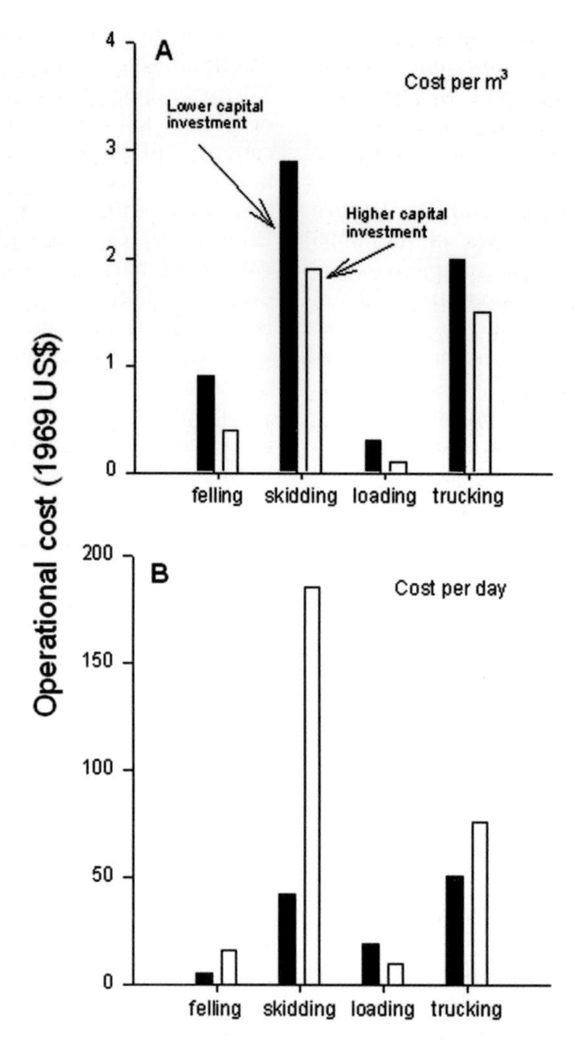

Fig. 9.8. The calculated cost of employing chainsaws and mechanized skidders and loaders in forest operations in central Guyana expressed as a function of volume extracted and operational time. Data source: Grayum (1971).

System have not precipitated the expected re-stocking necessary for another harvest after 20 years, even after liberal application of refinement treatments (de Graaf *et al.*, 1999).

Low, but highly concentrated, stocking of timber trees presents specific problems when heavy hardwood regeneration is optimized under relatively low levels of canopy openness (ter Steege *et al.*, 1995), but optimal harvest rates require extraction of a larger number of stems from these smaller areas. Forests throughout the Guiana Shield typically show relatively high dominance of 1–3 commercial taxa and their distribution is always concentrated at scales defined by edaphic or dispersal limits (see Chapter 7), a feature not as commonly encountered in other commercial forests in the neotropics. Thus, the optimal intensity of logging that makes best use of the investment in RIL depends to a great extent upon the distribution of commercial stems throughout the stand and how this interacts with the balance of financial considerations, such as achievable market price, amount of rent (taxes, royalties) payments and changing costs of operational inputs, such as fuel, parts and staff remuneration.

Fourthly, the absence of cost-effective silvicultural techniques also affects the relatively long term of investment required for sustainable timber production in the Guiana Shield. Most techniques, including liberation and enrichment planting, have been shown to yield comparatively better performance of target timber species, but have also proven far too costly (de Graaf *et al.*, 1999). Moreover, decision-making at the time of postharvest, silvicultural intervention may not adequately predict changing market demand and opportunities. Consequently treatment costs may be borne up-front, only to find they were incurred to eliminate species in demand at the time of the subsequent harvest. Non-commercial species may also play an important role in provisioning forest stands with important nutrient import (Perreijn, 2002) or dispersal services (Hammond *et al.*, 1996) (see Chapter 4). Vertebrate-dispersal of timber tree seeds is arguably greater in forests of the Guiana Shield than in any other tropical region (e.g. Jansen and Zuidema, 2001) and these services form an important part of natural regeneration approaches to timber management. The cost-ineffectiveness of postharvest intervention limits the range of tools that can be employed in stimulating faster growth, if that is the main objective in timber management, to the amount of canopy openness created during felling and extraction.

Quality vs. quantity as the timber management objective

A focus on timber quality would appear intuitive given the conditions structuring operational costs and constraints on forest productivity across the region (Hammond, 1999). Properties of high structural strength, resistance and durability define most timbers in demand overseas, principally for their load-bearing capacities (docks and wharves), and resistance to wear (flooring, decking) and maritime infestation (locks, docking), among other applications (Fig. 9.9). Higher harvesting rates, however, will inevitably liberate the canopy and stimulate faster growth. For many forests in the region, this will increase the rate of commercial volume accrual, but principally with faster-growing, light-wooded species. Heavy hardwood species recruited during these periods into harvestable size classes may also see a reduction in wood density as a consequence of growth stimulated by more open canopy conditions (but see factors affecting density in 'Wood density', Chapter 7 and 'Growth in relation to canopy openness' in Chapter 3). Consequently, future timber production may revolve around lower-density wood, a market that is open to a wide range of species grown principally in plantation forests throughout the world (compare across regions in Fig. 9.9). Given production levels within the Guiana Shield, this route would not appear to provide any comparative advantage over that already existing through production of increasingly scarce heavy hardwood material (Hammond, 1999). The trade-off required, however, is a focus on lower production volumes from forest managed for timber through natural regeneration. There are, however, other non-timber benefits attached to forests in the Guiana Shield and these will for the most part continue to accrue under low-impact systems of timber production. Compared with much higher rates of tropical forest loss elsewhere and remaining standing forest areas (see Table 9.1), continued low-level wood production from forests in the Guiana Shield should see increases in unit prices if a focus on wood product quality is maintained.

Operational waste and inefficiencies

RIL techniques measurably offset financial investments in planning by reducing operational waste and inefficiency. However, for the broader forest operation and forestry sector to achieve sustainability (after Goodland and Daly, 1996), other point

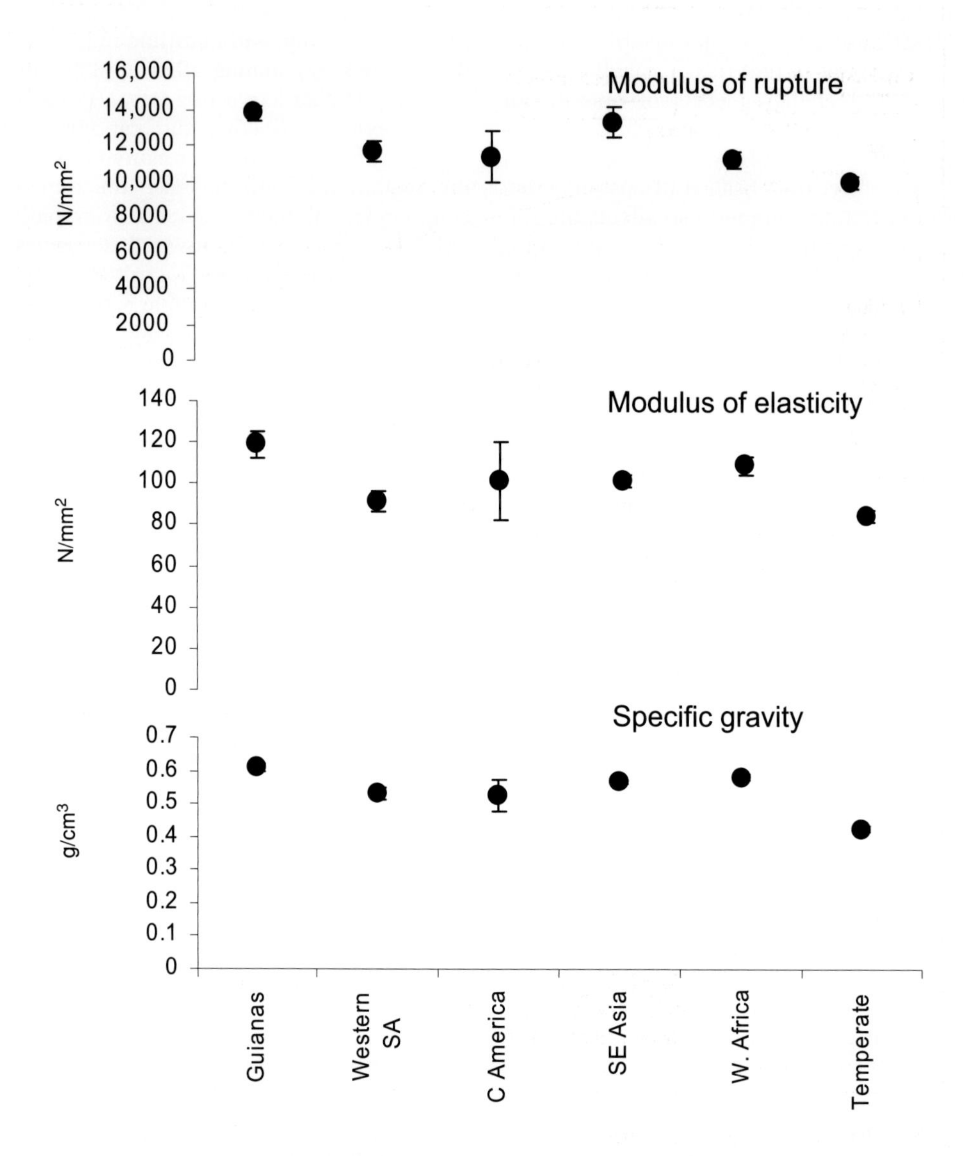

Fig. 9.9. Mean (± SE) wood specific gravity and two of its mechanical property correlates for timber species grouped by region. See Fig. 7.6 for data sources.

sources of waste and inefficiency must be wrangled and subdued if RIL is to achieve its measured objective of leaving a growing, non-depreciated forest without foregoing the immediate financial objective of doing business. Arguably, other factors may be spurring investment in marginal prospects of financial return from timber in the shield region in the first place, but these are too complex to be explored here (see Repetto and Gillis, 1988). It is clear, however, that operational waste and inefficiency in felling and extraction of timber form only the first links in a chain of waste (Fig. 9.10) that, depending on its length and strength, can terminate any prospect of positive

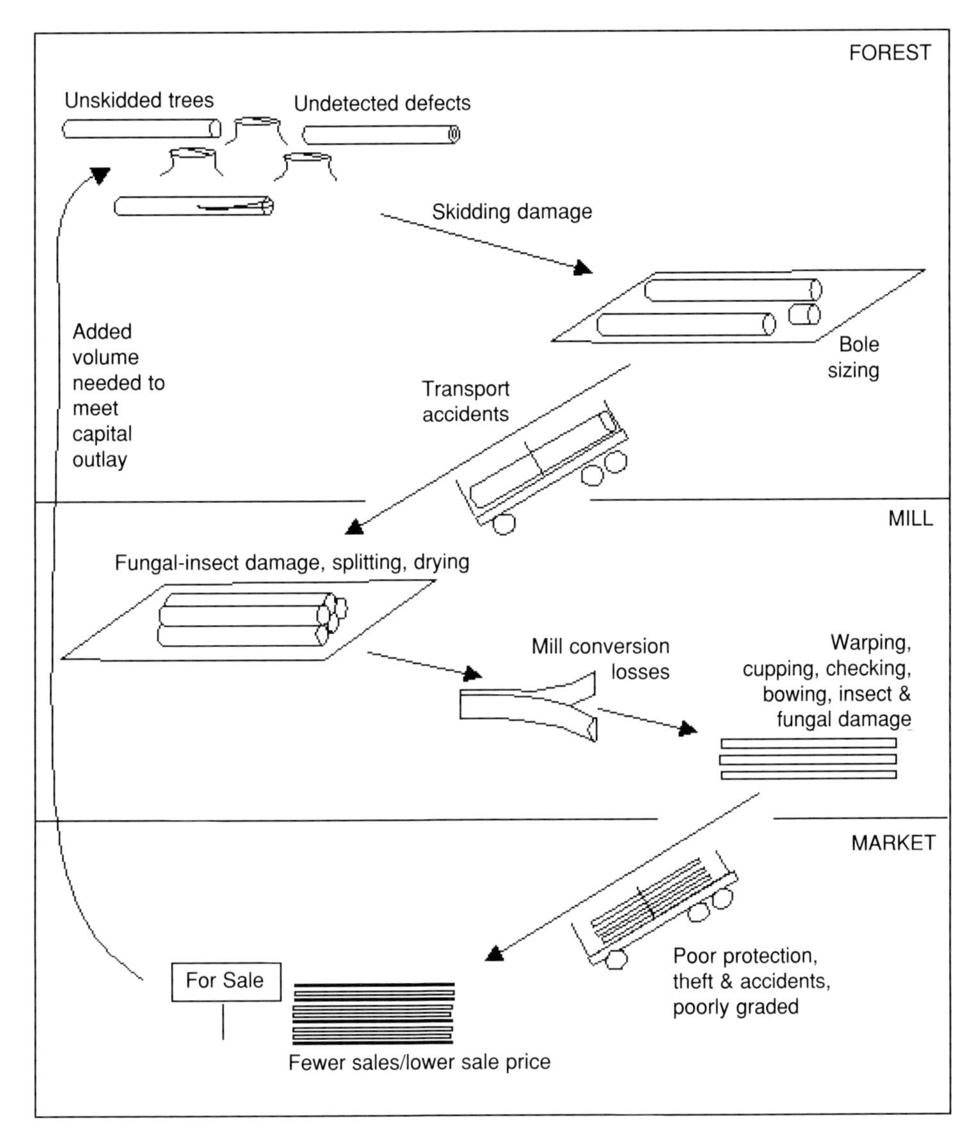

Fig. 9.10. Schematic describing the chain of waste that can exist in inefficient and poorly managed timber production companies. RIL techniques can reduce this waste up to the forest-to-mill phase (arrow). Cost savings of employing RIL can be rapidly eroded through inflated downstream waste.

financial return to timber operators after meeting reasonable tax burdens.[4] As a consequence, accumulated costs associated with small, but frequent, wastages downstream can accrue (Fig. 9.11A, B), putting pressure to sell finished products at lower prices. Ironically, high losses to waste and increasing operational costs can spark further demand for logs from the forest in an effort to offset initial capital outlay and mounting depreciation costs through economies of scale. Residual commercial stems in logged-over stands in particular can be targeted, particularly as transport distance between mill and unlogged forests increases unit costs and multiple re-entries, often within the same year, become commonplace (D. Hammond, personal observation).

Fig. 9.11. (A, B) Chasing waste. Extracting timber that is never processed creates unnecessary pressures on the forest resource since more logs must be extracted in order to meet financial targets. Reducing waste through transport, storage and milling phases will yield higher returns and improve government revenue. (C, D) Changing face. The changes to the forest system created by surface mining in the Guiana Shield can be severe when rehabilitation effort is low. Tall grass in (D) is the common non-flooded savanna species, *Andropogon bicornis.*

Discount rates and heavy hardwood growth and mortality

Defining appropriate discount rates to apply when assessing the net present value (NPV) of commercial forests remains a matter of almost arbitrary assignment. Yet NPV is a fundamental calculation influencing the economic viability of investment in tropical forests. Typically, discount rates are applied as brackets around typical lending rates when calculating forest NPV. However, the economic factors affecting lending rates are hardly sensitive to changes in the forestry sector, particularly across the Guiana Shield where it typically accounts for less than 5% of GDP (at factor cost). Presumably, when logged forests are viewed (solely) as timber assets, they appreciate based on the rate of increasing commercial stock, the cost of bringing the product to the market and the market price achieved. Therefore, ecological factors that influence the rate of increase should also shape the rate of asset appreciation, all else being equal. The primary determinants of this rate are based on basic population changes defined by growth and mortality rates. Yet, if growth is slow, as it is for many heavy hardwoods common to the Guiana Shield, then the fraction of these lost to mortality events while occupying commercially harvestable size classes will be considerably greater than faster growing species with typically shorter life spans (ter Steege and Hammond, 1996). This would suggest that discount rates based on appreciation of heavy hardwood timber stocks should be set higher than stands appreciating largely through growth of faster-growing commercial species. Since these invariably form a larger part of future commercial growing stocks under natural forest management regimes, discount rates applied to forest NPV should decrease with time, all else being equal. Factors that stimulate greater mortality rates, such as catastrophic disturbance events, and therefore greater investment risk and higher lending rates, would suggest spatial variation in the application of discount rates across geomorphographic regions.

Mainly mining

The extraction of gold, bauxite, diamonds and iron is arguably the most substantial non-timber use of forests in the Guiana Shield (see Chapter 8). In terms of biophysical and socio-economic impact, the production of minerals spreads across a much greater portion of the region, figures more prominently in the agenda of national economic development and conjures up greater visions of rapidly accruing wealth than virtually any other forest-based activity. Most mining operations open and shut over a 15-year period (see 'Commercial mining', Chapter 8), or are consumed by operational inefficiencies as global commodity markets fluctuate independent of extraction costs. The long planning horizon of timber production appears almost sluggish by comparison and mining does not need to carry with it any visage of sustainability comparable to the timber industry.

It does, however, increasingly need to deflect growing criticism over the way in which it is being managed across the region and the negative impacts that this is causing through:

1. Little or no management of sediment and mecury effluent;
2. Rapid spread of a wide variety of communicable diseases, including HIV and malaria (Palmer *et al.*, 2002);
3. Breakdown of family units structures (Heemskerk, 2001);
4. Negative impacts on neighbouring rural communities that may have very different forestland use priorities;
5. Extirpation of local wildlife through unrestricted hunting, fishing and live-animal collecting; and
6. Large-scale surface degradation with little or no post-closure restoration effort (Fig. 11C,D).

Mitigation of mining impacts

Although considerable attention has been focused traditionally on large, international mining consortia that are operating at various locations throughout the region (e.g. Colchester, 1997), small-scale miners are more numerous, less easily regulated and more often working illegally. When environmental and social impacts become unacceptable, it is virtually impossible to determine accountability in a way that is more transparent with larger, stationary operations. Among these impacts, mercury poisoning has arguably attracted the greatest attention because of the important public health consequences to people living in rural environments and dependent on aquatic resources for their livelihoods. Sing *et al.* (1996) measured significantly higher mean organic mercury levels (31.3 mg/l) in blood samples taken from Makuxi villagers along the Cotingo River with high exposure to mining operation affluent. Levels of total mercury in urine of Maroon gold workers in Suriname were significantly higher than those not involved in gold mining, although blood levels were generally low (de Kom *et al.*, 1998). In French Guiana, dietary habits accounted for a significant part of mercury-level variation in hair samples of rural residents, particularly among Amerindian children (Cordier *et al.*, 1998). The dietary connection is largely based on fish that readily sequester and concentrate mercury in their tissues. In Suriname, a survey of the freshwater fish community showed that predatory fish had significantly higher concentrations of mercury than non-predatory species and that these levels were often exceeding maximum permissible concentrations (Mol *et al.*, 2001). River bottom sediments in Guyana tend to concentrate mercury downstream from intensive mining districts (Miller *et al.*, 2003), where it is available for uptake by benthic feeders. Although significant background mercury levels have been detected throughout many parts of the region as a consequence of its long history of weathering and exposure to forest fires (Veiga *et al.*, 1994; Fostier *et al.*, 2000; Santos *et al.*, 2001), both important sources of naturally occurring mercury, the role of mining in catalysing widespread exposure to much higher concentrations of this toxic heavy metal is difficult to dispute (Nriagu *et al.*, 1992; Roulet *et al.*, 1999). Unfortunately, the hydrochemical properties of most rivers in the Guiana Shield make them ideal transporters of metallo-organic complexes of methylated mercury (Silva-Forsberg *et al.*, 1999).

Mainly non-timber forest products (NTFPs)

Non-timber approaches to forest use include a large basket of potential goods and services. Many studies do not discriminate commercially viable and subsistence-oriented materials, particularly when NTFP abundances are being inventoried. Frequently every plant has several useful purposes within the traditional use system of most Amerindian people, so assessing NTFP status alone does not allow for good discrimination of prospective income-earners. Here an emphasis is placed on NTFPs in the Guiana Shield that have or could generate sufficient market demand to provide some measure of income support to rural livelihoods.

Direct use forms include the production of plant oils, resins and latexes, fibres for furniture and handicraft manufacture, food plants, collection of wildlife for the pet trade, and fish and bushmeat for market sale. Biochemical components of plants and animals can also be surveyed for their prospective use in agro-chemical or pharmaceutical applications. Indirect uses include services provided by forests in the form of tourism, downstream water quality and flow regulation, carbon storage and sequestration and more complex connections associated with regulation of land–ocean–atmosphere fluxes forming the global biogeochemical cycles. Non-use forms include existence value concepts attached to philosophical or spiritual beliefs supporting forest persistence.

Non-timber forest plant products

Historically, a number of important NTFPs have been commercialized across the Guiana Shield (Table 8.2), including balata (*Manilkara*), rosewood oil (*Aniba*), copal resin (*Hymenaea*), sarrapia (cumaru) beans (*Dipteryx*), curare (*Strychnos*), among others. Depending on the part of the plant required, the harvest of many NTFPs led to commercial extinction across much of their range in a manner more commonly attributed to timber production (see Chapter 8). Thus, concepts of sustainable use and harvest intensity apply equally to both timber and non-timber products. In the Guiana Shield, this has become particularly important in those industries that have developed industrial-scale demand for raw materials.

Palm-heart sustainability

Among these, palm-heart (*Euterpe* spp.) production has received the greatest attention, both as a growing industry along the Atlantic margin of the Guiana Shield, but also with a concern towards the long-term sustainability of current production methods. The coastal Atlantic region of the Guiana Shield dominates the geographic distribution of *E. oleracea*, the multi-stemmed species that accounts for most palm-heart production from the region. In contrast, the single-stemmed *E. precatoria* is the primary source of material for the Bolivian palm-heart industry (Bojanic, 2001) and prior industries in southern Brazil centred on the single-stemmed *E. edulis* are now largely closed due to commercial extinction of the raw material source (Anderson and Jardim, 1989; Pollak *et al.*, 1995). In fact, sustainable management of palm-heart production from single-stemmed *Euterpe* appears largely unachievable at any reasonable timescale due to slow growth rates (Peña and Zuidema, 1999) and absence of root suckering (Anderson, 1988; Johnston, 1995; van Andel, 1998, 2000). Long-term production from *E. oleracea* is feasible if size-based harvesting limits, fractional shoot collection and adequate fallow periods are adhered to by palm-heart harvesters (Pollak *et al.*, 1995; van Andel, 2000). In most cases, a fallow period of 4–5 years appears adequate to ensure stability in the population size class distribution.

Thus, the industry across the Guiana Shield has a striking comparative advantage based on its *E. oleracea* stands if it can be managed sustainably. Competition in *Euterpe* heart production is restricted to areas in and around the Amazon estuary where palm-heart production must also compete with other profitable uses of *Euterpe* that do not currently generate any significant domestic demand across the Guianas. Competition from palm-heart production in *Bactris gasipaes,* a semi-domesticate found throughout the neotropics (Henderson *et al.*, 1995) is more substantial. According to Arkcoll and Clement (1989, cited in van Andel, 2000), this palm is capable of producing palm hearts at six times the rate of *E. oleracea.*

Palm-heart production appears capable of bring a modicum of livelihood support to rural communities across the northern shield perimeter if strict, but simple, harvesting rules are followed and enforced. It also provides a source of income that is close to their home communities, avoiding many of the social impacts associated with men leaving to work for extended periods away from home, a feature commonly associated with small-scale, gold-mining operations.

Wildlife use

The socio-economic issues attached to wildlife use in tropical forests can be exceptionally complex, as a wide variety of perspectives and objectives view forest animals in very different ways (Freese, 1998). Forest wildlife use is seen through a wide range of perspectives ranging from its value as a protein source, as the focus of an esoteric, but potentially important, culture of exotic pet-keeping through to their spiritual representations. The catalysts and consequent patterns of commercial wildlife use are, however, surprisingly similar (Fimbel *et al.*, 2001).

Wildlife is the most important commercial NTFP across the Guianas. Registered exports of wildlife, composed mainly of psittacines (parrots, macaws), reptiles (snakes, lizards, turtles), monkeys (*Cebus*, *Saguinas*, *Saimiri*) and aquarium fish constitute a significant foreign-exchange earner and source of income for people in Guyana, Suriname and the Rio Negro region of Brazil (Chao and Prang, 1997; De Souza, 1997; Duplaix, 2001; Ouboter, 2001).

Despite its prominent figuring in NTFP revenue generation across the region, wildlife use remains one of the least regulated of forest-based activities and little information is available regarding population sizes, rates of use or how long these can be sustained. All indications suggest that use is more targeted, less widespread and less intensive compared to other lowland neotropical regions that have been virtually defaunated, but productivity in the shield is also typically much lower and limits defined elsewhere are unlikely to be applicable across geomorphographic regions. Some species have already been clearly over-harvested historically, since their geographic ranges have contracted to more remote locations (e.g. black caiman, arapaima). Sociocultural conditions are also different. For example, monkeys are rarely hunted for bushmeat in Guyana, although consumptive rates across other parts of the neotropics are often as high or higher for primates than other taxa (Redford, 1993).

Unregulated collection of wildlife to meet a growing commercial trade in bushmeat, fish and live specimens is arguably the most important challenge facing vertebrate conservation and sustainable use across the Guiana Shield. Significant concerns remain regarding how best to establish functional quotas on the wildlife trade (Pilgrim, 1993) and the impact this trade has on rural societies (Forte, 1989), but the largest issue confronting the future of wildlife across the region is the virtual absence of any restrictions on hunting and fishing season, quantity or location (Duplaix, 2001). Unmitigated access to (mainly) public wildlife resources without any payment of taxes or rents similar to those placed on plant products invariably stimulates wasteful commercial consumption of wildlife because it creates opportunities to make a solid return on investment without the need to manage operations critically. Like logging, a chain of waste ensues from point of harvest to final market. Income derived from the use of wildlife that should bolster public revenues is instead taken as super-profit by operators or lost as wastage. Ultimately, local and national communities bear the cost of over-exploitation through a decreasing subsistence share and lost government revenues. At the national scale, the absence of any restrictions on hunting and fishing is difficult to resolve with interest in developing viable ecotourism industries, maintaining an efficient, profitable and legal export of live specimens and protecting local rural community subsistence use needs. All of these represent real comparative, but largely unmarketed, advantages held in the region due to the high degree of forest integrity that has been maintained relative to other regions that were capable of providing similar goods and services.

Managing forest roads and road-building

If lack of regulations, restrictions and their effective enforcement can critically impede progress towards sustainable forest management across the Guiana Shield, then an expanding road network is the main agent capable of ensuring that it fails altogether. This is not to engender the view that significant benefit cannot be derived from road establishment, but to emphasize that in the absence of effective road management, costs attached to their use will invariably outweigh the benefits roads can deliver at local, national and regional scales. The main cities of the region were connected by all-year roads by 2002 with the upgrading of the Georgetown to Lethem road in Guyana, although this remains unpaved (Fig. 9.12). The only other road prospectively connecting the main centres of population that remains unpaved is the Oiapoque to Macapa road in Amapá and this is earmarked for paving under the *Avança Brasil* programme (Carvalho *et al.*, 2001).

Pavement of the Georgetown to Lethem road would connect all of the major cities along the shield periphery, Manaus, Ciudad Bolivar, Georgetown, Paramaribo and Cayenne, in a way that would accommodate much larger volumes of traffic carrying much larger payloads. The feasibility of such an undertaking has been explored (Environmental Resources Management, 1995), although the advantage of upgrading the 585 km length of road to the resident Guyanese population of less than 750,000 remains largely unclear in comparison to the relative ease of access afforded Brazil's northernmost state, Roraima, with its population of 325,000 (Forte, 1990; Forte and Benjamin, 1993).

Roads across remote frontiers act as conduits for the legal transport of people and goods that can broaden cash income-earning opportunities for rural households through greater market access for their agricultural and timber products. But roads also can act as very effective thoroughfares for the transmission of disease, invasive and exotic pests, smuggling, poaching of wildlife, timber and other plant materials, unpermitted mining, violence and crime. Massive influxes of cash and working capital can rapidly overwhelm and then erode existing rural community social structures. For example, even modest increases in traffic along the Georgetown to Lethem road running through the Iwokrama Forest saw an increase in illegal hunting activities, timber poaching and conflicts regarding uses of roadside creeks that supplied water to adjoining communities (Ousman, 1999). More substantial impacts are well-established for roads built in other parts of the Amazon, largely due to an absence of any road management presence capable of organizing where and how roadside land is used and mitigating impacts created by road-users. The Guiana Shield remains one of the last forest frontiers largely due to its inaccessibility. Improving access will bring

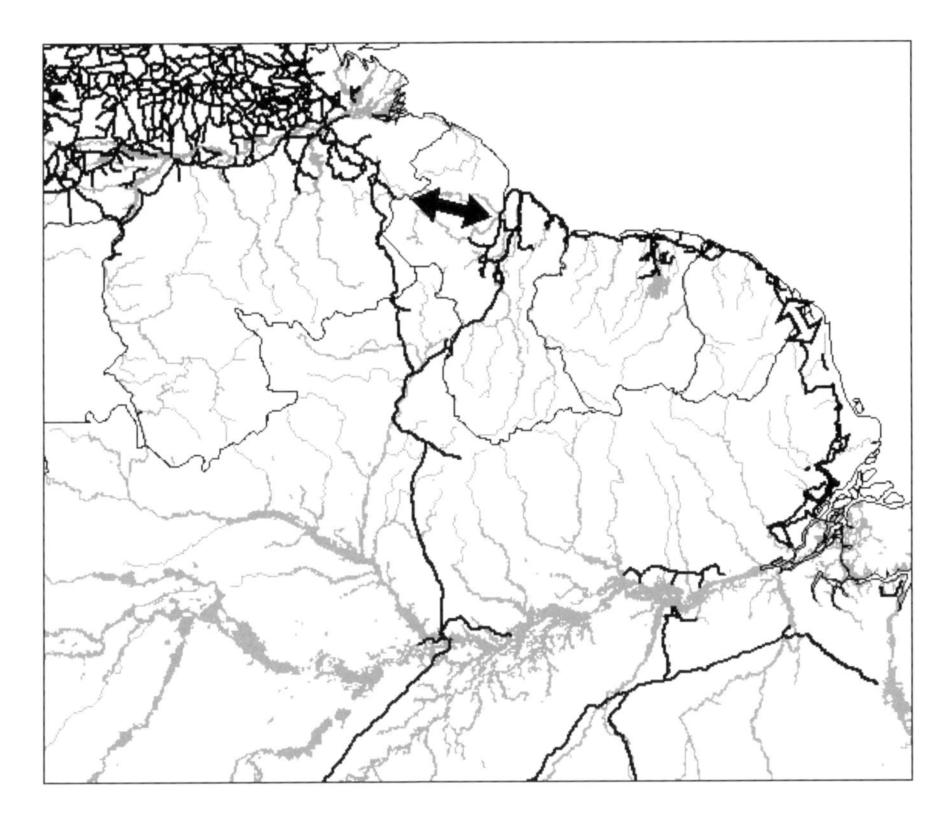

Fig. 9.12. Major road arteries connecting large towns and cities across the Guiana Shield and beyond. Note the pivotal position of Boa Vista in the overland flow of traffic between the Guianas, Brazil and Venezuela. Road development between Guyana and Venezuela (through Cuyuni-Nuria) (filled arrow) and French Guiana and Amapá (empty arrow) would diversify the transit access across the region.

further opportunities for economic development only if these roads are effectively managed to prevent or mitigate the wide range of negative impacts that can follow road development.

Management Scale, Focus and Objective – a Perspective

Scale

Environmental concerns over the use of tropical forests in many ways have been transfixed at the largest inclusive scales, but consistently studied at some of the smallest scales. In fact, scaling-up interpretation of many site- and time-specific results has unintentionally cemented the notion of underlying environmental uniformity and perpetuated an implicit belief that all tropical forests are the product of similar past histories and uniformly respond (negatively) to the same basket of destructive land-use practices. Selective extraction and forest clearance approaches to timber production get muddled into a single mass. Non-wood products equally begin to form a conglomerate of economic alternatives. Shifting agriculturalists are melded into a single group that ignores important variation in the marginal utility they receive, impacts they exact and core competencies through which they manage forest resources. The scale of mining operations and the realistic costs and benefits attached to these different levels are not always adequately considered. Thus studying the socio-economic and environmental consequences of commercial-scale use of *in situ* forestlands from a presumptive perspective of either net benefit or costs without faithfully assessing the other can further foster irrational decision-making and add a polemical coating to problem-solving efforts. In the end, these efforts should seek to bring the maximum net benefit to the widest cross-section of society. Implicit scaling-up of single case history examples of success and failure or research findings is often inappropriate when variation in biophysical and socio-economic features at these larger scales is considered. Getting the scale width of applicability correct is the first step in identifying the necessary tolerance (or robustness) of forest management focus in the Guiana Shield.

Focus

Management focus must be sufficiently robust to weather the vagaries of larger-scale socio-economic and biophysical fluctuations. Chronic boom-and-bust cycling of commodity production that has typified economies of the Guiana Shield creates enormous employment and inflationary ripples that deflect progress towards social and economic stability.

The resulting instabilities can drive further environmental degradation. Bust phases are dominated by large numbers of small operators seeking to meet their short-term subsistence needs with very little capacity or motivation to minimize collateral damage (e.g. Heemskerk, 2001). Their activities are difficult to regulate and manage. The sum total of their individual impacts creates a 'tragedy of the commons' scenario where compounded use of the same resources leads to rapid commercial extinction and options for sustainable management are limited to restoration.

Alternatively, large-scale industrial extraction of resources by a few corporate operators can ease and reduce the cost of monitoring for sustainable management if the political will is sufficient to enact legislation mandating appropriate forms of sustainable management and allow this legislation to be actively implemented. Historically, where large companies have dominated commodity economies, the excessive weight of their political leverage has led to large-scale devaluation of forestlands and minimized the flow of benefits as a result of their activities. A focus on either small- or large-scale routes to forest use can lead to similar outcomes, only through different routes to management failure.

How does management fail?

Few aspirations, few opportunities

The focus of forestland use is also shaped by human aspirations and these, in turn, by prevailing social norms at the time of decision-making. Education, health, communication, transport and the technologies that affect changes in living conditions can alter these social norms and aspirations attached to forest use. When opportunities are few, aspirations are limited and concepts of long-term management for future forest value rarely figure in the extremely short-term planning horizons of most people living in the commodity-driven economies of the Guiana Shield.

Extending the planning horizon is an important precondition to sustainable management, but this can only occur where economic and political processes culture a sense within both public and private sectors that forests deliver a wider and more permanent range of benefits. The distribution of benefit as a result of forest use also influences perspectives on acceptable levels of waste and extraction intensity. Those deriving the least direct benefit and bearing the greatest indirect costs from extraction are most likely to be the most conservative in their perspectives of acceptable levels. This implies that the focus can change with level of ownership, rather than necessarily with the scale of management.

Mismatched land-use approaches

Equally, the mismatch of land-use approach with biophysical features further drives gross leakages that slacken the inertia needed to achieve socio-economic development in commodity-driven economies. Large-scale forest clearance for agriculture or livestock on sand-dominated tropical soil facies leaves few livelihood alternatives. Once these are cleared and abandoned, an extremely slow forest re-establishment process modulated by nutrient evacuation, large-seed dispersal distances and inherently slow growth capacities (see Chapter 3) renders these areas of marginal economic potential. These sand-dominated soils occupy large areas of the region (and define several of the endemic ecoregional units) (see 'Ecoregions', this chapter, and 'Soils', Chapter 2).

In the Guiana Shield, biophysical trajectories are defined by low-energy systems (see Chapter 7). If the focus of forest land use is sustainable delivery of specific benefits without erosion of future opportunities, then adopted systems must also seek conservation measures that reduce their role as external importers of energy into the ecosystem. Across most land-use options, these conservation measures are a function of extraction intensity, operational inefficiencies or both. Whether seen as collateral stand damage, non-selective wildlife harvesting, uncontrolled anthropogenic fire, poor road-building or unmitigated mining effluent, the injection of energy into the system accelerates the slide down an intrinsic dampening slope of forest life-expectancy. At a geological timescale, forests of the Guiana Shield are already in a state of transitory decline and the application of management techniques adopted from other regions that are currently in a state of transitory ascent (e.g. western Amazonia, central America, SE Asia), rather than descent, will invariably accelerate this decline.

Objectives

If ownership influences acceptability of forest use and low-energy systems govern forest ecosystem change, then the objective of management at the largest scale should be an inclusive system that offers structured opportunities for a wide range of approaches tailored to different prospective types of ownership. At smaller scales, working approaches should integrate core conservation measures to mitigate energy imports and identify optimal harvesting intensities that will meet reasonable, short- and medium-term financial objectives without liquidating longer-term opportunities through accelerated forest decline. Establishing and enforcing allowable boundaries to the way in which the resource can be managed constitutes the main approach to achieving this objective. Public forest-use policy and the agencies responsible for its implementation constitute the main tools for ensuring that long-term opportunities are not unnecessarily liquidated and that the forest resource industry is not beset with unrealistic or uneconomic conservation measures. Maintaining this balance in the low-energy environment of the Guiana Shield represents one of the greatest challenges to sustainable management of tropical forests anywhere. Sustainable use and conservation of forests in the region will in the end depend on the international and national support for the work of these agencies and the commitment of their staff in making best use of this support in achieving objectives that benefit wider society. How efficiently natural capital is transformed into social capital through directed public and private sector investment will determine the role that regional forests will play in national development or, sadly, chronic under-development. How people in the Guiana Shield ultimately define a lost opportunity will and should determine how long and how much utility they and their future generations can expect to enjoy from their forests.

Notes

[1] This area incorporates the Jaú National Park (see Fig. 9.1 for location), Mamirauá and Amana Sustainable Development Reserves and the Anavilhanas Ecological Station (WCPA – World Database on Protected Areas).
[2] How the latter criterion was spatially discriminated is not clear to this author from the available literature.
[3] Explanatory variables used in the analysis appear to violate assumption of orthogonality and are subject to unknown error structure (Vanclay, 2001). One could just as easily argue that enforcement is proven most effective in maintaining tropical park integrity because other efforts to adequately pair protection with expanding livelihood opportunities and effort to demonstrate to people the purpose and benefits of habitat protection have been neglected or failed. In fact, degradation of adjoining habitat and effectiveness of guard density are not independent and may be strongly correlated. A more appropriate conclusion would have stated that individuals involved in the establishment, management and funding of tropical parks believe that guard density is the best predictor of park effectiveness. If local community residents did not form part of the survey, it is difficult to see how local support and participation variables were calculated independent of opinions expressed by park-associated respondents who may have quite different views on park performance and attribution of causality.
[4] This normal tax burden effectively equates to the appropriate level of economic rent sought by the owner. In the case of the Guiana Shield, where most commercial forests are owned by the public and stewarded on their behalf by the relevant government agency, this rent would be the difference between revenues and the sum of the production costs and a normal profit (often around 30%). In effect, it is a rent for use of the land in its mature, timber-bearing condition. Often rent capture has been well below this difference and is believed to have formed one of the more important incentives for otherwise inefficient logging practices to persist. Rents, through the various forms they can take, have traditionally been very low in most countries in the Guiana Shield (e.g. Oliphant, 1938; Palmer, 1996; Whiteman, 1999).

References

Agriconsulting (1993) *Preparatory Study for the Creation of a Protected Area in the Kanuku Mountains Region of Guyana.* Agriconsulting, Rome.

Anderson, A.B. (1988) Use and management of native forests dominated by açai palm (*Euterpe oleracea* Mart.) in the Amazon estuary. *Advances in Economic Botany* 6, 144–154.

Anderson, A.B. and Jardim, M.A.G. (1989) Costs and benefits of floodplain forest management by rural inhabitants in the Amazon estuary: a case study of Açai palm production. In: Browder, J.O. (ed.) *Fragile Lands of Latin America: Strategies for Sustainable Development.* Westview Press, London, pp. 114–129.

Arkcoll, D. B. and Clement, C. (1989) Potential new crops for the Amazon. In: Wickens, G.E., Haq, N. and Day, P. (eds) *New Crops for Food and Industry.* Chapman and Hall, London, pp. 150–165.

Armstrong, S. and Inglis, C.J. (2000) RIL for real: introducing reduced impact logging techniques into a commercial forestry operation in Guyana. *International Forestry Review* 2, 17–23.

Barbier, E.B., Burgess, J.C., Bishop, J. and Aylward, B. (1994) *The Economics of the Tropical Timber Trade.* Earthscan Publications, London.

Barreto, P., Amaral, P., Vidal, E. and Uhl, C. (1988) Costs and benefits of forest management for timber production in eastern Amazonia. *Forest Ecology and Management* 108, 9–26.

Bojanic, A.J. (2001) *Balance is Beautiful: Assessing Sustainable Development in the Rain Forests of the Bolivian Amazon.* PROMAB Scientific Series 4. PROMAB, Utrecht, The Netherlands.

Boltz, F., Holmes, T.P. and Carter, D.R. (2003) Economic and environmental impacts of conventional and reduced-impact logging in tropical South America: a comparative review. *Forest Policy and Economics* 5, 69–81.

British Guiana Forest Department (1940–1965) *Annual Report of the Forest Department.* Department of Lands and Mines, Georgetown, Guyana
Bruner, A.G., Gullison, R.E., Rice, R.E. and da Fonseca, G.A.B. (2001) Effectiveness of parks in protecting tropical biodiversity. *Science* 291, 125–128.
Bryant, D., Nielsen, D. and Tangley, L. (1997) *The Last Frontier Forests – Ecosystems and Economies on the Edge.* World Resources Institute, Washington, DC.
Carvalho, G., Barros, A.C., Moutinho, P. and Nepstad, D. (2001) Sensitive development could protect Amazonia instead of destroying it. *Nature* 409, 131.
Chao, N.L. and Prang, G. (1997) Project Piaba – towards a sustainable ornamental fishery in the Amazon. *Aquarium Sciences and Conservation* 1, 105–111.
Chape, S., Blyth, S., Fish, L., Fox, P. and Spalding, M. (2003) *2003 United Nations List of Protected Areas.* IUCN/UNEP-WCMC, Cambridge, UK.
Colchester, M. (1997) *Guyana: Fragile Frontier.* Earthscan Publications, London.
Colinvaux, P.A., De Oliveira, P.E., Moreno, J.E., Miller, M.C. and Bush, M.B. (1996) A long pollen record from lowland Amazonia: forest and cooling in glacial times. *Science* 274, 85–88.
Coomes, D.A. and Grubb, P.J. (1996) Amazonian caatinga and related communities at La Esmeralda, Venezuela: forest structure, physiognamy and floristics, and control by soil factors. *Vegetatio* 122, 167–191.
Cooper, A. (1979) Muri and white sand savannah in Guyana, Surinam and French Guiana. In: Specht, R.L. (ed.) *Heathlands and Related Shrublands.* Elsevier, Amsterdam, pp. 471–481.
Cordier, S., Grasmick, C., Paquier-Passelaigue, M., Mandereau, L., Weber, J.P. and Jouan, M. (1998) Mercury exposure in French Guiana: levels and determinants. *Archives of Environmental Health* 53, 299–303.
Davis, S.D., Heywood, V.H., Herrera-McBryde, O., Villa-Lobos, J. and Hamilton, A.C. (1997) *Centres of Plant Diversity. A Guide and Strategy for their Conservation. Volume 3 – the Americas.* WWF-IUCN, Gland, Switzerland.
Davis, T.A.W. and Richards, P.W. (1933) The vegetation of Moraballi creek, British Guiana: an ecological study of a limited area of tropical rain forest. I. *Journal of Ecology* 21, 350–384.
de Graaf, N.R. (1986) *A Silvicultural System for Natural Regeneration of Tropical Rain Forest in Suriname.* Pudoc/WAU, Wageningen, The Netherlands.
de Graaf, N.R. (2000) Reduced impact logging as part of the domestication of neotropical rainforest. *International Forestry Review* 2, 40–44.
de Graaf, N.R., Poels, R.L.H. and van Rompaey, R.S.A.R. (1999) Effect of silvicultural treatment on growth and mortality of rainforest in Surinam, over long periods. *Forest Ecology and Management* 124, 123–135.
de Kom, J.F., van der Voet, G.B. and de Wolff, F.A. (1998) Mercury exposure of maroon workers in the small scale gold mining in Suriname. *Environmental Research* 77, 91–97.
de Milde, R. (1970) *Assessment of Hidden Defects on Standing Trees During Inventory Work.* Technical Report No. 4 (SF/GUY 9). FAO/UNDP, Georgetown, Guyana.
de Milde, R. and Inglis, C.J. (1974) *Defect Assessment of Standing Trees.* FAO Project Working Document No. 14 FO/SF/SUR/71/506. FAO, Paramaribo, Suriname.
De Souza, B. (1997) *Towards a Methodology for the Economic Valuation of Non-timber Forest Products: A View from Guyana.* GNRA, Georgetown, Guyana.
Denevan, W.D. (1970) Aboriginal drained field cultivation in the Americas. *Science* 169, 647–654.
Dickinson, M.B., Dickinson, J.C. and Putz, F.E. (1996) Natural forest management as a conservation tool in the tropics: divergent views on possibilities and alternatives. *Commonwealth Forestry Review* 75, 309–315.
Duivenvoorden, J.F. and Lips, J.M. (1995) *A Land-Ecological Study of Soils, Vegetation and Plant Diversity in Colombian Amazonia.* Tropenbos Foundation, Wageningen, The Netherlands.
Duplaix, N. (2001) *Evaluation of the Animal and Plant Trade in the Guianas – Preliminary Findings.* WWF-Guianas, Cayenne, French Guiana.
Dykstra, D. and Heinrich, R. (eds) (1996) *Forest Codes of Practice – Contributing to Environmentally Sound Forest Operations.* FAO Forestry Paper 133. FAO, Rome.
Environmental Resources Management (1995) *Environmental and Social Impact Assessment: Linden-Lethem Road, Guyana.* ERM, London.
Evans, C. and Meggers, B.J. (1960) *Archeological Investigations in British Guiana.* Bulletin of the Bureau of American Ethnology, 177. Smithsonian Institution Press, Washington, DC.
FAO (2001) *Global Forest Resources Assessment 2000.* FAO Forestry Paper 140. FAO, Rome.

Fanshawe, D.B. (1952) *The Vegetation of British Guiana: A Preliminary Review.* Institute paper 29. Imperial Forestry Institute, Oxford, UK.
Fanshawe, D.B. (1954) Forest types of British Guiana. *Caribbean Forester* 15, 73–111.
Fimbel, R.A., Grajal, A. and Robinson, J.G. (eds) (2001) *The Cutting Edge – Conserving Wildlife in Logged Tropical Forests.* Biology and Resource Management Series. Columbia University Press, New York.
Forte, J. (1989) Wildlife, timber and balata. In: Forte, J. (ed.) *The Material Culture of the Wapishana People of the South Rupununi Savannahs in 1989.* ARU – University of Guyana, Turkeyen, pp. 41–44.
Forte, J. (1990) *A Preliminary Assessment of the Road to Brazil.* ARU – University of Guyana. Turekeyen.
Forte, J. and Benjamin, A. (1993) *The Road from Roraima State.* ARU – University of Guyana, Turkeyen.
Fostier, A.H., Forti, M.C., Guimaraes, J.R.D., Melfi, A.J., Boulet, R., Espirito Santo, C.M. and Krug, F.J. (2000) Mercury fluxes in a natural forested Amazonian catchment (Serra do Navio, Amapa state, Brazil). *Science of the Total Environment* 260, 201–211.
Freese, C.H. (1998) *Wild Species as Commodities.* Island Press, Washington, DC.
Funk, V.A. and Richardson, K.S. (2002) Systematic data in biodiversity studies: use it or lose it. *Systematic Biology* 51, 303–316.
Funk, V.A., Fernanda Zermoglio, M. and Nazir, N. (1999) Testing the use of specimen collection data and GIS in biodiversity exploration and conservation decision-making in Guyana. *Biodiversity and Conservation* 8, 727–751.
Furch, K. (1984) Water chemistry of the Amazon basin: the distribution of chemical elements among freshwaters. In: Sioli, H. (ed.) *The Amazon. Limnology and Landscape Ecology of a Mighty Tropical River and its Basin.* Dr. W. Junk, Dordrecht, The Netherlands, pp. 167–195.
Gibbs, A.K. and Barron, C.N. (1993) *The Geology of the Guiana Shield.* Oxford University Press, Oxford, UK.
Goodland, R. and Daly, H. (1996) Environmental sustainability: universal and non-negotiable. *Ecological Applications* 6, 1002–1017.
Goulding, M., Carvalho, M.L. and Ferreira, E.G. (1988) *Rio Negro: Rich Life in Poor Water.* SPB Academic, The Hague, The Netherlands.
Grayum, G. (1971) *Logging and Forest Management.* UNDP Technical Report 12 (Guyana). UNDP, Georgetown, Guyana.
Hammond, D.S. (1999) *Quality vs. Quantity in the Quest for Sustainable Forest Management in the Guiana Shield.* Sustainable Forest-Based Business Partnerships, Georgetown, Guyana, Iwokrama International Centre.
Hammond, D.S. and Brown, V.K. (1995) Seed size of woody plants in relation to disturbance, dispersal, soil type in wet neotropical forests. *Ecology* 76, 2544–2561.
Hammond, D.S. and ter Steege, H. (1998) Propensity for fire in the Guianan rainforests. *Conservation Biology* 12, 944–947.
Hammond, D.S., Gourlet-Fleury, S., van der Hout, P., ter Steege, H. and Brown, V.K. (1996) A compilation of known Guianan timber trees and the significance of their dispersal mode, seed size and taxonomic affinity to tropical rain forest management. *Forest Ecology and Management* 83, 99–106.
Hammond, D.S., van der Hout, P., Zagt, R., Marshall, G., Evans, J. and Cassells, D.S. (2000) Benefits, bottlenecks and uncertainties in the pantropical implementation of reduced impact logging techniques. *International Forestry Review* 2, 45–53.
Heemskerk, M. (2001) Do international commodity prices drive natural resource booms? An empirical analysis of small-scale gold mining in Suriname. *Ecological Economics* 39, 295–308.
Henderson, A., Galeano, G. and Bernal, R. (1995) *Field Guide to the Palms of the Americas.* Princeton University Press, Princeton, New Jersey.
Hendrison, J. (1989) *Damage-controlled Logging in Managed Tropical Rain Forest in Suriname.* WAU, Wageningen, The Netherlands.
Heyligers, P.C. (1963) Vegetation and soil of a white-sand savanna in Suriname. *Verhandelingen der Koninklijke Nederlandse Akademie van Wetenschappen, Afd. Natuurkunde* 54, 1–148.
Hohenkerk, L.S. (1922) A review of the timber industry of British Guiana. *Journal of the Board of Agriculture of British Guiana* 15, 2–22.
Holmes, T.P., Blate, G.M., Zweede, J.C., Pereira, R.J., Barreto, P., Boltz, F. and Bauch, R. (1999) *Financial Costs and Benefits of Reduced-impact Logging Relative to Conventional Logging in the Eastern Amazon.* Phase 1 final report. TFF/USFS, Belem.
Hoosein, M. (1996) *A Review of Protection Criteria in Previous Documentation for Protected Areas in Guyana.* Guyana Environmental Protection Agency, Turkeyen, Guyana.
Hubbell, S.P., Foster, R.B., O'Brien, S.T., Harms, K.E., Condit, R., Wechsler, B., Wright, S.J. and de Lao, S.L.

(1999) Light-gap disturbances, recruitment limitation, and tree diversity in a neotropical forest. *Science* 283, 554–557.
Huber, O. (1995) Conservation of the Venezuelan Guayana. In: Berry, P.E., Holst, B.K. and Yatskievych, K. (eds) *Flora of the Venezuelan Guayana.* Timber Press, Portland, Oregon, pp. 193–218.
Huber, O. (1997) Pantepui region of Venezuela. In: Davis, S.L.D., Heywood, V.H., Herrera-MacBryde, O., Villa-Lobos, J. and Hamilton, A.C. (eds) *Centres of Plant Diversity – A Guide and Strategy for their Conservation. Vol. 3. The Americas.* World Wildlife Fund for Nature (WWF) and The World Conservation Union-IUCN. IUCN Publications Unit, Cambridge, UK, pp. 308–311.
Huber, O. (2001) Conservation and environmental concerns in the Venezuelan Amazon. *Biodiversity and Conservation* 10, 1627–1643.
IBGE (2000) Cidades@. IBGE (http://www.ibge.gov.br/cidadesat).
IBGE (2002) Banco de dados agregados-extração vegetal. IBGE (http://www.sidra.ibge.gov.br).
Im Thurn, E.F. (1883) *Among the Indians of Guiana.* Kegan Paul, Trench and Co., London.
Jansen, P.A. and Zuidema, P. (2001) Logging, seed dispersal by vertebrates, and natural regeneration of tropical timber trees. In: Fimbel, R., Grajal, A. and Robinson, J.G. (eds) *The Cutting Edge – Conserving Wildlife in Logged Tropical Forests.* Columbia University Press, New York, pp. 35–59.
Johnston, D.V. (1995) Report on the palm cabbage industry in northwest Guyana. In: Forte, J. (ed.) *Indigenous Use of the Forest – Situation Analysis with Emphasis on Region 1.* ARU – University of Guyana, Georgetown, pp. 62–67.
Jonkers, W.B.J. and Schmidt, P. (1984) Ecology and timber production in tropical rain forest in Suriname. *Interciencia* 9, 290–298.
Kalliola, R.J., Jokinen, P. and Tuukki, E. (1999) Fluvial dynamics and sustainable development in upper Rio Amazonas, Peru. In: Padoch, C., Ayres, J.M., Pinedo-Vasquez, M. and Henderson, A. (eds) *Várzea – Diversity, Development and Conservation of Amazonia's Whitewater Floodplains.* NYBG Press, New York, pp. 271–282.
Kammesheidt, L., Köhler, P. and Huth, A. (2001) Sustainable timber harvesting in Venezuela: a modelling approach. *Journal of Applied Ecology* 38, 756–770.
Kennedy, L.M. and Horn, S.P. (1997) Prehistoric maize cultivation at the La Selva biological station, Costa Rica. *Biotropica* 29, 368–370.
King, K.F.S. (1963) The role of the forest in the future economy of British Guiana. *Commonwealth Forestry Review* 42, 237–241.
King, K F.S. (1965) The use of arboricides in the management of tropical high forest. *Turrialba* 15, 35–39.
Ledru, M.-P., Blanc, P., Charles-Dominique, P., Fournier, M., Martin, L., Riera, B. and Tardy, C. (1997) Reconstituion palynologique de la forêt guyanaise au cours des 3000 dernières années. *Comptes Rendu de l'Academie des Sciences, Paris* 324, 469–476.
Lescure, J.-P. and Boulet, R. (1985) Relationships between soil and vegetation in a tropical rain forest in French Guiana. *Biotropica* 17, 155–164.
Liu, K.-b. and Colinvaux, P.A. (1985) Forest changes in the Amazon basin during the last glacial maximum. *Nature* 318, 556–557.
Lugo, A.E. (1995) Management of tropical biodiversity. *Ecological Applications* 5, 956–961.
Mackay, M.S. (1926) The durability of Greenheart. *Journal of the Board of Agriculture of British Guiana* 19(1), 10–12.
McNabb, K.L. and Wadouski, L.H. (1999) Multiple rotation yields for intensively managed plantations in the Amazon basin. *New Forests* 18, 5–15.
McTurk, M. (1882) Notes on the forests of British Guiana. *Timehri* 1, 173–215.
Meggers, B.J. (1996) *Amazonia – Man and Culture in a Counterfeit Paradise.* Revised edition. Smithsonian Institution Press, London.
Miller, J.R., Lechler, P.J. and Bridge, G. (2003) Mercury contamination of alluvial sediments within the Essequibo and Mazaruni river basins, Guyana. *Water, Air and Soil Pollution* 148, 139–166.
Mittermeier, R.A., Mittermeier, C.G., Brooks, T.M., Pilgrim, J.D., Konstant, W.R., da Fonseca, G.A.B. and Kormos, C. (2001) Wilderness and biodiversity conservation. *Proceedings of the National Academy of Sciences USA* 100, 10309–10313.
Mol, J.H., Ramlal, J.S., Lietar, C. and Verloo, M. (2001) Mercury contamination in freshwater, estuarine, and marine fishes in relation to small-scale gold mining in Suriname, South America. *Environmental Research* 86, 183–197.
Moran, E.F. (1995) Rich and poor ecosystems of Amazonia: an approach to management. In: Nishizawa, T. and Uitto, J.I. (eds) *The Fragile Tropics of Latin America.* United Nations University Press, Tokyo, pp. 45–67.

Myers, N., Mittermeier, R.A., Mittermeier, C.G., Fonseca, G.A.B. and Kent, J. (2000) Biodiversity hotspots for conservation priorities. *Nature* 403, 853–858.

Nriagu, J.O., Pfeiffer, W.C., Malm, O., Souza, C.M.M. and Mierle, G. (1992) Mercury pollution in Brazil. *Nature* 356, 389.

OCEI/MARNR (2000) Densidad demográfica. OCEI (http://www.ocei.gov.ve).

Oliphant, F.M. (1938) *The Commercial Possibilities and Development of the Forests of British Guiana.* The Argosy Co. Ltd, Georgetown, Guyana.

Olson, D.M., Dinerstein, E., Wikramanayake, E.D., Burgess, N.D., Powell, G.V.N., Underwood, E.C., D'Amico, J.A., Itoua, I., Strand, H.E., Morrison, J.C., Loucks, C.J., Allnutt, T.F., Ricketts, T.H., Kura, Y., Lamoreux, J.F., Wettengel, W.W., Hedao, P. and Kassem, K.R. (2001) Terrestrial ecoregions of the world: a new map of life on Earth. *Bioscience* 51, 933–938.

Ouboter, P. (2001) *Assessment of the Traded Wildlife Species.* WWF-Guianas, Paramaribo, Suriname.

Ousman, S. (1999) A socio-economic and biophysical assessment of the Iwokrama road corridor, Guyana. MSc thesis, University of West Indies – Mona, Kingston.

Palmer, C.J., Validum, L., Loeffke, B., Laubach, H.E., Mitchell, C., Cummings, R. and Cuadrado, R.R. (2002) HIV prevalence in a gold mining camp in the Amazon region, Guyana. *Emerging Infectious Diseases* 8, 330–331.

Palmer, J.R. (1986) Lessons for land managers in the tropics. *Revue Bois et Forêts des Tropiques* 212, 16–27.

Palmer, J.R. (1996) *Report on Forest Revenue Systems.* GFC DfID Support Project, Oxford, UK.

Pearce, D., Putz, F.E. and Vanclay, J.K. (2003) Sustainable forestry in the tropics: panacea or folly? *Forest Ecology and Management* 172, 229–247.

Peña, M. and Zuidema, P. (1999) Limitaciones demográficas para el aprovechamiento sostenible de Euterpe precatoria para producción de palmito en dos tipos de bosque de Bolivia. *Ecología en Bolivia* 33, 3–21.

Perreijn, K. (2002) *Symbiotic Nitrogen Fixation by Leguminous Trees in Tropical Rain Forest in Guyana.* Tropenbos-Guyana Series 11. Tropenbos-Guyana, Georgetown.

Pilgrim, K. (1993) Guyana's wildlife trade. *Guyana Review* 5, 8–10.

Piperno, D. (1990) Fitolitos, arqueología y cambios prehistóricos de la vegetación en un lote de cincuenta hectáreas de la isla de Barro Colorado. In: Leigh, E.G.J., Rand, A.S. and Windsor, D.M. (eds) *Ecología de un Bosque Tropical.* STRI, Balboa, Panama, pp. 153–156.

Plumptre, A. (1995) The importance of 'seed trees' for the natural regeneration of selectively logged tropical forest. *Commonwealth Forestry Review* 74, 253–258.

Pollak, H., Mattos, M. and Uhl, C. (1995) A profile of palm heart extraction in the Amazon estuary. *Human Ecology* 23, 357–386.

Prince, A.J. (1971) The rate of growth of Greenheart (*Ocotea rodiaei* Schomb.). *Commonwealth Forestry Review* 52, 143–146.

Putz, F.E. (1991) Silvicultural effects of lianas. In: Putz, F.E. and Mooney, H.A. (eds) *The Biology of Vines.* Cambridge University Press, Cambridge, UK, pp. 493–501.

Pyne, S.J. (2001) *Year of the Fires – the Story of the Great Fires of 1910.* Viking Press, New York.

Ramdass, I. and Haniff, M. (1990) *A Definition of Priority Conservation Areas in Amazonia.* Guyana country paper. Biological priorities for conservation in Amazonia. Conservation International, Washington, DC.

Räsänen, M.E., Salo, J.S. and Kalliola, R.J. (1987) Fluvial perturbance in the western Amazon basin: regulation by long-term sub-Andean tectonics. *Science* 238, 1398–1401.

Redford, K.H. (1993) Hunting in neotropical forests: a subsidy from nature. In: Hladik, C.M., Hladik, A., Linares, O.F., Pagezy, H., Sample, A. and Hadley, M. (eds) *Tropical Forests, People and Food.* Parthenon Publishing, Paris, pp. 227–246.

Repetto, R. and Gillis, M. (eds) (1988) *Public Policies and the Misuse of Forest Resources.* Cambridge University Press, Cambridge, UK.

Richards, P.W. (1952) *The Tropical Rain Forest.* Cambridge University Press, Cambridge, UK.

Richards, P.W. (1996) *The Tropical Rain Forest: An Ecological Study.* Cambridge University Press, Cambridge, UK.

Rodríquez, J.P. and Rojas-Suárez, F. (1996) Guidelines for the design of conservation strategies for the animals of Venezuela. *Conservation Biology* 10, 1245–1252.

Rollet, B. (1980) Jari: succès ou échec? Un exemple de développement agro-sylvo-pastoral et industriel en Amazonie brésilienne. *Bois et Forêts des Tropiques* 192, 3–34.

Roulet, M., Lucotte, M., Farella, N., Serique, G., Coelho, H., Sousa Passos, C.J., Jesus da Silva, E., Scavone de Andrade, P., Mergler, D., Guimaraes, J.R.D. and Amorim, M. (1999) Effect of recent human colo-

nization on the presence of mercury in Amazonian ecosystems. *Water, Air and Soil Pollution* 112, 297–313.
Russell, C.E. (1987) Plantation forestry. In: Jordan, C.F. (ed.) *Amazonian Rain Forests – Ecosystem Disturbance and Recovery*. Springer, Berlin, pp. 76–89.
Santos, G.M., Cordeiro, R.C., Silva Filho, E.V., Turcq, B., Lacerda, L.D., Fifield, L.K., Gomes, P.R.S., Hausladen, P.A., Sifeddine, A. and Albuquerque, A.L.S. (2001) Chronology of the atmospheric mercury in Lagoa da Pata basin, upper Rio Negro region of Brazilian Amazon. *Radiocarbon* 43, 801–808.
Schulz, J.P. (1960) Ecological studies on rain forest in Northern Suriname. *Verhandelingen der Koninklijke Nederlandse Akademie van Wetenschappen, Afd. Natuurkunde* 163, 1–250.
Silva, J.M.N., de Carvalho, J.O.P., Lopes, J.C.A., de Almeida, B.F., Costa, D.H.M., de Oliviera, L.C., Vanclay, J.K. and Skovs.gaard, J.P. (1995) Growth and yield of a tropical rain forest in the Brazilian Amazon 13 years after logging. *Forest Ecology and Management* 71, 267–274.
Silva-Forsberg, M.C., Forsberg, B.R. and Zeidemann, V.K. (1999) Mercury contamination in humans linked to river chemistry in the Amazon Basin. *Ambio* 28, 519–521.
Sing, K.A., Hryhorczuk, D.O., Saffirio, G., Sinks, T., Paschal, D.C. and Chen, E.H. (1996) Environmental exposure to organic mercury among the Makuxi in the Amazon basin. *International Journal of Occupational and Environmental Health* 2, 165–171.
Sist, P., Nolan, T., Bertault, J.-G. and Dykstra, D. (1998) Logging intensity versus sustainability in Indonesia. *Forest Ecology and Management* 108, 251–260.
Spangenberg, A., Grimm, U., Sepeda da Silva, J.R. and Fölster, H. (1996) Nutrient store and export rates of *Eucalyptus urograndis* plantations in eastern Amazonia (Jari). *Forest Ecology and Management* 80, 225–234.
Spencer, C.S., Redmond, E.M. and Rinaldi, M. (1998) Prehispanic causeways and regional politics in the llanos of Barinas, Venezuela. *Latin American Antiquity* 9, 95–110.
Stattersfield, A.J., Crosby, M.J., Long, A.J. and Wege, D.C. (1998) *Endemic Bird Areas of the World – Priorities for Biodiversity Conservation.* Birdlife Conservation Series No. 7. Birdlife International, London.
ter Steege, H. (1998) The use of forest inventory data for a National Protected Area Strategy in Guyana. *Biodiversity Conservation* 7, 1457–1483.
ter Steege, H. (2000) *Plant Diversity in Guyana.* Tropenbos Series 18. Tropenbos Foundation, Wageningen, The Netherlands.
ter Steege, H. and Hammond, D.S. (1996) Forest management in the Guianas: ecological and evolutionary constraints on timber production. *BOS Nieuwsletter* 15, 62–69.
ter Steege, H. and Hammond, D.S. (2001) Character convergence, diversity, and disturbances in tropical rain forest in Guyana. *Ecology* 82, 3197–3212.
ter Steege, H. and Zondervan, G. (2000) A preliminary analysis of large-scale forest inventory data of the Guiana Shield. In: ter Steege, H. (ed.) *Plant Diversity in Guyana.* Tropenbos Foundation, Wageningen, The Netherlands, pp. 35–54.
ter Steege, H., Jetten, V.G., Polak, A.M. and Werger, M.J.A. (1993) Tropical rain forest types and soil factors in a watershed area in Guyana. *Journal of Vegetation Science* 4, 705–716.
ter Steege, H., Boot, R.G.A., Brouwer, L.C., Hammond, D.S., van der Hout, P., Jetten, V.G., Khan, Z., Polak, A.M., Raaimakers, D. and Zagt, R. (1995) Basic and applied research for sound rain forest management in Guyana. *Ecological Applications* 5, 904–910.
ter Steege, H., Sabatier, D., Castellanos, H., van Andel, T., Duivenvoorden, J.F., de Oliveira, A.A., Ek, R.C., Lilwah, R., Maas, P.J.M. and Mori, S.A. (2000) An analysis of Amazonian floristic composition, including those of the Guiana Shield. *Journal of Tropical Ecology* 16, 801–828.
ter Steege, H., Welch, I.A. and Zagt, R. (2002) Long-term effect of timber harvesting in the Bartica Triangle, Central Guyana. *Forest Ecology and Management* 170, 127–144.
Thompson, J., Proctor, J., Viana, V., Milliken, W., Ratter, J.A. and Scott, D.A. (1992) Ecological studies on a lowland evergreen rain forest on Maracá Island, Roraima, Brazil. I. Physical environment, forest structure and leaf chemistry. *Journal of Ecology* 80, 689–703.
USGS and CVGTM (1993) *Geology and Mineral Resource Assessment of the Venezuelan Guayana Shield.* USGS Bulletin 2062. US GPO, Washington, DC.
van Andel, T. (1998) Commercial exploitation of non-timber forest products in the north-west district of Guyana. *Caribbean Journal of Agriculture and Natural Resources* 2, 15–28.
van Andel, T. (2000) *Non-timber Forest Products of the North West District of Guyana.* Part I. Tropenbos-Guyana Series 8A. Tropenbos-Guyana Programme, Georgetown.
van der Hout, P. (1999) *Reduced Impact Logging in the Tropical Rain Forest of Guyana. Ecological,*

Economic and Silvicultural Consequences. Tropenbos-Guyana Series 6. Tropenbos-Guyana Programme, Georgetown.

van der Hout, P. (2000) Testing the applicability of reduced impact logging in greenheart forest in Guyana. *International Forestry Review* 2, 24–32.

Vanclay, J. (2001) The effectiveness of parks. *Science* 293, 1007.

Veiga, M.M., Meech, J.A. and Onate, N. (1994) Mercury inputs from forest fire in the Amazon. *Nature* 368, 816–817.

Veillon, J.P. (1985) El crecimiento de algunas bosques naturales de Venezuela en relación con los parámetros del medio ambiente. *Revista Forestal Venezolana* 29, 5–121.

Vink, A.T. (1970) *Forestry in Surinam – a Review of Policy, Planning, Projects, Progress.* Suriname Forest Service (LBB), Paramaribo.

Welch, I.A. (1975a) *A Short History of the Guyana Forest Department 1925–1975.* Guyana Forestry Commission, Georgetown.

Welch, I.A. (1975b) *The Timber Resources of Guyana.* Technical report. Guyana Forest Department, Georgetown.

Whiteman, A. (1999) *Economic Rent from Forest Operations in Suriname and a Proposal for Revising its Forest Revenue System.* FAO, Rome.

Williams, D. (1994) Annex C. Archaeology and anthropology. In: *Environmental Impact Assessment – the Linden Lethem Road.* ERM, Georgetown, Guyana

Winkler, N. (1997) *Environmentally Sound Forest Harvesting – Testing the Applicability of the FAO Model Code in the Amazon in Brazil.* Forest Harvesting Case Study No. 8. FAO, Rome.

Zagt, R.J. (1997) *Tree Demography in the Tropical Rain Forest of Guyana.* Tropenbos-Guyana Series 3. Tropenbos-Guyana Programme, Georgetown.

Index